INTERNATIONAL

INTERNATIONAL
CODE COUNCIL®

EXISTING

BUILDING

CODE®

COMMENTARY

2003

2003 International Existing Building Code® Commentary

First Printing: June 2005

ISBN # 1-58001-135-7

COPYRIGHT© 2005
by

INTERNATIONAL CODE COUNCIL, INC.

PRINTED IN THE U.S.A.

PREFACE

The principle purpose of the Commentary is to provide a basic volume of knowledge and facts relating to building construction as it pertains to the regulations set forth in the 2003 *International Existing Building Code*. The person who is serious about effectively designing, constructing and regulating existing buildings and structures will find the Commentary to be a reliable data source and reference to almost all components in the built environment.

Throughout all of this, strenuous effort has been made to keep the vast quantity of material accessible and its method of presentation useful. With a comprehensive yet concise summary of each section, the Commentary provides a convenient reference for regulations applicable to the construction of buildings and structures. In the chapters that follow, discussions focus on the full meaning of application and the consequences of not adhering to the code text. Illustrations and examples are provided to aid understanding; they do not necessarily illustrate the only methods of achieving code compliance.

The format of the Commentary includes the full text of each section, table and figure in the code, followed immediately by the commentary applicable to that text. At the time of printing, the Commentary reflects the most up-to-date text of the 2003 *International Existing Building Code*. As stated in the preface to the *International Existing Building Code*, the content of sections in the code that begin with a letter designation (i.e., Section 1301) are maintained by another code development committee. Each section's narrative includes a statement of its objective and intent, and usually includes a discussion about why the requirement commands the conditions set forth. Code text and commentary text are easily distinguished from each other. All code text is shown as it appears in the *International Existing Building Code*, and all commentary is indented below the code text and begins with the symbol ❖.

Readers should note that the Commentary is to be used in conjunction with the *International Existing Building Code* and not as a substitute for the code. **The Commentary is advisory only;** the code official alone possesses the authority and responsibility for interpreting the code.

Comments and recommendations are encouraged, for through your input, we can improve future editions. Please direct your comments to the Code and Standards Department in the Chicago District Office.

The International Code Council would like to extend its thanks to Susan Gentry, Melvyn Green, Wayne Jewell, Ken Schoonover, Mark Stimac and the Structural Engineers Association of California (SEAOC) whose help as contributing authors made this document possible.

TABLE OF CONTENTS

CHAPTER 1

ADMINISTRATION

General Comments

This chapter contains provisions for the application, enforcement and administration of subsequent requirements of the code. In addition to establishing the scope of the code, Chapter 1 identifies which buildings and structures come under its purview. Section 101 addresses the scope of the code and references the other *International Codes®* that are mentioned elsewhere in the code. Section 102 establishes the applicability of the code and addresses existing structures.

Section 103 establishes the department of building safety and the appointment of department personnel. Section 104 outlines the duties and authority of the code official with regard to permits, inspections and right of entry. It also establishes the authority of the code official to approve alternative materials, used materials and modifications. Section 105 states when permits are required and establishes the procedures for the review of applications and the issuance of permits. Section 106 describes the information that must be included on the construction documents submitted with the application. Section 107 authorizes the code official to issue permits for temporary structures and uses. Section 108 establishes requirements for a fee schedule. Section 109 includes inspection duties of the code official or an inspection agency that has been approved by the code official. Provisions for the issuance of certificates of occupancy are detailed in Section 110. Section 111 gives the code official the authority to approve utility connections. Section 112 establishes the board of appeals and the criteria for making applications for appeal. Administrative provisions for violations are addressed in Section 113, including provisions for unlawful acts, violation notices, prosecution and penalties. Section 114 describes procedures for stop work orders. Section 115 establishes the criteria for unsafe structures and equipment and the procedures to be followed by the code official for abatement and for notification to the responsible party. Section 116 describes the emergency measures that address structures in danger of collapse. Section 117 authorizes the code official to have structures demolished that are dangerous, unsafe, insanitary or otherwise unfit for human habitation or occupancy. Each state's building code enabling legislation, which is grounded within the police power of the state, is the source of all authority to enact building codes. In terms of how it is used, police power is the power of the state to legislate for the general welfare of its citizens. This power enables passage of such laws as building codes. If the state legislature has limited this power in any way, the municipality may not exceed these limitations. While the municipality may not further delegate its police power (e.g., by delegating the burden of determining code compliance to the building owner, contractor or architect), it may turn over the administration of the building code to a municipal official such as a code official, provided that sufficient criteria are given to the code official to clearly establish the basis for decisions as to whether or not a proposed building conforms to the code.

Chapter 1 is largely concerned with maintaining "due process of law" in enforcing the building performance criteria contained in the body of the code. Only through careful observation of the administrative provisions can the code official reasonably hope to demonstrate that "equal protection under the law" has been provided. While it is generally assumed the administration and enforcement section of a code is geared toward a code official, this is not entirely true. The provisions also establish the rights and privileges of the design professional, contractor and building owner. The position of the code official is merely to review the proposed and completed work and to determine if the construction conforms to the code requirements. The design professional is responsible for the design of a safe structure. The contractor is responsible for constructing the structure in conformance with the plans.

During the course of construction, the code official reviews the activity to ascertain that the spirit and intent of the law are being met and that the safety, health and welfare of the public will be protected. As a public servant, the code official enforces the code in an unbiased, proper manner. Every individual is guaranteed equal enforcement of the provisions of the code. Furthermore, design professionals, contractors and building owners have the right of due process for any requirement in the code.

Purpose

This code, as with any other code, is intended to be adopted as a legally enforceable document to safeguard health, safety, property and public welfare. A code cannot be effective without adequate provisions for its administration and enforcement. The official charged with the administration and enforcement of building regulations has a great responsibility, and with this responsibility goes authority. No matter how detailed the code may be, the code official must, to some extent, exercise his or her own judgement in determining code compliance. The code official has the responsibility to establish that the homes in which the citizens of the community reside and the buildings in which they work are designed and constructed to be structurally stable, with adequate means of egress, light and ventilation and to provide a minimum acceptable level of protection to life and property from fire.

A large number of existing buildings and structures do not comply with the current building code requirements for new construction. Although many of these buildings are potentially salvageable, rehabilitation is often cost prohibitive because they may not be able to comply with all the requirements for new construction. At the same time, it is necessary to regulate construction in existing buildings that undergo additions, alterations, renovations, extensive repairs or change of occupancy. Such activity represents an opportunity to ensure that new construction complies with the current building codes and that existing conditions are maintained, at a minimum, to their current level of compliance or are improved as required. To accomplish this objective, and to make the rehabilitation process easier, this chapter allows for a controlled departure from full compliance with the technical codes, without compromising the minimum standards for fire prevention and life safety features of the rehabilitated building.

SECTION 101
GENERAL

101.1 Title. These regulations shall be known as the *Existing Building Code* of [NAME OF JURISDICTION], hereinafter referred to as "this code."

❖ The purpose of this section is to identify the adopted regulations by inserting the name of the adopting jurisdiction into the code.

101.2 Scope. The provisions of the *International Existing Building Code* shall apply to the repair, alteration, change of occupancy, addition, and relocation of existing buildings. A building or portion of a building that has not been previously occupied or used for its intended purpose shall comply with the provisions of the *International Building Code* for new construction. Repairs, alterations, change of occupancy, existing buildings to which additions are made, historic buildings, and relocated buildings complying with the provisions of the *International Building Code*, *International Mechanical Code*, *International Plumbing Code*, and *International Residential Code* as applicable shall be considered in compliance with the provisions of this code.

❖ This section establishes when the regulations contained in the code must be followed, whether all or in part. For the code to be applicable, something must occur, such as modification to an existing building or allowing an existing building or structure to become unsafe. The code is not a maintenance document requiring periodic inspections and it does not result in enforcement action, although periodic inspections are addressed by the *International Fire Code®* (IFC®).

101.3 Intent. The purpose of this code is to establish the minimum requirements to safeguard the public health, safety, and welfare insofar as they are affected by the repair, alteration, change of occupancy, addition, and relocation of existing buildings.

❖ The intent of the code is to set forth regulations that establish the minimum acceptable level to safeguard public health, safety and welfare and to provide protection for fire fighters and emergency responders in building emergencies. The intent becomes important in the application of such sections as Sections 102, 104.11 and 113, as well as any enforcement-oriented interpretive action or judgement. Like any code, the written text is subject to interpretation. Interpretations should not be affected by economics or the potential impact on any party. The only considerations should be protection of public health, safety and welfare and emergency responder safety.

101.4 Existing buildings. The legal occupancy of any building existing on the date of adoption of this code shall be permitted to continue without change, except as is specifically covered in this code, the *International Fire Code*, or the *International Property Maintenance Code*, or as is deemed necessary by the code official for the general safety and welfare of the occupants and the public.

❖ An existing structure is generally "grandfathered" to be considered approved with code adoption, provided that the building meets a minimum level of safety. Frequently, the criteria for this level are the regulations (or code) under which the existing building was originally constructed. If there are no previous code criteria to apply, the code official must apply those provisions that are reasonably applicable to existing structures. A specific level of safety is dictated by provisions dealing with hazard abatement in existing buildings and maintenance provisions, as contained in this code, the *International Building Code®* (IBC®), the *International Property Maintenance Code®* (IPMC®) and the IFC. These codes are referenced (see Sections 101.4, 101.5, 1101.2 and 1201.3.2) and are applicable to existing buildings.

101.5 Maintenance. Buildings and parts thereof shall be maintained in a safe and sanitary condition. The provisions of the *International Property Maintenance Code* shall apply to the maintenance of existing buildings and premises; equipment and facilities; light, ventilation, space heating, sanitation, life and fire safety hazards; responsibilities of owners, operators, and occupants; and occupancy of existing premises and buildings. All existing devices or safeguards shall be maintained in all existing buildings. The owner or the owner's designated agent shall be responsible for the maintenance of the building. To determine compliance with this subsection, the code official shall have the authority to require a building to be reinspected. Except where specifically permitted by this code, the code shall not provide the basis for removal or abrogation of fire protection and safety systems and devices in existing buildings.

❖ This section establishes the owner's responsibility to keep the building maintained in accordance with the code and the other referenced *International Codes*.

The code official has the authority to rule on the performance of maintenance work when public health or safety is affected. In addition, he or she has the responsibility to require existing structures to be maintained in compliance with all health and safety provisions required by the code.

Fire protection and safety systems in existing structures are to remain in place and be maintained. The removal of an existing safety system is not permitted, even if the structure would meet the minimum requirements of Chapter 12 without that system. For example, an existing sprinkler system in part of a building must remain functional even if the system was not required to meet the compliance alternative provisions of Chapter 12.

101.5.1 Work on individual components or portions. Where the code official determines that a component or a portion of a building or structure is in need of repair, strengthening or replacement by provisions of this code, only that specific component or portion shall be required to be repaired, strengthened, or replaced unless specifically required by other provisions of this code.

❖ This section allows the repair, replacement or strengthening of a specific component to be limited to that specific component. For example, in a situation where a masonry chimney has become damaged due to deterioration, this section would only require the repair of the chimney. To require the entire masonry fireplace and chimney system to be repaired or replaced, whether or not it is needed, would be excessive.

101.5.2 Design values for existing materials and construction. The incorporation of existing materials, construction, and detailing into the structural system shall be permitted when approved by the code official. Minimum quality levels and maximum strength values shall comply with this code.

❖ This section specifically allows for the use of existing construction materials and detailing into the structural system of the building. It is anticipated that these materials are in good condition, structurally sound and capable of withstanding all loads in the same manner as when the materials were new. The code official still retains the right to approve the use of these existing materials.

101.6 Safeguards during construction. All construction work covered in this code, including any related demolition, shall comply with the requirements of Chapter 13.

❖ The fundamental rationale behind this section is to establish that reasonable safety precautions, in accordance with Chapter 13, be provided during all phases of construction and demolition. Chapter 13 also covers the protection of adjacent public and private properties.

101.7 Appendices. The code official is authorized to require rehabilitation and retrofit of buildings, structures, or individual structural members in accordance with the appendices of this code if such appendices have been individually adopted. When any of such appendices is specifically referenced in the text of this code, it becomes a part of this code without any special adoption by the local jurisdiction.

❖ This section describes one of the more unique aspects of the code in that any appendix referenced in this code becomes a part of the code without the jurisdiction having to specifically adopt the appendices. For example, in Chapter 7, voluntary alterations to lateral-force-resisting systems are allowed when conducted in accordance with Appendix A. Therefore, Appendix A, having been specifically referenced, is enforceable without having to be specifically adopted by the local jurisdiction. Any appendices not specifically referenced in the code must be individually adopted to be legally enforced.

101.8 Correction of violations of other codes. Repairs or alterations mandated by any property, housing, or fire safety maintenance code or mandated by any licensing rule or ordinance adopted pursuant to law shall conform only to the requirements of that code, rule, or ordinance and shall not be required to conform to this code unless the code requiring such repair or alteration so provides.

❖ This section is intended to keep the requirements of other codes or ordinances intact and separate from the requirements of the code.

SECTION 102
APPLICABILITY

102.1 General. Where in any specific case different sections of this code specify different materials, methods of construction, or other requirements, the most restrictive shall govern. Where there is a conflict between a general requirement and a specific requirement, the specific requirement shall be applicable.

❖ The most restrictive code requirement is to apply where there may be different requirements in the code for a specific installation. In cases where the code establishes a specific requirement for a certain condition, that requirement is applicable even if it is less restrictive than a general requirement elsewhere in the code.

102.2 Other laws. The provisions of this code shall not be deemed to nullify any provisions of local, state, or federal law.

❖ In some cases, other laws enacted by the jurisdiction or the state or federal government may be applicable to a condition that is also governed by a requirement in the code. In such circumstances, the requirements of the code are in addition to that other law that is still in effect, although the building official may not be responsible for its enforcement.

102.3 Application of references. References to chapter or section numbers or to provisions not specifically identified by number shall be construed to refer to such chapter, section, or provision of this code.

❖ In a situation where the code makes reference to a chapter or section number or to another code provision without specifically identifying its location in the code, then that referenced section, chapter or provision is in this code and not in a referenced code or standard.

102.4 Referenced codes and standards. The codes and standards referenced in this code shall be considered part of the requirements of this code to the prescribed extent of each such reference. Where differences occur between provisions of this code and referenced codes and standards, the provisions of this code shall apply.

❖ A referenced code, standard or portion thereof is an enforceable extension of the code as if the content of the standard were included in the body of the code. For example, Section 407.1.1.1 references ASCE 31 in its entirety for the seismic evaluation and design of an existing building; therefore, all of ASCE 31 is applicable. However, in those cases where the code references only portions of a standard, the use and application of the referenced standard is limited to those portions that are specifically identified. If conflicts occur between this code and a referenced code or standard, the text of this code governs.

102.4.1 Standards and guidelines for structural evaluation. The code official shall allow structural evaluation, condition assessment, and rehabilitation of buildings, structures, or individual structural members based on this code's appendix chapters, referenced standards, guidelines, or other approved standards and procedures.

❖ This section refers to the various tools a code official has at his or her disposal to properly evaluate a structure. The interdependence of codes and standards and the positive synergistic effect they have together cannot be overstated.

102.4.2 Compliance with other codes, standards, and guides. Compliance with the structural provisions of the 2000 *International Building Code,* 2003 *International Building Code,* 1999 *BOCA National Building Code,* 1997 *Standard Building Code* or 1997 *Uniform Building Code* shall be deemed exceeding or equivalent to compliance with the structural provisions of this code.

❖ This section establishes the equivalency of the listed codes with the structural provisions of the code.

102.5 Partial invalidity. In the event that any part or provision of this code is held to be illegal or void, this shall not have the effect of making void or illegal any of the other parts or provisions.

❖ Only invalid sections of the code (as established by the court of jurisdiction) can be set aside. This is essential to safeguard the application of the code text to situations whereby a provision is declared illegal or unconstitutional. This section preserves the legislative action that put the legal provisions in place.

SECTION 103
DEPARTMENT OF BUILDING SAFETY

103.1 Creation of enforcement agency. The Department of Building Safety is hereby created, and the official in charge thereof shall be known as the code official.

❖ This section creates the building department and describes its composition (see Section 109 for a discussion of the inspection duties of the department).

The executive official in charge of the building department is named the "code official" by this section. In actuality, the person who is in charge of the department may hold a different title, such as building commissioner, building inspector or construction official. For the purpose of the code, that person is referred to as the "code official."

103.2 Appointment. The code official shall be appointed by the chief appointing authority of the jurisdiction.

❖ This section establishes the code official as an appointed position of the jurisdiction.

103.3 Deputies. In accordance with the prescribed procedures of this jurisdiction and with the concurrence of the appointing authority, the code official shall have the authority to appoint a deputy code official, the related technical officers, inspectors, plan examiners, and other employees. Such employees shall have powers as delegated by the code official.

❖ This section provides the code official with the authority to appoint other individuals to assist with the administration and enforcement of the code. These individuals have the authority and responsibility as designated by the code official. Such appointments, however, may be exercised only with the authorization of the chief appointing authority.

SECTION 104
DUTIES AND POWERS OF CODE OFFICIAL

104.1 General. The code official is hereby authorized and directed to enforce the provisions of this code. The code official shall have the authority to render interpretations of this code and to adopt policies and procedures in order to clarify the application of its provisions. Such interpretations, policies, and procedures shall be in compliance with the intent and purpose of this code. Such policies and procedures shall not have the effect of waiving requirements specifically provided for in this code.

❖ The duty of the code official is to enforce the code, and he or she is the "authority having jurisdiction" for all matters relating to the code and its enforcement. It is the duty of the code official to interpret the code and to determine compliance. Code compliance will not always be easy to determine and will require judgement and expertise, particularly when enforcing the provisions of Sections 104.10 and 104.11. In exercising this authority, however, the code official cannot set aside or ignore any provision of the code.

104.2 Applications and permits. The code official shall receive applications, review construction documents, and issue permits for the repair, alteration, addition, demolition, change of occupancy, and relocation of buildings; inspect the premises for which such permits have been issued; and enforce compliance with the provisions of this code.

❖ The code enforcement process is normally initiated with an application for a permit. The code official is responsible for processing applications and issuing permits for the modification of buildings in accordance with the code.

104.2.1 Preliminary meeting. When requested by the permit applicant, the code official shall meet with the permit applicant prior to the application for a construction permit to discuss plans for the proposed work or change of occupancy in order to establish the specific applicability of the provisions of this code.

Exception: Repairs and Level 1 alterations.

❖ The preliminary meeting is an important aspect of any repair, alteration, change of occupancy, addition or relocation of any building. At this phase in a project it is considerably less expensive to make changes or corrections. Possible problem issues can be identified and solutions devised ahead of time, resulting in fewer correction notices and rework being required.

104.2.1.1 Building evaluation. The code official is authorized to require an existing building to be investigated and evaluated by a registered design professional based on the circumstances agreed upon at the preliminary meeting to determine the existence of any potential nonconformance with the provisions of this code.

❖ This section authorizes the code official to have an existing structure investigated and evaluated by a design professional based on agreements made at the preliminary meeting. Existing structures may have problems that are not immediately visible. The ability to call in an experienced design professional to aid in the proper evaluation of an existing structure is an invaluable tool.

104.3 Notices and orders. The code official shall issue all necessary notices or orders to ensure compliance with this code.

❖ An important element of code enforcement is the necessary advisement of deficiencies and corrections, which is accomplished through written notices and orders. The code official is required to issue orders to abate illegal or unsafe conditions. Section 115.3 contains additional information for these notices.

104.4 Inspections. The code official shall make all of the required inspections, or the code official shall have the authority to accept reports of inspection by approved agencies or individuals. Reports of such inspections shall be in writing and be certified by a responsible officer of such approved agency or by the responsible individual. The code official is authorized to engage such expert opinion as deemed necessary to report upon unusual technical issues that arise, subject to the approval of the appointing authority.

❖ The code official is required to make inspections as necessary to determine compliance with the code or to accept written reports of inspections by an approved agency. The inspection of the work in progress or accomplished to date is another significant element in determining code compliance. While a department does not have the resources to inspect every aspect of

all work, the required inspections are those that are dictated by administrative rules and procedures based on many parameters, including available inspection resources. In order to expand the available resources for inspection purposes, the code official may approve an agency which, in his or her opinion, is objective and competent, has adequate equipment to perform any required tests and employs experienced personnel educated in conducting, supervising and evaluating tests and/or inspections. When unusual, extraordinary or complex technical issues arise relative to building safety, the code official has the authority to seek the opinion and advice of experts. Since this usually involves the expenditure of funds, the approval of the jurisdiction's chief executive (or similar position) is required. A technical report from an expert requested by the code official can be used to assist in the approval process.

104.5 Identification. The code official shall carry proper identification when inspecting structures or premises in the performance of duties under this code.

❖ This section requires the code official (including by definition all authorized designees) to carry identification in the course of conducting the duties of the position. This removes any question as to the purpose and authority of the inspector.

104.6 Right of entry. Where it is necessary to make an inspection to enforce the provisions of this code, or where the code official has reasonable cause to believe that there exists in a structure or upon a premises a condition which is contrary to or in violation of this code which makes the structure or premises unsafe, dangerous, or hazardous, the code official is authorized to enter the structure or premises at reasonable times to inspect or to perform the duties imposed by this code, provided that if such structure or premises be occupied that credentials be presented to the occupant and entry requested. If such structure or premises be unoccupied, the code official shall first make a reasonable effort to locate the owner or other person having charge or control of the structure or premises and request entry. If entry is refused, the code official shall have recourse to the remedies provided by law to secure entry.

❖ The first part of this section establishes the right of the code official to enter the premises in order to make the permit inspections required by Section 109.3. Permit application forms typically include a statement in the certification signed by the applicant (who is the owner or owner's agent) granting the code official the authority to enter areas covered by the permit in order to enforce code provisions related to the permit. The right to enter other structures or premises is more limited. First, to protect the right of privacy, the owner or occupant must grant the code official permission before an interior inspection of the property can be conducted. Permission is not required for inspections that can be accomplished from within the public right-of-way. Second, such access may be denied by the owner or occupant. Unless the inspector has reasonable cause to believe that a violation of the code exists, access may be unattainable. Third, code officials must present proper identification (see Section 104.5) and request admittance during reasonable hours—usually the normal business hours of the establishment—to be admitted. Fourth, inspections must be aimed at securing or determining compliance with the provisions and intent of the regulations that are specifically within the established scope of the code official's authority.

Searches to gather information for the purpose of enforcing the other codes, ordinances or regulations are considered unreasonable and are prohibited by the Fourth Amendment to the U.S. Constitution. "Reasonable cause" in the context of this section must be distinguished from "probable cause," which is required to gain access to property in criminal cases. The burden of proof establishing reasonable cause may vary among jurisdictions. Usually, an inspector must show that the property is subject to inspection under the provisions of the code; that the interests of the public health, safety and welfare outweigh the individual's right to maintain privacy; and that such an inspection is required solely to determine compliance with the provisions of the code.

Many jurisdictions do not recognize the concept of an administrative warrant and may require the code official to prove probable cause in order to gain access upon refusal. This burden of proof is usually more substantial, often requiring the code official to stipulate in advance why access is needed (usually access is restricted to gathering evidence for seeking an indictment or making an arrest); what specific items or information is sought; its relevance to the case against the individual subject; how knowledge of the relevance of the information or items sought was obtained; and how the evidence sought will be used. In all such cases, the right to privacy must always be weighed against the right of the code official to conduct an inspection to verify that public health, safety and welfare are not in jeopardy. Such important and complex constitutional issues should be discussed with the jurisdiction's legal counsel. Jurisdictions should establish procedures for securing the necessary court orders when an inspection is deemed necessary following a refusal.

104.7 Department records. The code official shall keep official records of applications received, permits and certificates issued, fees collected, reports of inspections, and notices and orders issued. Such records shall be retained in the official records for the period required for retention of public records.

❖ In keeping with the need for an efficiently conducted business practice, the code official must keep official records pertaining to permit applications, permits, fees collected, inspections, notices and orders issued. Such documentation provides a valuable source of information if questions arise regarding the department's actions with respect to a building. The code does not require that construction documents be kept after the project is complete; however, it does require that other documents be kept for the length of time

mandated by laws or administrative rules of a state or jurisdiction for retaining public records.

104.8 Liability. The code official, member of the Board of Appeals, or employee charged with the enforcement of this code, while acting for the jurisdiction in good faith and without malice in the discharge of the duties required by this code or other pertinent law or ordinance, shall not thereby be rendered liable personally and is hereby relieved from personal liability for any damage accruing to persons or property as a result of any act or by reason of an act or omission in the discharge of official duties. Any suit instituted against an officer or employee because of an act performed by that officer or employee in the lawful discharge of duties and under the provisions of this code shall be defended by legal representative of the jurisdiction until the final termination of the proceedings. The code official or any subordinate shall not be liable for cost in any action, suit, or proceeding that is instituted in pursuance of the provisions of this code.

❖ The code official, other department employees and members of the appeals board are not intended to be held liable for actions performed in accordance with the code in a reasonable and lawful manner. The responsibility of the code official in this regard is subject to local, state and federal laws that may supersede this provision. This section further establishes that code officials (or subordinates) must not be liable for costs in any legal action instituted in response to the performance of lawful duties. These costs are to be borne by the state, county or municipality. The best way to be certain that the code official's action is a "lawful duty" is always to cite the applicable code section on which the enforcement action is based.

104.9 Approved materials and equipment. Materials, equipment, and devices approved by the code official shall be constructed and installed in accordance with such approval.

❖ The code is a compilation of criteria with which materials, equipment, devices and systems must comply to be suitable for a particular application. The code official has a duty to evaluate such materials, equipment, devices and systems for code compliance and, when compliance is determined, approve the same for use. The materials, equipment, devices and systems must be constructed and installed in compliance with, and all conditions and limitations considered as a basis for, that approval. For example, the manufacturer's instructions and recommendations are to be followed if the approval of the material was based even in part on those instructions and recommendations. The approval authority given the code official is a significant responsibility and is a key to code compliance. The approval process is first technical and then administrative and must be approached as such. For example, if data to determine code compliance are required, such data should be in the form of test reports or engineering analysis and not simply taken from a sales brochure.

104.9.1 Used materials and equipment. The use of used materials which meet the requirements of this code for new materials is permitted. Used equipment and devices shall not be reused unless approved by the code official.

❖ The code criteria for materials and equipment have changed over the years. Evaluation of testing and materials technology has permitted the development of new criteria that the old materials may not satisfy. As a result, used materials are required to be evaluated in the same manner as new materials. Used materials, equipment and devices must be equivalent to that required by the code if they are to be used again in a new installation.

104.10 Modifications. Wherever there are practical difficulties involved in carrying out the provisions of this code, the code official shall have the authority to grant modifications for individual cases upon application of the owner or owner's representative, provided the code official shall first find that special individual reason makes the strict letter of this code impractical and the modification is in compliance with the intent and purpose of this code, and that such modification does not lessen health, accessibility, life and fire safety, or structural requirements. The details of action granting modifications shall be recorded and entered in the files of the Department of Building Safety.

❖ The code official may amend or make exceptions to the code as needed where strict compliance is impractical. Only the code official has authority to grant modifications. Consideration of a particular difficulty is to be based on the application of the owner and a demonstration that the intent of the code is accomplished. This section is not intended to permit setting aside or ignoring a code provision; rather, it is intended to provide for the acceptance of equivalent protection. Such modifications do not, however, extend to actions that are necessary to correct violations of the code. In other words, a code violation or the expense of correcting one cannot constitute a practical difficulty.

104.11 Alternative materials, design and methods of construction, and equipment. The provisions of this code are not intended to prevent the installation of any material or to prohibit any design or method of construction not specifically prescribed by this code, provided that any such alternative has been approved. An alternative material, design, or method of construction shall be approved where the code official finds that the proposed design is satisfactory and complies with the intent of the provisions of this code, and that the material, method, or work offered is, for the purpose intended, at least the equivalent of that prescribed in this code in quality, strength, effectiveness, fire resistance, durability, and safety.

❖ The code is not intended to inhibit innovative ideas or technological advances. A comprehensive regulatory document, such as a building code, cannot envision and then address all future innovations in the industry. As a result, a performance code must be applicable to and provide a basis for the approval of an increasing number of newly developed, innovative materials, systems and methods for which no code text or referenced standards yet exist. The fact that a material, product or

method of construction is not addressed in the code is not an indication that such material, product or method is intended to be prohibited. The code official is expected to apply sound technical judgement in accepting materials, systems or methods that, while not anticipated by the drafters of the current code text, can be demonstrated to offer equivalent performance. By virtue of its text, the code regulates new and innovative construction practices while addressing the relative safety of building occupants. The code official is responsible for determining if a requested alternative provides the equivalent level of protection of public health, safety and welfare as required by the code.

104.11.1 Tests. Whenever there is insufficient evidence of compliance with the provisions of this code or evidence that a material or method does not conform to the requirements of this code, or in order to substantiate claims for alternative materials or methods, the code official shall have the authority to require tests as evidence of compliance to be made at no expense to the jurisdiction. Test methods shall be as specified in this code or by other recognized test standards. In the absence of recognized and accepted test methods, the code official shall approve the testing procedures. Tests shall be performed by an approved agency. Reports of such tests shall be retained by the code official for the period required for retention.

❖ To provide the basis on which the code official can make a decision regarding an alternative material or method, he or she must be provided with sufficient technical data, test reports and documentation for evaluation. If evidence satisfactory to the code official indicates that the alternative material or construction method is equivalent to that required by the code, he or she may approve it. Any such approval cannot have the effect of waiving any requirements of the code. The burden of proof of equivalence lies with the applicant who proposes the use of alternative materials or methods. The code official must require the submission of any appropriate information and data to assist in the determination of equivalency. This information should be submitted before a permit can be issued. The type of information required includes test data in accordance with referenced standards, evidence of compliance with the referenced standard specifications and design calculations. A research report issued by an authoritative agency is particularly useful in providing the code official with the technical basis for evaluation and approval of new and innovative materials and methods of construction. The use of authoritative research reports can greatly assist the code official by reducing the time-consuming engineering analysis necessary to review these materials and methods. Failure to substantiate adequately a request for the use of an alternative is a valid reason for the code official to deny a request. Any tests submitted in support of an application must have been performed by an agency approved by the code official based on evidence that the agency has the technical expertise, test equipment and quality assurance to properly conduct and report the necessary testing. The test reports submitted to the code official must be retained in accordance with the requirements of Section 104.7.

SECTION 105
PERMITS

105.1 Required. Any owner or authorized agent who intends to repair, add to, alter, relocate, demolish, or change the occupancy of a building or to repair, install, add, alter, remove, convert, or replace any electrical, gas, mechanical, or plumbing system, the installation of which is regulated by this code, or to cause any such work to be done, shall first make application to the code official and obtain the required permit.

❖ This section contains the administrative rules governing the issuance, suspension, revocation or modification of building permits. It also establishes how and by whom the application for a building permit is to be made, how it is to be processed, whether fees are involved and what information it must contain or have attached to it.

In general, a permit is required for all activities that are regulated by the code or its referenced codes (see Section 101.4), and these activities cannot begin until the permit is issued, unless the activity is specifically exempted by Section 105.2. Only the owner or a person authorized by the owner can apply for the permit. Note that this section indicates a need for a permit for a change in occupancy, even if no work is contemplated. Although the occupancy of a building or portion thereof may change and the new activity is still classified in the same group, different code provisions may be applicable. The means of egress, structural loads and light and ventilation provisions are examples of requirements that are occupancy sensitive. The purpose of the permit is to cause the work to be reviewed, approved and inspected to determine compliance with the code.

105.1.1 Annual permit. In lieu of an individual permit for each alteration to an already approved electrical, gas, mechanical, or plumbing installation, the code official is authorized to issue an annual permit upon application therefor to any person, firm, or corporation regularly employing one or more qualified trade persons in the building, structure, or on the premises owned or operated by the applicant for the permit.

❖ In some instances, such as large buildings or industrial facilities, the repair, replacement or alteration of electrical, gas, mechanical or plumbing systems occurs on a frequent basis and this section allows the code official to issue an annual permit for this work. This relieves both the building department and the owners of such facilities from the burden of filing and processing individual applications for this activity; however, there are restrictions on who is entitled to these permits. They can be issued only for work on a previously approved installation and only to an individual or corporation that employs persons specifically qualified in the trade for which the permit is issued. If tradespeople who perform the work involved are required to be li-

censed in the jurisdiction, then only those persons would be permitted to perform the work. If trade licensing is not required, then the code official needs to review and approve the qualifications of the persons who will be performing the work. The annual permit can apply only to the individual property that is owned or operated by the applicant.

105.1.2 Annual permit records. The person to whom an annual permit is issued shall keep a detailed record of alterations made under such annual permit. The code official shall have access to such records at all times, or such records shall be filed with the code official as designated.

❖ The work performed in accordance with an annual permit must be inspected by the code official, so it is necessary to know the location of such work and when it was performed. This can be accomplished by having records of the work available to the code official either at the premises or in his or her office, as determined by the official.

105.2 Work exempt from permit. Exemptions from permit requirements of this code shall not be deemed to grant authorization for any work to be done in any manner in violation of the provisions of this code or any other laws or ordinances of this jurisdiction. Permits shall not be required for the following:

Building:

1. Sidewalks and driveways not more than 30 inches (762 mm) above grade and not over any basement or story below and that are not part of an accessible route.

2. Painting, papering, tiling, carpeting, cabinets, counter tops, and similar finish work.

3. Temporary motion picture, television, and theater stage sets and scenery.

4. Shade cloth structures constructed for nursery or agricultural purposes, and not including service systems.

5. Window awnings supported by an exterior wall of Group R-3 or Group U occupancies.

6. Movable cases, counters, and partitions not over 69 inches (1753 mm) in height.

Electrical:

Repairs and maintenance: Minor repair work, including the replacement of lamps or the connection of approved portable electrical equipment to approved permanently installed receptacles.

Radio and television transmitting stations: The provisions of this code shall not apply to electrical equipment used for radio and television transmissions, but do apply to equipment and wiring for power supply, the installations of towers, and antennas.

Temporary testing systems: A permit shall not be required for the installation of any temporary system required for the testing or servicing of electrical equipment or apparatus.

Gas:

1. Portable heating appliance.

2. Replacement of any minor part that does not alter approval of equipment or make such equipment unsafe.

Mechanical:

1. Portable heating appliance.

2. Portable ventilation equipment.

3. Portable cooling unit.

4. Steam, hot, or chilled water piping within any heating or cooling equipment regulated by this code.

5. Replacement of any part that does not alter its approval or make it unsafe.

6. Portable evaporative cooler.

7. Self-contained refrigeration system containing 10 pounds (4.54 kg) or less of refrigerant and actuated by motors of 1 horsepower (746 W) or less.

Plumbing:

1. The stopping of leaks in drains, water, soil, waste, or vent pipe; provided, however, that if any concealed trap, drainpipe, water, soil, waste, or vent pipe becomes defective and it becomes necessary to remove and replace the same with new material, such work shall be considered as new work, and a permit shall be obtained and inspection made as provided in this code.

2. The clearing of stoppages or the repairing of leaks in pipes, valves, or fixtures, and the removal and reinstallation of water closets, provided such repairs do not involve or require the replacement or rearrangement of valves, pipes, or fixtures.

❖ Section 105.1 requires a permit for any activity involving work on a building and its systems and other structures. This section lists those activities that are permitted to take place without first obtaining a permit from the building department. Note that in some cases, such as Items 9, 10, 11 and 12, the work is exempt only for certain occupancies. Further, it is the intent of the code that even though work may be exempted for permit purposes, it must still comply with the code and the owner is responsible for proper and safe construction for all work being done. Work exempted by the codes adopted by reference in Section 101.4 is also included here.

105.2.1 Emergency repairs. Where equipment replacements and repairs must be performed in an emergency situation, the permit application shall be submitted within the next working business day to the code official.

❖ This section recognizes that in some cases emergency replacement and repair work must be done as quickly as possible, and it is not practical to take the time necessary to apply for and obtain approval. A permit for the work must be obtained the next day that the building department is open for business. Any work performed before the permit is issued must be done in accordance with the code and corrected if not approved by the code official.

105.2.2 Repairs. Application or notice to the code official is not required for ordinary repairs to structures and items listed in Section 105.2. Such repairs shall not include the cutting away of any wall, partition, or portion thereof, the removal or cutting of any structural beam or load-bearing support, or the removal or change of any required means of egress or rearrangement of parts of a structure affecting the egress requirements; nor shall ordinary repairs include addition to, alteration of, replacement, or relocation of any standpipe, water supply, sewer, drainage, drain leader, gas, soil, waste, vent, or similar piping, electric wiring, or mechanical or other work affecting public health or general safety.

❖ This section distinguishes between what might be termed by some as repairs, but are in fact alterations. The intent of the section is not to require permits for ordinary repair work.

105.2.3 Public service agencies. A permit shall not be required for the installation, alteration, or repair of generation, transmission, distribution, or metering or other related equipment that is under the ownership and control of public service agencies by established right.

❖ Utilities that supply electricity, gas, water, telephone, television cable, etc., are not required to obtain permits for work involving the transmission lines and metering equipment that they own and control, that is, to their point of delivery. They are typically regulated by other laws that give them specific rights and authority in this area. Any equipment or appliances installed or serviced by such agencies that are not owned by them and under their full control are not exempt from a permit.

105.3 Application for permit. To obtain a permit, the applicant shall first file an application therefor in writing on a form furnished by the Department of Building Safety for that purpose. Such application shall:

1. Identify and describe the work in accordance with Chapter 3 to be covered by the permit for which application is made.

2. Describe the land on which the proposed work is to be done by legal description, street address, or similar description that will readily identify and definitely locate the proposed building or work.

3. Indicate the use and occupancy for which the proposed work is intended.

4. Be accompanied by construction documents and other information as required in Section 106.3.

5. State the valuation of the proposed work.

6. Be signed by the applicant or the applicant's authorized agent.

7. Give such other data and information as required by the code official.

❖ This section requires that a written permit application be filed on forms provided by the building department and details the information required on the application. Permit forms will typically have sufficient space to write a very brief description of the work to be accomplished, which is sufficient for only small jobs. For larger projects, the description will be augmented by construction documents as indicated in Item 4. As required by Section 105.1, the applicant must be the owner of the property or an authorized agent of the owner, such as an engineer, architect, contractor, tenant or other. The applicant must sign the application and permit forms, which typically include a statement that if the applicant is not the owner, he or she has permission from the owner to make the application.

105.3.1 Action on application. The code official shall examine or cause to be examined applications for permits and amendments thereto within a reasonable time after filing. If the application or the construction documents do not conform to the requirements of pertinent laws, the code official shall reject such application in writing, stating the reasons therefor. If the code official is satisfied that the proposed work conforms to the requirements of this code and laws and ordinances applicable thereto, the code official shall issue a permit therefor as soon as practicable.

❖ This section requires the code official to act with reasonable speed on a permit application. In some instances this time period is set by state or local law. The code official must refuse to issue a permit when the application and accompanying documents do not conform to the code. In order to ensure effective communication and due process of law, the reasons for denial of a permit application are required to be in writing. Once the code official determines that the work described conforms with the code and other applicable laws, the permit must be issued upon payment of the fees required by Section 108.

105.3.2 Time limitation of application. An application for a permit for any proposed work shall be deemed to have been abandoned 180 days after the date of filing, unless such application has been pursued in good faith or a permit has been issued; except that the code official is authorized to grant one or more extensions of time for additional periods not exceeding 90 days each. The extension shall be requested in writing and justifiable cause demonstrated.

❖ Typically, a permit application is submitted and goes through a review process that ends with the issuance of a permit. If a permit has not been issued 180 days after the date of filing, however, the application is considered abandoned, unless the applicant was diligent in efforts to obtain the permit. The code official has the authority to extend this time limitation (in increments of 90 days), provided there is reasonable cause. This would cover delays beyond the applicant's control, such as prerequisite permits or approvals from other authorities within the jurisdiction or state. The intent of this section is to limit the time between the review process and the issuance of a permit.

105.4 Validity of permit. The issuance or granting of a permit shall not be construed to be a permit for, or an approval of, any violation of any of the provisions of this code or of any other or-

dinance of the jurisdiction. Permits presuming to give authority to violate or cancel the provisions of this code or other ordinances of the jurisdiction shall not be valid. The issuance of a permit based on construction documents and other data shall not prevent the code official from requiring the correction of errors in the construction documents and other data. The code official is also authorized to prevent occupancy or use of a structure where in violation of this code or of any other ordinances of this jurisdiction.

❖ This section states the fundamental premise that the permit is only a license to proceed with the work. It is not a license to violate, cancel or set aside any provisions of this code. This is significant because it means that despite any errors or oversights in the approval process, the permit applicant, not the code official, is responsible for code compliance. Also, the permit can be suspended or revoked in accordance with Section 105.6.

105.5 Expiration. Every permit issued shall become invalid unless the work on the site authorized by such permit is commenced within 180 days after its issuance, or if the work authorized on the site by such permit is suspended or abandoned for a period of 180 days after the time the work is commenced. The code official is authorized to grant, in writing, one or more extensions of time for periods not more than 180 days each. The extension shall be requested in writing and justifiable cause demonstrated.

❖ The permit becomes invalid under two distinct situations—both based on a 180-day period. The first situation is when no work has started 180 days from issuance of a permit. The second situation is when the authorized work has stopped for 180 days. The person who was issued the permit should be notified, in writing, that the permit is invalid as well as what steps must be taken to reinstate it and restart the work. The code official has the authority to extend this time limitation (in increments of 180 days), provided the extension is requested in writing and there is reasonable cause, which typically includes events beyond the permit holder's control.

105.6 Suspension or revocation. The code official is authorized to suspend or revoke a permit issued under the provisions of this code wherever the permit is issued in error or on the basis of incorrect, inaccurate, or incomplete information or in violation of any ordinance or regulation or any of the provisions of this code.

❖ A permit is in reality a license to proceed with the work. The code official, however, can suspend or revoke permits shown to be based, all or in part, on any false statement or misrepresentation of fact. A permit can also be suspended or revoked if it was issued in error, such as an omitted prerequisite approval or code violation indicated on the construction documents. An applicant may subsequently apply for a reinstatement of the permit with the appropriate corrections or modifications made to the application and construction documents.

105.7 Placement of permit. The building permit or copy shall be kept on the site of the work until the completion of the project.

❖ The building permit or copy shall be kept on the site of the work until the completion of the project and made available to the code official or representative to conveniently make required entries thereon.

SECTION 106
CONSTRUCTION DOCUMENTS

106.1 Submittal documents. Construction documents special inspection and structural observation programs, investigation and evaluation reports, and other data shall be submitted in one or more sets with each application for a permit. The construction documents shall be prepared by a registered design professional where required by the statutes of the jurisdiction in which the project is to be constructed. Where special conditions exist, the code official is authorized to require additional construction documents to be prepared by a registered design professional.

> **Exception:** The code official is authorized to waive the submission of construction documents and other data not required to be prepared by a registered design professional if it is found that the nature of the work applied for is such that reviewing of construction documents is not necessary to obtain compliance with this code.

❖ This section establishes the requirement to provide the code official with construction drawings, specifications and other documents that describe the structure or system for which a permit is sought. It describes the information that must be included in the documents, who must prepare the documents and procedures for approving these documents.

A detailed description of the work for which the permit application is made must be submitted. When the work can be briefly described on the application form and the services of a registered design professional are not required, the code official may utilize judgement in determining the need for detailed documents. An example of work that may not involve the submission of detailed construction documents is the replacement of an existing 60-amp electrical service with a 200-amp service. These provisions are intended to reflect the minimum scope of information needed to determine code compliance. Although this section specifies that "one or more" sets of construction documents be submitted, note that Section 106.3.1 requires one set of approved documents to be retained by the code official and one set to be returned to the applicant. The code official should establish a consistent policy of the number of sets required by the jurisdiction and make this information readily available to applicants.

This section also requires the code official to determine that any state professional registration laws be complied with as they apply to the preparation of construction documents.

106.1.1 Information on construction documents. Construction documents shall be dimensioned and drawn upon suitable material. Electronic media documents are permitted to be submitted when approved by the code official. Construction documents shall be of sufficient clarity to indicate the location, nature, and extent of the work proposed and show in detail that it will conform to the provisions of this code and relevant laws, ordinances, rules, and regulations, as determined by the code official. The work areas shall be shown.

❖ The construction documents are required to be of a quality and detail such that the code official can determine that the work conforms to the code and other applicable laws and regulations. General statements on the documents such as "all work must comply with the *International Existing Building Code*" are not an acceptable substitute for showing the required information. The following subsections specify the detailed information that must be shown on the submitted documents. When specifically allowed by the code official, documents can be submitted in electronic form.

106.1.1.1 Fire protection system shop drawings. Shop drawings for the fire protection system(s) shall be submitted to indicate conformance with this code and the construction documents and shall be approved prior to the start of system installation. Shop drawings shall contain all information as required by the referenced installation standards in Chapter 9 of the *International Building Code.*

❖ Since the fire protection contractor(s) may not have been selected at the time a permit is issued for construction of a building, detailed shop drawings for fire protection systems are not available. Because they provide the information necessary to determine code compliance, as specified in the appropriate referenced standard in Chapter 9 of the IBC, they must be submitted and approved by the code official before the contractor can begin installing the system. For example, the professional responsible for the design of an automatic sprinkler system should determine that the water supply is adequate, but will not be able to prepare a final set of hydraulic calculations if the specific materials and pipe sizes, lengths and arrangements have not been identified. Once the installing contractor is selected, specific hydraulic calculations can be prepared. Factors such as classification of the hazard, amount of water supply available and the density or concentration to be achieved by the system are to be included with the submission of the shop drawings. Specific data sheets identifying sprinklers, pipe dimensions, power requirements for smoke detectors, etc., should also be included with the submission.

106.1.2 Means of egress. The construction documents for alterations Level 2, alterations Level 3, additions, and changes of occupancy shall show in sufficient detail the location, construction, size, and character of all portions of the means of egress in compliance with the provisions of this code. The construction documents shall designate the number of occupants to be accommodated in every work area of every floor and in all affected rooms and spaces.

❖ The complete means of egress system is required to be indicated on the plans such that the code official is able to initiate a review and identify pertinent code requirements for each component. Additionally, requiring such information to be reflected in the construction documents requires the designer not only to become familiar with the code, but also to be aware of egress principles, concepts and purposes. The need to ensure that the means of egress leads to a public way is also a consideration during the plan review. Such an evaluation cannot be made without the inclusion of a site plan, as required by Section 106.2.

Information essential for determining the required capacity of the egress components (see Section 1005 of the IBC) and the number of egress components required from a space (see Sections 1014.1 and 1018.1 of the IBC) must be provided. The designer must be aware of the occupancy of a space and properly identify that, along with its resultant occupant load, on the construction documents.

106.1.3 Exterior wall envelope. Construction documents for all work affecting the exterior wall envelope shall describe the exterior wall envelope in sufficient detail to determine compliance with this code. The construction documents shall provide details of the exterior wall envelope as required, including windows, doors, flashing, intersections with dissimilar materials, corners, end details, control joints, intersections at roof, eaves, or parapets, means of drainage, water-resistive membrane, and details around openings.

The construction documents shall include manufacturer's installation instructions that provide supporting documentation that the proposed penetration and opening details described in the construction documents maintain the wind and weather resistance of the exterior wall envelope. The supporting documentation shall fully describe the exterior wall system which was tested, where applicable, as well as the test procedure used.

❖ This section specifically identifies details of exterior wall construction that are critical to the weather resistance of the wall, and requires those details to be provided on the construction documents. Where the weather resistance of the exterior wall assembly is based on tests, the submitted documentation is to describe the details of the wall envelope and the test procedure that was used. This provides the code official with the information necessary to determine code compliance.

106.2 Site plan. The construction documents submitted with the application for permit shall be accompanied by a site plan showing to scale the size and location of new construction and existing structures on the site, distances from lot lines, the established street grades, and the proposed finished grades; and it shall be drawn in accordance with an accurate boundary line survey. In the case of demolition, the site plan shall show construction to be demolished and the location and size of existing structures and construction that are to remain on the site or plot. The code official is authorized to waive or modify the requirement for a site plan when the application for permit is for alteration, repair, or change of occupancy.

❖ Certain code requirements are dependent on the structure's location on the lot and the topography of the site. As a result, a scaled site plan containing the data listed in this section is required to permit review for compliance. The code official can waive the requirement for a site plan when it is not required to determine code compliance, such as work involving only interior alterations or repairs.

106.3 Examination of documents. The code official shall examine or cause to be examined the construction documents and shall ascertain by such examinations whether the construction or occupancy indicated and described is in accordance with the requirements of this code and other pertinent laws or ordinances.

❖ The requirements of this section are related to those found in Section 105.3.1 regarding the action of the code official in response to a permit application. The code official can delegate review of the construction documents to subordinates as provided for in Section 103.3.

106.3.1 Approval of construction documents. When the code official issues a permit, the construction documents shall be approved in writing or by stamp as "Reviewed for Code Compliance." One set of construction documents so reviewed shall be retained by the code official. The other set shall be returned to the applicant, shall be kept at the site of work, and shall be open to inspection by the code official or a duly authorized representative.

❖ The code official must stamp or otherwise endorse as "Reviewed for Code Compliance" the construction documents on which the permit is based. One set of approved construction documents must be kept on the construction site to serve as the basis for all subsequent inspections. To avoid confusion, the construction documents on the site must be the documents that were approved and stamped. This is because inspections are to be performed with regard to the approved documents, not the code itself. Additionally, the contractor cannot determine compliance with the approved construction documents unless they are readily available. Unless the approved construction documents are available, the inspection should be postponed and work on the project halted.

106.3.2 Previous approvals. This code shall not require changes in the construction documents, construction, or designated occupancy of a structure for which a lawful permit has been heretofore issued or otherwise lawfully authorized, and the construction of which has been pursued in good faith within 180 days after the effective date of this code and has not been abandoned.

❖ If a permit is issued, and construction proceeds at a normal pace and a new edition of the code is adopted by the legislative body, then requiring the building to be constructed to conform to the new code is unreasonable. This section provides for the continuity of permits issued under previous codes, as long as such permits are being "actively prosecuted" subsequent to the effective date of the ordinance adopting the new edition of the code.

106.3.3 Phased approval. The code official is authorized to issue a permit for the construction of foundations or any other part of a building before the construction documents for the whole building or structure have been submitted, provided that adequate information and detailed statements have been filed complying with pertinent requirements of this code. The holder of such permit for the foundation or other parts of a building shall proceed at the holder's own risk with the building operation and without assurance that a permit for the entire structure will be granted.

❖ The code official has the authority to issue a partial permit to allow for the practice of "fast tracking" a job. Any construction under a partial permit is "at the holder's own risk" and "without assurance that a permit for the entire structure will be granted." The code official is under no obligation to accept work or issue a complete permit in violation of the code, ordinances or statutes simply because a partial permit has been issued. Fast tracking places an unusual administrative and technical burden on the code official. The purpose is to proceed with construction while the design continues for other aspects of the work. Coordinating and correlating the code aspects into the project in phases requires attention to detail and project tracking so that all code issues are addressed. The coordination of these submittals is the responsibility of the registered design professional in responsible charge.

106.3.4 Deferred submittals. For the purposes of this section, deferred submittals are defined as those portions of the design that are not submitted at the time of the application and that are to be submitted to the code official within a specified period.

Deferral of any submittal items shall have the prior approval of the code official. The registered design professional in responsible charge shall list the deferred submittals on the construction documents for review by the code official.

Submittal documents for deferred submittal items shall be submitted to the registered design professional in responsible charge who shall review them and forward them to the code official with a notation indicating that the deferred submittal documents have been reviewed and that they have been found to be in general conformance to the design of the building. The deferred submittal items shall not be installed until their design and submittal documents have been approved by the code official.

❖ Often, especially on larger projects, details of certain building parts are not available at the time of permit issuance because they have not yet been designed; for example, exterior cladding, prefabricated items such as trusses and stairs and the components of fire protection systems (see Section 106.1.1.1). The design professional in responsible charge must identify on construction documents the items to be included in any deferred submittals. Documents required for the approval of deferred items must be reviewed by the design professional in responsible charge for compatibility with the design of the building, forwarded to the code official with a notation that this is the case and approved by the code official before installation of the

items. Sufficient time must be allowed for the approval process. Note that deferred submittals differ from the phased permits described in Section 106.3.3 in that they occur after the permit for the building is issued, and are not for work covered by separate permits.

106.4 Amended construction documents. Work shall be installed in accordance with the reviewed construction documents, and any changes made during construction that are not in compliance with the approved construction documents shall be resubmitted for approval as an amended set of construction documents.

❖ Any amendments to the approved construction documents must be filed before constructing the amended item. In the broadest sense, amendments include all addenda, change orders, revised drawings and marked-up shop drawings. Code officials should maintain a policy that all amendments be submitted for review. Otherwise, a significant amendment may not be submitted because of misinterpretation, resulting in an activity that is not approved and causes a needless delay in obtaining approval of the finished work.

106.5 Retention of construction documents. One set of approved construction documents shall be retained by the code official for a period of not less than the period required for retention of public records.

❖ A set of approved construction documents must be kept by the code official as may be required by state or local laws, but for a period of not less than 180 days after the work is complete. Questions regarding an item shown on the approved documents may arise in the period immediately following completion of the work and documents should be available for review. See Section 104.7 for requirements to retain other records that are generated as a result of the work.

106.6 Design professional in responsible charge. When it is required that documents be prepared by a registered design professional, the code official shall be authorized to require the owner to engage and designate on the building permit application a registered design professional who shall act as the registered design professional in responsible charge. If the circumstances require, the owner shall designate a substitute registered design professional in responsible charge who shall perform the duties required of the original registered design professional in responsible charge. The code official shall be notified in writing by the owner if the registered design professional in responsible charge is changed or is unable to continue to perform the duties. The registered design professional in responsible charge shall be responsible for reviewing and coordinating submittal documents prepared by others, including phased and deferred submittal items, for compatibility with the design of the building. Where structural observation is required, the inspection program shall name the individual or firms who are to perform structural observation and describe the stages of construction at which structural observation is to occur.

❖ At the time of permit application and at various intervals during a project, the code requires detailed technical information to be submitted to the code official. This will vary depending on the complexity of the project, but typically includes the construction documents with supporting information, applications utilizing the phased approval procedure in Section 106.3.3 and reports from engineers, inspectors and testing agencies. Since these documents and reports are prepared by numerous individuals, firms and agencies, it is necessary to have a single person charged with responsibility for coordinating their submittal to the code official. This person is the point of contact for the code official for all information related to the project. Otherwise, the code official could waste time and effort attempting to locate the source of accurate information when trying to resolve an issue, such as a discrepancy in plans submitted by different designers.

The requirement that the owner engage a person to act as the design professional in responsible charge is applicable to projects where the construction documents are required by law to be prepared by a registered design professional (see Section 106.1) and when required by the code official. The person employed by the owner to act as the design professional in responsible charge must be identified on the permit application, but the owner can change the designated person at any time during the course of the review process or work, provided the code official is notified in writing.

SECTION 107
TEMPORARY STRUCTURES AND USES

107.1 General. The code official is authorized to issue a permit for temporary structures and temporary uses. Such permits shall be limited as to time of service but shall not be permitted for more than 180 days. The code official is authorized to grant extensions for demonstrated cause.

❖ In the course of construction or other activities, structures that have a limited service life are often necessary. This section contains the administrative provisions that permit such temporary structures without full compliance with the code requirements for permanently occupied structures.

This section allows the code official to issue permits for temporary structures or uses. The applicant must specify the time period desired for the temporary structure or use, but the approval period cannot exceed 180 days. Structures or uses that are "temporary" but are anticipated to be in existence for more than 180 days are required to conform to code requirements for permanent structures and uses. The code official is authorized to grant time extensions if the applicant can provide a valid reason for the extension. A typical example would be circumstances that have occurred beyond the applicant's control. This provision is not intended to be used to circumvent the 180-day limitation.

107.2 Conformance. Temporary structures and uses shall conform to the structural strength, fire safety, means of egress, ac-

cessibility, light, ventilation, and sanitary requirements of this code as necessary to ensure the public health, safety, and general welfare.

❖ This section prescribes those categories of the code that must be complied with, despite the fact that the structure will be removed or the use discontinued at some time in the future. These criteria are essential for measuring the safety of any structure or use, temporary or permanent. Therefore, the application of these criteria to a temporary structure cannot be waived.

107.3 Temporary power. The code official is authorized to give permission to temporarily supply and use power in part of an electric installation before such installation has been fully completed and the final certificate of completion has been issued. The part covered by the temporary certificate shall comply with the requirements specified for temporary lighting, heat, or power in the ICC *Electrical Code.*

❖ Commonly, the electrical service on most construction sites is installed and energized long before all of the wiring is completed. This procedure allows the power supply to be increased as construction demands; however, temporary permission is not intended to waive the requirements set forth in the ICC *Electrical Code®* (ICC EC™) or its referenced standard, NFPA 70. Construction power from the permanent wiring of the building does not require the installation of temporary ground-fault circuit-interrupter (GFCI) protection or the assured equipment grounding program, since the building wiring installed as required by the code should be as safe for construction use as it would be for use after completion of the building.

107.4 Termination of approval. The code official is authorized to terminate such permit for a temporary structure or use and to order the temporary structure or use to be discontinued.

❖ This section provides the code official with the necessary authority to terminate the permit for a temporary structure or use. The code official can order that a temporary structure be removed or a temporary use be discontinued if conditions of the permit have been violated or the structure or use poses an imminent hazard to the public, in which case the provisions of Section 115 become applicable. This text is important because it allows the code official to act quickly when time is of the essence in order to protect public health, safety and welfare.

SECTION 108
FEES

108.1 Payment of fees. A permit shall not be valid until the fees prescribed by law have been paid. Nor shall an amendment to a permit be released until the additional fee, if any, has been paid.

❖ The code anticipates that jurisdictions will establish their own fee schedules. It is the intent that the fees collected by the department for building permit issuance, plan review and inspection be adequate to cover the costs of the department in these areas.

This section requires that all fees be paid prior to permit issuance or release of an amendment to a permit. Since department operations are intended to be supported by fees paid by the user of department activities, it is important that these fees are received before incurring any expense. This philosophy has resulted in a common practice of paying fees prior to plan review and inspection.

108.2 Schedule of permit fees. On buildings, electrical, gas, mechanical, and plumbing systems or alterations requiring a permit, a fee for each permit shall be paid as required in accordance with the schedule as established by the applicable governing authority.

❖ The jurisdiction inserts its desired fee schedule at this location. The fees are established by law, such as in an ordinance adopting the code (see page v of the code for a sample), a separate ordinance or legally promulgated regulation, as required by state or local law. Fee schedules are often based on a valuation of the work to be performed. This concept is based on the proposition that the valuation of a project is related to the amount of work to be expended in plan review, inspections and administering the permit, plus an excess to cover the department overhead.

108.3 Building permit valuations. The applicant for a permit shall provide an estimated permit value at time of application. Permit valuations shall include total value of work including materials and labor for which the permit is being issued, such as electrical, gas, mechanical, plumbing equipment, and permanent systems. If, in the opinion of the code official, the valuation is underestimated on the application, the permit shall be denied unless the applicant can show detailed estimates to meet the approval of the code official. Final building permit valuation shall be set by the code official.

❖ As indicated in Section 108.2, jurisdictions usually base their fees on the value of the work being performed. This section, therefore, requires the applicant to provide this figure, which is to include the total value of the work, including materials and labor, for which the permit is sought. If the code official believes the value provided by the applicant is underestimated, the permit is to be denied unless the applicant can substantiate the value by providing detailed estimates of the work to the satisfaction of the code official. For the construction of new buildings, the building valuation data referred to in Section 108.2 can be used by the code official as a yardstick against which to compare the applicant's estimate.

108.4 Work commencing before permit issuance. Any person who commences any work on a building, electrical, gas, mechanical, or plumbing system before obtaining the necessary permits shall be subject to an additional fee established by the code official that shall be in addition to the required permit fees.

❖ The code official will incur certain costs (i.e., inspection time and administrative) when investigating and citing a person who has commenced work without having obtained a permit. The code official is therefore entitled to recover these costs by establishing a fee, in addition to that collected when the required permit is issued, imposed on the responsible party. Note that this is not a penalty, as described in Section 113.4, for which the person can also be liable.

108.5 Related fees. The payment of the fee for the construction, alteration, removal, or demolition of work done in connection to or concurrently with the work authorized by a building permit shall not relieve the applicant or holder of the permit from the payment of other fees that are prescribed by law.

❖ The fees for a building permit may be in addition to other fees required by the jurisdiction or for related items such as sewer connections, water service taps, driveways and signs. It cannot be construed that the building permit fee includes these other items.

108.6 Refunds. The code official is authorized to establish a refund policy.

❖ This section allows for a refund of fees, which may be full or partial, typically resulting from the revocation, abandonment or discontinuance of a construction project for which a permit has been issued and fees have been collected. The refund of fees should be related to the cost of enforcement services not provided because of the termination of the project. The code official, when authorizing a fee refund, is authorizing the disbursement of public funds. Therefore, the request for a refund must be in writing and for good cause.

SECTION 109
INSPECTIONS

109.1 General. Construction or work for which a permit is required shall be subject to inspection by the code official, and such construction or work shall remain accessible and exposed for inspection purposes until approved. Approval as a result of an inspection shall not be construed to be an approval of a violation of the provisions of this code or of other ordinances of the jurisdiction. Inspections presuming to give authority to violate or cancel the provisions of this code or of other ordinances of the jurisdiction shall not be valid. It shall be the duty of the permit applicant to cause the work to remain accessible and exposed for inspection purposes. Neither the code official nor the jurisdiction shall be liable for expense entailed in the removal or replacement of any material required to allow inspection.

❖ The inspection function is one of the more important aspects of building department operations. This section authorizes the code official to inspect the work for which a permit has been issued and requires that the work to be inspected remain accessible to the code official until inspected and approved. Any expense incurred in removing or replacing material that conceals an item to be inspected is not the responsibility of the code official or the jurisdiction. As with the issuance of permits (see Section 105.4), approval as a result of an inspection is not a license to violate the code, and an approval in violation of the code does not relieve the applicant from complying with the code and is, therefore, not valid.

109.2 Preliminary inspection. Before issuing a permit, the code official is authorized to examine or cause to be examined buildings and sites for which an application has been filed.

❖ The code official is granted authority to inspect the site before permit issuance. This may be necessary to verify existing conditions that impact on the plan review and permit approval. This section provides the code official with the right-of-entry authority that otherwise does not occur until after the permit is issued (see Section 104.6).

109.3 Required inspections. The code official, upon notification, shall make the inspections set forth in Sections 109.3.1 through 109.3.9.

❖ The code official is required to verify that the work is completed in accordance with the approved construction documents. It is the responsibility of the permit holder to notify the code official when the item is ready for inspection. The inspections that are necessary to provide such verification are listed in the following sections, with the caveat in Section 109.3.7 that inspections in addition to those listed here may be required depending on the work involved.

109.3.1 Footing or foundation inspection. Footing and foundation inspections shall be made after excavations for footings are complete and any required reinforcing steel is in place. For concrete foundations, any required forms shall be in place prior to inspection. Materials for the foundation shall be on the job, except where concrete is ready-mixed in accordance with ASTM C 94, the concrete need not be on the job.

❖ It is necessary for the code official to inspect the soil upon which the footing or foundation is to be placed. This inspection also includes any reinforcing steel, concrete forms and materials to be used in the foundation, except for ready-mixed concrete that is prepared off site.

109.3.2 Concrete slab or under-floor inspection. Concrete slab and under-floor inspections shall be made after in-slab or under-floor reinforcing steel and building service equipment, conduit, piping accessories, and other ancillary equipment items are in place but before any concrete is placed or floor sheathing installed, including the sub floor.

❖ The code official must be able to inspect the soil and any required under-slab drainage, waterproofing or dampproofing material, as well as reinforcing steel, conduit, piping and other service equipment embedded in or installed below a slab prior to placing the concrete. Similarly, items installed below a floor system other than concrete must be inspected before they are concealed by the floor sheathing or subfloor.

109.3.3 Lowest floor elevation. For additions and substantial improvements to existing buildings in flood hazard areas, upon placement of the lowest floor, including basement, and prior to further vertical construction, the elevation documentation required in the *International Building Code* shall be submitted to the code official.

❖ Where a structure is located in a flood hazard area, as established in Section 1612.3 of the IBC, the code official must be provided with certification that either the lowest floor elevation (for structures located in flood hazard areas not subject to high-velocity wave action) or the elevation of the lowest horizontal structural member (for structures located in flood hazard areas subject to high-velocity wave action) is in compliance with Section 1612 of the IBC. This certification must be submitted prior to any construction proceeding above this level.

109.3.4 Frame inspection. Framing inspections shall be made after the roof deck or sheathing, all framing, fire blocking, and bracing are in place and pipes, chimneys, and vents to be concealed are complete and the rough electrical, plumbing, heating wires, pipes, and ducts are approved.

❖ This section requires the code official be able to inspect the framing members, such as studs, joists, rafters and girders, and other items such as vents and chimneys that will be concealed by wall construction. Rough electrical work, plumbing, heating wires, pipes and ducts must have already been approved in accordance with the applicable codes prior to this inspection.

109.3.5 Lath or gypsum board inspection. Lath and gypsum board inspections shall be made after lathing and gypsum board, interior and exterior, is in place but before any plastering is applied or before gypsum board joints and fasteners are taped and finished.

Exception: Gypsum board that is not part of a fire-resistance-rated assembly or a shear assembly.

❖ In order to verify that lath and gypsum board is properly attached to framing members, it is necessary for the code official to be able to inspect before the plaster or joint finish material is applied. This is required only for gypsum board that is part of either a fire-resistant assembly or a shear wall.

109.3.6 Fire-resistant penetrations. Protection of joints and penetrations in fire-resistance-rated assemblies shall not be concealed from view until inspected and approved.

❖ The code official must have an opportunity to inspect joint protection and penetration protection for fire-resistance-rated assemblies before it is concealed from view.

109.3.7 Other inspections. In addition to the inspections specified above, the code official is authorized to make or require other inspections of any construction work to ascertain compliance with the provisions of this code and other laws that are enforced by the Department of Building Safety.

❖ Any item regulated by the code is subject to inspection by the code official to determine compliance with the applicable code provision, and no list can include all items in a given building. This section, therefore, gives the code official the authority to inspect any regulated items.

109.3.8 Special inspections. Special inspections shall be required in accordance with the *International Building Code.*

❖ Special inspections are to be provided by the owner for the types of work required in Section 1704 of the IBC. The code official is to approve special inspectors and verify that the required special inspections have been conducted.

109.3.9 Final inspection. The final inspection shall be made after all work required by the building permit is completed.

❖ Upon completion of the work for which the permit has been issued and before issuance of the certificate of occupancy required by Section 110.3, a final inspection is to be made. All violations of the approved construction documents and permit are to be noted and the holder of the permit is to be notified of the discrepancies.

109.4 Inspection agencies. The code official is authorized to accept reports of approved inspection agencies, provided such agencies satisfy the requirements as to qualifications and reliability.

❖ As an alternative to the code official conducting the inspection, he or she is permitted to accept inspections of and reports by approved inspection agencies. Appropriate criteria on which to base approval of inspection agencies can be found in Section 1703 of the IBC.

109.5 Inspection requests. It shall be the duty of the holder of the building permit or their duly authorized agent to notify the code official when work is ready for inspection. It shall be the duty of the permit holder to provide access to and means for any inspections of such work that are required by this code.

❖ It is the responsibility of the permit holder or other authorized person, such as the contractor performing the work, to arrange for the required inspections when completed work is ready and to allow for sufficient time for the code official to schedule a visit to the site to prevent work from being concealed prior to being inspected. Access to the work to be inspected must be provided, including any special means, such as a ladder.

109.6 Approval required. Work shall not be done beyond the point indicated in each successive inspection without first obtaining the approval of the code official. The code official, upon notification, shall make the requested inspections and shall either indicate the portion of the construction that is satisfactory as completed or shall notify the permit holder or an agent of the permit holder wherein the same fails to comply with this code. Any portions that do not comply shall be corrected and such portion shall not be covered or concealed until authorized by the code official.

❖ This section establishes that work cannot progress beyond the point of a required inspection without the code official's approval. Upon making the inspection, the code official must either approve the completed work or notify the permit holder or other responsible party of that which does not comply with the code. Approvals and notices of noncompliance must be in writing, as required by Section 104.4, to avoid any misunderstanding as to what is required. Any item not approved cannot be concealed until it has been corrected and approved by the building official.

SECTION 110
CERTIFICATE OF OCCUPANCY

110.1 Altered area use and occupancy classification change. No altered area of a building and no relocated building shall be used or occupied, and no change in the existing occupancy classification of a building or portion thereof shall be made until the code official has issued a certificate of occupancy therefor as provided herein. Issuance of a certificate of occupancy shall not be construed as an approval of a violation of the provisions of this code or of other ordinances of the jurisdiction.

❖ This section establishes that a building or structure that has been repaired, altered, experienced a change of occupancy or been relocated cannot be occupied until a certificate of occupancy is issued by the code official, which reflects the conclusion of the work allowed by the building permit. Also, no change in occupancy of an existing building is permitted without first obtaining a certificate of occupancy for the new use.

The tool that the code official uses to control the uses and occupancies of various buildings and structures within the jurisdiction is the certificate of occupancy. It is unlawful to use or occupy a building or structure unless a certificate of occupancy has been issued for that use. Its issuance does not relieve the building owner from the responsibility of correcting any code violation that may exist.

110.2 Certificate issued. After the code official inspects the building and finds no violations of the provisions of this code or other laws that are enforced by the Department of Building Safety, the code official shall issue a certificate of occupancy that shall contain the following:

1. The building permit number.
2. The address of the structure.
3. The name and address of the owner.
4. A description of that portion of the structure for which the certificate is issued.
5. A statement that the described portion of the structure has been inspected for compliance with the requirements of this code for the occupancy and division of occupancy and the use for which the proposed occupancy is classified.
6. The name of the code official.

7. The edition of the code under which the permit was issued.

8. The use and occupancy in accordance with the provisions of the *International Building Code.*

9. The type of construction as defined in the *International Building Code.*

10. The design occupant load and any impact the alteration has on the design occupant load of the area not within the scope of the work.

11. If an automatic sprinkler system is provided, whether the sprinkler system is required.

12. Any special stipulations and conditions of the building permit.

❖ The code official is required to issue a certificate of occupancy after a successful final inspection has been completed and all deficiencies and violations have been resolved. This section lists the information that must be included on the certificate. This information is useful to both the code official and owner because it indicates the criteria under which the structure was evaluated and approved at the time the certificate was issued. This is especially important when later applying Chapter 12 to existing structures.

110.3 Temporary occupancy. The code official is authorized to issue a temporary certificate of occupancy before the completion of the entire work covered by the permit, provided that such portion or portions shall be occupied safely. The code official shall set a time period during which the temporary certificate of occupancy is valid.

❖ The code official is permitted to issue a temporary certificate of occupancy for all or a portion of a building prior to the completion of all work. Such certification is to be issued only when the building or portion in question can be safely occupied prior to full completion. The certification is intended to acknowledge that some building features may not be completed even though the building is safe for occupancy, or that a portion of the building can be safely occupied while work continues in another area. This provision precludes the occupancy of a building or structure that does not contain all of the required fire protection systems and means of egress. Temporary certificates should be issued only when incidental construction remains, such as site and interior work that is not regulated by the code and exterior decoration not necessary to the integrity of the building envelope. The code official should view the issuance of a temporary certificate of occupancy as substantial an act as the issuance of the final certificate. Indeed, the issuance of a temporary certificate of occupancy offers a greater potential for conflict because once the building or structure is occupied, it is very difficult to remove the occupants through legal means. The certificate must specify the time period for which it is valid.

110.4 Revocation. The code official is authorized to, in writing, suspend or revoke a certificate of occupancy or completion

issued under the provisions of this code wherever the certificate is issued in error or on the basis of incorrect information supplied, or where it is determined that the building or structure or portion thereof is in violation of any ordinance or regulation or any of the provisions of this code.

❖ The code official is authorized to, in writing, suspend or revoke a certificate of occupancy or completion issued under the provisions of this code wherever the certificate is issued in error on the basis of incorrect information supplied or where it is determined that the building or structure, or portion thereof, is in violation of any ordinance, regulation or any of the provisions of this code.

This section is needed to give the code official the authority to revoke a certificate of occupancy for the reasons indicated in the code text. The code official may also suspend the certificate of occupancy until all of the code violations are corrected.

SECTION 111
SERVICE UTILITIES

111.1 Connection of service utilities. No person shall make connections from a utility, source of energy, fuel, or power to any building or system that is regulated by this code for which a permit is required, until approved by the code official.

❖ This section establishes the authority of the code official to approve utility connections to a building for items such as water, sewer, electricity, gas and steam, and to require their disconnection when hazardous conditions or emergencies exist.

The approval of the code official is required before a connection can be made from a utility to a building system that is regulated by the applicable code, including those referenced in Section 101.4. This includes utilities supplying water, sewer, electricity, gas and steam services. For the protection of building occupants, including workers, such systems must have had final inspection approvals, except as allowed by Section 111.2 for temporary connections.

111.2 Temporary connection. The code official shall have the authority to authorize the temporary connection of the building or system to the utility source of energy, fuel, or power.

❖ The code official is permitted to issue temporary authorization to make connections to the public utility system prior to the completion of all work. This acknowledges that, because of seasonal limitations, time constraints or the need for testing or partial operation of equipment, some building systems may be safely connected even though the building is not suitable for final occupancy. The temporary connection and utilization of connected equipment should be approved when the requesting permit holder has demonstrated to the code official's satisfaction that public health, safety and welfare will not be endangered.

111.3 Authority to disconnect service utilities. The code official shall have the authority to authorize disconnection of utility service to the building, structure, or system regulated by this code and the codes referenced in case of emergency where necessary to eliminate an immediate hazard to life or property. The code official shall notify the serving utility and, wherever possible, the owner and occupant of the building, structure, or service system of the decision to disconnect prior to taking such action. If not notified prior to disconnecting, the owner or occupant of the building, structure, or service system shall be notified in writing, as soon as practical thereafter.

❖ Disconnection of one or more of a building's utility services is the most radical method of hazard abatement available to the code official and should be reserved for cases in which all other lesser remedies have proven ineffective. Such an action must be preceded by written notice to the utility and the owner and occupants of the building. Disconnection must be accomplished within the time frame established by the code official in the notice. When the hazard to the public health, safety or welfare is so imminent as to mandate immediate disconnection, the code official has the authority and even the obligation to cause disconnection without notice. In such cases, the owner or occupants must be given written notice as soon as practical.

SECTION 112
BOARD OF APPEALS

112.1 General. In order to hear and decide appeals of orders, decisions, or determinations made by the code official relative to the application and interpretation of this code, there shall be and is hereby created a board of appeals. The board of appeals shall be appointed by the governing body and shall hold office at its pleasure. The board shall adopt rules of procedure for conducting its business.

❖ This section provides an aggrieved party with a material interest in the decision of the code official a process to appeal such a decision before a board of appeals. This provides a forum, other than the court of jurisdiction, in which to review the code official's actions.

This section literally allows any person to appeal a decision of the code official. In practice, this section has been interpreted to permit appeals only by those aggrieved parties with a material or definitive interest in the decision of the code official. An aggrieved party may not appeal a code requirement per se. The intent of the appeal process is not to waive or set aside a code requirement; rather it is intended to provide a means of reviewing a code official's decision on an interpretation or application of the code or to review the equivalency of protection to the code requirements. The members of the appeals board are appointed by the governing body of the jurisdiction (typically a council or administrator such as a mayor or city manager) and remain members until removed from office. The board must establish procedures for electing a chairperson, scheduling and conducting meetings and administration.

112.2 Limitations on authority. An application for appeal shall be based on a claim that the true intent of this code or the rules legally adopted thereunder have been incorrectly interpreted, the provisions of this code do not fully apply, or an equally good or better form of construction is proposed. The board shall have no authority to waive requirements of this code.

❖ This section establishes the grounds for an appeal, which claims that the code official has misinterpreted or misapplied a code provision. The board is not allowed to set aside any of the technical requirements of the code; however, it is allowed to consider alternative methods of compliance with the technical requirements (see Section 104.11).

112.3 Qualifications. The board of appeals shall consist of members who are qualified by experience and training to pass on matters pertaining to building construction and are not employees of the jurisdiction.

❖ It is important that the decisions of the appeals board are based purely on the technical merits involved in an appeal. It is not the place for policy or political deliberations. The members of the appeals board are therefore expected to have experience in building construction matters.

SECTION 113
VIOLATIONS

113.1 Unlawful acts. It shall be unlawful for any person, firm, or corporation to repair, alter, extend, add, move, remove, demolish, or change the occupancy of any building or equipment regulated by this code or cause same to be done in conflict with or in violation of any of the provisions of this code.

❖ Violations of the code are prohibited and form the basis for all citations and correction notices.

113.2 Notice of violation. The code official is authorized to serve a notice of violation or order on the person responsible for the repair, alteration, extension, addition, moving, removal, demolition, or change in the occupancy of a building in violation of the provisions of this code or in violation of a permit or certificate issued under the provisions of this code. Such order shall direct the discontinuance of the illegal action or condition and the abatement of the violation.

❖ The code official is required to notify the person responsible for the erection or use of a building found to be in violation of the code. The section that is allegedly being violated must be cited so that the responsible party can respond to the notice.

113.3 Prosecution of violation. If the notice of violation is not complied with promptly, the code official is authorized to request the legal counsel of the jurisdiction to institute the appropriate proceeding at law or in equity to restrain, correct, or abate such violation or to require the removal or termination of the unlawful occupancy of the building or structure in violation of the provisions of this code or of the order or direction made pursuant thereto.

❖ The code official must pursue, through the use of legal counsel of the jurisdiction, legal means to correct the violation. This is not optional.

Any extensions of time so that the violations may be corrected voluntarily must be for a reasonable, bona fide cause or the code official may be subject to criticism for "arbitrary and capricious" actions. In general, it is better to have a standard time limitation for correction of violations. Departures from this standard must be for a clear and reasonable purpose, usually stated in writing by the violator.

113.4 Violation penalties. Any person who violates a provision of this code or fails to comply with any of the requirements thereof or who repairs or alters or changes the occupancy of a building or structure in violation of the approved construction documents or directive of the code official or of a permit or certificate issued under the provisions of this code shall be subject to penalties as prescribed by law.

❖ Penalties for violating provisions of the code are typically contained in state law, particularly if the code is adopted at that level and the building department must follow those procedures. If there is no such procedure already in effect, one must be established with the aid of legal counsel.

SECTION 114
STOP WORK ORDER

114.1 Authority. Whenever the code official finds any work regulated by this code being performed in a manner contrary to the provisions of this code or in a dangerous or unsafe manner, the code official is authorized to issue a stop work order.

❖ This section provides for the suspension of work for which a permit was issued, pending the removal or correction of a severe violation or unsafe condition identified by the code official. Normally, correction notices, issued in accordance with Section 109.6, are used to inform the permit holder of code violations. Stop work orders are issued when there is no other way to accomplish enforcement or when a dangerous condition exists.

114.2 Issuance. The stop work order shall be in writing and shall be given to the owner of the property involved or to the owner's agent, or to the person doing the work. Upon issuance of a stop work order, the cited work shall immediately cease. The stop work order shall state the reason for the order and the conditions under which the cited work will be permitted to resume.

❖ Upon receipt of a violation notice from the code official, all construction activities identified in the notice must immediately cease, except as expressly permitted to correct the violation.

114.3 Unlawful continuance. Any person who shall continue any work after having been served with a stop work order, except such work as that person is directed to perform to remove a violation or unsafe condition, shall be subject to penalties as prescribed by law.

❖ This section states that work in violation must terminate, and that all work, except that which is necessary

to correct the violation or unsafe condition, must cease as well. As determined by the municipality or state, a penalty may be assessed for failure to comply with this section.

SECTION 115
UNSAFE BUILDINGS AND EQUIPMENT

115.1 Conditions. Buildings or existing equipment that are or hereafter become unsafe, insanitary, or deficient because of inadequate means of egress facilities, inadequate light and ventilation, or which constitute a fire hazard, or in which the structure or individual structural members exceed the limits established by the definition of Dangerous in Chapter 2, or that involve illegal or improper occupancy or inadequate maintenance, shall be deemed an unsafe condition. Unsafe buildings shall be taken down and removed or made safe, as the code official deems necessary and as provided for in this code. A vacant structure that is not secured against entry shall be deemed unsafe.

❖ This section describes the responsibility of the code official to investigate reports of unsafe structures and equipment and provides criteria for such determination. Unsafe structures are defined as buildings or structures that are insanitary, deficient in light and ventilation or adequate exit facilities, constitute a fire hazard or are otherwise dangerous to human life. This section establishes that unsafe buildings can result from illegal or improper occupancies. For example, prima facie evidence of an unsafe structure is an unsecured (open at door or window) vacant building. All unsafe buildings must either be demolished or made safe and secure as deemed appropriate by the code official.

115.2 Record. The code official shall cause a report to be filed on an unsafe condition. The report shall state the occupancy of the structure and the nature of the unsafe condition.

❖ The code official must file a report on each investigation of unsafe conditions, stating the occupancy of the structure and the nature of the unsafe condition. This report provides the basis for the notice described in Section 115.3.

115.3 Notice. If an unsafe condition is found, the code official shall serve on the owner, agent, or person in control of the structure a written notice that describes the condition deemed unsafe and specifies the required repairs or improvements to be made to abate the unsafe condition, or that requires the unsafe building to be demolished within a stipulated time. Such notice shall require the person thus notified to declare immediately to the code official acceptance or rejection of the terms of the order.

❖ When a building or structure is deemed unsafe, the code official is required to notify the owner or agent of the building as the first step in correcting the problem. Such notice must describe the necessary repairs and improvements to correct the deficiency or must require the unsafe building or structure to be demolished in a specified time in order to provide for public health, safety and welfare. Additionally, such notice requires the immediate response of the owner or agent. If the owner or agent is not available, public notice of such declaration should suffice for the purposes of complying with this section (see Section 115.4). The code official may also determine that immediate work is necessary to correct an unsafe condition and seek a lien against the building or structure to compensate the municipality for the cost of remedial action.

115.4 Method of service. Such notice shall be deemed properly served if a copy thereof is delivered to the owner personally; sent by certified or registered mail addressed to the owner at the last known address with the return receipt requested; or delivered in any other manner as prescribed by local law. If the certified or registered letter is returned showing that the letter was not delivered, a copy thereof shall be posted in a conspicuous place in or about the structure affected by such notice. Service of such notice in the foregoing manner upon the owner's agent or upon the person responsible for the structure shall constitute service of notice upon the owner.

❖ The notice must be delivered personally to the owner. If the owner or agent cannot be located, additional procedures are established, including posting the unsafe notice on the premises in question. Such action may be considered the equivalent of personal notice; however, it may or may not be deemed by the courts as representing a "good faith" effort to notify. Therefore, in addition to complying with this section, public notice through the use of newspapers and other postings in a prominent location at the government center should be used.

115.5 Restoration. The building or equipment determined to be unsafe by the code official is permitted to be restored to a safe condition. To the extent that repairs, alterations, or additions are made or a change of occupancy occurs during the restoration of the building, such repairs, alterations, additions, or change of occupancy shall comply with the requirements of this code.

❖ This section provides that unsafe structures may be restored to a safe condition. This means the cause of the unsafe structure notice can be abated without the structure being required to comply fully with the provisions for new construction. Any work done to eliminate the unsafe condition, as well as any change in occupancy that may occur, must comply with the code.

SECTION 116
EMERGENCY MEASURES

116.1 Imminent danger. When, in the opinion of the code official, there is imminent danger of failure or collapse of a building that endangers life, or when any building or part of a building has fallen and life is endangered by the occupation of the building, or when there is actual or potential danger to the building occupants or those in the proximity of any structure because of explosives, explosive fumes or vapors, or the presence of toxic fumes, gases, or materials, or operation of defec-

tive or dangerous equipment, the code official is hereby authorized and empowered to order and require the occupants to vacate the premises forthwith. The code official shall cause to be posted at each entrance to such structure a notice reading as follows: "This Structure Is Unsafe and Its Occupancy Has Been Prohibited by the Code Official." It shall be unlawful for any person to enter such structure except for the purpose of securing the structure, making the required repairs, removing the hazardous condition, or of demolishing the same.

❖ If the code official has determined that failure or collapse of a building or structure is imminent, failure has occurred that results in a continued threat to the remaining structure or adjacent properties or any other unsafe condition as described in this section exists in a structure, the code official is authorized to require the occupants to vacate the premises and to post such buildings or structures as unsafe and unoccupiable. Unless authorized by the code official to make repairs, secure or demolish the structure, it is illegal for anyone to enter the building or structure. This will minimize the potential for injury.

116.2 Temporary safeguards. Notwithstanding other provisions of this code, whenever, in the opinion of the code official, there is imminent danger due to an unsafe condition, the code official shall order the necessary work to be done, including the boarding up of openings, to render such structure temporarily safe whether or not the legal procedure herein described has been instituted; and shall cause such other action to be taken as the code official deems necessary to meet such emergency.

❖ This section recognizes the need for immediate and effective action in order to protect the public. This section empowers the code official to cause the necessary work to be done to minimize the imminent danger temporarily without regard for due process. This section has to be viewed critically insofar as the danger of structural failure to which the code official has responded must be "imminent;" that is, readily apparent and immediate.

116.3 Closing streets. When necessary for public safety, the code official shall temporarily close structures and close or order the authority having jurisdiction to close sidewalks, streets, public ways, and places adjacent to unsafe structures, and prohibit the same from being utilized.

❖ The code official is authorized to temporarily close sidewalks, streets and adjacent structures as needed to protect the public from the unsafe building or structure when an imminent danger exists. Since the code official may not have the direct authority to close sidewalks, streets and other public ways, the agency having such jurisdiction (e.g., the police or highway department) must be notified.

116.4 Emergency repairs. For the purposes of this section, the code official shall employ the necessary labor and materials to perform the required work as expeditiously as possible.

❖ The cost of emergency work may have to be paid initially by the jurisdiction. The important principle here is that the code official must act immediately to protect the public when warranted, leaving the details of costs and owner notification for a later date.

116.5 Costs of emergency repairs. Costs incurred in the performance of emergency work shall be paid by the jurisdiction. The legal counsel of the jurisdiction shall institute appropriate action against the owner of the premises where the unsafe structure is or was located for the recovery of such costs.

❖ The cost of emergency repairs is to be paid by the jurisdiction, with subsequent legal action against the owner to recover such costs. This does not preclude, however, reaching an alternative agreement with the owner.

116.6 Hearing. Any person ordered to take emergency measures shall comply with such order forthwith. Any affected person shall thereafter, upon petition directed to the appeals board, be afforded a hearing as described in this code.

❖ Anyone ordered to take an emergency measure or to vacate a structure because of an emergency condition must do so immediately. Thereafter, any affected party has the right to appeal the action to the appeals board to determine whether the order should be continued, modified or revoked. It is imperative that appeals to an emergency order occur after the hazard has been abated, rather than before, to minimize the risk to the occupants, employees, clients and the public.

SECTION 117
DEMOLITION

117.1 General. The code official shall order the owner of any premises upon which is located any structure that in the code official's judgment is so old, dilapidated, or has become so out of repair as to be dangerous, unsafe, insanitary, or otherwise unfit for human habitation or occupancy, and such that it is unreasonable to repair the structure, to demolish and remove such structure; or if such structure is capable of being made safe by repairs, to repair and make safe and sanitary or to demolish and remove at the owner's option; or where there has been a cessation of normal construction of any structure for a period of more than two years, to demolish and remove such structure.

❖ This section describes the conditions where the code official has the authority to order the owner to remove the structure. Conditions where the code official may give the owner the option of repairing the structure are also in this section. The code official should carefully document the condition of the structure prior to issuing a demolition order to provide an adequate basis for ordering the owner to remove the structure.

117.2 Notices and orders. All notices and orders shall comply with Section 113.

❖ Before the code official can pursue action to demolish a building in accordance with Section 117.1 or 117.3, it is imperative that all owners and any other persons with a recorded encumbrance on the property be given proper notice of the demolition plans. See Section 113 for notice and order requirements.

117.3 Failure to comply. If the owner of a premises fails to comply with a demolition order within the time prescribed, the code official shall cause the structure to be demolished and removed, either through an available public agency or by contract or arrangement with private persons, and the cost of such demolition and removal shall be charged against the real estate upon which the structure is located and shall be a lien upon such real estate.

❖ When the owner fails to comply with a demolition order, the code official is authorized to take action to have the building razed and removed. The costs are to be charged as a lien against the real estate. To reduce complaints regarding the validity of demolition costs, the code official should obtain competitive bids from several demolition contractors to raze the structure.

117.4 Salvage materials. When any structure has been ordered demolished and removed, the governing body or other designated officer under said contract or arrangement aforesaid shall have the right to sell the salvage and valuable materials at the highest price obtainable. The net proceeds of such sale, after deducting the expenses of such demolition and removal, shall be promptly remitted with a report of such sale or transaction, including the items of expense and the amounts deducted, for the person who is entitled thereto, subject to any order of a court. If such a surplus does not remain to be turned over, the report shall so state.

❖ The governing body may sell any valuables or salvageable materials for the highest price obtainable. The costs of demolition are then to be deducted from any proceeds from the sale of salvage. If a surplus of funds remains, it is to be remitted to the owner with an itemized expense and income account. If no surplus remains, this must also be reported.

Bibliography

The following resource materials are referenced in this chapter or are relevant to the subject matter addressed in this chapter.

Legal Aspects of Code Administration. Falls Church, VA: International Code Council, 2003.

NFPA 70-02, *National Electrical Code.* Quincy, MA: National Fire Protection Association, 2002.

Rhyne, Charles S. *Survey of the Law and Building Codes.* The American Institute of Architects and the National Association of Home Builders.

CHAPTER 2
DEFINITIONS

General Comments

All terms defined in the code are listed alphabetically in Chapter 2. The user should be familiar with the terms in this chapter because: (1) definitions are essential to the correct interpretation of the code; and (2) the user might not be aware that a particular term encountered in the text has the special definition found herein.

Section 201.1 contains the scope of the chapter. Section 201.2 establishes the interchangeability of the terms in the code. Section 201.3 establishes the use of terms defined in other codes. Section 201.4 establishes the use of undefined terms, and Section 202 lists terms and their definitions according to this code.

Purpose

Codes, by their very nature, are technical documents. As such, literally every word, term and punctuation mark can add to or change the meaning of the intended result. This is even more so with a performance-based code where the desired result often takes on more importance than the specific words. Furthermore, the code, with its broad scope of applicability, includes terms inherent in a variety of construction disciplines. These terms often have multiple meanings depending on the context or discipline being used at the time. For these reasons, it is necessary to maintain a consensus on the specific meaning of terms contained in the code. Chapter 2 performs this function by stating clearly what specific terms mean for the purpose of the code.

SECTION 201
GENERAL

❖ This section contains language and provisions that are supplemental to the use of Chapter 2. It gives guidance to the use of the defined words relevant to tense, gender and plurality. Finally, this section provides direction on how to apply terms that are not defined in the code.

201.1 Scope. Unless otherwise expressly stated, the following words and terms shall, for the purposes of this code, have the meanings shown in this chapter.

❖ The terms and definitions in this chapter relate to the use of the terms in the code, unless stated otherwise.

201.2 Interchangeability. Words used in the present tense include the future; words stated in the masculine gender include the feminine and neuter; the singular number includes the plural and the plural, the singular.

❖ While the definitions contained or referenced in Chapter 2 are to be taken literally, gender and tense are interchangeable; thus, any grammatical inconsistencies with the code text will not hinder the understanding or enforcement of the requirements.

201.3 Terms defined in other codes. Where terms are not defined in this code and are defined in the other *International Codes*, such terms shall have the meanings ascribed to them as in those codes.

❖ When a word or term appears in the code and that word or term is not defined in this chapter, other references may be used to find its definition, such as the ICC *Electrical Code*® (ICC EC™), the *International Building Code*® (IBC®), *International Residential Code*® (IRC®), *International Fire Code*® (IFC®), *International Plumbing Code*® (IPC®), *International Mechanical*

Code® (IMC®), *International Fuel Gas Code*® (IFGC®), ICC *Performance Code*™ (ICC PC™), *International Private Sewage Disposal Code*® (IPSDC®), *International Property Maintenance Code*® (IPMC®), *International Energy Conservation Code*® (IECC®), *International Urban Wildland Interface*™ (IUWIC™) and *International Zoning Code*® (IZC®). These codes contain additional definitions (some parallel and duplicative) that may be used in the enforcement of the code or in the enforcement of the other codes by reference.

201.4 Terms not defined. Where terms are not defined through the methods authorized by this chapter, such terms shall have ordinarily accepted meanings such as the context implies.

❖ Words or terms not defined within the *International Code*® series are intended to be applied based on their "ordinarily accepted meanings." The intent of this statement is that a dictionary definition may suffice, provided it is in context. Sometimes the construction terms used in the code are not specifically defined in the code or even in a dictionary. In such a case, the definitions contained in the referenced standards (see Chapter 14) and published textbooks on the subject in question are good resources.

SECTION 202
GENERAL DEFINITIONS

❖ This section contains definitions of terms that are used throughout the code. It is important to emphasize that these terms are used throughout the code and that these terms are applicable everywhere the term is used in the code. Definitions of terms can help in the understanding and application of the code requirements.

ADDITION. An extension or increase in floor area, number of stories, or height of a building or structure.

❖ This term is used to describe the condition when the floor area or height of an existing building or structure is increased. This term is only applicable to existing buildings, never new ones.

ALTERATION. Any construction or renovation to an existing structure other than repair or addition. Alterations are classified as Level 1, Level 2, and Level 3.

❖ The code utilizes this term to reflect construction operations intended for an existing building but not within the scope of an addition or repair (see the definitions of "Addition" and "Repair").

CHANGE OF OCCUPANCY. A change in the purpose or level of activity within a building that involves a change in application of the requirements of this code.

❖ When a change of occupancy occurs, the code provisions for new construction then apply to an existing structure having a new occupancy. Changing the occupancy classification in an existing structure may change the level of inherent hazards that the code was initially intended to address. For example, a change from mercantile occupancy to a business occupancy renders all Group B provisions applicable to all portions of the structure where the occupancy was changed. Change of occupancy is specifically addressed in Chapter 8 of this code.

DANGEROUS. Any building or structure or any individual member with any of the structural conditions or defects described below shall be deemed dangerous:

1. The stress in a member or portion thereof due to all factored dead and live loads is more than one and one third the nominal strength allowed in the *International Building Code* for new buildings of similar structure, purpose, or location.

2. Any portion, member, or appurtenance thereof likely to fail, or to become detached or dislodged, or to collapse and thereby injure persons.

3. Any portion of a building, or any member, appurtenance, or ornamentation on the exterior thereof is not of sufficient strength or stability, or is not anchored, attached, or fastened in place so as to be capable of resisting a wind pressure of two thirds of that specified in the *International Building Code* for new buildings of similar structure, purpose, or location without exceeding the nominal strength permitted in the *International Building Code* for such buildings.

4. The building, or any portion thereof, is likely to collapse partially or completely because of dilapidation, deterioration or decay; construction in violation of the *International Building Code*; the removal, movement or instability of any portion of the ground necessary for the purpose of supporting such building; the deterioration, decay or inadequacy of its foundation; damage due to

fire, earthquake, wind or flood; or any other similar cause.

5. The exterior walls or other vertical structural members list, lean, or buckle to such an extent that a plumb line passing through the center of gravity does not fall inside the middle one third of the base.

❖ This definition describes what is considered to be dangerous in building construction. There is a list of five specific criteria that will cause a structure to be classified as dangerous. These criteria take into consideration loading, deterioration and the ability of the structure to resist wind loads. This would also include any building element that could become detached and potentially result in the failure of one or more of the building's inherent systems.

EQUIPMENT OR FIXTURE. Any plumbing, heating, electrical, ventilating, air conditioning, refrigerating, and fire protection equipment, and elevators, dumb waiters, escalators, boilers, pressure vessels and other mechanical facilities or installations that are related to building services. Equipment or fixture shall not include manufacturing, production, or process equipment, but shall include connections from building service to process equipment.

❖ This definition outlines the type of building systems and devices that are considered to be categorized as equipment or fixtures. It is important to note that while the list is quite extensive, the definition does not specifically exclude the terms "manufacturing," "production" and "process equipment."

EXISTING BUILDING. A building erected prior to the date of adoption of the appropriate code, or one for which a legal building permit has been issued.

❖ This term is used to identify those structures or buildings that have been previously permitted, constructed and have received the applicable certificate of occupancy. Often erected under the provisions of an earlier edition of the code, the buildings are exempt from compliance with current code provisions, unless otherwise stated, when a hazardous condition is present or when alterations or changes in building height and areas are made.

[B] FLOOD HAZARD AREA. The greater of the following two areas:

1. The area within a flood plain subject to a 1-percent or greater chance of flooding in any year.

2. The area designated as a flood hazard area on a community's flood hazard map, or otherwise legally designated.

❖ The Federal Emergency Management Agency (FEMA) prepares Flood Insurance Rate Maps (FIRMs) that delineate the land area that is subject to inundation by the 1-percent annual chance flood. Some states and local jurisdictions develop and adopt maps of flood hazard areas that are more extensive than the areas shown on FEMA's maps. For the purpose of the code, the flood hazard area within which the requirements are to be applied is the greater of the two delineated areas.

HISTORIC BUILDING. Any building or structure that is listed in the State or National Register of Historic Places; designated as a historic property under local or state designation law or survey; certified as a contributing resource within a National Register listed or locally designated historic district; or with an opinion or certification that the property is eligible to be listed on the National or State Registers of Historic Places either individually or as a contributing building to a historic district by the State Historic Preservation Officer or the Keeper of the National Register of Historic Places.

❖ This definition specifies the criteria for consideration as an historic building. Chapter 10 contains provisions for buildings that qualify as historic.

LOAD-BEARING ELEMENT. Any column, girder, beam, joist, truss, rafter, wall, floor or roof sheathing that supports any vertical load in addition to its own weight or any lateral load.

❖ This term relates to all of the load-bearing elements in a structure. It is important to identify all such load-bearing elements to ensure that they can continue to accomplish their designed function of being able to transfer any vertical load and any lateral load in addition to its own weight effectively to the earth.

REHABILITATION. Any work, as described by the categories of work defined herein, undertaken in an existing building.

❖ This process of returning a property to a state of utility through repair or alteration makes it possible to effect a positive contemporary use while preserving those portions and features of the property that are significant to its historic, architectural and cultural values.

REHABILITATION, SEISMIC. Work conducted to improve the seismic lateral force resistance of an existing building.

❖ This definition relates specifically to the efforts designed to help improve and perhaps reestablish seismic lateral-force resistance of a property to an appropriate level.

REPAIR. The restoration to good or sound condition of any part of an existing building for the purpose of its maintenance.

❖ As indicated in Section 105.2.2, the repair of an item typically does not require a permit. This definition makes it clear that repair is limited to work on the item, and does not include complete or substantial replacement or other new work.

SEISMIC LOADING. The assumed forces prescribed herein, related to the response of the structure to earthquake motions, to be used in the analysis and design of the structure and its components.

❖ This definition refers to the forces to be used in the seismic design of structures and both structural and nonstructural components.

[B] SUBSTANTIAL DAMAGE. For the purpose of determining compliance with the flood provisions of this code, damage of any origin sustained by a structure whereby the cost of restoring the structure to its before-damaged condition would equal or exceed 50 percent of the market value of the structure before the damage occurred.

❖ The term is used in the definition of "Substantial improvement." Substantial damage is a special case of substantial improvement, and if the cost of restoring damage equals or exceeds 50 percent of the market value of the structure, then compliance of the existing building is required. It is notable that a substantial damage determination is to be made regardless of what causes the damage. Buildings have sustained substantial damage due to flood, fire, wind, earthquake, deterioration and other causes.

[B] SUBSTANTIAL IMPROVEMENT. For the purpose of determining compliance with the flood provisions of this code, any repair, alteration, addition, or improvement of a building or structure, the cost of which equals or exceeds 50 percent of the market value of the structure before the improvement or repair is started. If the structure has sustained substantial damage, any repairs are considered substantial improvement regardless of the actual repair work performed. The term does not, however, include either:

1. Any project for improvement of a building required to correct existing health, sanitary, or safety code violations identified by the code official and that is the minimum necessary to assure safe living conditions, or

2. Any alteration of a historic structure, provided that the alteration will not preclude the structure's continued designation as a historic structure.

❖ One of the long-range objectives of the National Flood Insurance Program (NFIP) is to reduce the exposure of older buildings that were built in flood hazard areas before local jurisdictions adopted flood hazard area maps and regulations. Section 105.3 directs the applicant to state the valuation of the proposed work as part of the information submitted to obtain a permit. To make a determination as to whether a proposed repair, reconstruction, rehabilitation, addition or improvement constitutes substantial improvement or damage, the cost of the proposed work is to be compared to the market value of the building or structure before the work is started. In order to determine market value, the code official may require the applicant to provide such information, as allowed under Section 105.3. For additional guidance, refer to FEMA 213 and FEMA 311.

SUBSTANTIAL STRUCTURAL DAMAGE. A condition where:

1. In any story, the vertical elements of the lateral-force-resisting system, in any direction and taken as a whole, have suffered damage such that the lateral load-carrying capacity has been reduced by more than 20 percent from its predamaged condition, or

2. The vertical load-carrying components supporting more than 30 percent of the structure's floor or roof area have suffered a reduction in vertical load-carrying capacity to below 75 percent of the *International Building Code* required strength levels calculated by either the strength or allowable stress method.

❖ This definition gives the specific parameters for evaluating when a building has sustained substantial structural damage. There are two separate criteria provided in the definition, either one of which will qualify a structure as being substantially damaged.

TECHNICALLY INFEASIBLE. An alteration of a building or a facility that has little likelihood of being accomplished because the existing structural conditions require the removal or alteration of a load-bearing member that is an essential part of the structural frame or because other existing physical or site constraints prohibit modification or addition of elements, spaces, or features that are in full and strict compliance with the minimum requirements for new construction and that are necessary to provide accessibility.

❖ This term is defined in order to provide a basis for the application of accessibility provisions to existing buildings. Bringing any given existing site or building that is altered into full compliance with all accessibility requirements applicable to new construction may require extraordinary effort because of existing physical characteristics. The code utilizes the concept of technical infeasibility to provide a basis for exceptions from strict compliance with the provisions for new construction in an existing building.

[B] UNSAFE BUILDINGS OR EQUIPMENT. Buildings or existing equipment that is insanitary or deficient because of inadequate means of egress facilities, inadequate light and ventilation, or that constitutes a fire hazard, or that is otherwise dangerous to human life or the public welfare or that involves illegal or improper occupancy or inadequate maintenance, shall be deemed an unsafe condition.

❖ This section describes the responsibility of the code official to investigate reports of unsafe buildings and equipment and provides criteria for such determination. Some unsafe buildings may be the result of unsafe or illegal occupancies. For example, prima facie evidence of an unsafe structure is an unsecured (open doors or windows) vacant structure. Because of the attractive nuisance they represent, all unsafe buildings must be either demolished or made safe and secure as deemed appropriate by the code official.

WORK AREA. That portion or portions of a building consisting of all reconfigured spaces as indicated on the construction documents. Work area excludes other portions of the building where incidental work entailed by the intended work must be performed and portions of the building where work not initially intended by the owner is specifically required by this code.

❖ This section specifically defines the area of all reconfigured spaces where work is expected to be occurring within the scope of a project. These areas are to be shown clearly on the construction documents. Incidental work areas are not required to be shown as work areas.

Bibliography

The following resource materials are referenced in this chapter or are relevant to the subject matter addressed in the this chapter.

FEMA 213, *Answers to Questions About Substantially Damaged Buildings*. Washington, DC: Federal Emergency Management Agency, 1991.

FEMA 311, *Guidance on Estimating Substantial Damage Using the NFIP Residential Substantial Damage Estimator*. Washington, DC: Federal Emergency Management Agency, 1998.

CHAPTER 3

CLASSIFICATION OF WORK

General Comments

This chapter provides an overview of the process for the repair, alteration and restoration of existing buildings. A brief description that identifies the differences between the three levels of alterations is provided. In addition, the topics of additions, historic buildings and relocated buildings are mentioned.

Purpose

This chapter enables the contractor, design professional or code official to easily identify the classification of and the associated chapter in the code for building alterations, additions and repairs.

SECTION 301
GENERAL

301.1 Scope. The work performed on an existing building shall be classified in accordance with this chapter.

❖ This section establishes when the regulations contained in the code must be followed, whether all or in part. Something must happen, such as modification to an existing building or allowing an existing building or structure to become unsafe, for the code to be applicable. The code is not a maintenance document requiring periodic inspections that will, in turn, result in an enforcement action. Periodic inspections are addressed by the *International Fire Code*® (IFC®).

301.2 Work area. The work area, as defined in Chapter 2, shall be identified on the construction documents.

❖ As defined in Chapter 2, "Work area" is the area of all reconfigured spaces where work is occurring within the scope of a project. These areas are to be shown clearly on the construction documents. Work areas exclude other portions of the building where incidental work is ongoing.

301.3 Compliance alternatives. The provisions of Chapters 4 through 10 are not applicable where the building complies with Chapter 12.

❖ In the interest of maintaining or increasing the level of public safety, health and general welfare, the code allows repairs, alterations, additions and change of occupancies of existing structures in these structures to comply with the compliance alternative requirements located in Chapter 12.

301.4 Occupancy and use. When determining the appropriate application of the referenced sections of this code, the occupancy and use of a building shall be determined in accordance with Chapter 3 of the *International Building Code*.

❖ In the early years of building code development, especially during the last century, the essence of regulatory safeguards from fire was to provide a reasonable level of protection to property. The idea was that if property was adequately protected from fire, then the building occupants would also be protected. From this outlook

on fire safety, the concept of equivalent risk has evolved in the code. This concept maintains that, in part, an acceptable level of risk against the damages of fire respective to a particular occupancy type (group) can be achieved by limiting the height and area of buildings containing such occupancies according to the building's construction type, i.e., its relative fire endurance.

The concept of equivalent risk involves three interdependent considerations: (1) the level of fire hazard associated with the specific occupancy of the facility; (2) the reduction of fire hazard by limiting the floor area(s) and the height of the building based on the fuel load (combustible contents and burnable building components); and (3) the level of overall fire resistance provided by the type of construction used for the building.

The interdependence of these fire safety considerations can be seen by first looking at Tables 601 and 602 of the *International Building Code*® (IBC®), which show the fire-resistance ratings of the principal structural elements comprising a building in relation to the five classifications for types of construction. Type I construction is the classification that generally requires the highest fire-resistance ratings for structural elements, whereas Type V construction, which is designated as a combustible type of construction, generally requires the least amount of fire-resistance-rated structural elements. If one then looks at Table 503 of the IBC, the relationship among group classification, allowable heights and areas and types of construction becomes apparent. Respective to each group classification, the greater the fire-resistance rating of structural elements, as represented by the type of construction, the greater the floor area and height allowances. The greater the potential fire hazards indicated as a function of the group, the lesser the height and area allowances for a particular construction type.

As a result of extensive research and advancements in fire technology, today's building codes are more comprehensive and complex regulatory instruments than they were in the earlier years of code development. While the principle of equivalent risk remains an important component in building codes, perspectives have changed and life safety is now the paramount fire issue. Even so, occupancy classification still plays a key part in organizing and prescribing the appropriate protection measures. As such, threshold requirements for fire protection and means of egress systems are based on occupancy classification.

SECTION 302
REPAIRS

302.1 Scope. Repairs, as defined in Chapter 2, include the patching or restoration of materials, elements, equipment, or fixtures for the purpose of maintaining such materials, elements, equipment, or fixtures in good or sound condition.

❖ This section describes repairs to existing structures, including the maintenance or restoration to a good or sound condition of any part of the building with any of the materials and methods listed in Section 302.1.

302.2 Application. Repairs shall comply with the provisions of Chapter 4.

❖ Chapter 4 provides the requirements for repairs to existing structures. It covers topics ranging from building elements and materials to fire protection and accessibility. The main focus of Chapter 4 is covered in the provisions of Section 407.

SECTION 303
ALTERATION—LEVEL 1

303.1 Scope. Level 1 alterations include the removal and replacement or the covering of existing materials, elements, equipment, or fixtures using new materials, elements, equipment, or fixtures that serve the same purpose.

❖ Level 1 alterations represent the most basic level of building alterations. This includes the removal and replacement or the covering of existing materials, elements, equipment or fixtures. An example would be the addition of a new roof to an existing building. Another example would be the removal of an aluminum siding exterior finish to be replaced with vinyl siding exterior finish.

303.2 Application. Level 1 alterations shall comply with the provisions of Chapter 5.

❖ Chapter 5 describes in detail the requirements for dealing with Level 1 alterations. It is important to note that historic buildings must also comply with this chapter unless there is a modification noted in Chapter 10.

SECTION 304
ALTERATION—LEVEL 2

304.1 Scope. Level 2 alterations include the reconfiguration of space, the addition or elimination of any door or window, the reconfiguration or extension of any system, or the installation of any additional equipment.

❖ Chapter 6 describes in detail the requirements for dealing with Level 2 alterations. The exception to Section 601 allows buildings undergoing alterations that are exclusively the result of compliance with the accessibility requirements of Section 506.2 to comply with Chapter 5.

304.2 Application. Level 2 alterations shall comply with the provisions of Chapter 5 for Level 1 alterations as well as the provisions of Chapter 6.

❖ In addition to the provisions listed in Chapter 6, Level 2 alterations are also required to meet all of the provisions of Chapter 5. This requirement effectively compounds the requirements for someone planning to alter an existing structure. For example, if during the process of replacing the aluminum siding on a building with vinyl siding, the building owner decides to eliminate one of four windows for a room. This project would be classified as a Level 2 alteration and would, therefore, be required to meet the provisions of Chapter 5 and 6.

SECTION 305
ALTERATION—LEVEL 3

305.1 Scope. Level 3 alterations apply where the work area exceeds 50 percent of the aggregate area of the building.

❖ Anytime the work area exceeds one-half of the aggregate building area the project is considered to be a Level 3 alteration and, therefore, has to meet the requirements of Chapter 7.

305.2 Application. Level 3 alterations shall comply with the provisions of Chapters 5 and 6 for Level 1 and 2 alterations, respectively, as well as the provisions of Chapter 7.

❖ Any project that qualifies as a Level 3 alteration project must meet all of the requirements for Chapters 5, 6 and 7.

SECTION 306
CHANGE OF OCCUPANCY

306.1 Scope. Change of occupancy provisions apply where the activity is classified as a change of occupancy as defined in Chapter 2.

❖ A change in occupancy in an existing structure may change the level of inherent hazards that the code was initially intended to address.

This is done so that the applicable code requirements adequately address the specific hazards of the new occupancy. For example, a change from an existing mercantile occupancy to a business occupancy renders all Group B provisions applicable to all portions of the structure where the occupancy has changed.

306.2 Application. Changes of occupancy shall comply with the provisions of Chapter 8.

❖ This section is one of the most frequently used provisions in the code for application to existing structures, since the occupancy in a building or structure is often subject to change during the life of the building.

SECTION 307
ADDITIONS

307.1 Scope. Provisions for additions shall apply where work is classified as an addition as defined in Chapter 2.

❖ Any project that would increase the floor area, the number of stories in a building or the height of a structure would qualify as an addition.

307.2 Application. Additions to existing buildings shall comply with the provisions of Chapter 9.

❖ Additions to existing structures are specifically covered in Chapter 9 of the IBC.

SECTION 308
HISTORIC BUILDINGS

308.1 Scope. Historic buildings provisions shall apply to buildings classified as historic as defined in Chapter 2.

❖ The most important criterion for application of this section is that the building must be certified as being of historic significance by a qualified party or agency. Usually this is done by a state or local authority after careful review of the historical value of the building. Most, if not all, states have such authorities, as do many local jurisdictions. The agencies with such authority can be located at the state or local government level or through the local chapter of the American Institute of Architects (AIA).

308.2 Application. Except as specifically provided for in Chapter 10, historic buildings shall comply with applicable provisions of this code for the type of work being performed.

❖ Chapter 10 covers the various aspects of existing historical structures to include specific sections on repairs, fire safety, alterations, change of occupancy and structure. In the absence of any specific requirements or provisions in Chapter 10 the remainder of this document shall apply.

SECTION 309
RELOCATED BUILDINGS

309.1 Scope. Relocated buildings provisions shall apply to relocated or moved buildings.

❖ Any structure that is relocated or moved to a different lot or on the same lot fall within the scope of this section.

309.2 Application. Relocated buildings shall comply with the provisions of Chapter 11.

❖ The requirements for relocated or moved buildings are found in Chapter 11 of the code.

CHAPTER 4

REPAIRS

General Comments

Chapter 4 discusses the repair of existing structures. Section 401.1 specifically references Chapter 10 for repair work on historic buildings. Original materials and methods can be utilized in the repair and maintenance of an historic structure. It is the intent of the code to allow buildings to be repaired and maintained without detracting from the historical significance of the structure.

Purpose

Repairs to an existing structure must be made with the proper materials in a manner that will safeguard the public and ensure the building does not become a hazard to life, health or property.

SECTION 401
GENERAL

401.1 Scope. Repairs as described in Section 302 shall comply with the requirements of this chapter. Repairs to historic buildings shall comply with this chapter, except as modified in Chapter 10.

❖ Repairs are described in Section 302 as the patching or restoration of materials and elements for the purpose of maintaining such materials and elements in good or sound condition. The scoping provisions of this section refer the user to Section 302 to be certain the work classification is "repair" so that Chapter 4 is the appropriate chapter to be used. There are additional provisions specific to repairs in historic buildings and the code user is referred to Chapter 10 for the possible applicability of those provisions in addition to the requirements of Chapter 4.

401.2 Permitted materials. Except as otherwise required herein, work shall be done using materials permitted by the applicable code for new construction or using like materials such that no hazard to life, health or property is created.

❖ There are two possible options for materials used in repair work on an existing building. Unless prohibited by other provisions of this section, it is acceptable to use materials consistent with those that are already present. This allowance follows the general concept that repair work is making the building no more unsafe or hazardous than it was prior to work being done. In lieu of using the same type of materials, this section permits the use of any materials presently permitted by the *International Codes*®.

401.3 Conformance. The work shall not make the building less conforming to the building, plumbing, mechanical, electrical or fire codes of the jurisdiction, or to alternative materials, design and methods of construction, or any previously approved plans, modifications, alternative methods, or compliance alternatives, than it was before the repair was undertaken.

❖ The general limitation on any work done under the provisions of this chapter is that the level of safety, health and public welfare of the existing building must not be reduced by any work being performed. In the case of structural stability, the existing degree of structural strength must be maintained or increased. In general terms, the structure is not to be made unsafe. This requirement can be broadly interpreted, as its applications vary on a case-by-case situation. As far as existing mechanical and plumbing systems are concerned, there must not be a reduced level of protection or sanitation.

401.4 Flood hazard areas. In flood hazard areas, repairs that constitute substantial improvement shall require that the building comply with Section 1612 of the *International Building Code.*

❖ If located in designated flood hazard areas, buildings and structures that are damaged by any means are to be examined to determine if the damage constitutes substantial damage, in which the cost of repairing/restoring the building or structure to its predamaged condition equals or exceeds 50 percent of its market value before the damage occurred. All substantial improvements and repairs of buildings and structures that are substantially damaged are to meet the flood-resistant provisions of the *International Building Code*® (IBC®).

SECTION 402
SPECIAL USE AND OCCUPANCY

402.1 General. Repair of buildings classified as special use or occupancy as described in the *International Building Code* shall comply with the requirements of this chapter.

❖ Special use and occupancy buildings described in Chapter 4 of the IBC, such as covered mall buildings and high-rise buildings, are treated the same as any other building when the work being performed is a repair. Please note that with Level 3 alterations or where there is a change of occupancy, some special use and occupancy buildings are required to comply with the IBC, *International Fire Code*® (IFC®) or *International Mechanical Code*® (IMC®).

SECTION 403
BUILDING ELEMENTS AND MATERIALS

403.1 Hazardous materials. Hazardous materials that are no longer permitted, such as asbestos and lead-based paint, shall not be used.

❖ It is generally possible to repair a structure , its components and its systems with materials consistent with those materials that were used previously. However, where materials that are now deemed hazardous are involved in the repair work, they may no longer be used. For example, the code identifies asbestos and lead-based paint as two common hazardous materials that cannot be used in the repair process. Certain materials previously considered acceptable for building construction are now a threat to the health of the occupants.

403.2 Glazing in hazardous locations. Replacement glazing in hazardous locations shall comply with the safety glazing requirements of the *International Building Code* or *International Residential Code* as applicable.

Exception: Glass block walls, louvered windows, and jalousies repaired with like materials.

❖ When glazing in an existing building is replaced within the same building, it must either comply with the current requirements and standards of the IBC or the *International Residential Code*® (IRC®), as applicable. This includes installing new glass in an existing window, door or other type of opening, even where the glass replaced did not comply with the standards of the code.

Glass block walls are described in Chapter 21 of the IBC, which eliminates the Consumer Product Safety Commission (CPSC) test requirement. Glass block walls are not required to meet the test requirements of CPSC 16 CFR for safety glazing; however, there are still safety requirements placed on the installation of the glass block.

Louvered and jalousie windows are exempt from safety glazing requirements in all applications, including those where a flat plane of glass is otherwise required to be safety glass. This exemption is based on records that show the injuries associated with this use of glass are primarily from persons impacting the glass edge with no cutting or piercing injuries resulting from glass breakage.

SECTION 404
FIRE PROTECTION

404.1 General. Repairs shall be done in a manner that maintains the level of fire protection provided.

❖ Any level of fire protection that currently exists in a building must not be adversely affected as a result of any repair. For example, repairing the existing ceiling and sprinkler heads or repair of the fire alarm system equipment must ultimately provide the same level of coverage and protection that existed prior to the repairs being undertaken.

SECTION 405
MEANS OF EGRESS

405.1 General. Repairs shall be done in a manner that maintains the level of protection provided for the means of egress.

❖ Any level of protection provided by the means of egress that currently exists in a building must not be adversely affected as a result of any repair. For example, repairing the walls and doors of a corridor must ultimately provide the same level of protection that existed prior to the repairs being undertaken.

SECTION 406
ACCESSIBILITY

406.1 General. Repairs shall be done in a manner that maintains the level of accessibility provided.

❖ The level of accessibility that currently exists in a building must not be adversely affected as a result of any repair. Continued compliance with the accessibility requirements of the code is dependent on the maintenance of such facilities throughout the life of the building. For example, drinking fountains that are required to be accessible are of little value if they malfunction through deterioration or failure of any of the working parts. In other cases, inoperable elevators, locked accessible doors and obstructed accessible routes must be maintained such that they are readily useable by individuals with disabilities.

SECTION 407
STRUCTURAL

407.1 General. Repairs of structural elements shall comply with this section.

❖ This section gives the requirements that pertain to structural materials and elements in need of repair. Specific requirements for repairing damaged items based on the extent of the damage are given in Section 407.3. Also note that Section 407.1.2 specifies acceptable methods of wind design, while Section 407.1.1 provides the methodologies for the seismic evaluation of existing buildings. The seismic evaluation requirements are also referred to by the other code sections that specifically require an evaluation.

Note that under Section 102.4.2 buildings that comply with the structural provisions of any of the following model codes are deemed to comply with this code:

International Building Code® (IBC®), 2003 or 2000 edition.

National Building Code (NBC), 1999 edition.

Standard Building Code (SBC), 1999 edition.

Uniform Building Code (UBC), 1997 edition.

Although the IRC is not explicitly listed, its use is required by the IBC for detached one- and two-family dwellings as well as townhouses with not more than three stories above grade plane and a separate means of egress.

407.1.1 Seismic evaluation and design. Seismic evaluation and design of an existing building and its components shall be based on the assumed forces related to the response of the structure to earthquake motions.

❖ Consideration of earthquake forces is largely aimed at the performance of the lateral-force-resisting system. Other earthquake hazards, such as the performance of nonstructural components, should also be considered. This section lists the various methods that are used for the seismic evaluation of existing buildings. It identifies the force levels or evaluation methods that are specifically required, or permitted, by other code sections. Portions of this section also apply to alterations or change in occupancy as well as to repairs.

Evaluation versus design.
The term "design" as it applies to new construction is readily understood as starting "with a blank sheet of paper" and the ability to select/control the materials and methods of construction. The term should be applied more loosely to existing construction, since the first step is evaluating what is already built against a specified loading or other performance criteria. Since older buildings often lack the desirable structural characteristics required in new buildings, such as ductile detailing, the seismic evaluation of an existing building requires that the design professional exercise considerable judgement. The minimum criteria that apply are established by the provisions of this code. The outcome of such an evaluation can be that the building, or portion thereof, being evaluated is either satisfactory or not. When an existing building, or portion thereof, is deemed unsatisfactory, options for corrective action must be considered. These "fixes" must be designed whether they involve adding a new member (or members), replacing an existing member with a new member or reinforcing an existing member. With respect to existing buildings, terms such as "upgrades," "retrofits" or "rehabilitation" are often used interchangeably with the term design.

407.1.1.1 Evaluation and design procedures. The seismic evaluation and design of an existing building shall be based on the procedures specified in the *International Building Code*, Appendix A of this code (GSREB), ASCE 31 or FEMA 356.

❖ This section lists four documents containing the procedures that are to be used for the seismic evaluation of an existing building as well as the design of any needed repairs. Since the scope of these documents varies considerably, brief descriptions are given below.

International Building Code® (IBC®)
The IBC is a comprehensive model building code with seismic provisions that are based, for the most part, on the National Earthquake Hazards Reduction Program

(NEHRP) *Recommended Provisions for Seismic Regulations for New Buildings and Other Structures*. The requirements are intended to minimize the hazard to life for all buildings, increase the expected performance of higher occupancy buildings as compared to ordinary buildings and improve the capability of essential facilities to function during and after an earthquake. In addition to minimum seismic loading criteria, the earthquake design provisions include requirements for special inspection and testing as well as material-specific design requirements. Achieving the intended performance depends on a number of factors, including, for example, the structural framing type, configuration and construction materials.

The significant earthquake load concepts include the following:

1. The ground motions are based on a maximum considered earthquake (MCE), which has an approximate average return period of 2,500 years in most of the United States. United States Geological Survey (USGS) ground motion maps [IBC Figures 1615(1) through 1615(10)] provide spectral response accelerations at short periods (S_s) and at one-second periods (S_1). These levels of ground motion are also used in FEMA 356 and ASCE 31. As an alternative to the code figures, the ground motion parameters can be determined by using the Seismic Design Parameter compact disc developed by USGS and available at no cost from the International Code Council® (ICC®). The USGS earthquake hazards program website also provides these ground motions.

2. Design for the effect of two-thirds of the MCE: Considering the margin of safety of 1.5 inherent in seismic design practice, this approach provides collapse prevention under MCE level ground motions. It is also intended that damage from the "design earthquake" ground motion would be repairable. For essential facilities (Occupancy Category IV with a Seismic Importance Factor of 1.5) it is intended that any damage from the design earthquake ground motion be relatively minor and allow the continued occupancy and function of the facility. For higher ground motions, the intent is that there be a low probability of structural collapse.

3. Occupancy category and importance factors: The importance factors of Table 1604.5 of the IBC directly impacts the calculation of seismic (as well as wind and snow) loads. The magnitude of the design load varies in proportion to the importance factor and a higher value is assigned to buildings with an occupancy that warrants a higher level of performance (or protection). The description of these occupancy categories and their relation to seismic use groups are summarized in Commentary Table 407.1.1.1.

TABLE 407.1.1.1 – FIGURE 407.1.1.1(1)   REPAIRS

Table 407.1.1.1
IBC OCCUPANCY CLASSIFICATIONS

OCCUPANCY CATEGORY	NATURE OF OCCUPANCY	IBC & GSREB SEISMIC USE GROUP	SEISMIC IMPORTANCE FACTOR
I	Buildings and other structures that represent a low hazard to human life in the event of failure.	I	1.0
II	Buildings and other structures except those listed in Categories I, III and IV.	I	1.0
III	Buildings and other structures that reqresent a substantial hazard to human life in the event of failure.	II	1.25
IV	Buildings and other structures designated as essential facilities.	III	1.5

4. Nonlinear seismic behavior is accounted for through the use of equivalent lateral forces that are reduced by a response modification factor (R). This approximates the internal forces under the design earthquake. The corresponding building displacements, however, must be increased by the deflection amplification factor (C_d) in meeting the drift requirements.

5. Detailing and limitations on the seismic-force-resisting system are a function of a structure's seismic design category classification, which considers the seismicity at the site, type of soil present at the site and the nature of the building occupancy. Several provisions of this code and *Guidelines for the Seismic Retrofit of Existing Buildings* (GSREB) use a structure's seismic design category as a trigger, but neither document contains the criteria for determining a building's seismic design category. Figure 407.1.1.1(1) provides step-by-step instructions on how to determine a structure's seismic design category using the IBC seismic criteria.

Two levels of IBC seismic forces are used as the basis for the code requirements for seismic analysis and design. These two levels are either the full seismic force required by the IBC (see IEBC Section 407.1.1.2) or the reduced seismic force level which is 75 percent of full seismic forces of the IBC (see IEBC Section 407.1.1.3).

Prestandard and Commentary for the Seismic Rehabilitation of Buildings, FEMA 356

FEMA 356 is essentially an updated version of FEMA 273, *NEHRP Guidelines for the Seismic Rehabilitation of Buildings,* written in mandatory language. It provides detailed guidance on performing a seismic rehabilitation analysis and design and includes an integrated commentary.

The significant concepts include the following:

1. Discrete rehabilitation objectives are established based on target-building performance levels at various levels of ground motion (earthquake risk). These are summarized in Figure 407.1.1.1(2), which is reproduced from the FEMA 356 commentary. Note that this entire array of rehabilitation objectives is only available under a voluntary upgrade to the seismic-force-resisting system in accordance with Section 707.7 of this code. Otherwise, the required performance level is established by the code.

2. Two specific levels of ground motion based on USGS ground motion maps are defined. Basic Safety Earthquake-2 (BSE-2) is the ground motion based on the MCE, and the mapped spectral response accelerations (S_s and S_1) are the same as those used in the IBC as well as ASCE 31. Basic Safety Earthquake-1 (BSE-1) is the *lesser* of 10 percent/50-year ground motion or two-thirds of the BSE-2. The latter is similar to the design

---

1. Determine the mapped MCE spectral response acceleration at short periods, S_s, and at 1-second period, S_1, for the site from code Figures 1615(1) through 1615(10) of the IBC. The Seismic Design Parameter CD is an alternate source for this information.
2. Determine the (soil) site class in accordance with Table 1615.1.1 of the IBC.
3. Determine the site coefficients F_a and F_v from Tables 1615.1.2(1) and 1615.1.2(2) of the IBC, respectively.
4. Determine the design spectral response acceleration at short periods, S_{DS}, and at 1-second period, S_{D1}, as follows:

$$S_{DS} = (2/3)\,(F_a)\,(S_s)$$

$$S_{D1} = (2/3)\,(F_v)\,(S_1)$$

5. Determine seismic use group in accordance with Section 1616.2 and Table 1604.5 of the IBC.
6. Determine the seismic design category as prescribed by Tables 1616.3(1) and 1616.3(2) of the IBC. The highest (most restrictive) of the seismic design categories from the two tables is the category assigned to the building.

Figure 407.1.1.1(1)
DETERMINATION OF SEISMIC DESIGN CATEGORY
USING THE *INTERNATIONAL BUILDING CODE*

		Target Building Performance Levels			
		Operational Performance Level (1-A)	Immediate Occupancy Performance Level (1-B)	Life Safety Performance Level (3-C)	Collapse Prevention Performance Level (5-E)
Earthquake Hazard Level	50%/50 year	a	b	c	d
	20%/50 year	e	f	g	h
	BSE-1 (approx. 10%/50 year)	i	j	k	l
	BSE-2 (approx. 2%/50 year)	m	n	o	p

Notes:

1. Each cell in the above matrix represents a discrete Rehabilitation Objective.

2. The Rehabilitation Objectives in the matrix above may be used to represent the three specific Rehabilitation Objectives defined in Sections 1.4.1, 1.4.2, and 1.4.3 as follows:

   k + p = Basic Safety Objective (BSO)

   k + p + any of a, e, i, b, f, j, or n = Enhanced Objectives

   o alone or n alone or m alone = Enchanced Objectives

   k alone or p alone = Limited Objectives

   c, g, d, h, l = Limited Objectives

**Figure 407.1.1.1(2)
FEMA 356 REHABILITATION OBJECTIVES**

ground motion level used under the IBC, ensuring that BSE-1 ground motions are not larger than the design ground motions used for design of new construction. BSE-1 and BSE-2 are identified in the left-hand (earthquake hazard level) column of Commentary Figure 407.1.1.1(2). FEMA 356 also provides procedures to establish other ground motion levels that may be of interest in rehabilitations. The USGS Seismic Design Parameter compact disc can be used to look up the required BSE-1 or BSE-2 spectral accelerations as shown in Commentary Figure 407.1.1.1(3). The USGS earthquake hazards program website also provides these ground motions.

3. Section 1.4.1 of FEMA 356 commentary, establishes a rehabilitation objective that is referred to as the basic safety objective (BSO). The BSO requires a life safety performance level for BSE-1 and a collapse prevention performance level for BSE-2. This dual objective corresponds to entries k and p as illustrated in Commentary Figure 407.1.1.1(4). The BSO approximates the risk to life safety that has traditionally been accepted in earth-

quake design and is comparable to the intended performance for new buildings under the IBC.

4. Systematic versus simplified rehabilitation methodologies: Commentary Figure 407.1.1.1(5) reprinted from the FEMA 356 commentary provides an overview of the rehabilitation process under FEMA 356.

Seismic Evaluation of Existing Buildings, ASCE 31

This document is a consensus standard that was developed as a replacement for FEMA 310, *Handbook for the Seismic Evaluation of Buildings – a Pre- standard*. It is an evaluation tool that provides a standardized process for identifying potential seismic deficiencies in existing buildings and includes an integrated commentary. It takes a three-tier approach to evaluation starting with a screening phase (Tier 1) and proceeding up to a detailed evaluation phase (Tier 3), if required.

The significant concepts include the following:

1. Seismic demand is based on the MCE. The MCE mapped spectral response accelerations (S_s and S_1) are based on ASCE 7 and are the same as those required under the IBC as well as FEMA 356 ground motion maps.

2. The level of seismicity for a building is defined as low, moderate or high, based on the design short period spectral response acceleration (S_{DS}) and design spectral response acceleration at a one-second period (S_{D1}). These design accelerations are identical to the IBC design level ground motions.

3. The level of performance is established as either life safety (LS) or immediate occupancy (IO). Note that in complying with the code, Table 407.1.1.2 establishes the performance level that applies based on a building's occupancy classification.

4. Accounts for nonlinear response to earthquake ground motions by applying pseudo static lateral forces representing the forces required to impose the expected actual deformations of the structure in its yielded state under the design ground motion.

Guidelines for the Seismic Retrofit of Existing Buildings (GSREB)

Seismic retrofit guidelines have been developed and utilized in the western United States for many years. The 1997 edition of the *Uniform Code for Building Conservation* (UCBC) included three appendix chapters (see A1, A2, and A3 below) dealing with seismic strengthening of specific building types. In 2000, these three chapters were combined with two new chapters (see A4 and A5 below) and published as a standalone document, GSREB. The code has incorporated the GSREB in Appendix A. These chapters provide remedies for the following areas of concern in existing buildings:

A1 *Unreinforced Masonry-Bearing Wall Buildings*: This chapter applies to buildings with one or

FIGURE 407.1.1.1(3)

REPAIRS

Figure 407.1.1.1(3)
USING SEISMIC DESIGN PARAMETERS FOR GROUND MOTIONS
Source: USGS Seismic Design Parameters on CD ROM Version 3.10

more unreinforced masonry-bearing walls. For Occupancy Category I or II in Table 1604.5 of the IBC, there are no limitations, but for Occupancy Category III or IV, this chapter is limited to buildings that are classified as Seismic Design Category A or B.

A2 *Reinforced Concrete and Reinforced Masonry Wall Buildings With Flexible Diaphragms*: This chapter has requirements for wall anchorage systems for reinforced concrete or masonry walls that are laterally supported by flexible diaphragms. It is intended for buildings classified as Seismic Design Category C, D, E or F.

A3 *Strengthening of Cripple Walls and Sill Plate Anchorage of Light, Wood-Frame Residential Buildings*: This chapter has requirements for perimeter foundations, sill plate connections and unbraced cripple walls. It is intended for light wood-frame residential buildings classified as Seismic Design Category D or E.

A4 *Wood-Frame Residential Buildings With Soft, Weak or Open-Front Walls*: This chapter has re-quirements for wood-frame multiple-unit resi-dential buildings that have soft, weak or open-front walls. It applies to buildings located where the design spectral acceleration at a one-second period (S_{D1}) is 0.3g or greater.

A5 *Concrete Buildings and Concrete With Masonry Infill Buildings*: This chapter has require-ments for concrete buildings and concrete frames with masonry infills. It exempts buildings classified as Seismic Design Category A.

Because the GSREB has its roots in the *Uniform Codes*, it also contains references to UBC seismic zones and continues to require the use of various UBC seismic provisions. The Structural Engineers Associa-tion of California (SEAOC) Bluebook provides exten-sive commentary on these UBC seismic provisions. As noted above, the scope and other thresholds include references to seismic criteria that correspond to the IBC as well (e.g., seismic design category). The GSREB is also based on building code forces, which require the use of a response modification coefficient (*R*) (see commentary, Section 407.1.1.2).

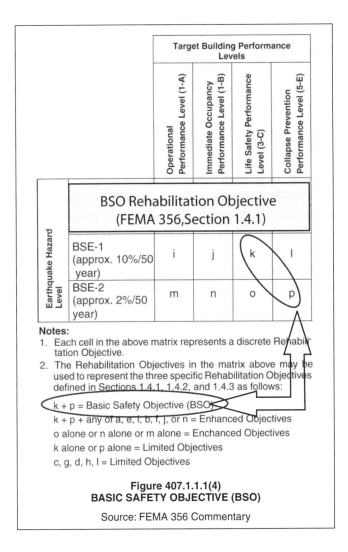

Notes:
1. Each cell in the above matrix represents a discrete Rehabilitation Objective.
2. The Rehabilitation Objectives in the matrix above may be used to represent the three specific Rehabilitation Objectives defined in Sections 1.4.1, 1.4.2, and 1.4.3 as follows:

k + p = Basic Safety Objective (BSO)

k + p + any of a, e, i, b, f, j, or n = Enhanced Objectives

o alone or n alone or m alone = Enchanced Objectives

k alone or p alone = Limited Objectives

c, g, d, h, l = Limited Objectives

**Figure 407.1.1.1(4)
BASIC SAFETY OBJECTIVE (BSO)**

Source: FEMA 356 Commentary

407.1.1.2 IBC level seismic forces. When seismic forces are required to meet the *International Building Code* level, they shall be based on 100 percent of the values in the *International Building Code* or FEMA 356. Where FEMA 356 is used, the FEMA 356 Basic Safety Objective (BSO) shall be used for buildings in Seismic Use Group I. For buildings in other Seismic Use Groups the applicable FEMA 356 performance levels shown in Table 407.1.1.2 for BSE-1 and BSE-2 Earthquake Hazard Levels shall be used.

❖ Where the code requires the use of IBC level seismic forces, it provides the same level of earthquake performance achieved in new construction. One difficulty in applying the seismic requirements intended for new buildings to existing structures is the use of the response modification coefficient (R) in calculating the design seismic force. Under the IBC, the value of R is directly linked to the level of detailing required for any seismic-force-resisting system. Systems are characterized as ordinary, intermediate or special, based on the extent of ductile detailing that is provided. In areas of moderate or higher seismicity (as reflected by a structure's seismic design category), the use of most ordinary systems–those with limited ductility–are typically restricted or prohibited. In an existing building, the sys-

tem detailing is in place and the problem is selecting an R-value that is consistent with the construction of that system. A good practice would be to limit R-values to no greater than those listed for a "comparable" ordinary system, unless there is clear evidence that a higher level of detailing has been provided. As an alternative, FEMA 356 and ASCE 31 avoid this dilemma by considering the ductility of individual components, rather than requiring the assumption of an overall R-value for the existing seismic-force-resisting system.

In this section, the code equates the use of full IBC seismic forces to the level of performance achieved by the FEMA 356 BSO for buildings classified as Seismic Use Group I (Occupancy Categories I and II). The BSO is described in the commentary to Section 407.1.1.1 and Commentary Figure 407.1.1.1(4). Also note that it corresponds directly to the dual requirements listed in Table 407.1.1.2. Where this section is cited by other code provisions, the GSREB is not permitted as an alternative, since it is primarily based on reduced force levels as described in Section 407.1.1.3.

**TABLE 407.1.1.2
IBC SEISMIC USE GROUP EQUIVALENTS TO FEMA 356 AND
ASCE 31 PERFORMANCE LEVELS[a]**

SEISMIC USE GROUP (BASED ON IBC TABLE 1604.5)	PERFORMANCE LEVELS OF ASCE 31 AND FEMA 356 BSE-1 EARTHQUAKE HAZARD LEVEL	PERFORMANCE LEVELS OF FEMA 356 BSE-2 EARTHQUAKE HAZARD LEVEL
I	Life Safety (LS)	Collapse Prevention (CP)
II	Life Safety (LS)	Collapse Prevention (CP)
III	Note b	Note b
IV	Immediate Occupancy (IO)	Life Safety (LS)

a. The charging provisions for Seismic Use Group equivalents to ASCE 31 and FEMA 356 BSE-1 for reduced *International Building Code* level seismic forces are located in Section 407.1.1.3.

b. Performance Levels for Seismic Use Group III shall be taken as halfway between the performance levels specified for Seismic Use Groups II and IV.

❖ The left-hand column heading should be "occupancy category" rather than "seismic use group." See Commentary Table 407.1.1.1 for the correct correlation of the occupancy category and seismic use group under the *International Codes*. The right-hand column of this table specifies the performance level that is to be achieved under the BSE-2 level of ground motion in FEMA 356. The center column specifies the performance level that is to be achieved in ASCE 31 or under the BSE-1 level of ground motion using FEMA 356.

Note b requires that the Occupancy Category III performance levels are midway between those required for Occupancy Category II and IV. This is analogous to the IBC's use of the seismic importance factor (see Commentary Table 407.1.1.1).

407.1.1.3 Reduced IBC level seismic forces. When seismic forces are permitted to meet reduced *International Building Code* levels, they shall be based on 75 percent of the assumed forces prescribed in the *International Building Code*, applica-

FIGURE 407.1.1.1(5)

REPAIRS

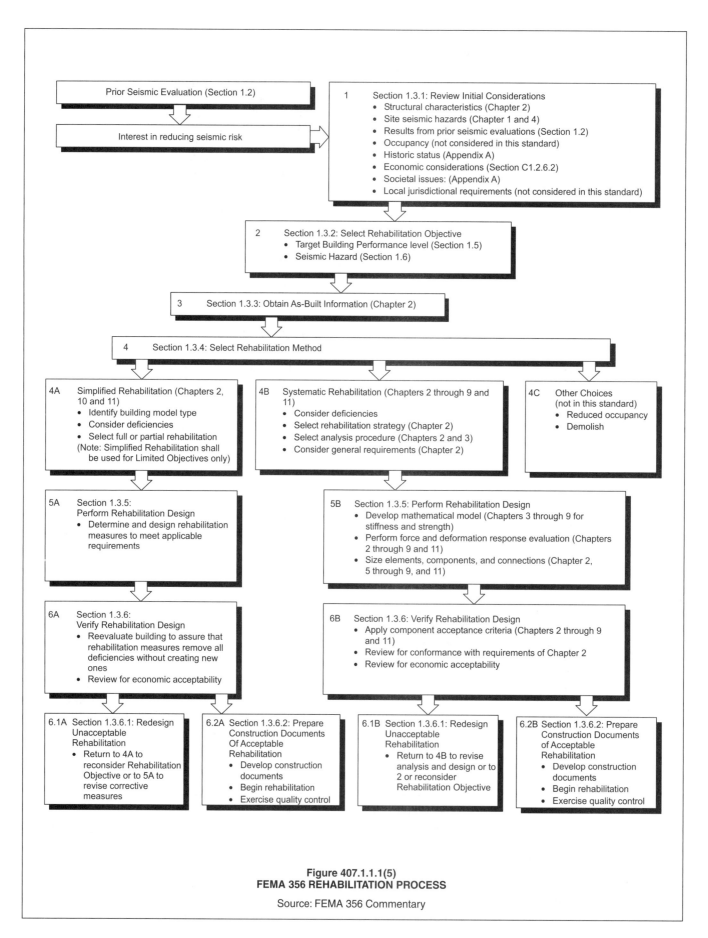

Figure 407.1.1.1(5)
FEMA 356 REHABILITATION PROCESS
Source: FEMA 356 Commentary

ble chapters in Appendix A of this code (GSREB), the applicable performance level of ASCE 31 as shown in Table 407.1.1.2, or the applicable performance level for the BSE-1 Earthquake Hazard Level of FEMA 356 shown in Table 407.1.1.2.

❖ The lateral-force-resisting systems in most older buildings are difficult, if not impossible, to upgrade to the same level of performance that is required of new construction. Where the code permits the use of this reduced seismic force level, it provides a means of achieving some improvement in earthquake performance of older buildings without making such efforts cost prohibitive as well as permitting a building that is relatively close to current earthquake standards to comply outright. The same approach is used in Appendix A (GSREB) and it is similar to the approach that has been used previously under FEMA 178. The same dilemma with the *R*-value determination discussed in the commentary under Section 407.1.1.2 occurs here as well.

The code equates these reduced seismic forces to the level of performance required by Table 407.1.1.2 for ASCE 31 or FEMA 356 for the BSE-1 earthquake hazard level.

407.1.2 Wind design. Wind design of existing buildings shall be based on the procedures specified in the *International Building Code* or *International Residential Code* as applicable.

❖ The IBC must be used for the analysis and design of existing buildings for the effects of wind. Detached one- and two-family dwellings as well as townhouses with no more than three stories above grade plane and having separate means of egress must comply with the wind requirements of the IRC.

The intent of the IBC wind criteria in Section 1609 is to establish minimum wind loading appropriate for the design and construction of new buildings. The wind criteria given in this section of the IBC generally reflect the wind load provisions of ASCE 7. Their use serves to reduce the potential for damage to property caused by windstorms, and to provide an acceptable level of protection to building occupants. The objective also includes the prevention of damage to adjacent properties because of the possible detachment of major building components (e.g., walls, roofs, etc.), structural collapse, wind-borne debris and for the safety of people in the immediate vicinity. Depending on the type of existing construction being evaluated, other material-specific wind provisions of the IBC will apply as well (e.g., requirements for wood diaphragms and shear walls in IBC Sections 2305 and 2306).

407.2 Reduction of strength. Repairs shall not reduce the structural strength or stability of the building, structure, or any individual member thereof.

Exception: Such reduction shall be allowed provided the capacity is not reduced to below the *International Building Code* levels.

❖ This section explicitly states that repairs must not reduce the strength of individual structural elements or the building structure as a whole. The exception to this

rule permits strength reductions to members that have "reserve capacity." In this case, however, the reduced strength must be equal to, or better than, that required by the IBC for a new structure.

407.3 Damaged buildings. Damaged buildings shall be repaired in accordance with this section.

❖ Buildings can suffer damage from numerous sources. Natural disasters such as earthquakes, floods, hurricanes and tornadoes can cause extensive damage over widespread areas, depending on the severity of the event. Water intrusion due to a failure in the building envelope, termite infestations or exposure to corrosive chemicals can all lead to the deterioration of structural members over time. Damage to a building's structure can also occur due to overloading portions of a floor or roof as well as an accidental impact by a vehicle or a fire.

For the most part, this section does not differentiate between the possible causes of the damage. Instead, the requirements for repairs to damaged buildings in this section are primarily based on the extent of the damage. Needless to say, determining the root cause of any damage would be necessary in order to accurately assess the extent of the damage, to ascertain the risk of recurrence and, if necessary, to develop a plan to address that risk.

407.3.1 New structural frame members. New structural frame members used in the repair of damaged buildings, including anchorage and connections, shall comply with the *International Building Code*.

Exception: For the design of new structural frame members connected to existing structural frame members, the use of reduced *International Building Code* level seismic forces as specified in Section 407.1.1.3 shall be permitted.

❖ The term "new structural frame member" is not specifically explained, but as it applies to the repair of damage, it refers to any new structural member that is installed in the course of repairing a structure. Any new member, including its connections to the existing structure, must comply with the IBC. The exception affects new structural frame members that are added for seismic resistance, since it permits the use of reduced IBC level seismic forces in the design of those members.

407.3.2 Substantial structural damage. Buildings that have sustained substantial structural damage shall comply with this section.

❖ Substantial structural damage is defined in Section 202. It establishes two thresholds beyond which the adequacy of the building structure must be evaluated by a registered design professional. One threshold is based on the extent of damage to the vertical elements of the lateral-force-resisting system in any story, and the other threshold is based on the extent of damage to the vertical load-carrying components. The determination of the extent of damage itself would necessitate some level of engineering evaluation.

407.3.2.1 Engineering evaluation and analysis. An engineering evaluation and analysis that establishes the structural adequacy of the damaged building shall be prepared by a registered design professional and submitted to the code official. The evaluation and analysis may assume that all damaged structural elements and systems have their original strength and stiffness. The seismic analysis shall be based on one of the procedures specified in Section 407.1.1.

❖ Once it has been determined that a building has sustained substantial structural damage, it is necessary to have an engineering evaluation of the entire building structure by a registered design professional. For the purpose of establishing the structure's adequacy, the evaluation can be performed using the original strength and stiffness of any damaged structural component. This suggests that where the original construction is determined to be adequate by this evaluation then the repairs may be limited to restoring the structure to its original state.

Since there are no limits stated, the structural evaluation must assess the entire structure for all types of loading. With the exception of wind and seismic provisions (see Section 407.3.2.1.1), the code provisions used to determine loading are not specified, leaving this area subject to interpretation. The procedures permitted for seismic analysis are those listed in Section 407.1.1.1.

407.3.2.1.1 Extent of repair. The evaluation and analysis shall demonstrate that the building, once repaired, complies with the wind and seismic provisions of the *International Building Code.*

> **Exception:** The seismic design level for the repair design shall be the higher of the Building Code in effect at the time of original construction or reduced *International Building Code* level seismic forces as specified in Section 407.1.1.3.

❖ The evaluation described in Section 407.3.2.1 must establish that the repaired building will meet the wind and seismic provisions of the IBC. As permitted in Section 407.3.2.1, this may be based upon the original construction. If the original construction is inadequate, any additional structural elements or strengthening of existing elements necessary to comply would need to be incorporated in this evaluation.

The exception permits the design of repairs using the reduced IBC seismic force as provided in Section 407.1.1.3. However, this force can be no lower than that required by the building code in effect at the time of the building's construction, thus preventing a reduction in seismic resistance to less than what should have existed prior to the damage. It is estimated that this limitation would apply primarily to buildings constructed under model building codes adopted after 1976. This generalization does not necessarily apply "across the board," since the seismic hazard mapping and the resulting seismic force under the IBC can be very different from that of its predecessors.

407.3.3 Below substantial structural damage. Repairs to buildings damaged to a level below the substantial structural damage level as defined in Section 202 shall be allowed to be made with the materials, methods, and strengths in existence prior to the damage unless such existing conditions are dangerous as defined in Chapter 2. New structural frame members as defined in Chapter 2 shall comply with Section 407.3.1.

❖ Damage that is determined to be below the threshold of substantial structural damage (see definition in Section 202 and commentary to Section 407.3.2), can be repaired with materials and construction methods that match those of the existing building. The exceptions are where the existing conditions are dangerous (see definition in Section 202) or where new structural frame members are installed.

407.3.4 Other uncovered structural elements. Where in the course of conducting repairs other uncovered structural elements are found to be unsound or otherwise structurally deficient, such elements shall be made to conform to the requirements of Section 407.3.2.1.1.

❖ In the course of repairing buildings, it not uncommon to discover additional structural elements that are also in need of repair. These elements must be checked and verified to be capable of supporting the wind and seismic loading required by the IBC.

407.3.5 Flood hazard areas. In flood hazard areas, damaged buildings that sustain substantial damage shall be brought into compliance with Section 1612 of the *International Building Code.*

❖ The definition of "Flood hazard area" provided in Section 202 is taken from the IBC and establishes where this provision is applicable. If located in designated flood hazard areas, buildings that are damaged must be examined to determine if the damage constitutes "substantial damage" (see definition in Section 202). Buildings are considered to have sustained substantial damage when the cost of repairing the building to its predamaged condition is 50 percent or more of its market value before the damage occurred. Buildings determined to be substantially damaged must meet the flood-resistant provisions of the IBC (see FEMA 213).

Section 1612 of the IBC addresses requirements for buildings in designated flood hazard areas. The design and construction is required to be in accordance with ASCE 24, which in turn references the flood loading given in ASCE 7. Through use of these IBC provisions, communities meet a significant portion of the floodplain management regulation requirements necessary to participate in the National Flood Insurance Program (NFIP).

SECTION 408
ELECTRICAL

408.1 Material. Existing electrical wiring and equipment undergoing repair shall be allowed to be repaired or replaced with like material.

Exceptions:

1. Replacement of electrical receptacles shall comply with the applicable requirements of Section 406.3(D) of NFPA 70.

2. Plug fuses of the Edison-base type shall be used for replacements only where there is no evidence of over fusing or tampering per applicable requirements of Section 240.51(B) of NFPA 70.

3. For replacement of nongrounding-type receptacles with grounding-type receptacles and for branch circuits that do not have an equipment grounding conductor in the branch circuitry, the grounding conductor of a grounding-type receptacle outlet shall be permitted to be grounded to any accessible point on the grounding electrode system, or to any accessible point on the grounding electrode conductor in accordance with Section 250.130(C) of NFPA 70.

4. Non-"hospital grade" receptacles in patient bed locations of Group I-2 shall be replaced with "hospital grade" receptacles, as required by NFPA 99 and Article 517 of NFPA 70.

5. Frames of electric ranges, wall-mounted ovens, counter-mounted cooking units, clothes dryers, and outlet or junction boxes that are part of the existing branch circuit for these appliances shall be permitted to be grounded to the grounded circuit conductor in accordance with Section 250.140 of NFPA 70.

❖ In essence, this section states that existing wiring systems can be maintained in the same manner in which they were installed. Repairs are to be made with materials and components that do not in any way make the existing wiring system less safe. Materials and components can be replaced with items of equal or superior quality and integrity. The intent is to allow necessary repairs without subjecting the system to new construction requirements. A wiring system material or component that is obsolete or no longer recognized by current codes is permitted to be used for the purpose of making repairs, provided that it is consistent with the existing materials and components and the system is not made any less safe.

Exception 1 ties the replacement of receptacle devices to the provisions of NFPA 70, which do not always allow replacement with like devices. For example, where a grounding means exists, ungrounded-type (2 conductor) receptacles must be replaced only with grounding-type (3 conductor) receptacles. Similarly, receptacles in locations where ground-fault circuit-interrupter (GFCI) protection is required must be replaced only with GFCI-type receptacles or the branch circuit must provide such GFCI protection. See NFPA 70 Sections 406.3(D)(1) through (3)(c) for more replacement provisions.

Exception 2 does not allow Edison-base (screw base)-type fuses to replace existing fuses except where there is no reason to believe the wrong size fuses have been or are being used or where there is no evidence of attempts to defeat the protection afforded by the fuses. Edison-base fuses are plug-style fuses with the same screw thread base as the common incandescent lamp. Such fuses have ampere ratings of 30 amps or less and are interchangeable, meaning that occupants are not prevented from inserting fuses that are rated higher than the capacity of the wiring they are intended to protect; therefore, fire hazards are likely to be created.

Evidence of oversized fuses being used or tampered with is the code's justification for prohibiting the installation of any new Edison-base plug fuses in existing fuseholders. Obviously, the hazardous condition of overfusing or tampering must not be allowed to continue to exist. In such cases, replacement fuses must be Type S fuses, which are designed to thwart overfusing and attempts at tampering or bypassing.

Exception 3 is not actually an exception to the provision of this section, but rather, is a recognition of the provisions of Section 250.130(C) of NFPA 70. Exception 1 addresses circumstances where existing receptacles cannot be replaced with like receptacles, and Exception 3 describes one of the options specified in Section 406.3(D) of NFPA 70.

Exception 4 reflects the intent of Section 517.18(B) of NFPA 70 and, again, is an exception to allowing devices to be replaced with like devices.

Exception 5 is not actually an exception to the provision of this section, but rather, is a recognition of a provision of Section 250.140 of NFPA 70 that parallels the intent of this code to allow repairs to be consistent with the original installation. Although allowed in the past, grounding of appliances to the grounded circuit conductor is considered to be unnecessarily risky and, therefore, is now allowed only for existing wiring installations under specified conditions that serve to limit the risk to an acceptable level.

SECTION 409
MECHANICAL

409.1 General. Existing mechanical systems undergoing repair shall comply with Section 401.1 and the scoping provisions of Chapter 1 where applicable.

❖ This section is essentially a reference section to direct the user's attention to the possibility of other sections that might be relevant to mechanical systems. Repair work must not alter the nature of appliances and equipment in a way that would invalidate the listing or conditions of approval.

[P] SECTION 410
PLUMBING

410.1 Materials. The following plumbing materials and supplies shall not be used:

1. Sheet and tubular copper and brass trap and tailpiece fittings less than the minimum wall thickness of .027 inch (0.69 mm).

2. Solder having more than 0.2-percent lead in the repair of potable water systems.

3. Water closets having a concealed trap seal or an unventilated space or having walls that are not thoroughly washed at each discharge in accordance with ASME A112.19.2M.

4. The following types of joints shall be prohibited:

    4.1. Cement or concrete joints.

    4.2. Mastic or hot-pour bituminous joints.

    4.3. Joints made with fittings not approved for the specific installation.

    4.4. Joints between different diameter pipes made with elastomeric rolling O-rings.

    4.5. Solvent-cement joints between different types of plastic pipe.

    4.6. Saddle-type fittings.

5. The following types of traps are prohibited:

    5.1. Traps that depend on moving parts to maintain the seal.

    5.2. Bell traps.

    5.3. Crown-vented traps.

    5.4. Traps not integral with a fixture and that depend on interior partitions for the seal, except those traps constructed of an approved material that is resistant to corrosion and degradation.

❖ 1. Tubular waste fittings utilized from the outlet of the bowl or tub to the trap can be stainless steel or copper alloy and are regulated by the minimum wall thickness of 0.027 inches (.68 mm).

2. A soldered joint is the most common method of joining copper pipe and tubing. Lead-based solders (containing more than 0.2-percent lead), however, are no longer permitted for joining copper tubing used for the distribution of potable water. Lead-based solders and fluxes have been implicated as a cause for an increased concentration of lead in drinking water.

3. Fixtures that do not thoroughly scour the used surface or area will harbor bacteria and germs resulting in these fixtures becoming a source of contamination to occupants. Fixtures either located in an unventilated space or not thoroughly washed at each discharge are prohibited. Any water closet that allows the content of the bowl to siphon up into the tank will subject occupants to the same type of health hazard as fixtures that do not thoroughly scour the used area.

4. Cement and concrete are not effective in sealing a pipe joint, as such materials are inflexible and susceptible to cracking displacement. A rolling O-ring has no resistance to being pushed or rolled out of a joint when exposed to pressure or the movement of pipe due to expansion and contraction. A solvent-cement joint is a homogeneous chemical bond made between a pipe and fitting. The bond is accomplished because the chemical composition of the joint surfaces is the same. If the materials are different, the joint would not form a proper chemical bond and the strength and integrity of the joint will be adversely affected. Saddle-type fittings could possibly be moved out of alignment, can weaken the pipe because of the pipe wall penetration and typically do not form a drainage pattern connection.

5. A trap is intended to be a simple U-shaped piping arrangement that offers minimal resistance to flow. Based on their design or configurations, prohibited traps are not a simple U-shape piping design. Such traps typically impede drainage flow by moving parts, design (tends to clog with debris), flow pattern and ability to lose a trap.

410.2 Water closet replacement. When any water closet is replaced, the replacement water closet shall comply with the *International Plumbing Code*. The maximum water consumption flow rates and quantities for all replaced water closets shall be 1.6 gallons (6 L) per flushing cycle.

Exception: Blowout-design water closets [3.5 gallons (13 L) per flushing cycle].

❖ Federal legislation mandates the design and use of water closets to have flow rates restricted to 1.6 gallons (6.1 L) per flushing cycle, except where blowout-type fixtures are utilized. Blowout-designed water closets are exempt from the 1.6-gallon (6.1 L) requirement because such fixtures depend on high-volume and high-velocity water flow to evacuate the contents of the bowl; therefore, such fixture designs may not be able to function at lower consumption rates.

Bibliography

The following resource materials are referenced in this chapter or are relevant to the subject matter addressed in this chapter.

ASCE 7-02, *Minimum Design Loads for Buildings and Other Structures*. Reston, VA: American Society of Civil Engineers, 2002.

ASCE 24-98, *Flood Resistant Design and Construction Standard*. Reston, VA: American Society of Civil Engineers, 1998.

ASCE 31-02, *Seismic Evaluation of Existing Buildings*. Reston, VA: American Society of Civil Engineers, 2002.

ATC 20, *Procedures for Postearthquake Safety Evaluation of Buildings*. Redwood City, CA: Applied Technology Council, 1989.

ATC 20-2, *Addendum to the ATC 20 Postearthquake Building Safety Evaluation Procedures*. Redwood City, CA: Applied Technology Council, 1995.

ATC 40, *Sesimic Evaluation and Retrofit of Concrete Buildings*. Redwood City, CA: Applied Technology Council for the California Seismic Safety Commission, 1996.

Chang, David and Hamid Naderi. "International Existing Building Code" *Structural Engineer*. October 2003, pp 22-25.

Chen, Wai-Fah and Scawthorn, Charles. *Earthquake Engineering Handbook*. Boca Raton, FL: CRC Press LLC, 2003.

FEMA 178, *NEHRP Handbook for the Seismic Evaluation of Existing Buildings*. Washington, DC: Federal Emergency Management Agency, 1992.

FEMA 213, *Answers to Questions About Substantially Damaged Buildings*. Washington, DC: Federal Emergency Management Agency, 1991.

FEMA 259, *Engineering Principles and Practices for Retrofitting Flood-Prone Residential Buildings*. Washington, DC: Federal Emergency Management Agency, 1995.

FEMA 273, *NEHRP Guidelines for the Seismic Rehabilitation of Buildings*. Washington, DC: Federal Emergency Management Agency, 1997.

FEMA 274, *NEHRP Commentary on the Guidelines for the Seismic Rehabilitation of Buildings*. Washington, DC: Federal Emergency Management Agency, 1997.

FEMA 310, *Handbook for the Seismic Evaluation of Buildings – a Prestandard*. Washington, DC: Federal Emergency Management Agency, 1998.

FEMA 311, *Guidance on Estimating Substantial Damage Using the NFIP Residential Substantial Damage Estimator*. Washington, DC: Federal Emergency Management Agency, 1998.

FEMA 312, *Homeowner's Guide to Retrofitting – Six Ways to Protect Your House From Flooding*. Washington, DC: Federal Emergency Management Agency, 1998.

FEMA 356, *Prestandard and Commentary for the Seismic Rehabilitation of Buildings*. Washington, DC: Federal Emergency Management Agency, 2000.

FEMA 368, *NEHRP Recommended Provisions for Seismic Regulations for New Buildings and Other Structures, Part 1*. Washington, DC: Building Seismic Safety Council for the Federal Emergency Management Agency, 2001.

FEMA 369, *NEHRP Recommended Provisions for Seismic Regulations for New Buildings and Other Structures Commentary, Part 2 Commentary*. Washington, DC: Building Seismic Safety Council, for the Federal Emergency Management Agency, 2001.

Ghosh, S. K. and Fanella, David A. *Seismic and Wind Design of Concrete Buildings (2000 IBC, ASCE 7-98, ACI 318-99)*. Falls Church, VA: International Code Council, 2003.

IBC-00, *International Building Code*. Falls Church, VA : International Code Council, 2000.

IBC-03, *International Building Code*. Falls Church, VA : International Code Council, 2003.

IRC-03, *International Residential Code*. Falls Church, VA: International Code Council, 2003.

Leyendecker, E. V., A. D. Frankel and K. S. Rustales. *Seismic Design Parameters*. U.S. Geological Survey Open-File Report 01-437. Washington, DC: United States Geological Survey, 2001.

Naeim, Farzad. *The Seismic Design Handbook*. Norwell, MA: Kluwer Academic Publishers, 2001.

NBC-99, *BOCA National Building Code*. Country Club Hills, IL: Building Officials and Code Administrators International, Inc. 1999.

Recommended Lateral Force Requirements and Commentary (SEAOC Blue Book). Sacramento, CA: Structural Engineers Association of California, 1999.

Reducing Flood Losses Through the International Code Series: Meeting the Requirements of the National Flood Insurance Program. Falls Church, VA: International Code Council, May 2000.

SBC-99, *Standard Building Code*. Birmingham, AL: Southern Building Code Congress International, 1999.

Taly, Narendra. *Loads and Loads Paths in Buildings Principles of Structural Design*. Falls Church, VA: International Code Council, 2003.

UBC-97, *Uniform Building Code*. Whittier, CA: International Conference of Building Officials, 1997.

UCBC-97, *Uniform Code for Building Conservation*. Whittier, CA: International Conference of Building Officials, 1997.

Williams, Alan. *Seismic and Wind Forces Structural Design Examples*. Falls Church, VA: International Code Council, 2003.

CHAPTER 5

ALTERATIONS—LEVEL 1

General Comments

This chapter provides the technical requirements for those existing buildings that undergo Level 1 alterations. Reference is made to Level 1 alterations in Section 303, which includes replacement or covering of existing materials, elements, equipment or fixtures using new materials for the same purpose.

This chapter, similar to other chapters of the code, covers all building-related subjects such as structural, mechanical, plumbing, electrical and accessibility as well as fire and life safety issues when alterations are classified as Level 1. Sections 501, 502 and 503 are related to scoping, special use and occupancy and building elements and materials. It should be noted that, in the interest of brevity and avoiding repetition of material, Section 601.2 requires that Level 2 alterations comply not only with Chapter 6, but also with Chapter 5. Similarly, Section 701.2 requires that Level 3 alterations comply with Chapters 5 and 6, as well as Chapter 7. As such, Chapter 5 is applicable to all levels of alteration work. Section 503 covers in detail elements such as interior finishes and carpeting. Section 503 also refers to the *International Building Code*® (IBC®), ICC *Electrical Code*® (ICC EC™), *International Energy Conservation Code*® (IECC®), *International Mechanical Code*® (IMC®) and *International Plumbing Code*® (IPC®) for new materi-

als. The remainder of the chapter is related to fire protection, means of egress, accessibility, structural, electrical, mechanical and plumbing provisions.

Alterations of Level 1 classification are considered the least drastic of alterations. The decision to divide the level typically known as "alterations" in the IBC and previous legacy building codes into three parts of Level 1, Level 2 and Level 3 was based on the fact that minor alterations, not including space reconfiguration, and extensive alterations that might range from relocation of walls and partitions in various floors (or in the entire building) should be treated differently and with different threshold levels for requiring upgrades or improvements to the building or spaces within the building.

Purpose

The purpose of this chapter is to provide detailed requirements and provisions to identify the required improvements in the existing building elements, building spaces and building structural system. This chapter is distinguished from Chapters 6 and 7 by involving only replacement of building components with new components. In contrast, Level 2 alterations involve space reconfiguration, and Level 3 alterations involve more extensive space reconfiguration, exceeding 50 percent of the building area.

SECTION 501
GENERAL

501.1 Scope. Level 1 alterations as described in Section 303 shall comply with the requirements of this chapter. Level 1 alterations to historic buildings shall comply with this chapter, except as modified in Chapter 10.

❖ Level 1 alterations are described in Section 303 as the type of alterations that include the removal and replacement or the covering of existing materials and elements. The scoping provisions of this section refer the user to Section 303 to be certain that the work classification is Level 1 alterations and that Chapter 5 is the appropriate chapter to be used. There are additional provisions specific to Level 1 alterations in historic buildings, and the code user is referred to Chapter 10 for the possible applicability of those provisions in addition to the requirements of Chapter 5.

501.2 Conformance. An existing building or portion thereof shall not be altered such that the building becomes less safe than its existing condition.

Exception: Where the current level of safety or sanitation is proposed to be reduced, the portion altered shall conform to the requirements of the *International Building Code.*

❖ The current level of safety or the level of compliance with regulatory provisions in a building are not, in general, allowed to be reduced, regardless of the type of work taking place in the building. For example, a Class B interior wall finish in an exit passageway will not be allowed to be replaced with a finish material that has a Class C rating. The exception describes the only situation where the reduction of the level of safety or the level of compliance is allowed. This is the condition where the existing level of compliance is above the level required by the IBC and the reduced level of compliance still meets or exceeds the building code requirements. Some other examples where reduction of level of compliance is not allowed might be to meet water supply requirements (see Figure 501.2), and to meet electrical loads or reduction in existing light and ventilation to below the levels required in the building code, or below the existing levels, whichever is lower.

501.3 Flood hazard areas. In flood hazard areas, alterations that constitute substantial improvement shall require that the building comply with Section 1612 of the *International Building Code.*

❖ When alterations to existing buildings that are located in flood hazard areas are proposed, determinations are to be made whether the proposed work is a sub-

FIGURE 501.2 – 503.3

ALTERATIONS—LEVEL 1

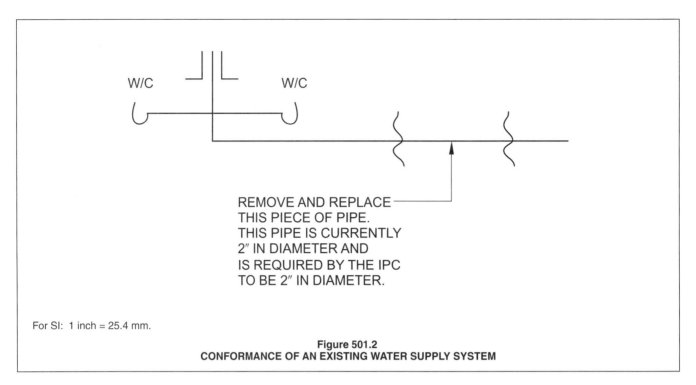

For SI: 1 inch = 25.4 mm.

Figure 501.2
CONFORMANCE OF AN EXISTING WATER SUPPLY SYSTEM

stantial improvement. If the proposed alterations are determined to be substantial improvements, then the existing building is to be brought into compliance with the flood-resistant provisions of the IBC.

SECTION 502
SPECIAL USE AND OCCUPANCY

502.1 General. Alteration of buildings classified as special use and occupancy as described in the *International Building Code* shall comply with the requirements of Section 501.1 and the scoping provisions of Chapter 1 where applicable.

❖ Special use and occupancy buildings described in the Chapter 4 of the IBC, such as covered mall buildings and high-rise buildings, are treated the same as any other building when the work taking place is a Level 1 alteration. This section specifically provides this information because at higher levels of alteration work, or where there is a change of occupancy, some special use and occupancy buildings are required to comply with the IBC (see Sections 702 and 802).

SECTION 503
BUILDING ELEMENTS AND MATERIALS

503.1 Interior finishes. All newly installed interior finishes shall comply with the flame spread requirements of the *International Building Code*.

❖ Newly installed interior finish materials are required to comply with the flame spread requirements of the IBC. The level of developed smoke, surface burning, flame spread and toxic byproducts of combustion are critical elements to be considered in fire situations. For this

reason, the code explicitly outlines the criteria for interior finishes. This criteria is consistent with the IBC in most cases. This is a requirement that has potentially dramatic effects on building and occupant safety in fire situations.

503.2 Carpeting. New carpeting used as an interior floor finish material shall comply with the radiant flux requirements of the *International Building Code*.

❖ The same description and reasoning provided in Section 503.1 applies here to carpeting. Carpeting is required to comply with the radiant flux requirements of the IBC. Even though Section 503.3 covers all materials used under Level 1 alterations, carpeting is specifically called out because of its potential contribution to flame spread under fire conditions.

503.3 Materials and methods. All new work shall comply with materials and methods requirements in the ICC *Electrical Code*, *International Building Code*, *International Energy Conservation Code*, *International Mechanical Code*, and *International Plumbing Code*, as applicable, that specify material standards, detail of installation and connection, joints, penetrations, and continuity of any element, component, or system in the building.

❖ Materials and methods requirements refer to requirements of various codes such as the IBC, IMC and IPC that specify material standards, details of installation and connection, joints, penetrations and continuity of any element, component or system in the building. This description, which is not presented in the form of a definition, is so broad that there are numerous sections in the *International Codes*® that are considered to be related to materials and methods. One way of dealing with this section might have been to list every section in every *International Code* that deals with materials and methods. As this would have been a long list of sections, the committee and the membership chose to address this issue in general terms rather than listing such sections from the *International Codes*, except for the materials and methods sections of the *International Fuel Gas Code®* (IFGC®) that have been specifically listed in Section 503.1.

All new work in Level 1 alterations must comply with materials and methods requirements of the *International Codes*.

For example, the sheetrock from one side of an existing corridor is to be removed and replaced with new sheetrock and finish paneling. Regardless of how the existing sheetrock was attached and what the characteristics of the finish material were, the new sheetrock must comply with material referenced standards in Section 2506 of the IBC and its installation must be in accordance with Table 2508.1 of the IBC. Accordingly, the interior finish paneling used must be of a class in compliance with Table 803.5 of the IBC.

[FG] 503.3.1 International Fuel Gas Code. The following sections of the *International Fuel Gas Code* shall constitute the fuel gas materials and methods requirements for Level 1 alterations.

1. All of Chapter 3, entitled "General Regulations," except Sections 303.7 and 306.
2. All of Chapter 4, entitled "Gas Piping Installations," except Sections 401.8 and 402.3.
   2.1. Sections 401.8 and 402.3 shall apply when the work being performed increases the load on the system such that the existing pipe does not meet the size required by code. Existing systems that are modified shall not require resizing as long as the load on the system is not increased and the system length is not increased even if the altered system does not meet code minimums.
3. All of Chapter 5, entitled "Chimneys and Vents."
4. All of Chapter 6, entitled "Specific Appliances."

❖ Any alteration of fuel gas equipment or piping that falls under the category of Level 1 alterations, must comply with materials and methods requirements of the IFGC. This section identifies those sections of the IFGC that are considered to be related to materials and methods. These, with the exception of four specific sections, include all of Chapters 3, 4, 5 and 6 of the IFGC.

SECTION 504
FIRE PROTECTION

504.1 General. Alterations shall be done in a manner that maintains the level of fire protection provided.

❖ Any level of fire protection that currently exists in a building must not be adversely affected or lessened as a result of any alteration. For example, removing and replacing the existing ceiling and rearrangement of

some fire sprinkler heads should ultimately provide the same level of sprinkler coverage and protection that existed prior to the alterations being undertaken. A fire protection feature that is already installed in a building cannot be removed even if the present code does not require the fire protection feature.

SECTION 505
MEANS OF EGRESS

505.1 General. Means of egress for buildings undergoing alteration shall comply with the requirements of Section 501.1 and the scoping provisions of Chapter 1 where applicable.

❖ This section is purely a reference section to direct the code user's attention to the possibility of other sections that might be relevant to the means of egress. Also, this section makes indirect reference to Sections 101.2 and 115 and Section 1003.

SECTION 506
ACCESSIBILITY

506.1 General. A building, facility, or element that is altered shall comply with the applicable provisions in Sections 506.1.1 through 506.1.12, Chapter 11 of the *International Building Code*, and ICC A117.1 unless technically infeasible. Where compliance with this section is technically infeasible, the alteration shall provide access to the maximum extent technically feasible.

Exceptions:
1. The altered element or space is not required to be on an accessible route unless required by Section 506.2.
2. Accessible means of egress required by Chapter 10 of the *International Building Code* are not required to be provided in existing buildings and facilities.
3. Type B dwelling or sleeping units required by Section 1107 of the *International Building Code* are not required to be provided in existing buildings and facilities.

❖ The purpose of Section 506.1 is to establish minimum criteria for accessibility when dealing with existing buildings and facilities that are being renovated or altered. The history and efforts involved are similar to that discussed in the commentary for Chapter 11 of the IBC. Briefly, access to buildings and structures for people with physical disabilities has been a subject that the building codes have regulated since the early 1970s. They have consistently relied on a consensus national standard, ICC/ANSI A117.1, *Accessible and Usable Buildings and Facilities*, as the technical basis for accessibility. There has been a great deal of emphasis and awareness recently placed on the subject of accessibility through the passage of two federal laws. The Americans with Disabilities Act (ADA) and the Fair Housing Amendment Act (FHAA) are federal regulations that affect building construction as it relates to accessibility. Efforts for coordination with the federal accessibility requirements are ongoing. Representatives from interested accessibility groups, the Department of Housing and Urban Development (HUD) and the Architectural and Transportation Barriers Compliance Board (ATBCB), commonly referred to as the Access Board, have been attending and participating in the code change process for the IBC and ICC/ANSI A117.1. In addition, the International Code Council® (ICC®) has participated in the public comment process on the development of federal regulations for accessibility. Appendix B includes information found in the new *American's with Disabilities Act Accessibility Guildelines* (ADAAG) that cannot be enforced through the typical code enforcement process, but would provide beneficial information for the designer/owner for full compliance purposes. At this time, HUD has stated that the 2000 IBC with the 2001 Supplement, and the ICC/ANSI A117.1-1998 are "safe harbor" for anyone wanting to comply with the *Fair Housing Administrative Guideline* (FHAG) scoping and technical requirements. For further information, see the commentary to Chapter 11 of the IBC.

The code approaches application of accessibility provisions to a facility that is altered by broadly requiring full conformance to new construction, meaning full accessibility is expected (see Section 506.1.12). Exceptions are then provided to indicate the conditions under which less than full accessibility is permitted.

When a facility or element is altered, it must meet new code requirements. For example, if a door and frame are removed and replaced, the door must meet the requirements for width, height, maneuvering clearances and hardware. If just the doorknob is being removed, it must be replaced with lever hardware.

The circumstance under which full compliance with accessibility provisions is not required is when it is deemed to be technically infeasible (see the commentary on the definition of "Technically infeasible" in Section 202). This is considered reasonable since, if not provided for, plans for alterations may be otherwise abandoned by the building owner. The opportunity to upgrade and increase the current level of accessibility in an existing building would then be lost. This concern is also embodied in the requirement that an altered element or space is expected to be made accessible to the extent to which it is technically feasible to do so. In this manner, the code accomplishes the greatest degree of accessibility while at the same time recognizing the justifiable difficulties that may be involved in providing full accessibility in existing buildings.

Per Exception 1, if the area undergoing alteration does not contain a primary function (see Section 506.2), there are no additional requirements. However, if the area contains a primary function, there are additional criteria to achieve accessibility that may require work not in the original scope of the project. This additional criteria is to provide an accessible route to the altered area, as well as improvements to any toilets and drinking fountains that serve the altered area. Requirements for an accessible route might specify that the door previously discussed be removed and replaced because it did not have adequate width or maneuvering clearances.

Exception 2 indicates that accessible means of egress are not required in existing buildings. In existing buildings, strict compliance with Section 1007 is often technically infeasible. The requirement for a 48 inches (1219 mm) clear width between handrails would require many stairways to be widened. This often would entail movement of major structural elements in order to accomplish this alteration. Note that this is not an exception for the accessible entrance requirements.

Exception 3 exempts existing buildings being altered from providing Type B dwelling and sleeping units. Since this section is referenced from Chapter 8, this would include existing buildings undergoing a change of occupancy. This is for consistency with the FHAA (see the definitions in Section 202 of the IBC for "Dwelling unit" and "Sleeping unit" and the definitions in Section 1102 of the IBC for "Dwelling unit or sleeping unit, Type B"). Additions that contain four or more dwelling and sleeping units required to be Type B in new construction would be required to provide Type B units. It should be noted that Accessible and Type A dwelling and sleeping units are required in existing residential and institutional buildings undergoing alterations (see Section 506.1.8 and the definitions in Section 1002 of the IBC for "Accessible unit" and "Dwelling or sleeping unit, Type A").

The specific provisions of the following subsections are intended to reflect conditions under which less than full accessibility, as would be required in new construction, is permitted in altered areas. As previously discussed, Section 506.1 requires altered areas to comply with the full range of accessibility-related provisions of the code for new construction. The exceptions and subsections reflect a reasonable set of conditions under which a different level of accessibility can be provided. Sections 506.1.1 through 506.1.12 are part of the IBC coordination effort with the ICC/ANSI A117.1 accessibility standard and the recommendations for the ADAAG Review Federal Advisory Committee.

506.1.1 Entrances. Where an alteration includes alterations to an entrance, and the building or facility has an accessible entrance on an accessible route, the altered entrance is not required to be accessible unless required by Section 506.2. Signs complying with Section 1110 of the *International Building Code* shall be provided.

❖ If the building already has the accessible entrances required by Section 1105 of the IBC, an entrance that is being altered is not required to be made accessible. An exception to this would be if the entrance was required to be made accessible as part of the route to the altered primary function area. If not all entrances are accessible, provide appropriate signage to notify persons with disabilities when an entrance is or is not accessible, and if not accessible, directs them to the nearest accessible entrance.

506.1.2 Elevators. Altered elements of existing elevators shall comply with ASME A17.1 and ICC A117.1. Such elements shall also be altered in elevators programmed to respond to the same hall call control as the altered elevator.

❖ Requirements for new construction state that all elevators on an accessible route must be fully accessible in accordance with ICC A117.1. If a passenger elevator is altered, the altered element must be accessible in accordance with the requirements for existing elevators in Section 407.5 of the ICC A117.1. If the altered elevator is part of a bank of elevators, the same element must be made accessible in every elevator that is part of that bank. The purpose of this requirement is to have consistency among elevators in a bank so that disabled people are not required to wait for a specific elevator, whereas the general population can take the first available elevator.

506.1.3 Platform lifts. Platform (wheelchair) lifts complying with ICC A117.1 and installed in accordance with ASME A18.1 shall be permitted as a component of an accessible route.

❖ This section provides for the use of platform (wheelchair) lifts in existing buildings. In order to create an accessible route where there are changes in floor levels, the provisions for new construction would most often require the installation of an elevator or ramp. Platform lifts are allowed in new construction for limited conditions (see Section 1109.7 of the IBC). If the space in an existing building precludes the installation of an elevator or ramp, a platform lift may be the only practical solution. Given the choice between no accessibility or accessibility by a platform lift, accessibility is prefered. Previously, platform lift requirements were addressed in the elevator standard, ASME A17.1, but they are now addressed in their own standard, ASME A18.1. One of the many changes was the removal of the requirement for key operation, which previously discouraged independent utilization of platform lifts. Note that in accordance with Section 1007.5 of the IBC, platform lifts are also permitted for an accessible means of egress in some limited locations. However, accessible means of egress are not required in existing buildings in accordance with Section 506.1, Exception 2.

506.1.4 Ramps. Where steeper slopes than allowed by Section 1010.2 of the *International Building Code* are necessitated by space limitations, the slope of ramps in or providing access to existing buildings or facilities shall comply with Table 506.1.4.

❖ This section recognizes the circumstances where, due to existing site or configuration constraints, a ramp with a slope of one unit vertical in 12 units horizontal (1:12) may not be feasible. A steeper slope is allowed where the elevation change does not exceed 6 inches (152 mm). The remainder of ramp requirements, such as width, landings, etc., is set forth in Section 1010 of the IBC.

TABLE 506.1.4
RAMPS

SLOPE	MAXIMUM RISE
Steeper than 1:10 but not steeper than 1:8	3 inches
Steeper than 1:12 but not steeper than 1:10	6 inches

For SI: 1 inch = 25.4 mm.

❖ In existing buildings, ramps that rise 3 inches (76 mm) or less may have a slope as steep as one unit vertical in eight units horizontal (1:8). In existing buildings, ramps that rise 6 inches (152 mm) or less may have a slope as steep as one unit vertical in 10 units horizontal (1:10). If it is possible to provide a lesser slope, it is desirable to do so. These steeper slopes should only be utilized when the one unit vertical in 12 units horizontal (2 percent) (1:12) slope is not possible.

506.1.5 Dining areas. An accessible route to raised or sunken dining areas or to outdoor seating areas is not required provided that the same services and decor are provided in an accessible space usable by any occupant and not restricted to use by people with a disability.

❖ The intent of this section is to provide equal access to dining services for a disabled individual without segregation. If equivalent dining services are available on an accessible level, an accessible route to other levels or to outside dining is not required as part of the alteration to an existing dining area. For example, where a snack bar is located on one level, while full dining is provided on two other levels, an accessible route would be required to the snack bar level and one of the full dining levels.

506.1.6 Performance areas. Where it is technically infeasible to alter performance areas to be on an accessible route, at least one of each type of performance area shall be made accessible.

❖ This section recognizes that, because of the existing arrangement and location of performing areas (e.g., stages, platforms, orchestra pits, etc.), it may be infeasible to alter all performing areas to be on an accessible route. In such cases, it is reasonable to require that a minimum of one of each type of performing area be made accessible. This is intended to include access to any supporting areas utilized by the performers, such as practice rooms, dressing rooms, green rooms, etc.

506.1.7 Jury boxes and witness stands. In alterations, accessible wheelchair spaces are not required to be located within the defined area of raised jury boxes or witness stands and shall be permitted to be located outside these spaces where ramp or lift access poses a hazard by restricting or projecting into a required means of egress.

❖ This exception for jury boxes and witness stands is consistent with Section 232 of ADAAG, Judicial facilities. The intent is that if ramp access to a jury box or witness stand would have the ramp limiting or blocking the means of egress for the general population in the space, alternative locations for potential jurors or witnesses is viable.

506.1.8 Dwelling or sleeping units. Where Group I-1, I-2, I-3, R-1, R-2, or R-4 dwelling or sleeping units are being altered, the requirements of Section 1107 of the *International Building Code* for accessible or Type A units and Chapter 9 of the *International Building Code* for accessible alarms apply only to the quantity of the spaces being altered.

❖ This section sets forth the rate for providing Accessible and Type A dwelling or sleeping units in Groups I-1, I-2, I-3, R-1, R-2 and R-4 when such facilities are altered (see also Section 606.3, the definitions in Section 202 for "Dwelling units" and "Sleeping units" and the definitions in Section 1102 for "Accessible units" and "Dwelling and sleeping unit, Type A"). Assuming that the required number of Accessible or Type A units are not already provided, the number of Accessible or Type A units to be incorporated into each alteration is based on the number being altered. For example, if a nursing home was being altered a portion at a time, 50 percent of the units being altered each time would be required to be wheelchair accessible. It is not the intent that all units being altered are required to be Accessible units until 50 percent of the units in the entire facility are Accessible units. The total number of Accessible units in the facility is not required to exceed that required for new construction, as indicated in Section 506.1.12. It is unreasonable to require a greater level of accessibility in an existing building than is required in new construction. The technical criteria for Accessible units is found in ICC/ANSI A117.1-1998, Chapters 3 through 9. The technical criteria for Type A units is found in ICC/ANSI A117.1-1998, Section 1002.

This section also references visible and audible alarm requirements in Chapter 9 of the IBC. Sleeping accommodations in Groups I-1 and R-1 are required to have visible alarms in accordance with Table 907.9.1.3 of the IBC. Section 907.9.1.4 of the IBC also contains requirements for alarms within Group R-2 units. In a repair or Level 1 alterations, if the alarm system is not part of the alteration, it is not the intent of this section to require the fire alarm system to be upgraded (see Sections 404.1 and 504.1). In Level 2 and 3 alterations, change of occupancy and additions, an upgrade of the system may be required (see Sections 604.4, 704.2, 804.1 and 904).

506.1.9 Toilet rooms. Where it is technically infeasible to alter existing toilet and bathing facilities to be accessible, an accessible unisex toilet or bathing facility is permitted. The unisex facility shall be located on the same floor and in the same area as the existing facilities.

❖ This section deals with circumstances in which it is technically infeasible to alter existing toilet or bathing facilities to be accessible. In new construction, both the men's and women's facilities would be required to be accessible. When it is technically infeasible to alter the existing toilet or bathing rooms to meet new construction requirements, an accessible unisex toilet or bathing room must be provided. If this alternative is selected, the unisex room must be located on the same floor and in the same area as the existing toilet or bathing room. Signage must be provided at the inaccessible toilet rooms to notify persons with a disability when a facility is not accessible and direct them to the nearest accessible facilities. It should be noted that this alternative is not offered as a choice between making the existing separate-sex toilet and bathing rooms accessible or providing a unisex accessible toilet room. The existing separate-sex toilet rooms must be altered when it is technically feasible. Consideration of a unisex facility is only available when alteration of the existing toilet rooms is technically infeasible (see the definition for "Technically infeasible" in Section 202).

506.1.10 Dressing, fitting, and locker rooms. Where it is technically infeasible to provide accessible dressing, fitting, or locker rooms at the same location as similar types of rooms, one accessible room on the same level shall be provided. Where separate sex facilities are provided, accessible rooms for each sex shall be provided. Separate sex facilities are not required where only unisex rooms are provided.

❖ This section takes a similar approach for dressing rooms as provided for in Section 506.1.9 for toilet and bathing facilities. If it is technically infeasible to alter existing dressing rooms to be accessible, then space elsewhere on the level must be committed to providing not less than one accessible dressing room. In this case, if the existing dressing rooms provide separate rooms for each sex, then not less than one accessible dressing room for each sex must be provided.

506.1.11 Thresholds. The maximum height of thresholds at doorways shall be $^3/_4$ inch (19.1 mm). Such thresholds shall have beveled edges on each side.

❖ Thresholds at doorways may be $^3/_4$ inch (19.1 mm) maximum in existing buildings. In new construction, a typical threshold is $^1/_2$ inch (12.7 mm) maximum in accordance with Section 1008.1.6 of IBC. This section recognizes that such things as differences in floor materials may create changes in elevation greater than that allowed in new construction. Edges of thresholds greater than $^1/_4$ inch (6.4 mm) must be beveled to allow passage of a wheelchair.

506.1.12 Extent of application. An alteration of an existing element, space, or area of a building or facility shall not impose a requirement for greater accessibility than that which would be required for new construction. Alterations shall not reduce or have the effect of reducing accessibility of a building, portion of a building, or facility.

❖ The purpose of this section is to clarify to which level the requrement of Sections 506.1.1 through 506.1.11

apply. The requirements are not intended to impose a higher level of accessibility than that required in new construction. At the same time, alterations cannot result in a lesser degree of accessibility than existed before alterations were undertaken.

506.2 Alterations affecting an area containing a primary function. Where an alteration affects the accessibility to, or contains an area of, primary function, the route to the primary function area shall be accessible. The accessible route to the primary function area shall include toilet facilities or drinking fountains serving the area of primary function. For the purposes of complying with this section, an area of primary function shall be defined by applicable provisions of 49 CFR Part 37.43(c) or 28 CFR Part 36.403.

Exceptions:

1. The costs of providing the accessible route are not required to exceed 20 percent of the costs of the alterations affecting the area of primary function.

2. This provision does not apply to alterations limited solely to windows, hardware, operating controls, electrical outlets, and signs.

3. This provision does not apply to alterations limited solely to mechanical systems, electrical systems, installation or alteration of fire protection systems, and abatement of hazardous materials.

4. This provision does not apply to alterations undertaken for the primary purpose of increasing the accessibility of an existing building, facility, or element.

❖ The text references documents from the Department of Transportation and the Department of Justice for the definition of "Primary function area."

A primary function area is defined by 49 CFR Part 37.43(c) as follows:

"A primary function is a major activity for which the facility is intended. Areas of transportation facilities that involve primary functions include, but are not necessarily limited to, ticket purchase and collection areas, passenger waiting areas, train or bus platforms, baggage checking and return areas and employment areas [except those involving nonoccupiable spaces accessed only by ladders, catwalks, crawl spaces, very narrow passageways, or freight (nonpassenger) elevators which are frequented only by repair personnel]."

A primary function area is defined by 28 CFR Part 36.430 as follows:

"Primary function. A primary function is a major activity for which the facility is intended. Areas that contain a primary function include, but are not limited to, the customer services lobby of a bank, the dining area of a cafeteria, the meeting rooms in a conference center, as well as offices and other work areas in which the activities of the public accommodation or other private entity using the facility are carried out. Mechanical rooms, boiler rooms, supply storage rooms, employee lounges or locker rooms, janitorial closets, entrances, corridors and restrooms are not areas containing a primary function."

While a primary function area is not defined in the IBC or the code, the intent is that an area containing a primary function is one in which a major activity for which the building or facility is intended is carried out. The key concept is that a primary function area is one that contains a major activity of the facility (e.g., dining area in a restaurant, meeting rooms in a conference center, sales and display areas on a store). Areas that contain activities not related to the main purpose of the facility would not be considered a primary function area (e.g., the kitchen in a restaurant, public bathrooms, employee breakrooms, storage areas, mechanical rooms). With this information, it is clear that areas containing a primary function are clearly more critical in terms of the purpose for which people enter and use the facility. Therefore, this section reflects that when such areas are altered or added, it is important to require that an accessible route to the primary function area be provided.

In addition, any toilet rooms and drinking fountains serving the primary function area must also be made accessible, even though such facilities and areas may not by themselves be considered primary function areas. This would include providing an accessible route to the toilet rooms and drinking fountains, as well as altering the existing toilet room, fixtures within the room and drinking fountains to meet accessibility requirements.

There are conditions under which it may not be reasonable to enforce strictly this requirement for an accessible route to an altered or added primary function area.

Exception 1 approaches this by utilizing the cost of the alterations or addition as a basis for determining if providing an accessible route is reasonable. The requirement for a complete accessible route does not apply when the cost of providing it exceeds 20 percent of the cost of the alterations or addition to the primary function area. These costs are intended to be based on the actual costs of the planned alterations or addition to the primary function area before consideration of the cost of providing an accessible route. For example, if the planned alterations will cost $100,000, not including the cost of an accessible route to a primary function area, this exception would apply if the additional cost of providing the accessible route would exceed $20,000.

It is not the intent to exempt all requirements for accessibility when the total cost for providing the accessible route exceeds the 20-percent threshold. Improvements to the accessible route are required to the extent that costs do not exceed 20 percent of the cost to the planned alteration or addition. It is not required that the full 20 percent be spent. If the accessible route (including accessible toilet rooms and drinking fountains) is already provided, no additional expenditure is required. Note that there is not a priority list given for where funds should be spent on improving the accessible route. The logical progression is access to the site, accessible exterior routes to accessible entrances, access throughout the facility, access to services within the facility, toilet and bathing rooms and, finally, drinking fountains. Evaluation on how and where the funds available should best be spent must be made on a case-by-case basis. For example, if an accessible route is not available to an upper level, and the cost of an elevator is more than 20 percent of the cost of the renovation, then other alternatives could be investigated, such as a platform lift or limited access elevator, or adding the elevator pit and shaft at this time, with elevator equipment added later. If all such items are in excess of the 20-percent limit, perhaps the available funds could be spent towards making the toilet rooms accessible. The idea is that existing buildings would become fully accessible over time.

Exceptions 2 and 3 identify certain alterations that are not intended to trigger the requirement for providing an accessible route to a primary function area. Alterations limited to such elements as windows, hardware, operating controls, electrical outlets, signage, mechanical, electrical and fire protection systems, including alterations for the purpose of abating a hazardous materials circumstance, do not affect the usability of a primary function area in the same manner as alterations that affect the floor plan or the configuration, location or size of rooms or spaces. It is therefore considered unreasonable to require the installation of an accessible route when the scope of alterations is limited to that reflected in these exceptions. Note that costs for these items are not "backed out" of the total cost for the alteration before applying Exception 1. Exceptions 2 and 3 are alterations limited to the specific items referenced.

Exception 4 is intended to avoid penalizing a building owner who is undertaking alterations or additions for the purpose of increasing accessibility. It is appropriate to encourage owners to make such alterations without requiring them to do more work simply because they chose to increase the accessibility of the space. This could otherwise have the opposite effect of discouraging such alterations to avoid the expense of undertaking more work and expense than was originally planned. For example, federal law (ADA) requires that owners of existing buildings remove certain existing barriers to accessibility. Removal of such barriers may require a permit from the code official. It would be unreasonable to have such activity trigger the mandatory requirement for further alterations to accomplish accessibility beyond the original planned work. In principle, the code takes the view that some extent of greater accessibility is positive progress and should be encouraged, not penalized.

SECTION 507
STRUCTURAL

507.1 General. Where alteration work includes replacement of equipment that is supported by the building or where a

reroofing permit is required, the structural provisions of this section shall apply.

❖ These structural requirements apply to alterations involving reroofing or the replacement of equipment that is supported by the building. Note that under Section 102.4.2 buildings that comply with the structural provisions of any of the following model codes are deemed to comply with this code:

International Building Code® (IBC®), 2003 or 2000 edition.
National Building Code (NBC), 1999 edition.
Standard Building Code (SBC), 1997 edition.
Uniform Building Code (UBC), 1997 edition.

Although the *International Residential Code®* (IRC®) is not explicitly listed, its use is required by the IBC for detached one- and two-family dwellings as well as townhouses not more than three stories above grade plane with a separate means of egress.

507.2 Design criteria. Existing structural components supporting alteration work shall comply with this section.

❖ For alterations involving reroofing or the replacement of equipment that is supported by the building, the criteria that applies to any affected structural member is given in this section.

507.2.1 Replacement of roofing or equipment. Where replacement of roofing or equipment results in additional dead loads, structural components supporting such re-roofing or equipment shall comply with the vertical load requirements of the *International Building Code*.

Exceptions:

1. Structural elements whose stress is not increased by more than 5 percent.

2. Buildings constructed in accordance with the *International Residential Code* or the conventional construction methods of the *International Building Code* and where the additional dead load from the equipment is not increased by more than 5 percent.

❖ Where reroofing or replacement of equipment results in a net increase in the supported dead load, the affected structural components must be checked to verify that the vertical load requirements of the IBC for a new structure are satisfied. The reference to only vertical loads is an indication that an analysis for the effects of lateral loads, such as seismic, would not be a necessity for this level of alteration.

There are two exceptions to satisfy full compliance with the IBC, one of which allows buildings constructed under the IRC or the conventional construction provisions of the IBC to increase dead loads from equipment by up to 5 percent. Another exception allows additional dead loads that do not increase stresses in affected structural elements by more than 5 percent. Allowing overstresses of up to 5 percent in existing structural members has been a long-standing rule of thumb used by structural engineers. These exceptions do not specifically address the cumulative affects from successive alterations; for instance, it would be prudent to limit cumulative increases under these exceptions to a total of 5 percent unless it is clearly documented that a structure has additional capacity.

507.2.2 Parapet bracing and wall anchors for reroof permits. Unreinforced masonry bearing wall buildings classified as Seismic Design Category D, E, or F shall have parapet bracing and wall anchors installed at the roof line whenever a reroofing permit is issued. Such parapet bracing and wall anchors shall be designed in accordance with the reduced *International Building Code* level seismic forces as specified in Section 407.1.1.3 and design procedures of Section 407.1.1.1.

❖ The failure of parapets in unreinforced masonry-(URM) bearing wall buildings has been a recurring problem in areas that experience significant earthquakes. Because this poses a very real risk, the code requires these elements to be braced where the seismic hazard is deemed to be relatively high as reflected in a building's seismic design category. Since the code does not provide the requirements for establishing the seismic design category, Section 1616.3 of the IBC must be used for this purpose. For an explanation of this process, see the commentary for Section 407.1.1.1. The code allows this parapet bracing to be designed using the reduced seismic forces of Section 407.1.1.3 and the design procedures listed in Section 407.1.1.1. See Section 11.9.1 of FEMA 356 or Section A113.6 of *Guidelines for Seismic Evaluation of Exist-*

ing Buildings (GSREB) in Appendix A of the code for specific requirements for parapets.

An example of the prescribed parapet bracing is illustrated schematically in Figure 507.2.2. The approach shown—a steel brace anchored to the parapet wall and the roof structure—is a commonly used method of providing lateral support for a parapet. Other methods of strengthening a parapet can be used to mitigate the hazard, provided their use is approved as alternative methods of design and construction under Section 104.11.

Note that only the parapet bracing for seismic loads is required. Some buildings, especially those east of the Rocky Mountains, were not designed for seismic loads. They were only designed to resist wind loads. At the present time the code does not require a review of the building for seismic loads.

507.3 Roof diaphragm. Where roofing materials are removed from more than 50 percent of the roof diaphragm of a building or section of a building where the roof diaphragm is a part of the main windforce-resisting system the integrity of the roof diaphragm shall be evaluated and if found deficient because of insufficient or deteriorated connections, such connections shall be provided or replaced.

❖ The removal of roofing provides an opportunity to inspect a portion of the structure that is otherwise concealed. In reroofing operations where more than 50 percent of the roof covering is removed, a roof diaphragm that is a part of the main windforce-resisting

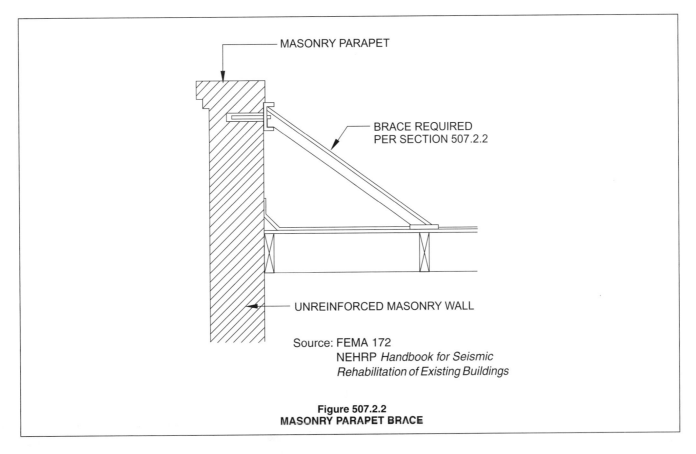

MASONRY PARAPET

BRACE REQUIRED
PER SECTION 507.2.2

UNREINFORCED MASONRY WALL

Source: FEMA 172
NEHRP *Handbook for Seismic
Rehabilitation of Existing Buildings*

**Figure 507.2.2
MASONRY PARAPET BRACE**

system, shall be evaluated for adequate strength to perform as a wind-resistant diaphragm. If found deficient because of insufficient or deteriorated connections, such connections shall be provided or replaced. Although not stated in the code text, the connections of interest are those needed for the roof sheathing to act as a diaphragm to transfer wind loads to the structure below.

Bibliography

The following resource materials are referenced in this chapter or are relevant to the subject matter addressed in this chapter.

FEMA 74, *Reducing the Risks of Nonstructural Earthquake Damage – a Practical Guide*. Washington, DC: Federal Emergency Management Agency, 1994.

FEMA 172, *NEHRP Handbook for the Seismic Rehabilitation of Existing Buildings*. Washington, DC: Federal Emergency Management Agency, 1997.

FEMA 356, *Prestandard and Commentary for the Seismic Rehabilitation of Buildings*. Washington, DC: Federal Emergency Management Agency, 2000.

IBC-00, *International Building Code*. Falls Church, VA: International Code Council, 2000.

IBC-03, *International Building Code*. Falls Church, VA: International Code Council, 2003.

IRC-03, *International Residential Code*. Falls Church, VA: International Code Council, 2003.

NBC-99, *BOCA National Building Code*. Country Club Hills, IL: Building Officials & Code Administrators International, Inc. 1999.

SBC-99, *Standard Building Code*. Birmingham, AL: Southern Building Code Congress International, 1999.

UBC-97, *Uniform Building Code*. Whittier, CA: International Conference of Building Officials, 1997.

CHAPTER 6

ALTERATIONS—LEVEL 2

General Comments

This chapter provides the technical requirements for those existing buildings that undergo Level 2 alterations. Reference is made to Level 2 alterations in Section 304, which includes reconfiguration of space. Installation of additional equipment that did not previously exist or the addition or elimination of doors and windows are also considered as Level 2 alterations, and hence covered by this chapter.

This chapter, similar to other chapters of the code, covers all building related subjects such as structural, mechanical, plumbing, electrical and accessibility as well as the fire and life safety issues when the alterations are classified as Level 2. Sections 601, 602 and 603 are related to scoping, special use and occupancy and building elements and materials. In the interest of brevity and avoiding repetition of materials, Section 601.2 requires that Level 2 alterations comply not only with Chapter 6, but also with Chapter 5. As such, any alteration work that is classified as Level 2 is assumed to include everything related to Level 1 alterations plus the reconfiguration of space. Section 603 covers in details elements such as existing vertical openings and when such openings might be required to be enclosed; stairway enclosures; interior finishes and guards. The remainder of the chapter is related to fire protection, means of egress, accessibility, structural, electrical, mechanical and plumbing provisions. While means of egress, Section 605, covers subjects such as the minimum number of exits, fire escape requirements and construction details, corridor openings and exit signs, the structural section, Section 607, addresses subjects such as new and existing structural members, gravity loads, lateral loads and snow drift loads.

The level of improvements required as a result of Level 2 alterations are extensive when compared to Level 1 al-terations because the extent of alterations are more drastic and space reconfiguration is involved. The decision to divide the level typically known as "alterations" in the *International Building Code®* (IBC®) and previous legacy building codes into three parts of Level 1, Level 2 and Level 3 was based on the fact that minor alterations that do not include space reconfiguration and extensive alterations that might range from relocation of walls and partitions in various floors to the entire building should be treated differently and with different threshold levels for requiring upgrades or improvements to the building or spaces within the building.

Purpose

The purpose of this chapter is to provide detailed requirements and provisions to identify the required improvements in the existing building elements, building spaces and building structural system. This chapter is distinguished from Chapters 5 and 7 by involving space reconfiguration that could be up to and including 50 percent of the area of the building. In contrast, Level 1 alterations do not involve space reconfiguration and Level 3 alterations involve extensive space reconfiguration that exceeds 50 percent of the building area. Depending on the nature of alteration work, its location within the building and whether it encompasses one or more tenants, improvements and upgrades could be required for the open-floor penetrations, sprinkler system or the installation of additional means of egress such as stairs or fire escapes. At times and under certain situations, this chapter also intends to improve the safety of certain building features beyond the work area and in other parts of the building where no alteration work might be taking place.

SECTION 601
GENERAL

601.1 Scope. Level 2 alterations as described in Section 304 shall comply with the requirements of this chapter.

Exception: Buildings in which the reconfiguration is exclusively the result of compliance with the accessibility requirements of Section 506.2 shall be permitted to comply with Chapter 5.

❖ Any alteration work that results in the reconfiguration of space or otherwise falls under the classification of Level 2 alterations as described in Section 304, must comply with Chapter 6 (see Figure 601.1). The exception is intended to encourage existing buildings to improve the accessibility for the disabled by allowing any such alterations (which are solely for compliance with accessibility provisions of Section 506) to only comply with Chapter 5

provisions, as if no space reconfiguration has taken place. This is an advantage for improving accessibility in buildings because Chapter 5 is related to alterations that do not encompass any space reconfiguration, and as such has a lower level of compliance requirement and higher thresholds for triggering other improvements.

Example 601.1:

An existing office building does not comply with current accessibility requirements. The owner wishes to make the building completely accessible to the disabled by installing ramps, enlarging toilet rooms and widening doorways.

Q: What, if any, additional code requirements are triggered?

A: All affected areas can be rebuilt with similar materials, except glass at human impact areas, which

FIGURE 601.1 – 601.3

ALTERATIONS—LEVEL 2

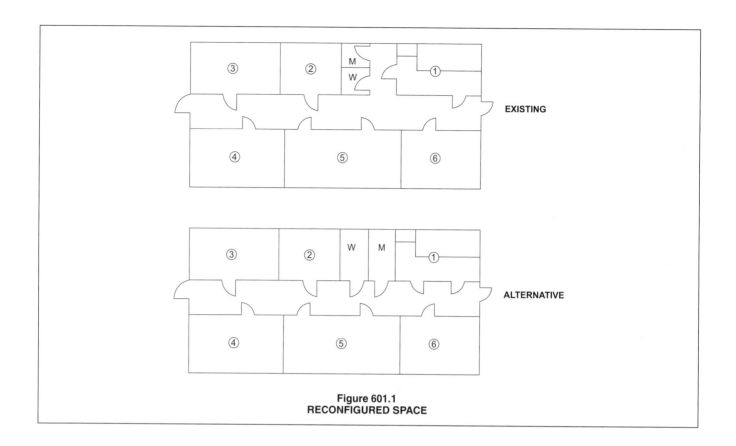

Figure 601.1
RECONFIGURED SPACE

must comply with safety glass requirements (see Section 403.2 and Figure 301.1).

601.2 Alteration Level 1 compliance. In addition to the requirements of this chapter, all work shall comply with the requirements of Chapter 5.

❖ Chapters 6 and 7 have cascading effects by requiring that each level of alteration must comply with the chapter for the lower level of alteration. Accordingly, an alteration project that is classified as Level 2 must comply with Chapters 5 and 6. Similarly, an alteration project classified as Level 3 must comply with Chapters 5, 6 and 7. This is to eliminate the repetition of various requirements from Chapter 5 in Chapters 6 and 7 and from Chapter 6 in Chapter 7, as a Level 2 alteration, in general, includes various aspects of a Level 1 alteration, and a Level 3 alteration, in general, includes various aspects of Level 1 and 2 alterations. The code drafting committee decided to make reference to these chapters rather than repeat all such provisions again.

601.3 Compliance. All new construction elements, components, systems, and spaces shall comply with the requirements of the *International Building Code.*

Exceptions:

1. Windows may be added without requiring compliance with the light and ventilation requirements of the *International Building Code.*

2. Newly installed electrical equipment shall comply with the requirements of Section 608.

3. The length of dead-end corridors in newly constructed spaces shall only be required to comply with the provisions of Section 605.6.

4. The minimum ceiling height of the newly created habitable and occupiable spaces and corridors shall be 7 feet (2134 mm).

❖ New elements, components and spaces created as a result of Level 2 alteration work must comply with the IBC or other applicable code, as Section 101.4 of the IBC makes reference to other *International Codes*® such as the *International Plumbing Code*® (IPC®), *International Mechanical Code*® (IMC®) and *International Fire Code*® (IFC®). Exception 1 allows new windows to be added without complying with the IBC light and ventilation requirements, so as not to discourage a building owner from making improvements in the current amount of natural light and ventilation. Exceptions 2 and 3 refer the user to Sections 605.6 and 608, which provide detailed provisions for electrical equipment and dead-end corridors. Exception 4 allows the minimum ceiling height of newly created spaces be lowered from the IBC requirement of 7 feet 6 inches (2289 mm) to 7 feet (2134 mm), recognizing that even though the space is newly created, it is within an existing building that might have framing members that will prohibit higher ceilings.

SECTION 602
SPECIAL USE AND OCCUPANCY

602.1 General. Alteration of buildings classified as special use and occupancy as described in the *International Building Code* shall comply with the requirements of Section 601.1 and the scoping provisions of Chapter 1 where applicable.

❖ Special use and occupancy buildings described in Chapter 4 of the IBC, such as covered mall buildings and high-rise buildings, are treated the same as any other building when the work taking place is a Level 2 alteration. This section specifically and clearly provides this information because at higher levels of alteration work, or where there is a change of occupancy, some special use and occupancy buildings are required to comply with the IBC, either completely or in particular alteration areas (see Sections 702 and 802). By reference to Section 601.1 and Chapter 1, this section makes indirect reference to Sections 101.2, 115 and applicable subsections of Section 1003.

SECTION 603
BUILDING ELEMENTS AND MATERIALS

603.1 Scope. The requirements of this section are limited to work areas in which Level 2 alterations are being performed, and shall apply beyond the work area where specified.

❖ This section provides the scoping provisions for vertical openings, smoke barriers, interior finish and guards by requiring that the work areas must comply with this section, and that at times there are supplementary requirements that apply to spaces beyond the work areas. Such supplementary requirements are found in Sections 603.2.2, 603.2.3 and 603.4.1 where the extent of compliance beyond the work area are identified.

603.2 Vertical openings. Existing vertical openings shall comply with the provisions of Sections 603.2.1, 603.2.2, and 603.2.3.

❖ The purpose of this section is to provide requirements for vertical openings in Level 2 alterations. Sections 603.2.1 and 603.2.2 provide these requirements.

603.2.1 Existing vertical openings. All existing interior vertical openings connecting two or more floors shall be enclosed with approved assemblies having a fire-resistance rating of not less than 1 hour with approved opening protectives.

Exceptions:

1. Where vertical opening enclosure is not required by the *International Building Code* or the *International Fire Code.*

2. Interior vertical openings other than stairways may be blocked at the floor and ceiling of the work area by installation of not less than 2 inches (51 mm) of solid wood or equivalent construction.

3. The enclosure shall not be required where:

3.1. Connecting the main floor and mezzanines; or

3.2. All of the following conditions are met:

3.2.1. The communicating area has a low hazard occupancy or has a moderate hazard occupancy that is protected throughout by an automatic sprinkler system.

3.2.2. The lowest or next to the lowest level is a street floor.

3.2.3. The entire area is open and unobstructed in a manner such that it may be assumed that a fire in any part of the interconnected spaces will be readily obvious to all of the occupants.

3.2.4. Exit capacity is sufficient to provide egress simultaneously for all the occupants of all levels by considering all areas to be a single floor area for the determination of required exit capacity.

3.2.5. Each floor level, considered separately, has at least one half of its individual required exit capacity provided by an exit or exits leading directly out of that level without having to traverse another communicating floor level or be exposed to the smoke or fire spreading from another communicating floor level.

4. In Group A occupancies, a minimum 30-minute enclosure shall be provided to protect all vertical openings not exceeding three stories.

5. In Group B occupancies, a minimum 30-minute enclosure shall be provided to protect all vertical openings not exceeding three stories. This enclosure, or the enclosure specified in Section 603.2.1, shall not be required in the following locations:

5.1. Buildings not exceeding 3,000 square feet (279 m^2) per floor.

5.2. Buildings protected throughout by an approved automatic fire sprinkler system.

6. In Group E occupancies, the enclosure shall not be required for vertical openings not exceeding three stories when the building is protected throughout by an approved automatic fire sprinkler system.

7. In Group F occupancies, the enclosure shall not be required in the following locations:

7.1. Vertical openings not exceeding three stories.

7.2. Special purpose occupancies where necessary for manufacturing operations and direct access is provided to at least one protected stairway.

7.3. Buildings protected throughout by an approved automatic sprinkler system.

8. In Group H occupancies, the enclosure shall not be required for vertical openings not exceeding three

stories where necessary for manufacturing operations and every floor level has direct access to at least two remote enclosed stairways or other approved exits.

9. In Group M occupancies, a minimum 30-minute enclosure shall be provided to protect all vertical openings not exceeding three stories. This enclosure, or the enclosure specified in Section 603.2.1, shall not be required in the following locations:

   9.1. Openings connecting only two floor levels.

   9.2. Occupancies protected throughout by an approved automatic sprinkler system.

10. In Group R-1 occupancies, the enclosure shall not be required for vertical openings not exceeding three stories in the following locations:

   10.1. Buildings protected throughout by an approved automatic sprinkler system.

   10.2. Buildings with less than 25 dwelling units or sleeping units where every sleeping room above the second floor is provided with direct access to a fire escape or other approved second exit by means of an approved exterior door or window having a sill height of not greater than 44 inches (1118 mm) and where:

      10.2.1. Any exit access corridor exceeding 8 feet (2438 mm) in length that serves two means of egress, one of which is an unprotected vertical opening, shall have at least one of the means of egress separated from the vertical opening by a 1-hour fire barrier; and

      10.2.2. The building is protected throughout by an automatic fire alarm system, installed and supervised in accordance with the *International Building Code*.

11. In Group R-2 occupancies, a minimum 30-minute enclosure shall be provided to protect all vertical openings not exceeding three stories. This enclosure, or the enclosure specified in Section 603.2.1, shall not be required in the following locations:

   11.1. Vertical openings not exceeding two stories with not more than four dwelling units per floor.

   11.2. Buildings protected throughout by an approved automatic sprinkler system.

   11.3. Buildings with not more than four dwelling units per floor where every sleeping room above the second floor is provided with direct access to a fire escape or other approved second exit by means of an approved exterior door or window having a sill height of not greater than 44 inches (1118 mm) and the building is protected throughout by an automatic fire alarm system complying with Section 604.4.

12. One- and two-family dwellings.

13. Group S occupancies where connecting not more than two floor levels or where connecting not more than three floor levels and the structure is equipped throughout with an approved automatic sprinkler system.

14. Group S occupancies where vertical opening protection is not required for open parking garages and ramps.

❖ All existing vertical openings connecting two or more floors must be enclosed with assemblies of 1-hour fire-resistance-rated construction and approved protected openings. Even without the following 14 exceptions, this provision is more relaxed than the IBC shaft enclosure requirements that require 2-hour fire-resistance-rated enclosures for buildings four stories or more in height. Additionally, the user should remember the scoping provisions of Section 603.1, which indicates that the enclosure requirements triggered under Level 2 alterations apply only to work areas. As such, if the alteration is on the first floor of a multistory building, and the vertical opening is within the work area, only the portion of the vertical opening within the first floor is required to be enclosed and not the entire vertical opening from the bottom to top floor. Exception 1 is provided so that the provisions of the code in relation to vertical openings are not more restrictive than those of the IBC and IFC. The IFC regulates vertical openings in existing buildings by Table 704.1. This table allows vertical openings without enclosures where they connect only two stories in other than hazardous or institutional occupancies.

Exception 2 allows vertical openings other than stairways to be blocked by solid wood of 2 inches (51 mm) in thickness or equivalent construction. The 2 inches (51 mm) used here is the nominal thickness of the solid wood members.

Exception 3.1 is a repetition of Exception 9 of Section 707.2 of the IBC. Exception 3.2 provides five conditions that, if all are met simultaneously, enclosure of vertical openings are not required. The combination of these five conditions is equivalent to a situation where the occupancy is a low hazard occupancy, fire sprinklers are present and there is sufficient means of egress capacity.

Exceptions 4 through 11 and Exception 13 allow either a 30-minute protection for vertical openings in lieu of 1 hour or allow the vertical opening to remain open for buildings three stories or less in height with certain features such as sprinklers or multiple remote means of egress.

Exceptions 12 and 14 allow existing vertical openings to remain open in single-family, duplex buildings and open parking garages. The IBC and IFC exception is listed under Exception 1. As can be seen, the exceptions within the IBC and this code allow for numerous methods of dealing with vertical openings so that the building owners and designers have many options to provide a minimum level of protection intended by the codes.

603.2.2 Supplemental shaft and floor opening enclosure requirements. Where the work area on any floor exceeds 50 percent of that floor area, the enclosure requirements of Section 603.2 shall apply to vertical openings other than stairways throughout the floor.

> **Exception:** Vertical openings located in tenant spaces that are entirely outside the work area.

❖ Unenclosed vertical openings and shafts, other than stairways that would typically be required to be enclosed by the IBC, are required to be enclosed throughout a floor where the work area on that floor exceeds 50 percent of the floor area. Stairways are not included here because they are specifically addressed in Section 603.2.3. This is the first of the supplemental requirements for vertical enclosures to which the code user is referred in Section 603.1. The proportionality philosophy of the code is applied here, as it is believed that a work area extending to more than 50 percent of the floor area of a story is large enough to trigger additional vertical openings, other than stairways, to be en-

closed. As the undertaking of such work in tenant spaces, which are completely outside of the work area and are not within the scope of the original proposed alteration, is disruptive and an unacceptable practice, the exception eliminates this requirement in such tenant spaces.

Example 603.2.2:

An existing four-story retail building is being altered on the third floor. The alterations involve reconfiguration of space in about 70 percent of the third floor. There are two enclosed stairways, an open escalator and an elevator without protected enclosures. The building is not sprinklered.

Q: Will the escalator and the elevator be required to be protected or enclosed in accordance with the IBC?

A: Since the work area on the third floor exceeds 50 percent of the area of the third floor, both the elevator and the escalator must be protected even though the elevator is outside of the work area. The protection is needed on the third floor only (see Section 603.1 and Figure 603.2.2).

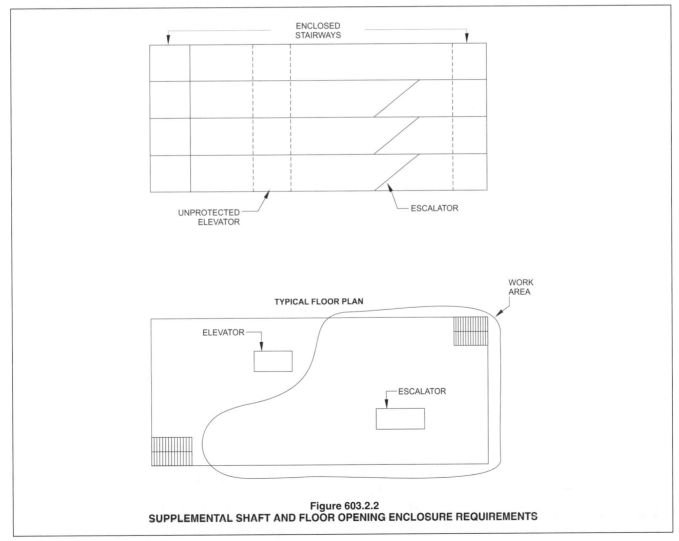

Figure 603.2.2
SUPPLEMENTAL SHAFT AND FLOOR OPENING ENCLOSURE REQUIREMENTS

603.2.3 Supplemental stairway enclosure requirements. Where the work area on any floor exceeds 50 percent of that floor area, stairways that are part of the means of egress serving the work area shall, at a minimum, be enclosed with smoke-tight construction on the highest work area floor and all floors below.

Exception: Where stairway enclosure is not required by the *International Building Code* or the *International Fire Code*.

❖ Stairways serving as part of the means of egress system must be enclosed where possible to increase the degree of protection for building occupants. The triggering mechanism in this section is the percentage of work area on any given floor. If the work area exceeds 50 percent of floor area on any floor, then stairways that are part of the means of egress and serve that work area must be enclosed, even though the stairways may not be located within the work area. This provision could affect several stairways, all of which might be located outside of the proposed work area. For example, on a multistory building that contains three required stairways, all of which are unenclosed. The work area on an upper floor exceeds 50 percent of that story area triggers the requirement for the enclosure of all three stairways if they each serve as one of the exits for that work area. The enclosure need only be smoke tight and need not provide any fire-resistance rating, and the openings into such a smoke-tight enclosure need not be protected with fire-resistant assemblies, only smoke-protected assemblies. The extent of enclosure is specified to be from the floor at which the work areas have triggered this requirement and all floors below it. The exception is a reminder that stairways not required to be enclosed by the IBC or IFC are not affected by this section.

603.3 Smoke barriers. Smoke barriers in Group I-2 occupancies shall be installed where required by Sections 603.3.1 and 603.3.2.

❖ This section is the charging paragraph for smoke barrier requirements in Group I-2 occupancies. Sections 603.3.1 and 603.3.2 give the requirements.

603.3.1 Compartmentation. Where the work area is on a story used for sleeping rooms for more than 30 patients, the story shall be divided into not less than two compartments by smoke barrier walls complying with Section 603.3.2 such that each compartment does not exceed 22,500 square feet (2093 m²), and the travel distance from any point to reach a door in the required smoke barrier shall not exceed 200 feet (60 960 mm).

Exception: Where neither the length nor the width of the smoke compartment exceeds 150 feet (45 720 mm), the travel distance to reach the smoke barrier door shall not be limited.

❖ The smoke barrier provisions in alterations classified as Level 2 apply only to Group I-2 occupancies, such as hospitals, nursing homes, mental hospitals and detoxification facilities. In such facilities, Section 603.3.1 requires that a story used for sleeping rooms for more

than 30 patients be divided into at least two compartments by the use of smoke barriers. No such compartment can be larger than 22,500 square feet (2093 m²). Realizing that a 22,500-square-foot (2093 m²) compartment could be a very long and narrow space, the travel distance from any point to reach a door in the smoke barrier is limited to 200 feet (60 960 mm). This requirement and the related compartment size and travel distance are taken directly from the IBC, with the exception that the trigger under the code is only for sleeping areas for more than 30 patients, as opposed to the IBC, which requires compartmentation in all sleeping areas regardless of occupant load, and in nonsleeping areas with an occupant load of 50 or more (see commentary, Section 407.4 of the IBC).

The exception is a reflection of space geometry where spaces do not have dimensions larger than 150 feet (45 720 mm) and, therefore, will most likely meet the maximum 200-foot (60 960 mm) travel distance limitation.

603.3.2 Fire-resistance rating. The smoke barriers shall be fire-resistance rated for 30 minutes and constructed in accordance with the *International Building Code*.

❖ This is a less-restrictive requirement than the requirement for a 1-hour fire-resistance rating given by the IBC to allow the possibility of the use of existing walls and partitions as smoke barriers that results in minimal disruption and construction in existing facilities while still improving the safety features for Group I-2 facilities. The construction requirements for smoke barriers, including requirements for continuity, openings, penetrations, joints and duct openings are given in Section 709 of the IBC.

603.4 Interior finish. The interior finish of walls and ceilings in exits and corridors in any work area shall comply with the requirements of the *International Building Code*.

Exception: Existing interior finish materials that do not comply with the interior finish requirements of the *International Building Code* shall be permitted to be treated with an approved fire-retardant coating in accordance with the manufacturer's instructions to achieve the required rating.

❖ The interior finish of corridor and exit walls and ceilings within the work areas must meet the flame spread and smoke-development limitations and requirements of the IBC. Corridors and exits are defined in Section 1002.10 of the IBC and flame spread and smoke-developed classification are determined from Table 803.5 of the IBC and based on ASTM E 84. To be able to maintain some interior finish materials that do not comply with the flame spread and smoke-development requirements of the code rather than remove and replace them, the exception provides the option of available technology for the use of fire-retardant treatment of such surfaces as long as the code official approves the coating and it is applied in accordance with the manufacturer's instructions to achieve the required rating.

603.4.1 Supplemental interior finish requirements. Where the work area on any floor exceeds 50 percent of the floor area, Section 603.4 shall also apply to the interior finish in exits and corridors serving the work area throughout the floor.

> **Exception:** Interior finish within tenant spaces that are entirely outside the work area.

❖ The interior finish requirements discussed above in Section 603.4 are also applicable in corridors and exits that are outside of the work area but serve that same work area, if the work area on any floor exceeds 50 percent of the floor area at that level. Similar to other supplemental requirements, this extended area of code application is not required in tenant spaces that are entirely outside of the work area.

603.5 Guards. The requirements of Sections 603.5.1 and 603.5.2 shall apply in all work areas.

❖ This is the charging paragraph for guard requirements. The detailed requirements are given in Sections 603.5.1 and 603.5.2.

603.5.1 Minimum requirement. Every portion of a floor, such as a balcony or a loading dock, that is more than 30 inches (762 mm) above the floor or grade below and is not provided with guards, or those in which the existing guards are judged to be in danger of collapsing, shall be provided with guards.

❖ To reduce the potential of an accidental fall, the IBC requires guards where open sides of walking surfaces, mezzanines, stairways, ramps, landings and other similar areas are located more than 30 inches (702 mm) above the floor or grade below. The code in this section uses the same criteria to require new guards where none existed before or to require that dangerous guards be replaced. This requirement is applicable only in the work areas and does not contain supplementary provisions. Replacement of existing guards is required if they are judged to be in danger of collapsing. This sentence makes it clear that the replacement of guards is triggered if anyone, including the owner, designer or code official, judge the guard to be in danger of collapsing. It needs to be stressed, though, that guards that could potentially collapse could be considered dangerous as defined in Section 202 and required to be replaced on the basis of Section 115.1, whether it is in the work area or not. Section 115 of the IBC and Section 110 of the IFC have the same effect and could require replacement of guards anywhere within the building, even if no alterations are planned.

Where Section 603.5.1 triggers installation of new guards or replacement of existing guards, the new guards must comply with the IBC for height, spacing between balusters and structural strength (see Sections 1012 and 1607.7 of the IBC).

Example 603.5:

An existing restaurant has an elevated platform for performances, an elevated dining area and an outside dining deck, all of which require guards under the IBC. The performance area has no guards and the elevated dining area has existing guards that are in acceptable physical condition, but the spacing between the rails is $5\frac{1}{2}$ inches (140 mm). The outside dining deck has guards that are deteriorated and might collapse under applied loads. The owner plans to make some rearrangements of the kitchen, toilets and the elevated dining, which are all in the same area.

Q: What requirements related to guards are applicable?

A: 1. The guards for the elevated dining area do not require any upgrading, as they are in acceptable physical shape.

2. The elevated performance area and the outside dining deck are not within the work area and as such are not covered by this section of the code. Both of these areas, however, could most likely be considered unsafe under the provisions of Section 115.1, and accordingly guards in full compliance with the IBC must be installed in both locations.

603.5.2 Design. Where there are no guards or where existing guards must be replaced, the guards shall be designed and installed in accordance with the *International Building Code*.

❖ The strength, load carrying capacity and maximum opening requirements for guards are given by the IBC. This is consistent with the requirements of Section 503.3 for new building materials and equipment.

SECTION 604
FIRE PROTECTION

604.1 Scope. The requirements of this section shall be limited to work areas in which Level 2 alterations are being performed, and where specified they shall apply throughout the floor on which the work areas are located or otherwise beyond the work area.

❖ The scoping provisions of fire protection are the same as all other subjects in Chapter 6. The requirements are triggered within work areas and, at times, there are supplemental requirements that apply to areas outside of the work area.

604.2 Automatic sprinkler systems. Automatic sprinkler systems shall be provided in accordance with the requirements of Sections 604.2.1 through 604.2.5. Installation requirements shall be in accordance with the *International Building Code*.

❖ The sprinkler requirements are grouped into four sections: high-rise buildings; Groups A, E, F-1, H, I, M, R-1, R-2, R-4, S-1 and S-2; windowless stories; and buildings and areas listed in Table 903.2.13 of the IBC.

604.2.1 High-rise buildings. In high-rise buildings, work areas that include exits or corridors shared by more than one tenant or that serve an occupant load greater than 30 shall be provided with automatic sprinkler protection where the work area is located on a floor that has a sufficient sprinkler water supply system from an existing standpipe or a sprinkler riser serving that floor.

❖ The requirement in high-rise buildings is dependent upon the availability of a sufficient sprinkler water supply from an existing standpipe, or a sprinkler riser serving the floor where the alteration work area is located. This criteria is included because of the complexity and extensive cost involved in constructing completely new sprinkler water mains, risers and associated piping and equipment in high-rise buildings. Where there are existing standpipe or sprinkler risers with a sufficient water supply serving a particular floor, work areas on that floor must be sprinklered if such work areas include corridors or exits shared by more than one tenant, or where such exits or corridors serve an occupant load greater than 30.

604.2.1.1 Supplemental automatic sprinkler system requirements. Where the work area on any floor exceeds 50 percent of that floor area, Section 604.2.1 shall apply to the entire floor on which the work area is located.

Exception: Tenant spaces that are entirely outside the work area.

❖ If the work area on a floor exceeds 50 percent of that floor area, sprinklers discussed above must be installed throughout the floor except in tenant areas that are entirely outside the work area. Note that the 50-percent rule applies to the floor area only, and therefore this does not become a Level 3 alteration until the work area exceeds 50 percent of the aggregate area of all floors.

604.2.2 Groups A, E, F-1, H, I, M, R-1, R-2, R-4, S-1, and S-2. In buildings with occupancies in Groups A, E, F-1, H, I, M, R-1, R-2, R-4, S-1, and S-2, work areas that include exits or corridors shared by more than one tenant or that serve an occupant load greater than 30 shall be provided with automatic sprinkler protection where all of the following conditions occur:

1. The work area is required to be provided with automatic sprinkler protection in accordance with the *International Building Code* as applicable to new construction;

2. The work area exceeds 50 percent of the floor area; and

3. The building has sufficient municipal water supply for design of a fire sprinkler system available to the floor without installation of a new fire pump.

Exception: Work areas in Group R occupancies three stories or less in height.

❖ While, in general, automatic sprinkler protection for various occupancies in the IBC is a function of "fire area" size, in the code it is a function of five conditions in select occupancies. This section establishes the criteria for all occupancies, except for Group B, F-2 and U occupancies, because these three are not required to be sprinklered in the IBC unless they are in special buildings, such as covered malls, high rises and the so-called windowless stories. First and foremost to be considered for the possibility of requiring a sprinkler system is whether such a work area would be required to be sprinklered under the provisions of the IBC for new construction. If the IBC does not require a sprin-

kler system for the space or building under consideration, neither will the code. Once this is established, the other conditions consisting of whether the work area includes corridors or exits that serve more than one tenant or that serve an occupant load greater than 30; whether the work area exceeds 50 percent of the floor area; and whether the building has a sufficient municipal water supply for a sprinkler system at the floor where the work area is located are evaluated. The availability of sufficient water at the floor being considered must be evaluated without consideration for a new fire pump.

An exception is provided for work areas of Group R occupancies three stories or less because the 2000 IBC and some previous legacy model codes allowed certain unsprinklered Group R occupancies under some specific conditions.

The 30-occupant load threshold in this section is taken from Table 1016.1 of the IBC, which uses this threshold in most occupancies to require either sprinklers or rated corridors.

604.2.2.1 Mixed uses. In work areas containing mixed uses, one or more of which requires automatic sprinkler protection in accordance with Section 604.2.2, such protection shall not be required throughout the work area provided that the uses requiring such protection are separated from those not requiring protection by fire-resistance-rated construction having a minimum 2-hour rating for Group H and a minimum 1-hour rating for all other occupancy groups.

❖ Designers and owners who do not wish to sprinkler entire work areas, just because one of a multiple number of mixed uses requires automatic sprinklers, are provided an option to separate the various occupancies. This concept is very similar to the concept in Section 302.3 of the IBC, where two options of nonseparated uses and separated uses are provided. Here the code user is given the option of separating the occupancies by a 1-hour fire-resistance rating (2-hour rating in Group H occupancies) and sprinkler only the occupancy that has been triggered to provide sprinklers, or provide the automatic sprinkler in the entire work area and multiple occupancies within such work areas, thus eliminating the need for separation of the various mixed occupancies.

Example 604.2.2.1:

Q: Which occupancies or tenants must be provided with a sprinkler system?

A: First, the three conditions of Section 604.2.2 must be verified. Conditions 2 and 3 are present. For Condition 1, Section 903 of the IBC requires sprinklers for tenants 2, 5, 7, and 10. Tenants 5 and 7, however, are outside of the work area and as such are not going to be required to provide sprinklers (see Section 604.1). Tenants 2 and 10 are within the work area. Accordingly, all of the tenants within the work area (tenants 1, 2, 3, 8, 9, 10 and 11) will be required to be sprinklered, even though tenants

1, 3, 8, 9 and 11 are not individually required to be sprinklered by the IBC. If the owner wishes to only sprinkler tenants 2 and 10, then full height 1-hour-rated partitions must be provided to separate tenants 2 and 10 from the adjacent tenants 1, 3, 9 and 11 (see Figure 604.2.2.1).

604.2.3 Windowless stories. Work located in a windowless story, as determined in accordance with the *International Building Code*, shall be sprinklered where the work area is required to be sprinklered under the provisions of the *International Building Code* for newly constructed buildings and the building has a sufficient municipal water supply available to the floor without installation of a new fire pump.

❖ In the context of the code, a windowless story is a reference to Section 903.2.10.1 of the IBC. Within this context, work areas within any story or basement larger than 1,500 square feet (139 m²) that do not have access to openings, as described in Items 1 and 2 of Section 903.2.10.1 of the IBC, are required to be sprinklered. The only exception is if there is not a sufficient municipal water supply available to such floor without the need for a new fire pump.

604.2.4 Other required suppression systems. In buildings and areas listed in Table 903.2.13 of the *International Building Code*, work areas that include exits or corridors shared by more than one tenant or serving an occupant load greater than 30

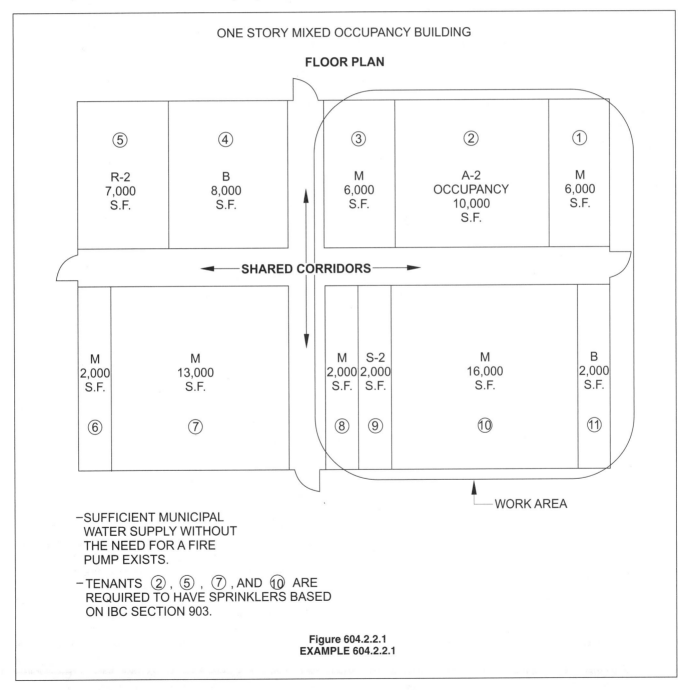

ONE STORY MIXED OCCUPANCY BUILDING

FLOOR PLAN

—SUFFICIENT MUNICIPAL WATER SUPPLY WITHOUT THE NEED FOR A FIRE PUMP EXISTS.

—TENANTS ②, ⑤, ⑦, AND ⑩ ARE REQUIRED TO HAVE SPRINKLERS BASED ON IBC SECTION 903.

Figure 604.2.2.1
EXAMPLE 604.2.2.1

shall be provided with sprinkler protection under the following conditions:

1. The work area is required to be provided with automatic sprinkler protection in accordance with the *International Building Code* applicable to new construction; and

2. The building has sufficient municipal water supply for design of a fire sprinkler system available to the floor without installation of a new fire pump.

❖ Special buildings and occupancies such as covered malls, atriums, stages and others listed in Table 903.2.13 of the IBC that use suppression systems such as smoke-protected seating, dry cleaning plants, spray booths and others listed in Table 903.2.13 of the IFC must also be separately considered for the possibility of automatic sprinklers or other suppression system requirements. The first condition to be considered for an area to require a sprinkler system is whether such a work area would be required to be sprinklered under the provisions of the IBC for new construction. If the IBC does not require a sprinkler system for the space or building under consideration, neither will the code. Once this is established, the other conditions consisting of whether the work area includes corridors or exits that serve more than one tenant or that serve an occupant load greater than 30 and whether the building has a sufficient municipal water supply for a sprinkler system at the floor where the work area is located are evaluated. The availability of sufficient water at the floor being considered must be evaluated without consideration for a new fire pump. The 30-occupant load threshold in this section is taken from Table 1016.1 of the IBC, which uses this threshold in most occupancies to require either sprinklers or rated corridors.

604.2.5 Supervision. Fire sprinkler systems required by this section shall be supervised by one of the following methods:

1. Approved central station system in accordance with NFPA 72;

2. Approved proprietary system in accordance with NFPA 72;

3. Approved remote station system of the jurisdiction in accordance with NFPA 72; or

4. Approved local alarm service that will cause the sounding of an alarm in accordance with NFPA 72.

Exception: Supervision is not required for the following:

1. Underground gate valve with roadway boxes.

2. Halogenated extinguishing systems.

3. Carbon dioxide extinguishing systems.

4. Dry and wet chemical extinguishing systems.

5. Automatic sprinkler systems installed in accordance with NFPA 13R where a common supply main is used to supply both domestic and automatic sprinkler systems and a separate shutoff valve for the automatic sprinkler system is not provided.

❖ If a sprinkler system is required based on Sections 604.2.1, 604.2.2, 604.2.3 and 604.2.4, then such a system must be supervised in accordance with NFPA 72, the same standard used for supervision of sprinklers in new buildings constructed under the IBC. This section identifies the four acceptable methods described in NFPA 72: central station system, proprietary systems, remote station system of the jurisdiction or local alarm service that will cause the sounding of an alarm. Five exceptions are provided where supervision is not required; these five exceptions are similar to the exceptions found in Sections 903.4 and 903.4.1 of the IBC.

604.3 Standpipes. Where the work area includes exits or corridors shared by more than one tenant and is located more than 50 feet (15 240 mm) above or below the lowest level of fire department access, a standpipe system shall be provided. Standpipes shall have an approved fire department connection with hose connections at each floor level above or below the lowest level of fire department access. Standpipe systems shall be installed in accordance with the *International Building Code*.

Exceptions:

1. No pump shall be required provided that the standpipes are capable of accepting delivery by fire department apparatus of a minimum of 250 gallons per minute (gpm) at 65 pounds per square inch (psi) (946 L/m at 448KPa) to the topmost floor in buildings equipped throughout with an automatic sprinkler system or a minimum of 500 gpm at 65 psi (1892 L/m at 448KPa) to the topmost floor in all other buildings. Where the standpipe terminates below the topmost floor, the standpipe shall be designed to meet (gpm/psi) (L/m/KPa) requirements of this exception for possible future extension of the standpipe.

2. The interconnection of multiple standpipe risers shall not be required.

❖ Once a work area encompasses corridors that are shared by more than one tenant or exits, the standpipe requirement is triggered only by work area height criteria, similar to building height criteria in Section 905.3.1 of the IBC. The standpipe triggers in the IBC for new buildings contain additional criteria based on Group A occupancies (see Section 905.3.2), covered mall buildings (see Section 905.3.3), underground buildings (see Section 905.3.5) and helistops and heliports (see Section 905.3.6), none of which are used as the basis for standpipe triggers in the code. Where the work area that includes exits or corridors serving more than one tenant is located more than 50 feet (15 240 mm) above or below the lowest level of fire department vehicle access, a standpipe system is required. Such standpipes must have a fire department connection with hose connection at each floor level up to the level where the work area is located in accordance with the IBC. Even though the class of the standpipe system is not specified in this section, because the trigger is based on work area height, the standpipe intended is a

Class III standpipe. The four exceptions in Section 905.3.1 of the IBC that allow Class I standpipe in lieu of Class III would also be allowed here. Exception 1 allows the installation of a standpipe system without a fire pump if sufficient water with sufficient pressure could be delivered by means of fire department apparatus pumps to the topmost floor of the building. Under this exception, the location of the work area that has triggered the standpipe requirement is irrelevant, and the topmost floor must be considered whether the work area is located at the topmost floor or not. Exception 2 recognizes the fact that over the life of a building, and as a result of various alterations work in different parts of the building, multiple independent standpipe risers might be present and these are allowed to remain independent and function separately as needed.

604.4 Fire alarm and detection. An approved fire alarm system shall be installed in accordance with Sections 604.4.1 through 604.4.3. Where automatic sprinkler protection is provided in accordance with Section 604.2 and is connected to the building fire alarm system, automatic heat detection shall not be required.

An approved automatic fire detection system shall be installed in accordance with the provisions of this code and NFPA 72. Devices, combinations of devices, appliances, and equipment shall be approved. The automatic fire detectors shall be smoke detectors, except that an approved alternative type of detector shall be installed in spaces such as boiler rooms, where products of combustion are present during normal operation in sufficient quantity to actuate a smoke detector.

❖ Fire alarm system requirements are solely based on the nature of occupancy. Fire detection system requirements are based on specific features to perform specific functions as described in the IBC, such as elevator recall (see Section 3003.2), smokeproof enclosure ventilation (see Section 909.20) or when detectors are installed to comply with the permitted alternative, such as door-closing devices (see Section 714.2.7.3). These requirements apply within the work area only. Such devices, where required, much the same as the IBC, must be installed in accordance with NFPA 72 and need not be anything other than smoke detectors with certain minor exceptions. The requirements in this section are the same as the provisions found in Section 907.3 of the IFC.

604.4.1 Occupancy requirements. A fire alarm system shall be installed in accordance with Sections 604.4.1.1 through 604.4.1.7. Existing alarm-notification appliances shall be automatically activated throughout the building. Where the building is not equipped with a fire alarm system, alarm-notification appliances within the work area shall be provided and automatically activated.

Exceptions:

1. Occupancies with an existing, previously approved fire alarm system.

2. Where selective notification is permitted, alarm-notification appliances shall be automatically activated in the areas selected.

❖ As mentioned above, fire alarm system requirements are based on occupancy classification as determined by Sections 604.4.1.1 through 604.4.1.7. If the automatic alarm system is required in the work area in a certain occupancy, and there also happens to be alarm notification devices already existing in other parts of the building, all such existing devices must be connected to the new alarm system and be automatically activated upon activation of the new alarm system. If there are no such existing alarms or notification devices, then the new alarm and its notification devices, are only required within the work area. Unless the supplemental fire alarm system requirements of Section 604.4.2 are applicable, fire alarm systems are required in work areas only, in Groups E, I-1 residential care/assisted living, I-2, I-3, R-1 R-2 and R-4 residential care/assisted-living facilities. The triggers are the same as IFC retroactive triggers for the same occupancies.

604.4.1.1 Group E. A fire alarm system shall be installed in work areas of Group E occupancies as required by the *International Fire Code* for existing Group E occupancies.

❖ A building with a maximum of 1,000 square feet (93 m²) in area that contains a single classroom and is located 50 feet (15 240 mm) or more from another building, and any building with an occupant load less than 50 are not required to have an alarm system. Work areas in Group E buildings are required to be provided with a manual fire alarm system. It must be noted, however, that in this case the requirements of the IFC are more stringent than this code. If a jurisdiction enforces the retroactive provisions of the IFC, the entire Group E occupancy (and not just the work area) is required to be provided with a manual fire alarm system based on the size and occupant load criteria just discussed.

604.4.1.2 Group I-1. A fire alarm system shall be installed in work areas of Group I-1 residential care/assisted living facilities as required by the *International Fire Code* for existing Group I-1 occupancies.

❖ A manual fire alarm system must be installed in work areas of all Group I-1 residential care/assisted living facilities. Again, Section 907.2.6 of the IFC requires this same system retroactively in the entire Group I-1 occupancy, which, if enforced, is more comprehensive than the code's requirement.

604.4.1.3 Group I-2. A fire alarm system shall be installed in work areas of Group I-2 occupancies as required by the *International Fire Code* for existing Group I-2 occupancies.

❖ A manual fire alarm system is required to be provided in the work areas. Here again, the IFC retroactive provisions are more stringent than the code and require the entire existing Group I-2 occupancies to be provided with a manual fire alarm system (see Sections 907.3.1.3 and 907.2.6 of the IFC).

604.4.1.4 Group I-3. A fire alarm system shall be installed in work areas of Group I-3 occupancies as required by the *International Fire Code* for existing Group I-3 occupancies.

❖ The work areas must be provided with a manual and automatic fire alarm system.

604.4.1.5 Group R-1. A fire alarm system shall be installed in Group R-1 occupancies as required by the *International Fire Code* for existing Group R-1 occupancies.

❖ The fire alarm requirement is again referred to the retroactive provisions of the IFC. The IFC deals with Group R-1 hotels and motels and Group R-1 boarding and rooming houses differently. Boarding and rooming houses are required to be provided with a fire alarm system within the work area, regardless of the size of the building, height or number of stories. For hotels and motels, the trigger is set for buildings that are more than three stories in height or with more than 20 guestrooms. There is also an exception that is related to buildings of less than two stories in height (see Sections 907.3.1.5 and 907.3.1.6 of the IFC).

604.4.1.6 Group R-2. A fire alarm system shall be installed in work areas of Group R-2 apartment buildings as required by the *International Fire Code* for existing Group R-2 occupancies.

❖ The fire alarm system trigger for apartments is similar to that of hotels and motels and is set at buildings of more than three stories in height or containing more than 16 dwelling units. Section 907.3.1.7 of the IFC includes three exceptions that are also applicable under the code for Group R-2 apartment buildings.

604.4.1.7 Group R-4. A fire alarm system shall be installed in work areas of Group R-4 residential care/assisted living facilities as required by the *International Fire Code* for existing Group R-4 occupancies.

❖ Other than the two exceptions found under Section 907.3.1.8 of the IFC, work areas in Group R-4 residential care/assisted living facilities must be provided with a fire alarm system.

604.4.2 Supplemental fire alarm system requirements. Where the work area on any floor exceeds 50 percent of that floor area, Section 604.4.1 shall apply throughout the floor.

Exception: Alarm-initiating and notification appliances shall not be required to be installed in tenant spaces outside of the work area.

❖ The fire alarm requirements discussed above for various occupancies apply to the work area only (except where the retroactive provisions of the IFC govern and are enforced by the jurisdiction, as discussed above), unless the work area on any floor exceeds 50 percent of that floor area, in which case the fire alarm system

must be provided beyond the work area throughout that floor. Where this provision applies and there are multiple tenants on the floor that is required to be equipped with a fire alarm system, the extension of the alarm system into tenant areas, whose space is not part of the alteration work is very interruptive and, in most cases, acts as a major deterrent to undertaking certain levels of alteration. For this reason, the exception eliminates the requirement for the alarm-initiating and notification appliances to be installed within such tenant spaces that are not part of the alteration work.

604.4.3 Smoke alarms. Individual sleeping units and individual dwelling units in any work area in Group R-1, R-2, R-3, R-4, and I-1 occupancies shall be provided with smoke alarms in accordance with the *International Fire Code*.

Exception: Interconnection of smoke alarms outside of the rehabilitation work area shall not be required.

❖ Smoke detectors in sleeping units and in dwelling units that are within the work area must be provided as required by the IFC. Sections 907.3.2 and 907.2.10 of the IFC are the applicable sections for smoke detectors. According to these sections, smoke detectors are required in the sleeping rooms, in every room in the path of the means of egress from the sleeping areas and in each story and basement. Interconnection of such smoke detectors is still required, unless such smoke alarms are located outside of the work area or qualify under Exception 2 of Section 907.3.2.2 of the IFC for Groups R-2 and R-4 only.

SECTION 605
MEANS OF EGRESS

605.1 Scope. The requirements of this section shall be limited to work areas that include exits or corridors shared by more than one tenant within the work area in which Level 2 alterations are being performed, and where specified they shall apply throughout the floor on which the work areas are located or otherwise beyond the work area.

❖ Section 605 is entirely devoted to the means of egress requirements in existing buildings that are undergoing Level 2 or 3 alterations. The requirements of this section apply to the work area only, unless there are specific sections addressing supplemental requirements, in which case the requirements will apply throughout the floor under consideration or beyond. The provisions of Section 605 are applicable only when the alteration work area includes exits or corridors shared by more than one tenant. As such, in multitenant buildings where reconfiguration of space takes place within one of the tenant spaces and does not include corridors or exits that affect others, the tenant space undergoing alterations need only comply with the means of egress requirements of Section 505. In essence, if there are any egress conditions that are considered unsafe, the conditions must be remedied. It is not necessary to confirm that the level of egress safety has not been reduced compared to the current conditions.

605.2 General. The means of egress shall comply with the requirements of this section.

Exceptions:

1. Where the work area and the means of egress serving it complies with NFPA 101.

2. Means of egress conforming to the requirements of the *International Building Code* under which the building was constructed shall be considered compliant means of egress if, in the opinion of the code official, they do not constitute a distinct hazard to life.

❖ The provisions of Section 605 are intended to address improvements in the means of egress that are crucial for the safe egress of occupants. With the exception of very old buildings, most existing buildings designed under the building codes of a jurisdiction provide a certain degree of safe egress. Accordingly, unless the code official finds that all or parts of the means of egress system constitute a distinct hazard, the means of egress does not need to comply with Section 605 as long as it complies with the requirements of the building code under which it was built, including NFPA 101. Where the code official finds a distinct hazard, the provisions of Section 605 could be applied or whatever other remedy determined by the code official must be followed. This provision is also found in Section 1026.1 of the IFC.

605.3 Number of exits. The number of exits shall be in accordance with Sections 605.3.1 through 605.3.3.

❖ The number of exits in any building is one of the most important factors in safe egress of occupants. The IBC and IFC make it unlawful to reduce the number of exits in a building.

605.3.1 Minimum number. Every story utilized for human occupancy on which there is a work area that includes exits or corridors shared by more than one tenant within the work area shall be provided with the minimum number of exits based on the occupancy and the occupant load in accordance with the *International Building Code*. In addition, the exits shall comply with Sections 605.3.1.1 and 605.3.1.2.

❖ Given that the number of exits is so important, this section requires that every story with a work area that falls within the limitations discussed in Section 605.1 be provided with a minimum number of exits in accordance with the IBC. This code does, however, provide exceptions that allow single exit construction, many of which are common to the IBC, as listed in Section 605.3.1.1. In addition, fire escapes can be used as a solution where more than one exit is required, as provided in Section 605.3.1.2.

605.3.1.1 Single-exit buildings. Only one exit is required from buildings and spaces of the following occupancies:

1. In Group A, B, E, F, M, U, and S occupancies, a single exit is permitted in the story at the level of exit discharge when the occupant load of the story does not exceed 50 and the exit access travel distance does not exceed 75 feet (22 860 mm).

2. Group B, F-2, and S-2 occupancies not more than two stories in height that are not greater than 3,000 square feet per floor (279 m²), when the exit access travel distance does not exceed 75 feet (22 860 mm). The minimum fire-resistance rating of the exit enclosure and of the opening protection shall be 1 hour.

3. Open parking structures where vehicles are mechanically parked.

4. Groups R-1 and R-2, except that in community residences for the developmentally disabled, the maximum occupant load excluding staff is 12.

5. Groups R-1 and R-2 not more than two stories in height, when there are not more than four dwelling units per floor and the exit access travel distance does not exceed 50 feet (15 240 mm). The minimum fire-resistance rating of the exit enclosure and of the opening protection shall be 1 hour.

6. In multilevel dwelling units in buildings of Occupancy Group R-1 or R-2, an exit shall not be required from every level of the dwelling unit provided that one of the following conditions is met:

    6.1. The travel distance within the dwelling unit does not exceed 75 feet (22 860 mm); or

    6.2. The building is not more than three stories in height and all third-floor space is part of one or more dwelling units located in part on the second floor; and no habitable room within any such dwelling unit shall have a travel distance that exceeds 50 feet (15 240 mm) from the outside of the habitable room entrance door to the inside of the entrance door to the dwelling unit.

7. In Group R-2, H-4, H-5, and I occupancies and in rooming houses and childcare centers, a single exit is permitted in a one-story building with a maximum occupant load of 10 and the exit access travel distance does not exceed 75 feet (22 860 mm).

8. In buildings of Group R-2 occupancy that are equipped throughout with an automatic fire sprinkler system, a single exit shall be permitted from a basement or story below grade if every dwelling unit on that floor is equipped with an approved window providing a clear opening of at least 5 square feet (0.47 m²) in area, a minimum net clear opening of 24 inches (610 mm) in height and 20 inches (508 mm) in width, and a sill height of not more than 44 inches (1118 mm) above the finished floor.

9. In buildings of Group R-2 occupancy of any height with not more than four dwelling units per floor; with a smokeproof enclosure or outside stair as an exit; and with such exit located within 20 feet (6096 mm) of travel to the entrance doors to all dwelling units served thereby.

10. In buildings of Group R-3 occupancy equipped throughout with an automatic fire sprinkler system, only one exit shall be required from basements or stories below grade.

❖ There are 10 specific conditions described where the building or a certain story could be provided with only one single exit. Several of these 10 conditions are either the same or similar to various provisions in the IBC. For example, Item 1 in this section is similar to the first row in Table 1018.2 of the IBC. The only difference is that Item 1 of this section allows the condition at the level of exit discharge in multistory buildings as opposed to Table 1018.2 of the IBC that allows this condition in a one-story building only. Item 3 of this section is the same as Section 1018.1.1 of the IBC, and Item 7 is the same as the third row in Table 1018.2 of the IBC.

605.3.1.2 Fire escapes required. When more than one exit is required, an existing or newly constructed fire escape complying with Section 605.3.1.2.1 shall be accepted as providing one of the required means of egress.

❖ The use of fire escapes as an element of the required means of egress has not been allowed by building codes in many years. The typical problems with fire escapes are the difficulty of access to it in conditions of panic, the minimal width and their location outside the building where they can be subjected to ice, snow and slippery conditions. In existing buildings, however, where the construction of a new exit stair is either impossible due to proximity of the exterior walls to property lines; the infeasibility of penetrating or rearranging interior or exterior structural elements occurs; when it is extremely cost prohibitive due to the need for major restructuring of building elements; or when affecting other relevant occupancy issues, the utilization of fire escapes is very valuable. As such, this section provides for the use of existing or newly constructed fire escapes to be used as a required element of the means of egress system in cases where more than one exit is required and only if it meets the construction requirements of this section. The fire escapes can be counted only as one of the required means of egress and not both. Specific detail requirements and construction requirements are given in Sections 605.3.1.2.1 and 605.3.1.2.2.

605.3.1.2.1 Fire escape access and details. Fire escapes shall comply with all of the following requirements:

1. Occupants shall have unobstructed access to the fire escape without having to pass through a room subject to locking.

2. Access to a new fire escape shall be through a door, except that windows shall be permitted to provide access from single dwelling units or sleeping units in Group R-1, R-2, and I-1 occupancies or to provide access from spaces having a maximum occupant load of 10 in other occupancy classifications.

3. Newly constructed fire escapes shall be permitted only where exterior stairs cannot be utilized because of lot

lines limiting the stair size or because of the sidewalks, alleys, or roads at grade level.

4. Openings within 10 feet (3048 mm) of fire escape stairs shall be protected by fire assemblies having minimum $^3/_4$-hour fire-resistance ratings.

   Exception: Opening protection shall not be required in buildings equipped throughout with an approved automatic sprinkler system.

5. In all buildings of Group E occupancy, up to and including the 12th grade, buildings of Group I occupancy, rooming houses, and childcare centers, ladders of any type are prohibited on fire escapes used as a required means of egress.

❖ This section provides the access and detail conditions that must be followed in order to make a fire escape an acceptable second exit.

   Newly constructed fire escapes are allowed only if construction of exterior stairs is impossible due to property lines, sidewalks, alleys or roads. If allowed to be used by this section, then a fire escape must be readily accessible without any obstructions, and access to it must be through a door with the exception of a single dwelling unit or sleeping unit and spaces with a maximum occupant load of 10, in which case access through a window is acceptable. The code does not have a minimum size limitation for such windows and does not have any additional criteria in this regard. It might be reasonable to expect such windows to comply with the same requirements for a window opening area, minimum opening height, minimum opening width and maximum opening height above floor and other criteria as required for emergency escape provisions of Section 1025 of the IBC or Section R310 of the *International Residential Code®* (IRC®).

   All openings within 10 feet (3048 mm) of fire escape stairs used for a required means of egress must be protected by fire assemblies having a minimum of $^3/_4$-hour fire-resistance rating. This is consistent with protection of openings adjacent to exterior exit stairs required by Section 1019.1.4 of the IBC. Much the same as Section 1019.1.4 of the IBC, the protection of openings within 10 feet (3048 mm) of fire escapes is not required for buildings equipped throughout with an approved automatic sprinkler system. Based on the type of occupancy involved, the appropriate sprinkler system, complying with NFPA 13 or 13R, is appropriate.

Example 605.3.1.2:

An existing two-story downtown building is used for shops on the first floor and the second floor is used for offices of several tenants. The owner wishes to rearrange and add interior partitions and alter about 80 percent of the second floor. The second floor has only one stairway and due to its current uses and the building location on the lot, is not able to add a second stairway. The area of the second floor is such that it would require two stairways under the IBC.

Q: Would the alteration trigger the requirement of the second stairway?

A: This would require the construction of a second stairway, however it would allow fire escapes to be utilized in this case (see Sections 605.3 and 605.3.1.2).

605.3.1.2.2 Construction. The fire escape shall be designed to support a live load of 100 pounds per square foot (4788 Pa) and shall be constructed of steel or other approved noncombustible materials. Fire escapes constructed of wood not less than nominal 2 inches (51 mm) thick are permitted on buildings of Type V construction. Walkways and railings located over or supported by combustible roofs in buildings of Types III and IV construction are permitted to be of wood not less than nominal 2 inches (51 mm) thick.

❖ Sections 605.3.1.2.2 and 605.3.1.2.3 provide the minimum structural and the dimensional requirements for new fire escapes. Existing fire escapes are not restricted to compliance with Sections 605.3.1.2.2 and 605.3.1.2.3 but must be inspected and evaluated to be sure no dangerous conditions exist and that they will function as intended for a required means of egress.

605.3.1.2.3 Dimensions. Stairs shall be at least 22 inches (559 mm) wide with risers not more than, and treads not less than, 8 inches (203 mm). Landings at the foot of stairs shall not be less than 40 inches (1016 mm) wide by 36 inches (914 mm) long and located not more than 8 inches (203 mm) below the door.

❖ Sections 605.3.1.2.2 and 605.3.1.2.3 provide the minimum structural and the dimensional requirements for new fire escapes. Existing fire escapes are not restricted to compliance with Sections 605.3.1.2.2 and 605.3.1.2.3 but must be inspected and evaluated to be sure no dangerous conditions exist and that they will function as intended for a required means of egress.

605.3.2 Mezzanines. Mezzanines in the work area and with an occupant load of more than 50 or in which the travel distance to an exit exceeds 75 feet (22 860 mm) shall have access to at least two independent means of egress.

Exception: Two independent means of egress are not required where the travel distance to an exit does not exceed 100 feet (30 480 mm) and the building is protected throughout with an automatic sprinkler system.

❖ The means of egress requirements for mezzanines are almost exactly the same as the provisions found in the IBC. This applies to those mezzanines that are in the work area. The triggers for requiring two independent means of egress from mezzanines is the occupant load being greater than 50 or the travel distance being greater than 75 feet (22 860 mm), with an exception for sprinklered buildings allowing travel distance up to 100 feet (30 480 mm). Similar provisions in the IBC are found in Sections 505.3, 1013.3 and 1014.1. The code is somewhat less restrictive than the IBC since the occupant load trigger is set at over 50 regardless of occupancy classification, whereas Table 1014.1 of the IBC has a lower level of threshold trigger for hazardous, in-

dustrial, residential and storage occupancies. Additionally, the code limitation of 75 feet (22 860 mm) [or 100 feet (30 480 mm) in sprinklered buildings] is the measurement of travel distance, whereas IBC's similar provisions in Section 1013.3 of the IBC is the measurement of the common path of egress travel.

605.3.3 Main entrance—Group A. All buildings of Group A with an occupant load of 300 or more shall be provided with a main entrance capable of serving as the main exit with an egress capacity of at least one half of the total occupant load. The remaining exits shall be capable of providing one half of the total required exit capacity.

Exception: Where there is no well-defined main exit or where multiple main exits are provided, exits shall be permitted to be distributed around the perimeter of the building provided that the total width of egress is not less than 100 percent of the required width.

❖ Assembly occupancy egress, particularly the capacity of the main entrance/main exit, is so critical for high occupant load assembly occupancies that the code, in this section provides the same exact requirements found in Section 1024.2 of the IBC for new construction. Accordingly, when Level 2 or 3 alterations work is taking place in assembly occupancies with an occupant load greater than 300, the main entrance/exit must provide sufficient width to accommodate one-half of the total occupant load.

Example 605.3.3:

An existing community hall shown in the figure is to undergo some reconfiguration of its interior spaces. There are currently several 3-foot (914 mm) doors that provide the needed exit width and separation of exit requirements based on the IBC.

Q: Are there any additional requirements related to the exterior doors that must be complied with?

A: Yes. A main entrance must be provided that is wide enough to accommodate one-half of the total occupant load (450/2 = 225). This is a width of 225 x 0.2 = 45 inches (1143 mm) (see Table 1005.1 of the IBC and Figure 605.3.3).

605.4 Egress doorways. Egress doorways in any work area shall comply with Sections 605.4.1 through 605.4.5.

❖ This is the charging paragraph for the requirements for egress doorways in Level 2 or 3 alterations. The specific requirements are given in Sections 605.4.1 through 605.4.5, relating the number of egress doorways, the door swing, the door closing, panic hardware and emergency power in detention facilities.

605.4.1 Two egress doorways required. Work areas shall be provided with two egress doorways in accordance with the requirements of Sections 605.4.1.1 and 605.4.1.2.

❖ The triggers for requiring two egress doorways are similar in some respects to the IBC triggers. Section 605.4.1.1 relates these triggers for all occupancies, and Section 605.4.1.2 deals with Group I-2 occupancies.

FIGURE 605.3.3 – 605.4.2

ALTERATIONS—LEVEL 2

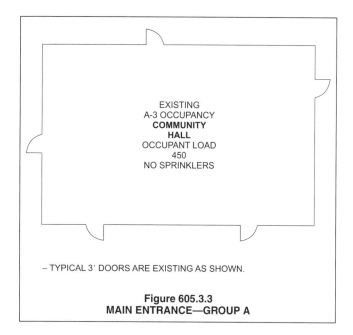

– TYPICAL 3′ DOORS ARE EXISTING AS SHOWN.

Figure 605.3.3
MAIN ENTRANCE—GROUP A

605.4.1.1 Occupant load and travel distance. In any work area, all rooms and spaces having an occupant load greater than 50 or in which the travel distance to an exit exceeds 75 feet (22 860 mm) shall have a minimum of two egress doorways.

Exceptions:

1. Storage rooms having a maximum occupant load of 10.

2. Where the work area is served by a single exit in accordance with Section 605.3.1.1.

❖ The discussion in Section 605.3 demonstrated how several criteria for requiring only one means of egress is also found in the IBC for new construction. The same is true here; the occupant load being greater than 50 and travel distance exceeding 75 feet (22 860 mm) are the triggers used to require at least two egress doorways for all occupancies.

Exception 1 under Section 605.4.1.1 addresses storage rooms with small occupant loads that might be of such dimensions that travel distance might exceed 75 feet (22 860 mm). Realizing that many storage rooms are generally not occupied regularly, and when occupied, such occupants are either employees or individuals who are familiar with the surroundings, this exception allows storage rooms of any size and any length of travel distance to have only one egress doorway as long as the occupant load does not exceed 10.

Example 605.4.1.1:

An existing grocery store as shown is undergoing Level 2 or 3 alterations. The storage area has a travel distance of 90 feet (27 432 mm) within the storage room and an overall travel distance of 10 feet (3048 mm) to exit the building. There is currently one egress doorway from the storage room to the retail area.

Q: Is a second egress doorway required from the storage area?

A: No. The storage room has an occupant load of 10 (3000/300 = 10) and falls under Exception 1.

- Storage room occupant load = 10 (see Table 1004.1.2 of the IBC).
- Maximum travel distance in storage room = 90 feet (27 432 mm).
- One egress doorway from storage room acceptable (see Exception 1 to Section 605.4.1.1 and Figure 605.4.1.1).

Exception 2 refers the user back to Section 605.3.1.1 so that a building or space that qualifies under any of the 10 items listed would still be required to have one egress doorway.

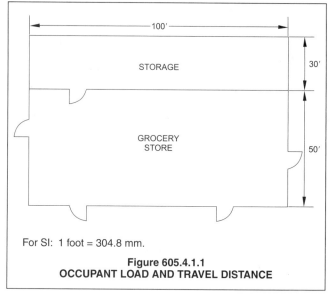

For SI: 1 foot = 304.8 mm.

Figure 605.4.1.1
OCCUPANT LOAD AND TRAVEL DISTANCE

605.4.1.2 Group I-2. In buildings of Group I-2 occupancy, any patient sleeping room or suite of patient rooms greater than 1,000 square feet (93 m²) within the work area shall have a minimum of two egress doorways.

❖ Group I-2 occupancies have an additional criteria for patient sleeping rooms or suites in this section. This additional criteria for Group I-2 patient sleeping areas is taken directly from a part of Section 1013.2.2 of the IBC.

605.4.2 Door swing. In the work area and in the egress path from any work area to the exit discharge, all egress doors serving an occupant load greater than 50 shall swing in the direction of exit travel.

❖ Door swings must be considered not only in the work area, but all the way from the work area along the path of egress to the exit discharge. Here again, door swing is such a critical element of safe egress that existing doors must comply just as new doors do under the IBC. The threshold of "serving an occupant load greater than 50" of Section 605.4.2 is taken directly from Section 1008.1.2 of the IBC.

Example 605.4.2:

An existing three-story office building as shown is undergoing Level 2 or 3 alterations on the third floor. Stairway 1 is the most obvious, the closest and the most likely path of egress for the occupants in the work area under consideration.

Q: Which doors are required to swing in the direction of egress travel?

A: Doors 2, 3, 4 and 7 in Figure 605.4.2 must swing in the direction of egress travel as they serve an occupant load of greater than 50 and are either in the work area or along the egress path to the exit discharge (door 1 serves an occupant load less than 50, and doors 5 and 6 are not along the path of egress of the work area and, as such, are not required to comply with Section 605.4.2) (see Figure 605.4.2).

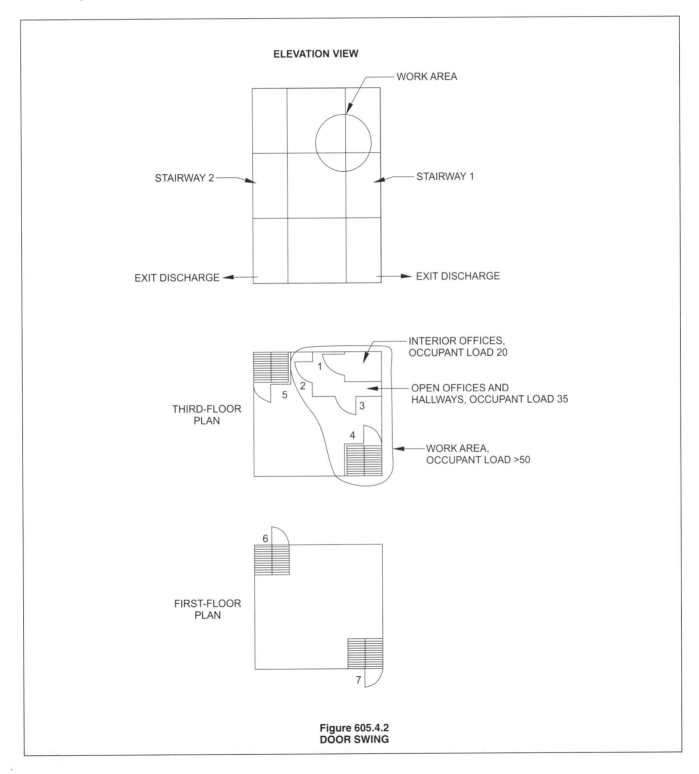

**Figure 605.4.2
DOOR SWING**

605.4.2.1 Supplemental requirements for door swing.
Where the work area exceeds 50 percent of the floor area, door swing shall comply with Section 605.4.2 throughout the floor.

Exception: Means of egress within or serving only a tenant space that is entirely outside the work area.

❖ Door swings in the entire floor, except doors that are within a tenant space located entirely outside the work area, must swing in the direction of egress travel if the work area on that floor exceeds 50 percent of the area of that floor and if the occupant load they serve is greater than 50. Accordingly, all doors along the means of egress to the exit discharge must also swing in the direction of egress.

Example 605.4.2.1:

An existing two-story building has three tenants on the third floor as shown in Figure 605.4.2.1. The owner plans some alterations of the second floor. The work area will be more than 50 percent of the second floor area.

Q: Which doors must be brought into compliance with the door swing in the direction of egress travel?

A: Doors 4, 5, 6, 7, 10, 11, 12 and 13 in Figure 605.4.2.1 must swing in the direction of egress travel. Doors 6, 7 and 10 serve an occupant load of more than 50 and are in the work area. Door 11 serves an occupant load of more than 50 and is outside of the work area; however, because the work area is greater than 50 percent of the second floor area, this door must also swing in the direction of egress. Doors 4, 5, 12 and 13 are on the egress path from the work area to exit discharge.

Doors 1, 2, 3, 8 and 9 are not required to change their swing to the direction of egress if they currently swing inward. Doors 1, 2 and 3 serve an occupant load greater than 50, but are in a tenant space completely outside of the work area and fall under the exception. Door 8 serves an occupant load less than 50, and door 9 is a convenience door connecting the two parts of tenant 3 (see Figure 605.4.2.1).

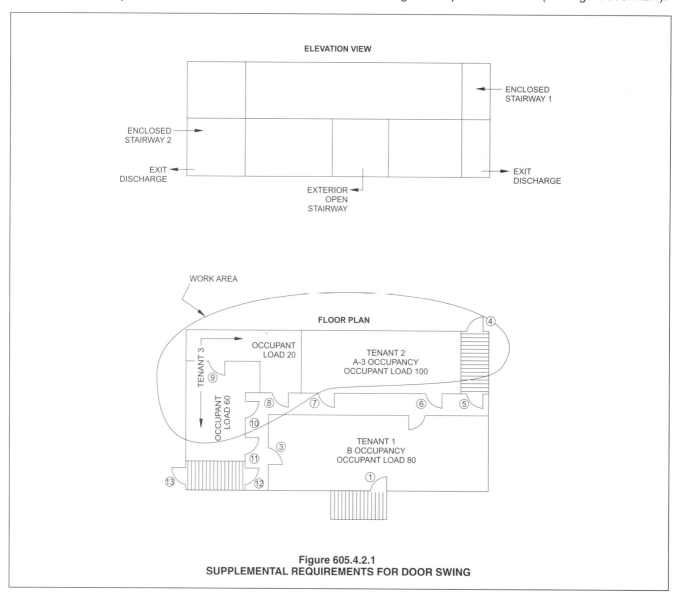

Figure 605.4.2.1
SUPPLEMENTAL REQUIREMENTS FOR DOOR SWING

605.4.3 Door closing. In any work area, all doors opening onto an exit passageway at grade or an exit stair shall be self-closing or automatically closing by listed closing devices.

Exceptions:

1. Where exit enclosure is not required by the *International Building Code.*
2. Means of egress within or serving only a tenant space that is entirely outside the work area.

❖ The same concepts discussed for door swing in the work area and beyond the work area when supplemental provisions are applicable are true for door closings. Doors opening into exit passageways and exit stair enclosures are required to be self-closing or automatic closing by means of listed devices.

605.4.3.1 Supplemental requirements for door closing. Where the work area exceeds 50 percent of the floor area, doors shall comply with Section 605.4.3 throughout the exit stair from the work area to the level of exit discharge.

❖ If the work area on the floor exceeds 50 percent of the area of that floor and if the occupant load it serves is greater than 50, all doors along the means of egress to the level of exit discharge must also be self-closing or automatic closing.

605.4.4 Panic hardware. In any work area, and in the egress path from any work area to the exit discharge, in buildings or portions thereof of Group A assembly occupancies with an occupant load greater than 100, all required exit doors equipped with latching devices shall be equipped with approved panic hardware.

❖ The panic hardware provisions are only applicable to Group A occupancies and are much the same as Section 1008.1.9 of the IBC. The triggers for requiring panic hardware are the presence of latching devices on the door and the occupant load of the assembly occupancy being greater than 100. The panic hardware requirements are required in the work areas and the path of egress from the work area to the exit discharge. Panic hardware requirements could be applicable in other parts of the space or building beyond the work area and the path of egress when the supplemental provisions apply.

605.4.4.1 Supplemental requirements for panic hardware. Where the work area exceeds 50 percent of the floor area, panic hardware shall comply with Section 605.4.4 throughout the floor.

Exception: Means of egress within a tenant space that is entirely outside the work area.

❖ Similar to the supplemental requirements for door swing and door closing, work exceeding 50 percent of the floor area triggers a more restrictive requirement for panic hardware as well. Except for means of egress within an unaffected tenant space, the 50-percent threshold triggers a requirement for panic hardware throughout the work area floor.

605.4.5 Emergency power source in Group I-3. Work areas in buildings of Group I-3 occupancy having remote power unlocking capability for more than 10 locks shall be provided with an emergency power source for such locks. Power shall be arranged to operate automatically upon failure of normal power within 10 seconds and for a duration of not less than 1 hour.

❖ This threshold for emergency power for doors with remote power unlocking equipment is the same as given in Section 408.4.2 in the IBC for Group I-3 facilities.

605.5 Openings in corridor walls. Openings in corridor walls in any work area shall comply with Sections 605.5.1 through 605.5.4.

Exception: Openings in corridors where such corridors are not required to be rated in accordance with the *International Building Code.*

❖ This is the charging paragraph for the requirements for openings in corridor walls in Level 2 or 3 alterations. The specific requirements are given in Sections 605.5.1 through 605.5.4, relating the requirements for corridor doors, transoms, corridor openings, such as windows, and supplemental requirements. The protections prescribed in these three sections are required only if the corridor under consideration is required to be a fire-resistance-rated corridor in accordance with the IBC. The overall goal is to eliminate certain unsuitable doors and openings within the rated corridors located in a Level 2 or 3 work area, even if there was no intention by the scope of the alteration work to do anything to these elements.

605.5.1 Corridor doors. Corridor doors in the work area shall not be constructed of hollow core wood and shall not contain louvers. All dwelling unit or sleeping unit corridor doors in work areas in buildings of Groups R-1, R-2, and I-1 shall be at least 1³/₈-inch (35 mm) solid core wood or approved equivalent and shall not have any glass panels, other than approved wired glass or other approved glazing material in metal frames. All dwelling unit or sleeping unit corridor doors in work areas in buildings of Groups R-1, R-2, and I-1 shall be equipped with approved door closers. All replacement doors shall be 1³/₄-inch (45 mm) solid bonded wood core or approved equivalent, unless the existing frame will accommodate only a 1³/₈-inch (35 mm) door.

Exceptions:

1. Corridor doors within a dwelling unit or sleeping unit.
2. Existing doors meeting the requirements of *HUD Guideline on Fire Ratings of Archaic Materials and Assemblies* (IEBC Resource A) for a rating of 15 minutes or more shall be accepted as meeting the provisions of this requirement.
3. Existing doors in buildings protected throughout with an approved automatic sprinkler system shall be required only to resist smoke, be reasonably tight fitting, and shall not contain louvers.
4. In group homes with a maximum of 15 occupants and that are protected with an approved automatic detection system, closing devices may be omitted.

5. Door assemblies having a fire-protection rating of at least 20 minutes.

❖ In cases where doors are being replaced, these provisions require that a certain level of fire or smoke protection be provided. As such, no hollow core wood doors and no louvers are allowed within a rated corridor in the work area. In addition to this prohibition, if the occupancy of the building undergoing alteration is Group R-1, R-2 or B-1 and the corridor under consideration serves the dwelling or sleeping units of these occupancies, all doors within the rated corridors must be equipped with approved closers and be at least $1^3/_8$-inches (35 mm) solid core wood or approved equivalent. Further, these doors cannot contain any glass panels unless tested assemblies or approved wired glass is used. Approved wired glass is generally considered to be $^1/_4$-inch (6.4 mm) wired glass.

Once the doors that are to be replaced have been identified, then the replacement doors must be at least $1^3/_4$-inch (45 mm) solid bonded wood core or approved equivalent. One-and-three-eights-inch (35 mm) solid wood doors are allowed for such replacement doors if the existing frame can only accommodate a $1^3/_8$-inch-thick (35 mm) door. Five exceptions have been provided for corridor door requirements that are self-explanatory. The replacement doors have a greater protection requirement compared to those doors that are not being replaced but happen to be in the work area, because it is unreasonable to require every existing door that provides a certain level of protection to be replaced within the entire work even though they provide a certain level of protection with a $1^3/_8$-inch (35 mm) solid core door or equivalent.

605.5.2 Transoms. In all buildings of Group I-1, R-1, and R-2 occupancy, all transoms in corridor walls in work areas shall either be glazed with $^1/_4$-inch (6.4 mm) wired glass set in metal frames or other glazing assemblies having a fire-protection rating as required for the door and permanently secured in the closed position or sealed with materials consistent with the corridor construction.

❖ Transoms in corridors are specifically regulated only in Group I-1, R-1 and R-2 occupancies. It should be noted again that the applicability here is only in corridors that would be required to be rated. The IBC requires that glass sidelites and transoms perform the same as fire windows and does not consider these elements part of the door assembly. Based on this approach, sidelites and transoms in occupancies other than Group I-1, R-1 and R-2 would be covered in Section 605.5.3, which addresses sash, grilles and windows. There are three options in dealing with transoms:

1. Use $^1/_4$-inch (6.4 mm) wired glass in metal frame,

2. A minimum of 20-minute fire-resistant-protected glazing fixed assembly, or

3. Do away with the transom and seal it with wall construction similar to the existing corridor wall.

605.5.3 Other corridor openings. In any work area, any other sash, grille, or opening in a corridor and any window in a corridor not opening to the outside air shall be sealed with materials consistent with the corridor construction.

❖ Openings other than doors such as sash, grille, louver and windows that open into a fire-resistance-rated corridor and that do not open to the outside air must be covered and sealed the same as the existing corridor wall construction. This is required for the protection of the corridor from the potential of smoke contamination from adjacent rooms. The reference to such openings "not opening to the outside air" is consistent with Section 708.5 of the IBC that allows corridor exterior walls and their openings to comply with the fire-resistance rating based primarily on proximity to property lines, otherwise known as fire separation distance. Two questions arising from this provision are why windows in a rated corridor are required to be sealed with a wall construction, and is there a way to allow such windows to remain in place and not be covered with wall construction. The code in this section intends to improve the safety and protection of corridors and does not intend to necessarily eliminate all windows opening into a corridor. As such, if it is desired for such windows to remain, then the window and its glazing must comply with fire window and fire-protection-rated glazing requirements of Section 715.4 of the IBC. It must be clarified that sidelites and transoms are not considered part of the door assembly and are regulated by this section. If the total area of windows exceeds 25 percent of the area of a common wall within any room, then some of the area of such windows must be sealed to bring the total percentage of openings to below 25 percent (see Section 715.4.7.2 of the IBC).

605.5.3.1 Supplemental requirements for other corridor opening. Where the work area exceeds 50 percent of the floor area, Section 605.5.3 shall be applicable to all corridor windows, grills, sashes, and other openings on the floor.

Exception: Means of egress within or serving only a tenant space that is entirely outside the work area.

❖ The supplemental provisions, where the work area exceeds 50 percent of the floor area, are the same here as they are in all previous sections discussed. As in all other areas, the 50-percent threshold brings a more serious concern regarding the alteration, and thus more restrictive requirements are applied throughout the floor, even in areas where no alteration is being performed. For other corridor openings, however, the tenant spaces that are unaffected do not require these alterations.

605.5.4 Supplemental requirements for corridor openings. Where the work area on any floor exceeds 50 percent of the floor area, the requirements of Sections 605.5.1 through 605.5.3 shall apply throughout the floor.

❖ The supplemental provisions, where the work area exceeds 50 percent of the floor area, are the same here as they are in all previous sections discussed. As in all other areas, the 50-percent threshold brings a more serious concern regarding the alteration, and thus more restrictive requirements are applied throughout the floor, even in areas where no alteration is being performed. This paragraph does not exclude tenant spaces that are unaffected by the alterations. Therefore, the doors, openings and transoms must be modified according to the applicable section. This paragraph supersedes the exception given in Section 605.5.3.1.

605.6 Dead-end corridors. Dead-end corridors in any work area shall not exceed 35 feet (10 670 mm).

Exceptions:

1. Where dead-end corridors of greater length are permitted by the *International Building Code*.

2. In other than Group A and H occupancies, the maximum length of an existing dead-end corridor shall be 50 feet (15 240 mm) in buildings equipped throughout with an automatic fire alarm system installed in accordance with the *International Building Code*.

3. In other than Group A and H occupancies, the maximum length of an existing dead-end corridor shall be 70 feet (21 356 mm) in buildings equipped throughout with an automatic sprinkler system installed in accordance with the *International Building Code*.

4. In other than Group A and H occupancies, the maximum length of an existing, newly constructed, or extended dead-end corridor shall not exceed 50 feet (15 240 mm) on floors equipped with an automatic sprinkler system installed in accordance with the *International Building Code*.

❖ The dead-end corridor limitations are only applicable where more than one exit or exit access doorway is required. The typical existing dead-end length allowed is 35 feet (10 670 mm) with exceptions that allow length up to 70 feet (21 336 mm). Existing dead ends in Group A and H occupancies are limited to 35 feet (10 670 mm), unless Exception 3 of Section 1016.3 of the IBC could apply. This is the exception that allows unlimited dead-end length as long as the length of the corridor is less than two and one-half times the least width of the dead-end corridor.

The installation of an automatic fire alarm system would increase the allowed length of an existing dead end to as much as 50 feet (15 240 mm), and the installation of an automatic sprinkler system throughout the building would increase the allowed length to 70 feet (21 336 mm). These two allowances are only applicable to existing corridors that happen to be longer in length than allowed by the IBC and are intended to encourage the installation of fire alarms and fire sprinklers throughout the building where such a corridor exists. Existing dead-end corridors must be brought into compliance with the 35-foot (10 670 mm) length limitation if these exceptions are not used.

605.7 Means-of-egress lighting. Means-of-egress lighting shall be in accordance with this section, as applicable.

❖ This is the charging paragraph for the requirements for means of egress lighting in Level 2 or 3 alterations. The specific requirements are given in Sections 605.7.1 and 605.7.2, relating the requirements for artificial lighting and supplemental requirements, respectively.

605.7.1 Artificial lighting required. Means of egress in all work areas shall be provided with artificial lighting in accordance with the requirements of the *International Building Code*.

❖ The means of egress in the work area and beyond the work area in the entire floor, if the work area exceeds 50 percent of the floor area, must be illuminated in accordance with the requirements of the IBC. This illumination must be provided by artificial lighting and must be a minimum of 1 foot-candle (11 lux) at the floor level. There are some exceptions to this requirement that can be found in Sections 1006.1 and 1006.2 of the IBC, which would apply here.

605.7.2 Supplemental requirements for means-of-egress lighting. Where the work area on any floor exceeds 50 percent of that floor area, means of egress throughout the floor shall comply with Section 605.7.1.

Exception: Means of egress within or serving only a tenant space that is entirely outside the work area.

❖ The supplemental provisions, where the work area exceeds 50 percent of the floor area, are the same here as they are in all previous sections discussed. As in all other areas, the 50-percent threshold brings a more serious concern regarding the alteration, and thus more restrictive requirements are applied throughout the floor, even in areas where no alteration is being performed. However, the tenant spaces that are unaffected do not require these alterations.

605.8 Exit signs. Exit signs shall be in accordance with this section, as applicable.

❖ This is the charging paragraph for the requirements for exit signs in Level 2 or 3 alterations. The specific requirements are given in Sections 605.8.1 and 605.8.2.

605.8.1 Work areas. Means of egress in all work areas shall be provided with exit signs in accordance with the requirements of the *International Building Code*.

❖ Exit signs are a critical element of directing occupants to safety outside the building. Exit signs must be installed in all Level 2 or 3 alterations work areas and beyond, where supplemental rules apply, in accordance with the IBC as if this was new construction. The exit sign provisions of the IBC are found in Section 1011, and it is important to note that such requirements as illumination and power source would apply to exit signs being installed in an alteration.

605.8.2 Supplemental requirements for exit signs. Where the work area on any floor exceeds 50 percent of that floor area, means of egress throughout the floor shall comply with Section 605.8.1.

Exception: Means of egress within a tenant space that is entirely outside the work area.

❖ The supplemental provisions, where the work area exceeds 50 percent of the floor area, are the same here as they are in all previous sections discussed. As in all other areas, the 50-percent threshold brings a more serious concern regarding the alteration, and thus more restrictive requirements are applied throughout the floor, even in areas where no alteration is being performed. However, the tenant spaces that are unaffected do not require these alterations.

605.9 Handrails. The requirements of Sections 605.9.1 and 605.9.2 shall apply to handrails from the work area floor to the level of exit discharge.

❖ This is the charging paragraph for the requirements for means of egress lighting in Level 2 or 3 alterations. The specific requirements are given in Sections 605.9.1 and 605.9.2, relating the minimum requirements for handrails and design requirements, respectively.

605.9.1 Minimum requirement. Every required exit stairway that is part of the means of egress for any work area and that has three or more risers and is not provided with at least one handrail, or in which the existing handrails are judged to be in danger of collapsing, shall be provided with handrails for the full length of the run of steps on at least one side. All exit stairways with a required egress width of more than 66 inches (1676 mm) shall have handrails on both sides.

❖ Complying handrails must be installed in all required exit stairways that serve a work area. If additional stairways are provided that are not required by IBC as an element of means of egress, then the handrail provisions of this section is not applicable to such stairways. The degree of compliance required is limited to a handrail on one side of the stairway only until the occupant load served is so high that the required width of the stairway would be a minimum of 66 inches (1676 mm). At this level of occupant load, complying handrails must be provided on both sides of the stairway. Intermediate handrails are not required as an improvement criteria under the code. This section intends to require complying handrails for those stairs that have three or more risers. This trigger level is an attempt to be compatible with an exception in some of the legacy building codes that did not require a handrail for stairways containing less than four risers in certain occupancies. The handrail requirement is applicable to required exit stairways that serve the work area, starting at the work area all the way to the level of exit discharge, but does not include the exit discharge itself.

605.9.2 Design. Handrails required in accordance with Section 605.9.1 shall be designed and installed in accordance with the provisions of the *International Building Code.*

❖ Once a handrail is required by this section, its structural strength, height, continuity and other design features must comply with the IBC.

Example 605.9:

An existing two-story office building has two enclosed stairways, one convenience stairway, a mezzanine stair and a stair outside the building as shown in Figure 605.9. Stairs either have no handrails or have handrails that are not in compliance with the IBC.

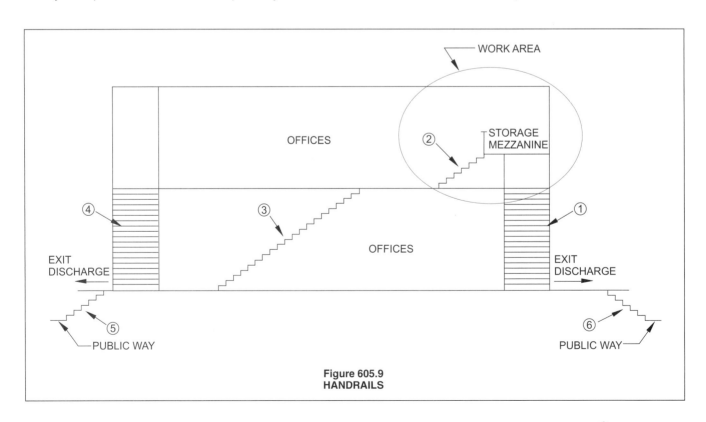

Figure 605.9
HANDRAILS

Q: Which stairs are required to install new complying handrails or bring the current handrails up to the IBC requirements?

A: Stairs 1 and 2 must be provided with new complying handrails to bring their existing handrails up to IBC requirements. These stairs serve the work area and are both part of the required means of egress pathway (see Figure 605.9).

605.10 Guards. The requirements of Sections 605.10.1 and 605.10.2 shall apply to guards from the work area floor to the level of exit discharge but shall be confined to the egress path of any work area.

❖ This is the charging paragraph for the requirements for means of egress lighting in Level 2 or 3 alterations. The specific requirements are given in Sections 605.10.1 and 605.10.2, relating the minimum requirements for handrails and design requirements, respectively.

605.10.1 Minimum requirement. Every open portion of a stair, landing, or balcony that is more than 30 inches (762 mm) above the floor or grade below and is not provided with guards, or those portions in which existing guards are judged to be in danger of collapsing, shall be provided with guards.

❖ The criteria and triggers for guards is much the same as those for handrails. Along the path of egress from the work area to the level of exit discharge, anywhere at stairs, landings or balconies where an elevation difference of more than 30 inches (762 mm) exists and a complying guard is not present, then a guard complying with details and structural requirements of the IBC must be installed.

605.10.2 Design. Guards required in accordance with Section 605.10.1 shall be designed and installed in accordance with the *International Building Code*.

❖ Once a guard is required by this section, its structural strength, height, size of openings and other design features must comply with the IBC.

SECTION 606
ACCESSIBILITY

606.1 General. A building, facility, or element that is altered shall comply with Section 506.

❖ A Level 2 or 3 alteration would include the reconfiguration of space, while a Level 1 alteration would not. Therefore, alterations covered under this section would have all accessibility requirements in Section 506 as well as two additional requirements that address reconfiguration in a structure.

606.2 Stairs and escalators in existing buildings. In alterations where an escalator or stair is added where none existed previously, an accessible route shall be provided in accordance

with Sections 1104.4 and 1104.5 of the *International Building Code*.

❖ If a stair or escalator is added as part of an alteration in a location where one did not previously exist, the alteration must also include an accessible route between the same two levels. If an accessible route is already available between the two levels, this requirement is not applicable. If the stair or escalator is replacing an existing stair or escalator, this requirement is not applicable. In conjunction with Section 506.1.12, if the requirement for the accessible route would be in excess of what is required for new construction, such as an accessible route to an area that was exempted by Section 1103.2, 1104, 1107, or 1108 of the IBC, this requirement is not applicable. The intent is that if a route is provided between accessible levels for a nondisabled person to use, it is reasonable to also expect an accessible route.

606.3 Dwelling units and sleeping units. Where Group I-1, I-2, I-3, R-1, R-2, or R-4 dwelling units or sleeping units are being added, the requirements of Section 1107 of the *International Building Code* for accessible units or Type A units and Chapter 9 of the *International Building Code* for accessible alarms apply only to the quantity of spaces being added.

❖ This section sets forth the rate for providing Accessible and Type A dwelling or sleeping units in Groups I-1, I-2, I-3, R-1, R-2 and R-4 when such facilities are added as part of the alteration (see also Section 506.1.8). Assuming that the required number of Accessible or Type A units are not already provided, the number of Accessible or Type A units to be incorporated into each alteration is based on the number being added. For example, if a nursing home was being altered a portion at a time, 50 percent of the units being added each time would be required to be wheelchair accessible. It is not the intent that all units being added are required to be Accessible units until 50 percent of the units in the entire facility are Accessible units. The total number of Accessible units in the facility is not required to exceed that required for new construction, as indicated in Section 506.1.12. It is unreasonable to require a greater level of accessibility in an existing building than is required in new construction. The technical criteria for Accessible units is found in ICC/ANSI A117.1-1998, Chapters 3 through 9. The technical criteria for Type A units is found in ICC/ANSI A117.1-1998, Section 1002.

This section also references visible and audible alarm requirements in Chapter 9 of the IBC. Sleeping accommodations in Groups I-1 and R-1 are required to have visible alarms in accordance with Table 907.9.1.3 of the IBC. Section 907.9.1.4 of the IBC also contains requirements for alarms within Group R-2 units. It is not the intent of this section to require the fire alarm system to be upgraded unless it is required by Section 604.4, 704.2, 804.1 or 904, as applicable.

SECTION 607
STRUCTURAL

607.1 General. Where alteration work includes installation of additional equipment that is structurally supported by the building or reconfiguration of space such that portions of the building become subjected to higher gravity loads as required by Tables 1607.1 and 1607.6 of the *International Building Code*, the provisions of this section shall apply.

❖ These structural requirements apply to alterations involving the addition of equipment or where a reconfiguration of building space results in an increased minimum live load required by Section 1607 of the IBC. Note that the structural requirements for Level 1 alterations in Section 507 are applicable according to Section 304.2.

Under Section 102.4.2, buildings that comply with the structural provisions of any of the following model codes are deemed to comply with this code:

International Building Code® (IBC®), 2003 or 2000 edition.

National Building Code (NBC), 1999 edition.

Standard Building Code (SBC), 1999 edition.

Uniform Building Code (UBC), 1997 edition.

Although the IRC is not explicitly listed, its use is required by the IBC for detached one- and two-family dwellings as well as townhouses with not more than three stories above grade plane and separate means of egress.

607.2 Reduction of strength. Alterations shall not reduce the structural strength or stability of the building, structure, or any individual member thereof.

Exception: Such reduction shall be allowed as long as the strength and the stability of the building are not reduced to below the *International Building Code* levels.

❖ This section explicitly states that alterations must not reduce the strength of individual structural elements or the building structure as a whole. The exception to this rule permits strength reductions to members that have "reserve capacity." In this case, however, the reduced strength must be equal to, or better than, that required by the IBC for a new structure.

607.3 New structural members. New structural members in alterations, including connections and anchorage, shall comply with the *International Building Code*.

❖ Any new structural element that is added in the course of alteration work, including its connections to the existing structure, must comply with the IBC requirements for new construction. There is no exception to this requirement.

607.4 Existing structural members. Existing structural components supporting additional equipment or subjected to additional loads based on *International Building Code* Tables 1607.1 and 1607.6 as a result of a reconfiguration of spaces shall comply with Sections 607.4.1 through 607.4.3.

❖ In the course of reconfiguring building spaces it is often necessary to relocate or add equipment. Sometimes a change in use (without actually changing the occupancy classification) necessitates use of a higher design live load. Consequently, existing structural members that are subjected to increased loads must comply with requirements in this section for gravity loads, seismic loads and snow drifts.

607.4.1 Gravity loads. Existing structural elements supporting any additional gravity loads as a result of additional equipment or space reconfiguration shall comply with the *International Building Code*.

Exceptions:

1. Structural elements whose stress is not increased by more than 5 percent.

2. Buildings of Group R occupancy with not more than five dwelling units or sleeping units used solely for residential purposes where the existing building and its alteration comply with the conventional light-frame construction methods of the *International Building Code* or the provisions of the *International Residential Code*.

❖ Where the alterations result in a net increase in the gravity load that is supported by the existing structure, the affected structural components must be checked to verify they satisfy the requirements of the IBC for a new structure. There are two exceptions to full compliance with the IBC, one of which permits smaller Group R buildings and their alterations that comply with either the IRC or the conventional light-frame construction provisions of the IBC. Another exception allows additional gravity loads that do not increase stresses in affected structural elements by more than 5 percent. Allowing overstresses of up to 5 percent in existing structural members has been a long-standing rule of thumb used by structural engineers. This exception does not specifically address the cumulative affects from successive alterations, but it would be prudent to limit cumulative increases under this exceptions to a total of 5 percent unless it is clearly documented that a structure has additional capacity.

607.4.2 Lateral loads. Buildings in which Level 2 alterations increase the seismic base shear by more than 5 percent shall comply with the structural requirements specified in Section 707.

❖ When Level 2 or 3 alterations increase the seismic base shear by more than 5 percent, the structural requirements for Level 3 alterations are applicable. The seismic base shear depends on several variables such as: the mapped spectral accelerations at the site; the site soil coefficients; the importance factor based on the nature of the occupancy; as well as the structure's effective seismic weight and response modification factor. For a Level 2 or 3 alteration, it is primarily the structure's effective seismic weight that is of interest, as the other variables would remain fixed. The effective seismic weight (W) used to determine the base shear consists of the total weight of both the structure, all equipment and other permanently attached items as well as a percentage of the design live load for storage uses. It reflects the weight of all dead loads and contents that might reasonably be expected to be attached to the structure at the time the design earthquake occurs. Therefore, the threshold of a 5-percent increase in the seismic base shear in turn permits a 5-percent increase in the effective seismic weight due to the alteration without requiring compliance with Section 707.

The 5-percent rule assumes that the existing building has a load path capable of transforming seismic load effects to the foundation for the existing seismic lateral load. This may or may not be the case since many existing buildings, especially those east of the Rocky Mountains, were not designed for seismic effects.

No mention of any wind lateral load analysis of the building is made in this section. Level 2 alterations would not increase the wind area of the building and it is assumed that the original building was adequately designed for wind load.

607.4.3 Snow drift loads. Any structural element of an existing building subjected to additional loads from the effects of snow drift as a result of additional equipment shall comply with the *International Building Code.*

Exceptions:

1. Structural elements whose stress is not increased by more than 5 percent.

2. Buildings of Group R occupancy with no more than five dwelling units or sleeping units used solely for residential purposes where the existing building and its alteration comply with the conventional light-frame construction methods of the *International Building Code* or the provisions of the *International Residential Code.*

❖ When a piece of equipment is added to the roof of an existing building, Section 607.4.1 requires consideration of the equipment load(s) on the supporting structure. Under certain conditions, roof projections cause additional accumulation of snow, or drifting, in their immediate vicinity. Therefore, adding (or relocating) roof-mounted equipment necessitates the consideration of the effects of snow drift loads on the existing structure. Section 1608.7 of the IBC requires the use of ASCE 7 procedures for snow drifts. These procedures use a triangular loading to approximate the drift (see "After" portion of Figure 607.4.3). This loading is combined with the balanced snow load (see "Before" portion of Figure 607.4.3) in evaluating the existing structure. The magnitude of this drift load depends on the dimensions of the equipment, its location on the roof, as well the ground snow load. The permitted exceptions are similar to those of Section 607.4.1.

SECTION 608
ELECTRICAL

608.1 New installations. All newly installed electrical equipment and wiring relating to work done in any work area shall comply with the materials and methods requirements of Chapter 5.

Exception: Electrical equipment and wiring in newly installed partitions and ceilings shall comply with all applicable requirements of the ICC *Electrical Code.*

❖ In the course of alterations that involve reconfiguration of space, there is sometimes a need to do certain rewiring of electrical equipment or fixtures or a need for replacement of such equipment or fixtures. Depending on whether the electrical work is taking place in existing walls, partitions, ceilings or floors or in new walls, partitions, ceilings or floors, the requirements are slightly different. In the case of new walls, partitions and other elements, all electrical equipment, fixtures and all electrical work must comply with the provisions of the electrical code. This is reasonable since the building element, being a wall, ceiling or other element is new, and all the electrical-related equipment and wiring in it are new. For example, in such situations, the required separation of outlets from sinks, maximum separation of outlets, minimum or maximum distances from the floor and all related wiring methods, wiring materials and wire gauge must completely comply with the electrical code. On the other hand, if such electrical work is taking place in existing walls, partitions, ceilings and floors, then only the materials and methods of the electrical code must be complied with. In such a case, for example, where the number of outlets, their proximity to wet locations and other dimensional criteria do not meet the current electrical code, then the replacement and rewiring of such outlets is allowed at these same locations as long as the equipment, wiring and wiring methods are in compliance with the electrical code.

608.2 Existing installations. Existing wiring in all work areas in Group A-1, A-2, A-5, H, and I occupancies shall be upgraded to meet the materials and methods requirements of Chapter 5.

❖ Faulty and damaged wiring or electrical wiring that might not be suitable for certain applications are contributors to many fires that might have electrical origins. To reduce the risk of such fires, this section requires upgrading electrical wiring in high occupant load, hazardous and institutional occupancies. The electrical wiring in the entire work area, in Group A-1, A-2, A-5, H and I occupancies, must be examined, and where such wiring does not comply with the materials and methods provisions of the electrical code, it must be upgraded to comply. It should be noted that electrical wiring in partitions and other elements that are not being altered or relocated must also comply as long as they are within the work area. The wiring materials and methods are mostly covered in Chapter 3 of the *National Electrical Code* (NEC).

FIGURE 607.4.3 – 608.3

ALTERATIONS—LEVEL 2

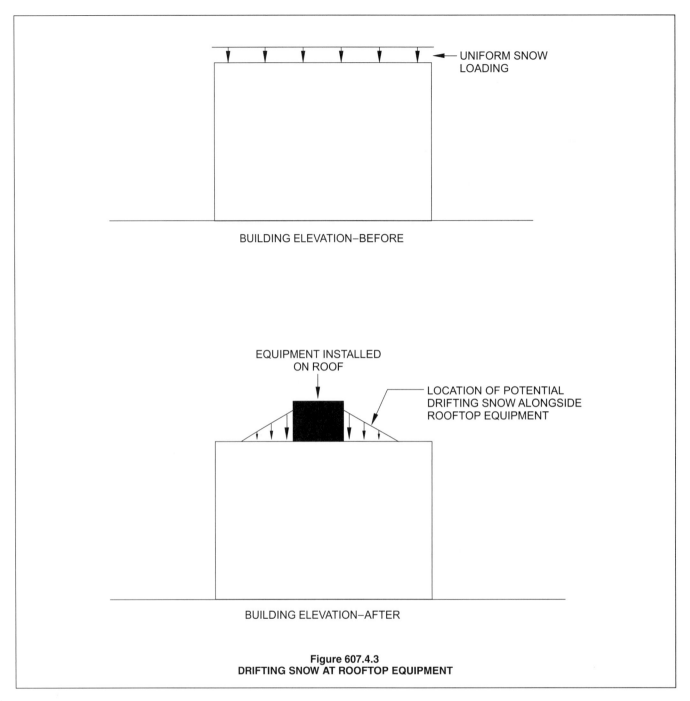

Figure 607.4.3
DRIFTING SNOW AT ROOFTOP EQUIPMENT

608.3 Residential occupancies. In Group R-2, R-3, and R-4 occupancies and buildings regulated by the *International Residential Code*, the requirements of Sections 608.3.1 through 608.3.7 shall be applicable only to work areas located within a dwelling unit.

❖ This is the charging paragraph for the requirements for electrical work in residential occupancies in Level 2 or 3 alterations. The specific requirements are given in Sections 608.3.1 through 608.3.7.

The work areas within a dwelling unit in Group R-2, R-3 and R-4 occupancies are required to meet a certain minimum number of receptacle outlets, ground-fault circuit interruptors (GFCI) and lighting outlets.

This is true regardless of whether the listed occupancies are being regulated under the IBC or IRC. Kitchens, laundry areas, sleeping rooms, study rooms or other similar rooms are required to provide at least one or two duplex receptacle outlets.

Closets, storage areas, hallways, garages and basements are exempt from this requirement even if they are within the work area. Even though receptacle outlets are needed and used in these areas, their use is not as frequent or as necessary compared to areas that are required to provide such receptacles under this section.

Where new receptacle outlets are installed, and if the electrical code requires GFCI for the location under consideration, then such GFCI must also be installed.

Detached garages that are served by electric power, attached garages, bathrooms, hallways, stairways, utility rooms, basements used for storage or mechanical equipment, exits and outdoor entrances that are within the work area must be provided with lighting outlets.

Example 608.3:

Q: A home owner wishes to do some remodeling that includes reconfiguration of some interior spaces including a bathroom, laundry and the living room. There is currently no light fixture outside the front door. Is a light fixture required to be installed outside of the front door?

A. Yes. Even though the exterior wall at the front door is not being reconfigured, it is located at the wall, which is part of the alteration work area and, as such, is required to be provided with a light fixture (see Figure 608.3).

Electrical service equipment that is located within the work area and within the dwelling unit must be adjusted, relocated or its surrounding obstructions removed or relocated such that the minimum clearances required by the electrical code are provided. Providing this clearance in compliance with the electrical code addresses the safety of repair personnel.

608.3.1 Enclosed areas. All enclosed areas, other than closets, kitchens, basements, garages, hallways, laundry areas, utility areas, storage areas, and bathrooms shall have a minimum of two duplex receptacle outlets or one duplex receptacle outlet and one ceiling or wall-type lighting outlet.

❖ See the commentary for Section 608.3.

608.3.2 Kitchens. Kitchen areas shall have a minimum of two duplex receptacle outlets.

❖ See the commentary for Section 608.3.

608.3.3 Laundry areas. Laundry areas shall have a minimum of one duplex receptacle outlet located near the laundry equipment and installed on an independent circuit.

❖ See the commentary for Section 608.3.

608.3.4 Ground fault circuit interruption. Newly installed receptacle outlets shall be provided with ground fault circuit interruption as required by the ICC *Electrical Code.*

❖ See the commentary for Section 608.3.

608.3.5 Minimum lighting outlets. At least one lighting outlet shall be provided in every bathroom, hallway, stairway, attached garage, and detached garage with electric power, and to illuminate outdoor entrances and exits.

❖ See the commentary for Section 608.3.

608.3.6 Utility rooms and basements. At least one lighting outlet shall be provided in utility rooms and basements where such spaces are used for storage or contain equipment requiring service.

❖ See the commentary for Section 608.3.

608.3.7 Clearance for equipment. Clearance for electrical service equipment shall be provided in accordance with the ICC *Electrical Code.*

❖ See the commentary for Section 608.3.

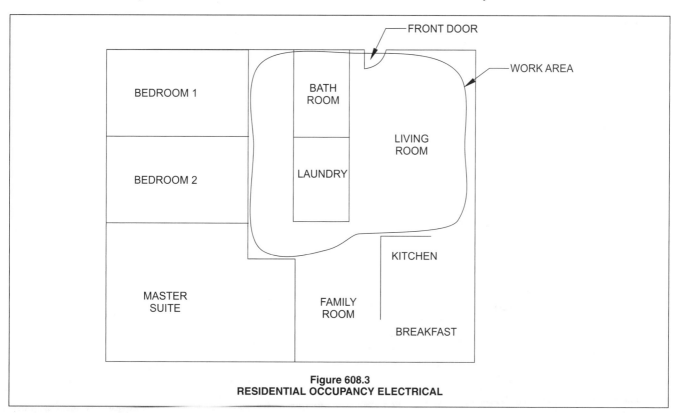

Figure 608.3
RESIDENTIAL OCCUPANCY ELECTRICAL

SECTION 609
MECHANICAL

609.1 Reconfigured or converted spaces. All reconfigured spaces intended for occupancy and all spaces converted to habitable or occupiable space in any work area shall be provided with natural or mechanical ventilation in accordance with the *International Mechanical Code.*

> **Exception:** Existing mechanical ventilation systems shall comply with the requirements of Section 609.2.

❖ This section is consistent with the basic premise for new building elements and materials given in Chapter 5, Section 503.3, that replacing elements required to comply with the applicable code for new construction. The amount of mechanical ventilation must be provided in accordance with the IMC as required for new construction. This could be that air-moving equipment would need to be altered as well. However, the exception allows special minimum provisions for existing mechanical equipment that would not require increasing capacities for existing air-moving equipment, as long as minimum ventilation rates are provided.

609.2 Altered existing systems. In mechanically ventilated spaces, existing mechanical ventilation systems that are altered, reconfigured, or extended shall provide not less than 5 cubic feet per minute (cfm) (0.0024 m³/s) per person of outdoor air and not less than 15 cfm (0.0071 m³/s) of ventilation air per person; or not less than the amount of ventilation air determined by the Indoor Air Quality Procedure of ASHRAE 62.

❖ This section essentially gives alternative minimum requirements to the IBC for ventilation in work areas of Level 2 or 3 alterations. However, this does not really reduce the amount of engineering work that needs to be done in an altered space. The referenced ASHRAE 62 must be consulted for the overarching ventilation requirements, and then the minimums stated in this section must be provided. Therefore, even in a situation where the intent was not to change the air moving equipment, but only the configuration of the ductwork or air channels, the minimum requirements could require changes to air-moving equipment capabilities.

609.3 Local exhaust. All newly introduced devices, equipment, or operations that produce airborne particulate matter, odors, fumes, vapor, combustion products, gaseous contaminants, pathogenic and allergenic organisms, and microbial contaminants in such quantities as to affect adversely or impair health or cause discomfort to occupants shall be provided with local exhaust.

❖ This is a straightforward, common-sense requirement that will readily be observed, and probably already dealt with in manufacturers' installation instructions for the equipment. However, the code official should carefully review the particular details for all such equipment to ensure that appropriate measures are taken for local exhaust.

SECTION 610
PLUMBING

610.1 Minimum fixtures. Where the occupant load of the story is increased by more than 20 percent, plumbing fixtures for the story shall be provided in quantities specified in the *International Plumbing Code* based on the increased occupant load.

❖ Alterations involving reconfiguration of spaces could at times result in an increased occupant load. The minimum number of plumbing fixtures is not affected and does not need to be reviewed or revised as long as the increased occupant load is 20 percent or less than the existing occupant load. This criteria is applied story by story and not to the building as a whole, because in most cases building occupants normally use the toilet facilities on the floor where they work or live and rarely travel to another floor to use the plumbing facilities. Once the 20-percent increased occupant load threshold is passed, then the entire occupant load of the story after alterations must be considered and the minimum plumbing fixtures as required by the IPC must be provided.

Example 610.1:

An existing mixed occupancy building as shown in Figure 610.1 is to undergo alterations to reconfigure certain areas and increase the occupant load as shown. There are four water closets currently provided in each restroom.

Q: What is the impact of the alterations on the required number of water closets?

A: Current occupant load is 650 (based on Table 2902.1 of the IBC).

Offices: 100 occupant load, two water closets for male and two for female.

Restaurant: 300 occupant load, two water closets for male and two for female.

Retail: 250 occupant load, one water closet for male and one for female.

Total number of water closets in each restroom is 5

Case 1: Office occupant load (100) + restaurant (375) + 292 = 767

Percent change of occupant load for the story = (767-650)/650 = 18 percent

The number of water closets does not need to be increased; there can remain four water closets in each male and female restroom.

Case 2: Office occupant load (100) + restaurant (395) + 292 = 787

Percent change of occupant load for the story = (787-650)/650 = 21 percent

The number of water closets need to be increased:

Offices: 100 occupant load, two water closets for male and two for female.

Restaurant: 395 occupant load, three water closets for male and three for female.

Retail: 292 occupant load, one water closet for male and one for female.

The number of water closets must be increased to six in each male and female restroom (see Figure 610.1).

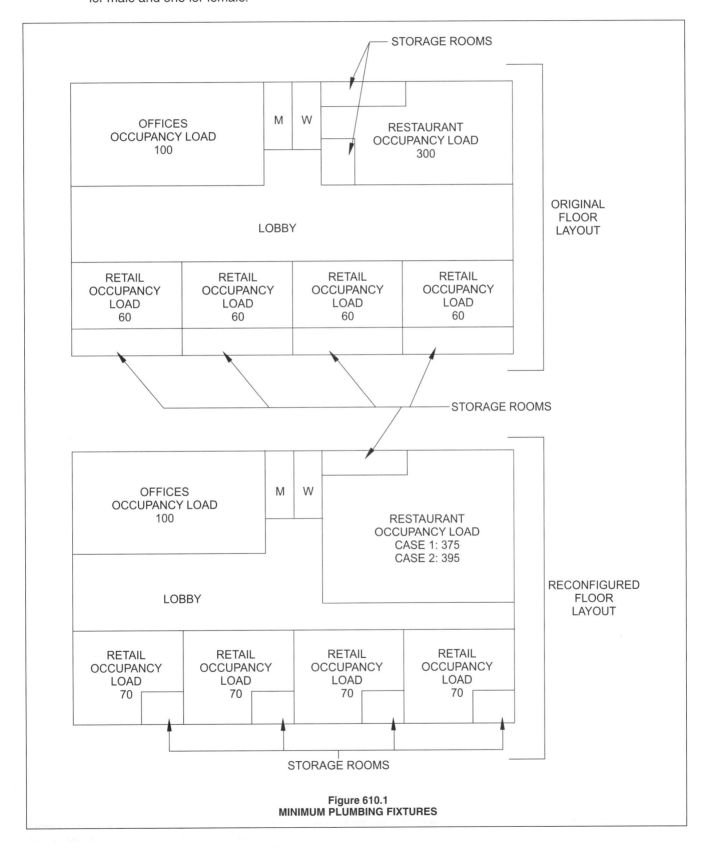

Figure 610.1
MINIMUM PLUMBING FIXTURES

Bibliography

The following resource materials are referenced in this chapter or are relevant to the subject matter addressed in this chapter.

ASCE 7-02, *Minimum Design Loads for Buildings and Other Structures*. Reston, VA: American Society of Civil Engineers, 2002.

ASHRAE 62-89, *Ventilation for Acceptable Indoor Air Quality.* Atlanta, GA: American Society of Heating and Air-conditioning Engineers, 1989.

IBC-03, *International Building Code*. Falls Church, VA: International Code Council, 2003.

IMC-03, *International Mechanical Code*. Falls Church, VA: International Code Council, 2003.

IRC-03, *International Residential Code*. Falls Church, VA: International Code Council, 2003.

NBC-99, *BOCA National Building Code*. Country Club Hills, IL: Building Officials & Code Administrators International, Inc. 1999.

SBC-99, *Standard Building Code*. Birmingham, AL: Southern Building Code Congress International, 1999.

UBC-97, *Uniform Building Code*. Whittier, CA: International Conference of Building Officials, 1997.

CHAPTER 7

ALTERATIONS—LEVEL 3

General Comments

This chapter provides the technical requirements for those existing buildings that undergo Level 3 alterations. Reference is made to Level 3 alterations in Section 305, which includes any alteration that involves more than 50 percent of the aggregate area of the building.

This chapter, similar to other chapters of this code, covers all building-related subjects such as: structural, mechanical, plumbing, electrical and accessibility as well as the fire and life safety issues when the alterations are classified as Level 3. Sections 701, 702 and 703 are related to scoping, special use and occupancy and building elements and materials. In the interest of brevity and avoiding repetition of materials, Section 701.2 requires that Level 3 alterations comply not only with Chapter 7, but also with Chapters 5 and 6. As such, any alteration work that is classified as Level 3 is assumed to include everything related to Level 1 and 2 alterations plus the provisions of this chapter. Section 703 covers in detail elements such as existing shafts and vertical openings, including when such openings might be required to be enclosed, fire partitions and interior finishes. The remainder of the chapter is related to fire protection, means of egress, accessibility, structural, electrical, mechanical and plumbing provisions. Section 705 requires additional means of egress lighting and exit signs beyond what is required already based upon Chapter 6 requirements for means of egress. Keep in mind that Section 605 covers a wide range of means of egress subjects such as the minimum number of exits, fire escape requirements and construction details, corridor openings and exit signs, which are also applicable here. The structural provisions for Level 3 alterations are contained in Section 607.

The decision to divide the level typically known as "alterations" in the *International Building Code*® (IBC®) and previous legacy building codes into three parts of Level 1, Level 2 and Level 3 was based on the fact that minor alterations that do not include space reconfiguration and extensive alterations that might range from relocation of walls and partitions in various floors or in the entire building should be treated differently and with different threshold levels for requiring upgrades or improvements to the building or spaces within the building.

Purpose

The purpose of this chapter is to provide detailed requirements and provisions to identify the required improvements in the existing building elements, building spaces and building structural system. This chapter is distinguished from Chapters 5 and 6 by involving alterations that are over 50 percent of the aggregate area of the building. In contrast, Level 1 alterations do not involve space reconfiguration and Level 2 alterations involve extensive space reconfiguration that do not exceed 50 percent of the building area. Depending on the nature of alteration work, its location within the building and whether it encompasses one or more tenants, improvements and upgrades could be required for the open floor penetrations, sprinkler system or the installation of additional means of egress, such as stairs or fire escapes. At times and under certain situations, this chapter also intends to improve the safety of certain building features beyond the work area and in other parts of the building where no alteration work might be taking place.

SECTION 701
GENERAL

701.1 Scope. Level 3 alterations as described in Section 305 shall comply with the requirements of this chapter.

❖ Any alteration work that results in reconfiguration of space or otherwise falls under the classification of Level 3 alterations, as described in Section 305, must comply with Chapter 7. The exception of Section 701.2 is intended to encourage existing buildings to improve the accessibility for the disabled by allowing any such alterations (which are solely for compliance with accessibility provisions of Section 506) to only comply with Chapter 5 provisions as if no space reconfiguration has taken place. This is an advantage for improving accessibility of buildings since Chapter 5 is related to alterations that do not encompass any space reconfiguration and, as such, has a lower level of compliance requirement and lower threshold for triggering other improvements.

701.2 Compliance. In addition to the provisions of this chapter, work shall comply with all of the requirements of Chapters 5 and 6. The requirements of Sections 603, 604, and 605 shall apply within all work areas whether or not they include exits and corridors shared by more than one tenant and regardless of the occupant load.

Exception: Buildings in which the reconfiguration of space affecting exits or shared egress access is exclusively the result of compliance with the accessibility requirements of Section 506.2 shall not be required to comply with this chapter.

❖ Chapters 6 and 7 have cascading effects by requiring that each level of alteration must comply with the chapter for the lower level of alteration. Accordingly, an alteration project that is classified as Level 2 must comply with Chapter 6 as well as with Chapter 5. Similarly, an alteration project classified as Level 3 must comply with Chapters 5, 6 and 7. This is to eliminate the repetition of various requirements from Chapter 5 in Chapters 6 and 7 and from Chapter 6 in Chapter 7, as a Level 2 alteration, in general, includes various aspects of a Level 1 alteration, and a Level 3 alteration, in general, includes various aspects of Level 1 and 2 alterations. The code drafting committee decided to make reference to these chapters rather than repeat all such provisions again.

Keep in mind that Section 601.3 is applicable in this chapter as well. Section 601.3 requires that new elements, components and spaces created as a result of Level 2 alteration work must comply with the IBC or other applicable code, just as Section 101.4 of the IBC makes reference to other *International Codes*® such as *International Plumbing Code*® (IPC®), *International Mechanical Code*® (IMC®) and *International Fire Code*® (IFC®). Please review the requirements and commentary in Section 601.3, including the exceptions, which also apply.

SECTION 702
SPECIAL USE AND OCCUPANCY

702.1 High-rise buildings. Any building having occupied floors more than 75 feet (22 860 mm) above the lowest level of fire department vehicle access shall comply with the requirements of Sections 702.1.1 and 702.1.2.

❖ This is the charging paragraph for high-rise requirements. The specific requirements contained herein pertain to recirculating or exhaust of air (see Section 702.1.1) or elevators (see Section 702.1.2). A high-rise building is defined as a building having an occupied floor located more than 75 feet (22 860 mm) above the lowest level of fire department vehicle access, which is the same as the IBC criteria for identifying a high-rise building. In high-rise buildings undergoing a Level 3 alteration, the subjects of recirculating air and elevators must specifically be addressed. All other provisions of Chapter 7 still apply to high-rise buildings, but these two are specific only to high-rises. The notable difference between this section and Section 403 of the IBC, is that Section 403.1 of the IBC has five exceptions. The five exceptions are buildings that do not fall under the high-rise provisions of Section 403 of the IBC. The code does not contain any of these five exceptions, so whether or not the building under consideration is an airport traffic control tower (see Section 403.1 of the IBC, Exception 1), an open parking garage (see Section 403.1 of the IBC, Exception 2), a Group A-5 occupancy (see Section 403.1 of the IBC, Exception 3), a low-hazard special industrial occupancy (see Section 403.1 of the IBC, Exception 4) or a Group H-1, H-2 or H-3 occupancy (see Section 403.1 of the IBC, Exception 5) the recirculating air and elevator recall requirements of Section 702.1.1 and 702.1.2 are applicable. Even though this appears to be more

restrictive than the IBC, it really is not, because in most buildings the five exceptions in Section 403.1 of the IBC have a special section within the IBC that they must comply with (see Sections 406.3, 412, and 415 of the IBC).

702.1.1 Recirculating air or exhaust systems. When a floor is served by a recirculating air or exhaust system with a capacity greater than 15,000 cubic feet per minute (701 m³/s), that system shall be equipped with approved smoke and heat detection devices installed in accordance with the *International Mechanical Code*.

❖ This section applies to high-rise building undergoing a Level 3 alteration, as described in Section 305, that has a floor served by a recirculating air or exhaust system with a capacity of greater than 15,000 cubic feet per minute (701 m/s³). Such a system is required to be equipped with approved smoke and heat detection devices. This is clearly a condition where large mechanical devices have the potential of spreading products of combustion quickly through a particular floor and beyond. Over the years, it has been recognized that many injuries and deaths in high-rise fires have been the result of smoke inhalation due to smoke penetrating the floors above the floor of fire origin. The smoke and heat detection devices that are required to be installed in accordance with the IMC are intended to shut down the mechanical circulating air or exhaust system and to reduce the potential spread of smoke.

702.1.2 Elevators. Where there is an elevator or elevators for public use, at least one elevator serving the work area shall comply with Section 607.1 of the *International Fire Code*.

❖ The elevator recall provisions of Section 607.1 of the IFC are intended to accommodate the needs of emergency personnel for fire-fighting or rescue purposes. Elevators are often used as a tool by emergency responders when responding to fires and other emergencies. Due to these needs, the elevators must be capable of providing certain functions such as recall and emergency operation. The following sections simply denote when such controls and necessary signage are required to clarify to building occupants that elevators are not to be used during a fire.

Section 607.1 of the IFC sets out requirements for both new and existing elevators. Existing elevators that travel 25 feet (7620 mm) or more above or below the main level should, as a minimum, be equipped with emergency operation capabilities that comply with ASME A17.3. New elevator installations are held to more restrictive requirements for increased cost effectiveness and must have both emergency recall and emergency in-car operation to comply with ASME A17.1 for any amount of travel distance. The ASME standards are safety codes for elevators and escalators: ASME A17.3 for existing elevators and ASME A17.1 for new elevator installations.

702.2 Boiler and furnace equipment rooms. Boiler and furnace equipment rooms adjacent to or within the following facilities shall be enclosed by 1-hour fire-resistance-rated construction: day nurseries, children's shelter facilities, residential childcare facilities, and similar facilities with children below the age of $2^1/_2$ years or that are classified as Group I-2 occupancies, shelter facilities, residences for the developmentally disabled, group homes, teaching family homes, transitional living homes, rooming and boarding houses, hotels, and multiple dwellings.

Exceptions:

1. Furnace and boiler equipment of low-pressure type, operating at pressures of 15 pounds per square inch gauge (psig) (103.4 KPa) or less for steam equipment or 170 psig (1171 KPa) or less for hot water equipment, when installed in accordance with manufacturer recommendations.

2. Furnace and boiler equipment of residential R-3 type with 200,000 British thermal units (Btu) (2.11×108 J) per hour input rating or less is not required to be enclosed.

3. Furnace rooms protected with automatic sprinkler protection.

❖ The boiler and furnace room separation requirements are based on the separation concept of Section 302.1.1 of the IBC but the trigger mechanism is different from that of the IBC. The IBC requires any boiler or furnace room with a single piece of equipment larger than 400,000 Btu per hour input be separated from the main areas by a fire barrier, whereas the code requires any such room with high-pressure-type equipment be separated from certain uses. High pressure is considered 15 pounds per square inch (psig) (103.4 kPa) for steam equipment and 170 psig (1171 kPa) for hot water equipment. The uses from which the separation is required are those that are occupied by infants, certain institutional-type facilities, hotels and multiple-family occupancies. Furnace rooms protected with an automatic sprinkler system are not required to be separated.

702.2.1 Emergency controls. Emergency controls for boilers and furnace equipment shall be provided in accordance with the *International Mechanical Code* in all buildings classified as day nurseries, children's shelter facilities, residential childcare facilities, and similar facilities with children below the age of $2^1/_2$ years or that are classified as Group I-2 occupancies, and in group homes, teaching family homes, and supervised transitional living homes in accordance with the following:

1. Emergency shutoff switches for furnaces and boilers in basements shall be located at the top of the stairs leading to the basement; and

2. Emergency shutoff switches for furnaces and boilers in other enclosed rooms shall be located outside of such room.

❖ The boiler and furnace equipment must have emergency controls in accordance with the IMC.

SECTION 703
BUILDING ELEMENTS AND MATERIALS

703.1 Existing shafts and vertical openings. Existing stairways that are part of the means of egress shall be enclosed in accordance with Section 603.2.1 between the highest work area floor and the level of exit discharge and all floors below.

❖ This section addresses the enclosure of existing stairways only. Other shafts and floor openings are regulated exactly the same as under Level 2 alterations. The code user is referred to Section 603.2.1 for stairway enclosure methods and triggers. Such enclosures, though, must be provided from the highest work area floor all the way to the level of exit discharge and all floors below the level of exit discharge. This requirement is only applicable to stairways that are part of the means of egress and not other stairways used for convenience, attending equipment or those that have other specific purposes.

703.2 Fire partitions in Group R-3. Fire separation in Group R-3 occupancies shall be in accordance with Section 703.2.1.

❖ This is the charging paragraph for fire separations between dwelling units in Group R-3 occupancies, sending the user of this code to Section 703.2.1. It should be noted that the title of this section is somewhat misleading, referring to "fire partitions," which are defined and detailed specifically in the IBC. However, the regulatory text describes the required separation between the dwelling units, never actually requiring construction of fire partitions as described in the IBC.

703.2.1 Separation required. Where the work area is in any attached dwelling unit in Group R-3 or any multiple single family dwelling (townhouse), walls separating the dwelling-units that are not continuous from the foundation to the underside of the roof sheathing shall be constructed to provide a continuous fire separation using construction materials consistent with the existing wall or complying with the requirements for new structures. All work shall be performed on the side of the dwelling unit wall that is part of the work area.

Exception: Where alterations or repairs do not result in the removal of wall or ceiling finishes exposing the structure, walls are not required to be continuous through concealed floor spaces.

❖ The separation of dwelling units in duplex and townhouse buildings is found in Section R317 of the *International Residential Code* (IRC) for new construction. The code in this section addresses the vertical separation aspect only. If the alteration includes the removal of wall or ceiling finishes that expose the wall structure, then the separation wall must be evaluated for continuity from foundation to roof deck. The existing wall construction is allowed to remain and be continued to the roof deck even if it does not provide the IRC's required 1-hour fire-resistance rating.

It is intended that this requirement not create a burden for the residents of adjacent dwellings if they are not involved in the alteration project; as such, it is allowed that extension of such walls to roof deck be placed on the side where an alteration project is taking place and not intrude in the neighboring units.

703.3 Interior finish. Interior finish in exits serving the work area shall comply with Section 603.4 between the highest floor on which there is a work area to the floor of exit discharge.

❖ The interior finish materials in Level 3 alterations are regulated the same as Level 2 alterations. The reader is referred to Section 603.4 where the code requires interior finishes in exits and corridors serving the work area to comply with the IBC interior finish requirements. The only additional requirement in Level 3 alterations is the extent of coverage. Exits serving the work area must comply with the interior finish requirements of Section 603.4 starting at the highest floor where there is any work area to the floor of exit discharge. Any corridors within the work area must also comply with Section 603.4, as Section 701.2 requires all Level 3 alterations to comply with Chapters 5, 6 and 7.

SECTION 704
FIRE PROTECTION

704.1 Automatic sprinkler systems. Automatic sprinkler systems in accordance with Section 604.2 shall be provided in all work areas.

❖ Automatic sprinkler requirements are referenced to Section 604.2 where the requirements for various occupancies, mixed uses and windowless stories are provided. This section has specific requirements for high-rise buildings and rubbish and linen chutes.

704.1.1 High-rise buildings. In high-rise buildings, work areas shall be provided with automatic sprinkler protection where the building has a sufficient municipal water supply system to the site. Where the work area exceeds 50 percent of floor area, sprinklers shall be provided in the specified areas where sufficient municipal water supply for design and installation of a fire sprinkler system is available at the site.

❖ Where there is sufficient municipal water supply to the site to provide an automatic sprinkler system, then all of the work areas within the high-rise building must be sprinklered. "Available to the site" refers to the availability of water at the property line and not at the building itself. In certain high-rise Level 3 alteration projects this could have the effect of requiring design and installation of a new fire line main and associated vaults, valves and other related components from the municipal water main to the building. Sprinklers could be required beyond the work areas on any floor where the work area exceeds 50 percent of that floor area. This extension of sprinklers is not required in tenant spaces that are entirely outside of the work area (see Figure 704.1.1).

Example:

Sprinklers required in:

Level 1: Tenants 1 and 3 (more than 50 percent of floor – tenant 2 is entirely outside the work area).

Level 2: Parts of tenants 1, 2 and 3 that are within the work area (work areas only – less than 50 percent of floor area).

Level 3: Tenant 2 (work area not more than 50 percent – tenant 1 is entirely outside of work area).

Level 4: Tenants 1, 2, 3 and 4 (work area is more than 50 percent of floor area – tenants 2 and 4 are not entirely outside the work area).

Level 5: Tenants 1, 2, 3 and 4 (work area is more than 50 percent of the area – tenants 2 and 4 are not entirely outside the work area).

704.1.2 Rubbish and linen chutes. Rubbish and linen chutes located in the work area shall be provided with sprinklered protection where protection of the rubbish and linen chute would be required under the provisions of the *International Building Code* for new construction and the building has sufficient municipal water supply available to the site.

❖ In any building undergoing a Level 3 alteration, the rubbish and linen chutes within the work areas must be provided with sprinklers if there is sufficient municipal water supply to the site (for a discussion of sufficient municipal water available at the site, see the commentary for Section 704.1.1). The IBC requires an automatic sprinkler system at the top and in the terminal rooms of the rubbish and linen chutes, and when chutes extend through three or more floors additional sprinkler heads are required at alternate floors. Accordingly, in a four-story building where there is a linen chute from fourth floor to the first and there are alterations on all floors, the code sprinkler requirement for the chute will be exactly the same as the IBC. Chute sprinklers must be available for servicing (see Figure 704.1.2).

704.2 Fire alarm and detection systems. Fire alarm and detection systems complying with Sections 604.4.1 and 604.4.3 shall be provided throughout the building in accordance with the *International Building Code*.

❖ This section refers back to Chapter 6, Sections 604.4.1 and 604.4.3, except that the significant phrase in this section is "throughout the building." The code user is referred to Sections 604.4.1 and 604.4.3 for the fire alarm and detection system requirements, where the triggering mechanism for various occupancies are provided. Section 604.4.2, the supplemental fire alarm system requirements, is not referenced because in Level 3 alterations the alarm system installation is not limited just to the work areas; rather, it is required throughout the building.

704.2.1 Manual fire alarm systems. In Group A, B, E, F, H, I, M, R-1, and R-2 occupancies a manual fire alarm system shall be provided on all floors in the work area. Alarm notification appliances shall be provided on such floors and shall be automatically activated as required by the *International Building Code*.

Exceptions:

1. Where the *International Building Code* does not require a manual fire alarm system.

2. Alarm-initiating and notification appliances shall not be required to be installed in tenant spaces outside of the work area.

3. Visual alarm notification appliances are not required, except where an existing alarm system is upgraded or replaced or where a new fire alarm system is installed.

❖ Manual fire alarm systems are required within the work areas in certain occupancies such as Groups A, E and M. This is required within the work areas on any floor regardless of the size of the area on a particular floor. There are three exceptions for this requirement. Exception 1 is a referral to the IBC to make sure the manual alarm requirements in this section would not be applied where such a system is not required for new construction. Exception 2 addresses the typical exception that we have seen in many sections of this code, which is an exemption for tenant spaces that are completely outside of the work area. Exception 3 is an exemption for visual alarm notification when upgrading of an existing system or installation of a new system are not involved.

SECTION 705
MEANS OF EGRESS

705.1 General. The means of egress shall comply with the requirements of Section 605 except as specifically required in Sections 705.2 and 705.3.

❖ The means of egress requirements for Level 3 alterations are exactly the same as Level 2 alterations found in Section 605, with the exception that the means of egress lighting and exit signs must meet an additional criteria.

705.2 Means-of-egress lighting. Means of egress from the highest work area floor to the floor of exit discharge shall be provided with artificial lighting within the exit enclosure in ac-

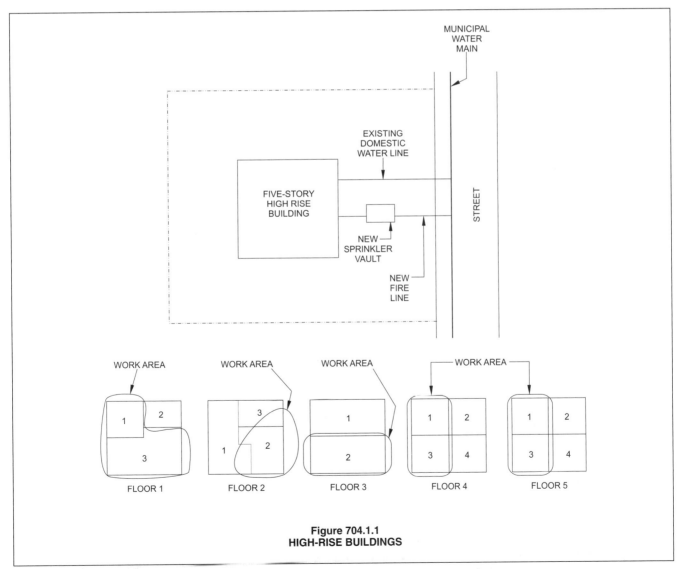

Figure 704.1.1
HIGH-RISE BUILDINGS

FIGURE 704.1.2 – 705.3

ALTERATIONS—LEVEL 3

cordance with the requirements of the *International Building Code*.

❖ The additional criteria for means of egress lighting requires that within exit enclosures, illumination be provided from the highest floor where there is any work area down to the level of exit discharge. The intensity of illumination, emergency power and performance of the system is required based on the IBC.

705.3 Exit signs. Means of egress from the highest work area floor to the floor of exit discharge shall be provided with exit signs in accordance with the requirements of the *International Building Code*.

❖ The additional criteria for exit signs requires that the means of egress from the highest floor where there is any work area to the level of exit discharge be provided with exit signs. The location, height, illumination and other requirements of such exit signs must comply with the IBC.

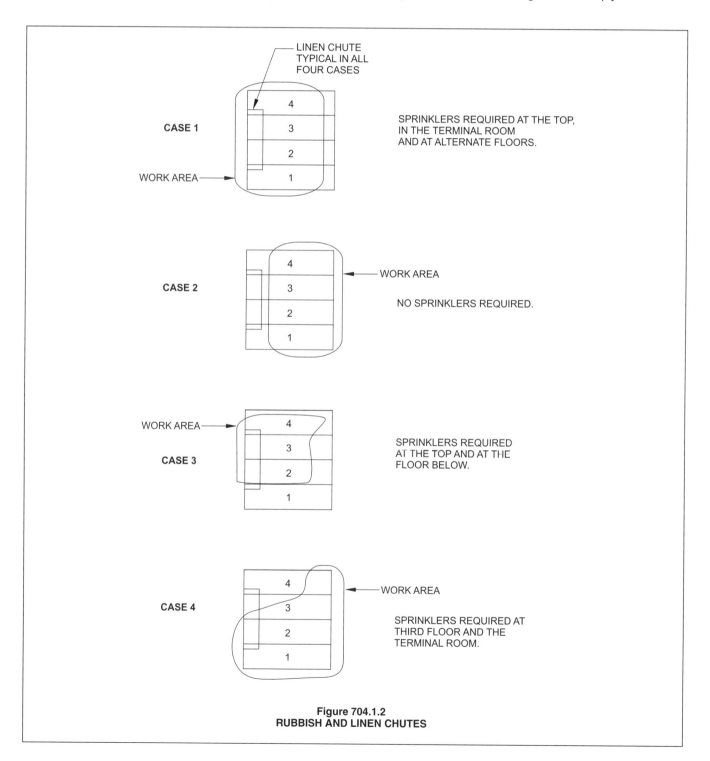

Figure 704.1.2
RUBBISH AND LINEN CHUTES

SECTION 706
ACCESSIBILITY

706.1 General. A building, facility, or element that is altered shall comply with Section 506.

❖ While this section only references Section 506, Section 701.2 specifically states that Level 3 alterations shall comply with Chapters 5, 6 and 7. Therefore, alterations covered under this section would have to meet all accessibility requirements in Section 506 as well as two additional requirements covered in Section 606.

SECTION 707
STRUCTURAL

707.1 General. Where buildings are undergoing Level 3 alterations including structural alterations, the provisions of this section shall apply.

❖ These structural requirements apply to alterations where the work area exceeds 50 percent of the aggregate area of the building. Note that the structural requirements for Level 1 and 2 alterations in Sections 507 and 607, respectively, are also applicable to Level 3 alterations according to Section 305.2.

Under Section 102.4.2, buildings that comply with the structural provisions of any of the following model codes are deemed to comply with the code:

International Building Code® (IBC®), 2003 or 2000 edition.

National Building Code (NBC), 1999 edition.

Standard Building Code (SBC), 1997 edition.

Uniform Building Code (UBC), 1997 edition.

Although the IRC is not explicitly listed, its use is required by the IBC for detached one- and two-family dwellings as well as townhouses with not more than three stories above grade plane and separate means of egress.

707.2 Reduction of strength. Alterations shall not reduce the structural strength or stability of the building, structure, or any individual member thereof.

Exception: Such reduction shall be allowed provided that the structural strength and the stability of the building are not reduced to below the *International Building Code* levels.

❖ This section explicitly states that alterations must not reduce the strength of individual structural elements or the building structure as a whole. The exception to this rule permits strength reductions to members that have "reserve capacity." In this case, however, the reduced strength must be equal to, or better than, that required by the IBC for a new structure.

707.3 New structural members. New structural members in alterations, including connections and anchorage, shall comply with the *International Building Code.*

❖ Any new structural element that is added in the course of alteration work, including its connections to the existing structure, must comply with the IBC requirements for new construction. There is no exception to this requirement.

707.4 Minimum design loads. The minimum design loads on existing elements of a structure that do not support additional loads as a result of an alteration shall be the loads applicable at the time the building was constructed.

❖ Unless an alteration adds load to an existing structural member, there is no need to reevaluate it against current code loading criteria. This is in contrast to the treatment of existing members that have increased loading in accordance with Section 607.4. The applicable loads of the code at the time the building was built apply.

707.5 Structural alterations. Buildings and structures undergoing structural alterations or buildings in which the seismic base shear is increased by more than 5 percent because of alterations shall comply with this section.

❖ This section applies to both Level 2 and 3 alterations that increase the seismic base shear by more than 5 percent. Where this occurs, it is necessary to have an engineering evaluation of the building. The loading criteria to be used in the evaluation is dependent upon the extent of the alteration work that is undertaken. The code assumes that the structural members were properly designed for the applicable loads at the time of construction.

The seismic base shear depends on several variables such as the mapped spectral accelerations at the site, the site soil coefficients, the importance factor based on the nature of the occupancy as well as the structure's effective seismic weight and response modification factor. For work that is considered an alteration, it is primarily the structure's effective seismic weight that is of interest, as the other variables would likely remain fixed. The effective seismic weight (W) used to determine the base shear consists of the total weight of the structure, all equipment and other permanently attached items as well as a percentage of the design live load for storage uses. It reflects the weight of all dead loads and contents that might reasonably be expected to be attached to the structure at the time the design earthquake occurs. Therefore, the threshold of a 5-percent increase in the seismic base shear in turn permits a 5-percent increase in the effective seismic weight due to the alteration without requiring compliance with this section.

707.5.1 Evaluation and analysis. An engineering evaluation and analysis that establishes the structural adequacy of the altered structure shall be prepared by a registered design professional and submitted to the code official. Where more than 30 percent of the total floor and roof areas of the building or structure has been or is proposed to be involved in structural alteration within a 12-month period, the evaluation and analysis shall demonstrate that the altered building or structure complies with the *International Building Code* for wind loading and with reduced *International Building Code* level seismic forces as specified in Section 407.1.1.3 for seismic loading. For seismic considerations, the analysis shall be based on one of the procedures specified in Section 407.1.1.1. The areas to be counted toward the 30 percent shall be those areas tributary to the vertical load-carrying components such as joists, beams, columns, walls, and other structural components that have been or will be removed, added, or altered, as well as areas such as mezzanines, penthouses, roof structures, and in-filled courts and shafts.

Exceptions:

1. Buildings of Group R occupancy with no more than five dwelling units or sleeping units used solely for residential purposes that are altered based on the conventional light-frame construction methods of the *International Building Code* or in compliance with the provisions of the *International Residential Code*.

2. Where such alterations involve only the lowest story of a building and the change of occupancy provisions of Chapter 8 do not apply, only the lateral-force-resisting components in and below that story need comply with this section.

❖ A building undergoing alterations that result in a seismic base shear increase greater than 5 percent and that also exceed the area threshold established in this section requires an engineering evaluation of the structure as described in this section. The evaluation must demonstrate that the altered structure complies with the IBC wind-loading criteria and with reduced IBC seismic forces in Section 407.1.1.3. The latter permits the use of the *Guidelines for Seismic Evaluation of Existing Buildings* (GSREB) in Appendix A of the code, ASCE 31 and FEMA 356 (see commentary, to Section 407.1.1.1).

The area threshold of 30 percent of the floor and roof areas would likely include all Level 3 alterations, since Section 305.1 establishes that these are in excess of 50 percent of the aggregate area of the building. Level 2 alterations that exceed the 30-percent threshold are also covered by this section. The additional stipulation that all structural alterations occurring over a 12-month period must be considered prevents a building owner from breaking a large alteration project into a series of smaller alterations in hopes of complying with less-stringent requirements. The clarification of exactly which areas contribute to meeting the area threshold is important. For instance, mezzanines complying with Section 505 of the IBC do not contribute to the building area.

There are two exceptions to this evaluation requirement, one of which permits smaller Group R buildings that are altered to comply with either the IRC or the conventional light-frame construction provisions of the IBC. Exception 2 pertains to multistory buildings, where alterations are made to only the lowest story. This permits compliance with the IBC's wind and seismic criteria to be limited to the lateral-force-resisting

elements in that story and below. While the impact of wind and seismic loading contributed by the stories above must be considered, no evaluation of those stories is required. The code assumes the upper stories have been designed for wind and earthquake according to the code in effect at the time of construction.

707.5.2 Limited structural alteration. Where not more than 30 percent of the total floor and roof areas of the building is involved in structural alteration within a 12-month period, the evaluation and analysis shall demonstrate that the altered building or structure complies with the loads applicable at the time the building was constructed.

❖ A building undergoing alterations not exceeding the threshold established in Section 707.5.1, but resulting in a seismic base shear increase greater than 5 percent, need only be evaluated for the loads applicable at the time the building was constructed rather than current code requirements. Depending on the wind and seismic criteria that were originally applicable to the structure, this may or may not offer an actual advantage in the structural evaluation. There are no exceptions given, but at the very least Exception 1 to Section 707.5.1 could be applied to this more limited amount of alteration work as well. Interestingly, this section requires a verification of the prior design for loads at the time of construction while the upper floors need not be verified according to Section 707.5.1, Exception 2.

707.6 Additional vertical loads. Where gravity loading is increased on the roof or floor of a building or structure, all structural members affected by such increase shall meet the gravity load requirements of the *International Building Code*.

Exceptions:

1. Structural elements whose stress is not increased by more than 5 percent.

2. Buildings of Group R occupancy with no more than five dwelling units or sleeping units used solely for residential purposes that are altered based on the conventional light-frame construction methods of the *International Building Code* or in compliance with the provisions of the *International Residential Code*.

❖ Alterations that result in a net increase in gravity loading on the structure require the affected structural components be checked to verify they satisfy the gravity load requirements of the IBC for a new structure. There are two exceptions to full compliance with the IBC, one of which allows smaller Group R buildings to have alterations conforming to either the IRC or the conventional light-frame construction provisions of the IBC. Another exception allows additional gravity loads that do not increase stresses in affected structural elements by more than 5 percent. Allowing overstresses of up to 5 percent in existing structural members has been a long-standing rule of thumb used by structural engineers and provides latitude by allowing the addition of minor loads without the burden of strengthening or replacing portions of a structure. The exception does not specifically address the cumulative affects from successive alterations, but it would be prudent to limit cumulative increases under this exception to a total of 5 percent unless it is clearly documented that a structure has additional capacity. This code section assumes a code compliant original design for gravity loads.

707.7 Voluntary lateral-force-resisting system alterations. Alterations of existing structural elements that are initiated for the purpose of increasing the lateral-force-resisting strength or stiffness of an existing structure and that are not required by other sections of this code shall not be required to be designed for forces conforming to the *International Building Code* provided that an engineering analysis is submitted to show that:

1. The capacity of existing structural elements required to resist forces is not reduced;

2. The lateral loading to existing structural elements is not increased beyond their capacity;

3. New structural elements are detailed and connected to the existing structural elements as required by the *International Building Code*;

4. New or relocated nonstructural elements are detailed and connected to existing or new structural elements as required by the *International Building Code*; and

5. A dangerous condition as defined in this code is not created.

Voluntary alterations to lateral-force-resisting systems conducted in accordance with Appendix A and the referenced standards of this code shall be permitted.

❖ This section addresses the issue of upgrading a building's lateral-force-resisting system voluntarily for improved resistance to wind and seismic forces. It does not apply in situations where other code sections require compliance with the IBC or other specific minimum wind or seismic design loads. This section allows an owner to initiate an improvement to the lateral-force-resisting system to the extent it is viable to do so and provided the requisite engineering analysis is furnished. Since no minimum load requirement is established by the code, the building owner and the design professional have the latitude to establish performance goals and objectives. Thus, an owner can do something to mitigate the hazard of future earthquakes or wind storms in an existing building without being discouraged from doing so by incurring prohibitive costs.

The retrofit of an existing building for improved performance in an earthquake, for example, to a level of resistance less than that required for a new building considers the overall benefit of hazard mitigation. It is an often used approach to managing risk in areas that are susceptible to frequent earthquakes. This is the thrust of program's such as FEMA's Project Impact. The GSREB in Appendix A and the other referenced standards for seismic evaluation and rehabilitation have been developed for this purpose, and this section of the code recognizes their use. A voluntary seismic upgrade of a building structure can, therefore, consider the full range of rehabilitation objectives available under FEMA 356, for instance (see commentary, Section 407.1.1.1).

Bibliography

The following resource materials are referenced in this chapter or are relevant to the subject matter addressed in this chapter.

FEMA 356, *Prestandard and Commentary for the Seismic Rehabilitation of Buildings.* Washington, DC: Federal Emergency Management Agency, 2000.

IBC-03, *International Building Code.* Falls Church, VA: International Code Council, 2003.

IMC-03, *International Mechanical Code.* Falls Church, VA: International Code Council, 2003.

IRC-03, *International Residential Code.* Falls Church, VA: International Code Council, 2003.

NBC-99, *BOCA National Building Code.* Country Club Hills, IL: Building Officials & Code Administrators International, Inc., 1999.

SBC-99, *Standard Building Code.* Birmingham, AL: Southern Building Code Congress International, 1999.

UBC-97, *Uniform Building Code.* Whittier, CA: International Conference of Building Officials, 1997.

CHAPTER 8

CHANGE OF OCCUPANCY

General Comments

This chapter deals with the special situations involved in an existing building when a change of occupancy occurs. A change of occupancy is not to be confused with a change of occupancy classification. The *International Building Code®* (IBC®) defines different occupancy classifications in Chapter 3, and special occupancy requirements in Chapter 4. Within specific occupancy classifications, there can be many different types of actual activities that can take place. For instance, a Group A-3 occupancy classification deals with a wide variation of types of activities, including bowling alleys and courtrooms, indoor tennis courts and dance halls. When a facility is changed from use as, say, a bowling alley to use as a dance hall, the occupancy classification remains as Group A-3, but the different uses could lead to drastically different code requirements. Therefore, this chapter deals with special circumstances that are associated with a change in the use of a building within the same occupancy classification as well as a change of occupancy classification.

As a general rule, when a change in occupancy classification occurs, the requirements of Chapter 7 for Level 3 alterations apply, along with all of the provisions of Section 812. However, there are exceptions to this rule as provided in Section 812.3, which deals with a change in occupancy where the change is clearly to a lesser hazard.

Purpose

The purpose of this chapter is to provide regulations for the circumstances when an existing building is subject to a change in occupancy or a change in occupancy classification. As discussed above, a change in occupancy is defined in Chapter 2 of this code as:

CHANGE OF OCCUPANCY. A change in the purpose or level of activity within a building that involves a change in application of the requirements of this code.

In either case, some scenarios exist that are not dealt with in Chapters 5, 6, or 7 that must be addressed nevertheless. Changes in occupancy or occupancy classification change the activities or activity levels with a building, which can have implications on the life safety hazards associated with this change in activity. A change from a dance hall to a nightclub is a change in occupancy classification that has clear implications with regard to risks to life. A change from a restaurant to a nightclub is not a change in occupancy classification, but is, nevertheless, a significant change in activity levels with different related life safety risks.

SECTION 801
GENERAL

801.1 Repair and alteration with no change of occupancy classification. Any repair or alteration work undertaken in connection with a change of occupancy that does not involve a change of occupancy classification as described in the *International Building Code* shall conform to the applicable requirements for the work as classified in Chapter 3 and to the requirements of Sections 802 through 811.

Exceptions:

1. Compliance with all of the provisions of Chapter 7 is not required where the change of occupancy classification complies with the requirements of Section 812.3.

2. As modified in Section 1005 for historic buildings.

3. As permitted in Chapter 12.

❖ A change of occupancy without a change in occupancy classification is a change in the level or type of activity in a building that could have life safety implications. For instance, imagine that a community center is sold and converted to a dance hall. Both of these are considered by the IBC as being an Occupancy Classification A-3, but the requirements of this code for each facility could be considerably different. The dance hall might have a higher occupant load, and, therefore, require different egress facilities.

This code intends that a change in occupancy can be dealt with as a Level 1, 2, or 3 alteration in accordance with Chapters 5, 6, or 7, depending upon the extent of the remodel, and with the particular requirements contained in Sections 807, 808, 809 and 810. This code further intends that, for a change in occupancy classification, the alteration must be treated as a Level 3 alteration, making Chapter 7 apply, along with the special provisions of Section 812. Exception 1 deals with a change in occupancy classification where the new change is clearly a reduction in hazard as outlined in Section 812.3. Exception 2 provides for special considerations for historic buildings, as given in Section 1005. Finally, Chapter 12 provides compliance alternatives exactly as provided for in Chapter 34 of the IBC. This involves reevaluation of a building that was built before a community began utilizing some comprehensive form of a building code.

801.2 Partial change of occupancy group. Where a portion of an existing building is changed to a new occupancy group, Section 812 shall apply.

❖ Circumstances occur where only a portion of a building is changed to a different occupancy classification.

This section refers to Section 812, where Sections 812.1.1 and 812.1.2 address partial change of occupancies for two options of separated and nonseparated conditions.

801.3 Certificate of occupancy required. A certificate of occupancy shall be issued where a change of occupancy occurs that results in a different occupancy classification as determined by the *International Building Code*.

❖ A change in occupancy has historically been treated just as severely as if the construction were new. In this code, the change in occupancy is treated as the most extensive alteration (Level 7); therefore, the occupancy requires that another certificate of occupancy be issued as if the construction were new.

SECTION 802
SPECIAL USE AND OCCUPANCY

802.1 Compliance with the Building Code. Where the character or use of an existing building or part of an existing building is changed to one of the following special use or occupancy categories as defined in Chapter 4 of the *International Building Code*, the building shall comply with all of the applicable requirements of the *International Building Code*.

1. Covered mall buildings.
2. Atriums.
3. Motor vehicle related occupancies.
4. Aircraft related occupancies.
5. Motion picture projection rooms.
6. Stages and platforms.
7. Special amusement buildings.
8. Incidental use areas.
9. Hazardous materials.

❖ The IBC contains some special types of constructions or occupancies. These are examples where the alteration to the building does not actually involve a change in occupancy classification, but involves a significant change to these special occupancies that are regulated by very specific provisions in Chapter 4 of the IBC. As such, this code cannot be applied to these specific provisions, because such a change is outside the scope of this code. For instance, a large retail store could be converted to a multiple-tenant facility where the owner would like to use a covered mall concept. The utilization of the general provisions of this code would not adequately deal with the specific considerations for life safety envisioned as necessary in the special occupancy provisions of the IBC. Therefore, the designer must simply follow the provisions of the IBC as if the construction were new.

802.2 Underground buildings. An underground building in which there is a change of use shall comply with the requirements of the *International Building Code* applicable to underground structures.

❖ Similar to the discussion for special occupancies given in Section 802.1, underground buildings are dealt with in a specific manner in Chapter 4 of the IBC. The difference here is that, wherever there is a change in occupancy in an underground building, only the IBC requirements applicable to underground buildings apply. The building is otherwise dealt with as a change in occupancy in accordance with this code.

SECTION 803
BUILDING ELEMENTS AND MATERIALS

803.1 General. Building elements and materials in portions of buildings undergoing a change of occupancy classification shall comply with Section 812.

❖ Chapters 5, 6, 7 and 8 are all organized in the same manner: Section 802, Special use and occupancy; Section 803, Building elements and materials; Section 804, Fire protection; Section 805, Means of egress; Section 806, Accessibility, and so on. Consistent with the stated requirements of Section 801.1, a change in occupancy not involving a change in occupancy classification must be dealt with as a Level 1, 2 or 3 alteration. Therefore, the requirements contained in those chapters for building elements and materials would apply, depending upon the applicable requirements based upon the classification of the work.

SECTION 804
FIRE PROTECTION

804.1 General. Fire protection requirements of Section 812 shall apply where a building or portions thereof undergo a change of occupancy classification.

❖ Chapters 5, 6, 7 and 8 are all organized in the same manner: Section 802, Special use and occupancy; Section 803, Building elements and materials; Section 804, Fire protection; Section 805, Means of egress; Section 806, Accessibility, and so on. Consistent with the stated requirements of Section 801.1, a change in occupancy not involving a change in occupancy classification must be dealt with as a Level 1, 2 or 3 alteration. Therefore, the requirements contained in those chapters for fire protection would apply, depending upon the applicable requirements based upon the classification of the work.

SECTION 805
MEANS OF EGRESS

805.1 General. Means of egress in portions of buildings undergoing a change of occupancy classification shall comply with Section 812.

❖ Chapters 5, 6, 7 and 8 are all organized in the same manner: Section 802, Special use and occupancy; Section 803, Building elements and materials; Section 804, Fire protection; Section 805, Means of egress;

Section 806, Accessibility, and so on. Consistent with the stated requirements of Section 801.1, a change in occupancy not involving a change in occupancy classification must be dealt with as a Level 1, 2 or 3 alteration. Therefore, the requirements contained in those chapters for means of egress would apply, depending on the applicable requirements based on the classification of the work.

SECTION 806
ACCESSIBILITY

806.1 General. Accessibility in portions of buildings undergoing a change of occupancy classification shall comply with Section 812.5.

❖ When the use in a structure or a portion of a structure changes, the space must comply with new construction unless technically infeasible. For additional information, see Section 812.5.

Chapters 5, 6, 7, and 8 are all organized in the same manner: Section 802, Special use and occupancy; Section 803, Building elements and materials; Section 804, Fire protection; Section 805, Means of egress; Section 806, Accessibility, and so on. Consistent with the stated requirements of Section 801.1 a change in occupancy not involving a change in occupancy classification must be dealt with as a Level 1, 2 or 3 alteration. Therefore, the requirements contained in those chapters for accessibility would apply, depending upon the applicable requirements based upon the classification of the work.

SECTION 807
STRUCTURAL

807.1 Gravity loads. Buildings or portions thereof subject to a change of occupancy where such change in the nature of occupancy results in higher uniform or concentrated loads based on Tables 1607.1 and 1607.6 of the *International Building Code* shall comply with the gravity load provisions of the *International Building Code*.

Exception: Structural elements whose stress is not increased by more than 5 percent.

❖ These structural requirements apply where the change of occupancy results in an increased minimum live load required by Section 1607 of the IBC. The exception allows increased gravity design loads that do not increase stresses in affected structural elements by more than 5 percent. Allowing overstresses of up to 5 percent in existing structural members has been a long-standing rule of thumb used by structural engineers.

Under Section 102.4.2, buildings that comply with the structural provisions of any of the following model codes are deemed to comply with this code:

International Building Code® (IBC®), 2003 or 2000 edition.

BOCA *National Building Code* (NBC), 1999 edition.

Standard Building Code (SBC), 1999 edition.

Uniform Building Code (UBC), 1997 edition.

807.2 Snow and wind loads. Buildings and structures subject to a change of occupancy where such change in the nature of occupancy results in higher wind or snow importance factors based on Table 1604.5 of the *International Building Code* shall be analyzed and shall comply with the applicable wind or snow load provisions of the *International Building Code*.

Exception: Where the new occupancy with a higher importance factor is less than or equal to 10 percent of the total building floor area. The cumulative effect of the area of occupancy changes shall be considered for the purposes of this exception.

❖ This section addresses a change of occupancy that results in higher wind or snow importance factors than required for the current occupancy. The design snow and wind loads required under the IBC are directly proportional to the importance factors for snow and wind that are listed in Table 1604.5 of the IBC (also see Section 1609.5 for wind and Section 1608.3.3 for snow). These importance factors are a function of the building's occupancy. The values of importance factors are listed in Table 807.2 of the commentary. The intent is to achieve a higher level of performance for the facilities that have higher importance factors; thus, the importance factor generally increases with the importance of the facility. The facilities listed as Category IV are considered essential facilities and are assigned the highest importance factors. The facilities are essential in that their continuous use is needed. For example, fire, rescue and police stations and emergency vehicle garages are needed to be in service during and immediately after a major windstorm or earthquake event.

Where this applies, the structure needs to be analyzed for snow and wind loads and shown to be adequate under the provisions of the IBC for new buildings. The impact this may have on the existing structure and the minimum design loads depends on the magnitude of the importance factors for the new (proposed) occupancy category relative to those required for the current occupancy category. Other less apparent impacts will result due to the fact that the basic design loads prior to considering the importance factor (e.g., wind pressure) required by the IBC will often be different than the predecessor model code used in the building's original design.

Table 807.2 IBC TABLE 1604.5 IMPORTANCE FACTORS			
OCCUPANCY CATEGORY	SEISMIC	SNOW	WIND
I	1.0	0.8	0.87
II	1.00	1.0	1.00
III	1.25	1.1	1.15
IV	1.50	1.2	1.15

The exception allows a change of occupancy to an occupancy requiring a higher importance factor without a structural analysis where this occupancy comprises only a relatively small portion of the entire building.

807.3 Seismic loads. Existing buildings with a change of occupancy shall comply with the seismic provisions of Sections 807.3.1 and 807.3.2.

❖ A change of occupancy must comply with the seismic provisions of Sections 807.3.1 and 807.3.2.

807.3.1 Compliance with the *International Building Code*. When a building or portion thereof is subject to a change of occupancy such that a change in the nature of the occupancy results in a higher seismic factor based on Table 1604.5 of the *International Building Code* or where such change of occupancy results in a reclassification of a building to a higher hazard category as shown in Table 812.4.1 and a change of a Group M occupancy to a Group A, E, I-1 R-1, R-2, or R-4 occupancy with two-thirds or more of the floors involved in Level 3 alteration work, the building shall conform to the seismic requirements of the *International Building Code* for the new seismic use group.

Exceptions:

1. Group M occupancies being changed to Group A, E, I-1, R-1, R-2, or R-4 occupancies for buildings less than six stories in height and in Seismic Design Category A, B, or C.

2. Specific detailing provisions required for a new structure are not required to be met where it can be shown that an acceptable level of performance and seismic safety is obtained for the applicable seismic use group using reduced *International Building Code* level seismic forces as specified in Section 407.1.1.3. The rehabilitation procedures shall be approved by the code official and shall consider the regularity, overstrength, redundancy, and ductility of the lateral-load-resisting system within the context of the existing detailing of the system.

3. Where the area of the new occupancy with a higher hazard category is less than or equal to 10 percent of the total building floor area and the new occupancy is not classified as Seismic Use Group IV. For the purposes of this exception, where a structure is occupied for two or more occupancies not included in the same seismic use group, the structure shall be assigned the classification of the highest seismic use group corresponding to the various occupancies. Where structures have two or more portions that are structurally separated in accordance with Section 1620 of the *International Building Code*, each portion shall be separately classified. Where a structurally separated portion of a structure provides required access to, required egress from, or shares life safety components with another portion having a higher seismic use group, both portions shall be assigned the higher seismic use group. The cumulative effect of the area of occupancy changes shall be considered for the purposes of this exception.

4. Where the new occupancy with a higher hazard category is within only one story of a building or structure, only the lateral-force-resisting elements in that story and all lateral-force-resisting elements below that story shall be required to comply with Section 807.3.1 and Exception 2. The lateral forces generated by masses of such upper floors shall be included in the analysis and design of the lateral-force-resisting systems for the strengthened floor. Such forces may be applied to the floor level immediately above the topmost strengthened floor and be distributed in that floor in a manner consistent with the construction and layout of the exempted floor.

5. Unreinforced masonry bearing wall buildings in Seismic Use Group II and in Seismic Use Groups II and III when in Seismic Design Categories A, B, and C shall be allowed to be strengthened to meet the requirements of Appendix A of the code (GSREB).

❖ The three types of occupancy change that require compliance with IBC seismic provisions for new construction are as follows:

- The change of occupancy results in the new occupancy requiring a higher importance factor (see Commentary Table 807.2). The explanation is the same as that given under snow and wind loads in Section 807.2.

- The change of occupancy results in the new occupancy having a higher hazard category in accordance with Table 812.4.1.

- A Group M occupancy that is changed to either a Group A, E, I-1 R-1, R-2 or R-4 occupancy and at least two-thirds of the floors have alterations classified as Level 3. However, Exception 1 exempts buildings less than six stories in height that are classified as Seismic Design Category A, B or C.

The code recognizes that the risk to safety is likely to be greater in the above situations and conformance to the IBC seismic provisions is mandated, unless compliance with any of the exceptions can be demonstrated.

Exception 2 is based on a similar exception to Section 1614.2 of the IBC and recognizes the difficulty of making an older structure comply with the seismic requirements for a new structure. It provides general guidance to the code official and designer on areas that need to be investigated when compliance with the seismic requirements for new buildings is not feasible. See Section 407.1.1.3 for the permitted evaluation and rehabilitation procedures.

Exception 3 contains requirements as well as exceptions that apply to buildings that house two or more occupancies classified in different occupancy categories. The general requirement for a change in occupancy resulting in a mixed-use building is that the entire structure must be designed in accordance with the requirements of the most critical occupancy (higher hazard classification) category, because it resists the earthquake forces as a unit. This is consistent with Section 1616.2.4 of the IBC. The exception to this rule is where a new occupancy having a higher hazard category comprises no more than 10 percent of the total building area and it is not considered an essential facility. A second portion (also consistent with Section 16161.2.4 of the IBC) of this exception allows multiple occupancies that are structurally separated to be classified and designed independently using the importance factors applicable to each occupancy category.

Exception 4 pertains to multistory buildings, where the new (higher hazard) occupancy occurs within a single story. This permits the seismic evaluation to be limited to the lateral-force-resisting elements in that story and below. While the loading contributed by the stories above must, of course, be considered, no seismic rehabilitation of the stories above would be necessary due to the change of occupancy.

Exception 5 refers to Chapter A1 of Appendix A *Guidelines for Seismic Evaluation of Existing Buildings* (GSREB), which is applicable to unreinforced masonry (URM)-bearing wall buildings. These buildings would likely fall into the category of buildings already exempt under Exception 2. Nevertheless, this exception provides a direct link to the GSREB requirements for URM buildings that undergo a change in occupancy. Section A102.2 limitations make Chapter A1 applicable to Occupancy Categories I and II in Table 1604.5 of the IBC as well as Occupancy Categories III or IV that are classified as Seismic Design Category A or B.

807.3.2 Access to Seismic Use Group IV. Where the change of occupancy is such that compliance with Section 807.3.1 is required and the seismic use group is a Category IV, the operational access to such Seismic Use Group IV existing structure shall not be through an adjacent structure.

Exception: Where the adjacent structure conforms to the requirements for Seismic Use Group IV structures.

Where operational access is less than 10 feet (3048 mm) from an interior lot line or less than 10 feet (3048 mm) from another structure, access protection from potential falling debris shall be provided by the owner of the Seismic Use Group IV structure.

❖ This section provides for access to the occupancy of essential facilities after an earthquake by prohibiting access through an adjacent structure that does not comply with the seismic criteria for essential facilities. These requirements are the result of lessons learned from previous earthquakes in which essential facilities have been rendered out of service because of the failure of an adjacent structure or a structure's exterior components. These requirements are based on Section 1616.2.3 of the IBC as well as FEMA 368.

SECTION 808
ELECTRICAL

808.1 Special occupancies. Where the occupancy of an existing building or part of an existing building is changed to one of the following special occupancies as described in the ICC *Electrical Code*, the electrical wiring and equipment of the building or portion thereof that contains the proposed occupancy shall

comply with the applicable requirements of the ICC *Electrical Code* whether or not a change of occupancy group is involved:

1. Hazardous locations.
2. Commercial garages, repair, and storage.
3. Aircraft hangars.
4. Gasoline dispensing and service stations.
5. Bulk storage plants.
6. Spray application, dipping, and coating processes.
7. Health care facilities.
8. Places of assembly.
9. Theaters, audience areas of motion picture and television studios, and similar locations.
10. Motion picture and television studios and similar locations.
11. Motion picture projectors.
12. Agricultural buildings.

❖ Similar to the discussion in Section 802.1 of this code regarding the special occupancy provisions of Chapter 4 of the IBC, the ICC *Electrical Code*® (ICC EC™) also contains provisions for special types of constructions or occupancies. The examples in this section discuss where the alteration to the building does not actually involve a change in occupancy classification, but involves a significant change to these special occupancies that are regulated by very specific provisions in the ICC EC. As such, this code cannot be applied to these specific provisions because such a change is outside the scope of this code. For instance, a large hospital could be renovated to change some of the spaces to a new emergency room or new testing facility. The utilization of the general provisions of this code would not adequately deal with the specific considerations for life safety envisioned as necessary in the special occupancy provisions of the ICC EC. Therefore, the designer must simply follow the provisions of the ICC EC without modification.

808.2 Unsafe conditions. Where the occupancy of an existing building or part of an existing building is changed, all unsafe conditions shall be corrected without requiring that all parts of the electrical system be brought up to the current edition of the ICC *Electrical Code*.

❖ Buildings, structures or premises that constitute a fire hazard or are otherwise dangerous to human life, or which in relation to existing use, constitute a hazard to safety, health or public welfare, by reason of inadequate maintenance, dilapidation, obsolescence, fire hazard, disaster damage or abandonment as specified in this code or any other ordinance, are unsafe conditions.

The intent of this section is to address unsafe electrical conditions before they can lead to injury or death. The code specifically requires only the unsafe wiring condition to be brought up to the current ICC EC. Any additional existing wiring that is isolated from the damaged area is not required to be repaired or replaced.

For example, on a typical downtown street, above

the main street commercial occupancies, there are some older existing apartments. Prior to the rental of one particular unit, an inspector notices a damaged receptacle with wiring exposed. According to Section 808.2 of the code, this outlet must be properly repaired or replaced as per the current ICC EC. The remainder of the existing electrical system would not be required to be changed unless it was damaged.

808.3 Service upgrade. Where the occupancy of an existing building or part of an existing building is changed, electrical service shall be upgraded to meet the requirements of the ICC *Electrical Code* for the new occupancy.

❖ If an existing structure has a portion or all of the building undergo a change of occupancy, this section requires the wiring supporting the changed area to meet the ICC EC.

808.4 Number of electrical outlets. Where the occupancy of an existing building or part of an existing building is changed, the number of electrical outlets shall comply with the ICC *Electrical Code* for the new occupancy.

❖ The number of outlets that are inherent to an area undergoing a change of occupancy must meet the ICC EC requirements for the new occupancy.

For example, consider a room that undergoes a change from meeting room usage to a kitchen. The old rule of being able to go in any direction along the wall no more than 6 feet (1829 mm) and reach an outlet is no longer sufficient. There would have to be outlets added such that no point along the counter wall line is more than 24 inches (610 mm) from an outlet.

SECTION 809
MECHANICAL

809.1 Mechanical requirements. Where the occupancy of an existing building or part of an existing building is changed such that the new occupancy is subject to different kitchen exhaust requirements or to increased mechanical ventilation requirements in accordance with the *International Mechanical Code*, the new occupancy shall comply with the intent of the respective *International Mechanical Code* provisions.

❖ This section deals with two specific scenarios that could occur in a change in occupancy that do not qualify as a change in occupancy classification.

The first scenario is a change in kitchen equipment that would require different exhaust requirements. A simple change in the size of a grill would require a different hood and different levels of exhaust. This type of change meets the definition of a change in occupancy.

The second scenario is a change in ventilation requirements. Many changes in occupancy could lead to changes in the occupant load of the building. An increase in the occupant load could mean an increase in ventilation requirements. Therefore, ventilation levels would need to be based upon the intent of the *International Mechanical Code*® (IMC®).

SECTION 810
PLUMBING

810.1 Increased demand. Where the occupancy of an existing building or part of an existing building is changed such that the new occupancy is subject to increased or different plumbing fixture requirements or to increased water supply requirements in accordance with the *International Plumbing Code*, the new occupancy shall comply with the intent of the respective *International Plumbing Code* provisions.

❖ Many changes in occupancy could lead to changes in the occupant load of the building. An increase in the occupant load could mean an increase in needed plumbing fixtures. Therefore, fixture levels would need to be based on the intent of the *International Plumbing Code*® (IPC®).

810.2 Food handling occupancies. If the new occupancy is a food handling establishment, all existing sanitary waste lines above the food or drink preparation or storage areas shall be panned or otherwise protected to prevent leaking pipes or condensation on pipes from contaminating food or drink. New drainage lines shall not be installed above such areas and shall be protected in accordance with the *International Plumbing Code*.

❖ When a change in occupancy involves new food handling facilities, the health safety issues associated with the integrity of the existing plumbing equipment and facilities must be addressed. Again, this code does not contain special provisions for such a specific issue. Therefore, the facilities must be modified in accordance with the intent of the IPC for new facilities.

810.3 Interceptor required. If the new occupancy will produce grease or oil-laden wastes, interceptors shall be provided as required in the *International Plumbing Code*.

❖ When a change in occupancy involves new facilities that produce grease or oil-laden wastes, the issues associated with the integrity of the existing plumbing equipment and facilities must be addressed. Again, this code does not contain special provisions for such a specific issue. Therefore, the facilities must be modified in accordance with the intent of the IPC for new facilities.

810.4 Chemical wastes. If the new occupancy will produce chemical wastes, the following shall apply:

1. If the existing piping is not compatible with the chemical waste, the waste shall be neutralized prior to entering the drainage system, or the piping shall be changed to a compatible material.

2. No chemical waste shall discharge to a public sewer system without the approval of the sewage authority.

❖ When a change in occupancy involves new facilities that produce chemical wastes, the issues associated with the integrity of the existing plumbing equipment and facilities must be addressed. In this case, the performance requirements stated essentially address the intent of the IPC for health, life and environmental safety by requiring that the piping be made to handle the particular chemicals or the chemicals be treated such as to not harm the piping, and that the impact on the public sewer system be addressed.

810.5 Group I-2. If the occupancy group is changed to Group I-2, the plumbing system shall comply with the applicable requirements of the *International Plumbing Code*.

❖ Similar to the provisions given in Section 802.1 for special occupancies in the IBC, the utilization of the general provisions of the code would not adequately deal with the specific considerations for life and health safety envisioned as necessary in the specific provisions of the IPC for the plumbing facilities in a Group I-2 occupancy. Therefore, the designer must simply follow the provisions of the IPC as if the construction were new.

SECTION 811
OTHER REQUIREMENTS

811.1 Light and ventilation. Light and ventilation shall comply with the requirements of the *International Building Code* for the new occupancy.

❖ Many changes in occupancy could lead to changes in the occupant load of the building. An increase in the occupant load could mean an increase in needed lighting or ventilation. Therefore, the requirements of the IBC applicable to new construction are applied to light and ventilation.

SECTION 812
CHANGE OF OCCUPANCY CLASSIFICATION

812.1 Compliance with Chapter 7. The occupancy classification of an existing building may be changed, provided that the building meets all of the requirements of Chapter 7 applied throughout the building for the new occupancy group and complies with the requirements of Sections 802 through 812.

❖ This section contains provisions that allow a change of occupancy classificiation within an existing building as defined in Section 302.1 of the IBC. Conditions to allow that change of occupancy classification require that a review of the existing building be conducted to determine compliance with the requirements of Chapter 7. This essentially triggers a review of all the building's elements against the requirements of Chapters 5, 6, 7 and Sections 802 through 812 of Chapter 8. All deficiencies disclosed by the review are required to be brought into compliance with the provisions of the code. Design and detail of the corrective construction shall be incorporated into the scope of work and identified in the construction documents. The basic philosophy of the code continues to be that as the level of alteration activity increases in scope or area of the building, the level of compliance to new construction code requirements also increases.

812.1.1 Change of occupancy group without separation. Where a portion of an existing building is changed to a new occupancy group and that portion is not separated from the remainder of the building with fire barriers having a fire-resistance rating as required in the *International Building Code* for the separate occupancy, the entire building shall comply with all of the requirements of Chapter 7 applied throughout the building for the most restrictive occupancy group in the building and with the requirements of this chapter.

Exception: Compliance with all of the provisions of Chapter 7 is not required when the change of occupancy group complies with the requirements of Section 812.3.

❖ Requirements of this subsection and that of Section 812.1.2 provide alternatives for the evaluation process when the change of occupancy occurs in only a portion of an existing building. These two sections incorporate two of the philosophies for mixed-use occupancies found in Section 302.3 of the IBC.

If the portion of the building that is undergoing the change in occupancy classification is not separated from the remainder of the building by fire barriers, this section allows two or more portions to remain unseparated from each other. However, the entire building shall comply with the requirements of Chapter 7 and those of this chapter for the most restrictive occupancy group. This philosophy is consistent with the approach for nonseperated mixed uses, as given in Section 302.3.1 of the IBC. Note that this could require construction activity within existing occupied spaces or portions of the building not otherwise impacted by the change in occupancy classification.

An exception to mandatory compliance with the provisions of Chapter 7 is compliance with the provisions of Section 812.3. Those provisions require that the change of occupancy classification is an equal or lesser hazard in all three hazard classifications as determined in Sections 812.4.1, 812.4.2 and 812.4.3. If the change is an equal or lesser hazard category, then the mandatory compliance requirements for building components and/or systems are identified in Sections 812.3.1 through 812.3.5. Section 812.3 contains five subsections that contain both minimum requirements that would be applicable to all occupancy groups and some that are occupancy specific. This exception expresses the fact that if the previous occupancy had been legally occupying the building, it would have been permitted to continue. It then takes that fact and extends the logic to allow a new occupancy that is of an equal or lesser relative hazard as the occupancy that it has replaced. However, it is not a blanket exception. The subsections of Section 812.3 still require compliance with minimum standards for egress capacity and interior finish. Additionally, there are further requirements for some specific occupancies that include the installation of fire alarms, fire suppression systems and fire-resistance- rated assemblies to separate rooms and spaces.

812.1.2 Change of occupancy group with separation. A portion of an existing building that is changed to a new occupancy group and that is separated from the remainder of the building with fire barriers having a fire-resistance rating as required in the *International Building Code* for the separate occupancy shall comply with all the requirements of Chapter 7 for the new occupancy group and with the requirements of this chapter.

> **Exception:** Compliance with all of the provisions of Chapter 7 is not required when the change of use complies with the requirements of Section 812.3.

❖ When the change of occupancy classification is wholly contained in a portion of the building completely separated from other uses occupying the building by fire barriers as required by Table 302.3.2 of the IBC, compliance with the requirements of Chapter 7 and Sections 802 through 812 are only incorporated into that portion of the building occupied by the new occupancy classification. The remaining portion(s) of the building legally occupied are permitted to continue occupancy without change, until activities specifically covered in the code occur.

Again, an exception is provided for when a change in occupancy classification is to an equal or lesser hazard classification. See the commentary to Section 812.1.1 for a discussion on the provisions of this exception.

812.2 Hazard category classifications. The relative degree of hazard between different occupancy groups shall be as set forth in the hazard category classifications specified in Tables 812.4.1, 812.4.2, and 812.4.3 of Sections 812.4.1, 812.4.2, and 812.4.3.

❖ All the classified occupancy groups were evaluated in three areas in which the degree of inherent hazards of an occupancy group can differ. These evaluations were reviewed and like results were grouped and then assigned a relative rating number, 1 through 4 or 5, depending on the element being evaluated. The lower the number, the higher the relative hazard of the occupancy classification. Aspects of a building that were considered were: fire and life safety, including the ability to exit; the height and area of the building; and the need to protect surrounding buildings from fire conflagration. Occupancies with equivalent relative hazard numbers were then grouped together and formed into Tables 812.4.1, 812.4.3 and 812.4.4 of Sections 812.4.1, 812.4.3 and 812.4.4, respectively. These tables are used to determine whether a change in occupancy classification is to a higher, lesser or equivalent occupancy classification, in relation to the relative hazard of the use.

812.2.1 Change of occupancy classification to an equal or lesser hazard. An existing building or portion thereof may have its use changed to an occupancy group within the same hazard classification category or to an occupancy group within a lesser hazard classification category (higher number) in all four hazard category classifications, provided it complies with the provisions of Chapter 7 for the new occupancy group, applied throughout the building or portion thereof.

> **Exception:** Compliance with all the provisions of Chapter 7 is not required where the change of occupancy group complies with the requirements of Section 812.3.

❖ When a change of occupancy to a building or portion thereof is changed to an occupancy group within the same or lesser hazard classification category for all three hazard category classifications, compliance with the provisions of Chapter 7 are required for the building, or portion thereof, that is undergoing a change in occupancy.

An exception is provided to not require compliance with the provisions of Chapter 7 when the hazard classification change is to an equivalent or lesser hazard for all three hazard categories and the building, or portion there,of is brought into compliance with all the minimum requirements of Section 812.3.1 and the specific use group requirements of Sections 812.3.2 through 812.3.5.

812.2.2 Change of occupancy classification to a higher hazard. An existing building shall comply with all of the applicable requirements of this chapter when a change in occupancy group places it in a higher hazard category or when the occupancy group is changed within Group H.

❖ This section applies to circumstances where the change of occupancy classification places it in a higher relative hazard (lower number) or when the occupancy is changed within Group H (e.g., H-1 to H-3 or H-4 to H-2). When either of these conditions occurs, all the provisions of this chapter shall apply throughout the building.

While it would seem that the changes within Group H are simply changes to equivalent or lesser hazards, Group H occupancies have very separate and distinct hazards and, therefore, must be treated as increases in hazard classification.

812.2.3 Change of occupancy classification to a higher hazard in all three hazard classifications. An existing building may have its use changed to a higher hazard rating (lower number) in all three hazard category classifications designated in Tables 812.4.1, 812.4.2, and 812.4.3, provided it complies with this chapter or with Chapter 12.

❖ When a change of occupancy classification is to a higher relative hazard (lower number) in all three hazard classifications, the building is required to comply with all the provisions of Chapter 8 or 12. Chapter 12 contains provisions for an alternative method of evaluation for the existing building based on a numerical scoring system involving 18 various safety issues. Existing conditions are reviewed for the degree of compliance with several descriptions of conditions that could exist for each of the safety issues. Each description is called a "category" and is assigned a numerical value, which is used to generate the final score of the building. Final building scores are calculated for three different aspects of building safety—fire safety, means of egress and general safety. Mandatory minimum safety scores have been determined for each of these three aspects to establish an equivalent level of safety.

812.3 Change of occupancy classification to an equal or lesser hazard in all three hazard classifications. A change of use to an occupancy group within the same hazard classification category or to an occupancy group within a lesser hazard classification category (higher number) in the three hazard category classifications addressed by Tables 812.4.1, 812.4.2, and 812.4.3 shall be permitted in an existing building or portion thereof, provided the provisions of Sections 812.3.1 through 812.3.5 are met.

❖ If the occupancy classification change is to an equal or lesser hazard category in all three hazard classifications as determined by Tables 812.4.1, 812.4.2 and 812.4.3, requirements for building components and/or systems are identified in Sections 812.3.1 through 812.3.5. These subdivisions contain both minimum requirements that are applicable to all occupancy groups and some that are occupancy group specific. Occupancy group specific requirements are referenced to a section of compliance required by Level 1 or 2 alterations. This is a logical link, as some level of alteration is probably required to accomodate a change of occupancy occurring within a building or portion thereof.

This concept is originally found in Section 102.6 of the IBC, which states that an occupancy that has been legally occupying a building would be permitted to continue without requiring any compliance to current code provisions, unless specifically identified in the IBC, *International Property Maintenance Code®* (IPMC®) or the *International Fire Code®* (IFC®), or as deemed necessary by the code official for the general safety and welfare of the occupants and public. Section 812.3 then extends the logic to allow a new occupancy that is of an equal or lesser relative hazard than the occupancy it has replaced. However, just as in Section 102.6 of the IBC, it is not a blanket exception to all code compliance. Following the initial charging language of this section are subdivisions that mandate some specific requirements and help to remove what could be seen as very subjective decisions made by a code official regarding the general safety and welfare of the occupants and public. This is not to say it eliminates the discretion of the code official, but provides more guidance and direction. Section 812.3.1 mandates compliance with minimum standards for egress capacity and interior finish. Additionally, there are further mandatory requirements in Sections 812.3.2, 812.3.3, 812.3.4 and 812.3.5 for specific occupancies that include the installation or upgrading of fire alarm, fire suppression and fire-resistance assemblies to separate rooms and spaces.

812.3.1 Minimum requirements. Regardless of the occupancy group involved, the following requirements shall be met:

1. The capacity of the means of egress shall comply with *International Building Code*.

2. The interior finish of walls and ceilings shall comply with the requirements of the *International Building Code* for the new occupancy group.

❖ Two areas are addressed in this section: means of egress capacity and interior finish requirements. These are two requirements that must be met in all cases. The criteria for means of egress capacity reverts to the IBC for new construction, and the flame spread limitations for interior finish materials must also be as required for new construction.

The requirement to meet minimum capacity requirements for means of egress is consistent with the basic statement made in Section 1001.2 of the IBC and IFC, making it unlawful to reduce the capacity and number of means of egress below that given in those codes.

812.3.2 Groups I-1, R-1, R-2 or R-4. Where the new use is classified as a Group I-1, R-1, R-2 or R-4 occupancy the following requirements shall be met.

1. Corridor doors and transoms shall comply with the requirements of Sections 605.5.1 and 605.5.2.

2. Automatic sprinkler systems shall comply with the requirements of Section 604.2.

3. Fire alarm and detection systems shall comply with the requirements of Section 604.4.

❖ Essentially, fundamental requirements for these specific occupancy groups are based on some specific provisions given in Chapter 6 or 7. In the case of Item 1, the call out for requirements for corridor doors and transoms is to Sections 605.5.1 and 605.5.2, which give specific requirements for Level 2 and 3 alterations in these same occupancies. The concern is to provide a minimum level of protection against spread of smoke and fire in these residential occupancies.

Item 2 requires the use of Level 2 alteration requirements for automatic sprinkler systems in these occupancies, as well as Group I-2 (Section 812.3.3 item 3) and I-3 occupancies (Section 812.3.4, Item 3) (see commentary, Section 604.2.2).

Item 3 requires the use of Level 2 alteration requirments for fire alarm and detection systems, consistent with the same requirements for these occupancies given in Chapter 6. The same requirements apply to Group I-2 (Section 812.3.3, Item 5) and I-3 (Section 812.3.4, Item 4), which is consistent with the requirments of Chapter 6 (see commentary, Section 604.4).

812.3.3 Group I-2. Where the new use is classified as a Group I-2 occupancy, the following requirements shall be met:

1. Egress doorways from patient sleeping rooms and from suites of rooms shall comply with the requirements of Section 605.4.1.2.

2. Shaft enclosures shall comply with the requirements of Section 703.1.

3. Smoke barriers shall comply with the requirements of Section 603.3.

4. Automatic sprinkler systems shall comply with the requirements of Section 604.2.

5. Fire alarm and detection systems shall comply with the requirments of Section 604.4.

❖ Like Groups I-1, R-1, R-2 or R-4 listed in Section 812.3.2 above, this section requires automatic sprinkler systems and fire alarm systems consistent with the requirements for Level 2 and 3 alterations given in Chapter 6. In addition, Level 3 alterations are required for enclosure of stairways (see commentary, Section 703.1) and smoke barriers consistent with the requirements for Level 2 and 3 alterations (see commentary, Section 603.3). Finally, Group I-2 occupancies require two egress doors for patient rooms greater than 1,000 square feet (92.9 m²), consistent with the same requirements for Group I-2 given in Chapter 6 for Level 2 and 3 alterations (see commentary, Section 605.4.1.2).

812.3.4 Group I-3. Where the new use is classified as a Group I-3 occupancy, the following requirements shall be met:

1. Locking of egress doors shall comply with the requirements of Section 605.4.5.

2. Shaft enclosures shall comply with the requirements of Section 703.1.

3. Automatic sprinkler systems shall comply with the requirements of Section 604.2.

4. Fire alarm and detection systems shall comply with the requirements of Section 604.4.

❖ Like Groups I-1, R-1, R-2 or R-4 listed in Section 812.3.2 above, this section requires automatic sprinkler systems and fire alarm systems consistent with the requirements for Level 2 and 3 alterations given in Chapter 6. Like Group I-2, Level 3 alterations are required for enclosure of stairways (see commentary, Section 703.1). Finally, the special locking arrangements given for Level 2 and 3 alterations in Chapter 6 for Group I-3 are imposed here (see commentary, Section 604.4.5).

812.3.5 Group R-3. Where the new use is classified as a Group R-3 occupancy, the following requirements shall be met:

1. Dwelling unit separation shall comply with the requirements of Section 703.2.1.

2. The smoke alarm requirements of Section 604.4.3 shall be met.

❖ Group R-3 occupancies require dwelling unit separation consistent with the requirements of Chapter 7 for Level 3 alterations to Group R-3 occupancies (see commentary, Section 703.2.1). In addition, smoke alarms are required for Group R-3, just as they are for Level 2 and 3 alterations in Group R-3 occupancies given in Chapter 6 (see commentary, Section 604.4.3).

812.4 Fire and life safety. The fire and life safety provisions of this section shall be applicable to buildings or portions of buildings undergoing a change of occupancy classification.

❖ Sections 812.4.1, 812.4.2 and 812.4.3 contain provisions specific to elements of a building for changes of occupancy. These mandates are not specifically related to the occupancy classification, but are to be applied to all changes in occupancy classification that meet the charg-

ing language of the provisions. Some of the mandates are applicable to changes that are of an equal or lower hazard category, while others are applicable when the change involves a use classification to a higher (lower number) category. These mandates are correlated with the provisions of Section 812.3 as demonstrated by inclusion of exceptions for conditions that comply with Section 812.3 or simple acceptance of existing conditions (these mandates are in addition to the general requirements of Section 812.3, as applicable). The difference is that only requiring compliance with the minimal provisions of Section 812.3 requires the consistent rating of an equal or lesser value in all three hazard classification categories. Mandates required in these four subsections of Section 812.4 are applied individually, based on a review against the hazard category and classification tables contained within the respective subsections. These provisions are meant to address existing conditions in buildings that may have differences in the ratings established in the three hazard category and classification tables. The possibility exists that a change in occupancy results in an equal or lesser hazard change in one table and an increase in hazard in another table, or the inverse situation is possible. For example, a change from business to an assembly group is an increase in Tables 812.4.1 and 812.4.2 and a equal rating in Table 812.4.3. Further, a change from storage Group S-1 to residential Group R-2 is a decrease in Tables 812.4.1 and 812.4.2 and increase in Table 812.4.3.

812.4.1 Means of egress, general. Hazard categories in regard to life safety and means of egress shall be in accordance with Table 812.4.1.

TABLE 812.4.1
HAZARD CATEGORIES AND CLASSIFICATIONS:
LIFE SAFETY AND EXITS

RELATIVE HAZARD	OCCUPANCY CLASSIFICATIONS
1 (Highest Hazard)	H
2	I-2, I-3, I-4
3	A, E I-1, M, R-1, R-2, R-4
4	B, F-1, R-3, S-1
5 (Lowest Hazard)	F-2, S-2, U

❖ As can be seen, the means of egress provisions for hazard classification contains five levels. These levels are established based upon IBC requirements for different occupancies for means of egress, fire protection and fire-resistance-rated construction.

812.4.1.1 Means of egress for change to higher hazard category. When a change of occupancy group is made to a higher hazard category (lower number) as shown in Table 812.4.1, the means of egress shall comply with the requirements of Chapter 10 of the *International Building Code*.

Exceptions:

1. Stairways shall be enclosed in compliance with the applicable provisions of Section 703.1.

2. Existing stairways including handrails and guards complying with the requirements of Chapter 7 shall be permitted for continued use subject to approval of the code official.

3. Any stairway replacing an existing stairway within a space where the pitch or slope cannot be reduced because of existing construction shall not be required to comply with the maximum riser height and minimum tread depth requirements.

4. Existing corridor walls constructed of wood lath and plaster in good condition or $^1/_2$-inch-thick (12.7 mm) gypsum wallboard shall be permitted.

5. Existing corridor doorways, transoms, and other corridor openings shall comply with the requirements in Sections 605.5.1, 605.5.2, and 605.5.3.

6. Existing dead-end corridors shall comply with the requirements in Section 605.6.

7. An existing operable window with clear opening area no less than 4 square feet (0.38 m²) and with minimum opening height and width of 22 inches (559 mm) and 20 inches (508 mm), respectively, shall be accepted as an emergency escape and rescue opening.

❖ The basic premise here is for means of egress requirements to be the same as required for new construction when the change of occupancy group is to a higher (lower number) classification. The exceptions relate to common problem areas where it is impractical to fully comply with requirements for new construction. Exception 1 allows the same provision for limited enclosures of stairways as allowed for Level 3 alterations. Exception 2 allows handrails and guards to meet the limited requirements of Chapter 7. The reader will note that Chapter 7 contains no specific requirements for guards and handrails. However, it must be remembered that the provisions of Chapter 6 also apply to Chapter 7; therefore, the provisions of Sections 605.9 and 605.10 are applicable. Exceptions 3 through 7 give allowances for construction problems commonly found in existing construction that make full compliance with the present IBC impractical.

812.4.1.2 Means of egress for change of use to equal or lower hazard category. When a change of occupancy group is made to an equal or lesser hazard category (higher number) as shown in Table 812.4.1, existing elements of the means of egress shall comply with the requirements of Section 705 for the new occupancy group. Newly constructed or configured means of egress shall comply with the requirements of Chapter 10 of the *International Building Code*.

Exception:

1. Any stairway replacing an existing stairway within a space where the pitch or slope cannot be reduced because of existing construction shall not be required to comply with the maximum riser height and minimum tread depth requirements.

2. Compliance with Section 705 is not required where the change of occupancy group complies with the requirements of Section 812.3.

❖ Where the change of occupancy is to a lower hazard category (higher number), the code allows for the modifications to means of egress provided for Level 3 alterations in Chapter 7 (see commentary, Section 705). Detailed requirements are actually provided in Section 605, which are applicable to Level 3 alterations in Chapter 7 as well. Section 705 gives additional requirements regarding means of egress lighting and exit signs, requiring these to be installed in accordance with the IBC.

Exception 1 deals with a practical problem common in existing buildings, where the space provided for the stairway cannot accommodate shallower pitch of steps required in the IBC.

Exception 2 reminds the user that changes of occupancy classifications to equal or lesser values in all three major categories would then not require compliance with this section.

812.4.1.3 Egress capacity. Egress capacity shall meet or exceed the occupant load as specified in the *International Building Code* if the change of occupancy classification is to an equal or lesser hazard category when evaluated in accordance with Table 812.4.1.

❖ This provision is consistent throughout the code for all alterations and changes of occupancy. Reduction of egress capacity is strictly prohibited by Section 1001.2 of the IBC.

812.4.1.4 Handrails. Existing stairways shall comply with the handrail requirements of Section 605.9 in the area of the change of occupancy classification.

❖ The special provisions of Section 605.9 for handrails are allowed for this category of change of use, which is similar to that allowed for change of occupancy to a higher classification in Exception 2 of Section 812.4.1.1, except that the code official's specific approval is required in Section 812.4.1.1. The provisions of Section 605.9 boil down to a simple requirement that a single handrail is required for all stairways with a width less than 66 inches (1676.4 mm), at which point two handrails are required. The design of the handrails are otherwise regulated by the applicable provisions of the IBC.

812.4.1.5 Guards. Existing guards shall comply with the requirements in Section 605.10 in the area of the change of occupancy classification.

❖ The provisions of Section 605.10 apply to guards for changes of occupancy in this category. The provisions are the same as that of the IBC; however, guards are only required to modify the egress path from work area to exit.

812.4.2 Heights and areas. Hazard categories in regard to height and area shall be in accordance with Table 812.4.2.

❖ The provisions for hazard categories and requirements related to change of occupancy classification related to height and areas are considerably simpler

TABLE 812.4.2 – 812.4.3.1

and more straightforward than the life safety and exit hazard category. Essentially, the change to a higher hazard category requires compliance with height and area limitations for new construction as given by the IBC, and the change to an equal or lesser hazard category allows for the height and area of the existing building to be deemed acceptable.

TABLE 812.4.2
HAZARD CATEGORIES AND CLASSIFICATIONS:
HEIGHTS AND AREAS

RELATIVE HAZARD	OCCUPANCY CLASSIFICATIONS
1 (Highest Hazard)	H
2	A-1, A-2, A-3, A-4, I, R-1, R-2, R-4
3	E, F-1, S-1, M
4 (Lowest Hazard)	B, F-2, S-2, A-5, R-3, U

❖ Table 812.4.2 groups the relative hazards based upon allowable heights and areas given by the IBC.

812.4.2.1 Height and area for change to higher hazard category. When a change of occupancy group is made to a higher hazard category as shown in Table 812.4.2, heights and areas of buildings and structures shall comply with the requirements of Chapter 5 of the *International Building Code* for the new occupancy group.

Exception: A one-story building changed to Group E shall not be required to meet the area limitations of the *International Building Code*.

❖ As stated above, the change to a higher hazard category requires compliance with height and area limitations for new construction as given by the IBC, with a single exception regarding Group E one-story buildings. This approach is consistent with the premise that any alteration to a building cannot increase the hazard in the building. By definition, utilization of a building for a use in which the height or area for the occupancy is exceeded is an increase in the hazard of the building.

The exception facilitates an economic solution for use of existing facilities to house an educational facility in a rapidly growing school district. One should remember, however, that this section would not allow an expansion of a facility to be unlimited in area. Chapter 9 would clearly prohibit that. Ostensibly, the existing facility was built under some area limitation, which required sprinklers, or some other special provisions to allow for a very large building.

812.4.2.2 Height and area for change to equal or lesser hazard category. When a change of occupancy group is made to an equal or lesser hazard category as shown in Table 812.4.2, the height and area of the existing building shall be deemed acceptable.

❖ When the change of occupancy is an equal or lesser hazard, the height and area of the existing building is not at issue. This can, however, create an interesting scenario where the new occupancy is housed in a fa-

cility with height or area that would exceed that allowable height and area for new construction for that use, such as in a case where a Type IIB, Group S-1 building constructed to be 17,500 square feet (1626 m²) could be converted entirely to a Group M occupancy without any special provisions, even though the maximum allowable area for Type IIB construction for Group M is 12,500 square feet (1161 m²).

812.4.2.3 Fire barriers. When a change of occupancy group is made to a higher hazard category as shown in Table 812.4.2, fire barriers in separated mixed-use buildings shall comply with the fire resistance requirements of the *International Building Code*.

Exception: Where the fire barriers are required to have a 1-hour fire-resistance rating, existing wood lath and plaster in good condition or existing $^1/_2$-inch-thick (12.7 mm) gypsum wallboard shall be permitted.

❖ This provision is consistent with the provisions of Section 812.1.2, requiring fire barriers in compliance with the IBC when a mixed-use separation is necessary. The exception relates to a form of construction using plaster walls in existing buildings.

812.4.3 Exterior wall fire-resistance ratings. Hazard categories in regard to fire-resistance ratings of exterior walls shall be in accordance with Table 812.4.3.

❖ The third hazard classification area is exposure of exterior walls, relating to the need to protect surrounding buildings from fire conflagration. Table 812.4.3 groups the uses in the relative hazard levels based primarily on the requirements of Section 704, and Table 602 of the IBC.

TABLE 812.4.3
HAZARD CATEGORIES AND CLASSIFICATIONS:
EXPOSURE OF EXTERIOR WALLS

RELATIVE HAZARD	OCCUPANCY CLASSIFICATION
1 (Highest Hazard)	H
2	F-1, M, S-1
3	A, B, E, I, R
4 (Lowest Hazard)	F-2, S-2, U

812.4.3.1 Exterior wall rating for change of occupancy classification to a higher hazard category. When a change of occupancy group is made to a higher hazard category as shown in Table 812.4.3, exterior walls shall have fire resistance and exterior opening protectives as required by the *International Building Code*. This provision shall not apply to walls at right angles to the property line.

Exception: A 2-hour fire-resistance rating shall be allowed where the building does not exceed three stories in height and is classified as one of the following groups: A-2 and A-3 with an occupant load of less than 300, B, F, M, or S.

❖ Consistent with the general rule throughout this chapter for a change of occupancy to a higher hazard category, the exterior wall rating is required to meet those called for by the IBC for new construction. These provisions apply to exterior walls where the separation distance is a distance measured perpendicular from the building face to the interior lot line, the center line of a public street or to an imaginary line between two buildings. The exception is somewhat academic. These occupancy groups never are required to have a 3-hour fire-resistance rating on exterior walls.

812.4.3.2 Exterior wall rating for change of occupancy classification to an equal or lesser hazard category. When a change of occupancy group is made to an equal or lesser hazard category as shown in Table 812.4.3, existing exterior walls, including openings, shall be accepted.

❖ The exterior wall fire-resistance rating is allowed to remain unaltered when the change in occupancy classification is to an equal or lesser hazard category. Again, this is in keeping with the philosophy that no change is needed when the hazard to the building is not increased by the change or alteration of the building.

812.4.3.3 Opening protectives. Openings in exterior walls shall be protected as required by the *International Building Code*. Where openings in the exterior walls are required to be protected because of their distance from the property line, the sum of the area of such openings shall not exceed 50 percent of the total area of the wall in each story.

Exceptions:

1. Where the *International Building Code* permits openings in excess of 50 percent.

2. Protected openings shall not be required in buildings of Group R occupancy that do not exceed three stories in height and that are located not less than 3 feet (914 mm) from the property line.

3. Where exterior opening protectives are required, an automatic sprinkler system throughout may be substituted for opening protection.

4. Exterior opening protectives are not required when the change of occupancy group is to an equal or lower hazard classification in accordance with Table 812.4.3.

❖ At first glance, this provision would seem to contradict the provision in Section 812.4.3.2, which states that opening protectives need not be altered for change of occupancy to an equal or lesser hazard classification. This is, however, reflected in Exception 4 at any rate. It should be remembered, however, that this would apply in all cases where change is to a higher hazard category as described in Section 812.4.3.1. The significance of this section is that it imposes a maximum area of openings to be 50 percent of the wall area when the fire separation distance requires opening protectives, contrary to the percentages given by Table 704.8 of the IBC.

Exception 1 allows a higher percentage when the separation distance is such that the IBC would otherwise allow greater than 50 percent. Exception 2 relates

to the separation distances and requirements allowed by the *International Residential Code*® (IRC®). Exception 3 is consistent with, albeit more liberal than, the trade-off provided for automatic sprinkler systems given by Section 704.8.1 of the IBC.

812.4.4 Enclosure of vertical shafts. Enclosure of vertical shafts shall be in accordance with Sections 812.4.4.1 through 812.4.4.4.

❖ Section 812.4.4 outlines the requirements for the enclosure of vertical shafts within the building.

812.4.4.1 Minimum requirements. Vertical shafts shall be designed to meet the *International Building Code* requirements for atriums or the requirements of this section.

❖ Minimum requirements are set forth in Section 812.4.4.1, which offers two options for compliance. First is that vertical shafts shall be designed to meet the requirements for atriums as prescribed in Section 404.5 of the IBC. As an alternative, vertical shafts can comply with the requirements contained in the remaining subsections.

812.4.4.2 Stairways. When a change of occupancy group is made to a higher hazard category as shown in Table 812.4.1, interior stairways shall be enclosed as required by the *International Building Code*.

Exceptions:

1. In other than Group I occupancies, an enclosure shall not be required for openings serving only one adjacent floor and that are not connected with corridors or stairways serving other floors.

2. Unenclosed existing stairways need not be enclosed in a continuous vertical shaft if each story is separated from other stories by 1-hour fire-resistance-rated construction or approved wired glass set in steel frames and all exit corridors are sprinklered. The openings between the corridor and the occupant space shall have at least one sprinkler head above the openings on the tenant side. The sprinkler system shall be permitted to be supplied from the domestic water-supply systems, provided the system is of adequate pressure, capacity, and sizing for the combined domestic and sprinkler requirements.

3. Existing penetrations of stairway enclosures shall be accepted if they are protected in accordance with the *International Building Code*.

❖ Specific requirements for stairways are found in Section 812.4.4.2. When the change in occupancy is to a higher hazard category as shown in Table 812.4.1, compliance to the IBC is required. Three exceptions are provided. The first two of these exceptions are providing equivalencies to the requirements for atrium enclosures, and the third exception allows existing penetrations into the stairway if they are protected in accordance with the IBC.

812.4.4.3 Other vertical shafts. Interior vertical shafts other than stairways, including but not limited to elevator hoistways and service and utility shafts, shall be enclosed as required by the *International Building Code* when there is a change of use to a higher hazard category as specified in Table 812.4.1.

Exceptions:

1. Existing 1-hour interior shaft enclosures shall be accepted where a higher rating is required.

2. Vertical openings, other than stairways, in buildings of other than Group I occupancy and connecting less than 6 stories shall not be required to be enclosed if the entire building is provided with an approved automatic sprinkler system.

❖ Section 812.4.4.3 outlines compliance requirements for all other vertical shafts other than stairways. Again, there are minimum requirements when an increase in hazard category occurs. Again, there are two exceptions that are found in the IBC. The first is found in the provisions for atriums, Section 404, and the second is found in the exceptions of Section 707.2 for shafts and vertical exit enclosures.

812.4.4.4 Openings. All openings into existing vertical shaft enclosures shall be protected by fire assemblies having a fire-protection rating of not less than 1 hour and shall be maintained self-closing or shall be automatic closing by actuation of a smoke detector. All other openings shall be fire protected in an approved manner. Existing fusible link-type automatic door-closing devices shall be permitted in all shafts except stairways if the fusible link rating does not exceed 135°F (57°C).

❖ Section 812.4.4.4 prescribes a minimum fire protection rating of 1 hour for all openings into vertical shaft enclosures. All openings, such as those created by doors, archways or window-type openings, are to be self-closing or automatic closing by actuation of a smoke detector. Any other openings are required to be protected by an approved method. The use of fusible links are acceptable to control the closing of doors in all vertical shafts, except for stairways. This acceptance is limited to fusible links that have a temperature rating not exceeding 135°F (57°C).

812.5 Accessibility. Existing buildings or portions thereof that undergo a change of group or occupancy classification shall have all of the following accessible features:

1. At least one accessible building entrance.

2. At least one accessible route from an accessible building entrance to primary function areas.

3. Signage complying with Section 1110 of the *International Building Code.*

4. Accessible parking, where parking is provided.

5. At least one accessible passenger loading zone, where loading zones are provided.

6. At least one accessible route connecting accessible parking and accessible passenger loading zones to an accessible entrance.

Where it is technically infeasible to comply with the new construction standards for any of these requirements for a change of group or occupancy, the above items shall conform to the requirements to the maximum extent technically feasible. Changes of group or occupancy that incorporate any alterations or additions shall comply with this section and Sections 506.1 and 905.1 as applicable.

Exception: Type B dwelling or sleeping units required by Section 1107 of the *International Building Code* are not required to be provided in existing buildings and facilities.

❖ This section establishes that when an existing building, or a portion thereof, undergoes a change of occupancy, full compliance with accessibility requirements is expected and reasonable for the portion undergoing the change of occupancy, except where technical infeasibility can be demonstrated. If full compliance is technically infeasible, the element must be made accessible to the fullest extent that is feasible. This is consistent with the general approach that has always been taken relative to other matters regulated by the code.

In addition to accessibility requirements within the space, an accessible route is required to that space. Six items that make up that accessible route are listed. The intent is to provide the bare minimum to get people from a point of arrival into the building and to the area of the new use. This is not based on specific provisions of the *Americans with Disabilities Act Accessibility Guidelines* (ADAAG), but parallels the intent of requirements for removal of barriers. If the altered area is not required to be served by an accessible route in new construction, an accessible route would not be required for a change of occupancy (see Section 506.1.12).

If the area undergoing a change of occupancy is being altered and contains a primary function area by reference to Sections 506.1 and 905.1, the accessible route provisions in Section 506.2 are also applicable. This would not only require an accessible route to the change of occupancy area, but would also include possible upgrades to toilet rooms and drinking fountains that serve the area. See the commentary to Section 506.2 for additional information on the exceptions.

Type B dwelling or sleeping units are not required in a change of occupancy in residential and institutional facilities. This is consistent with the Fair Housing Act requiring compliance only in new construction.

812.6 Seismic loads. Existing buildings with a change of occupancy classification shall comply with the seismic provisions of Section 807.3.

❖ The seismic provisions of this chapter do not change for the special circumstances stated in Section 812 (see commentary, Section 807.3).

Bibliography

The following resource materials are referenced in this chapter or are relevant to the subject matter addressed in this chapter.

ASCE 7-02, *Minimum Design Loads for Buildings and Other Structures*. Reston, VA: American Society of Civil Engineers, 2002.

FEMA 368, *NEHRP Recommended Provisions for Seismic Regulations for New Buildings and Other Structures, Part 1*. Washington, DC: Federal Emergency Management Agency, 2001.

FEMA 369, *NEHRP Recommended Provisions for Seismic Regulations for New Buildings and Other Structures Commentary, Part 2 Commentary*. Washington, DC: Federal Emergency Management Agency, 2001.

IBC-00, *International Building Code*. Falls Church, VA: International Code Council, 2000.

IBC-03, *International Building Code*. Falls Church, VA: International Code Council, 2003.

NBC-99, *BOCA National Building Code*. Country Club Hills, IL: Building Officials & Code Administrators International, Inc. 1999.

SBC-99, *Standard Building Code*. Birmingham, AL: Southern Building Code Congress International, 1999.

UBC-97, *Uniform Building Code*. Whittier, CA: International Conference of Building Officials, 1997.

CHAPTER 9

ADDITIONS

General Comments

An "Addition" is defined in Chapter 2 as "an extension or increase in the floor area, number of stories or height of a building or structure." Chapter 9 contains the minimum requirements for an addition, which is not separated from the existing building by a fire wall.

Purpose

Chapter 9 provides the requirements for additions that correlate to the code requirements for new construction. There are, however, some exceptions that are specifically stated within this chapter.

SECTION 901
GENERAL

901.1 Scope. An addition to a building or structure shall comply with the building, plumbing, electrical, and mechanical codes, without requiring the existing building or structure to comply with any requirements of those codes or of these provisions.

Exception: In flood hazard areas, the existing building is subject to the requirements of Section 903.5.

❖ This section states the scope of Chapter 9 and states that the addition, but not the existing, unaltered, portion of the building or structure, must meet the requirements of the current editions of the building, plumbing, mechanical and electrical codes. The exception imposes the flood requirements from Section 903.5 on the existing building, regardless of whether work is planned for the remainder of the existing building.

When a new building is erected immediately adjacent to an existing building and they are separated by a fire wall, then they are considered as separate buildings, not an addition to the existing structure. The new building must be designed to comply with the technical provisions of *International Building Code*® (IBC®), not with the provisions of this chapter. The existing building must be evaluated taking, into consideration the elimination of the adjacent open space now occupied by the new building.

901.2 Creation or extension of nonconformity. An addition shall not create or extend any nonconformity in the existing building to which the addition is being made with regard to accessibility, structural strength, fire safety, means of egress, or the capacity of mechanical, plumbing, or electrical systems.

❖ If an existing building is noncompliant in any way that may be affected by the addition, or if the addition causes a noncompliant situation, the addition is not allowed to be constructed without correction of the noncompliance. For example, if the existing building has a travel distance that is greater than allowed to an exit and the addition adds to that distance, the addition is not allowed unless the total travel distance is brought to within that which is required by the code. Another example is that an addition that blocks an existing required exit would not be allowed.

An existing structure that is of a type of construction that does not comply with the height and area limitations of Table 503 of IBC may not be added to, unless the type of construction is upgraded or the allowable building height and area are increased through the addition of a sprinkler system throughout both the existing building and the addition as allowed by Chapter 5 of the IBC.

901.3 Other work. Any repair or alteration work within an existing building to which an addition is being made shall comply with the applicable requirements for the work as classified in Chapter 3.

❖ This section clearly requires that any structure having an addition added will have to comply with the provisions of Chapter 3 for the type of work performed as defined in Chapter 3. In other words, the provisions of Chapters 5, 6, 7, 8 and 10 would also apply depending on whether the work is a Level 1, 2 or 3 alteration, a change of occupancy or a renovation to an historic building, etc.

SECTION 902
HEIGHTS AND AREAS

902.1 Height limitations. No addition shall increase the height of an existing building beyond that permitted under the applicable provisions of Chapter 5 of the *International Building Code* for new buildings.

❖ See the commentary to Section 902.2.

902.2 Area limitations. No addition shall increase the area of an existing building beyond that permitted under the applicable provisions of Chapter 5 of the *International Building Code* for new buildings unless fire separation as required by the *International Building Code* is provided.

Exception: In-filling of floor openings and nonoccupiable appendages such as elevator and exit stair shafts shall be permitted beyond that permitted by the *International Building Code*.

❖ The purpose of Sections 902.1 and 902.2 are to establish that, with one exception, the height and area of an existing building plus its addition must comply with the requirements for a new building of its same occupancy and type of construction. A building with a proposed addition is to be evaluated based on the type of construction of the existing building or the addition, whichever is the lower type.

When reviewing for compliance with Table 503 of the IBC, Section 602.1.1 of the IBC may also be applicable. If a sprinkler system is not planned for both the existing building and the addition, the proposed addition does not meet the requirements of Table 503. The area limitation in Table 503 for the existing construction Type VB, Group F-1 is 8,500 square feet (790 m^2) and the total proposed area is 11,000 square feet (1,022 m^2). The area evaluation is based on Type VB construction, even though the proposed addition is Type VA construction, because the allowable area in Table 503 for Type VB construction is less than the allowable area for Type VA construction. One way to satisfy the requirements of Table 503 is to upgrade the type of construction of the existing building to Type VA, which has an allowable area of 14,000 square feet (1301 m^2). Another solution is to install an automatic sprinkler system that complies with NFPA 13 throughout the building. The revised allowable area of the building would be:

8,500 square feet + (3 x 8,500 square feet) = 34,000 square feet (3159 m^2) in accordance with the sprinkler area increase, which is more than the proposed building area of 11,000 square feet (1022 m^2).

902.3 Fire protection systems. Existing fire areas increased by the addition shall comply with Chapter 9 of the *International Building Code.*

❖ If an addition increases an existing building's fire area or areas to a level that is required to have fire protection systems by Chapter 9 of the NCBC, those fire areas must comply with the new construction requirements, both in the addition and the existing building. The code provides no exceptions to the sprinkler threshold limits given in Chapter 9 of the IBC.

For example, a restaurant is being increased from 4,000 square feet (372 m^2) to 6,000 square feet (557 m^2), with no intervening fire barriers in the facility. Section 903.2.1.2 of the IBC requires sprinklers for this Group A-2 occupancy when the fire area is greater than 5,000 square feet (465 m^2). Therefore, with the new addition, the threshold of 5,000 square feet (465 m^2) is exceeded, and sprinklers are required.

SECTION 903
STRUCTURAL

903.1 Compliance with the *International Building Code.* Additions to existing buildings or structures are new construction and shall comply with the *International Building Code.*

❖ An addition, by its very nature, is new construction, and the addition itself must comply with the IBC requirements for new construction.

903.2 Additional gravity loads. Existing structural elements supporting any additional gravity loads as a result of additions shall comply with the *International Building Code.*

Exceptions:

1. Structural elements whose stress is not increased by more than 5 percent.

2. Buildings of Group R occupancy with no more than five dwelling units or sleeping units used solely for residential purposes where the existing building and the addition comply with the conventional light-frame construction methods of the *International Building Code* or the provisions of the *International Residential Code.*

❖ Unless the existing building was initially designed with future expansion in mind, it is very likely that an addition will impose loads on some portion of the existing structure that exceed the member's capacity and require reinforcement or other mitigation. In any event, where an addition results in an added gravity load supported by an existing structure, the affected structural components must be checked to verify that they satisfy the requirements of the IBC for a new structure. Verification for gravity loading indicates compliance with the majority of Chapter 16 of the IBC except for seismic and wind loads.

There are two exceptions to full compliance with the IBC, one of which permits smaller Group R buildings and their additions that comply with either the *International Residential Code*® (IRC®) or the conventional light-frame construction provisions of the IBC. This is a prescriptive alternative to Chapter 16 loading provisions also permitted for new construction. See the limitations on these provisions in the IRC or Chapter 23 of the IBC.

903.3 Lateral-force-resisting system. The lateral-force-resisting system of existing buildings to which additions are made shall comply with Sections 903.3.1, 903.3.2, and 903.3.3.

Exceptions:

1. In Type V construction, Group R occupancies where the lateral-force story shear in any story is not increased by more than 10 percent.

2. Buildings of Group R occupancy with no more than five dwelling units or sleeping units used solely for residential purposes where the existing building and the addition comply with the conventional light-frame construction methods of the *International Building Code* or the provisions of the *International Residential Code.*

3. Additions where the lateral-force story shear in any story is not increased by more than 5 percent.

❖ An addition to an existing building requires that the lateral-force-resisting system is to comply with either Section 903.3.1 if a story is added or the height is increased; Section 903.3.2 if the addition is horizontal such as an extension of the floor area; or Section 903.3.3 if there is a voluntary addition of structural elements to the lateral-force-resisting system.

There are three exceptions to the above, one of which permits smaller Group R buildings and their additions to comply with either the IRC or the conventional light-frame construction provisions of the IBC. Exception 3 allows additions that do not increase the story shear in any story by more than 5 percent, equating roughly to allowing overstresses of up to 5 percent for gravity loads (see Section 903.2). Exception 1 is similar to Exception 3, but permits an increased story shear in any story of up to 10 percent in buildings of Type V construction that house Group R occupancies.

The *SEAOC Bluebook* and the *Uniform Building Code* define story shear as "the summation of design lateral forces above the story under consideration."

The term "story shear" is not defined in either the code or IBC but rather is expressed in mathematical terms by a formula included under the seismic load provisions that is equivalent to the preceding definition and illustration in Figure 903.3.

903.3.1 Vertical addition. Any element of the lateral-force-resisting system of an existing building subjected to an increase in vertical or lateral loads from the vertical addition shall comply with the lateral load provisions of the *International Building Code.*

❖ Where the addition of a story or an increase in height impose additional loads, either vertical or lateral, on portions of the existing lateral-force-resisting system, this section requires those members meet the lateral load requirements of the IBC. This would mean checking those elements for wind and seismic loading in accordance with applicable provisions of the IBC. Any element not meeting these IBC provisions requires replacement, reinforcement or other measures.

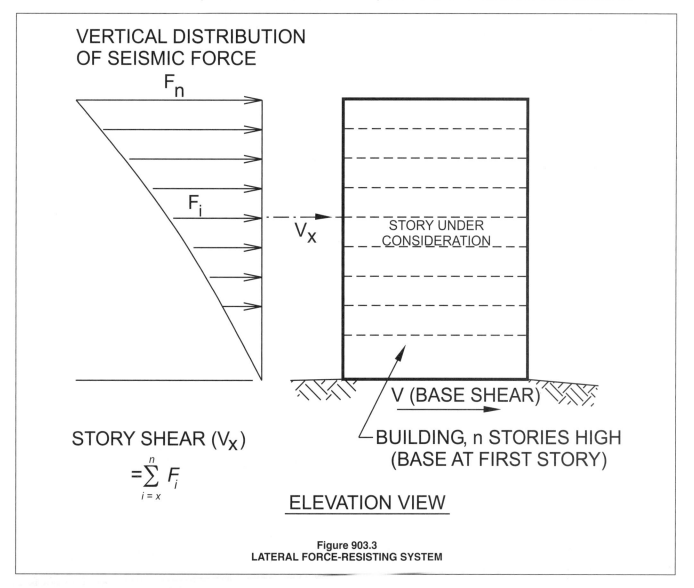

VERTICAL DISTRIBUTION OF SEISMIC FORCE

F_n

F_i

V_x

STORY UNDER CONSIDERATION

V (BASE SHEAR)

BUILDING, n STORIES HIGH (BASE AT FIRST STORY)

STORY SHEAR (V_x)

$$=\sum_{i=x}^{n} F_i$$

ELEVATION VIEW

**Figure 903.3
LATERAL FORCE-RESISTING SYSTEM**

FIGURE 903.3.2 – 903.3.3

ADDITIONS

Meeting the seismic provisions of the IBC requires consideration of more than just the magnitude of the loading. For example, it would be necessary to confirm that the seismic-force-resisting system of the combined structure does not exceed the limitations on these systems (such as height). These limitations are established in the IBC based on seismic design category (for an explanation of seismic design category see the commentary to Section 407.1.1.1).

903.3.2 Horizontal addition. Where horizontal additions are structurally connected to an existing structure, all lateral-force-resisting elements of the existing structure affected by such addition shall comply with the lateral load provisions of the *International Building Code*. Lateral loads imposed on the elements of the existing structure and the addition shall be determined by a relative stiffness analysis of the combined structure including torsional effects.

❖ A horizontal addition that is isolated from the existing structure is self-supporting and, therefore, has no impact on the existing structure. Where this is not the case, the resulting combined structure must be investigated for wind and seismic loading in accordance with the IBC, and any affected elements of the existing structure must also be shown to comply with these wind and seismic load requirements.

Compliance with the seismic load provisions of the IBC necessitates consideration of many aspects of the new and existing construction. One of these is that in rigid diaphragm structures, the story shear is distributed according to the relative stiffness of the vertical elements of the lateral force system (rather than merely using tributary area as is permitted for flexible diaphragms). This requires consideration of torsional effects as well as direct shear. This is illustrated in Figure 903.3.2 and, as can be seen, a horizontal addition is likely to effect the center of mass as well as the structure's center of rigidity. It is important, therefore, that the net effect of the resulting torsion on the combined structure be considered, and this section simply reiterates what is already a required step in the seismic analysis of any rigid diaphragm structure.

903.3.3 Voluntary addition of structural elements to improve the lateral-force-resisting system. Voluntary addition of structural elements to improve the lateral-force-resisting system of a building shall comply with Section 707.7.

❖ Strictly speaking, the definition of "Addition" in Section 202 would not include simply adding structural elements voluntarily in order to improve a building's resistance to lateral forces. This is more likely to be considered an alteration. For good measure, this section cross-references the provision under alterations that apply to voluntary upgrades.

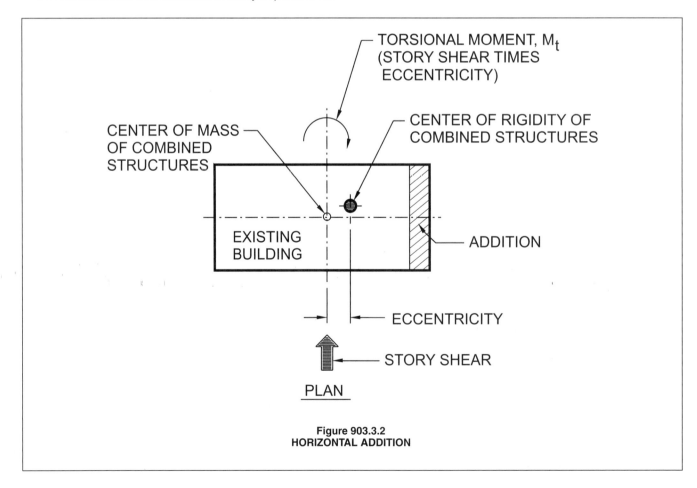

Figure 903.3.2
HORIZONTAL ADDITION

903.4 Snow drift loads. Any structural element of an existing building subjected to additional loads from the effects of snow drift as a result of an addition shall comply with the *International Building Code.*

Exceptions:

1. Structural elements whose stress is not increased by more than 5 percent.

2. Buildings of Group R occupancy with no more than five dwelling units or sleeping units used solely for residential purposes where the existing building and the addition comply with the conventional light-frame construction methods of the *International Building Code* or the provisions of the *International Residential Code.*

❖ An addition with a roof level that is different than that of an adjoining existing roof necessitates the consideration of the potential for drifting snow. Figure 903.4 illustrates this condition. If the addition is the portion of the building having the higher roof (side A in Figure 903.4), then the potential impact on the existing roof structure (side B) is the greatest. If the addition is the lower portion of the building (side B), then the impact of drifting snow on the existing structure is limited only to those members, if any, of the existing structure that support the new roof. The magnitude of the drift loading that must be included is a function of building geometry. All things being equal, the larger the difference in roof heights, the higher the code-required drift loading.

Section 1608.7 of the IBC requires the use of ASCE 7 procedures for snow drifts. The exceptions in this section of the code are similar to those of Section 903.2 and could be construed as already requiring compliance with all the snow load provisions of Section 1608 of the IBC, including snow drifts. For good measure, this section makes it clear that the impact of this commonly occurring condition of existing elements must be considered.

903.5 Flood hazard areas. Additions and foundations in flood hazard areas shall comply with the following requirements:

1. For horizontal additions that are structurally interconnected to the existing building:

    1.1. If the addition and all other proposed work, when combined, constitute substantial improvement, the existing building and the addition shall comply with Section 1612 of the *International Building Code.*

    1.2. If the addition constitutes substantial improvement, the existing building and the addition shall comply with Section 1612 of the *International Building Code.*

2. For horizontal additions that are not structurally interconnected to the existing building:

    2.1. The addition shall comply with Section 1612 of the *International Building Code.*

    2.2. If the addition and all other proposed work, when combined, constitute substantial improvement, the existing building and the addition shall comply with Section 1612 of the *International Building Code.*

3. For vertical additions and all other proposed work that, when combined, constitute substantial improvement, the existing building shall comply with Section 1612 of the *International Building Code.*

4. For a new, replacement, raised, or extended foundation, if the foundation work and all other proposed work, when combined, constitute substantial improvement, the

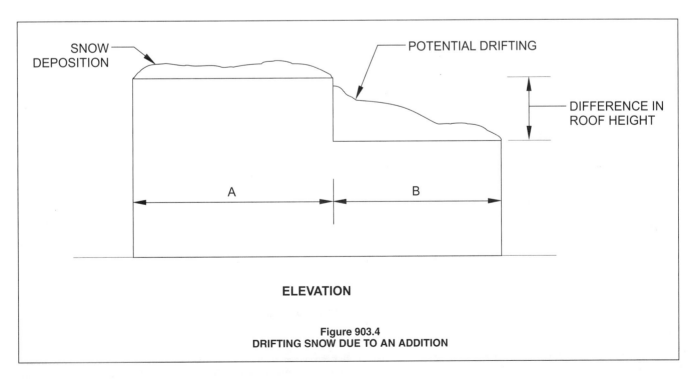

SNOW DEPOSITION

POTENTIAL DRIFTING

DIFFERENCE IN ROOF HEIGHT

A

B

ELEVATION

Figure 903.4
DRIFTING SNOW DUE TO AN ADDITION

existing building shall comply with Section 1612 of the *International Building Code.*

❖ The definition of "Flood hazard area" provided in Section 202 is taken from the IBC and establishes where this provision applies. In these areas, additions that separately or in combination with other repairs or alterations are considered to be "substantial improvements" (see definition in Section 202) then require the entire structure to meet the flood-resistant provisions of Section 1612 of the IBC. Buildings are considered to have undergone substantial improvement when the cost of the improvement is 50 percent or more of its market value before the improvement is undertaken. A horizontal addition that is structurally separated must always comply with the flood-resistant provisions of the IBC even if it does not qualify as a substantial improvement. This is merely a restatement of the requirement in Section 903.1 for additions to comply with the IBC.

Section 1612 of the IBC addresses requirements for buildings in designated flood hazard areas. The design and construction is required to be in accordance with ASCE 24, which in turn references the flood loading given in ASCE 7. Through use of these IBC provisions, communities meet a significant portion of the floodplain management regulation requirements necessary to participate in the National Flood Insurance Program (NFIP). [See the definition in Chapter 2 for "Substantial improvements." The requirements of Section 903.5 correlate to the Federal Emergency Management Agency (FEMA) requirements for additions and existing buildings.]

SECTION 904
SMOKE ALARMS IN OCCUPANCY
GROUPS R-3 AND R-4

904.1 Smoke alarms in an addition. Whenever an addition is made to a building or structure of a Group R-3 or R-4 occupancy, hardwired, interconnected smoke alarms meeting the requirements of the *International Building Code* or *International Residential Code* as applicable shall be installed and maintained in the addition.

❖ The installation of smoke alarms in areas remote from the sleeping area will be of minimal value if the alarm is not heard by the occupants. Interconnection of multiple smoke alarms within an individual dwelling unit or guestroom (or suite) is required in order to alert a sleeping occupant of a remote fire within the unit before the products reach the smoke alarm in the sleeping area, thus providing additional time for evacuation of the structure. It should be noted that the term "interconnection" is intended to allow the use of not only hard-wired systems, but also those that use radio signals (wireless systems). Underwriters Laboratories (UL) has listed smoke alarms that use this technology. It is presumed that on safely evacuating the unit or

room of fire origin, an occupant will notify other occupants by actuating the manual fire alarm system or other means available.

904.2 Smoke alarms in existing portions of a building. Whenever an addition is made to a building or structure of a Group R-3 or R-4 occupancy, the existing building shall be provided with smoke alarms as required by the *International Building Code* or *International Residential Code* as applicable.

❖ Sections 904.1 and 904.2 require that, for buildings containing Group R-3 and R-4 occupancies, both the addition and the existing building to which it is added must comply with the requirements of Chapter 9 of the IBC or Section 317 of the IRC (if single family) for new construction with respect to smoke alarms.

SECTION 905
ACCESSIBILITY

905.1 Minimum requirements. Accessibility provisions for new construction shall apply to additions. An addition that affects the accessibility to, or contains an area of, primary function shall comply with the requirements of Section 506.2 for accessible routes.

❖ Additions must comply with new construction; however, an addition is also an alteration to an existing building; therefore, accessible route provisions for existing buildings are applicable (see commentary, Section 506.2). For example, a new dining area is added onto a restaurant. All accessible elements within the parameter of the addition must be constructed accessible. If the accessible route to the addition is through the existing building, the route must be evaluated to see if elements need to be altered to meet accessibility requirements. In addition, any toilet rooms or drinking fountains that serve the addition have to be evaluated for accessibility. This would include the route to these elements, the toilet rooms themselves, the fixtures in the toilet room and the drinking fountains.

SECTION 906
ENERGY CONSERVATION

906.1 Minimum requirements. Additions to existing buildings or structures may be made to such buildings or structures without making the entire building or structure comply with the requirements of the *International Energy Conservation Code.* The addition shall conform to the requirements of the *International Energy Conservation Code* as they relate to new construction only.

❖ This section requires that only the addition meet the *International Energy Conservation Code®* (IECC®) requirements for new construction. The existing building to which it is added is exempt.

Bibliography

The following resource materials are referenced in this chapter or are relevant to the subject matter addressed in this chapter.

ASCE 7-02, *Minimum Design Loads for Buildings and Other Structures*. Reston, VA: American Society of Civil Engineers, 2002.

ASCE 24-98, *Flood Resistant Design and Construction Standard*. Reston, VA: American Society of Civil Engineers, 1998.

IBC-03, *International Building Code*. Falls Church, VA: International Code Council, 2003.

IRC-03, *International Residential Code*. Falls Church, VA: International Code Council, 2003.

Recommended Lateral Force Requirements and Commentary (SEAOC Blue Book). Sacramento, CA: Structural Engineers Association of California, 1999.

Reducing Flood Losses Through the International Code Series: Meeting the Requirements of the National Flood Insurance Program. Falls Church, VA: International Code Council, May 2000.

Taly, Narendra. *Loads and Loads Paths in Buildings Principles of Structural Design*. Falls Church, VA: International Code Council, 2003.

UBC-97, *Uniform Building Code*. Whittier, CA: International Conference of Building Officials, 1997.

Williams, Alan. *Seismic and Wind Forces Structural Design Examples*. Falls Church, VA: International Code Council, 2003.

CHAPTER 10

HISTORIC BUILDINGS

General Comments

This section provides some blanket exception from code requirements when the building in question has historic value. The most important criterion for application of this section is that the building must be essentially accredited as being of historic significance by a qualified third party or agency. Usually this is done by a state or local authority after careful review of the historical value of the building. Most, if not all, states have such authorities, as do many local jurisdictions. The agencies with such authority can be located at the state or local government level or through the local chapter of the American Institute of Architects (AIA). Other considerations include the structural condition of the building (i.e., is the building structurally sound), its proposed use, its impact on life safety and how the intent of the code, if not the letter, will be achieved.

Purpose

A large number of existing buildings and structures do not comply with the current building code requirements for new construction. Although many of these buildings are potentially salvageable, rehabilitation is often cost prohibitive because they may not be able to comply with all the requirements for new construction. At the same time, it is necessary to regulate construction in existing buildings that undergo additions, alterations, renovations, extensive repairs or change of occupancy. Such activity represents an opportunity to ensure that new construction complies with the current building codes and that existing conditions are maintained, at a minimum, to their current level of compliance or are improved as required. To accomplish this objective, and to make the rehabilitation process easier, this chapter allows for a controlled departure from full compliance with the technical codes, without compromising the minimum standards for fire prevention and life safety features of the rehabilitated building.

SECTION 1001
GENERAL

1001.1 Scope. It is the intent of this chapter to provide means for the preservation of historic buildings. Historical buildings shall comply with the provisions of this chapter relating to their repair, alteration, relocation and change of occupancy.

❖ This section establishes that this chapter deals with the requirements for the repair, alteration, relocation and change of occupancy of buildings previously defined by the code as historic. Note that the code would not regard reconstruction of an historic building as qualifying for treatment under this chapter, even if the building is reconstructed entirely from original stored materials and assemblies.

1001.2 Report. A historic building undergoing repair, alteration, or change of occupancy shall be investigated and evaluated. If it is intended that the building meet the requirements of this chapter, a written report shall be prepared and filed with the code official by a registered design professional when such a report is necessary in the opinion of the code official. Such report shall be in accordance with Chapter 1 and shall identify each required safety feature that is in compliance with this chapter and where compliance with other chapters of these provisions would be damaging to the contributing historic features. In high seismic zones, a structural evaluation describing, at minimum, a complete load path and other earthquake-resistant features shall be prepared. In addition, the report shall describe each feature that is not in compliance with these provisions and shall demonstrate how the intent of these provisions is complied with in providing an equivalent level of safety.

❖ This section provides for a report that acts as a technical backup for the code official's judgment. The intent is to be able to retain a record of the decisions and the basis for those decisions for future reference.

Since the definition of "Historic building" refers to listing/eligibility for national register, state designation, inclusion in a survey or contributing to a national or state historic district, contributing features may be limited to those features designated or mentioned as contributing in nominations, historic structure reports or other such documents. This represents a systematized approach for evaluating proposals of equivalent performance.

1001.3 Special occupancy exceptions—museums. When a building in Group R-3 is also used for Group A, B, or M purposes such as museum tours, exhibits, and other public assembly activities, or for museums less than 3000 square feet (279 m²), the code official may determine that the occupancy is Group B when life-safety conditions can be demonstrated in accordance with Section 1001.2. Adequate means of egress in such buildings, which may include a means of maintaining doors in an open position to permit egress, a limit on building occupancy to an occupant load permitted by the means of egress capacity, a limit on occupancy of certain areas or floors, or supervision by a person knowledgeable in the emergency exiting procedures, shall be provided.

❖ This section acknowledges that dwellings used as museums and other public assembly activities are a special case.

The provision allows a small museum or historic house used as a museum to demonstrate life safety using operational controls that would not be permitted in most other uses. This is likely to be effective because of the high degree of supervision that would be necessary for security reasons in this occupancy. In this case, the building use group shall be classified as Use Group B even though it may contain assembly or mercantile uses. It would also permit the occasional use of a residential building for tours and museum use.

1001.4 Flood hazard areas. In flood hazard areas, if all proposed work, including repairs, work required because of a change of occupancy, and alterations, constitutes substantial improvement, then the existing building shall comply with Section 1612 of the *International Building Code.*

> **Exception:** If a historic building will continue to be a historic building after the proposed work is completed, then the proposed work is not considered a substantial improvement. For the purposes of this exception, a historic building is:
>
> 1. Listed or preliminarily determined to be eligible for listing in the National Register of Historic Places;
>
> 2. Determined by the Secretary of the U.S. Department of Interior to contribute to the historical significance of a registered historic district or a district preliminarily determined to qualify as a historic district; or
>
> 3. Designated as historic under a state or local historic preservation program that is approved by the Department of Interior.

❖ For the purpose of application of the provisions of this code that pertain to structures within flood areas, historic buildings that meet one of the exceptions are not required to comply with the flood provisions, including the substantial improvement provision, provided the building will continue to be designated as historic. However, if plans to substantially improve an historic building would result in loss of its designation as such, the building would be required to be brought into compliance with the requirements for flood hazard areas. The code official may require applicants to consult with the appropriate historic preservation authority to determine whether proposed work will jeopardize a structure's historic designation. To the extent practicable, owners of historic buildings may wish to incorporate flood-resistant provisions when such buildings undergo repairs and alterations. If the work on an historic structure does not cause the loss of its eligibility as an historic building as a result of the proposed work, then the work would not be considered as a substantial improvement; therefore, the building would not have to comply with Section 1612 of the code. This exception, however, does not apply to work constituting substantial improvements that are a result of a change of use in an historic building.

SECTION 1002
REPAIRS

1002.1 General. Repairs to any portion of a historic building or structure shall be permitted with original or like materials and original methods of construction, subject to the provisions of this chapter.

❖ Repair with like materials may be needed for aesthetic reasons, or to meet material conservation needs.

1002.2 Dangerous buildings. When a historic building is determined to be dangerous, no work shall be required except as necessary to correct identified unsafe conditions.

❖ This exception makes it possible to repair identified hazardous conditions without triggering secondary requirements.

1002.3 Relocated buildings. Foundations of relocated historic buildings and structures shall comply with the *International Building Code.* Relocated historic buildings shall otherwise be considered a historic building for the purposes of this code. Relocated historic buildings and structures shall be sited so that exterior wall and opening requirements comply with the *International Building Code* or with the compliance alternatives of this code.

❖ The foundation of a relocated building is a new structure; therefore, there is no reason why it should not comply with provisions for new construction. However, compared with code provisions that treat a relocated building as a new building, the rest of the building is still treated as an historic structure. In relocation, siting is to some degree a new design decision. Relationship of walls and openings to property lines are, therefore, reasonably able to be brought into compliance with the fire separation requirements of the code.

1002.4 Chapter 4 compliance. Historic buildings undergoing repairs shall comply with all of the applicable requirements of Chapter 4, except as specifically permitted in this chapter.

❖ This section refers to the fact that the general code provisions of Chapter 4 will still apply except for the exceptions in this chapter.

1002.5 Replacement. Replacement of existing or missing features using original materials shall be permitted. Partial replacement for repairs that match the original in configuration, height, and size shall be permitted. Such replacements shall not be required to meet the materials and methods requirements of Section 401.2.

> **Exception:** Replacement glazing in hazardous locations shall comply with the safety glazing requirements of Chapter 24 of the *International Building Code.*

❖ In some cases, historic materials have not been tested for fire resistance or structural performance, but have been shown adequate in use. An excellent example would be a railing, where spacing, height or end conditions would not meet modern requirements. If sections of a railing are missing, or if repairs are needed, the railing is not required to meet the code provisions for new construction. The exception gives an example of one case in which an established hazard exists and the cost to fabricate an historically correct reproduction of a proper material is negligible.

SECTION 1003
FIRE SAFETY

1003.1 Scope. Historic buildings undergoing alterations, changes of occupancy, or that are moved shall comply with Section 1003.

❖ This section recognizes the unique aspects associated with older historical structures. It reinforces the importance of the means of egress pathway while granting the code official some latitude in accepting some degree of variance in the egress components (for example, direction of exit door swing, egress path width and height).

1003.2 General. Every historic building that does not conform to the construction requirements specified in this code for the occupancy or use and that constitutes a distinct fire hazard as defined herein shall be provided with an approved automatic fire-extinguishing system as determined appropriate by the code official. However, an automatic fire-extinguishing system shall not be used to substitute for, or act as an alternative to, the required number of exits from any facility.

❖ Fire-extinguishing systems are effective substitutes for some requirements that are typical for new construction, particularly passive systems such as rated doors and corridors. By slowing or terminating the development of a fire, a sprinkler system will make passive fire resistance unnecessary.

1003.3 Means of egress. Existing door openings and corridor and stairway widths less than those specified elsewhere in this code may be approved, provided that, in the opinion of the code official, there is sufficient width and height for a person to pass through the opening or traverse the means of egress. When approved by the code official, the front or main exit doors need not swing in the direction of the path of exit travel, provided that other approved means of egress having sufficient capacity to serve the total occupant load are provided.

❖ This provision would permit the continuance of a structure no more hazardous than before rehabilitation, with minimum standards of usability. Provisions for new construction would require that exit doors swing in the direction of exit travel for an occupant load exceeding 50.

1003.4 Transoms. In fully sprinklered buildings of Group R-1, R-2 or R-3 occupancy, existing transoms in corridors and other fire-resistance-rated, walls may be maintained if fixed in the closed position. A sprinkler shall be installed on each side of the transom.

❖ This permits the retention of nonwired/nonrated glass in historic transoms within rated walls in residential occupancies when protected by an automatic fire sprinkler system throughout the building.

1003.5 Interior finishes. The existing finishes of walls and ceilings shall be accepted when it is demonstrated that they are the historic finishes.

❖ Flame spread requirements for walls or ceilings may not be met by historic wood paneling or other finishes.

1003.6 Stairway enclosure. In buildings of three stories or less, exit enclosure construction shall limit the spread of smoke by the use of tight-fitting doors and solid elements. Such elements are not required to have a fire-resistance rating.

❖ Enclosure of stairs to control smoke would provide an improvement, but permitting the enclosure to be nonrated would allow for use of traditional materials. Example enclosures include plain or wired glass, smoke-activated doors and similar assemblies.

1003.7 One-hour fire-resistant assemblies. Where 1-hour fire-resistance-rated construction is required by these provisions, it need not be provided, regardless of construction or occupancy, where the existing wall and ceiling finish is wood or metal lath and plaster.

❖ The substitution of standard, old-fashioned lath and plaster for 1-hour-rated wall construction is a well-established alternative and is considered as meeting the intent of the code to provide a safe path for exit.

1003.8 Glazing in fire-resistance-rated systems. Historic glazing materials in interior walls required to have a 1-hour fire-resistance rating may be permitted when provided with approved smoke seals and when the area affected is provided with an automatic sprinkler system.

❖ Glazing of interior partitions can be vulnerable because of potential leakage around the edge of the glazing or because of heat-induced glass breakage. This provision addresses both concerns. The sprinkler system should be provided on both sides of the wall containing the glazing.

1003.9 Stairway railings. Grand stairways shall be accepted without complying with the handrail and guard requirements. Existing handrails and guards at all stairs shall be permitted to remain, provided they are not structurally dangerous.

❖ Requirements for handrail and guardrail height have increased over the years. While desirable, heights as required by current code are not essential for safety. Since a railing is very difficult to modify, particularly without significant architectural change, the code permits historic railings that are essentially safe to remain in use. This also deals with handrail profiles that may not meet current grip requirements.

1003.10 Guards. Guards shall comply with Sections 1003.10.1 and 1003.10.2.

❖ This section establishes that guards in existing historic structures can remain in the configuration in which they were originally constructed. Guards would still have to conform with the restrictions in effect at the time the building was built.

1003.10.1 Height. Existing guards shall comply with the requirements of Section 405.

❖ Section 405 requires that repairs maintain the existing level of protection.

1003.10.2 Guard openings. The spacing between existing intermediate railings or openings in existing ornamental patterns shall be accepted. Missing elements or members of a guard may be replaced in a manner that will preserve the historic appearance of the building or structure.

❖ Requirements for voids in railings and grilles have become more stringent over the years. The spacing of a balustrade or ornamental pattern of a grille is frequently a significant contributor to the building's historical character. This provision enables retention and repair of these elements.

1003.11 Exit signs. Where exit sign or egress path marking location would damage the historic character of the building, alternative exit signs are permitted with approval of the code official. Alternative signs shall identify the exits and egress path.

❖ Signs and exit path identification are still required, but location is permitted to be flexible within the limits of functionality.

1003.12 Automatic fire-extinguishing systems.

❖ This term is the generic name for all types of automatic fire-extinguishing systems, including the most common type: the automatic sprinkler system. See Section 904 of the *International Building Code®* (IBC®) for requirements for particular alternative automatic fire-extinguishing systems, such as wet-chemical, dry-chemical, foam, carbon dioxide, halon and clean-agent systems.

1003.12.1 General. Every historical building that cannot be made to conform to the construction requirements specified in the *International Building Code* for the occupancy or use and that constitutes a distinct fire hazard shall be deemed to be in compliance if provided with an approved automatic fire-extinguishing system.

Exception: When the code official approves an alternative life-safety system.

❖ Note that the alternative is specifically to the construction requirements, i.e., height and area limitations, construction type and passive fire resistance.

SECTION 1004
ALTERATIONS

1004.1 Accessibility requirements. The provisions of Section 506 shall apply to buildings and facilities designated as historic structures that undergo alterations, unless technically infeasible. Where compliance with the requirements for accessible routes, ramps, entrances, or toilet facilities would threaten or destroy the historic significance of the building or facility, as determined by the code official, the alternative requirements of Sections 1004.1.1 through 1004.1.5 for that element shall be permitted.

❖ The regulations of individual adopting jurisdictions may provide additional guidance or limitations on the use of this section. The basic Americans with Disabilities Act (ADA) requirements permit exceptions and alternatives for historic buildings. These minimum provisions pro-

vide reasonable accommodation for building users.

For this section to be applicable, the building must be registered as historic. Historic buildings are treated much the same as provided for in Section 506, in that an historic building that is altered is expected to comply with accessibility requirements, unless technical infeasibility can be demonstrated. However, this section also goes on to acknowledge that the historical character of a building may be adversely affected by strict compliance with accessibility provisions. For example, compliance with door width requirements may necessitate the removal of an existing set of doors that is critical to the historical character of the building. To assist the code official in determining if the required provisions are detrimental to the historical significance, recommended guidelines have been incorporated into Appendix B, Section B101.

This section is intended to exempt such conditions in order to maintain the historical character of the building. Because limited extent of accessibility is desired in all facilities, Sections 1004.1.1 through 1004.1.5 allow for alternatives.

In an effort to inform designers and owners of federal regulations that are not typically handled through the code enforcement process, information on the *Americans with Disabilities Act Accessibility Guidelines* (ADAAG) requirements for displays associated with historical buildings have been included in Appendix B, Section B101.5.

1004.1.1 Site arrival points. At least one main entrance shall be accessible.

❖ Full compliance would require an accessible route from all site arrival points. If this requirement would adversely affect the historical significance of the building, the available alternative is to provide an accessible route from one site arrival point to an accessible entrance.

1004.1.2 Multilevel buildings and facilities. An accessible route from an accessible entrance to public spaces on the level of the accessible entrance shall be provided.

❖ It is not required in a building alteration that accessibility to spaces above or below the level of an accessible entrance be provided. Full compliance for new construction might require an accessible route to levels above or below, as well as throughout, the entrance level. If this requirement would adversely affect the historical significance of the building, the alternative is to provide an accessible route from the accessible entrance to all spaces open to the public on the entrance level. If elevators are provided, but are not accessible, signage in accordance with Section 1110 of the IBC is required.

1004.1.3 Entrances. At least one main entrance shall be accessible.

Exceptions:

1. If a main entrance cannot be made accessible, an accessible nonpublic entrance that is unlocked while the building is occupied shall be provided; or

2. If a main entrance cannot be made accessible, a locked accessible entrance with a notification system or remote monitoring shall be provided.

❖ Although uniform building access is the rule for new construction, older buildings may have main entrances that are both inaccessible and that maintain a significant element of the buildings' character. In these cases, providing access by an alternative route is deemed to meet the primary intent of the accessibility regulations.

Full compliance would require 50 percent of the public entrances to be accessible, as well as any of the special types of entrances listed in Section 1105. If this requirement would adversely affect the historical significance of the building, only one main entrance is required to be made accessible. If a main entrance cannot be made accessible, then an employee or service entrance may serve as the accessible entrance, provided that it remains unlocked when the building is open. Alternatively, if the entrance is locked, some type of notification system (e.g., doorbell, intercom) or remote monitoring (e.g., security camera) is provided so that a person inside the facility would know to go to that entrance to admit someone. Signage must be provided at accessible and inaccessible entrances in accordance with Section 1110 of the IBC.

1004.1.4 Toilet and bathing facilities. Where toilet rooms are provided, at least one accessible toilet room shall be provided for each sex, or a unisex toilet room complying with Section 1109.2.1 of the *International Building Code* shall be provided.

❖ This provision contains several options. One is to not offer toilet rooms for anyone. Another is to provide an accessible toilet room for each sex. A third is to provide an accessible unisex toilet room. These exceptions are offered to help the designer deal with a major problem in dealing with the size of accessible toilet rooms.

Full compliance would require an accessible toilet/bathing facility at each location where toilet/bathing facilities are provided. If altering the existing facilities to be accessible would adversely affect the historical significance of the building, only one accessible unisex toilet/bathing facility is required. Signage must be provided at accessible and inaccessible toilet rooms in accordance with Section 1110 of the IBC.

1004.1.5 Ramps. The slope of a ramp run of 24 inches (610 mm) maximum shall not be steeper than one unit vertical in eight units horizontal (12-percent slope).

❖ Full compliance would allow a maximum slope for ramps of one unit vertical in 12 units horizontal (1:12) (12-percent slope). If providing a fully compliant ramp would adversely affect the historical significance of the building, for an elevation change of 3 inches (76 mm) or less, a slope of one unit vertical in eight units horizontal (1:8) (12.5-percent slope) maximum is permitted.

SECTION 1005
CHANGE OF OCCUPANCY

1005.1 General. Historic buildings undergoing a change of occupancy shall comply with the applicable provisions of Chapter 8, except as specifically permitted in this chapter. When Chapter 8 requires compliance with specific requirements of Chapter 4, Chapter 5, or Chapter 6 and when those requirements are subject to the exceptions in Section 1002, the same exceptions shall apply to this section.

❖ Occupancy choice ideally minimizes code conflicts and the necessity for change. According to the Secretary of the Interior's *Standards for Treatment of Historic Properties*, "A property will be used as it was historically, or be given a new use that maximizes the retention of distinctive materials, features, spaces and spatial relationships." However, economic pressures may dictate that a structure be used in a different occupancy than that for which it was intended.

1005.2 Building area. The allowable floor area for historic buildings undergoing a change of occupancy shall be permitted to exceed by 20-percent the allowable areas specified in Chapter 5 of the *International Building Code*.

❖ This section permits an allowable floor area in excess of the 20-percent restriction listed in the IBC.

1005.3 Location on property. Historic structures undergoing a change of use to a higher hazard category in accordance with Section 812.4.3 may use alternative methods to comply with the fire resistance and exterior opening protective requirements. Such alternatives shall comply with Section 1001.2.

❖ When changing to a higher hazard category, a building should be evaluated for means to manage the new or increased hazards. Literal code compliance may not be the only way, or the best way, to accomplish this objective in an historic building.

1005.4 Occupancy separation. Required occupancy separations of 1 hour may be omitted when the building is provided with an approved automatic sprinkler system throughout.

❖ Substitution of automatic extinguishing systems for passive protection is established as an acceptable procedure.

1005.5 Roof covering. Regardless of occupancy or use group, roof-covering materials not less than Class C shall be permitted where a fire-retardant roof covering is required.

❖ Roof-covering requirements are a spread-of-fire issue: first, to prevent fire spread to the building in question; second, to keep burning brands generated by a fire in that building from igniting other buildings in a general conflagration. However, if a limited number of buildings do not meet this requirement, the hazard is not significantly increased.

1005.6 Means of egress. Existing door openings and corridor and stairway widths less than those that would be acceptable for nonhistoric buildings under these provisions shall be approved, provided that, in the opinion of the code official, there is sufficient width and height for a person to pass through the opening or traverse the exit and that the capacity of the exit system is adequate for the occupant load, or where other operational controls to limit occupancy are approved by the code official.

❖ This provision would permit the continuance of a structure no more hazardous than before rehabilitation, with minimum standards of usability.

1005.7 Door swing. When approved by the code official, existing front doors need not swing in the direction of exit travel, provided that other approved exits having sufficient capacity to serve the total occupant load are provided.

❖ Provisions for new construction would require that exit doors swing in the direction of exit travel for an occupant load exceeding 50.

1005.8 Transoms. In corridor walls required by these provisions to be fire-resistance rated, existing transoms may be maintained if fixed in the closed position, and fixed wired glass set in a steel frame or other approved glazing shall be installed on one side of the transom.

Exception: Transoms conforming to Section 1003.4 shall be accepted.

❖ These alternatives would be used primarily by nonresidential occupancies, as residential occupancies would use Section 1003.4.

1005.9 Finishes. Where finish materials are required to have a flame-spread classification of Class III or better, existing nonconforming materials shall be surfaced with an approved fire-retardant paint or finish.

Exception: Existing nonconforming materials need not be surfaced with an approved fire-retardant paint or finish where the building is equipped throughout with an automatic fire-suppression system installed in accordance with the *International Building Code* and the nonconforming materials can be substantiated as being historic in character.

❖ The fire-retardant paint would have the effect of slowing the development and spread of fire. The exception can be used where conditions are unsuitable for use of intumescent paint or varnish or other fire-retardant finishes.

1005.10 One-hour fire-resistant assemblies. Where 1-hour fire-resistance-rated construction is required by these provisions, it need not be provided, regardless of construction or occupancy, where the existing wall and ceiling finish is wood lath and plaster.

❖ The substitution of standard, old-fashioned lath and plaster for 1-hour-rated wall construction is a well-established alternative. Pressed tin ceilings are an example of this issue. Some state historic preservation

offices (SHPO) may provide additional guidance on this subject and alternatives.

1005.11 Stairs and railings. Existing stairways shall comply with the requirements of these provisions. The code official shall grant alternatives for stairways and railings if alternative stairways are found to be acceptable or are judged to meet the intent of these provisions. Existing stairways shall comply with Section 1003.

Exception: For buildings less than 3000 square feet (279 m²), existing conditions are permitted to remain at all stairs and rails.

❖ This provision gives an opportunity to analyze the stairs functionality as an exit, and to alter only those elements that are judged to be unsafe or inadequate.

1005.12 Exit signs. The code official may accept alternative exit sign locations where such signs would damage the historic character of the building or structure. Such signs shall identify the exits and exit path.

❖ Signs and exit path identification are still required, but location is permitted to be flexible within limits of functionality.

1005.13 Exit stair live load. Existing historic stairways in buildings changed to a Group R-1 or R-2 occupancy shall be accepted where it can be shown that the stairway can support a 75-pounds-per-square-foot (366 kg/m²) live load.

❖ In the case of a building changed to a Group R occupancy from another occupancy, the likelihood is that stairs will be designed to a higher capacity than use would require.

1005.14 Natural light. When it is determined by the code official that compliance with the natural light requirements of Section 811.1.1 will lead to loss of historic character or historic materials in the building, the existing level of natural lighting shall be considered acceptable.

❖ At the time the provisions requiring natural light in residential occupancies were enacted, they were a necessary public health reform. At this time, deviations from these requirements would probably not have health consequences.

1005.15 Accessibility requirements. The provisions of Section 812.5 shall apply to buildings and facilities designated as historic structures that undergo a change of occupancy, unless technically infeasible. Where compliance with the requirements for accessible routes, ramps, entrances, or toilet facilities would threaten or destroy the historic significance of the building or facility, as determined by the authority having jurisdiction, the alternative requirements of Sections 1004.1.1 through 1004.1.5 for those elements shall be permitted.

❖ For this section to be applicable, the building must be registered as historic. Historic buildings are treated much the same as provided for in Sections 812.5, in that an historic building undergoing a change of occupancy is expected to comply with accessibility requirements, unless technical infeasibility can be demon- strated. However, this section also goes on to acknowledge that the historical character of a building may be adversely affected by strict compliance with accessibility provisions. For example, compliance with door width requirements may necessitate the removal of an existing set of doors that is critical to the historical character of the building. To assist the code official in determining if the required provisions are detrimental to the historical significance, recommended guidelines have been incorporated into Appendix B, Section B101.

This section is intended to exempt such conditions in order to maintain the historical character of the building. Because limited extent of accessibility is desired in all facilities, Sections 1004.1.1 through 1004.1.5 allow for alternatives.

In an effort to inform designers and owners of federal regulations that are not typically handled through the code enforcement process, information on the ADAAG requirements for displays associated with historical buildings have been included in Appendix B, Section B101.5.

SECTION 1006
STRUCTURAL

1006.1 General. Historic buildings shall comply with the applicable structural provisions for the work as classified in Chapter 3.

> **Exception:** The code official shall be authorized to accept existing floors and approve operational controls that limit the live load on any such floor.

❖ Any proposed work on historic buildings must comply with the structural provisions that apply based on the nature of that work. In other words, the classification of any proposed work is the first consideration. The only exception to this authorizes the code official to accept existing floors and approve operational controls that limit the live load on any such floor. This could be achieved by posting load limit signs in conspicuous locations. Actual loads on floors are typically less than the code required loads. This section permits consideration of the actual loads as opposed to the probabilistic loads used for new design.

1006.2 Unsafe structural elements. Where the code official determines that a component or a portion of a building or structure is dangerous as defined in this code and is in need of repair, strengthening, or replacement by provisions of this code, only that specific component or portion shall be required to be repaired, strengthened, or replaced.

❖ The definition of "Dangerous" in Section 202 gives five conditions under which a building, a portion of a building or an individual structural component is considered dangerous. Any building or structural component that is determined to be dangerous presents an unacceptable risk to public safety. Section 115.1 clarifies that a building considered dangerous is deemed to be un-

safe, and in order for an unsafe building to remain in place, the building must be made safe.

This section clarifies that, in an historic building any structural repair, strengthening or replacement applies only to the specific component(s) or portion(s) necessary to correct the dangerous condition(s).

CHAPTER 11

RELOCATED OR MOVED BUILDINGS

General Comments

Chapter 11 is applicable to any building that is moved or relocated. The relocation of a building will automatically cause an inspection and evaluation process that enables the jurisdiction to determine the level of compliance with the *International Fire Code®* (IFC®) and the *International Property Maintenance Code®* (IPMC®). These two codes by their scope are applicable to existing buildings. This is the case regardless of any repair, remodeling, alteration work or change of occupancy occurring (see Section 101.2 of the IPMC and IFC).

Section 1102 addresses the building location of lot and structural considerations. Note that electrical, plumbing and heating, ventilating and air conditioning (HVAC) are not addressed for relocated buildings. This means that relocated buildings are not necessarily required to improve the existing electrical, plumbing or HVAC systems beyond what is necessary for the appropriate interface with the public utilities at the new location, unless the code official discovers an unsafe or hazardous condition.

Section 1102.7 contains an inspection requirement that helps to determine the structural integrity of the structure's components and connections. The inspection will help the code official to properly evaluate a building to ensure it is safe for occupancy.

If the project scope involves repairs, alterations or a change of occupancy in addition to moving or relocating the building, then it will also have to comply with the provisions of the chapters related to repair, alteration or change of occupancy. Any new construction of accessory elements to include decks, stairs, fences or accessory buildings must comply with the applicable code requirements for new construction.

Purpose

A large number of existing buildings and structures are not be able to comply with the current building code requirements for new construction. Although many of these buildings are potentially salvageable, relocation and rehabilitation is often cost prohibitive. It is necessary to regulate construction in existing buildings that undergo additions, alterations, renovations, extensive repairs or change of occupancy to ensure public safety.

SECTION 1101
GENERAL

1101.1 Scope. This chapter provides requirements for relocated or moved structures.

❖ This section states the scope of Chapter 11 and references the requirements for relocated or moved structures.

1101.2 Conformance. The building shall be safe for human occupancy as determined by the *International Fire Code* and the *International Property Maintenance Code*. Any repair, alteration, or change of occupancy undertaken within the moved structure shall comply with the requirements of this code applicable to the work being performed. Any field-fabricated elements shall comply with the requirements of the *International Building Code* or the *International Residential Code* as applicable.

❖ Moved structures are required to comply with the provisions of this code for any repair, alteration or change of occupancy. The moved structure may comply with the alternative provisions of Chapter 12 instead of the code requirements for new structures, which may be particularly useful if the moved structure is older than the effective date of the adoption of building codes within the jurisdiction. The IFC and IPMC provisions for fire separation distance, safety and zoning requirements of the moved structure must comply with requirements for new structures.

SECTION 1102
REQUIREMENTS

1102.1 Location on the lot. The building shall be located on the lot in accordance with the requirements of the *International Building Code* or the *International Residential Code* as applicable.

❖ Exterior walls of buildings may need fire-resistance ratings in accordance with either the *International Building Code®* (IBC®) (see Section 704.5 of the IBC) or the *International Residential Code®* (IRC®) (see Section R302 of the IRC), as applicable. The required rating in either code is based on the fire separation distance, and under the IBC, the occupancy and type of construction must also be considered. Since the relocated building's exterior wall construction and its fire-resistance rating is fixed, the building must be situated so that the exterior wall's fire-resistance rating is in compliance.

1102.2 Foundation. The foundation system of relocated buildings shall comply with the *International Building Code* or the *International Residential Code* as applicable.

❖ The foundation for a relocated building is constructed at the building's new location. As new construction, the foundation, as well as the building's connection to it, is therefore required to comply with either the IBC or the IRC, as applicable.

1102.2.1 Connection to the foundation. The connection of the relocated building to the foundation shall comply with the *International Building Code* or the *International Residential Code* as applicable.

❖ See the commentary to Section 1102.2.

1102.3 Wind loads. Buildings shall comply with *International Building Code* or *International Residential Code* wind provisions as applicable.

Exceptions:

1. Detached one- and two-family dwellings and Group U occupancies where wind loads at the new location are not higher than those at the previous location.

2. Structural elements whose stress is not increased by more than 5 percent.

❖ When a building is moved, it can be subjected to different design wind loading at the new location. The structure must be checked, therefore, to verify that it satisfies the wind load criteria for a new structure using either the IBC or IRC as applicable. There are two exceptions to full compliance, one of which exempts detached one- and two-family dwellings as well as Group U occupancies if the design wind loads at the new location are no greater than those at the former location. Exception 2 allows additional wind loads that do not increase stresses in affected structural elements by more than 5 percent. Allowing overstresses of up to 5 percent in existing structural members has been a long-standing rule of thumb used by structural engineers.

1102.4 Seismic loads. Buildings shall comply with *International Building Code* or *International Residential Code* seismic provisions at the new location as applicable.

Exceptions:

1. Structures in Seismic Design Categories A and B and detached one- and two-family dwellings in Seismic Design Categories A, B, and C where the seismic loads at the new location are not higher than those at the previous location.

2. Stuctural elements whose stress is not increased by more than 5 percent.

❖ When a building is moved, the structure must satisfy the seismic load criteria for a new structure using either the IBC or IRC as applicable at the new location. There are two exceptions to full compliance. Exception 1 exempts any building classified as Seismic Design Category A or B in addition to detached one- and two-family dwellings classified as Seismic Design Category C, provided the design seismic loads at the new location are no greater than those at the former location. See the commentary to Section 407.1.1.1 for an explanation of seismic design category classification. While the design seismic loads are a function of several variables, with a relocated building that is not also undergoing a change in occupancy, the difference in the seismic load will merely be a function of two of

those variables, namely the mapped spectral accelerations (e.g., S_s) and the site soil coefficients (e.g., F_a). In other words, comparing the design spectral accelerations (e.g., $S_{DS} = 2 F_a S_s/3$) calculated at each site provides an indication of the difference in seismic loads. These parameters need to be determined using the IBC and are explained in the commentary to Section 407.1.1.1 (see determination of seismic design category under Section 1616.3 of the IBC). Exception 2 allows additional seismic loads that do not increase stresses in affected structural elements by more than 5 percent. Allowing overstresses of up to 5 percent in existing structural members has been a long-standing rule of thumb used by structural engineers.

1102.5 Snow loads. Structures shall comply with *International Building Code* or *International Residential Code* snow loads as applicable where snow loads at the new location are higher than those at the previous location.

Exception: Structural elements whose stress is not increased by more than 5 percent.

❖ When a building is moved to a location that requires an increased design snow load, the structure must be checked to verify that it satisfies the snow load requirements for a new structure using either the IBC or IRC, as applicable. The exception to full compliance allows additional snow loading that does not increase stresses in affected structural elements by more than 5 percent. Allowing overstresses of up to 5 percent in existing structural members has been a long-standing rule of thumb used by structural engineers.

1102.6 Flood hazard areas. If relocated or moved into a flood hazard area, structures shall comply with Section 1612 of the *International Building Code*.

❖ When a building is moved into or relocated within a flood hazard area, the structure must satisfy the flood hazard criteria for a new structure under the IBC.

1102.7 Required inspection and repairs. The code official shall be authorized to inspect, or to require approved professionals to inspect at the expense of the owner, the various structural parts of a relocated building to verify that structural components and connections have not sustained structural damage. Any repairs required by the code official as a result of such inspection shall be made prior to the final approval.

❖ In addition to the inspections required by Section 109, this section gives the code official the authority to inspect the structure of a relocated building to verify that structural components and connections have not been damaged. This is important because in the course of being moved, the building can be subjected to movements and displacements that were never anticipated in the original structural design. In the event more specific expertise is required, the code official may require such inspections to be made by an approved professional at the owner's expense.

Bibliography

The following resource materials are referenced in this chapter or are relevant to the subject matter addressed in this chapter.

ASCE 7-02, *Minimum Design Loads for Buildings and Other Structures*. Reston, VA: American Society of Civil Engineers, 2002.

IBC-03, *International Building Code*. Falls Church, VA: International Code Council, 2003.

IFC-03, *International Fire Code*. Falls Church, VA: International Code Council, 2003.

IPMC-03, *International Property Maintenance Code*. Falls Church, VA: International Code Council, 2003.

IRC-03, *International Residential Code*. Falls Church, VA: International Code Council, 2003.

CHAPTER 12

COMPLIANCE ALTERNATIVES

General Comments

Neither the designer nor the code official can physically inspect and evaluate every aspect of an existing building or structure, because many of its features may be concealed within the construction. It is therefore necessary to emphasize those items that can be evaluated. There are 19 critically important elements that can be quantified and evaluated to determine the level of safety for an existing building.

This type of analysis provides the designer and the code official with a rational basis for establishing the safety of an existing building or structure without having physical access to every part of the building, documentation of the original design or the construction history of a building.

Purpose

A large number of existing buildings and structures do not comply with the current code requirements for new construction. Although many of these buildings are potentially salvageable, rehabilitation is often cost prohibitive because they may not be able to comply with all the requirements for new construction. At the same time, it is necessary to regulate construction in existing buildings that undergo additions, alterations, renovations, extensive repairs or change of occupancy. Such activity represents an opportunity to ensure that new construction complies with the current code requirements and that existing conditions are maintained, at a minimum, to their current level of compliance or are improved as required. To accomplish this objective, and to make the rehabilitation process easier, this chapter allows for a controlled departure from full compliance with the technical codes, without compromising the minimum standards for fire prevention and life safety features of the rehabilitated building.

SECTION 1201
GENERAL

1201.1 Scope. The provisions of this chapter are intended to maintain or increase the current degree of public safety, health, and general welfare in existing buildings while permitting repair, alteration, addition, and change of occupancy without requiring full compliance with Chapters 4 through 10, except where compliance with other provisions of this code is specifically required in this chapter.

❖ This section states the scope of Chapter 12 and references alternative methods of code compliance for alteration, repair, addition and change of occupancy of existing structures. Chapter 12 also describes the responsibilities for maintenance, repairs, compliance with other codes and periodic testing.

1201.2 Applicability. Structures existing prior to [DATE TO BE INSERTED BY THE JURISDICTION]. Note: it is recommended that this date coincide with the effective date of building codes within the jurisdiction], in which there is work involving additions, alterations, or changes of occupancy shall be made to conform to the requirements of this chapter or the provisions of Chapters 4 through 10. The provisions of Sections 1201.2.1 through 1201.2.5 shall apply to existing occupancies that will continue to be, or are proposed to be, in Groups A, B, E, F, M, R, and S. These provisions shall not apply to buildings with occupancies in Group H or Group I.

❖ The adopting jurisdiction is to insert the desired date for applicability as indicated in this section; therefore, Chapter 12 applies only to structures existing prior to the established date. The date that construction was first regulated through a comprehensive building code in the jurisdiction is recommended because buildings predating any building regulation are often not equipped with the types of systems and features that modern codes require. Newer buildings are more likely to be in closer compliance with contemporary code requirements for new construction. These older buildings are assumed to face more difficulty in achieving a minimum level of life safety and are more likely to need the greater flexibility provisions of Chapter 12. The occupancies that qualify for the provisions of Chapter 12 are listed in Section 1201.2. Chapter 12 does not apply to Groups H (high hazard) and I (institutional).

1201.2.1 Change in occupancy. Where an existing building is changed to a new occupancy classification and this section is applicable, the provisions of this section for the new occupancy shall be used to determine compliance with this code.

❖ When a building undergoes a change of occupancy classification and Chapter 12 is applied, the evaluation method in Chapter 12 must be applied to the new occupancy for determining whether the existing building meets the compliance alternative in the code. This recognizes that it is the proposed conditions and relative hazards that will exist and must be determined to be acceptable. It is also consistent with how changes of occupancy are regulated in the absence of the Chapter 12 alternative.

1201.2.2 Partial change in occupancy. Where a portion of the building is changed to a new occupancy classification and that portion is separated from the remainder of the building with fire barrier wall assemblies having a fire-resistance rating as required by Table 302.3.3 of the *International Building Code* or Section R317 of the *International Residential Code* for the separate occupancies, or with approved compliance alternatives, the portion changed shall be made to conform to the provisions of this section.

Where a portion of the building is changed to a new occupancy classification and that portion is not separated from the remainder of the building with fire separation assemblies having a fire-resistance rating as required by Table 302.3.3 of the *International Building Code* or Section R317 of the *International Residential Code* for the separate occupancies, or with approved compliance alternatives, the provisions of this section which apply to each occupancy shall apply to the entire building. Where there are conflicting provisions, those requirements which secure the greater public safety shall apply to the entire building or structure.

❖ Where a portion of the building is changed to a new occupancy classification, the following options may be employed, dependent upon the fire separation of the portion of the building from the remainder of the existing building.

Where a portion of the building is changed to a new occupancy classification and that portion is separated from the remainder of the building by a fire barrier that complies with the requirements for new construction, the new occupancy portion must be evaluated with the existing or proposed building design to be in full compliance with the provisions of Chapter 12. The remainder of the existing building must also be evaluated in accordance with Chapter 12. The mandatory safety scores for the new occupancy portion of the building and the existing occupancy are obtained from those listed in Table 1201.8 and are incorporated in the building's final evaluation score (see Table 1201.7).

Where a portion of the building is changed to a new occupancy classification and that portion is not separated by a fire barrier that complies with the requirements for new construction, the provisions of Chapter 12 for each occupancy must apply to the entire building. The requirements offering the greater public safety are applied to the entire building. Any proposed or existing building attributes and modifications must be reviewed in light of these requirements. See the examples described in Sections 1201.6.16 and 1201.9.1 for mixed occupancies.

1201.2.3 Additions. Additions to existing buildings shall comply with the requirements of the *International Building Code*, *International Residential Code*, and this code for new construction. The combined height and area of the existing building and the new addition shall not exceed the height and area allowed by Chapter 5 of the *International Building Code*. Where a fire wall that complies with Section 705 of the *International Building Code* is provided between the addition and the existing building, the addition shall be considered a separate building.

❖ Additions effectively represent new construction, and it is therefore reasonable to require additions to comply with the code requirements for new construction. There are not assumed to be any existing conditions or other factors that would make full compliance impractical or difficult.

The requirements in this section are applied in the same way as those in Section 1201.1. This section is included in Chapter 12 so that provisions for additions are included for both optional methods of code compliance indicated in Section 1201.1 (see the commentary to Section 1201.1 for a discussion of these options).

The evaluation method in Chapter 12 may be used for an existing building that has been modified by an addition, provided that it meets the applicability requirements of Section 1201.2.

1201.2.4 Alterations and repairs. An existing building or portion thereof that does not comply with the requirements of this code for new construction shall not be altered or repaired in such a manner that results in the building being less safe or sanitary than such building is currently. If, in the alteration or repair, the current level of safety or sanitation is to be reduced, the portion altered or repaired shall conform to the requirements of Chapters 2 through 12 and Chapters 14 through 33 of the *International Building Code*.

❖ An existing building that is altered or repaired may be designed and evaluated in accordance with Chapter 12, provided it meets the applicability provisions of Section 1201.2.

When an existing building is altered or repaired, materials or methods consistent with the original construction must be used. This is described in Chapter 3 of the IEBC. The alteration or repair must not cause the building to be less safe or sanitary than it is currently; that is, before the alterations are undertaken. This is true even if the existing condition exceeds the minimum requirements of the code under which it was originally built. Should the alteration or repair cause a reduction in safety or sanitation, the resulting condition must meet the requirements of Chapters 2 through 12 and Chapters 14 through 33 of the *International Building Code®* (IBC®). This effectively allows existing conditions that exceed new construction requirements to be reduced to, but not below, new construction requirements.

1201.2.5 Accessibility requirements. All portions of the buildings proposed for change of occupancy shall conform to the accessibility provisions of Chapter 11 of the *International Building Code*.

❖ Any building or part of a building that has a change of occupancy must meet the requirements for accessibility in Chapter 11 of the IBC. This is substantially more restrictive than Chapter 8 of the code, which does not require full compliance with new construction accessibility requirements for a change of occupancy (see Section 812.5). Accessibility is generally independent of safety and sanitation requirements with which Chapter 12 is primarily concerned. The extent of compliance with accessibility requirements does not affect how the building performs with respect to the categories addressed in Sections 1201.6.1 through

1201.6.19. Therefore, compliance with Chapter 11 of the IBC could be considered an equivalent alternative, if approved by the code official, since that would establish the same level of accessibility that would be required by the code if one did not choose to follow the Chapter 12 alternative for a change of occupancy.

1201.3 Acceptance. For repairs, alterations, additions, and changes of occupancy to existing buildings that are evaluated in accordance with this section, compliance with this section shall be accepted by the code official.

❖ The *International Codes®* regulate safety in existing buildings by establishing the appropriate minimum levels of safety and sanitation deemed necessary for the safe occupancy of buildings. This is accomplished in several different portions of the family of *International Codes.* The *International Fire Code®* (IFC®) addresses fire safety issues in existing buildings and the *International Property Maintenance Code®* (IPMC®) addresses matters of health and sanitation in existing buildings and sites. Codes traditionally give broad authority to the code official to abate unusually hazardous conditions or operations that may be encountered in existing buildings. Although these fire safety, health and sanitation requirements are addressed in different codes, they together make up the package of requirements that represent the minimum conditions which any existing building must meet to be considered acceptable for occupancy under the *International Codes.* Chapter 12 works in conjunction with the other codes in the family of *International Codes* that regulate existing buildings by requiring compliance with those same minimum provisions. Without such references, Chapter 12 would be incomplete as an alternative tool for regulating existing buildings because it would otherwise allow conditions to exist that would not be permitted for any other existing building.

When an owner or designer of an existing building decides to apply Chapter 12 and complies with all the provisions of the chapter, including the applicability requirements in Section 1201.2, the code official must accept for review the proposed work or change of occupancy.

1201.3.1 Hazards. Where the code official determines that an unsafe condition exists as provided for in Section 115, such unsafe condition shall be abated in accordance with Section 115.

❖ When the code official finds an unsafe condition in the building that is not being corrected by the proposed work, he or she must order the abatement or correction of the unsafe condition or hazard just as would be ordered in an existing building that is not being renovated, as stipulated by Section 115. This section sets forth a required comprehensive performance objective of abating any condition that is unsafe, insanitary, illegal or of improper occupancy, egress deficient, fire hazardous, poorly maintained or otherwise dangerous to human life or the public welfare. Guidelines for abatement are provided in the code and in Chapter 1 of

both the IPMC (see Sections 108 and 110) and the IFC.

1201.3.2 Compliance with other codes. Buildings that are evaluated in accordance with this section shall comply with the *International Fire Code* and *International Property Maintenance Code.*

❖ This section requires an existing building that is subjected to the evaluation scoring process of Section 1201.6 to also comply with the IFC and IPMC. Those codes provide minimum requirements for health and safety that all existing buildings are expected to meet, regardless of whether there are any changes being made to the building or occupancy. Regardless of an existing building's final safety scores, the requirements of these referenced codes must be followed so occupants are safeguarded from the hazards addressed by those codes. This provision is similar to the requirements of Section 101.5.

1201.3.3 Compliance with flood hazard provisions. In flood hazard areas, buildings that are evaluated in accordance with this section shall comply with Section 1612 of the *International Building Code* if the work covered by this section constitutes substantial improvement.

❖ Regardless of how compliance with the provisions of this code are evaluated, buildings that are in flood hazard areas must meet the flood resistant provisions of the IBC if the proposed work is determined to be a substantial improvement.

1201.4 Investigation and evaluation. For proposed work covered by this chapter, the building owner shall cause the existing building to be investigated and evaluated in accordance with the provisions of Sections 1201.4 through 1201.9.

❖ This section and the subsequent subsections address what must be done by the owner or designer who elects to employ Chapter 12 for a proposed rehabilitation program and the corresponding action by the code official to assess the program objectively for approval or disapproval. The following actions must be taken:

- Fully investigate the building using both on-site inspections and research of all available building construction documents;
- Evaluate the building for conformance to Sections 1201.5 through 1201.9;
- Perform the required structural analysis of the structure;
- Determine whether the existing building, proposed work or change in occupancy complies with the accessibility provisions of Section 1201.2.5 (see commentary, Section 1201.5); and
- Submit to the code official documented results of the investigation and evaluation plus any proposed compliance alternatives.

Thus, the election and implementation of Chapter 12 by the designer or owner is part of a code compliance option that may be used for existing buildings that the code official has the responsibility to review and assess. A proper and satisfactorily prepared submission by the owner or designer offering qualitative and quantitative data requires review by the code official for approval. If the submission is disapproved, the code official must specifically cite any deficiencies and violations.

The owner is required to have an existing building and any proposed work therein investigated and evaluated for compliance with the 19 parameters of the evaluation process as specified in Section 1201.6.

1201.4.1 Structural analysis. The owner shall have a structural analysis of the existing building made to determine adequacy of structural systems for the proposed alteration, addition, or change of occupancy. The existing building shall be capable of supporting the minimum load requirements of Chapter 16 of the *International Building Code.*

❖ The owner is required to have a complete structural analysis of the building performed to ensure that it can support the required loads. This requires that all interior loads meet the minimum load requirements of Chapter 16 of the IBC. Existing and altered buildings must be shown to be capable of supporting the expected loading. Any existing exterior member not affected by either interior loading or additional exterior loading need not be evaluated, provided that the structural member has, over a period of time, proven its ability to withstand the forces that normally create stress. Loads imposed on existing structural members by alterations, additions or a change of occupancy must be shown to sustain the requirements of Chapter 16 of the IBC, as stated in Sections 1201.2 through 1201.2.2 of the code. This structural analysis provides the owner and the code official with reasonable assurance that the building is structurally safe.

1201.4.2 Submittal. The results of the investigation and evaluation as required in Section 1201.4, along with proposed compliance alternatives, shall be submitted to the code official.

❖ The results of the investigation, including structural analysis and evaluation, must be submitted to the code official. If alternative methods, materials or equivalent concepts are proposed, these must also be submitted to the code official for review and approval.

1201.4.3 Determination of compliance. The code official shall determine whether the existing building, with the proposed addition, alteration, or change of occupancy, complies with the provisions of this section in accordance with the evaluation process in Sections 1201.5 through 1201.9.

❖ When the results of the investigation and evaluation are submitted to the code official, he or she must determine whether the proposed work conforms to the provisions of Chapter 12 and whether the evaluation was performed in accordance with Sections 1201.5 through 1201.6.19.

1201.5 Evaluation. The evaluation shall be comprised of three categories: fire safety, means of egress, and general safety, as described in Sections 1201.5.1 through 1201.5.3.

❖ This section and the subsequent subsections address three general areas of safety to be evaluated: fire safety (FS), means of egress (ME) and general safety (GS). Section 1201.6 and the subsequent subsections address 19 safety parameters that reflect on those areas. Each of the 19 safety parameters indicated in Sections 1201.6.1 through 1201.6.19 must be carefully reviewed and assigned a numerical value that signifies the degree of safety influence on the three overall general safety categories. The three categories or areas are defined in Sections 1201.5.1 through 1201.5.3.

1201.5.1 Fire safety. Included within the fire safety category are the structural fire resistance, automatic fire detection, fire alarm, and fire-suppression system features of the facility.

❖ A partial list of the items used to evaluate fire safety in a building is given in this section.

1201.5.2 Means of egress. Included within the means of egress category are the configuration, characteristics, and support features for means of egress in the facility.

❖ The means of egress features that are evaluated by Chapter 12 fall into the general areas of configuration, characteristics and support features. The specific features include travel distance, dead ends, emergency lighting and exit capacity and number.

1201.5.3 General safety. Included within the general safety category are the fire safety parameters and the means-of-egress parameters.

❖ This category includes every item that is used in either the fire safety or means of egress evaluation.

1201.6 Evaluation process. The evaluation process specified herein shall be followed in its entirety to evaluate existing buildings. Table 1201.7 shall be utilized for tabulating the results of the evaluation. References to other sections of this code indicate that compliance with those sections is required in order to gain credit in the evaluation herein outlined. In applying this section to a building with mixed occupancies, where the separation between the mixed occupancies does not qualify for any category indicated in Section 1201.6.16, the score for each occupancy shall be determined, and the lower score determined for each section of the evaluation process shall apply to the entire building.

Where the separation between the mixed occupancies qualifies for any category indicated in Section 1201.6.16, the score for each occupancy shall apply to each portion of the building based on the occupancy of the space.

❖ This section is the key to understanding the entire evaluation process. The first sentence of this section clearly states that every one of the 19 safety parameters indicated in Sections 1201.6.1 through 1201.6.19 must be evaluated and nothing may be omitted. Although the evaluation process does not specifically evaluate all of the many issues that are regulated by the IBC for new construction, these 19 safety parameters have been determined to be the most critical factors related to the minimum degree of life safety and property protection needed in an existing building. The 19 safety parameters that must be evaluated are:

Building height (Section 1201.6.1)
Building area (Section 1201.6.2)
Compartmentation (Section 1201.6.3)
Tenant and dwelling unit separations (Section 1201.6.4)
Corridor walls (Section 1201.6.5)
Vertical openings (Section 1201.6.6)
HVAC systems (Section 1201.6.7)
Automatic fire detection (Section 1201.6.8)
Fire alarm systems (Section 1201.6.9)
Smoke control (Section 1201.6.10)
Means of egress capacity and number (Section 1201.6.11)
Dead ends (Section 1201.6.12)
Maximum exit access travel distance to an exit (Section 1201.6.13)
Elevator control (Section 1201.6.14)
Means of egress emergency lighting (Section 1201.6.15)
Mixed occupancies (Section 1201.6.16)
Automatic sprinklers (Section 1201.6.17)
Standpipes (Section 1201.6.18)
Incidental use (Section 1201.6.19)

The assigning of numerical values to each of the 19 safety parameters establishes a measurable quantity of what each of the parameters contributes to the overall safety of the building. Some evaluated parameters have a negative influence; others have a positive one. In total, the parameters may or may not result in an acceptable building score. The evaluation will determine whether the existing building has enough positive parameters to overcome the negative parameters, or will indicate the negative factors that must be overcome. Modifications can be made to any aspect of the building that will accrue sufficient additional positives points to achieve the mandatory safety scores in each of the three evaluation categories. In other words, if for example -10 points was scored because of the absence of a mixed occupancy separation under Section 1201.6.16.1 (assume for example purposes that the evaluation failed by 10 points in the fire safety category), it does not necessarily mean that a fire barrier must be constructed to separate the mixed occupancies. While that would be an acceptable solution, modifications could be made that accrue at least 10 additional points in any one or more of the other parameters in the fire safety category.

After the 19 safety parameters have been evaluated and assigned a numerical value, the values are entered in Table 1201.7. The values must be tabulated to obtain the building score for each of the three evaluation categories.

Some of the 19 safety parameters listed require mandatory compliance with other sections of the code and establish the foundation for determining a proper

evaluation of those parameters, regardless of whether the result is positive or negative. This involves a coordination of mandatory basic requirements with the respective existing building conditions to arrive at the numerical evaluations prescribed in Sections 1201.6.1 through 1201.6.19.

Section 1201.6.16 also addresses how mixed occupancies are handled in the evaluation. To apply this section, it is necessary to understand Section 302 of the IBC. When mixed occupancies in an existing building are not separated by fire barriers or fire walls complying with the requirements for new construction, the entire evaluation must be based on the occupancy with the most restrictive requirements. The evaluation process considers the score for the various occupancies and applies the lowest score to the entire building. When the mixed occupancies are separated by fire barriers in compliance with Section 302.3.3 of the IBC, they are to be evaluated separately and the score for each occupancy will apply to each portion based on its occupancy classification. For instance, if there are four different occupancies in the building, the values must be computed for each of the four occupancies. Each occupancy is required to meet the applicable mandatory building score for its occupancy classification. When mixed occupancies are separated by fire walls complying with the requirements of Section 705 of the IBC, separate buildings are created and must be evaluated separately for all 19 safety parameters. See the commentary to Sections 1201.2.2 and 1201.6.16 for a further discussion of the application of the evaluation procedure for mixed occupancies.

1201.6.1 Building height. The value for building height shall be the lesser value determined by the formula in Section 1201.6.1.1. Chapter 5 of the *International Building Code*, including allowable increases due to automatic sprinklers as provided for in Section 504.2, shall be used to determine the allowable height of the building. Subtract the actual building height from the allowable height and divide by $12^{1}/_{2}$ feet (3810 mm). Enter the height value and its sign (positive or negative) in Table 1201.7 under Safety Parameter 1201.6.1, Building Height, for fire safety, means of egress, and general safety. The maximum score for a building shall be 10.

❖ As a starting point for the actual evaluation process, this section and Section 1201.6.1.1 define in detail how to perform the building height evaluation. The exact values are to be computed and compared against the mandatory safety scores. The values are not to be rounded; for calculation purposes, two decimal places would be appropriate. It is not necessary to build in any inaccuracy that occurs through the rounding process.

The maximum number of points that can be scored for a building, regardless of the building's allowable height as compared to its actual height, is 10. Building height, as well as building area, which is covered in Section 1201.6.2, determines the type of construction classification. Limiting the number positive points that can be scored for height limits the weight that the type of classification could otherwise have on the overall evaluation. A low-rise building that is built of a high type of construction could accrue a disproportionately high number of points for height as compared to the number of points that are counted for other safety parameters. The intent is to avoid a situation in which many deficiencies in other critical safety parameters can be overcome simply by a high type of construction classification.

1201.6.1.1 Height formula. The following formulas shall be used in computing the building height value.

$$\text{Height value, feet} = \frac{(AH) - (EBH) \times CF}{12.5}$$

$$\text{Height value, stories} = (AS - EBS) \times CF$$

where: **(Equation 12-1)**

AH = Allowable height in feet (mm) from Table 503 of the *International Building Code.*

EBH = Existing building height in feet (mm).

AS = Allowable height in stories from Table 503 of the *International Building Code.*

EBS = Existing building height in stories.

CF = 1 if *(AH) – (EBH)* is positive.

CF = Construction type factor shown in Table 1201.6.6(2) if *(AH) – (EBH)* is negative.

Note: Where mixed occupancies are separated and individually evaluated as indicated in Section 1201.6, the values *AH, AS, EBH,* and *EBS* shall be based on the height of the fire area of the occupancy being evaluated.

❖ Two height formulas for calculating the score to be entered in Table 1201.7 are given in this section. One formula determines a height value based on the height in feet and the other formula determines a height value in number of stories. Both are based on the height limitations from Table 503 of the IBC and the height of the existing building. The denominator in the formula for height in feet, 12.5, represents an average story height in feet. Both formulas must be calculated, and the lesser of the two calculated values is the value that must be used in the evaluation.

The actual story height and the overall existing building height are to be directly compared to Table 503. Table 503 serves as a datum level that allows for establishing a numerical height value for the existing building by comparing its actual height and its type of construction as represented by a construction factor (*CF*). To determine the *CF*, if the existing building's actual height in feet is less than or equal to the allowable height of Table 503, then the *CF* value is 1 (no negative or positive multiplier is factored into the calculation). If the actual height in feet exceeds the Table 503 allowable height, the building is not in compliance with Table 503. Consequently, the *CF*, which is determined from Table 1201.6.6(2), must be factored into both equations and increases the negative points accrued be-

cause of the noncompliant building height condition. The penalty for noncompliant height is proportionally greater for lower types of construction. This can be readily seen in Table 1201.6.6(2) where, as the type of construction classification goes down, the *CF* value increases.

When a building is not in compliance with Table 503, it is considered less safe than a building that does comply with Table 503. As a result, deficiency points are assessed. Additional safeguards must be provided to compensate for this condition.

Example 1:

A six-story building of Type IB construction is 60 feet (18 288 mm) tall. The building is not sprinklered. It contains a Group B business occupancy.

The allowable height from Table 503 is 11 stories, 160 feet (48 800 mm).

AH = 160 feet

AS = 11 stories

EBH = 60 feet

EBS = 6 stories

CF = 1 from Table 1201.6.6(2) (because 160 - 60 is a positive number)

$$\text{Height value, feet} = \frac{(160 - 60) \times 1}{12.5} = 8$$

$$\text{Height value, stories} = (11 - 6) \times 1 = 5$$

The governing building height value is 5, because it is the lesser of the two height values.

Example 2:

A seven-story building of Type IIIB construction is 80 feet (24 400 mm) tall. The building is sprinklered. It contains a Group M mercantile occupancy.

The allowable height from Table 503 is four stories, 55 feet (16 775 mm).

AH = 55 + 20 (for sprinklers) = 75 feet

AS = 4 + 1 (for sprinklers) = 5 stories

EBS = 7 stories

EBH = 80 feet

CF = 3.5 from Table 1201.6.6(2) (because 75 - 80 is a negative number)

$$\text{Height value, feet} = \frac{(75 - 80) \times 3.5}{12.5} = -1.4$$

$$\text{Height value, stories} = (5 - 7) \times 3.5 = -7$$

The governing building height value is -7, because it is the lesser of the two height values.

The above values determine the entries for the "Summary Sheet—Building Score" of Table 1201.7 for the "Safety Parameter of Building Height" column under line 1201.6.1.

In Example 1, the height value of 5 is entered into the columns for fire safety (FS), means of egress (ME) and general safety (GS).

In Example 2, the height value of -7 is entered into the columns for fire safety (FS), means of egress (ME) and general safety (GS).

The assessed height value parameter is only one of the 19 safety parameters that need to be evaluated. Example 1 shows a positive contribution; however, this may not be enough to result in a sufficient overall building score. Similarly, the negative value in Example 2 may not in itself result in an overall insufficient building score.

1201.6.2 Building area. The value for building area shall be determined by the formula in Section 1201.6.2.2. Section 503 of the *International Building Code* and the formula in Section 1201.6.2.1 shall be used to determine the allowable area of the building. The allowable area shall be the lesser value calculated by Equations 12-2 and 12-3. This shall include any allowable increases due to open perimeter and automatic sprinklers as provided for in Section 506 of the *International Building Code*. Subtract the actual building area from the allowable area and divide by 1,200 square feet (112 m²). Enter the area value and its sign (positive or negative) in Table 1201.7 under Safety Parameter 1201.6.2, Building Area, for fire safety, means of egress, and general safety. In determining the area value, the maximum permitted positive value for area is 50 percent of the fire safety score as listed in Table 1201.8, Mandatory Safety Scores.

❖ Consideration of building area is similar in principle to that for building height discussed above. The contribution of building area to the overall evaluation is limited in the same manner and for the same reason as discussed in Section 1201.6.1. It is not uncommon for the allowable area of existing buildings to greatly exceed the actual area, which would make the area value calculated in this section disproportionately high. As such, the maximum area value that can be scored is 50 percent of the mandatory fire safety score listed in Table 1201.8 (see commentary, Section 1201.6.2.2). This maximum score is applicable and equal for the three building score categories—fire safety (FS), means of egress (ME) and general safety (GS).

1201.6.2.1 Allowable area formula. The following formula shall be used in computing allowable area:

$$A_a = \frac{(100 + I_f + I_s) \times A_t}{100}$$

(Equation 12-2)

$A_{max.} = 3 \times A_a$, as calculated in accordance with Section 503.3 of the *International Building Code*.

$$A_{a,max.} = \frac{A_{max}}{\text{Number of stories}}$$

(Equation 12-3)

where:

A_a = Allowable area per floor.

I_s = Area increase due to sprinkler protection, percent as calculated in accordance with Section 506.3 of the *International Building Code*.

I_f = Area increase due to frontage, percent as calculated in accordance with Section 506.2 of the *International Building Code*.

A_t = Tabular area per floor in accordance with Table 503 of the *International Building Code*, square feet (m²).

A_{max} = Total area of the entire building.

$A_{a,max.}$ = Allowable area per floor based on the limitations of Section 506.4 of the *International Building Code*.

❖ The formula used to calculate the allowable area of the building is a direct correlation of the area increases and reductions allowed for new buildings in Section 506 of the IBC. The use of these increases and reductions is addressed in Chapter 5 of the IBC.

1201.6.2.2 Area formula. The following formula shall be used in computing the area value. Determine the area value for each occupancy fire area on a floor-by-floor basis. For each occupancy, choose the minimum area value of the set of values obtained for the particular occupancy.

(Equation 12-4)

$$\text{Area value } i = \frac{\text{Allowable area}_i}{1200 \text{ square feet}} \left[1 - \left(\frac{\text{Actual area}_i}{\text{Allowable area}_i} + \dots + \frac{\text{Actual area}_n}{\text{Allowable area}_n} \right) \right]$$

where:

i = Value for an individual separated occupancy on a floor.

n = Number of separated occupancies on a floor.

❖ The area formula provides a numerical value for the actual building area to be entered into Table 1201.7 under the "Safety Parameter" column next to "1201.6.2 Building Area." If the area of the existing building is less than the allowable area, it is considered to be safer and the building receives a positive score. If the area of the existing building is larger than the allowable area, it represents a condition that is judged to be less safe and the building receives a negative score.

To use the formula for existing buildings that contain mixed occupancies, an area value must be determined for each story of a building. If a building contains just one occupancy or just one occupancy on a particular floor separated from other floors, the formula simply reduces to the allowable area minus the actual area divided by the constant of 1,200. The formula is also applicable to a story containing several separated occupancies. In such a situation, the formula requires each actual area to be divided by its respective allowable area. The resulting fractions are added together, subtracted from the constant of 1 and multiplied by the ratio of the allowable area of the particular use divided by the constant 1,200. This method of determining area values for sev-

FIGURE 1201.6.2.2(1)   COMPLIANCE ALTERNATIVES

eral occupancies that are separated on a story is directly comparable to the unity formula in Section 302.3.2 of the IBC.

Example 1: Figure 1201.6.2.2(1) illustrates a Group B building of Type IIA construction.

The building is five stories in height, unsprinklered.

The overall building area is 200 feet (60 960 mm) × 200 feet (60 960 mm).

$$\text{Area value} = \frac{75,000}{1,200}\left[1 - \left(\frac{40,000}{75,000}\right)\right]$$

$$= 62.5\,[1 - 0.533]$$

$$= 29.2$$

Allowable area   = 75,000 square feet

Actual area   = 40,000 square feet

In Table 1201.7, enter the following values under the "Safety Parameter" column next to "1201.6.2 Building Area":

Fire safety (FS)   =   12

Means of egress (ME)   =   12

General safety (GS)   =   12

These values are equal to 50 percent of the value in Table 1201.8 for the mandatory fire safety score (MFS) of 30 for Group B occupancy. This is the maximum value credit permitted for the building area parameter of Section 1201.6.2, even though the area value has been computed as 29.2 in Section 1201.6.2.2.

The positive values entered in Table 1201.7 under the "Safety Parameter" column next to "1201.6.2 Building Area" (FS = 12, ME = 12 and GS = 12), represents the evaluation for only one of the 19 safety parameters that must be assessed in arriving at the overall building score evaluation of Table 1201.7. As previously described for the building height parameter values under line 1201.6.1 of Table 1201.7, a single parameter of the total 19 safety parameters to be assessed does not, in itself, determine the ultimate acceptable building score.

Example 2: Figure 1201.6.2.2(2) illustrates a separated mixed occupancy building (Groups B, M and S-1) of Type IIIB construction.

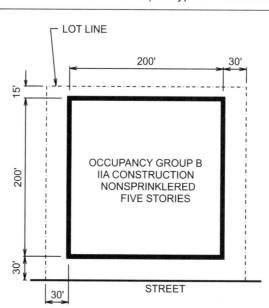

LOT LINE

200'   30'

15'

200'

OCCUPANCY GROUP B
IIA CONSTRUCTION
NONSPRINKLERED
FIVE STORIES

30'

30'   STREET

ACTUAL AREA = 200' × 200' = 40,000 SQ.FT.

SP = 0 (FROM SECTION 506.3)
OP = 100% (FROM SECTION 506.2)
ALLOWABLE AREA IN TABLE 503 = 37,500 SQ.FT.

$$AA = \frac{(0 + 100 + 100) \times 37,500}{100} = \frac{200 \times 37,500}{100} = 75,000 \text{ SQ.FT.}$$

For SI: 1 foot = 304.8 mm, 1 square foot = 0.0929 m².

Figure 1201.6.2.2(1)
ALLOWABLE AREA—SINGLE OCCUPANCY BUILDING

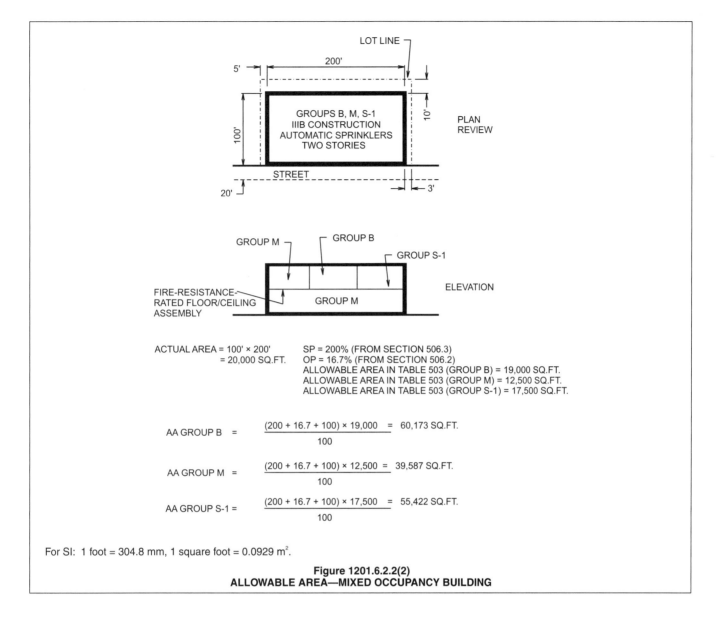

ACTUAL AREA = 100' × 200'
 = 20,000 SQ.FT.

SP = 200% (FROM SECTION 506.3)
OP = 16.7% (FROM SECTION 506.2)
ALLOWABLE AREA IN TABLE 503 (GROUP B) = 19,000 SQ.FT.
ALLOWABLE AREA IN TABLE 503 (GROUP M) = 12,500 SQ.FT.
ALLOWABLE AREA IN TABLE 503 (GROUP S-1) = 17,500 SQ.FT.

$$\text{AA GROUP B} = \frac{(200 + 16.7 + 100) \times 19,000}{100} = 60,173 \text{ SQ.FT.}$$

$$\text{AA GROUP M} = \frac{(200 + 16.7 + 100) \times 12,500}{100} = 39,587 \text{ SQ.FT.}$$

$$\text{AA GROUP S-1} = \frac{(200 + 16.7 + 100) \times 17,500}{100} = 55,422 \text{ SQ.FT.}$$

For SI: 1 foot = 304.8 mm, 1 square foot = 0.0929 m².

Figure 1201.6.2.2(2)
ALLOWABLE AREA—MIXED OCCUPANCY BUILDING

The building is two stories in height, fully sprinklered.

The overall building area is 100 feet (30 480 mm) by 200 feet (60 960 mm)

Actual area for Group M on first story = 20,000 square feet

Actual area for Group M on second story = 4,400 square feet

Actual area for Group B on second story = 10,000 square feet

Actual area for Group S-1 on second story = 5,600 square feet

SP = 200% (from Section 506.3)

OP = 16.7% (from Section 506.2)

Tabular areas from Table 503:

Group B = 19,000 square feet

Group M = 12,500 square feet

Group S-1 = 17,500 square feet

Allowable Areas:

$$AA = \frac{(200 + 16.7 + 100) \times \text{Tabular Area}}{100}$$

AA for Group B = 60,173 square feet

AA for Group M = 39,587 square feet

AA for Group S-1 = 55,422 square feet

Area Values:

For Group M, first story:

$$\text{Area value} = \frac{39,587}{1,200}\left[1 - \left(\frac{20,000}{39,587}\right)\right]$$

$$= 33\left[1 - 0.505\right]$$

$$= 16.3$$

For Group B, second story:

$$\text{Area value} = \frac{60,173}{1,200}\left[1-\left(\frac{10,000}{60,173}+\frac{4,400}{39,587}+\frac{5,600}{55,422}\right)\right]$$
$$= 50\left[1-0.505\right]$$
$$= 50\left[1-0.378\right]$$
$$\text{Area value} = 31$$

For Group M, second story:

$$\text{Area value} = \frac{39,587}{1,200}\left[1-0.378\right]$$
$$= 32.9\left[1-0.378\right]$$
$$\text{Area value} = 20.4$$

For Group S-1, second story:

$$\text{Area value} = \frac{55,422}{1,200}\left[1-0.378\right]$$
$$= 46\left[1-0.378\right]$$
$$\text{Area value} = 28.6$$

Since these mixed occupancies are being separated so that one of the categories indicated in Section 1201.6.16 is applicable, a separate score must be computed for each occupancy.

In this example, the area value for the Group B occupancy is calculated to be 31, but the maximum area value permitted is 50 percent of the MFS score listed in Table 1201.8. For a Group B occupancy, the MFS score is 30; therefore, the maximum positive value that can be entered into Table 1201.7 is 15.

When an occupancy is located in more than one story, a separate area value must be calculated for each story. For input into Table 1201.7, the area value to be used is the lesser of all the individual area values for that occupancy group, but not greater than 50 percent of the MFS score. In Figure 1201.6.2.2(2), there is an area classified as a Group M occupancy on both the first and second stories. The area value for the Group M occupancy on the first story is 16.3, and the area value for the second story is 20.4. For a Group M occupancy, the MFS score is 23; therefore, the maximum positive value that can be entered into Table 1201.7 is 11.5.

The area value for the Group S-1 occupancy portion of the building is 28.6. This value, just as the values for the other occupancies in this example, exceeds the 50 percent maximum permitted by the MFS score. For a Group S-1 occupancy, the MFS score is 19; therefore, the maximum positive value that can be entered into Table 1201.7 is 19.

1201.6.3 Compartmentation. Evaluate the compartments created by fire barrier walls which comply with Sections 1201.6.3.1 and 1201.6.3.2 and which are exclusive of the wall elements considered under Sections 1201.6.4 and 1201.6.5. Conforming compartments shall be figured as the net area and do not include shafts, chases, stairways, walls, or columns. Using Table 1201.6.3, determine the appropriate compartmentation value (CV) and enter that value into Table 1201.7 under Safety Parameter 1201.6.3, Compartmentation, for fire safety, means of egress, and general safety.

❖ This section establishes and evaluates the compartments contained within an existing building by the effectiveness of the enclosing fire barrier walls and fire-resistant floor/ceiling assemblies. Larger compartments are considered to be a greater safety risk than smaller compartments because the entire compartment is assumed to be involved when a fire incident occurs and, therefore, a single fire incident affects a greater portion of the building at one time.

Fire barriers must comply with Sections 1201.6.3.1 and 1201.6.3.2. Fire barriers are exclusive of the other separations or enclosures that are evaluated in Sections 1201.6.4 and 1201.6.5.

The evaluation of the compartments contained within an existing building is a linear function allowing interpolation between the various categories. This approach allows the compartmentation value to increase or decrease consistent with the actual changes in compartment sizes. Such an adjustment removes the previously built-in bias against smaller-sized buildings. Higher compartmentation values are assigned to buildings with smaller compartments.

TABLE 1201.6.3
COMPARTMENTATION VALUES

OCCUPANCY	CATEGORIES				
	a Compartment size equal to or greater than 15,000 square feet	b Compartment size of 10,000 square feet	c Compartment Size of 7,500 square feet	d Compartment size of 5,000 square feet	e Compartment size of 2,500 square feet or less
A-1, A-3	0	6	10	14	18
A-1	0	4	10	14	18
A-4, B, E, S-2	0	5	10	15	20
F, M, R, S-1	0	4	10	16	22

For SI: 1 square foot = 0.0929 m^2.

1201.6.3.1 Wall construction. A wall used to create separate compartments shall be a fire barrier conforming to Section 706 of the *International Building Code* with a fire-resistance rating of not less than 2 hours. Where the building is not divided into more than one compartment, the compartment size shall be taken as the total floor area on all floors. Where there is more than one compartment within a story, each compartmented area on such story shall be provided with a horizontal exit conforming to Section 1021 of the *International Building Code*. The fire door serving as the horizontal exit between compartments shall be so installed, fitted, and gasketed that such fire door will provide a substantial barrier to the passage of smoke.

❖ This section states that the walls determining the boundary of the compartment need to have fire barrier ratings of not less than 2 hours. These assemblies must be constructed in accordance with Section 706 of the IBC. For an existing building, this may need to be evaluated by both analyzing available plans and on-site investigations with a professional engineer's determination of the required 2-hour fire-resistance rating. If the fire-resistance rating is less than 2 hours or cannot be reasonably determined, such walls should not be considered as creating a compartment. The entire story of an existing building must then be considered the compartment. If 2-hour-rated floor/ceiling assemblies are not present, the compartment size becomes the total area on all stories.

Opening protection and continuity requirements of Section 706 of the IBC must be followed to maintain the integrity of the fire-resistance-rated wall assemblies, and thus the compartments. Horizontal exits and their fire doors must comply with Section 1021 of the IBC.

The evaluation of an existing door condition requires an investigation of available engineering data and on-site inspections. Any uncertainty as to the performance of such doors may result in the openings being considered unprotected, which results in a larger assumed compartment area or requires modification of the door to meet the current requirements of Section 1021.3 of the IBC.

1201.6.3.2 Floor/ceiling construction. A floor/ceiling assembly used to create compartments shall conform to Section 711 of the *International Building Code* and shall have a fire-resistance rating of not less than 2 hours.

❖ The building features that provide the horizontal boundaries of the compartment need to provide effective fire-resistive integrity between floors. The existing floor/ceiling assemblies must be rated for 2 hours and be tight against exterior walls. Penetrations in the floor/ceiling assemblies must be protected in accordance with Section 711 of the IBC to maintain their fire-resistant integrity. The floor/ceiling assemblies must conform to all of the requirements of Section 711 of the IBC to create the level of compartmentation required for this evaluation parameter.

1201.6.4 Tenant and dwelling unit separations. Evaluate the fire-resistance rating of floors and walls separating tenants, including dwelling units, and not evaluated under Sections

1201.6.3 and 1201.6.5. Under the categories and occupancies in Table 1201.6.4, determine the appropriate value and enter that value in Table 1201.7 under Safety Parameter 1201.6.4, Tenant and Dwelling Unit Separation, for fire safety, means of egress, and general safety.

❖ This parameter is used to evaluate partitions in an existing building other than those used for the creation of compartments in Section 1201.6.3 or the enclosure of corridors in Section 1201.6.5. This section examines the level of separation between tenant spaces and dwelling units. The listed categories specifically reference Sections 706, 708 and 710 of the IBC not only for fire-resistance ratings but also for continuity and opening protection purposes. Further credit is provided for existing buildings that have a 2-hour-rated separation between adjacent tenant spaces or dwelling units, which exceeds the requirement for new construction.

TABLE 1201.6.4
SEPARATION VALUES

OCCUPANCY	CATEGORIES				
	a	b	c	d	e
A-1	0	0	0	0	1
A-2	-5	-3	0	1	3
R	-4	-2	0	2	4
A-3, A-4, B, E, F, M, S-1	-4	-3	0	2	4
S-2	-5	-2	0	2	4

❖ Table 1201.6.4 provides values for tenant space and dwelling unit separations. The rationale for considering a nonfire-resistance-rated or incomplete separation as a safety deficiency is that even though the assembly may have some limited fire-resistant capability, it cannot be assumed to provide the level of fire performance expected of a fully complying assembly. Buildings containing Group R occupancies have separation values based on the assumption that dwelling unit separations are more critical than tenant separations in other occupancies.

1201.6.4.1 Categories. The categories for tenant and dwelling unit separations are:

1. Category a—No fire partitions; incomplete fire partitions; no doors; doors not self-closing or automatic closing.

2. Category b—Fire partitions or floor assembly less than 1-hour fire-resistance rating or not constructed in accordance with Sections 708 or 711 of the *International Building Code*, respectively.

3. Category c—Fire partitions with 1-hour or greater fire-resistance rating constructed in accordance with Section 708 of the *International Building Code* and floor assemblies with 1-hour but less than 2-hour fire-resistance rating constructed in accordance with Section 711 of the *International Building Code* or with only one tenant within the fire area.

FIGURE 1201.6.4.1(1)

COMPLIANCE ALTERNATIVES

4. Category d—Fire barriers with 1-hour but less than 2-hour fire-resistance rating constructed in accordance with Section 706 of the *International Building Code* and floor assemblies with 2-hour or greater fire-resistance rating constructed in accordance with Section 711 of the *International Building Code.*

5. Category e—Fire barriers and floor assemblies with 2-hour or greater fire-resistance rating and constructed in accordance with Sections 706 and 711 of the *International Building Code,* respectively.

❖ Tenant space and dwelling unit separations are categorized by the partitions being evaluated. The values of each category are listed in Table 1201.6.4 by occupancy classifications. Typical illustrations of the types of partitions are shown in Figures 1201.6.4.1(1) and 1201.6.4.1(2). The listed categories provide a graduated level of separation when compared to new construction requirements.

Category a addresses the situation where there is no separation or there are gaps in the separation provided between tenant spaces or dwelling units.

Category b accounts for those tenant spaces or dwelling units that are separated from one another with less than a 1-hour rating.

Category c represents what is required for new construction for tenant space and dwelling unit separation. An existing building meeting this level of compliance gains no benefit or penalty, therefore making the separation value zero.

Category d is given additional credit because the tenant space or dwelling unit separation has a fire-resistance rating that marginally exceeds the minimum required for new construction. Additionally, the walls are required to meet the fire barrier requirements of Section 706 of the IBC.

Category e provides increased credit for existing buildings that have walls and floor/ceiling assemblies with fire-resistance ratings exceeding the new construction requirement by at least 1 hour. This increased level of fire-resistant separation in an existing building is credited accordingly with high separation values.

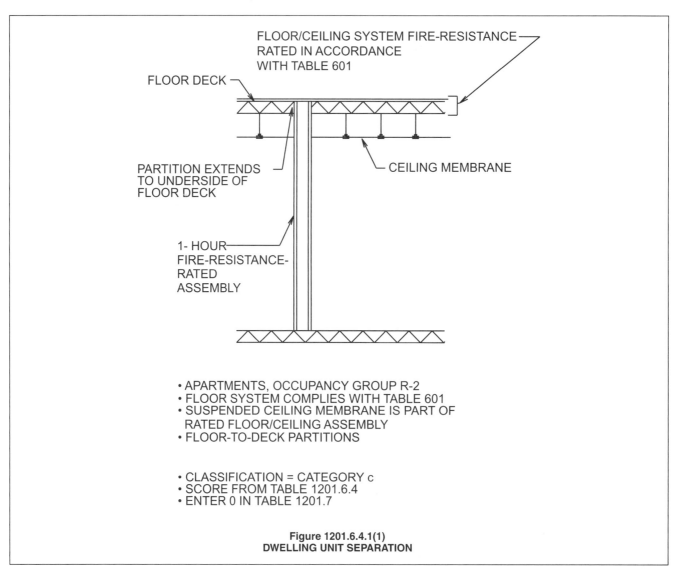

- APARTMENTS, OCCUPANCY GROUP R-2
- FLOOR SYSTEM COMPLIES WITH TABLE 601
- SUSPENDED CEILING MEMBRANE IS PART OF RATED FLOOR/CEILING ASSEMBLY
- FLOOR-TO-DECK PARTITIONS

- CLASSIFICATION = CATEGORY c
- SCORE FROM TABLE 1201.6.4
- ENTER 0 IN TABLE 1201.7

Figure 1201.6.4.1(1)
DWELLING UNIT SEPARATION

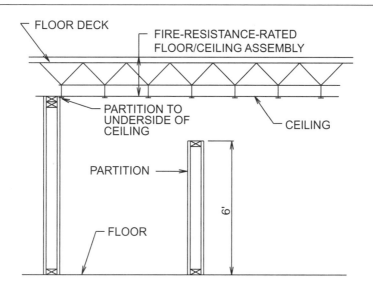

- FOOD COURT RESTAURANTS, OCCUPANCY GROUP A-3
- SOME PARTITIONS EXTEND TO CEILING OF FIRE-RESISTANCE-RATED
  FLOOR/CEILING ASSEMBLY
- NO CLOSERS ON DOORS BETWEEN ADJACENT TENANT SPACES
- SOME 6' PRIVACY PARTITIONS (PARTIAL PARTITIONS)

- CLASSIFICATION = CATEGORY a
- SCORE FROM TABLE 1201.6.4
- ENTER -4 IN TABLE 1201.7

For SI: 1 foot = 304.8 mm.

Figure 1201.6.4.1(2)
TENANT SEPARATION

1201.6.5 Corridor walls. Evaluate the fire-resistance rating and degree of completeness of walls which create corridors serving the floor and that are constructed in accordance with Section 1013 of the *International Building Code*. This evaluation shall not include the wall elements considered under Sections 1201.6.3 and 1201.6.4. Under the categories and groups in Table 1201.6.5, determine the appropriate value and enter that value into Table 1201.7 under Safety Parameter 1201.6.5, Corridor Walls, for fire safety, means of egress, and general safety.

❖ Corridor walls are evaluated as fire partitions possessing an adequate fire-resistance rating and completeness to restrict the spread of fire into the corridor. Various categories require compliance with prescribed requirements in Sections 708.4, 715 and 1016.1 of the IBC. Existing corridor walls require investigation and analysis to determine equivalency to code requirements. Figures 1201.6.5(1) and 1201.6.5(2) illustrate various corridor wall values. Corridor walls contrast with the compartmentation and tenant dwelling unit separations of Sections 1201.6.3 and 1201.6.4 by requiring an appropriate fire-resistance rating and continuity (see commentary, Section 708.4 of the IBC). The corridor wall evaluations do not include partitions required to establish a compartment (see Section

1201.6.3) or tenant and dwelling unit separations (see Section 1201.6.4). If a corridor wall serves as more than one of these elements, for example, where it is both a corridor wall and it defines a compartment under Section 1201.6.3 and a corridor under this subsection, it may be evaluated under either one parameter or the other at the designer's option, but not both.

TABLE 1201.6.5
CORRIDOR WALL VALUES

OCCUPANCY	CATEGORIES			
	a	b	c[a]	d[a]
A-1	-10	-4	0	2
A-2	-30	-12	0	2
A-3, F, M, R, S-2	-7	-3	0	2
A-4, B, E, S-2	-5	-2	0	2

a. Corridors not providing at least one-half the travel distance for all occupants on a floor shall use Category b.

❖ Table 1201.6.5 assigns values to the various occupancies based on the fire-resistance rating and continuity of the corridor wall construction. Since corridors are

FIGURE 1201.6.5(1) – 1201.6.5.1   COMPLIANCE ALTERNATIVES

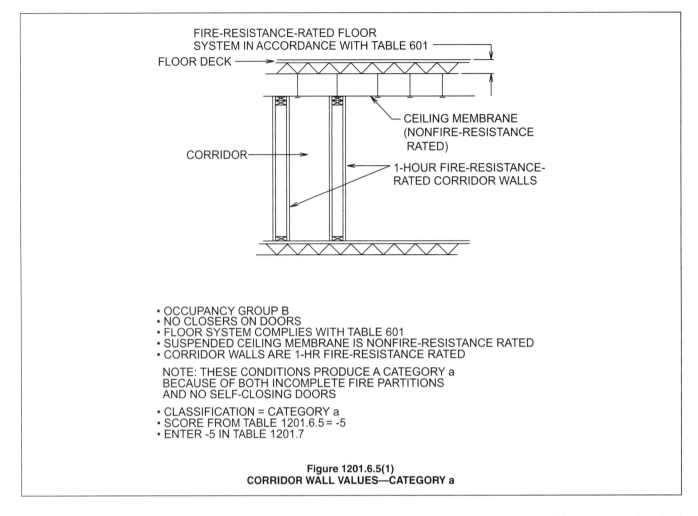

· OCCUPANCY GROUP B
· NO CLOSERS ON DOORS
· FLOOR SYSTEM COMPLIES WITH TABLE 601
· SUSPENDED CEILING MEMBRANE IS NONFIRE-RESISTANCE RATED
· CORRIDOR WALLS ARE 1-HR FIRE-RESISTANCE RATED

NOTE: THESE CONDITIONS PRODUCE A CATEGORY a
BECAUSE OF BOTH INCOMPLETE FIRE PARTITIONS
AND NO SELF-CLOSING DOORS

· CLASSIFICATION = CATEGORY a
· SCORE FROM TABLE 1201.6.5 = -5
· ENTER -5 IN TABLE 1201.7

Figure 1201.6.5(1)
CORRIDOR WALL VALUES—CATEGORY a

enclosed (confined) spaces subject to the rapid buildup of smoke and heat, a degree of protection of the exit access route is necessary for occupants. The table reflects this emphasis through substantial negative scores for buildings without properly enclosed corridors. Both Section 1201.6.5 and this table place an emphasis on corridor walls that differs from that expressed elsewhere in the code. For example, Chapter 10 does not require corridors to be provided except in Group I-2 occupancies. The emphasis here is that when corridors are already present, they must provide a minimum level of protection because of the confined path of travel that they create. The table is an assessment of the relative risk represented by the corridor.

Note a to the table further controls the application of Categories c and d to existing buildings with certain means of egress arrangements. Although an existing building may have corridors with significant fire-resistance ratings and protected openings, little credit can be granted when the building occupants are protected for only a short period of time. Very short corridors in large floor plans or corridors located in just one tenant space of a floor plan provide a benefit to only a portion of the occupant load for a small portion of the overall exit access travel and, thus, do not accrue any positive points. Unless rated corridors are available for all occupants of

that particular floor or they provide a protected path of travel for at least one-half of the occupants' overall travel length, negative corridor wall values are assigned. In existing buildings with very short or limited-use corridors, the code user is directed to use Category b, regardless of the corridors' fire-resistance ratings and opening protectives.

1201.6.5.1 Categories. The categories for corridor walls are:

1. Category a—No fire partitions; incomplete fire partitions; no doors; or doors not self-closing.
2. Category b—Less than 1-hour fire-resistance rating or not constructed in accordance with Section 708.4 of the *International Building Code*.
3. Category c—1-hour to less than 2-hour fire-resistance rating, with doors conforming to Section 715 of the *International Building Code* or without corridors as permitted by Section 1013 of the *International Building Code*.
4. Category d—2-hour or greater fire-resistance rating, with doors conforming to Section 715 of the *International Building Code*.

❖ Corridor walls are categorized by the partitions being evaluated. The values of each category are listed in Table 1201.6.5 by occupancy classification.

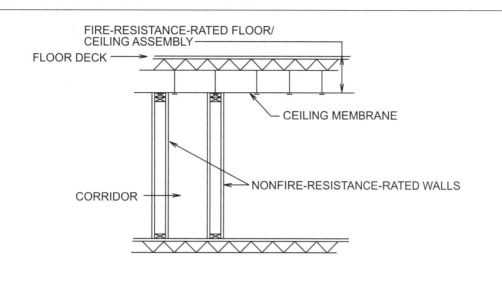

- OCCUPANCY GROUP B
- CLOSERS ON DOORS
- FIRE-RESISTANCE-RATED FLOOR/CEILING ASSEMBLY
- CORRIDOR WALLS ARE NONFIRE-RESISTANCE RATED
- PARTITIONS EXTEND TO UNDERSIDE OF CEILING MEMBRANE

NOTE: THESE CONDITIONS PRODUCE A CATEGORY b
AS THE PARTITIONS HAVE LESS THAN A 1-HR RATING

- CLASSIFICATION = CATEGORY b
- SCORE FROM TABLE 1201.6.5 = -2
- ENTER -2 IN TABLE 1201.7

Figure 1201.6.5(2)
CORRIDOR WALL VALUES—CATEGORY b

1201.6.6 Vertical openings. Evaluate the fire-resistance rating of vertical exit enclosures, hoistways, escalator openings, and other shaft enclosures within the building, and openings between two or more floors. Table 1201.6.6(1) contains the appropriate protection values. Multiply that value by the construction type factor found in Table 1201.6.6(2). Enter the vertical opening value and its sign (positive or negative) in Table 1201.7 under Safety Parameter 1201.6.6, Vertical Openings, for fire safety, means of egress, and general safety. If the structure is a one-story building, enter a value of 2. Unenclosed vertical openings that conform to the requirements of Section 707 of the *International Building Code* shall not be considered in the evaluation of vertical openings.

❖ This section evaluates the fire-resistant enclosure of openings between floors of a building and shaft enclosures, such as stairs, elevator hoistways and escalator openings. This section also gives the formula for determining the score to be entered in Table 1201.7. This section does not apply to a floor opening that is allowed by the requirements for new construction to be unenclosed. Section 707.2 of the IBC provides a number of exceptions to the requirement for enclosure of floor openings. This acknowledges that in new construction, unenclosed floor openings in the circumstances addressed in these exceptions do not present

an undue fire safety risk and, therefore, need not accrue negative points in this evaluation method.

1201.6.6.1 Vertical opening formula. The following formula shall be used in computing vertical opening value.

$$VO = PV \times CF \qquad \textbf{(Equation 12-5)}$$

where:

VO = Vertical opening value.

PV = Protection value from Table 1201.6.6(1).

CF = Construction type factor from Table 1201.6.6(2).

TABLE 1201.6.6(1)
VERTICAL OPENING PROTECTION VALUE

PROTECTION	VALUE
None (unprotected opening)	-2 times number of floors connected
Less than 1 hour	-1 times number of floors connected
1 to less than 2 hours	1
2 hours or more	2

TABLE 1201.6.6(2) – 1201.6.7.1

COMPLIANCE ALTERNATIVES

❖ Table 1201.6.6(1) assigns relative protection values based on the fire-resistance ratings of the vertical openings in a building. The lower the fire-resistance rating, the greater the hazard to the rest of the building. The table also reflects the varying levels of impact to an existing building based on the number of stories that are connected by unprotected openings. The greater the number of floors that are interconnected by unprotected openings, the greater the number of negative points assessed in the existing building's evaluation scores. The closer the building comes to meeting new construction requirements for shaft protection, the greater the number of positive points that can be accrued. Noncomplying unenclosed vertical openings consistently show up as contributing factors in unsuccessful fires. Consequently, a substantial number of points are at risk in this parameter, which is intended to be an incentive to bring noncomplying situations into compliance with new construction requirements.

TABLE 1201.6.6(2)
CONSTRUCTION-TYPE FACTOR

FACTOR	TYPE OF CONSTRUCTION								
	IA	IB	IIA	IIB	IIIA	IIIB	IV	VA	VB
	1.2	1.5	2.2	3.5	2.5	3.5	2.3	3.3	7

❖ Relative values for each type of construction are assigned in Table 1201.6.6(2). These represent the relative degree of fire hazard of each type of construction when compared to other types of construction. Similar factors were considered in the original development of Table 503 for height and area limitations. When one building has two different opening circumstances that individually result in different values, the lower value must be used.

Example 1:

Assume a building of Type IIB construction, three stories in height. The building has a 2-hour fire-resistance-rated exhaust shaft with $1^1/_2$-hour fire dampers (complying with UL 555).

VO = 2 x 3.5 = 7.0

Enter 7.0 in Table 1201.7 under the "Safety Parameter" column next to "1201.6.6 Vertical Openings."

1201.6.7 HVAC systems. Evaluate the ability of the HVAC system to resist the movement of smoke and fire beyond the point of origin. Under the categories in Section 1201.6.7.1, determine the appropriate value and enter that value into Table 1201.7 under Safety Parameter 1201.6.7, HVAC Systems, for fire safety, means of egress, and general safety.

❖ This section evaluates the HVAC system's potential for resisting the movement and spread of fire and smoke. This section does not address HVAC systems that are used exclusively for smoke control in the building. The systems evaluated in this section are those that use either supply, return or exhaust air. For example, a typical building might have supply air ducts or shafts, return air ducts or shafts, toilet exhaust ducts or shafts and kitchen exhaust ducts or shafts, all of which are considered HVAC systems. All systems in the building are evaluated and the lowest score obtained by any of the systems is the score that must be assigned to the entire building.

These provisions include two other safety aspects that are also applicable for new construction: plenums and air movement in egress elements, such as exit access corridors and exit stairways. These factors can significantly affect the relative safety of the occupants of the existing building in a fire condition. In some cases, these safety aspects can be more important than just the number of stories connected by an HVAC system.

1201.6.7.1 Categories. The categories for HVAC systems are:

1. Category a—Plenums not in accordance with Section 602 of the *International Mechanical Code.* -10 points.

2. Category b—Air movement in egress elements not in accordance with Section 1016.4 of the *International Building Code.* -5 points.

3. Category c—Both Categories a and b are applicable. -15 points.

4. Category d—Compliance of the HVAC system with Section 1016.4 of the *International Building Code* and Section 602 of the *International Mechanical Code.* 0 points.

5. Category e—Systems serving one story; or a central boiler/chiller system without ductwork connecting two or more stories. +5 points.

❖ The five categories that must be used in the evaluation process are defined in this section, along with their applicable values. The applicable value is entered in Table 1201.7 under the "Safety Parameter" column next to "1201.6.7 HVAC Systems." These values are not occupancy sensitive, since the spread of fire is dependent on the HVAC system present in the existing building, not on its occupancy classification.

Category a requires a value of -10 for existing buildings that contain plenums not in compliance with the requirements of Section 602 of the *International Mechanical Code*® (IMC®). Locations of plenums within the existing buildings, the materials they are built of relative to the building's type of construction and the materials exposed to plenum air must all be evaluated.

Category b corresponds to existing buildings that have exit access corridors or exit stairways that are used for supply, return or exhaust air or for other ventilation purposes. That type of layout puts the existing building's occupants at greater risk and must therefore be penalized with a value of -5. An existing building complying with one of the exceptions of Section 1004.3.2.4 is not considered as Category b.

Category c is applicable when an existing building has both noncomplying plenums and corridors or stairways used for air movement. A value of -15 must be assigned to such a building because the movement of smoke and fire throughout would pose an even greater hazard to the occupants.

Category d represents the base value of zero. A newly constructed building is required to meet all the provisions of Section 1016.4 of the IBC and Section 602 of the IMC. An existing building that complies with these provisions is meeting this same minimum compliance level and, therefore, is neither penalized nor given benefit for that compliance.

Category e specifies a value of 5 points for HVAC systems that serve only one story of a building. The hazards of a fire spreading laterally through a story of a building via the HVAC system are minimal; therefore, the code assigns a positive value. A boiler/chiller system also does not lend itself to fire spread as long as there is no air movement in ducts.

Example 1:

A Group R-2 apartment building is six stories in height. Each apartment has its own HVAC equipment located within the dwelling unit. Bathrooms are ventilated by fans connected to a central exhaust shaft; kitchen exhaust hoods connect to a central exhaust shaft serving all floors. Corridors have a direct outside supply-air system on each floor with no exhaust. The HVAC systems are classified as Category b because the corridors are being used as the make-up air source for the bathroom and kitchen exhausts. A score of -5 is therefore entered in Table 1201.7 under the "Safety Parameter" column next to "1201.6.7 HVAC Systems."

Example 2:

A Group S-2 low-hazard storage occupancy is located in a one-story building. The building is preengineered steel Type IIB construction, and has a suspended gypsum board ceiling. Gypsum board has also been attached to the underside of the ceiling joists to serve as the upper membrane of a return air plenum. The plenum also contains plastic fire sprinkler piping for the automatic fire suppression system that is installed throughout the building.

The plenum is classified as noncombustible in accordance with Section 602 of the IMC. As long as the plastic fire sprinkler piping meets the optical density and flame spread limits, the plenum is in compliance with the code requirements for new construction. Therefore, the HVAC system will be classified as Category d. A value of zero is therefore entered in Table 1201.7 under the "Safety Parameter" column next to "1201.6.7 HVAC Systems."

1201.6.8 Automatic fire detection. Evaluate the smoke detection capability based on the location and operation of automatic fire detectors in accordance with Section 907 of the *International Building Code* and the *International Mechanical Code*. Under the categories and occupancies in Table 1201.6.8, determine the appropriate value and enter that value into Table 1201.7 under Safety Parameter 1201.6.8, Automatic Fire Detection, for fire safety, means of egress, and general safety.

❖ This section considers the use of smoke detectors in a building. To receive credit for the smoke detectors, they must be connected to audible alarms and installed in accordance with Section 907 of the IBC and Section 606 of the IMC.

TABLE 1201.6.8
AUTOMATIC FIRE DETECTION VALUES

OCCUPANCY	CATEGORIES				
	a	b	c	d	e
A-1, A-3, F, M, R, S-1	-10	-5	0	2	6
A-2	-25	-5	0	5	9
A-4, B, E, S-2	-4	-2	0	4	8

❖ Table 1201.6.8 assigns values for each occupancy. The use of detectors increases the safety in a building by providing early warning of a fire condition to occupants. The lack of detectors and the associated early warning is considered less than optimum for safety in occupancies with high population densities and high combustible loads. Large deficiency points are accrued if adequate detection systems are not provided.

Example:

A four-story Group R-1 hotel has corridors and an elevator lobby on each story. Smoke detectors are installed throughout the corridors, closets, rooms and elevator lobbies. Single-station detectors are installed in the guestrooms. There are smoke detectors in the HVAC return air system and a fire alarm system is provided. An analysis of these building characteristics results in the selection of Category d. To be classified in Category e, guestrooms must have detectors connected to the building's emergency electrical system and annunciated room by room at a constantly attended location, such as the front desk. Additionally, the fire alarm system must be capable of being manually activated by the front desk when a smoke detector operates.

Category d value from Table 1201.6.8 = 2.
Enter 2 in Table 1201.7 under the "Safety Parameter" column next to "1201.6.8 Automatic Fire Detection."

1201.6.8.1 Categories. The categories for automatic fire detection are:

1. Category a—None.
2. Category b—Existing smoke detectors in HVAC systems and maintained in accordance with the *International Fire Code*.
3. Category c—Smoke detectors in HVAC systems. The detectors are installed in accordance with the requirements for new buildings in the *International Mechanical Code*.
4. Category d—Smoke detectors throughout all floor areas other than individual sleeping units, tenant spaces, and dwelling units.
5. Category e—Smoke detectors installed throughout the fire area.

❖ The categories are based on the location and completeness of the smoke detection system.

The categories represent a graduation in the levels of smoke detection that ranges from no detectors in Category a to full detection throughout all fire area spaces in Category e.

Category a is a facility that has no automatic smoke detection system.

Category b acknowledges that the IMC currently requires HVAC system detectors in more locations that may have been required in less contemporary model codes. This category, therefore, assumes that there are some smoke detectors in an existing building's HVAC system, but not to the extent required by the IMC for new construction. If the limited HVAC system detectors are maintained in accordance with the IFC, the detection system qualifies as Category b.

Category c addresses existing buildings that have upgraded HVAC systems with duct detectors installed in accordance with the IMC. The detection values for Category c are zero, since this is the level of protection required for new construction.

Category d is the classification for existing buildings that have full detection coverage throughout the public and common use spaces.

Category e has smoke detectors throughout fire areas in compliance with the provisions for new construction.

1201.6.9 Fire alarm systems. Evaluate the capability of the fire alarm system in accordance with Section 907 of the *International Building Code.* Under the categories and occupancies in Table 1201.6.9, determine the appropriate value and enter that value into Table 1201.7 under Safety Parameter 1201.6.9, Fire Alarm System, for fire safety, means of egress, and general safety.

❖ This section evaluates the capabilities of building fire alarm systems that are separate from the automatic fire detection system evaluated in Section 1201.6.8. A fire alarm system that is manually operated or activated by smoke detectors or sprinkler waterflow devices alerts the occupants to a fire condition. The fire alarm system will notify the occupants with visible or audible alarms so they may begin to take appropriate action. These systems are of particular importance in assembly, business, educational or residential occupancies, which can have large numbers of occupants in rooms with concentrated seating or people who are sleeping.

TABLE 1201.6.9
FIRE ALARM SYSTEM VALUES

OCCUPANCY	CATEGORIES			
	a	bª	c	d
A-1, A-2, A-3, A-4, B, E, R	-10	-5	0	5
F, M, S	0	5	10	15

a. For buildings equipped throughout with an automatic sprinkler system, add 2 points for activation by a sprinkler water-flow device.

❖ Table 1201.6.9 gives values for each occupancy and type of fire alarm system provided. It reflects the idea that the presence of an alarm system in a building usually creates a safer condition for the occupants when compared to a building without a fire alarm system.

Deficiency points are assigned to occupancies without a fire alarm system but which have a high occupant load of sleeping occupants, such as Groups A-1, A-2, A-3, A-4, B, E and R, which are included in Categories a and b.

Example:

A one-story building with an assembly occupancy has a complete manual fire alarm system, a voice alarm system, a public address system and a fire command station that does not contain status indicators and controls for the air-handling system. The fire command station does not have emergency power or lighting system controls. It also does not have a fire department communication panel. As a result, the building is classified in Category c.

Category c value from Table 1201.6.9 = 0.
Enter 0 in Table 1201.7 under the "Safety Parameter" column next to "1201.6.9 Fire Alarm System."

1201.6.9.1 Categories. The categories for fire alarm systems are:

1. Category a—None.

2. Category b—Fire alarm system with manual fire alarm boxes in accordance with Section 907.3 of the *International Building Code* and alarm notification appliances in accordance with Section 907.9 of the *International Building Code.*

3. Category c—Fire alarm system in accordance with Section 907 of the *International Building Code.*

4. Category d—Category c plus a required emergency voice/alarm communications system and a fire command station that conforms to Section 403.8 of the *International Building Code* and contains the emergency voice/alarm communications system controls, fire department communication system controls, and any other controls specified in Section 911 of the *International Building Code* where those systems are provided.

❖ These categories pertain to the fire alarm system that is provided within the existing building.

Category a means that there is no fire alarm system in the building or it does not conform to all the requirements of Section 907. This includes a system that does not have a secondary power supply in accordance with Section 907 or one where the zones of a floor exceed 22,500 square feet (2090 m²) as noted in Section 907.8.

Category b applies when the manual fire alarm boxes comply with Section 907.3.1 and NFPA 72. The alarm-notification appliances, specifically the audible alarms, are in accordance with Section 907.9.2. Location, height and color of the manual fire alarm boxes must be specifically evaluated for compliance, along with the sound levels of the audible alarms when compared to the normal sound levels within the existing building.

Category c requires that the fire alarm system comply with Section 907 of the IBC and NFPA 72. This category indicates that a complete fire alarm system is present in the existing building and that the system complies with all the requirements for new construction.

Category d is applicable for a building that is provided with a fire command station for fire department operations in accordance with Section 911 of the IBC. This type of fire command station is only required in new construction for high-rise buildings (see Section 403.8 of the IBC). However, for purposes of this evaluation parameter, Category d is applicable for any building that provides the fire command station that complies with Section 911, including low-rise buildings. The elements required in the fire command station include:

- The emergency voice/alarm communication system unit;
- The fire department communication unit;
- Fire detection and alarm system annunciator unit;
- An annunciator unit that visually indicates the location of elevators and whether they are operational;
- Status indicators and controls for air-handling systems;
- Fire-fighter's control panel required by Section 909.16 of the IBC, if the building has a smoke control system;
- Controls for unlocking stairway doors simultaneously;
- Sprinkler valve and waterflow detector display panels;
- Emergency and standby power status indicators;
- A telephone for fire department use with controlled access to the public telephone system;
- Fire pump status indicators, if the building has one or more fire pumps;
- Schematic building plans indicating the typical floor plan and detailing the building core, means of egress, fire protection systems, fire-fighting equipment and fire department access;
- A worktable;
- Generator supervision devices, manual start and transfer features if the building has one or more generators; and
- A public address system if such a system is required by other provisions of the code.

1201.6.10 Smoke control. Evaluate the ability of a natural or mechanical venting, exhaust, or pressurization system to control the movement of smoke from a fire. Under the categories and occupancies in Table 1201.6.10, determine the appropriate value and enter that value into Table 1201.7 under Safety Parameter 1201.6.10, Smoke Control, for means of egress and general safety.

❖ This section is used to evaluate characteristics that could limit smoke migration in the building, including operable windows, mechanical exhaust systems, pressurized stairways or smokeproof enclosures.

**TABLE 1201.6.10
SMOKE CONTROL VALUES**

OCCUPANCY	CATEGORIES					
	a	b	c	d	e	f
A-1, A-2, A-3	0	1	2	3	6	6
A-4, E	0	0	0	1	3	5
B, M, R	0	2a	3a	3a	3a	4a
F, S	0	2a	2a	3a	3a	3a

a. This value shall be 0 if compliance with Category d or e in Section 1201.6.8.1 has not been obtained.

❖ Table 1201.6.10 assigns values for each occupancy and category of smoke control in the building. The table represents the relative benefits to the building occupants due to various levels of smoke control methods provided in the building.

There are no negative points accrued under any circumstances, but zero positive points are indicated if no smoke control or only limited smoke control is provided. The table also indicates the value of smoke control in the exit stairs of a building and of automatic sprinkler systems as a means of limiting the volume of smoke production from a fire. Note a of the table assigns zero credit points for Groups B, M, R, F and S unless the building complies with Category d or e in Section 1201.6.8.1, which are the automatic fire detection system requirements.

Example 1:

A Group A-4 church sanctuary is located in a one-story building. It has no operable windows and no smoke control system.

Category a is applicable and the value from Table 1201.6.10 = 0.
Enter 0 in Table 1201.7 under the "Safety Parameter" column next to "1201.6.10 Smoke Control" for only means of egress (ME) and general safety (GS) parameters. No entry is made under the fire safety (FS) parameter column.

Example 2:

A three-story Group E high school has operable windows throughout the building. The stairways are interior without windows or a pressurization system.

Category b is applicable and the value from Table 1201.6.10 = 0.
Enter 0 in Table 1201.7 under the "Safety Parameter" column next to "1201.6.10 Smoke Control" for means of egress (ME) and general safety (GS) parameters.

Example 3:

A two-story Group A-2 nightclub is sprinklered throughout and has a smoke control system that meets the requirements for Category e.

Category e is applicable and the value from Table 1201.6.10 = 6.

Enter 6 in Table 1201.7 under the "Safety Parameter" column next to "1201.6.10 Smoke Control" for means of egress (ME) and general safety (GS) parameters.

Example 4:

A six-story Group B office building has three stairways: one is a smokeproof enclosure conforming to Section 909.20; one is pressurized in accordance with Section 909.20.5; and one has operable exterior windows. The building has smoke detectors throughout all floor spaces and fire areas.

Category f is applicable and the value from Table 1201.6.10 = 4.

Enter 4 in Table 1201.7 under the "Safety Parameter" column next to "1201.6.10 Smoke Control" for means of egress (ME) and general safety (GS).

If detectors are omitted from some of the offices of the buildings, then a score of 0 must be entered in Table 1201.7. This is determined from Note a in Table 1201.6.10.

1201.6.10.1 Categories. The categories for smoke control are:

1. Category a—None.

2. Category b—The building is equipped throughout with an automatic sprinkler system. Openings are provided in exterior walls at the rate of 20 square feet (1.86 m²) per 50 linear feet (15 240 mm) of exterior wall in each story and distributed around the building perimeter at intervals not exceeding 50 feet (15 240 mm). Such openings shall be readily openable from the inside without a key or separate tool and shall be provided with ready access thereto. In lieu of operable openings, clearly and permanently marked tempered glass panels shall be used.

3. Category c—One enclosed exit stairway, with ready access thereto, from each occupied floor of the building. The stairway has operable exterior windows, and the building has openings in accordance with Category b.

4. Category d—One smokeproof enclosure and the building has openings in accordance with Category b.

5. Category e—The building is equipped throughout with an automatic sprinkler system. Each fire area is provided with a mechanical air-handling system designed to accomplish smoke containment. Return and exhaust air shall be moved directly to the outside without recirculation to other fire areas of the building under fire conditions. The system shall exhaust not less than six air changes per hour from the fire area. Supply air by mechanical means to the fire area is not required. Containment of smoke shall be considered as confining smoke to the fire area involved without migration to other fire areas. Any other tested and approved design that will adequately accomplish smoke containment is permitted.

6. Category f—Each stairway shall be one of the following: a smokeproof enclosure in accordance with Section 1019.1.8 of the *International Building Code*; pressurized in accordance with Section 909.20.5 of the *International Building Code*; or shall have operable exterior windows.

❖ The six categories to be evaluated are compared to the occupancies to determine the score to be entered in Table 1201.7 under the "Safety Parameter" column next to "1201.6.10 Smoke Control."

Category a means there is no method of controlling smoke in the building.

Category b means the existing building is sprinklered and there are exterior windows that can be readily opened without the use of keys or tools. This category recognizes the benefit of manual venting capability taken from the high-rise provisions of the IBC for new construction, which includes automatic sprinkler protection. Operable panels or windows in the exterior walls must be provided at the rate of 20 square feet per 50 lineal feet (1.85 m² per 15 240 mm) of exterior wall in each story. The openings must be distributed around the perimeter of the building at intervals not more than 50 feet (15 240 mm) between windows. A story with very high ceilings may have a strip of windows located well above the floor with latches or controls that are not easily reachable. Such openings would not meet Category b standards, even if they are provided and distributed at the required rate.

Category c requires at least one enclosed exit stairway to have operable windows that open to the exterior. Additionally, the building must have operable windows complying with the size and spacing requirements of Category b.

Category d requires a minimum of one smokeproof enclosure and operable window in the building. These operable windows are the same as those addressed in Category b. By definition, a smokeproof enclosure refers to an enclosed interior exit stairway that conforms to Sections 1019.1 and 909.20 of the IBC.

Section 909.20 contains three methods for obtaining a satisfactory smokeproof enclosure: natural ventilation design (see Section 909.20.3), mechanical ventilation design (see Section 909.20.4) and stair pressurization design (see Section 909.20.5).

Additional requirements that must be considered when evaluating a smokeproof enclosure are: access (see Section 909.20.1), construction (see Section 909.20.2), ventilating equipment (see Section 909.20.6) and standby power (see Section 909.20.6.2).

Category e recognizes the use of a mechanical smoke control system designed in accordance with the provisions of the code in a fully sprinklered building. Each fire area within the building must have a mechanical smoke control system. Return and exhaust air from the system must be discharged directly to the exterior to achieve the necessary level of protection. A specific air change requirement is provided, independent of the fire area's volume or size.

Category f recognizes the merits of smokeproof enclosures, pressurized stairs and stairs with operable exterior windows. To be classified in this category, all stairs in the building must comply with the requirements of any one, or a combination of, the three types of stairs. It is possible for a single building to have a smokeproof enclosure, a pressurized stairway and a stairway with operable exterior windows.

Both the categories and the occupancy of the building are used in determining the score that will be entered in Table 1201.7. This score is determined from Table 1201.6.10. Note a in Table 1201.6.10 can have a significant impact on the scores. The note states that, even if the building has some level of smoke control, it is to receive no credit if it does not have an automatic fire detection system complying with Category d or e in Section 1201.6.8.1.

1201.6.11 Means-of-egress capacity and number. Evaluate the means-of-egress capacity and the number of exits available to the building occupants. In applying this section, the means of egress are required to conform to Sections 1013 of the *International Building Code* (with the exception of Section 1015), 1003 of the *International Building Code* (except that the minimum width required by this section shall be determined solely by the width for the required capacity in accordance with Table 1005.1 of the *International Building Code*), 1017, and 1023 of the *International Building Code*. The number of exits credited is the number that is available to each occupant of the area being evaluated. Existing fire escapes shall be accepted as a component in the means of egress when conforming to Section 605.3.1.2. Under the categories and occupancies in Table 1201.6.11, determine the appropriate value and enter that value into Table 1201.7 under Safety Parameter 1201.6.11, Means-of-Egress Capacity, for means of egress and general safety.

❖ This section addresses the exit capacity and number of existing exits available to the building occupants. Before a building can be evaluated in this category, and Section 1201 in general, the building must comply with several provisions of the IBC. This reflects the principle that the means of egress system is a critical aspect of safety in existing buildings and must meet certain baseline requirements for new construction in order for the building to utilize the evaluation method of Chapter 12. If these conditions are not met, then this parameter cannot be evaluated and, consequently, the overall evaluation under Chapter 12 cannot be completed.

A mandatory requirement is not placed on the length of travel in accordance with Section 1003 of the IBC, because travel length does not directly impact exit capacity. Overall exit safety is influenced by travel length, however, and is evaluated in Section 1201.6.13.

The list of mandatory requirements includes Sections 1003, 1013, 1017 and 1023. Sections 1003, 1013 and 1017 in turn cross-reference virtually the entire means of egress chapter. It was not the intent of the legacy model code from which Chapter 12 originated to mandate that,

as a minimum, all aspects of the means of egress fully comply with the requirements for new construction. Fully upgrading the means of egress to meet all requirements for new construction is not always practical and is not necessary to achieve an acceptable level of safety in an existing building. One example of why these references may be questionable is that they encompass Section 1016.3 on dead ends. Requiring compliance with the new construction dead-end limitations conflicts with the evaluation parameter in Section 1201.6.12. Category a in that parameter allows dead-end lengths consistent with the dead-end provisions for existing buildings in the IFC, which are longer than permitted in new construction. The origin of this parameter required compliance with requirements in the following subject areas, for which the corresponding provisions of the IBC are given as follows:

- Type and arrangement of means of egress: Sections 1003.7, 1013.2, 1013.3, 1014.2, 1023.1, 1023.2, 1023.6 and 1024.2 through 1024.4;
- Occupant load determination: Section 1004;
- Capacity of exits: Section 1005.1;
- Number of exits: Section 1018;
- Emergency escape windows: Section 1025.

Evaluation of the means of egress capacity involves all egress components, including exit access, exits and exit discharge. This evaluation is correlated to the requirements for the means of egress for new construction.

TABLE 1201.6.11
MEANS OF EGRESS VALUES

OCCUPANCY	CATEGORIES				
	a[a]	b	c	d	e
A-1, A-2, A-3, A-4, E	-10	0	2	8	10
M	-3	0	1	2	4
B, F, S	-1	0	0	0	0
R	-3	0	0	0	0

a. The values indicated are for buildings six stories or less in height. For buildings over six stories in height, add an additional -10 points.

❖ Table 1201.6.11 assigns values for each occupancy and category from Section 1201.6.11.1. The table gives credit for providing additional numbers of exits and additional exit capacity beyond the minimum required for new construction.

This additional evaluation contributes to the overall assessment of the building's safety and may offset other safety deficiencies [see Figures 1201.6.11(1) and 1201.6.11(2)]. Note a of the table provides a significant penalty for high-rise buildings that use fire escapes as part of the means of egress. In this case, if the building is seven stories or greater in height and uses fire escapes, -10 points must be added to the values already listed for Category a buildings.

FIGURE 1201.6.11(1) – FIGURE 1201.6.11(2)

COMPLIANCE ALTERNATIVES

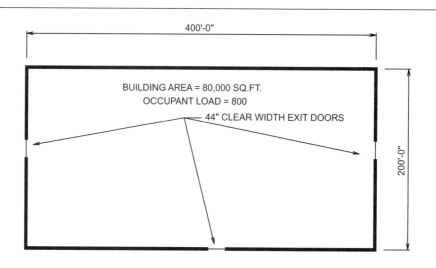

• OCCUPANCY GROUP F
• COMPLETE SPRINKLER SYSTEM
• EGRESS WIDTH PER SECTION 1003.2.3 AND TABLE 1003.2.3
• THE EXIT CAPACITY OF THE THREE 44"
  CLEAR WIDTH EXIT DOORS = $\frac{3 \times 44"}{0.15}$ = 880 OCCUPANTS

• CLASSIFICATION = CATEGORY b. THE
  EXIT CAPACITY OF 880 EXCEEDS
  THE ALLOWABLE CAPACITY OF 800, COMPLYING
  WITH SECTION 1003.2.3

• SCORE = 0 FROM TABLE 1201.6.11
• ENTER 0 IN TABLE 1201.7

For SI: 1 inch = 25.4 mm, 1 foot = 304.8 mm, 1 square foot = 0.0929 m².

Figure 1201.6.11(1)
MEANS OF EGRESS VALUES—CATEGORY b

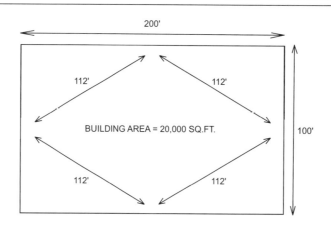

• OCCUPANCY GROUP A-3
• BUILDING FULLY SPRINKLERED
• NUMBER OF OCCUPANTS = 400
• TABLE 1005.2.1 REQUIRES TWO EXITS, FOUR
  EXTERIOR DOORS (EXITS) ARE PROVIDED
• THE 1/3 DIAGONAL = $\frac{224'}{3}$ = 75'
• EACH DOOR IS SPACED 112' APART
• CLASSIFICATION = CATEGORY d FROM TABLE 1201.6.11
• ENTER 8 IN TABLE 1201.7

For SI: 1 foot = 304.8 mm, 1 square foot = 0.0929 m².

Figure 1201.6.11(2)
MEANS OF EGRESS VALUES—CATEGORY b

1201.6.11.1 Categories. The categories for means-of-egress capacity and number of exits are:

1. Category a—Compliance with the minimum required means-of-egress capacity or number of exits is achieved through the use of a fire escape in accordance with Section 605.3.1.2.

2. Category b—Capacity of the means of egress complies with Section 1003 of the *International Building Code*, and the number of exits complies with the minimum number required by Section 1017 of the *International Building Code*.

3. Category c—Capacity of the means of egress is equal to or exceeds 125 percent of the required means-of-egress capacity, the means of egress complies with the minimum required width dimensions specified in the *International Building Code*, and the number of exits complies with the minimum number required by Section 1017 of the *International Building Code*.

4. Category d—The number of exits provided exceeds the number of exits required by Section 1017 of the *International Building Code*. Exits shall be located a distance apart from each other equal to not less than that specified in Section 1014.2 of the *International Building Code*.

5. Category e—The area being evaluated meets both Categories c and d.

❖ Five categories must be considered. These categories and the occupancy of the building will determine the score entered in Table 1201.7.

Category a is applicable to buildings that comply with either the means of egress capacity or the number of exits, including the use of fire escapes. Although the code allows fire escapes to be used as an egress element in existing buildings, this is the least desirable option in any type of building. The code requires a building using fire escapes to use Category a and its corresponding negative points.

Category b is applicable to buildings that meet the minimum requirements of Sections 1003 and 1007. These represent the code requirements for new construction and, consequently, there are no positive or negative points awarded.

Category c is applicable to buildings that meet the minimum width and number of exits but exceed the egress capacity requirements for new construction. This category provides small positive points for existing buildings that meet all of the following requirements:

- The capacity of all the means of egress components is greater than 125 percent of the required capacity.
- All of the means of egress components comply with the minimum required widths [e.g., 32 inch (813 mm) clear for doors, corridors 44 inches (1118 mm) wide and stairways].
- The minimum number of exits is provided based on the number of occupants in each floor level.

By providing oversized egress capacity and minimum egress width requirements, an existing building has added safety, which is rewarded with a small number of positive points.

Category d is applicable to buildings that provide a greater number of exits than required by Section 1007, which contributes a positive factor based on providing occupants with more available routes for exiting the building. Before credit can be given to an existing building with a greater number of exits, the exits must meet the remoteness criteria for new construction set forth in Section 1014.2 of the IBC.

Category e provides additional credit for existing buildings that have the characteristics of both Categories c and d. A building that has oversized capacities in its egress elements, meets the minimum required clear widths for the egress components and has additional exits that are all remotely located from one another merits additional points in occupancies with high occupant loads. In Group B, F, S and R occupancies, any additional capacity or increased number of exits is not as significant a factor and no positive points are awarded for these occupancies.

1201.6.12 Dead ends. In spaces required to be served by more than one means of egress, evaluate the length of the exit access travel path in which the building occupants are confined to a single path of travel. Under the categories and occupancies in Table 1201.6.12, determine the appropriate value and enter that value into Table 1201.7 under Safety Parameter 1201.6.12, Dead Ends, for means of egress and general safety.

❖ This section is used to evaluate dead-end exit access conditions within the building. This section uses the terminology "confined to a single path of travel." Another way to illustrate the meaning of this is "a single direction of travel to reach an exit." A corridor is typically of a width that effectively limits the occupants' route to the exits to one path of travel. A dead end may be a single path, but the key feature of a dead end is that only one direction is available to reach an exit. When building occupants have only one direction of travel available, a potentially hazardous condition is created because they may become trapped if the direction of travel to the exit is blocked by fire or smoke.

This section addresses only dead ends that are a component of exit access travel, which is "that portion of a means of egress that leads to an entrance to an exit." Although a room containing only one egress door has only one way out, it is not considered a dead-end condition. The code takes this condition into account in establishing the limitations under which a room is allowed to have only one means of egress. Dead ends are a concern in passageways and corridors where occupants may not realize that the corridor deadends, causing them to retrace their steps in order to remain on a path toward an exit.

TABLE 1201.6.12 - 1201.6.14

COMPLIANCE ALTERNATIVES

TABLE 1201.6.12
DEAD-END VALUES

OCCUPANCY	CATEGORIES[a]		
	a	b	c
A-1, A-3, A-4, B, F, M, R, S	-2	0	2
A-2, E	-2	0	2

a. For dead-end distances between categories, the dead end value shall be obtained by linear interpolation.

❖ The table reflects the relative degree of hazard associated with dead-end passageways and corridors. This is shown by the deficiency points that apply where a dead end exceeds 35 or 70 feet (10 668 or 21 336 mm) in an occupancy with a relatively high occupant load, such as an assembly occupancy. Note a allows for the interpolation for actual dead-end lengths between the distances specified in the categories. For example, if a building contains a corridor that has a dead-end length of 10 feet (3048 mm) at each end beyond the exits, the value of 1 is used; this is the midpoint between Categories b and c.

1201.6.12.1 Categories. The categories for dead ends are:

1. Category a — Dead end of 35 feet (10 670 mm) in nonsprinklered buildings or 70 feet (21 340 mm) in sprinklered buildings.

2. Category b — Dead end of 20 feet (6096 mm); or 50 feet (15 240 mm) in Group B in accordance with Section 1016.3, Exception 2 of the *International Building Code*.

3. Category c — No dead ends; or ratio of length to width (l/w) is less than 2.5:1.

❖ This section defines the categories for dead-end exit access conditions.

Category a allows dead-end conditions of up to 35 feet (10 668 mm) for existing buildings that are not fully sprinklered, and up to 70 feet (21 336 mm) for existing buildings that are fully sprinklered. These distances correspond to the absolute maximum allowable dead-end lengths allowed in existing buildings by the IFC. Dead ends greater than these distances are considered an unsafe condition because they are not allowed by the IFC to occur in an existing building; therefore, it would be inconsistent to allow greater dead- end lengths under this evaluation method. Since these lengths far exceed the allowable lengths permitted for new construction, negative values are associated with this category.

Category b is the classification for buildings that comply with the requirements for new construction. This zero-based category provides no extra credit for complying with new construction requirements.

Category c represents conditions that exceed the requirements for new construction. If there are no dead-end corridors or the corridor is more of a "space," Category c can be used. In Category c the length-to-width ratio of 2.5 to define a space/corridor is the same as Exception 3 in Section 1004.3.2.3 of the IBC.

Such a space gives the building user a more circular route and full view of the space, and therefore does not represent as great a potential hazard as a classic dead-end corridor.

These categories and occupancies of the building are used in Table 1201.6.12 to determine the score to be entered in Table 1201.7 under the "Safety Parameter" column next to "1201.6.12 Dead Ends" [see Figures 1201.6.12.1(1), 1201.6.12.1(2) and 1201.6.12.1(3)].

1201.6.13 Maximum exit access travel distance to an exit. Evaluate the length of exit access travel to an approved exit. Determine the appropriate points in accordance with the following equation and enter that value into Table 1201.7 under Safety Parameter 1201.6.13, Maximum Exit Access Travel Distance for means of egress and general safety. The maximum allowable exit access travel distance shall be determined in accordance with Section 1015 of the *International Building Code*.

$$\text{Points} = 20 \times \frac{\begin{array}{c}\text{Maximum allowable}\\\text{travel distance}\end{array} - \begin{array}{c}\text{Maximum actual}\\\text{travel distance}\end{array}}{\text{Maximum allowable travel distance}}$$

(Equation 12-6)

❖ The length of exit access travel distance is evaluated in comparison to the travel distance allowed by Table 1015.1 of the IBC for a particular occupancy. The exit access travel distance is measured as set forth in Section 1015.1 of the IBC and the notes to Table 1015.1 of the IBC.

To determine the value to be assigned for maximum travel distances to an exit, Equation 12-6 must be used. This equation allows a graduated scale to be used to evaluate compliance of an existing situation with new construction travel distance requirements. With the equation, a more definite evaluation can occur with a broader range of scores. Any existing building having overall travel distances less than those specified in Table 1015.1 of the IBC will achieve a positive credit. Existing buildings with travel distances greater than those allowed for new construction will be assigned negative points based on how much the travel distance exceeds the allowable limit.

For example, an existing Group B business building has a travel distance of 150 feet (45 720 mm). The distance is measured from the most remote corner of a partitioned office, down a corridor and to an exit door. Table 1015.1 of the IBC permits an unsprinklered Group B building to have a travel distance of 200 feet (60 960 mm). The equation yields the following:

$$20 \times \left(\frac{200 - 150}{200}\right) = 5 \text{ points}$$

See Figures 1201.6.13(1), 1201.6.13(2) and 1201.6.13(3) for other examples.

1201.6.14 Elevator control. Evaluate the passenger elevator equipment and controls that are available to the fire department to reach all occupied floors. Elevator recall controls shall be

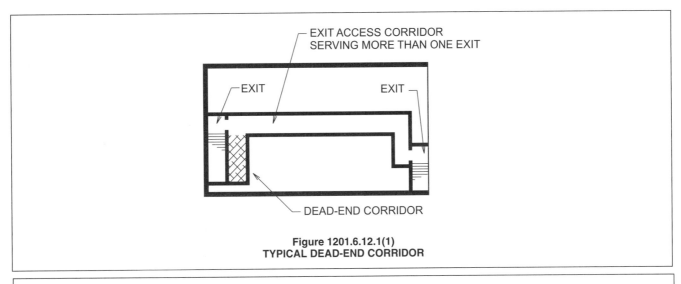

Figure 1201.6.12.1(1)
TYPICAL DEAD-END CORRIDOR

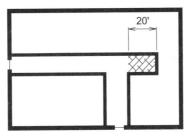

• OCCUPANCY GROUP B WITH MOVABLE PARTITIONS
  AND FURNITURE 6'-6" HIGH
• DEAD-END PASSAGEWAY IS 20'-0"
• CLASSIFICATION = CATEGORY b
• VALUE = 0 FROM TABLE 1201.6.12
• ENTER 0 IN TABLE 1201.7

For SI: 1 inch = 25.4 mm, 1 foot = 304.8 mm.

Figure 1201.6.12.1(2)
DEAD-END VALUES—CATEGORY b

provided in accordance with the *International Fire Code*. Under the categories an occupancies in Table 1201.6.14, determine the appropriate value and enter that value into Table 1201.7 under Safety Parameter 1201.6.14, Elevator Control, for fire safety, means of egress, and general safety. The values shall be zero for a single story building.

❖ The availability of elevators in the building and their capability for fire department use in an emergency must be evaluated. This section addresses four different categories of elevator capability in existing buildings. This section does not address the requirements of Chapter 30. Access must be provided to all occupied floors by passenger elevators for the purpose of this evaluation. Freight elevators cannot be considered, since they may be in locations not readily accessible for fire department use. Elevator recall controls must comply with the IFC for either a Phase I or II category. One-story buildings, including those with mezzanines served by an elevator, are awarded zero value.

TABLE 1201.6.14
ELEVATOR CONTROL VALUES

ELEVATOR	CATEGORIES			
	a	b	c	d
Less than 25 feet of travel above or below the primary level of elevator access for emergency fire-fighting or rescue personnel	-2	0	0	+2
Travel than 25 feet or more above or below the primary level of elevator access for emergency fire-fighting or rescue personnel	-4	NP	0	+4

For SI: 1 foot = 304.8 mm.

NP = Not permitted.

❖ The table assigns values based on the types of controls the elevators have, and the distance the elevators must travel to reach the floors they serve. The 25-foot

FIGURE 1201.6.12.1(3) – 1201.6.14.1

COMPLIANCE ALTERNATIVES

(7620 mm) threshold of elevator travel is based on the IFC requirements and ASME A17.1. Elevator travel distance is based on the level where the elevator is accessed by the fire department. Usually, this is the ground floor or grade-level floor and corresponds to the level where fire command stations or fire alarm system annunciator panels are located. Any elevator that travels 25 feet (7620 mm) or more must always be provided with Phase I and II recall capabilities.

Example:

Assume a Group B business occupancy, three stories in height. The elevator lobbies are equipped with smoke detectors that recall the elevator to the main floor, or to an alternate floor if the main floor detector is activated. The elevator travels more than 25 feet (7620 mm) above the main floor.

This building must be placed in Category c. Because the elevator travels more than 25 feet (7620 mm) above the main floor, it must be brought into compliance with the IFC, which requires Phase I or II recall. When the elevators are brought into compliance with the IFC, they are placed automatically in Category c. The value from Table 1201.6.14 is 0; therefore, 0 must be entered in Table 1201.7 under the "Safety Parameter" column next to "1201.6.14 Elevator Control."

1201.6.14.1 Categories. The categories for elevator controls are:

1. Category a—No elevator.
2. Category b—Any elevator without Phase I and II recall.
3. Category c—All elevators with Phase I and II recall as required by the *International Fire Code*.
4. Category d—All meet Category c; or Category b where permitted to be without recall; and at least one elevator

that complies with new construction requirements serves all occupied floors.

❖ The categories that a building may be placed into range from buildings with no elevators to buildings with elevators complying with new construction requirements and total elevator recall capability.

Category a is applicable to buildings with no elevators. Without an elevator in a multistory building, fire department personnel are required to use the stairs for rescuing people and accessing fire floors. Category a is assigned a negative value.

Category b is applicable to buildings in which elevators are present but have no recall controls. Without recall or fire department control, the elevators are allowed vertical travel distances of less than 25 feet (7620 mm). This approach is consistent with the IFC, which requires controls for elevators reaching 25 feet (7620 mm) or more.

Category c is applicable to buildings with elevators that have automatic recall as required by the IFC.

Category d is applicable to buildings when the elevator controls comply with either Category b or c, and at least one of the elevators in the existing building serves all occupied floor levels. The controls must also comply with all the requirements for new construction. This category recognizes the benefits of having all elevators with Phase I and II controls, as well as having an elevator with all the features required in new construction to facilitate fire-fighting and rescue operations.

Based on the controls provided, or the lack of elevators, the appropriate category is determined. This category and the distance the elevator travels are then used along with Table 1201.6.14 to determine the value score the building will receive for this item. The value is then entered in Table 1201.7.

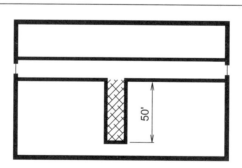

- OCCUPANCY GROUP R
- UNSPRINKLERED BUILDING
- DEAD-END CORRIDOR LENGTH IS 50'-0"
- CLASSIFICATION = CATEGORY a
- VALUE = -2 FROM TABLE 1201.6.12
- ENTER -2 IN TABLE 1201.7

For SI: 1 inch = 25.4 mm, 1 foot = 304.8 mm.

Figure 1201.6.12.1(3)
DEAD-END VALUES—CATEGORY a

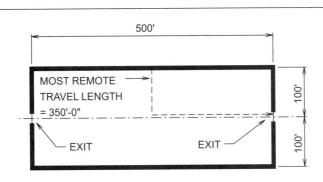

- COMBUSTIBLE STORAGE, OCCUPANCY GROUP S-1 WITH AUTOMATIC FIRE SUPPRESSION SYSTEM
- MOST REMOTE LENGTH OF EXIT ACCESS TRAVEL (SECTION 1004.2.4)
- TRAVEL DISTANCE IN BUILDING = 350'-0"
- TRAVEL DISTANCE LIMIT FROM TABLE 1004.2.4 = 250'-0"
- POINTS $= 20 \times \frac{250 - 350}{250} = -8$

For SI: 1 inch = 25.4 mm, 1 foot = 304.8 mm.

Figure 1201.6.13(1)
EXIT ACCESS TRAVEL DISTANCE VALUES—OCCUPANCY GROUP S-1

- OCCUPANCY GROUP B WITHOUT FIRE SUPPRESSION SYSTEM
- MOST REMOTE TRAVEL LENGTH = 200'-0"
- TRAVEL DISTANCE IN BUILDING = 200'-0"
- TRAVEL DISTANCE LIMIT FROM TABLE 1004.2.4 = 200'-0"
- POINTS $= 20 \times \frac{200 - 200}{200} = 0$

For SI: 1 inch = 25.4 mm, 1 foot = 304.8 mm.

Figure 1201.6.13(2)
EXIT ACCESS TRAVEL DISTANCE VALUES—OCCUPANCY GROUP B

1201.6.15 Means-of-egress emergency lighting. Evaluate the presence of and reliability of means-of-egress emergency lighting. Under the categories and occupancies in Table 1201.6.15, determine the appropriate value and enter that value into Table 1201.7 under Safety Parameter 1201.6.15, Means-of-Egress Emergency Lighting, for means of egress and general safety.

❖ Lighting throughout the entire means of egress is evaluated in this section. Illumination of the means of egress is essential in a building during normal occupancy. During an emergency, illumination becomes even more important, because occupants may be under more stress when seeking to evacuate the building and visibility may be reduced by the buildup of smoke from a fire. The relative value of means of egress lighting is represented by the positive points assigned by

Table 1201.6.15 when reliability of egress lighting is provided beyond the minimum required by the code.

TABLE 1201.6.15
MEANS-OF-EGRESS EMERGENCY LIGHTING VALUES

NUMBER OF EXITS REQUIRED BY SECTIONS 1018.1 AND 1018.2 OF THE *INTERNATIONAL BUILDING CODE*	CATEGORIES		
	a	b	c
Two or more exits	NP	0	4
Minimum of one exit	0	1	1

NP = Not permitted.

❖ This table is used to assess the relative risk to the occupants of the building when reliable sources of power

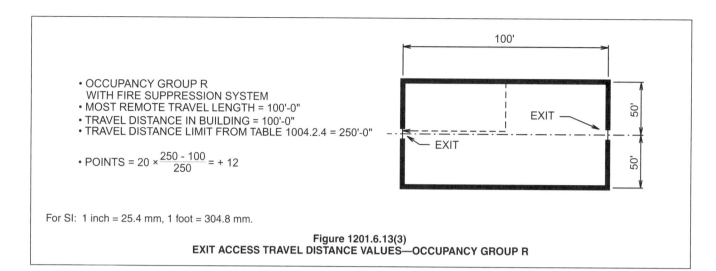

- OCCUPANCY GROUP R
  WITH FIRE SUPPRESSION SYSTEM
- MOST REMOTE TRAVEL LENGTH = 100'-0"
- TRAVEL DISTANCE IN BUILDING = 100'-0"
- TRAVEL DISTANCE LIMIT FROM TABLE 1004.2.4 = 250'-0"

- POINTS = $20 \times \dfrac{250 - 100}{250}$ = + 12

For SI: 1 inch = 25.4 mm, 1 foot = 304.8 mm.

Figure 1201.6.13(3)
EXIT ACCESS TRAVEL DISTANCE VALUES—OCCUPANCY GROUP R

are provided for egress lighting. The table correlates with the power source requirements of the IBC for new construction.

Example:

A Group A-3 church building with a number of classrooms that will accommodate up to 60 people has at least 1 footcandle (11 lux) of illumination at the floor level throughout the entire means of egress. The means of egress lighting in the sanctuary, corridors and stairs is provided with battery backup power that is sized to provide lighting for 1 hour. The classroom lighting is wired to the main switch panel in the building and has no emergency power source.

Category a is applicable because the classrooms are part of the means of egress (exit access) and the lack of emergency power does not comply with Section 1006 of the IBC. The classroom will accommodate up to 60 occupants; therefore, in accordance with the requirements for new construction in Section 1018.2 and Table 1018.2 of the IBC, the occupancy would be required to have two means of egress. The lighting must be provided with an emergency power source complying with Section 2702 of the IBC. The value determined from Table 1201.6.15 is "NP." Emergency power must be added to the lighting in the classrooms. The resulting Category b has a value of 0; therefore, enter 0 in Table 1201.7.

1201.6.15.1 Categories. The categories for means-of-egress emergency lighting are:

1. Category a—Means-of-egress lighting and exit signs not provided with emergency power in accordance with Section 2702 of the *International Building Code*.

2. Category b—Means-of-egress lighting and exit signs provided with emergency power in accordance with Section 2702 of the *International Building Code*.

3. Category c—Emergency power provided to means-of-egress lighting and exit signs, which provides protection in the event of power failure to the site or building.

❖ There are three categories of egress lighting. These categories are consistent with the requirements established by the IFC.

Category a is applicable when there is no emergency power source provided for the lighting in the means of egress. Without an emergency power source to ensure continued illumination of the means of egress, the means of egress lighting and exit signs would not provide a reliable level of safety to the building occupants. For buildings in which only exit is required, this category represents the zero-based criteria, which is the level required for new construction.

Category b is applicable to buildings that are provided with an emergency power source for means of egress illumination. For buildings in which a minimum of two exits are required, this category represents the zero-based criteria, which is the level required for new construction.

Category c is applicable when emergency power is provided for means of egress lighting and exit signs in excess of the minimum requirements for new construction. The emergency power requirements for new construction assume a power failure occurs within the building or somewhere within the building site. It does not assume that the power failure occurs at the source of power to the site (i.e. from the provider). If the emergency power provides full protection to the site or building during power failure, Category c is applicable. Campus-type complexes or buildings that require extra security may have a power plant available to provide complete back-up power for indefinite periods. This will qualify for the positive points of this category.

These categories, along with the number of required exits, are used to determine the appropriate value from Table1201.6.15. The value determined from Table1201.6.15 is then entered in Table1201.7.

1201.6.16 Mixed occupancies. Where a building has two or more occupancies that are not in the same occupancy classification, the separation between the mixed occupancies shall be evaluated in accordance with this section. Where there is no separation between the mixed occupancies or the separation between mixed occupancies does not qualify for any of the categories indicated in Section 1201.6.16.1, the building shall be evaluated as indicated in Section 1201.6, and the value for mixed occupancies shall be zero. Under the categories and occupancies in Table 1201.6.16, determine the appropriate value and enter that value into Table 1201.7 under Safety Parameter 1201.6.16, Mixed Occupancies, for fire safety and general safety. For buildings without mixed occupancies, the value shall be zero.

❖ This section is used to evaluate mixed occupancies within an existing building and whether the method of separating mixed occupancies conforms to the requirements of Section 302.3 of the IBC. If Section 302.3 of the IBC is not completely understood, see the associated commentary before proceeding with the evaluation of the existing building in this section. Also, refer to the commentary for Section 1201.6.

This section is applicable only to separated mixed occupancies. If a building is a single occupancy, the applicable value for this section is zero. If an existing mixed occupancy building has no fire-resistance-rated separation between the different uses, or the separation is fire-resistance rated for less than 1 hour, the applicable value is also zero. The building must also be evaluated in accordance with Section 1201.6. The zero-based category for this section is equivalent to full compliance with Section 302.3 of the IBC, which is the requirement for new construction.

**TABLE 1201.6.16
MIXED OCCUPANCY VALUES[a]**

OCCUPANCY	CATEGORIES		
	a	b	c
A-1, A-2, R	-10	0	10
A-3, A-4, B, E, F, M, S	-5	0	5

a. For fire-resistance ratings between categories, the value shall be obtained by linear interpolation.

❖ This table addresses the relative risk of a building in or close to compliance with the provisions for separated mixed occupancies. When mixed occupancies are not separated from each other, the risk from hazards is greater in high-density occupancies, such as Groups A-1 and A-2. This risk is also greater in residential occupancies, because occupants may be sleeping and not fully alert. For this reason, inadequate separation is given greater negative values. In buildings with lower occupant loads, and where the occupants are alert, the risks are relatively lower.

Note a permits linear interpolation of the corresponding values in each category. For example, an unsprinklered, Group B/Group F-1 mixed occupancy building is separated with a 2-hour fire-resistance-rated assembly. From Table 302.3.2 of the IBC, the required fire-resistance rating of the separation between a Group B and a Group F-1 occupancy for Category b is 3 hours. Since the rating that is provided is only 2 hours,

an interpolation halfway between Category a and Category b can be used. The resulting points are -2.5.

Example 1:

A three-story building has a first floor that is a Group M mercantile occupancy and the upper two stories are Group R residential occupancies. The building is Type VB construction, 9,000 square feet (836.1 m²) per floor. It is fully sprinklered, and only 25 percent of the perimeter is accessible. Although Type VB construction is not required to be protected, the Group M occupancy on the first floor is separated from the Group R occupancy on the second floor with a 2-hour fire-resistance-rated floor/ceiling assembly. In this case, the exception to Section 302.3.2 of the IBC permits the required rating of 2 hours from Table 302.3.2 of the IBC to be reduced to 1 hour because the building is equipped throughout with an automatic sprinkler system in accordance with Section 903 of the IBC. The 2-hour rating provided is twice that required; therefore, this building qualifies for Category c. A value of 10 points is assigned to the residential portion of the building, and a value of 5 points is assigned to the mercantile portion (see Section 1201.6 for commentary about mixed uses).

Example 2:

A four-story building is Type IIB construction and is fully sprinklered. It has 15,000 square feet (1393.5 m²) per floor with 25-percent open perimeter. The building is to be a Group R-1 hotel, except for a portion of the first floor which will be used as a Group A-2 nightclub, occupying 1,000 square feet (92.9 m²). Although the building is unprotected Type IIB construction, the floor/ceiling assemblies of the first and second floors are rated for 3 hours and the fire barrier assembly at the first floor between the R-1 occupancy and the Group A-2 occupancy is also rated for 3 hours. Considering these building characteristics, the following determinations are made:

The exception to Section 302.3.2 of the IBC is applicable in this example just as it was applicable in the previous example. Because of the presence of an automatic sprinkler system, the exception permits the rating of the separation required in Table 302.3.2 of the IBC to be reduced by 1 hour. Although this building is required to have 1-hour separation, it has a 3-hour separation, which is more than double the minimum. This building is classified as Category c. A value of 10 points is awarded to both occupancies.

1201.6.16.1 Categories. The categories for mixed occupancies are:

1. Category a—Minimum 1-hour fire barriers between occupancies.

2. Category b—Fire barriers between occupancies in accordance with Section 302.3.2 of the *International Building Code.*

3. Category c—Fire barriers between occupancies having a fire-resistance rating of not less than twice that required by Section 302.3.2 of the *International Building Code.*

❖ This section addresses three different conditions:

Category a is applicable when the rating between mixed occupancies is less than specified in Section 302.3.2 of the IBC but is a fire barrier with a minimum 1-hour fire-resistance rating.

Category b is applicable when the separation of mixed occupancies is a fire barrier that conforms to Section 302.3.2 of the IBC for fire-resistance ratings.

Category c is applicable when the fire-resistance rating that separates occupancies in an existing building is no less than twice the rating required by Section 302.3.2 of the IBC. Category c gives bonus points or increased credits to existing buildings that have higher fire-resistance ratings.

Section 1201 does not require full compliance with all of Section 302.3.2 of the IBC. Section 1201.6.16 evaluates whether the existing building's compliance with Section 302.3 of the IBC provides relative safety within the building. If the building does comply with Section 302.3.3 of the IBC, Section 1201.6.16 acknowledges that there is no basis for assigning negative points, nor is there any basis for assigning positive points for safety; therefore, a neutral zero value is assigned. The appropriate value is entered in Table 1201.7 under the "Safety Parameter" column next to "1201.6.16 Mixed Occupancies."

1201.6.17 Automatic sprinklers. Evaluate the ability to suppress a fire based on the installation of an automatic sprinkler system in accordance with Section 903.3.1.1 of the *International Building Code.* "Required sprinklers" shall be based on the requirements of this code. Under the categories and occupancies in Table 1201.6.17, determine the appropriate value and enter that value into Table 1201.7 under Safety Parameter 1201.6.17, Automatic Sprinklers, for fire safety, means of egress divided by 2, and general safety. High-rise buildings defined in Section 403.1 of the *International Building Code* that undergo a change of occupancy to Group R shall be equipped throughout with an automatic sprinkler system in accordance with Section 403.2 of the *International Building Code* and Chapter 9 of the *International Building Code.*

❖ These provisions are used to determine the amount of credit that can be applied to the evaluation for the installation of an automatic sprinkler system in an existing building.

The value of sprinkler protection is reflected in the the additional credit available in this parameter over and above the basic credits that are provided for building height in Section 1201.6.1, building area in Section 1201.6.2 and maximum travel distance to an exit in Section 1201.6.13. Throughout the code, additional credits are offered for the use of an automatic sprinkler system.

The evaluation of sprinklers in an existing building is based on whether an automatic sprinkler system is both required and installed. The criteria used to determine when an automatic sprinkler system is required are tied to the same requirements for new construction in Section 903.2 of the IBC. The thresholds listed in Sections 903.2.1 through 903.2.12 of the IBC must be

used to evaluate whether those characteristics and occupancies are present and whether a sprinkler system would be required for new construction. The exception to Section 903.2 of the IBC, and the additional requirements cross-referenced in Section 903.2.13 of the IBC must also be considered. The determination of whether sprinklers are required in a building or a portion of a building must be done to correctly determine the category applicable to the existing building.

These parameters for sprinklers allow for a more equitable evaluation of the contribution of that feature to overall building safety. This factor encourages the installation of automatic sprinkler systems in existing buildings by providing substantial negative and positive points.

The values for sprinklers are given in Table 1201.6.17. The appropriate credit values from Table 1201.6.17 are entered in Table 1201.7 under the "Safety Parameter" column next to "1201.6.17 Automatic Sprinklers" for both fire safety (FS) and general safety (GS), but only one-half the value is entered under the means of egress (ME). The one-half credit for egress is allowed because some credits for sprinklers are incorporated into the parameters for means of egress capacity (see Section 1201.6.11), dead ends (see Section 1201.6.12) and maximum travel distance to an exit (see Section 1201.6.13).

TABLE 1201.6.17
SPRINKLER SYSTEM VALUES

OCCUPANCY	CATEGORIES					
	a[a]	b[a]	c	d	e	f
A-1, A-3, F, M, R, S-1	-6	-3	0	2	4	6
A-2	-4	-2	0	1	2	4
A-4, B, E, S-2	-12	-6	0	3	6	12

a. These options cannot be taken if Category a in Section 1201.6.18 is used.

❖ This table lists the credit values for the respective categories of Section 1201.6.17.1, based on the occupancy being evaluated in the existing building. The assembly occupancies containing large combustible fuel loads with large occupant loads are included with those occupancies containing large fuel loads. These occupancies represent buildings where experience has shown that adequate on-site sprinkler systems save lives and are necessary to supplement the local fire department capabilities.

Group A-2 buildings are contained in a separate line in the table for determining sprinkler system values. This occupancy with its densely packed high occupant loads, ill-defined seating and aisle arrangements and facilities services must be separately evaluated to adequately match its hazards with sprinkler system requirements. Churches and schools, which have high-occupant loads but low fuel loads, are combined with other low fuel-load occupancies in the last line item in the table. Category c is the zero-based category. The categories to the left of this column contain negative values that define buildings and occupancies that would be required to be sprinklered as new construction, but are not sprinklered or are provided with inadequate sprinkler systems. The three categories to the right of this column contain positive values that represent buildings and occupancies with sprinkler systems that are deemed adequate or comply with current standards and requirements.

1201.6.17.1 Categories. The categories for automatic sprinkler system protection are:

1. Category a—Sprinklers are required throughout; sprinkler protection is not provided or the sprinkler system design is not adequate for the hazard protected in accordance with Section 903 of the *International Building Code.*

2. Category b—Sprinklers are required in a portion of the building; sprinkler protection is not provided or the sprinkler system design is not adequate for the hazard protected in accordance with Section 903 of the *International Building Code.*

3. Category c—Sprinklers are not required; none are provided.

4. Category d—Sprinklers are required in a portion of the building; sprinklers are provided in such portion; the system is one that complied with the code at the time of installation and is maintained and supervised in accordance with Section 903 of the *International Building Code.*

5. Category e—Sprinklers are required throughout; sprinklers are provided throughout in accordance with Chapter 9 of the *International Building Code.*

6. Categoryf—Sprinklers are not required throughout; sprinklers are provided throughout in accordance with Chapter 9 of the *International Building Code.*

❖ Six categories are defined in this section for evaluating the automatic sprinkler system in an existing building. These categories address all aspects, from existing buildings that are unsprinklered but are required to be sprinklered if new construction, to existing buildings that are sprinklered and are not required to be sprinklered if new construction. The range of points provided between various categories increases the flexibility in the use of this evaluation method.

Category a includes buildings or occupancies that would be required by the new construction criteria of Section 903 of the IBC to be sprinklered throughout the building. Category a buildings are provided with either no sprinkler system or one that is inadequate and does not provide the required level of protection. To evaluate the existing sprinkler system, a trained fire protection engineer should evaluate the system's design against the applicable referenced standards. This is not the same type of consideration needed in Category d, where the existing system actually meets all of the requirements of any earlier edition of the applicable referenced standards. This category is considered the lowest acceptable level of compliance and is, therefore,

associated with the largest negative values in Table 1201.6.17.

Category b is applicable when only a portion of the existing building is required by the provisions of Section 903 of the IBC to be sprinklered. For example, a multistory building may have two of its stories that qualify as windowless stories that require sprinklers, or a mercantile building may have one fire area exceeding 12,000 square feet (1114.8 m²) that requires sprinklers. In both cases, the buildings are without an automatic sprinkler system, or the sprinkler systems that are in place are inadequate and do not comply with the technical provisions of the code. Since only portions of the buildings are required to be sprinklered, the negative sprinkler system values for this category in Table 1201.6.17 are exactly half of the Category a values.

Category c is the zero-based category for this safety parameter. The existing building is not required by Section 903 of the IBC to be sprinklered and no sprinkler system is provided. Such buildings are neither penalized nor rewarded by Table 1201.6.17.

Category d is similar to Category b because only a portion of the existing building is required by Section 903 of the IBC to be sprinklered. In this case, there is a sprinkler system in place in that portion of the building and the system was designed at the time of its installation to comply with the requirements of an earlier edition of the applicable standards as stated in Section 903 of the IBC. An example of this is a sprinkler system designed to comply with the hydraulic design criteria of the 1989 edition of NFPA 13 for an ordinary Group 2 hazard. As long as this existing sprinkler system is still properly maintained and supervised, the building qualifies for the small positive values listed in Table 1201.6.17.

Category e includes existing buildings that are required by the provisions of Chapter 9 of the IBC to be sprinklered throughout and are protected throughout by a properly designed, installed and supervised sprinkler system. Although this category more closely represents the requirements for new construction, moderate positive values are awarded by Table 1201.6.17. The evaluation rewards existing buildings that are sprinklered with higher points.

Category f is the highest level of protection and is rewarded with maximum positive points in the accompanying table. Existing buildings, which are not required by Section 903 of the IBC to be sprinklered but are voluntarily provided with a fully designed, installed and supervised system, provide an added level of protection that justifies substantial bonus points that may ultimately determine whether or not an existing building meets its mandatory safety scores.

1201.6.18 Standpipes. Evaluate the ability to initiate attack on a fire by making supply of water available readily through the installation of standpipes in accordance with Section 905 of the *International Building Code.* "Required Standpipes" shall be based on the requirements of the *International Building Code.* Under the categories and occupancies in Table 1201.6.18, de-

TABLE 1201.6.18 – 1201.6.18.1   COMPLIANCE ALTERNATIVES

termine the appropriate value and enter that value into Table 1201.7 under Safety Parameter 1201.6.18, Standpipes, for fire safety, means of egress, and general safety.

❖ These provisions are used to determine the amount of credit that can be applied to the evaluation for the installation of a standpipe system in an existing building. Standpipes are provided as a tool for use by fire fighters to aid their fire-fighting operations. The general consensus is that they are not recommended or intended for use by untrained building occupants.

The evaluation of standpipe systems in an existing building, much like that for automatic sprinker systems, is based on whether a standpipe system is both required and installed. The criteria used to determine when a standpipe system is required are tied to the same requirements for new construction in Section 905.3 of the IBC. The thresholds listed in Sections 905.3.1 through 905.3.6 of the IBC must be used to evaluate whether those characteristics and occupancies are present and whether a standpipe system would be required for new construction. These parameters for standpipes encourage the installation of standpipe systems in existing buildings by providing substantial negative and positive points.

The values for standpipe systems are given in Table 1201.6.18. The appropriate values from Table 1201.6.18 are entered in Table 1201.7 under the "Safety Parameter" column next to "1201.6.18 Standpipes" for fire safety (FS), means of egress (ME) and general safety (GS).

TABLE 1201.6.18
STANDPIPE SYSTEM VALUES

OCCUPANCY	CATEGORIES			
	a[a]	b	c	d
A-1, A-3, F, M, R, S-1	-6	0	4	6
A-2	-4	0	2	4
A-4, B,	-12	0	6	12

a. This option cannot be taken if Category a or Category b in Section 1201.6.17 is used.

❖ This table lists the credit values for the respective categories of Section 1201.6.18.1, based on the occupancy being evaluated in the existing building. The grouping of occupancies is the same as provided for the sprinkler system parameter since the need for and value of a standpipe system, like sprinkler systems, is a function of the likelihood and potential severity of a fire in the various occupancies.

Category b is the zero-based category. Category a to the left of this column contains negative values that define buildings and occupancies that would be required to have a standpipe system as new construction but are not provided with one, or are provided with a standpipe system that is not designed and installed in accordance with the design and installation requirements of Section 905 of the IBC for new construction. The three categories to

the right of this column contain positive values that represent buildings and occupancies with a standpipe system that are deemed adequate or comply with current design standards and installation requirements.

1201.6.18.1 Standpipe. The categories for standpipe systems are:

1. Category a—Standpipes are required; standpipe is not provided or the standpipe system design is not in compliance with Section 905.3 of the *International Building Code.*

2. Category b—Standpipes are not required; none are provided.

3. Category c—Standpipes are required; standpipes are provided in accordance with Section 905 of the *International Building Code.*

4. Category d—Standpipes are not required; standpipes are provided in accordance with Section 905 of the *International Building Code.*

❖ Four categories are defined in this section for evaluating the standpipe system in an existing building. These categories address all aspects from existing buildings that do not have standpipes but would be required to have standpipes if new construction, to existing buildings that have standpipes and would not be required to have standpipes if new construction.

Category a is applicable to buildings or occupancies that would be required by the new construction criteria to be provided with a Class I, Class II or Class III standpipe system. Category a buildings are provided with either no standpipe system, or one that is inadequate and does not comply with the applicable design and installation criteria of Section 905 of the IBC. To evaluate the existing standpipe system, a trained fire protection engineer should evaluate the system's design against the applicable referenced standards. This category is considered the lowest acceptable level of compliance and is, therefore, associated with the largest negative values in Table 1201.6.18. Category a is not an option in unsprinklered or inadequately sprinklered buildings that would be required to be partially or fully sprinklered as new construction. This is reflected in Note a to Table 1201.6.18, which indicates that Category a cannot be used if Category a or b is used in the automatic sprinkler parameter in Section 1201.6.17. The effect of this provision is that the absence of both sprinklers and standpipes is unacceptable in buildings that are required to provide both sprinklers and standpipes. One or the other system must be provided in order to complete this evaluation method. If only one of the required systems is provided, negative points will be accrued in the parameter for the other system, but it is still possible to have an overall passing score for the building.

Category b is the zero-based category for this safety parameter. The existing building is not required by Section 905 of the IBC to have a standpipe system and no standpipe system is provided. Such buildings are neither penalized nor rewarded by Table 1201.6.18.

Category c is applicable to buildings that are required by Section 905 of the IBC to have a standpipe system, and are provided with the class standpipe system that would be required for new construction and is designed and installed in accordance with Section 905.2 of the IBC. Although this category more closely represents the requirements for new construction, moderate positive values are awarded by Table 1201.6.17. The evaluation rewards existing buildings that have a contemporary, reliable standpipe system with higher points.

Category d is the highest level of protection and is rewarded with maximum positive points in the accompanying table. Existing buildings, which are not required by Section 905 of the IBC to have a standpipe system but are voluntarily provided with a fully designed and installed system, provide an added level of protection that justifies substantial bonus points.

1201.6.19 Incidental use. Evaluate the protection of incidental use areas in accordance with Section 302.1.1 of the *International Building Code*. Do not include those where this code requires suppression throughout the building, including covered mall buildings, high-rise buildings, public garages, and unlimited area buildings. Assign the lowest score from Table 1201.6.19 for the building or fire area being evaluated. If there are no specific occupancy areas in the building or fire area being evaluated, the value shall be zero.

❖ This section includes an evaluation system for the separation and protection requirements indicated for incidental use areas in an existing building. This evaluation is based on the requirements for new construction in Table 302.1.1 of the IBC. The designer is required to comply with this section or to treat these use areas as a mixed occupancy. If the building designer chooses to separate or protect these rooms or areas in accordance with this table, the building is classified according to its main use. Because some existing buildings may have been designed in this fashion, an evaluation procedure has been added to account for the level of protection provided. The lowest score must be assigned to the building or fire area for the specific occupancy areas. For example, an existing Group B building has six separate storage rooms. Each room is more than 100 square feet (9.3 m²) in area. If five of the storage rooms are protected with an automatic fire-extinguishing system and separated with smoke partitions complying with Section 302.1.1.1 of the IBC, and the sixth room is not protected in the same manner, the lowest score for all the storage rooms is determined by the single unprotected storage room. If the building designer chooses to treat the incidental use areas as a separate occupancy, the provisions of Section 1201.9.1 are applicable and a zero is inserted in Table 1201.7 under the "Safety Parameter" column next to "1201.6.19 Incidental Use Area Protection."

TABLE 1201.6.19
INCIDENTAL USE AREA VALUES[a]

PROTECTION REQUIRED BY TABLE 302.1.1 OF THE *INTERNATIONAL BUILDING CODE*	PROTECTION PROVIDED						
	None	1 hour	AFSS	AFSS with SP	1 hour and AFSS	2 hours	2 hours and AFSS
2 hours and AFSS	-4	-3	-2	-2	-1	-2	0
2 hours, or 1 hour and AFSS	-3	-2	-1	-1	0	0	0
1 hour and AFSS	-3	-2	-1	-1	0	-1	0
1 hour	-1	0	-1	-1	0	0	
1 hour, or AFSS with SP	-1	0	-1	-1	0	0	0
AFSS with SP	-1	-1	-1	-1	0	-1	0
1 hour or AFSS	-1	0	0	0	0	0	0

a. AFSS = Automatic fire suppression system; SP = Smoke partitions (See IBC Section 302.1.1.1).

Note: For Table 1201.7, see page 63.

❖ This table provides a matrix of characteristics arranged in a format of rows and columns for determining values for incidental use areas. The left-hand column of the table is arranged for the separation or protection specified by Table 302.1.1 of the IBC for a particular occupancy room or area. Table 302.1.1 of the IBC must first be consulted to determine the level of separation or protection required by the code for new construction. This same entry is then found in the left-hand column of Table 1201.6.19. The top row represents the actual level of separation or protection that is provided in the existing building. The corresponding occupancy area value is read and then inserted into Table 1201.7 under the "Safety Parameter" column next to "1201.6.19 Incidental Use Area Protection." Values of zero are assigned to all arrangements that represent compliance with the requirements for new construction. Negative values are assigned based on the degree of noncompliance with the requirements for new construction.

1201.7 Building score. After determining the appropriate data from Section 1201.6, enter those data in Table 1201.7 and total the building score.

❖ This section is the tally sheet for all of the 19 safety parameters evaluated in Sections 1201.6.1 through 1201.6.19, which determine the building's overall safety profile for fire safety (FS), means of egress (ME) and general safety (GS).

This section also directs the data and values of the 19 safety parameters of Sections 1201.6.1 through 1201.6.19 to be entered into Table 1201.7 for totaling the building scores.

1201.8 Safety scores. The values in Table 1201.8 are the required mandatory safety scores for the evaluation process listed in Section 1201.6.

❖ This section lists the minimum scores for fire safety (FS), means of egress (ME) and general safety (GS) that must be obtained from the evaluation of the 19 safety parameters that qualifies a building as acceptable in meeting the code's objectives for public safety and health. This section summarizes the mandatory values of Table 1201.8 that must be met from the evaluation program.

1201.9 Evaluation of building safety. The mandatory safety score in Table 1201.8 shall be subtracted from the building score in Table 1201.7 for each category. Where the final score for any category equals zero or more, the building is in compliance with the requirements of this section for that category. Where the final score for any category is less than zero, the building is not in compliance with the requirements of this section.

❖ The sections and tables that follow are the final steps in the evaluation process. This section also discusses how mixed occupancies must to be treated during the final step of the evaluation process. This section compares the three building scores from Table 1201.7 to the three mandatory safety scores from Table 1201.8. If the values in all three categories from Table 1201.7 exceed the corresponding mandatory safety scores in Table 1201.8, the building passes and is in compliance with the code. If the score in any one category is less than the mandatory safety score, the building is deemed to have failed and additional measures must be taken to bring the scores to a point that will at least equal the mandatory safety scores.

1201.9.1 Mixed occupancies. For mixed occupancies, the following provisions shall apply:

1. Where the separation between mixed occupancies does not qualify for any category indicated in Section 1201.6.16, the mandatory safety scores for the occupancy with the lowest general safety score in Table 1201.8 shall be utilized. (See Section 1201.6.)

2. Where the separation between mixed occupancies qualifies for any category indicated in Section 1201.6.16, the mandatory safety scores for each occupancy shall be placed against the evaluation scores for the appropriate occupancy.

❖ This section explains how to determine whether a mixed occupancy building passes or fails the process of evaluation. It restates the information in Sections 1201.6 and 1201.6.16. The mixed occupancy evaluation is based on the requirements of Sections 302.3.1 and 302.3.2 of the IBC. The two procedures described are:

Procedure 1:

For unseparated occupancies in accordance with Section 302.3.1, or for occupancies that are separated with a fire-resistance rating of less than 1 hour, mandatory safety scores for the occupancy with the lowest general safety score of Table 1201.8 apply.

Procedure 2:

For separated occupancies in accordance with Section 302.3.2, or one of the categories listed in Section 1201.6.16, mandatory safety scores for each occupancy are compared to the evaluation scores for the appropriate occupancy. The total building score of Table 1201.7 is computed for each appropriate occupancy.

This section does not include a category or condition for a third option for mixed occupancies, and that is the separation of multiple occupancies within a single building or structure with one or more fire walls. Likewise, this option is not addressed in Section 302.3. A fire wall creates separate and independent buildings; therefore, a separate evaluation must be done for each building that is created by fire walls.

Example:

A 40-foot (12 192 mm), three-story building has an open perimeter of 25 percent. It is Type IIB unprotected construction, unsprinklered. The building has 9,000 square feet (836.1 m^2) per floor. A Group M mercantile is on the first floor with Group B business occupancies on the second and third floors. Although the building is unprotected Type IIB construction, there is a 2-hour fire-resistance-rated floor/ceiling assembly separating the first and second floors.

This building qualifies for Category b in Section 1201.6.16.1. A separate summary sheet for tabulating the building score in Table 1201.7 has to be completed for each of the two occupancies. The values of the building scores for fire safety (FS), means of egress (ME) and general safety (GS) for the Group M mercantile occupancy and the Group B business occupancy must be calculated. The mercantile building scores are compared to the mandatory safety scores of 23, 40 and 40, and the business building scores are compared to the mandatory safety scores of 30, 40 and 40. If any one of the three building scores from either of the two occupancies does not equal or exceed the applicable safety score, the entire building fails the evaluation. For the building to pass, the safety parameters that are less than the mandatory safety score must be upgraded until a passing score is achieved in each of the three categories.

TABLE 1201.7
SUMMARY SHEET—BUILDING CODE

Proposed occupancy _____ Existing occupancy _____

Year building was constructed _____ Number of stories _____ Height in feet _____

Type of construction _____

Percentage of frontage increase _____ % Percentage of height reduction _____ %

Completely suppressed: Yes _____ No _____ Corridor wall rating _____

Compartmentation: Yes _____ No _____ Required door closers: Yes _____ No _____

Fire-resistance rating of vertical opening opening enclosures _____

Type of HVAC system _____ Serving number of floors _____

Automatic fire detection: Yes _____ No _____ Type of location _____

Fire alarm system: Yes _____ No _____ Type _____

Smoke control: Yes _____ No _____ Type _____

Adequate exit routes: Yes _____ No _____ Dead ends: Yes _____ No _____

Maximum exit access travel distance _____ Elevator controls: Yes _____ No _____

Means-of-egress emergency lighting: Yes _____ No _____ Mixed occupancies: Yes _____ No _____

SAFETY PARAMETERS	FIRE SAFETY (FS)	MEANS OF EGRESS (ME)	GENERAL SAFETY (GS)
1201.6.1 Building Height 1201.6.2 Building Area 1201.6.3 Compartmentation			
1201.6.4 Tenant and Dwelling Unit Separations 1201.6.5 Corridor Walls 1201.6.6 Vertical Openings			
1201.6.7 HVAC Systems 1201.6.8 Automatic Fir Detection 1201.6.9 Fire Alarm System			
1201.6.10 Smoke Control 1201.6.11 Means-of-Egress Capacity 1201.6.12 Dead Ends	**** **** ****		
1201.6.13 Maximum Exit Access Travel Distance 1201.6.14 Elevator Control 1201.6.15 Means-of-Egress Emergency Lighting	**** ****		
1201.6.16 Mixed Occupancies 1201.6.17 Automatic Sprinklers 1201.6.18 Standpipes 1201.6.19 Incidental Use Area Protection		**** Divide by 2	
Building Score—Total Value			

****No applicable value to be inserted

❖ Table 1201.7 is the summary sheet containing all the relative attributes of the building. The summary sheet also contains a complete listing of the 19 safety parameters that have been evaluated. The upper portion of the summary sheet serves as a guide to the user to catalog and highlight existing building elements that relate to the 19 safety parameters and to the evaluation. The lower portion of the summary sheet is used to record the results of the 19 safety parameters that have been evaluated. These are added to produce the building score total values for fire safety (FS), means of egress (ME) and general safety (GS).

TABLE 1201.8 – TABLE 1201.9

COMPLIANCE ALTERNATIVES

TABLE 1201.8
MANDATORY SAFETY SCORES[a]

OCCUPANCY	FIRE SAFETY (MFS)	MEANS OF EGRESS (MME)	GENERAL SAFETY (MGS)
A-1	20	31	31
A-2	21	32	32
A-3	22	33	33
A-4, E	29	40	40
B	30	40	40
F	24	34	34
M	23	40	40
R	21	38	38
S-1	19	29	29
S-2	29	39	39

a.　MFS　=　Mandatory Fire Safety
　　MME = Mandatory Means of Egress
　　MGS = Mandatory General Safety

❖ The table lists the minimum mandatory safety scores for the evaluation of various occupancies for the three major mandatory safety scores of fire safety (MFS), means of egress (MME) and general safety (MGS). The mandatory safety values are based on the scores considered to provide an overall acceptable level of safety in an existing building upon which approval of the alterations, repairs, change of occupancy or addition can be based. This is the zero-based concept. The scores have been determined as representing one level of compliance higher than the code's minimum requirements for new construction. The mandatory safety scores are consistent with the idea of establishing an equivalent level of safety, even though the existing building is only evaluated for the 19 safety parameters.

TABLE 1201.9
EVALUATION FORMULAS[a]

FORMULA			T1201.7	T1201.8		SCORE	PASS	FAIL
FS - MFS	>	0	—— (FS) -	—— (MFS)	=	——	——	——
ME	≥	0	—— (ME) -	—— (MME)	=	——	——	——
GS - MGS	≥	0	—— (GS -	—— (MGS)	=	——	——	——

a.　FS　=　Fire Safety　　　MF　=　Mandatory Fire Safety
　　ME = Means of Egress　　MME = Mandatory Means of Egress
　　GS = General Safety　　　MGS = Mandatory General Safety

❖ Table 1201.9 shows in simple equations whether a building passes the evaluation by subtracting the mandatory safety score from Table 1201.7. This is done for each of the three general categories of evaluation: fire safety (FS), means of egress (ME) and general safety (GS). If the difference for each category is zero or greater, the existing building passes and is considered to comply with the code objectives for public safety.

Example:

A Group M mercantile occupancy receives evaluations for the 19 safety parameters, which results in total building scores in Table 1201.7 as follows:

Fire safety (FS) = 21
Means of egress (ME) = 40
General safety (GS) = 36

These scores are compared to the mandatory safety scores for a Group M occupancy from Table 1201.8.

Mandatory fire safety (MFS) = 23
Mandatory means of egress (MME) = 40
Mandatory general safety (MGS) = 40

The mandatory score is subtracted from the building score.

Conclusion:

The building fails the overall evaluation because, in the fire safety category, the building fire safety (FS) score of 21 is less than the MFS score of 23. The means of egress category was just barely satisfactory because the building score (ME) of 40 equals the mandatory score (MME) of 40. The passage of the general safety category by four points does not compensate for the failure by two points in the fire safety category. Each category must individually have a building score equal to or greater than the respective mandatory safety score for the building to pass the overall evaluation. In this case, modification of the design must be made so that at least two additional points are accrued in any one or more of the 14 parameters that affect the building fire safety (FS) score.

Bibliography

The following resource materials are referenced in this chapter or are relevant to the subject matter addressed in this chapter.

36 CFR Parts 1190 and 1191 Draft, *The Americans with Disabilities Act Accessibility Guidelines* (ADAAG). Washington, DC: Architectural and Transportation Barriers Compliance Board, April 2, 2002.

42 USC 3601-88, *Fair Housing Amendments Act* (FHAA). Washington, DC: United States Code, 1988.

"Accessibility and Egress for People with Physical Disabilities." CABO Board for the Coordination of the Model Codes Report, October 5, 1993.

ASME A17.1-00, *Safety Code for Elevators and Escalators*. New York: American Society of Mechanical Engineers, 2000.

ASME A17.3-96, *Safety Code for Existing Elevators and Escalators*. New York: American Society of Mechanical Engineers, 1996.

ASME 18.1-99, *Safety Standard for Platform Lifts and Stairway Chairlifts–with Addenda A18.1a-2001*. New York: American Society of Mechanical Engineers, 1999.

DOJ 28 CFR, Part 36-91, *Americans with Disabilities Act* (ADA). Washington, DC: Department of Justice, 1991.

DOJ 28 CFR, Part 36-91 (Appendix A), *ADA Accessibility Guidelines for Buildings and Facilities* (ADAAG). Washington, DC: Department of Justice, 1991.

FED-STD-795-88, *Uniform Federal Accessibility Standards*. Washington, DC: General Services Administration; Department of Defense; Department of Housing and Urban Development; U.S. Postal Service, 1988.

"Final Report, Recommendations for a New ADAAG." ADAAG Review Federal Advisory Committee, September 30, 1996.

ICC A117.1-98, *Accessible and Usable Buildings and Facilities*. Falls Church, VA: International Code Council, 1998.

IMC-03, *International Mechanical Code*. Falls Church, VA: International Code Council, 2003.

IPMC-03, *International Property Maintenance Code*. Falls Church, VA: International Code Council, 2003.

NFPA 13-99, *Installation of Sprinkler Systems*. Quincy, MA: National Fire Protection Association, 1999.

NFPA 13D-99, *Installation of Sprinkler Systems in One- and Two-Family Dwellings and Manufactured Homes*. Quincy, MA: National Fire Protection Association, 1999.

NFPA 13R-99, *Installation of Sprinkler Systems in Residential Occupancies Up to and Including Four Stories in Height*. Quincy, MA: National Fire Protection Association, 1999.

NFPA 72-99, *National Fire Alarm Code*. Quincy, MA: National Fire Protection Association, 1999.

NFPA 101-00, *Alternative Approaches to Life Safety*. Quincy, MA: National Fire Protection Association, 2000.

Rehabilitation Guidelines/1980. Washington, DC: National Institute of Building Science for the U.S. Department of Housing and Urban Development, 1980.

1. Guidelines for Setting and Adopting Standards for Building Rehabilitation.
2. Guideline for Approval of Building Rehabilitation.
3. Statutory Guideline for Building Rehabilitation.
4. Guideline for Managing Official Liability Associated with Building Rehabilitation.
5. Egress Guideline for Residential Rehabilitation.
6. Electrical Guideline for Residential Rehabilitation;
7. Plumbing DWV Guideline for Residential Rehabilitation; and
8. Guideline on Fire Ratings of Archaic Materials and Assemblies.

UL 555-96, *Fire Dampers*. Northbrook, IL: Underwriters Laboratories Inc., 1996.

CHAPTER 13

CONSTRUCTION SAFEGUARDS

General Comments

The building construction process involves a number of known and unanticipated hazards. Chapter 13 establishes specific regulations in order to minimize the risk to the public and adjacent properties. Some construction failures have resulted during the initial stages of grading, excavation and demolition. During these early stages, poorly designed and installed sheeting and shoring have resulted in ditch and embankment cave-ins. Also, inadequate underpinning of adjoining existing structures or careless removal of existing structures has produced construction failures.

The most critical period in building construction related to the physical safety of those on the job site is during the ongoing process when all building components have not yet been completed. Compounding this incomplete state is the use of dangerous construction methods, materials and equipment. The importance of reasonable precautions is evidenced by the federal Occupational Safety and Health Act and related state and local regulations.

Purpose

The purpose of Chapter 13 is to cite safety requirements during construction or demolition of buildings and structures. These requirements are intended to protect the public from injury and adjoining property from damage.

SECTION 1301
GENERAL

[B] 1301.1 Scope. The provisions of this chapter shall govern safety during construction that is under the jurisdiction of this code and the protection of adjacent public and private properties.

❖ The fundamental rationale behind this section is to establish that reasonable safety precautions in accordance with this chapter be provided during the construction or demolition process to protect the public from injury and adjoining property from damage.

[B] 1301.2 Storage and placement. Construction equipment and materials shall be stored and placed so as not to endanger the public, the workers or adjoining property for the duration of the construction project.

❖ This section requires that construction materials and equipment must be located and protected pursuant to the governing provisions of this chapter so that the public and adjoining property are safeguarded at all times during construction or demolition processes.

1301.3 Alterations, repairs, and additions. Required exits, existing structural elements, fire protection devices, and sanitary safeguards shall be maintained at all times during alterations, repairs, or additions to any building or structure.

Exceptions:

1. When such required elements or devices are being altered or repaired, adequate substitute provisions shall be made.

2. When the existing building is not occupied.

❖ Demolition and construction operations must not create a hazard for the occupants of a building during an alteration or addition. As such, the existing fire protection, means of egress elements and safety systems must remain in place and be functional. However, two options are provided in this section that permit the fire protection, means of egress elements and safety systems to be changed, modified or removed. The first option provides an alternative method that must be approved by the code official. The second option is simply to have an unoccupied building, which means that residents of a dwelling unit or employees of a business are not to be present during the construction or demolition process.

[B] 1301.4 Manner of removal. Waste materials shall be removed in a manner which prevents injury or damage to persons, adjoining properties, and public rights-of-way.

❖ Safe and sanitary procedures for the removal of building construction and demolition waste must be provided. The method of waste removal must be controlled such that debris will not pose a hazard, eyesore or nuisance to the public and neighboring properties. Examples of acceptable practices include: evidence that a professional disposal service will haul away the debris; limiting areas used for storing and handling demolished materials; enclosing storage areas so that only authorized personnel can gain access; establishing routes that waste removal vehicles are permitted to use; covering or tarping debris to prevent flying objects; providing fully enclosed chutes to control falling objects and dust; and scheduling waste removal when adjoining property and the public will be least exposed to unusual and possibly dangerous situations caused by fumes, noise, dust and unfamiliar events involved during a demolition process. Note that the accepted practice of waste removal is subject to the approval of the code official.

[B] 1301.5 Facilities required. Sanitary facilities shall be provided during construction or demolition activities in accordance with the *International Plumbing Code*.

❖ Construction employees must have plumbing facilities available during the construction or demolition process of a building. The facilities must conform to the requirements set forth in the *International Plumbing Code®* (IPC®).

[B] 1301.6 Protection of pedestrians. Pedestrians shall be protected during construction and demolition activities as required by Sections 1301.6.1 through 1301.6.7 and Table 1301.6. Signs shall be provided to direct pedestrian traffic.

❖ Safeguards are required to be in place during construction or demolition operations in accordance with Table 1301.6 and this chapter. In addition, since construction operations alter the familiar setting and path of travel, it is necessary to provide some form of visible directional signage to lead the public toward safety and away from potential hazards.

   This table establishes the type of protection required based on the location to overhead hazards and distance in relation to ground hazards. Once the type of protection is determined from the table, the applicable code section, such as Section 1301.6.3 for construction railings, Section 1301.6.4 for barriers and Section 1301.6.5 for covered walkways, must be provided as regulated.

[B] 1301.6.1 Walkways. A walkway shall be provided for pedestrian travel in front of every construction and demolition site unless the authority having jurisdiction authorizes the sidewalk to be fenced or closed. Walkways shall be of sufficient width to accommodate the pedestrian traffic, but in no case shall they be less than 4 feet (1219 mm) in width. Walkways shall be provided with a durable walking surface. Walkways shall be accessible in accordance with Chapter 11 of the *International Building Code* and shall be designed to support all imposed loads and in no case shall the design live load be less than 150 psf (7.2 kN/m²).

❖ Construction operations must not narrow or impede the normal flow of pedestrian traffic along a walkway by the placement of a fence or other enclosure. The

authority having jurisdiction must approve any construction operations that will cause the narrowing or impedance of a walkway, such as the sidewalk. If a walkway is narrowed or enclosed, the construction of another or wider walkway is required for all pedestrians. The walkway must be able to handle the normal anticipated flow of pedestrian traffic and must not be less than the minimum 4-foot (1219 mm) width. In addition, the walkway must be a stable surface that is capable of supporting all imposed loads. The minimum design load of the walkway must not be less than 150 pounds per square feet (psf) (7.2 kN/m²).

[B] 1301.6.2 Directional barricades. Pedestrian traffic shall be protected by a directional barricade where the walkway extends into the street. The directional barricade shall be of sufficient size and construction to direct vehicular traffic away from the pedestrian path.

❖ Similar to the local, federal and state guidelines that govern provisions to protect employees while doing road work, this section establishes that a barrier must be erected. The barrier must be capable of redirecting and regulating the flow of vehicular traffic where constructed or projected into a street. An example of how to regulate traffic is to provide the necessary warnings, notice of caution and instructions to the operators of motor vehicles with signs and colors that are common to the local, federal and state guidelines. Note that the intent of this visual barrier is to keep pedestrians out of vehicular traffic and prevent vehicles from encroaching on or using the pedestrian walkway as a roadway.

[B] 1301.6.3 Construction railings. Construction railings shall be at least 42 inches (1067 mm) in height and shall be sufficient to direct pedestrians around construction areas.

❖ A barrier consisting of a horizontal rail and supports must be constructed with a minimum height of 42 inches (1067 mm). In addition, this barrier must be capable of controlling the flow of pedestrian traffic by

[B] TABLE 1301.6
PROTECTION OF PEDESTRIANS

HEIGHT OF CONSTRUCTION	DISTANCE OF CONSTRUCTION TO LOTLINE	TYPE OF PROTECTION REQUIRED
8 feet or less	Less than 5 feet	Construction railings
	5 feet or more	None
More than 8 feet	Less than 5 feet	Barrier and covered walkway
	5 feet or more, but not more than one-fourth the height of construction	Barrier and covered walkway
	5 feet or more, but between one-fourth and one-half the height of construction	Barrier
	5 feet or more, but exceeding one-half the height of construction	None

For SI: 1 foot = 304.8 mm.

routing travel away from the hazards associated with a construction area.

[B] 1301.6.4 Barriers. Barriers shall be a minimum of 8 feet (2438 mm) in height and shall be placed on the side of the walkway nearest the construction. Barriers shall extend the entire length of the construction site. Openings in such barriers shall be protected by doors which are normally kept closed.

❖ When a barrier is required by Table 3306.1, it must be constructed to impede, separate and obstruct the passage of pedestrians onto a construction site. The structure must be a minimum of 8 feet (2438 mm) in height and located continuously along the walkway on the side where construction activities are being performed. Note that openings such as doors or gates are permitted as long as the only time they are open is during use by authorized personnel. As such, when the doors or gates are not being used, they must remain closed.

[B] 1301.6.4.1 Barrier design. Barriers shall be designed to resist loads required in Chapter 16 of the *International Building Code* unless constructed as follows:

1. Barriers shall be provided with 2 × 4 top and bottom plates.

2. The barrier material shall be a minimum of $^3/_4$ inch (19.1 mm) inch boards or $^1/_4$ inch (6.4 mm) wood structural use panels.

3. Wood structural use panels shall be bonded with an adhesive identical to that for exterior wood structural use panels.

4. Wood structural use panels $^1/_4$ inch (6.4 mm) or $^1/_{16}$ inch (23.8 mm) in thickness shall have studs spaced not more than 2 feet (610 mm) on center.

5. Wood structural use panels $^3/_8$ inch (9.5 mm) or $^1/_2$ inch (12.7 mm) in thickness shall have studs spaced not more than 4 feet (1219 mm) on center, provided a 2 inch by 4 inch (51 mm by 102 mm) stiffener is placed horizontally at the mid height where the stud spacing exceeds 2 feet (610 mm) on center.

6. Wood structural use panels $^5/_8$ inch (15.9 mm) or thicker shall not span over 8 feet (2438 mm).

❖ This section establishes two methods that can be used to design barriers. The first method establishes that the barrier can be designed using the same materials and assembly instructions as indicated in Items 1 through 6. The second method provides for a barrier that will resist the loads imposed on it to be part of a designed assembly. The loads, which need to be resisted, are the same as those established in Chapter 16 of the *International Building Code*® (IBC®).

[B] 1301.6.5 Covered walkways. Covered walkways shall have a minimum clear height of 8 feet (2438 mm) as measured from the floor surface to the canopy overhead. Adequate lighting shall be provided at all times. Covered walkways shall be designed to support all imposed loads. In no case shall the design live load be less than 150 psf (7.2 kN/m²) for the entire structure.

Exception: Roofs and supporting structures of covered walkways for new, light-frame construction not exceeding two stories in height are permitted to be designed for a live load of 75 psf (3.6 kN/m²) or the loads imposed on them, whichever is greater. In lieu of such designs, the roof and supporting structure of a covered walkway are permitted to be constructed as follows:

1. Footings shall be continuous 2 × 6 members.

2. Posts not less than 4 × 6 shall be provided on both sides of the roof and spaced not more than 12 feet (3658 mm) on center.

3. Stringers not less than 4 × 12 shall be placed on edge upon the posts.

4. Joists resting on the stringers shall be at least 2 × 8 and shall be spaced not more than 2 feet (610 mm) on center.

5. The deck shall be planks at least 2 inches (51 mm) thick or wood structural panels with an exterior exposure durability classification at least 23/32 inch (18.3 mm) thick nailed to the joists.

6. Each post shall be knee-braced to joists and stringers by 2 × 4 minimum members 4 feet (1219 mm) long.

7. A 2 × 4 minimum curb shall be set on edge along the outside edge of the deck.

❖ Walkways that are provided with a covering that extends out over the pedestrian path of travel, pursuant to Table 1301.6, must be at least 8 feet (2438 mm) in height. The measurement must be taken from the top of the walking surface vertically upward to the underside of the roof covering. The walkway must maintain a satisfactory level of light to the space that will be equal to or better than that provided prior to the installation of the protective covering. The covered walkway must also be structurally capable of resisting the design loads established in Chapter 16 of the IBC, but not less than 150 psf (7.2 kN/m²) for the live load.

Note that this section provides two alternative design options. One option permits the covered walkway to resist the design loads established in Chapter 16 of the IBC, but not less than the minimum 75 psf (3.6 kN/m²) live load provision if nearby construction does not exceed two stories in height and the construction materials are lightweight, such as light-frame construction. The second option is to construct the covered walkway in accordance with the seven criteria set forth in this section.

[B] 1301.6.6 Repair, maintenance and removal. Pedestrian protection required by Section 1301.6 shall be maintained in place and kept in good order for the entire length of time pedestrians may be endangered. The owner or the owner's agent, upon the completion of the construction activity, shall immediately remove walkways, debris and other obstructions and leave such public property in as good a condition as it was before such work was commenced.

❖ Any safeguards required by this chapter must be kept in good functional condition for the entire duration of the construction or demolition activity so that the public will not be placed in harm's way. Once a building or structure is occupiable and the site is properly graded, the protection must be removed. As such, public property affected by the construction activity must be restored to or left in the condition that existed prior to the work.

[B] 1301.6.7 Adjacent to excavations. Every excavation on a site located 5 feet (1524 mm) or less from the street lot line shall be enclosed with a barrier not less than 6 feet (1829 mm) high. Where located more than 5 feet (1524 mm) from the street lot line, a barrier shall be erected when required by the code official. Barriers shall be of adequate strength to resist wind pressure as specified in Chapter 16 of the *International Building Code.*

❖ This section establishes that whenever excavation is to take place less than 5 feet (1524 mm) from the edge of a roadway, a barrier must be erected with a minimum height of 6 feet (1829 mm). If the excavation is located greater than 5 feet (1524 mm) from the edge of the roadway, the code official must evaluate the level of hazard for the public and the necessary precautions to take. As such, the code official is the authority to order the construction of a structural barrier. Note that any barrier erected must maintain other provisions of the code such that it is capable of handling and resisting design wind loads denoted in Chapter 16 of the IBC.

[B] SECTION 1302
PROTECTION OF ADJOINING PROPERTY

1302.1 Protection required. Adjoining public and private property shall be protected from damage during construction and demolition work. Protection must be provided for footings, foundations, party walls, chimneys, skylights and roofs. Provisions shall be made to control water run-off and erosion during construction or demolition activities. The person making or causing an excavation to be made shall provide written notice to the owners of adjoining buildings advising them that the excavation is to be made and that the adjoining buildings should be protected. Said notification shall be delivered not less than 10 days prior to the scheduled starting date of the excavation.

❖ This section emphasizes the need to protect all existing public and private property bordering the proposed construction or demolition operations. The term "property" only alludes to existing buildings. As such, any building element or system must be provided with a safeguard that will limit the damage that could be caused from the processes that involve the equipment and materials used. Additionally, soil erosion and land disbursement control resulting from the construction or demolition operations must be provided to prevent spillage and spread of disturbed soil debris. The owner or owner's agent has the responsibility to provide a written notice 10 days in advance for any demolition or construction activities that may warrant bordering lots to be protected from damage.

[B] SECTION 1303
TEMPORARY USE OF STREETS, ALLEYS AND PUBLIC PROPERTY

1303.1 Storage and handling of materials. The temporary use of streets or public property for the storage or handling of materials or of equipment required for construction or demolition, and the protection provided to the public shall comply with the provisions of the authority having jurisdiction and Section 1303.

❖ In addition to jurisdictional regulations, this chapter establishes procedures for the storage of construction materials and equipment to protect the access to public safety equipment, utilities and public transportation facilities.

1303.2 Obstructions. Construction materials and equipment shall not be placed or stored so as to obstruct access to fire hydrants, standpipes, fire or police alarm boxes, catch basins or manholes, nor shall such material or equipment be located within 20 feet (6.1 m) of a street intersection, or placed so as to obstruct normal observations of traffic signals or to hinder the use of public transit loading platforms.

❖ This section indicates that precautions for material and equipment storage and placement must be provided for so as not to block or obstruct access to fire hydrants; standpipes (including fire department siamese connections for sprinklers and standpipes); fire or police alarm boxes; utility boxes and meters; catch basins or manholes; or any other vital facility whose function contributes to the health, safety and welfare of the public. Also, storage must not be placed within 20 feet (6096 mm) of a street intersection if it obstructs the normal observation of traffic signals or hinders the use of any mass-transit loading platforms, such as sidewalk bus stops, taxi waiting areas, etc.

1303.3 Utility fixtures. Building materials, fences, sheds or any obstruction of any kind shall not be placed so as to obstruct free approach to any fire hydrant, fire department connection, utility pole, manhole, fire alarm box, or catch basin, or so as to interfere with the passage of water in the gutter. Protection against damage shall be provided to such utility fixtures during the progress of the work, but sight of them shall not be obstructed.

❖ Utility fixtures such as those used by electrical, telephone, water, gas or sewer companies as well as fire protection devices must be hidden from view and not be blocked from access and use. When construction operations are near a utility fixture, precautions must be provided so as not to cause damage. Note that the utility companies and the authority having jurisdiction (including the fire department) may have specific requirements and guidelines to follow that limit the possibility of damage and obstruction.

SECTION 1304
FIRE EXTINGUISHERS

[F] 1304.1 Where required. All structures under construction, alteration, or demolition shall be provided with not less than one approved portable fire extinguisher in accordance with Section 906 of the *International Fire Code* and sized for not less than ordinary hazard as follows:

1. At each stairway on all floor levels where combustible materials have accumulated.

2. In every storage and construction shed.

3. Additional portable fire extinguishers shall be provided where special hazards exist, such as the storage and use of flammable and combustible liquids.

❖ A means to provide fire protection during the construction and demolition process is required. As such, this section indicates provisions for portable fire extinguishers. In addition to provisions of this section, the regulations of Section 906 of the *International Fire Code®* (IFC®) must be maintained.

[B] 1304.2 Fire hazards. The provisions of this code and of the *International Fire Code* shall be strictly observed to safeguard against all fire hazards attendant upon construction operations.

❖ Methods, procedures and construction materials each contribute, in some way, to creating a fire hazard. Therefore, in addition to the provisions of the code, the IFC must be used to regulate the proper means of safety that must be provided for a building that is being demolished, altered or constructed.

[B] SECTION 1305
EXITS

1305.1 Stairways required. Where an existing building exceeding 50 feet (15 240 mm) in height is altered, at least one temporary lighted stairway shall be provided unless one or more of the permanent stairways is available for egress as the construction progresses.

❖ Temporary stairways must be constructed in accordance with Section 1003.3.3 of the IBC, including riser, tread, guard and handrail requirements.

1305.2 Maintenance of exits. Required means of egress shall be maintained at all times during alterations, repairs and additions to any building.

❖ Access to all required existing exits must remain unobstructed and useable by any occupants within the existing building while construction or demolition work is being done. Note that depending upon the extent of work, an exit may become obstructed. Therefore, an alternative exit must be provided if the blocked exit is a required exit.

[F] SECTION 1306
STANDPIPE SYSTEMS

1306.1 Where required. Buildings required to have a standpipe system in accordance with this code shall be provided with not less than one standpipe for use during construction. Such standpipes shall be installed where the progress of construction is not more than 40 feet (12 192 mm) in height above the lowest level of fire department access. Such standpipe shall be provided with fire department hose connections at accessible locations adjacent to usable stairs. Such standpipes shall be extended as construction progresses to within one floor of the highest point of construction having secured decking or flooring.

❖ The scope of this section is to provide fire safety procedures during the construction operation in accordance with the code and the IFC. Standpipes that are required by Section 905 of the IFC are to be a permanent part of the building and must be installed and remain functional as construction or demolition progresses. Functional standpipes are required so that fire-fighting capability is available at all times within a reasonable proximity to all potential fire locations. Note that standpipes must be in place when a building or structure under construction exceeds 40 feet (12 192 mm) in height and thereafter as it progresses to its completed height. During demolition or construction, the standpipe must be operational no lower than one floor below the highest point of construction.

1306.2 Buildings being demolished. Where a building or portion of a building is being demolished and a standpipe is existing within such a building, such standpipe shall be maintained in an operable condition so as to be available for use by the fire department. Such standpipe shall be demolished with the building but shall not be demolished more than one floor below the floor being demolished.

❖ Standpipes in buildings under demolition must remain in service during the demolition process so that fire-fighting capability is maintained. Similar to buildings under construction, the standpipe is to remain in operation up to one floor level below the floor being demolished. The purpose of this section is to establish the requirement for fire safety procedures during the demolition process in accordance with the code and the IFC.

1306.3 Detailed requirements. Standpipes shall be installed in accordance with the provisions of Chapter 9 of the *International Building Code*.

Exception: Standpipes shall be either temporary or permanent in nature, and with or without a water supply, provided that such standpipes conform to the requirements of Section 905 of the *International Building Code* as to capacity, outlets and materials.

❖ During the construction or demolition process, standpipes complying with Section 1306.1 must be provided in a temporary or permanent location so fire fighters will have sufficient means to supply water to their connection. These standpipes must either be connected to a permanent water supply source or have the capabilities of being connected to one in accordance with Section 1306.4. Note that all standpipes must comply with Section 905 of the IFC and must be installed and provided in a building pursuant to the regulations of Chapter 9 of the IFC.

1306.4 Water supply. Water supply for fire protection, either temporary or permanent, shall be made available as soon as combustible material accumulates.

❖ This section is to provide a fire safety provision in accordance with the code and the IFC. As such, when a standpipe system is present during the construction process pursuant to Section 1306.1, a water source must be readily available to provide fire-fighting capabilities at all times.

[F] SECTION 1307
AUTOMATIC SPRINKLER SYSTEM

1307.1 Completion before occupancy. In portions of a building where an automatic sprinkler system is required by this code, it shall be unlawful to occupy those portions of the building until the automatic sprinkler system installation has been tested and approved, except as provided in Section 110.3.

❖ A certificate of occupancy must not be issued by the code official if a required automatic sprinkler system is not approved. However, a temporary occupancy may be granted at the discretion of the code official.

1307.2 Operation of valves. Operation of sprinkler control valves shall be permitted only by properly authorized personnel and shall be accompanied by notification of duly designated parties. When the sprinkler protection is being regularly turned off and on to facilitate connection of newly completed segments, the sprinkler control valves shall be checked at the end of each work period to ascertain that protection is in service.

❖ The scope of this section is to provide fire safety procedures during construction operations in accordance with the code and the IFC. Note that a sprinkler system must remain operable unless work is being done on it. In such a case, the water must only be shut off by authorized personnel coupled with the notification of the proper authorities so that a form of check and balance is achieved to make certain the water is turned back on.

SECTION 1308
ACCESSIBILITY

1308.1 Construction sites. Structures, sites, and equipment directly associated with the actual process of construction, including but not limited to scaffolding, bridging, material hoists, material storage, or construction trailers are not required to be accessible.

❖ This section exempts structures directly associated with the construction process because the need for accessibility on a continuous or regular basis in those circumstances is unlikely to arise. Note that all structures that may be involved during a construction project are not exempt, only those specifically involved in the actual process of construction. For example, if mobile units are brought into house classrooms during a school addition project, those mobile units would have to comply with accessibility provisions. This is consistent with the exception permitted for new construction in Section 1103.2.6 of the IBC.

Bibliography

The following resource materials are referenced in this chapter or are relevant to the subject matter addressed in this chapter.

DOL 29 CFR, Part 1910-74, *Occupational Safety and Health Standards*. Washington, DC: U.S. Department of Labor–Occupational Safety and Health Administration, 1974.

IBC-2003, *International Building Code.* Falls Church, VA: International Code Council, 2003.

IFC-2003, *International Fire Code.* Falls Church, VA: International Code Council, 2003.

IPC-2003, *International Plumbing Code.* Falls Church, VA: International Code Council, 2003.

CHAPTER 14

REFERENCED STANDARDS

General Comments

Chapter 14 contains a comprehensive list of all standards that are referenced in the code. It is organized in a manner that makes it easy to locate specific document references.

This chapter lists the standards that are referenced in various sections of this document. The standards are listed herein by the promulgating agency of the standard; the standard identification; the date and title; and the section or sections of this document that reference the standard. The application of the referenced standards shall be as specified in Section 102.4.

It is important to understand that not every document related to building design and construction is qualified to be a "referenced standard." The International Code Council® (ICC®) has adopted a criterion that standards referenced in the *International Codes*® and standards intended for adoption into the *International Codes* must meet in order to qualify as a referenced standard. The policy is summarized as follows:

- *Code references*: The scope and application of the standard must be clearly identified in the code text.

- *Standard content*: The standard must be written in mandatory language and appropriate for the subject covered. The standard shall not have the effect of requiring proprietary materials or prescribing a proprietary testing agency.

- *Standard promulgation*: The standard must be readily available and developed and maintained in a consensus process, such as ASTM or ANSI.

It should be noted that the ICC Code Development Procedures, of which the standards policy is a part, are updated periodically. A copy of the latest version can be obtained from the ICC offices.

Once a standard is incorporated into the code through the code development process, it becomes an enforceable part of the code. When the code is adopted by a jurisdiction, the standard is also a part of that jurisdiction's adopted code. It is for this reason that the criteria were developed. Compliance with this policy provides that documents incorporated into the code are, among others, developed through the use of the consensus process, written in mandatory language and do not mandate the use of proprietary materials or agencies. The requirement for a standard to be developed through a consensus process is vital, as it means that the standard will be representative of the most current body of available knowledge on the subject as determined by a broad spectrum of interested or affected parties without dominance by any single interest group. A true consensus process has many attributes, including but not limited to:

- An open process that has formal (published) procedures that allow for the consideration of all viewpoints;

- A definitive review period that allows for the standard to be updated or revised;

- A process of notification to all interested parties; and

- An appeals process.

Many available documents related to design, installation and construction, though useful, are not "standards" and are not appropriate for reference in the code. Often, these documents are developed or written with the intention of being used for regulatory purposes and are unsuitable for use as a regulation due to extensive use of recommendations, advisory comments and nonmandatory terms. Typical examples of such documents include installation instructions, guidelines and practices.

The objective of ICC's standards policy is to provide regulations that are clear, concise and enforceable–thus the requirement for standards to be written in mandatory language. This requirement is not intended to mean that a standard cannot contain informational or explanatory material that will aid the user of the standard in its application. When it is the desire of the standard's promulgating agency for such material to be included, however, the information must appear in a nonmandatory location, such as an annex or appendix, and be clearly identified as not being part of the standard.

Overall, standards referenced by the code must be authoritative, relevant, up to date and, most important, reasonable and enforceable. Standards that comply with ICC's standards policy fulfill these expectations.

Purpose

As a performance-oriented code, the code contains numerous references to documents that are used to regulate materials and methods of construction. The references to these documents within the code text consist of the promulgating agency's acronym and its publication designation (e.g., ASME A17.1) and a further indication that the document being referenced is the one that is listed in Chapter 14. Chapter 14 contains all the information that is necessary to identify the specific referenced document. Included is the following information on a document's promulgating agency (see Figure 14):

- The promulgating agency (i.e. the agency's title);

- The promulgating agency's acronym; and

- The promulgating agency's address.

For example, a reference to an ASME standard within the code indicates that the document is promulgated by

FIGURE 14   REFERENCED STANDARDS

the American Society of Mechanical Engineers (ASME), which is located in New York City. This chapter lists the standards agencies alphabetically for ease of identification. This chapter also includes the following information on the referenced document itself (see Figure 14):

- The document's publication designation;
- The document's edition year;
- The document's title;
- Any addenda or revisions to the document that are applicable; and
- Every section of the code in which the document is referred.

For example, a reference to ASME A17.1 indicates that this document can be found in Chapter 14 under the heading ASME. The specific standards designation is A17.1. For convenience, these designations are listed in alphanumeric order. Chapter 14 identifies that ASME A17.1 is titled *Safety Code for Elevators and Escalators*, the applicable edition (i.e., its year of publication) is 2000 and is referenced in numerous sections of the code.

This chapter also indicates when a document has been discontinued or replaced by its promulgating agency. When a document is replaced by a different one, a note will appear to tell the user the designation and title of the new document.

The key aspect of the manner in which standards are referenced by the code is that a specific edition of a specific standard is clearly identified. In this manner, the requirements necessary for compliance can be readily determined. The basis for code compliance is, therefore, established and available on an equal basis to the code official, contractor, designer and owner.

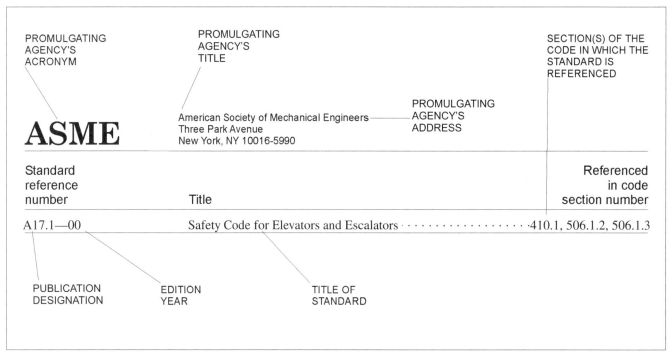

Figure 14
REFERENCED STANDARDS

This chapter lists the standards that are referenced in various sections of this document. The standards are listed herein by the promulgating agency of the standard, the standard identification, the effective date and title, and the section or sections of this document that reference the standard. The application of the referenced standards shall be as specified in Section 102.4.

ASCE

American Society of Civil Engineers
1801 Alexander Bell Drive
Reston, VA 20191-4400

Standard reference number	Title	Referenced in code section number
31—02	Seismic Evaluation of Existing Buildings	407.1.1, Table 407.1.1.2, 407.1.1.3

ASHRAE

American Society of Heating, Refrigerating and Air Conditioning Engineers
1791 Tullie Circle, NE
Atlanta, GA 30329

Standard reference number	Title	Referenced in code section number
62—01	Ventilation for Acceptable Indoor Air Quality	609.2

ASME

American Society of Mechanical Engineers
3 Park Avenue
New York, NY 10016

Standard reference number	Title	Referenced in code section number
A17.1—2000	Safety Code for Elevators and Escalators	506.1.2
A18.1—1999	Safety Standard for Platform Lifts and Stairway Chair Lifts—with A18.1a—2001 Addenda	506.1.3
A112.19.2M—1998	Vitreous China Plumbing Fixtures	410.1

DOJ

Department of Justice
950 Pennsylvania Avenue NW
Washington, DC 20530-0001

Standard reference number	Title	Referenced in code section number
28 CFR Part 36.403	Nondiscrimination on the Basis of Disability by Public Accommodations and in Commercial Facilities—New Construction and Alterations—Alterations: Path of Travel	506.2

DOTn

Department of Transportation
400 7th Street SW
Room 8102
Washington, DC 20590-0001

Standard reference number	Title	Referenced in code section number
49 CFR Part 37.43 (c)	Alteration of Transportation Facilities by Public Entities	506.2

FEMA

Federal Emergency Management Agency
Federal Center Plaza
500 C Street SW
Washington, DC 20472

Standard reference number	Title	Referenced in code section number
PUB 356	Pre-standard and Commentary for the Seismic Rehabilitation of Buildings	407.1.1.1, Table 407.1.1.2, 407.1.1.3

ICC

International Code Council, Inc.
5203 Leesburg Pike, Suite 600
Falls Church, VA 22041

Standard reference number	Title	Referenced in code section number
IBC—00	International Building Code®	102.4.2
IBC—03	International Building Code®	101.2, 102.4.2, 106.1.1.1, 109.3.3, 109.3.8, 110.2, 202, 301.4, 401.4, 402.1, 403.2, 407.1.1.1, 407.1.1.3, Table 407.1.1.2, 407.1.2, 407.2, 407.3.1, 407.3.2.1.1, 407.3.5, 501.3, 502.1, 503.1, 503.3, 506.1, 506.1.1, 506.1.7, 506.1.9, 507.2.1, 507.2.2, 601.3, 602.1, 603.2.1, 603.2.3, 603.3.2, 603.4, 603.5.2, 604.2, 604.2.2, 604.2.3, 604.2.4, 604.3, 605.2, 605.3.1, 605.4.3, 605.5, 605.6, 605.7.1, 605.8.1, 605.9.2, 605.10.2, 606.2, 606.3, 607.1, 607.2, 607.3, 607.4, 607.4.1, 607.4.3, 704.1.2, 704.2, 704.2.1, 705.2, 705.3, 707.2, 707.3, 707.5.1, 707.6, 707.7, 801.1, 801.3, 802.1, 802.2, 807.1, 807.2, 807.3.1, 811.1.1.1, 812.1.1, 812.1.2, 812.3.1, 812.4.1.1, 812.4.1.2, 812.4.1.3, 812.4.2.1, 812.4.2.3, 812.4.3.1, 812.4.3.3, 812.4.4.1, 812.4.4.3, 812.5, 902.1, 902.2, 903.1, 903.2, 903.3, 903.3.1, 903.3.2, 903.4, 903.5, 904.1, 904.2, 1001.4, 1002.3, 1002.5, 1005.2, 1005.9, 1101.2, 1102.1, 1102.2, 1102.2.1, 1102.3, 1102.4, 1102.5, 1102.6, 1201.2.2, 1201.2.3, 1201.2.4, 1201.2.5, 1201.3.3, 1201.4.1, 1201.6.1, 1201.6.1.1, 1201.6.2, 1201.6.2.1, 1201.6.3.1, 1201.6.3.2, 1201.6.13, 1201.6.4.1, 1201.6.5, 1201.6.5.1, 1201.6.6, 1201.6.7.1, 1201.6.8, 1201.6.9.1, 1201.6.10.1, 1201.6.11, 1201.6.11.1, 1201.6.12.1, 1201.6.15.1, Table 1201.6.15, 1201.6.16.1, 1201.6.17, 1201.6.17.1, 1201.6.18, 1201.6.18.1, 1201.6.19, 1301.6.1, 1301.6.4.1, 1301.6.7, 1306.3
ICC/ANSI A117.1—98	Guidelines for Accessible and Usable Buildings and Facilities	Table 407.1.1.2, 506.1.2, 506.1.3
ICC EC—03	ICC Electrical Code™	107.3, 503.3, 608.1, 608.3.4, 608.3.7, 808.1, 808.2, 808.3, 808.4
IECC—03	International Energy Conservation Code®	503.3, 906.1
IFC—03	International Fire Code®	101.4, 603.2.1, 603.2.3, 604.4.1.1, 604.4.1.2, 604.4.1.3, 604.4.1.4, 604.4.1.5, 604.4.1.6, 604.4.1.7, 604.4.3, 702.1.2, 1101.2, 1201.3.2, 1201.6.8.1, 1201.6.14, 1201.6.14.1, 1304.1, 1304.2
IFGC—03	International Fuel Gas Code®	503.3.1
IMC—03	International Mechanical Code®	101.2, 503.3, 609.1, 702.1.1, 702.2.1, 809.1, 1201.6.7.1, 1201.6.8, 1201.6.8.1
IPC—03	International Plumbing Code®	101.2, 410.2, 503.3, 610.1, 810.1, 810.2, 810.3, 810.5, 1301.5
IPMC—03	International Property Maintenance Code®	101.4, 101.5, 1101.2, 1201.3.2
IRC—03	International Residential Code®	101.2, 403.2, 407.1.2, 507.2.1, 607.4.1, 607.4.3, 608.3, 703.2.1, 707.5.1, 707.6, 903.2, 903.3, 903.4, 904.1, 904.2, 1102.1, 1201.2.2, 1201.2.3
NBC—99	BOCA National Building Code®	102.4.2
SBC—99	Standard Building Code®	102.4.2
UBC—97	Uniform Building Code™	102.4.2

NFPA

National Fire Protection Agency
1 Batterymarch Park
Quincy, MA 02269-9101

Standard reference number	Title	Referenced in code section number
NFPA 13R—99	Installation of Sprinkler Systems in Residential Occupancies up to and Including Four Stories in Height	604.2.5
NFPA 70—99	National Electrical Code	408.1
NFPA 72—99	National Fire Alarm Code	604.2.5, 604.4
NFPA 99—99	Health Care Facilities	408.1
NFPA 101—00	Life Safety Code	605.2

Appendix A: Guidelines for the Seismic Retrofit of Existing Buildings
CHAPTER A1

SEISMIC STRENGTHENING PROVISIONS FOR UNREINFORCED MASONRY BEARING WALL BUILDINGS

[Commentary portions of Chapters A1 through A5 are produced by the Structural Engineers Association of California (SEAOC). This commentary is jointly copyrighted by SEAOC and ICC.]

General

The provisions in Chapter A1 call for the selection of one of two procedures–either the general procedure or the special procedure. The special procedure is based on research on the behavior of URM buildings at their limit state and on the experience of engineers who have upgraded buildings in the Los Angeles hazard reduction program; use of the special procedure is expected for most buildings. The general procedure is a procedure for buildings with rigid diaphragms that do not qualify for the special procedure; it is a combination of a conventional code approach and some portions of the special procedure. The commentary discusses the general and special procedures individually and in detail at appropriate places in the document.

"Building code" means the current code adopted by the governing jurisdiction. Notations and definitions as given in the *Uniform Building Code* (UBC) 1997 edition and in the *International Building Code®* (IBC®) 2003 edition apply to Chapter A1.

Unreinforced masonry (URM)-bearing wall buildings have shown poor performance in past earthquakes: 1868 Hayward, California; 1906 San Francisco, California; 1925 Santa Barbara, California; 1933 Long Beach, California; 1952 Kern County, California; 1935 Helena, Montana; 1964 Olympic, Washington; 1971 San Fernando, California; 1983 Coalinga, California; 1987 Whittier, California; 1989 Loma Prieta, California; 1994 Northridge, and 2001 Nisqually, Washington.

California building codes changed after the 1933 Long Beach earthquake and few URM buildings have been built in California since then; however, there are large numbers of URM buildings that remain in the United States. Following the lead of the cities of Long Beach in the 1970s and Los Angeles in the 1980s, the state of California declared through Senate Bill 547 (Sec. 8875 et seq. of the Government Code) that the hazard posed by this class of building is unacceptable and that communities must identify them. The Senate bill does not specify the level of performance required or expected by communities. Although it was the Senate bill that prompted the predecessor to this document, the hazard of URM buildings is not limited to UBC Seismic Zones 3 and 4 (which are roughly equivalent to IBC Seismic Design Categories C and D) in California. The current edition of the *Guidelines for Seismic Retrofit of Existing Buildings* (GSREB) is cor-related to the mapped seismic hazards of the United States. The loss of lives and property damage in future earthquakes will be heaviest in buildings that exist today. The short-term impact of improved knowledge, codes and design practices will be limited to those relatively few recent buildings that take advantage of this improved knowledge. Thus, the dominant policy issues posed by earthquakes involve not new but existing buildings, particularly those structures that have obvious weaknesses and do not comply with the general intent and necessary requirements of current regulations. The issue before the public and the profession is determining how to set standards for these noncompliant existing buildings that are consistent with both the desire for safety and the limited resources available to achieve improved safety.

It is generally accepted that the intensity of earthquakes reasonably expected to occur in the moderate and high seismic zones of the United States would be sufficient to cause buildings with minimal seismic resistance characteristics to be seriously damaged, causing injury or death to occupants or passers-by.

It is reasonable, when a real hazard exists, to take steps to significantly reduce the hazard. The objective of Chapter A1 is the reduction or elimination of seismic hazards associated with URM buildings.

The goal should be to reduce life safety hazards as best as possible with the available resources. The efforts should be directed toward ensuring a coherent load path for lateral loads, reduction of out-of-plane wall failures, reduction of loss of support for floors and roofs and reduction of falling parapets or ornamentation. Application of this chapter will decrease the probability of loss of life, but this cannot be prevented. In order to provide a basis for designing lateral-force-resisting systems and details for URM buildings it is necessary to prescribe analysis and retrofit design procedures that are consistent with the desire to reduce the chances for failures of the types noted. We should be willing to accept some major and irreparable damage as long as there is a decrease in the likelihood of falling building elements or loss of support for horizontal framing.

The goal of this recommended strengthening is lower than the goal set for new construction. Chapter A1 recognizes that the economic difficulty of strengthening existing buildings necessitates reliance on building components with seismic performance characteristics that are less than ideal.

SECTION A101
PURPOSE

The purpose of this chapter is to promote public safety and welfare by reducing the risk of death or injury that may result from the effects of earthquakes on existing unreinforced masonry bearing wall buildings.

The provisions of this chapter are intended as minimum standards for structural seismic resistance, and are established primarily to reduce the risk of life loss or injury. Compliance with these provisions will not necessarily prevent loss of life or injury, or prevent earthquake damage to rehabilitated buildings.

❖ Chapter A1 is intended for use as a hazard reduction program. It represents minimum standards required to reduce risk of life loss or injury. The objective is accomplished by reducing the possibility of damage and the extent of damage. However, compliance with this chapter will not necessarily prevent earthquake damage to rehabilitated buildings. Risk of life loss or injury is significantly reduced but not eliminated with compliance of this chapter.

The risk reduction applies to the overall average performance of rehabilitated URM buildings. An individual building may have damage levels above or below the average depending on its structural characteristics and the local ground motion. If it is desired to further limit damage to any specific building or preserve its post-earthquake function, the engineer should consider additional measures for strengthening the building.

Chapter A1 is not intended to be applicable in locations of the United States where new construction using plain or URM is allowed.

SECTION A102
SCOPE

A102.1 General. The provisions of this chapter shall apply to all existing buildings having at least one unreinforced masonry bearing wall. The elements regulated by this chapter shall be determined in accordance with Table A1-A. Except as provided herein, other structural provisions of the Building Code shall apply. This chapter does not require alteration of existing electrical, plumbing, mechanical or fire-safety systems.

❖ This chapter is applicable to all existing buildings that have least one URM-bearing wall. The provisions are not applicable to buildings having only nonbearing unreinforced masonry walls. The definition of a bearing wall in the 1997 *Uniform Building Code* (UBC) is given in Section 1629.6.2 as a wall providing support for gravity loads. Section 2106.2.8 provides a definition of a nonbearing wall that explains the difference between a bearing wall and a nonbearing wall. The GSREB defines a URM-bearing wall as a wall that supports more than 200 pounds per linear foot (2919 N/M) of vertical load in addition to its own weight. URM walls constructed within concrete or steel frames are not bearing walls. These infilled frame buildings with URM

walls are considered as potentially hazardous and they are not covered by this chapter. Chapter 5 of the GSREB provides analysis procedures for nonductile concrete frame buildings with URM walls infilled into the frames.

Table A1-A lists elements of buildings that are regulated by this chapter. This listing of elements is related to observed damage or partial collapse of these elements and spectral intensity of ground shaking. Separation of unbraced parapets has been reported for cities distant from a large magnitude earthquake source. The measure of spectral velocity, S_{D1} (or C_v for the 1997 UBC) is used instead of short period spectral acceleration because this is better related to potential instability of the parapet. The upper bound value of S_{D1} given in Table A1-A corresponds to descriptions of Seismic Design Categories B, C and D. The contours were generally the boundaries of UBC Seismic Zones 2A, 2B, 3 and 4.

Chapter A1 does not regulate nonstructural systems. Requirements of the building code will apply to reconstructed, nonstructural systems.

A102.2 Essential and hazardous facilities. The provisions of this chapter are not intended to apply to the strengthening of buildings or structures in Occupancy Categories 1 and 2 of Table 16-K of the 1997 *Uniform Building Code* when located in Seismic Zones 2B, 3 and 4, or in Seismic Use Groups II and III, where Seismic Design Categories C, D, E and F as defined in the 2003 *International Building Code* are required. Such buildings or structures shall be strengthened to meet the requirements of the Building Code for new buildings of the same occupancy category or other such criteria that have been established by the jurisdiction.

❖ The degree of earthquake hazard reduction anticipated in Chapter A1 is not considered acceptable for the 1997 UBC occupancy categories given in Table 16-K of the 1997 UBC other than for standard occupancies, or 2003 IBC Seismic Use Groups II and III where Seismic Design Categories C, D, E and F are required. These occupancies or use groups require special detailing considerations, increased seismic loading and, in the 2003 IBC, reduced allowable story drift.

SECTION A103
DEFINITIONS

For the purpose of this chapter, the applicable definitions in the Building Code shall also apply.

❖ In general, the definitions given in this section are self-explanatory. The commentary expands on several of the definitions that are critical to the use of Chapter A1.

BUILDING CODE. The code currently adopted by the jurisdiction for new buildings.

COLLAR JOINT. The vertical space between adjacent wythes. A collar joint may contain mortar or grout.

❖ The collar joint is the space between wythes, space that may be empty or may be filled with mortar or grout. Although the condition of the collar joint may not be critical for in-plane wall shear stress, the procedure for determining allowable stress from in-place shear tests takes into account the probable effect of mortar in the collar joint. The condition of the joint is of greater importance for out-of-plane forces where it is necessary to have the wythes of the wall act integrally. A visual examination of the collar joint to determine its mortar coverage is required to allow the use of the increased height-to-thickness ratios given in Table A1-B for buildings with crosswalls in all stories.

CROSSWALL. A new or existing wall that meets the requirements of Section A111.3 and the definition of Section A111.3. A crosswall is not a shear wall.

❖ A crosswall is a light-frame wall sheathed with new or existing materials. Light-frame walls with shear panels have a desirable hysteretic behavior and can, by their coupling strength, cause adjacent diaphragms of different stiffness to have nearly the same relative displacement. They function as energy dissipaters when connecting diaphragms or a diaphragm to grade within the span of the diaphragm. They act similar to shear walls to the extent that they diminish the displacement of a floor or roof relative to the building base, but are not true shear walls. Their in-plane stiffness is not comparable with shear walls of masonry, concrete or lateral-load-resisting elements of structural steel. Moment-resisting frames may also be designed as crosswalls.

CROSSWALL SHEAR CAPACITY. The unit shear value times the length of the crosswall, $v_c L_c$.

DIAPHRAGM EDGE. The intersection of the horizontal diaphragm and a shear wall.

DIAPHRAGM SHEAR CAPACITY. The unit shear value times the depth of the diaphragm, $v_u D$.

INTERNATIONAL BUILDING CODE. The *2003 International Building Code* (IBC).

NEHRP RECOMMENDED PROVISIONS. The 1997 edition of NEHRP *Recommended Provisions for Seismic Regulations for New Buildings*, issued by the Federal Emergency Management Agency.

NORMAL WALL. A wall perpendicular to the direction of seismic forces.

❖ A normal wall is a URM-bearing wall with loads perpendicular to the wall. When the earthquake loads are parallel to an URM-bearing wall, the wall is considered a shear wall.

OPEN FRONT. An exterior building wall line without vertical elements of the lateral-force-resisting system in one or more, stories.

❖ "Open front" is a term for the side of a building that does not have a shear wall, frame or braced frame at the exterior wall line. The open front may be at the first story or at all stories. The open front is defined as "on one side only." The definition is intended to be a limitation. Open-front buildings on street corners must have lateral-load-resisting elements, such as shear walls installed on one of the streetfront sides.

POINTING. The partial reconstruction of the bed joints of an unreinforced masonry wall as defined in UBC Standard 21-8.

❖ Pointing is equivalent to repointing.

RIGID DIAPHRAGM. A diaphragm of reinforced concrete construction supported by concrete beams and columns or by structural steel beams and columns.

❖ A rigid diaphragm is a floor or roof diaphragm of reinforced concrete construction. This floor or roof could be supported by reinforced concrete beams and girders or by structural steel members. The floor or roof could be a self-supporting slab or a pan-joist system. Chapter A1 does not consider semirigid diaphragms. It uses only two categories of the relationship of diaphragm in-plane shear stiffness to masonry wall in-plane shear stiffness. The engineer must consider the relative stiffness and strength limit states of other materials, such as gypsum concrete roofs, to determine a rational analysis procedure for the existing structural materials. An alternative procedure would be to use the building code definition of a flexible diaphragm that is based on the shear stiffness of only the diaphragm and the shear walls below the diaphragm level.

UNIFORM BUILDING CODE. The 1997 *Uniform Building Code* (UBC).

UNREINFORCED MASONRY. Includes burned clay, concrete or sand-lime brick; hollow clay or concrete block; plain concrete; and hollow clay tile. These materials shall comply with the requirements of Section A106 as applicable.

UNREINFORCED MASONRY BEARING WALL. A URM wall that provides the vertical support for the reaction of floor or roof-framing members.

UNREINFORCED MASONRY (URM) WALL. A masonry wall that relies on the tensile strength of masonry units, mortar and grout in resisting design loads, and in which the area of reinforcement is less than 25 percent of the minimum ratio required by the Building Code for reinforced masonry.

❖ A URM wall is a wall constructed of solid clay or concrete units, with or without cores, or a masonry wall constructed of hollow units, either clay or concrete. If the wall is constructed of hollow units, the net area of the masonry wall must be used for plane shear calculations. The allowable height-to-thickness ratios given in Table A1-B are not affected by the solidity of the masonry unit. Only the height-to-thickness ratio affects the stability of the wall. The weight per square foot of the wall is not a parameter for out-of-plane wall stability.

The in-plane bed joint shear strength or tensile splitting strength of the masonry must be determined by testing. Field stone or adobe masonry does not have a reliable test method for determination of in-plane shear strength. The engineer should, with the concurrence of the building official, determine strength limit state materials properties.

Chapter A1 has been developed for URM. The tensile strength of bed joints is considered as zero. A wall is considered unreinforced if the amount of reinforcing is less than 25 percent of the minimum amount that would be required by the building code. The intention is to exclude engineered systems that have small but definite amounts of horizontal and vertical reinforcing.

In most cases, a wall will either contain reinforcing at or near the building code requirements, or it will have no reinforcing at all. The reinforcing must be both vertical and horizontal. Steel straps laid along bed joints exist in some older types of construction; these should not be considered horizontally reinforced since these straps could actually weaken the bed joint.

YIELD STORY DRIFT. The lateral displacement of one level relative to the level above or below at which yield stress is first developed in a frame member.

SECTION A104
SYMBOLS AND NOTATIONS

For the purpose of this chapter, the following notations supplement the applicable symbols and notations in the Building Code.

a_n = Diameter of core multiplied by its length or the area of the side of a square prism.

A = Cross-sectional area of unreinforced masonry pier or wall, square inches (10^{-6} m^2).

A_b = Total area of the bed joints above and below the test specimen for each in-place shear test, square inches (10^{-6} m^2).

D = In-plane width dimension of pier, inches (10^{-3} m), or depth of diaphragm, feet (m).

DCR = Demand-capacity ratio specified in Section A111.4.2.

$f'm$ = Compressive strength of masonry.

f_{sp} = Tensile-splitting strength of masonry.

F_{wx} = Force applied to a wall at level x, pounds (N).

H = Least clear height of opening on either side of a pier, inches (10^{-3} m).

h/t = Height-to-thickness ratio of URM wall. Height, h, is measured between wall anchorage levels and/or slab-on-grade.

L = Span of diaphragm between shear walls, or span between shear wall and open front, feet (m).

L_c = Length of crosswall, feet (m).

L_i = Effective span for an open-front building specified in Section A111.8, feet (m).

P = Applied force as determined by standard test method of ASTM C 496 or ASTM E 519, pounds (N).

P_D = Superimposed dead load at the location under consideration, pounds (kN). For determination of the rocking shear capacity, dead load at the top of the pier under consideration shall be used.

p_{D+L} = Stress resulting from the dead plus actual live load in place at the time of testing, pounds per square inch (kPa).

P_w = Weight of wall, pounds (N).

R = Response modification factor for Ordinary plain masonry shear walls in Bearing Wall System from Table 1617.6 of the IBC, where $R = 1.5$.

S_{DS} = Design spectral acceleration at short period, in g units. $S_{DS} = 2.5C_a$ for use in UBC.

S_{D1} = Design spectral acceleration at 1-second period, in g units. $S_{D1} = Cv$ for use in UBC.

v_a = The shear strength of any URM pier, $v_m A/1.5$ pounds (N).

v_c = Unit shear capacity value for a crosswall sheathed with any of the materials given in Table A1-D or A1-E, pounds per foot (N/m).

v_m = Shear strength of unreinforced masonry, pounds per square inch (kPa).

V_a = The shear strength of any URM pier or wall, pounds (N).

V_{ca} = Total shear capacity of crosswalls in the direction of analysis immediately above the diaphragm level being investigated, $v_c L_c$, pounds (N).

V_{cb} = Total shear capacity of crosswalls in the direction of analysis immediately below the diaphragm level being investigated, $v_c L_c$, pounds (N).

V_p = Shear force assigned to a pier on the basis of its relative shear rigidity, pounds (N).

V_r = Pier rocking shear capacity of any URM wall or wall pier, pounds (N).

v_t = Mortar shear strength as specified in Section A106.3.3.5, pounds per square inch (kPa).

V_{test} = Load at incipient cracking for each in-place shear test per UBC Standard 21-6, pounds (kN).

v_{to} = Mortar shear test values as specified in Section A106.3.3.5, pounds per square inch (kPa).

v_u = Unit shear capacity value for a diaphragm sheathed with any of the materials given in Table A1-D or A1-E, pounds per foot (N/m).

V_{wx} = Total shear force resisted by a shear wall at the level under consideration, pounds (N).

W = Total seismic dead load as defined in the Building Code, pounds (N).

W_d = Total dead load tributary to a diaphragm level, pounds (N).

W_w = Total dead load of a URM wall above the level under consideration or above an open-front building, pounds (N).

W_{wx} = Dead load of a URM wall assigned to level x halfway above and below the level under consideration, pounds (N).

$\Sigma v_u D$ = Sum of diaphragm shear capacities of both ends of the diaphragm, pounds (N).

$\Sigma\Sigma v_u D$ = For diaphragms coupled with crosswalls, $v_u D$ includes the sum of shear capacities of both ends ofdiaphragms coupled at and above the level under consideration, pounds (N).

ΣW_d = Total dead load of all the diaphragms at and above the level under consideration, pounds (N).

❖ Extended definitions are given to clarify the use of the following symbols:

P_D = The tributary dead load at the top of the pier under consideration. For uniformly loaded bearing walls, it consists of the dead load on a width of wall extending between the centers of the openings adjacent to an interior pier, or from the wall edge to the center of the adjacent opening for the exterior piers, or from wall edge to wall edge. For use in Equations A1-4, A1-5 and A1-6, P_D is the superimposed dead load at the test location.

P_{D+L} = Stress resulting from the dead load plus actual live load in place at the time of testing, in pounds per square inch. For use in Equation A1-3, the quantity P_{D+L} is the stress obtained by dividing the axial load on the wall by the wall area, not the area of the test bed joints.

$\Sigma v_u D$ = The sum of the diaphragm shear capacities of both ends of the diaphragm that are parallel to the direction of the lateral load.

$\Sigma\Sigma v_u D$ = The sum of the $\Sigma v_u D$ values for the diaphragms at and above the level under consideration.

SECTION A105
GENERAL REQUIREMENTS

A105.1 General. Buildings shall have a seismic-resisting system conforming to the Building Code, except as modified by this chapter.

❖ Chapter 16 of the IBC provides the basic requirements for a seismic resisting system. Distribution of lateral loads shall consider the relative rigidity of horizontal diaphragms. Flexible diaphragms have inadequate relative stiffness to redistribute force between those elements that are considered shear walls by the special procedure. Diaphragms in buildings that use the general procedure are capable of redistributing the lateral loads, causing increased forces due to torsional response and coupling the weight at each story level with the shear walls.

A105.2 Alterations and repairs. Alterations and repairs required to meet the provisions of this chapter shall comply with

applicable structural requirements of the Building Code unless specifically provided for in this chapter.

❖ The building code applies unless explicitly excluded or modified herein.

A105.3 Requirements for plans. The following construction information shall be included in the plans required by this chapter:

1. Dimensioned floor and roof plans showing existing walls and the size and spacing of floor and roof-framing members and sheathing materials. The plans shall indicate all existing and new crosswalls and shear walls and their materials of construction. The location of these walls and their openings shall be fully dimensioned and drawn to scale on the plans.

2. Dimensioned wall elevations showing openings, piers, wall classes as defined in Section A106.3.3.8, thickness, heights, wall shear test locations, cracks or damaged portions requiring repairs, the general condition of the mortar joints, and if and where pointing is required. Where the exterior face is veneer, the type of veneer, its thickness and its bonding and/or ties to the structural wall masonry shall also be noted.

3. The type of interior wall and ceiling materials, and framing.

4. The extent and type of existing wall anchorage to floors and roof when used in the design.

5. The extent and type of parapet corrections that were previously performed, if any.

6. Repair details, if any, of cracked or damaged unreinforced masonry walls required to resist forces specified in this chapter.

7. All other plans, sections and details necessary to delineate required retrofit construction.

8. The design procedure used shall be stated on both the plans and the permit application.

9. Details of the anchor prequalification program required by UBC Standard 21-7, if used, including location and results of all tests.

❖ These are minimum requirements for the plans to be submitted. Although this may seem like an administrative requirement, this section is needed to record that a thorough investigation of the building has been made and is shown on required plans that are submitted to the building official.

The plans should include openings and their dimensions. Each diaphragm should be investigated at several locations. The lay-up of the sheathing should be observed and properly described in the plans. However, the continuity of the finish floor sheathing under partitions need not be verified.

A105.4 Structural observation. Structural observation shall be provided if required by the building official for structures regulated by this chapter. The owner shall employ the engineer or architect responsible for the structural design, or another engineer or architect designated by the engineer or architect responsible for the structural design, to perform structural observation as defined in the Building Code. Observed deficiencies shall be reported in writing to the owner's representative, special inspector, contractor and the building official. The structural observer shall submit to the building official a written statement that the site visits have been made and shall identify any reported deficiencies that, to the best of the structural observer's knowledge, have not been resolved.

❖ The construction phase of the retrofit program will discover unforeseen conditions. The engineer or architect responsible for the design must provide a more extensive observation of the project than that required for new construction.

SECTION A106
MATERIALS REQUIREMENTS

A106.1 General. Materials permitted by this chapter, including their appropriate strength design values and those existing configurations of materials specified herein, may be used to meet the requirements of this chapter.

❖ Existing materials as described in Table A1-D, new materials as described in Table I-E and masonry materials when tested as described in this section or categorized by f'_m may be used as a part of the seismic resisting structural system. New materials permitted by the building code may be used to supplement the strength of existing materials.

A106.2 Existing materials. Existing materials used as part of the required vertical-load-carrying or lateral-force-resisting system shall be in sound condition, or shall be repaired orremoved and replaced with new materials. All other unreinforced masonry materials shall comply with the followingrequirements:

1. The lay-up of the masonry units shall comply with Section A106.3.2, and the quality of bond between the units has been verified to the satisfaction of the building official;

2. Concrete masonry units are verified to be load-bearing units complying with UBC Standard 21-4 or such other standard as is acceptable to the building official; and

3. The compressive strength of plain concrete walls shall be determined based on cores taken from each class of concrete wall. The location and number of tests shall be the same as those prescribed for tensile-splitting strength tests in Sections A106.3.3.3 and A106.3.3.4, or in Section A108.1.

The use of materials not specified herein or in Section A108.1 shall be based on substantiating research data or engineering judgment, with the approval of the building official.

❖ Existing materials shall be in a sound condition or repaired. If replaced, they must be replaced with new materials that are permitted by the building code for seismic resistance.

A106.3 Existing unreinforced masonry.

A106.3.1 General. Unreinforced masonry walls used to carry vertical loads or seismic forces parallel and perpendicular to

the wall plane shall be tested as specified in this section. All masonry that does not meet the minimum standards established by this chapter shall be removed and replaced with new materials, or alternatively, shall have its structural functions replaced with new materials and shall be anchored to supporting elements.

❖ It is essential for the engineer to make a proper evaluation of the existing masonry. Allowable shear values for use in the analysis of shear walls and allowable height-to-thickness values for normal walls that are not analyzed for out-of-plane loading are based on the assumption of adequate strengths and lay-up of the masonry. Nonconforming masonry must be removed or treated as veneer.

A106.3.2 Lay-up of walls.

A106.3.2.1 Multiwythe solid brick. The facing and backing shall be bonded so that not less than 10 percent of the exposed face area is composed of solid headers extending not less than 4 inches (102 mm) into the backing. The clear distance between adjacent full-length headers shall not exceed 24 inches (610 mm) vertically or horizontally. Where the backing consists of two or more wythes, the headers shall extend not less than 4 inches (102 mm) into the most distant wythe, or the backing wythes shall be bonded together with separate headers whose area and spacing conform to the foregoing. Wythes of walls not bonded as described above shall be considered veneer. Veneer wythes shall not be included in the effective thickness used in calculating the height-to-thickness ratio and the shear capacity of the wall.

> **Exception:** In other than Seismic Zone 4, or where S_{D1} exceeds 0.3g, veneer wythes anchored as specified in the Building Code and made composite with backup masonry may be used for calculation of the effective thickness.

❖ The concern is for the integrity of the multiwythe wall acting as a whole in resisting out-of-plane forces. "Facing" means the wythe at the face of the wall, whether interior or exterior; "backing" means the inner wythes that are not normally visible. "Lay-up" means the pattern of masonry units in each wythe and in the interlocking of wythes. The primary requirement is that all wythes be adequately interlocked by header bricks. The quantity of mortar in the collar joints between wythes is also important, having an affect on the allowable height-to-thickness ratio for the wall as specified in the notes to Table A1-B.

An exception to the requirements of the common lay-up of multiwythe walls is permitted for lower seismic hazard zones. This exception is applicable in UBC seismic zones other than Zone 4 and where S_{D1} as defined and mapped in the IBC is 0.3g or less.

This exception permits veneer wythes with anchorage as specified by the building code and made composite with the backing masonry to be considered as a part of the structural wall. The wythes may be bonded by a combination of additional ties or the combination of grout and ties and should be checked to determine if they can transfer the calculated stress between wythes.

A106.3.2.2 Grouted or ungrouted hollow concrete or clay block and structural hollow clay tile. Grouted or ungrouted hollow concrete or clay block and structural hollow clay tile shall be laid in a running bond pattern.

A106.3.2.3 Other lay-up patterns. Lay-up patterns other than those specified in Sections A106.3.2.1 and A106.3.2.2 above are allowed if their performance can be justified.

A106.3.3 Testing of masonry.

A106.3.3.1 Mortar tests. The quality of mortar in all masonry walls shall be determined by performing in-place shear tests in accordance with the following:

1. The bed joints of the outer wythe of the masonry should be tested in shear by laterally displacing a single brick relative to the adjacent bricks in the same wythe. The head joint opposite the loaded end of the test brick should be carefully excavated and cleared. The brick adjacent to the loaded end of the test brick should be carefully removed by sawing or drilling and excavating to provide space for a hydraulic ram and steel loading blocks. Steel blocks, the size of the end of the brick, should be used on each end of the ram to distribute the load to the brick. The blocks should not contact the mortar joints. The load should be applied horizontally, in the plane of the wythe. The load recorded at first movement of the test brick as indicated by spalling of the face of the mortar bed joints is V_{test} in Equation (A1-3).

2. Alternative procedures for testing shall be used where in-place testing is not practical because of crushing or other failure mode of the masonry unit (see Section A106.3.3.2).

❖ **Item 1.** Mortar tests are performed by shearing of the mortar joints above and below the test masonry unit. The test load is recorded at the first sign of movement of the test brick. This can be detected when sand grains are detaching from the mortar joint.

Item 2. Many masonry walls in the United States have been constructed with mortars that have shear strength such that the mortar shear strength test is not possible (e.g., hollow masonry). The failure may be the bearing on the end of the masonry units at the jack. Where this occurs, an alternative testing method described in Section 106.3.3.2 determines tensile splitting strength by the standard ASTM procedure.

A106.3.3.2 Alternative procedures for testing maonry. The tensile-splitting strength of existing masonry, f_{sp}, or the prism strength of existing masonry, f'_m, may be determined in accordance with one of the following procedures:

1. Wythes of solid masonry units shall be tested by sampling the masonry by drilled cores of not less than 8 inches (203 mm) in diameter. A bed joint intersection with a head joint shall be in the center of the core. The tensile-splitting strength of these cores should be determined by the standard test method of ASTM C 496. The core should be placed in the test apparatus with the bed joint 45 degrees from the horizontal. The tensile-splitting strength should be determined by the following equation:

$$f_{sp} = \frac{2P}{\pi a_n} \qquad \textbf{(Equation A1-1)}$$

2. Hollow unit masonry constructed of through-the-wall units shall be tested by sampling the masonry by a sawn square prism of not less than 18 inches square (11 613 mm²). The tensile-splitting strength should be determined by the standardtest method of ASTM E 519. The diagonal of the prism should be placed in a vertical position. The tensile-splitting strength should be determined by the following equation:

$$f_{sp} = \frac{0.494P}{a_n} \qquad \textbf{(Equation A1-2)}$$

3. An alternative to material testing is estimation of the f'_m of the existing masonry. This alternative should be limited to recently constructed masonry. The determination of f'_m requires that the unit correspond to a specification of the unit by an ASTM standard and classification of the mortar by type.

❖ These procedures determine the tensile splitting strength of masonry. The tensile splitting strength is equivalent to the peak shear strength when pier shear strength, V_m, is calculated using $v_m A_n/1.5$.

Recently constructed masonry (post-1950) may be categorized by unit strength and mortar mix, Table 21-D of the UBC or Tables 2105.2.2.1.1 and 2105.2.2.1.2 of the IBC. These tables may be used to estimate f'_m of the masonry. The estimate will be a conservative prism strength, which may be used to calculate peak shear stress. The shear strength of a pier is calculated by Equation A1-20.

The shear strength is an expected shear strength; no load factors or capacity reduction factors are used in Chapter A1.

A106.3.3.3 Location of tests. The shear tests shall be taken at locations representative of the mortar conditions throughout the entire building, taking into account variations in workmanship at different building height levels, variations in weathering of the exterior surfaces, and variations in the condition of the interior surfaces due to deterioration caused by leaks and condensation of water and/or by the deleterious effects of other substances contained within the building. The exact test locations shall be determined at the building site by the engineer or architect in responsible charge of the structural design work. An accurate record of all such tests and their locations in the building shall be recorded, and these results shall be submitted to the building department for approval as part of the structural analysis.

❖ The test locations should be uniformly distributed over the wall surface. The exact location of each test shall be determined at the site by the engineer. Relatively small areas that are pointed due to local deterioration should not be used for test areas since pointed areas will produce higher values that can skew the test results.

A106.3.3.4 Number of tests. The minimum number of tests per class shall be as follows:

1. At each of both the first and top stories, not less than two tests per wall or line of wall elements providing a common line of resistance to lateral forces.

2. At each of all other stories, not less than one test per wall or line of wall elements providing a common line of resistance to lateral forces.

3. In any case, not less than one test per 1,500 square feet (139.4 m²) of wall surface and not less than a total of eight tests.

❖ Numbers and locations are specified to ensure a representative sample of existing masonry; considering that the walls were not necessarily built with the same time with the same workmanship, workmanship may vary in a given wall, often being poorer near the top, and may not have been subjected to the same environmental influences. These and other conditions may make it desirable to divide walls into classes according to their relative overall quality (see Section A106.3.3.8). The number of tests in each class of wall should be increased to the maximum number specified in Section A106.3.3.4.3 to provide a body of data that is adequate to establish a 20-percent value, v_t.

A common line of resistance is defined as a set of one or more masonry walls at an edge of a diaphragm. The wall may have offsets in that common line. A minimum of eight tests for a building is specified. This would be the minimum number for a single-story building. The area of 1,500 square feet (139 m²) is gross area, including wall openings.

A106.3.3.5 Minimum quality of mortar.

1. Mortar shear test values, v_{to}, in pounds per square inch (kPa) shall be obtained for each in-place shear test in accordance with the following equation:

$$v_{to} = (V_{test}/A_b) - p_{D+L} \qquad \textbf{(Equation A1-3)}$$

2. Individual unreinforced masonry walls with v_{to} consistently less than 30 pounds per square inch (207 kPa) shall be entirely pointed prior to retesting.

3. The mortar shear strength, v_t, is the value in pounds per square inch (kPa) that is exceeded by 80 percent of the mortar shear test values, v_{to}.

4. Unreinforced masonry with mortar shear strength, v_t, less than 30 pounds per square inch (207 kPa) shall be removed, pointed and retested or shall have its structural function replaced, and shall be anchored to supporting elements in accordance with Sections A106.3.1 and A113.8. When existing mortar in any wythe is pointed to increase its shear strength and is retested, the condition of the mortar in the adjacent bed joints of the inner wythe or wythes and the opposite outer wythe shall be examined for extent of deterioration. The shear strength of any wall class shall be no greater than that of the weakest wythe of that class.

❖ A mortar shear test value, v_{to}, is obtained for each test location by dividing the test force, V_{test}, by the sum of the areas of the upper and lower surfaces of the brick, and subtracting the axial stress in the wall at the time of testing.

The axial stress is defined to include actual live load where it is significant; ordinarily the use of the dead weight of the superimposed masonry is close enough. A set of values of v_{to} is assembled for the whole building.

When an individual wall has values of v_{to} that are consistently below 30 pounds per square inch (207 kPa), the wall may be upgraded by pointing. Where this is done, the whole wall must be pointed. Then the upgraded wall is retested and improved values of v_{to} will be obtained. These improved values of v_{to} may be used in place of the original values for that wall, and a new set of values of v_{to} is assembled for calculation of v_{to}, since this wall now represents a class of masonry.

Once a set of values of v_{to} is obtained for the building or a class of masonry, the mortar shear strength, v_t, for the building or the class of masonry is determined as the value of v_{to} that is exceeded by 80 percent of the values of v_{to}.

When v_t is less than 30 psi (207 kPa) for the building or a common line of resistance, all of the masonry in the building, or line of resistance, must be either removed or upgraded by repointing, and then retested.

A106.3.3.6 Minimum quality of masonry.

1. The minimum average value of tensile-splitting strength determined by Equation (A1-1) or (A1-2) shall be 50 pounds per square inch (344.7 kPa). The minimum value of f'_m determined by categorization of the masonry units and mortar should be 1,000 pounds per square inch (6895 kPa).

2. Individual unreinforced masonry walls with average tensile-splitting strength of less than 50 pounds per square inch (344.7 kPa) shall be entirely pointed prior to retesting.

3. Hollow unit unreinforced masonry walls with estimated prism compressive strength of less than 1,000 pounds per square inch (6895 kPa) shall be grouted to increase the average net area compressive strength.

❖ A minimum quality of an assemblage of units and mortar is required to have confidence in the determination of strength limit values. Hollow unit masonry can have its tensile splitting strength increased by pointing. Hollow unit masonry commonly has only the face shell supported on a mortar bed joint. Hollow unit masonry may also be grouted to increase its net area and tensile splitting strength.

A106.3.3.7 Collar joints. The collar joints shall be inspected at the test locations during each in-place shear test, and estimates of the percentage of adjacent wythe surfaces that are covered with mortar shall be reported along with the results of the in-place shear tests.

❖ When a masonry unit is removed for the in-place shear test, the collar joint between the exterior wythe and the interior wythe is exposed. The percentage of mortar coverage of the collar joint is estimated and reported with the results of the in-place shear testing. Fifty percent of the collar joint must be filled to meet the requirements of the notes to Table A1-B. Table A1-B specifies the allowable height-thickness ratios of URM walls.

A106.3.3.8 Unreinforced masonry classes. Existing unreinforced masonry shall be categorized into one or more classes based on shear strength, quality of construction, state of repair, deterioration and weathering. A class shall be characterized by the allowable masonry shear stress determined in accordance with Section A108.2. Classes shall be defined for whole walls, not for small areas of masonry within a wall.

❖ This section allows categorization of masonry into classes. A single wall may be defined as a class. The provision that 80 percent of the tests made in this wall exceed the test value v_{to} would require a minimum of five tests in that wall. A three-story or higher building will have a minimum of five tests in any wall. If categorization of a wall or walls as a class is contemplated in a building of fewer than three stories, additional tests will be required in each of the walls or line of walls.

A106.3.3.9 Pointing. Deteriorated mortar joints in unreinforced masonry walls shall be pointed according to UBC Standard 21-8. Nothing shall prevent pointing of any deteriorated masonry wall joints before the tests are made, except as required in Section A107.1.

❖ Deteriorated mortar joints must be pointed. Mortar joints may be pointed before testing to increase the shear test values, but this pointing must be performed with special inspection as required by Section A107.1.

Pointing is maintenance work. Old mortars are subject to deterioration, and if so, they need to be renewed. Deterioration is most often due to moisture penetration and evaporation: water carries dissolved salts into the wall, or dissolves them from the masonry itself; upon evaporation of the water, crystals are left behind. The crystals break up the mortar matrix (and sometimes the brick itself), leaving it weak and powdery. The process of weathering propagates into the wall. In many cases, sound material will be found within an inch [or $1^1/_2$ inches (38 mm) at the most] of the original surface.

SECTION A107
QUALITY CONTROL

A107.1 Pointing. Preparation and mortar pointing shall be performed with special inspection.

Exception: At the discretion of the building official, incidental pointing may be performed without special inspection.

❖ Preparation and pointing of mortar joints shall have special inspection. UBC 21-8 provides guidance for the pointing work. The building official may allow pointing of small, deteriorated areas at his or her discretion without special inspection.

A107.2 Masonry shear tests. In-place masonry shear tests shall comply with Section A106.3.3.1. Testing of masonry for

determination of tensile-splitting strength shall comply with Section A106.3.3.2.

❖ See the commentary for Section A106.

A107.3 Existing wall anchors. Existing wall anchors used as all or part of the required tension anchors shall be tested in pull-out according to UBC Standard 21-7. The minimum number of anchors tested shall be four per floor, with two tests at walls with joists framing into the wall and two tests at walls with joists parallel to the wall, but not less than 10 percent of the total number of existing tension anchors at each level.

A107.4 New bolts. All new embedded bolts shall be subject to periodic special inspection in accordance with the Building Code, prior to placement of the bolt and grout or adhesive in the drilled hole. Five percent of all bolts that do not extend through the wall shall be subject to a direct-tension test, and an additional 20 percent shall be tested using a calibrated torque wrench. Testing shall be performed in accordance with UBC Standard 21-7. New bolts that extend through the wall with steel plates on the far side of the wall need not be tested.

Exception: Special inspection in accordance with the Building Code may be provided during installation of new anchors in lieu of testing.

All new embedded bolts resisting tension forces or a combination of tension and shear forces shall be subject to periodic special inspection in accordance with the Building Code, prior to placement of the bolt and grout or adhesive in the drilled hole. Five percent of all bolts resisting tension forces shall be subject to a direct-tension test, and an additional 20 percent shall be tested using a calibrated torque wrench. Testing shall be performed in accordance with UBC Standard 21-7. New through-bolts need not be tested.

SECTION A108
DESIGN STRENGTHS

A108.1 Values.

1. Strength values for existing materials are given in Table A1-D, and for new materials, in Table A1-E.

2. Strength values not specified herein or in the Building Code may be as specified in the NEHRP *Recommended Provisions*.

3. Capacity reduction factors need not be used.

4. The use of new materials not specified herein shall be based on substantiating research data or engineering judgment, with the approval of the building official.

❖ The materials resistance values used in Chapter A1 are strength values but are not used with capacity reduction factors. Strength values obtained by experimental testing and in-place testing of URM are expected strength values. The strength values obtained by the AKB (Agbabian, Barnes, Kariotis) dynamic testing of ductile materials, such as sheathed diaphragms, are strength limit values. It is recommended that strength values needed for retrofit materials not included in this chapter be obtained from the latest edition of the NEHRP *Recommended Provisions*.

Design of new elements that are intended to supplement the existing material should be designed using strength procedures, but not with capacity reduction factors.

When designing new elements, special attention should be paid to the relative stiffness of the existing structural elements and the new elements. For example, if new shear walls or braced frames supplement the strength of a URM wall, the loading should not be reduced by an increased R-factor. However, if the existing structural system is strengthened so that the system's ductility is increased, a higher R-factor could be justified. Such an example would be where new concrete shear walls are added and are designed to carry the full lateral load and URM piers are adequate to accommodate the expected sum of linear and non-linear displacements without shear failure.

A108.2 Masonry shear strength. The unreinforced masonry shear strength, v_m, shall be determined for each masonry class from one of the following equations:

1. The unreinforced masonry shear strength, vm, shall be determined by Equation (A1-4) when the mortar shear strength has been determined by Section A106.3.3.1.

$$v_m = 0.56v_t + \frac{0.75P_D}{A} \qquad \text{(Equation A1-4)}$$

The mortar shear strength values, v_t, shall be determined in accordance with Section 106.3.3.5 and shall not exceed 100 pounds per square inch (689.5 kPa) for the determination of v_m.

2. The unreinforced masonry shear, v_m, shall be determined by Equation (A1-5) when tensile-splitting strength has been determined in accordance with Section A106.3.3.2, Item 1 or 2.

$$v_m = 0.8f_{sp} + 0.5\frac{P_D}{A} \qquad \text{(Equation A1-5)}$$

3. When f'_m has been estimated by categorization of the units and mortar in accordance with IBC Section 2105.2.2.1 or UBC Section 2105.3.4, the unreinforced masonry shear strength, v_m, shall not exceed 200 pounds per square inch (1380 kPa) or the lesser of the following:

   a) $25\sqrt{f'_m}$ or

   b) 200 psi or

   c) $v + 0.75\frac{P_D}{A}$ **(Equation A1-6)**

For SI: 1 psi = 6.895 kPa.

Where:

    v = 62.5 psi (430 kPa) for running bond masonry not grouted solid.

    v = 100 psi (690 kPa) for running bond masonry grouted solid.

    v = 25 psi (170 kPa) for stack bond grouted solid.

❖ **Item 1.** The correlation of v_t and v_m was obtained by physical testing made by the ABK joint venture. Equation A1-4 is an empirical formula.

Item 2. Equation A1-5 is a theoretical formula adjusted for a probable coefficient of variation of the test data.

Item 3. Equation A1-6 are similar to the shear values given in the building code. They are allowable strength values adjusted upward by about 1.7.

A108.3 Masonry compression. Where any increase in dead plus live compression stress occurs, the compression stress in unreinforced masonry shall not exceed 300 pounds per square inch (2070 kPa).

❖ This maximum allowable stress is a conservative estimate of the compressive strength of a minimum acceptable masonry. No strength increase for seismic loading is permitted.

A108.4 Masonry tension. Unreinforced masonry shall be assumed to have no tensile capacity.

A108.5 Existing tension anchors. The resistance values of the existing anchors shall be the average of the tension tests of existing anchors having the same wall thickness and joist orientation.

A108.6 Foundations. For existing foundations, new total dead loads may be increased over the existing dead load by 25 percent. New total dead load plus live load plus seismic forces may be increased over the existing dead load plus live load by 50 percent. Higher values may be justified only in conjunction with a geotechnical investigation.

❖ Foundation loads may be increased over existing loads for several reasons. First, consolidation of foundation materials has decreased the probability of settlement due to added loads. Second, the dynamic loading of soils by earthquakes can be tolerated by foundation materials without additional settlement. Third, in many cases, the dead load added by seismic strengthening is usually relatively insignificant.

The restrictions may be illustrated by the following example:

    Given original dead load = 100
    Added dead load = 20
    Live load = 50
    Seismic = 30
    The new dead load is 100 + 20 = 120.

By the first sentence in Section A108.6, new dead load may be 1.25 multiplied by the existing dead load: in the example, 120 is less than 1.25 x 100 = 125; therefore, the added dead load is acceptable. By the second sentence, new dead plus live plus seismic may be 1.50 multiplied by the existing dead plus live: in the example, new dead plus live plus seismic is 120 + 50 + 30 = 200, which is less than 1.5 x (100 + 150) = 225; therefore the added dead load is acceptable.

Every structure should provide adequate resistance to resist overturning effects. This is less important for existing unreinforced masonry elements since their lateral load carrying capacity is limited by their rocking capacity. However, where new elements are added, overturning can be a controlling issue. In these cases the new elements should be designed to resist overturning in general accordance with the building code.

SECTION A109
ANALYSIS AND DESIGN PROCEDURE

A109.1 General. The elements of buildings hereby required to be analyzed are specified in Table A1-A.

❖ Chapter A1 uses analysis and design procedures that are specific for unreinforced masonry buildings. The chapter is for analysis and retrofit for existing buildings. The elements of the masonry buildings that are regulated by Chapter A1 are given in Table A1-A. This table exempts certain items from regulation in low-to-moderate seismic hazard regions.

A109.2 Selection of procedure. Buildings with rigid diaphragms shall be analyzed by the general procedure of Section A110, which is based on the Building Code. Buildings with flexible diaphragms shall be analyzed by the general procedure or, when applicable, may be analyzed by the special procedure of Section A111.

❖ URM buildings with concrete floors and roofs (rigid diaphragms) must be analyzed using the general procedure. The general procedure follows the building code in that the analysis base shear is calculated using the total mass of the building. The rigid diaphragms couple the shear walls of each level with the weight of the story level and normal walls. URM buildings with flexible diaphragms have an entirely different dynamic response characteristic. The mass of the floor, roofs and normal walls are coupled with the shear wall parallel to the direction of the seismic loading by a shear beam element that has a fundamental elastic period in the range where dynamic response is related to *T*, the elastic period of the diaphragm. This shear beam has significant ductility potential and the coupling of the story and normal wall mass with the shear wall will be limited to the shear capacity of the diaphragm at each story level. Section 111.1 provides the limits a building must satisfy in order for the special procedure to be used.

SECTION A110
GENERAL PROCEDURE

A110.1 Minimum design lateral forces. Buildings shall be analyzed to resist minimum lateral forces assumed to act nonconcurrently in the direction of each of the main axes of the structure in accordance with the following:

$$V = \frac{0.75\, S_{DS}\, W}{R} \qquad \textbf{(Equation A1-7)}$$

❖ The base shear of the URM building with rigid diaphragms is given by Equation A1-7. This formula is recognizable as being related to the seismic design formula in the building code. It is a simplified form of the usual base shear formula.

The fundamental elastic period of the building is taken as being in the constant acceleration range, i.e., S_{DS} for the IBC or $2.5C_a$ for the UBC. The base shear is at the maximum value for the seismic hazard zone.

This base shear is reduced by 25 percent to be consistent with the policy that the analysis forces for existing buildings is three-fourths of the design forces specified for new buildings. The base shear is also divided by an *R*-value of 1.5. This is consistent with Table 1617.6 of the IBC for bearing wall systems having ordinary plain (unreinforced) masonry shear walls. There is no comparable *R*-factor in the UBC for bearing wall buildings with URM walls. It is recommended that this base shear be applied in a uniform loading pattern on a multidegree of freedom model, if the shear wall has substantial perforations at each story level. The multidegree of freedom model is a shear-yielding model, not a flexural beam model. Story yield mechanisms may be only a single floor level if the yield mechanism is in-plane shear in the piers. If rocking of the piers is the yield mechanism, it is a possibility that yield mechanisms may form at several story levels. If the wall is basically solid, a triangular loading should be applied, as per the building code. The engineer should make preliminary calculations as to probable story strengths and yield mechanisms to determine a rational distribution of the base shear over the height of a multistory URM shear wall.

Global overturning is not a critical issue for URM shear walls. Flexural tensile forces due to global overturning cannot be carried downward without a tension load path. Story height overturning moments cause a modification of the axial load on the pier, but equilibrium requires that the reduction in vertical pier loading be equal to the increase in vertical pier loading. The in-plane shear capacity and rocking shear capacity is affected by pier loading, but the total interstory shear capacity is not significantly changed.

A110.2 Lateral forces on elements of structures. Parts and portions of a structure not covered in Sections A110.3 and A110.4 shall be analyzed and designed per the current Building Code, using force levels defined in Section A110.1.

Exceptions:

1. Unreinforced masonry walls for which height-to-thickness ratios do not exceed ratios set forth in Table A1-B need not be analyzed for out-of-plane loading. Unreinforced masonry walls that exceed the allowable *h/t* ratios of Table A1-B shall be braced according to Section A113.5.

2. Parapets complying with Section A113.6 need not be analyzed for out-of-plane loading.

❖ The failure of building elements, such as parapets and portions of walls, represents a major source of hazards posed by URM buildings. All vulnerable elements should be identified and analyzed.

Diaphragms are analyzed for the loadings calculated by the distribution of base shear at the level of consideration. The allowable shear values of existing and new materials applied to existing materials given in Tables A1-D and A1-E are applicable to the general and special procedures. These provisions require the same force factors on parts and portions as are required for a new building.

Although not specifically stated in buildings being analyzed by the general procedure and having flexible diaphragms and crosswalls conforming to Section A111.3, use of the crosswalls to provide partial lateral support for the diaphragm loading has been used by some engineers and permitted by several jurisdictions.

Exception 1 states that normal walls of unreinforced masonry need not be analyzed for flexural capacity but may be deemed adequate if the height-to-thickness ratios specified in Table A1-B are not exceeded. Use of this table is very conservative for rigid diaphragms. The rigid diaphragm does not amplify the story level acceleration like a flexible diaphragm. In addition to this effect, the overburden/wall-weight ratio used in the development of Table A1-B uses the weight of wood floors and only one additional story for multistory buildings. The probable vertical load of concrete floors and roofs would greatly increase the acceptable height-to-thickness ratio for normal walls.

Exception 2 states that parapets that have height-to-thickness ratios less than specified in Section A113.6 need not be analyzed. Parapets with height-to-thickness ratios in excess of this limit shall be braced.

A110.3 Out-of-plane loading for URM walls. Unreinforced masonry walls for which height-to-thickness ratios do not exceed the ratios set forth in Table A1-B need not be analyzed for out-of-plane loading. Unreinforced masonry walls that exceed the allowable h/t ratios of Table A1-B shall be braced according to Section A113.5. Parapets of such walls that comply with Section A113.6 need not be analyzed for out-of-plane loading. Walls shall be anchored to floor and roof diaphragms in accordance with Section A113.1.

A110.4 In-plane loading of URM shear walls and frames. Vertical lateral-load-resisting elements shall be analyzed in accordance with Section A112.

❖ Once the story shear forces are determined by the methods of the general procedure, the analysis of unreinforced masonry shear walls follows the same procedures as specified for the special procedure. The masonry is tested as prescribed by Section A106; the allowable shear stress is determined in accordance with Section 108; and the piers in the shear wall are analyzed in accordance with Section A112.2.3.

A110.5 Redundancy and overstrength factors. Any redundancy or overstrength factors contained in the Building Code may be taken as unity. The vertical component of earthquake load (E_v) may be taken as zero.

SECTION A111
SPECIAL PROCEDURE

A111.1 Limits for the application of this procedure. The special procedures of this section may be applied only to buildings having the following characteristics:

1. Flexible diaphragms at all levels above the base of the structure.

2. Vertical elements of the lateral-force-resisting system consisting predominantly of masonry or concrete shear walls.

3. Except for single-story buildings with an open front on one side only, a minimum of two lines of vertical elements of the lateral-force-resisting system parallel to each axis of the building. (See Section A111.8 for open-front buildings.)

❖ **Item 1.** The building must have flexible diaphragms at all story levels to qualify for this procedure, and the seismic loading used in this section is exclusively for flexible diaphragms. The definition of "rigid diaphragms" given in Section A103 is purposely very narrow. Investigation of existing URM wall buildings will find materials that do not fit into this narrow definition. Such a building would have to be modeled with the unique stiffness characteristics of the materials. Stiffness degradation should be assessed if a linear elastic analysis is made. It is outside the scope of Chapter A1 to provide guidance for analysis of unique buildings. The engineer and the building official should make the decisions for an analysis procedure that is specific for a unique structural system.

Item 2. The vertical lateral-force-resisting system should have stiffness such that the relative stiffness ratio of the diaphragms and shear walls assumed for this procedure is maintained. A steel or reinforced concrete moment frame does not have the story stiffness of a reinforced concrete or masonry shear wall designed for the same lateral loading. A moment frame may be adequate to act as a shear wall in the middle portion of a flexible diaphragm or at an open front of an unreinforced masonry building. Section A111.3.5 has special limits on yield drift of a moment frame used as a crosswall. Section A111.6.4 requires use of the same design forces as for a shear wall and further limits the elastic drift of this moment frame. Use of an R-factor in the design of the moment frame is prohibited.

Item 3. A minimum of two lines of vertical lateral-load-resisting elements on each axis of the building is required. Section A111.1.2 requires that each of these elements be "predominantly of masonry or concrete shear walls." If structural steel or reinforced concrete bracing is used for one or both of these vertical elements, a relative rigidity check of the bracing versus a shear wall should be made to confirm that the concept of a "rigid wall-flexible diaphragm" response to earthquake is probable.

A111.2 Lateral forces on elements of structures. With the exception of the diaphragm provisions in Section A111.4, elements of structures shall comply with Sections A110.2 through A110.5.

❖ This section specifies that floor and roof diaphragms are only checked for strength limit state and stiffness. The requirement for the check is given in Section A111.4. In addition, requirements for diaphragm shear, transfer shear walls and out-of-plane forces are given in Sections A111.5, A111.6 and A111.7, respectively. Other elements of structures are analyzed as prescribed by the building code, but the forces are those given in Section A110, and the material's resistance values are those given in Section A108. These design strengths are calculated using the principle of load and resistance factor design (LRFD) (strength) methods. However, load is not factored and capacity reduction factors are not used for resistance values.

A111.3 Crosswalls. Crosswalls shall meet the requirements of this section.

A111.3.1 Crosswall definition. A crosswall is a wood-framed wall sheathed with any of the materials described in Table A1-D or A1-E or other system as defined in Section A111.3.5. Crosswalls shall be spaced no more than 40 feet (12 192 mm) on center measured perpendicular to the direction of consideration, and shall be placed in each story of the building. Crosswalls shall extend the full story height between diaphragms.

Exceptions:

1. Crosswalls need not be provided at all levels when used in accordance with Section A111.4.2, Item 4.

2. Existing crosswalls need not be continuous below a wood diaphragm at or within 4 feet (1219 mm) of grade, provided:

   2.1 Shear connections and anchorage requirements of Section A111.5 are satisfied at all edges of the diaphragm.

   2.2 Crosswalls with total shear capacity of $0.5S_{D1}\Sigma W_d$ interconnect the diaphragm to the foundation.

   2.3 The demand-capacity ratio of the diaphragm between the crosswalls that are continuous to their foundations does not exceed 2.5, calculated as follows:

$$DCR = \frac{(2.1S_{D1}W_d + V_{ca})}{Dv_u D} \quad \textbf{(Equation A1-8)}$$

❖ The definition of a "Crosswall" is given and fully defined in this commentary in Sections A103 and A111.3.1. A crosswall is a light-frame sheathed wall parallel to the direction of earthquake loading. A crosswall is not considered as a shear wall in that it need not be designed for the tributary loads of the flexible diaphragm or diaphragms. The crosswall decreases the displacement of the center of the diaphragm relative to the shear walls and provides damping of the response of the diaphragm to earthquake shaking.

The spacing of the crosswalls along the diaphragm span length is limited to 40 feet (12 192 mm). This re-

striction is intended to limit the higher mode response of the diaphragm between crosswalls. In general, if crosswalls are used to increase the allowable height-to-thickness ratio of normal walls, the crosswalls must be in each story of the building. Existing and new crosswalls must extend the full story height between diaphragm levels.

Exception 1 allows the use of crosswalls that extend only between the roof diaphragm and a floor diaphragm to couple these into a combined diaphragm for its DCR analysis.

Exception 2 allows a special condition that commonly occurs in residential occupancies. Often in this occupancy, many crosswalls interconnect the roof and floors. However, the first floor is commonly constructed over a crawl space without crosswalls. This exception allows the use of special crosswalls spaced as far apart as 40 feet (12 192 mm) to stiffen the lowest diaphragm. The crosswalls must have adequate strength to limit the deformation of the diaphragm at the special crosswalls. The loading of the first floor diaphragm is its tributary load and the entire capacity of the crosswalls that are above the first floor.

A111.3.2 Crosswall shear capacity. Within any 40 feet (12 192 mm) measured along the span of the diaphragm, the sum of the crosswall shear capacities shall be at least 30 percent of the diaphragm shear capacity of the strongest diaphragm at or above the level under consideration.

❖ The stiffness of sheathed systems such as flexible diaphragms and crosswalls is directly related to their peak strength; therefore, the minimum strength of a crosswall is related to the strength of the diaphragm that it is restraining. The minimum strength of a crosswall must be 30 percent of the strength of the diaphragm shear capacity of the stronger diaphragm at or above the level under consideration.

A111.3.3 Existing crosswalls. Existing crosswalls shall have a maximum height-to-length ratio between openings of 1.5 to 1. Existing crosswall connections to diaphragms need not be investigated as long as the crosswall extends to the framing of the diaphragms above and below.

❖ Existing crosswalls generally are perforated by door openings. The effective length of the crosswall used for calculations of the required crosswall strength cannot include portions that have a height-to-length ratio less than 1.5. If the openings were 7-foot-high (2134 mm) doors, a section between door or doors and the end of the wall of less than $4\frac{1}{2}$ feet (1372 mm) cannot be used for the calculation of crosswall capacity.

The capacities of the connections of the existing crosswall to the diaphragms above or below the wall need not be investigated if the crosswall extends to the diaphragm level above. If the crosswall only extends to ceiling joists that are separate from the floor or roof framing above, then continuity of the crosswall must be provided. Note 2 of Table A1-D limits the total capacity to 900 pounds per foot (408.6 kg/304.8mm) regardless of the combined capacity of the existing materials on the crosswall. This limitation was provided in lieu of an investigation of the capacity of the connection of the crosswall to the diaphragm.

A111.3.4 New crosswalls. New crosswall connections to the diaphragm shall develop the crosswall shear capacity. New crosswalls shall have the capacity to resist an overturning moment equal to the crosswall shear capacity times the story height. Crosswall overturning moments need not be cumulative over more than two stories.

❖ Connections of new crosswalls to diaphragms shall be designed by the applicable strength design sections of the building code. This includes the design of a vertical connection of the end of the crosswall to the foundation of the crosswall or support below the level of the crosswall. This connection is required to protect the floor framing from wall overturning effects. The dead load of the floor or roof framing supported by the crosswall cannot be used to resist the calculated overturning due to lateral forces. However, the calculated force at the end of the crosswall need not use an overturning moment accumulated from more than two stories, including the level of calculation.

A111.3.5 Other crosswall systems. Other systems, such as moment-resisting frames, may be used as crosswalls provided that the yield story drift does not exceed 1 inch (25.4 mm) in any story.

❖ Other systems may be used as crosswalls. A steel frame may be more desirable than a crosswall in a commercial space. Such a frame should be a moment frame rather than a braced frame, so as to have the damping effect of a crosswall rather than the rigid bracing effect of a shear wall. The prescription of yield story drift not to exceed 1 inch (25 mm) is intended to provide stiffness, an inelastic threshold and load deformation behavior similar to that of existing crosswalls.

The frame used as a crosswall is intended to yield, and the frame sections at which yielding will occur must be braced to building elements that can provide adequate restraint against lateral buckling.

Considering the cost of adding a frame, regardless of the design criteria, it may be better to provide a moment frame designed as a shear wall. When used as a shear wall, the moment frame is designed to resist the loading of the tributary area of the diaphragm or diaphragms, and its drift under this loading is limited by Section A111.6.4 to 0.015 times the story height.

A111.4 Wood diaphragms.

A111.4.1 Acceptable diaphragm span. A diaphragm is acceptable if the point (L, DCR) on Figure A1-1 falls within Region 1, 2 or 3.

❖ Conventional diaphragm analysis is not required by the special procedure. The strength and stiffness of a flexible diaphragm is acceptable if its span length and demand capacity ratio (DCR) falls within Regions 1, 2, or 3 of Figure A1-1. Figure A1-1 is a plot of the estimated dynamic displacement of the center of the dia-

phragm of 5 inches (127 mm) measured relative to the shear walls that are shaking the diaphragm. The ground motions used for these calculations were scaled to the ATC-3 5-percent damped response spectrum having 0.40g effective peak acceleration and 12 inches per second (305 mm/s) peak ground velocity. The analysis of the DCR of diaphragms is only required in UBC Seismic Zones 3 and 4, or in IBC seismic areas where S_{D1} is 0.20g or greater. The probable ground motions of UBC Seismic Zone 3 are accommodated by the substitution of C_v in lieu of S_{D1} in the calculation of DCR.

A111.4.2 Demand-capacity ratios. Demand-capacity ratios shall be calculated for the diaphragm at any level according to the following formulas:

1. For a diaphragm without qualifying crosswalls at levels immediately above or below:

$$DCR = 2.1 S_{D1} W_d / \Sigma v_u D \qquad \textbf{(Equation A1-9)}$$

2. For a diaphragm in a single-story building with qualifying crosswalls, or for a roof diaphragm coupled by crosswalls to the diaphragm directly below:

$$DCR = 2.1 S_{D1} W_d / (\Sigma v_u D + V_{cb}) \qquad \textbf{(Equation A1-10)}$$

3. For diaphragms in a multistory building with qualifying crosswalls in all levels:

$$DCR = 2.1 S_{D1} \Sigma W_d / (\Sigma \Sigma v_u D + V_{cb}) \quad \textbf{(Equation A1-11)}$$

DCR shall be calculated at each level for the set of diaphragms at and above the level under consideration. In addition, the roof diaphragm shall also meet the requirements of Equation (A1-10).

4. For a roof diaphragm and the diaphragm directly below, if coupled by crosswalls:

$$DCR = 2.1 S_{D1} \Sigma W_d / \Sigma \Sigma v_u D \qquad \textbf{(Equation A1-12)}$$

❖ All of the DCR formulas assume that the resistance of each end of the diaphragm is nearly equal. If a large inequality exists, the total capacity of the diaphragm should be adjusted. A conservative assumption would be to use twice the capacity of the end with the least depth. The analysis of a flexible diaphragm for a DCR assumes that the diaphragm shape in any span is relatively rectangular. Buildings with plan irregularities but parallel walls can be analyzed by this method. The span of the diaphragm is the distance between shear walls when ties or collectors develop the load transfer to the shear wall.

DCRs are calculated for the following four conditions:

Item 1. If there are no qualifying crosswalls above or below the diaphragm level analyzed, the total lateral load tributary to the diaphragm, multiplied by a lateral load coefficient and spectral response factor is divided by the total shear capacity of both ends of the diaphragm.

Item 2. The crosswalls in a single-story building are also usually an additional load path between the roof diaphragm and the ground. The capacity of both ends of the diaphragm and the crosswalls below the diaphragm are added together to determine the capacity of the diaphragm.

Item 3. If the building has crosswalls in all levels of the building, the calculation of DCR begins with the roof level and is made at each story level. The sum of the diaphragm loads above the level under consideration is used as the demand. The total capacity of all the diaphragms above the level under consideration and the capacity of the crosswalls below the diaphragm level analyzed is the combined capacity. The crosswalls at each level have a minimum capacity of 30 percent of the diaphragm capacity above and cause a nearly common displacement of the diaphragms relative to the shear walls.

Item 4. A special treatment of the roof diaphragm is provided in this paragraph. Roof diaphragms are generally more flexible than adjacent floors. This provision allows the calculation of a DCR for the combination of a roof and the adjacent floor when only these two diaphragms are coupled by crosswalls. Crosswalls may or may not exist in the stories below the floor adjacent to the roof diaphragm.

A111.4.3 Chords. An analysis for diaphragm flexure need not be made, and chords need not be provided.

❖ Diaphragms are checked using the demand-capacity formulas and Figure A1-1. The conventional analysis for diaphragm chords is not part of the special procedure and the analysis is explicitly excluded in the provisions. The reason is that there is not a significant flexural response in flexible wood diaphragms; the primary effect is shear yielding in the end zones. The source of flexural capacity of the diaphragm is the continuity inherent in the diaphragm construction. Since close life-threatening damage related to lack of diaphragm chords has not been observed and virtually every diaphragm has some form of edge restraint and/or internal continuity capable of providing adequate flexural capacity, no rules for formal analysis of chords are included in Chapter A1.

A111.4.4 Collectors. An analysis of diaphragm collector forces shall be made for the transfer of diaphragm edge shears into vertical elements of the lateral-force-resisting system. Collector forces may be resisted by new or existing elements.

❖ The transfer of shear forces at the edge of a diaphragm to the shear resisting elements at that edge may require a collector at the edge of the diaphragm for distribution of the diaphragm shear to the elements of the shear wall. If the shear wall is the full depth of the diaphragm, no need for a collector would exist as the uniformly distributed shear load of the diaphragm would be uniformly resisted along the length of the shear wall. If the shear wall strength and stiffness were at one end of that line of resistance, the collector would accumulate load from the diaphragm and act as a tie to the shear wall element. The calculated collector force may

be resisted by existing or new elements. The relative rigidity of new and existing elements should be considered in apportioning the loads to the various elements.

A pier adjacent to a corner of the two intersecting shear walls is much stiffer than the usual pier analysis indicates. The effect of the flange, which is the other wall around the corner, is significant when the flange is in tension. The stiffness of this pier may cause the spandrel to have a significant collector stress. This should be considered in the design of the collector. Because of this increased collector force, a minimum required spacing of the first shear bolt from the corner is specified in Section A113.1.4 to provide an attachment of the collector immediately adjacent to the corner.

A111.4.5 Diaphragm openings.

1. Diaphragm forces at corners of openings shall be investigated and shall be developed into the diaphragm by new or existing materials.

2. In addition to the demand-capacity ratios of Section A111.4.2, the demand-capacity ratio of the portion of the diaphragm adjacent to an opening shall be calculated using the opening dimension as the span.

3. Where an opening occurs in the end quarter of the diaphragm span, the calculation of $v_u D$ for the demand-capacity ratio shall be based on the net depth of the diaphragm.

❖ **Item 1.** The analysis of a diaphragm for its DCR assumes a rectangular shape. The presence of a large opening can increase the flexibility of the diaphragm and cause localized stresses. The diaphragm shear is assumed to be an average shear stress and this assumption implies that the uniform shear in the diaphragm adjacent to the opening must be transferred by collectors to the portion of the diaphragm that is continuous past the opening. Section A111.4.9 describes both the stiffness and strength check; the reduction in stiffness caused by an opening is checked as indicated in Item 3. If the strength of the existing diaphragm with the opening is adequate for the strength analysis, the transfer of uniform shear to the remaining section of the diaphragm could be assumed to be accomplished by the diaphragm itself, and the diaphragm edge adjacent to the opening could be assumed to have no shear stress.

Item 2. When the length of an opening is a significant part of the diaphragm span length and it is in the center half of the diaphragm, the DCR of the diaphragm that is continuous shall be checked. The demand is that part of the diaphragm load, W_o, within the length of opening and the span is the length of the opening measured parallel to the diaphragm span.

Item 3. Openings within the length of the diaphragm significantly affect its stiffness. The reduction of stiffness is approximated by using a capacity of the diaphragm that is calculated for the net depth rather than the full depth.

A111.5 Diaphragm shear transfer. Diaphragms shall be connected to shear walls with connections capable of developing the diaphragm-loading tributary to the shear wall given by the lesser of the following formulas:

$$V = 1.2S_{D1} C_p W_d \qquad \textbf{(Equation A1-13)}$$

using the C_p values in Table A1-C, or

$$V = v_u D \qquad \textbf{(Equation A1-14)}$$

❖ The calculation of the capacity of the shear connection of the diaphragm to the shear wall is based on the coefficients for spectral acceleration, C_v for the UBC and S_{D1} for the IBC, and a C_p factor that is based on the materials used for construction of the diaphragm. The values of the C_p range from 0.5 to 0.75. The least value is for a single layer of sheathing. The greatest value of C_p is for double or multiple layers of sheathing. A single layer of plywood with all edges supported on the framing or blocked is considered as equivalent to a double layer system. Steel decks without fill material have C_p values in between the single- and the double-sheathed diaphragms. The stronger and stiffer diaphragms amplify the ground motions transmitted to the ends of the diaphragm by the shear walls more than single-sheathed systems and are assigned a high C_p coefficient.

The evaluation of the required capacity of the shear connection of a diaphragm to the shear wall is not a calculation of the reaction of a beam with a loading that may be variable within its span length. The required capacity of the connection is one-half of the total dead load of the diaphragm and normal walls, multiplied by the spectral response and construction coefficients. The analysis of diaphragms by the special procedure accepts nonlinear behavior of the diaphragm and limits the required capacity of the shear connection to the capacity of the diaphragm. The check for the acceptable DCR of the diaphragm is for displacement control. The limitation of the shear connection capacity to the shear capacity of the diaphragm provides a connection strength that causes shear yielding in the diaphragm.

A111.6 Shear walls (In-plane loading).

A111.6.1 Wall story force. The wall story force distributed to a shear wall at any diaphragm level shall be the lesser value calculated as:

$$F_{wx} = 0.8S_{D1}(W_{wx} + W_d/2) \qquad \textbf{(Equation A1-15)}$$

but need not exceed

$$F_{wx} = 0.8S_{D1}W_{wx} + v_u D \qquad \textbf{(Equation A1-16)}$$

❖ The special procedure differs from the general procedure in that the loading of the shear wall is determined by the diaphragm capacity rather than an arbitrary distribution of a base shear to the levels of the diaphragm. If there are no crosswalls, the story force of a diaphragm level is equal to the spectral response coefficient (C_v for UBC, S_{D1} for IBC) times the tributary weight of the wall at the level of calculation, plus one-half of the dead load of

the diaphragm and normal walls at the level of consideration. However, the story force need not exceed the spectral response coefficient times the tributary weight of the shear wall, plus the shear capacity of the diaphragm at the level of consideration.

A111.6.2 Wall story shear. The wall story shear shall be the sum of the wall story forces at and above the level of consideration.

$$V_{wx} = \Sigma F_{wx} \qquad \textbf{(Equation A1-17)}$$

❖ The design shear load in a shear wall at any level is the total of the wall story forces calculated at the stories above by Equations A1-15 through A1-16. The least value of calculated story shear is to be used for analysis of the existing shear wall.

A111.6.3 Shear wall analysis. Shear walls shall comply with Section A112.

❖ The forces calculated in this section are used to analyze the existing shear walls by the procedures specified in Section A112.

A111.6.4 Moment frames. Moment frames used in place of shear walls shall be designed as required by the Building Code, except that the forces shall be as specified in Section A111.6.1, and the story drift ratio shall be limited to 0.015, except as further limited by Section A112.4.2.

❖ Moment frames may be used as shear walls in locations where a shear wall does not exist. A common use for moment frames is at the open front of a commercial building. The loading of the moment frame is calculated as if it were an existing shear wall. The interstory drift of the moment frame is limited to 0.015 of the story height. If the moment frame is used in a line of resistance with a URM wall, more severe limitations on interstory drift are prescribed by Section A112.4.2.

A111.7 Out-of-plane forces—unreinforced masonry walls.

A111.7.1 Allowable unreinforced masonry wall height-to-thickness ratios. The provisions of Section A110.2 are applicable, except the allowable height-to-thickness ratios given in Table A1-B shall be determined from Figure A1-1 as follows:

1. In Region 1, height-to-thickness ratios for buildings with crosswalls may be used if qualifying crosswalls are present in all stories.

2. In Region 2, height-to-thickness ratios for buildings with crosswalls may be used whether or not qualifying crosswalls are present.

3. In Region 3, height-to-thickness ratios for "all other buildings" shall be used whether or not qualifying crosswalls are present.

❖ URM walls need not be analyzed for forces normal to the wall surface. Instead of a usual flexural analysis, acceptable height-to-thickness ratios for the URM walls are specified. Walls conforming to these height-to-thickness ratios are stable due to dynamic rocking stability, not because of the tensile capacity of the bed joints. The concept of dynamic stability assumes that the bed joints are cracked at the wall anchorage levels and near the center of the wall between wall anchorage levels. If the height-to-thickness ratio of the URM wall exceeds these prescribed ratios, the wall must be braced to increase its stability.

The allowable height-to-thickness ratios are given in Table A1-B. The height-to-thickness ratios were determined by data obtained by dynamic testing of full size walls. The walls at the uppermost story level are most vulnerable because there is no overburden. Walls at the first story of the buildings have less vulnerability as the unamplified ground motion shakes one end of the wall. The increased ratios for buildings with crosswalls are due to the effect of damping by the crosswalls of the diaphragms. These height-to-thickness ratios, with the exception of buildings with crosswalls, are applicable to buildings analyzed by either the general or special procedure. Buildings with crosswalls analyzed by the special procedure may have larger height-to-thickness ratios if several special limitations are met.

Item 1. In Region 1, the increased height-to-thickness ratios are applicable only to buildings with crosswalls in all story levels and if the following conditions are met:

- The tested mortar shear strength shall not be less than 100 psi (690 kPa) or where the visual examination of the collar joints indicates 50 percent or more of the collar joint is filled with mortar and the tested mortar shear strength is 60 psi (414 mm) or more.

- The separation of the building from an adjacent building shall not be less than 5 inches (127 mm) (see Section A113.10).

Item 2. A special condition exists when the DCR of the diaphragms falls in Region 2 of Figure A1-1. In this special condition, the height-to-thickness ratios for crosswalls may be used even though qualifying crosswalls are not present.

Item 3. This special provision allowing ratios of buildings "with crosswalls" when no crosswalls are present is due to the presence of hysteretic damping of diaphragms that have DCRs in excess of 2.5.

A111.7.2 Walls with diaphragms in different regions. When diaphragms above and below the wall under consideration have demand-capacity ratios in different regions of Figure A1-1, the lesser height-to-thickness ratio shall be used.

❖ The DCR of both the upper and lower diaphragms that shake the out-of-plane wall shall be checked. If the DCR of either diaphragm falls into Region 3 of Figure A1-1, the height-to-thickness ratios for "all other" buildings shall be used even if qualifying crosswalls are not present. The DCR of both diaphragms must fall within Region 2 for the height-to-thickness ratios for "with crosswalls" to be used, even though crosswalls are not present. In the calculation of DCR for determination of height-to-thickness ratios, the effect of crosswalls is excluded.

A111.8 Open-front design procedure. A single-story building with an open front on one side and crosswalls parallel to the open front may be designed by the following procedure:

1. Effective diaphragm span, L_i, for use in Figure A1-1 shall be determined in accordance with the following formula:

$$L_i = 2 \left[(W_w/W_d)L + L \right]$$ **(Equation A1-18)**

2. Diaphragm demand-capacity ratio shall be calculated as:

$$DCR = 2.12 S_{D1}(W_d + W_w)/[(v_u D) + V_{cb}]$$

(Equation A1-19)

❖ It should be noted that "open-front" has an ordinary meaning, referring to a building face that is open, i.e., free from significant solid walls, usually at one story. Most commonly, an open-front building is the ground floor of a commercial building. Buildings on corners of blocks often have open fronts on two sides.

If a moment frame complying with Section A111.6.4 is placed in the open front, then the building meets the requirement of Section A111.1.3 for two lines of vertical elements in the direction parallel to the front; therefore, the open-front procedure is not applicable.

Under the exception of Section A111.1.3, the open front of a single-story building is permitted to remain open if the building satisfies the requirements of the open-front procedure in this section and the building contains crosswalls.

The open-front procedure is permitted only in one direction: if there are two open fronts, one of them must be provided with a braced frame or shear wall.

In case the open-front procedure is tried but cannot be satisfied, the solution would be to provide a braced frame or a moment frame in the open front. The open-front procedure uses Equation A1-18 to determine an effective span, L_i; the distance from the nearest shear wall to the open front is the length L. An additional length is calculated in order to account for the weight of masonry at the open front. This weight, which is a concentrated load at the end of the diaphragm, is assumed to be distributed uniformly along a length, $(W_w/W_D) L$. The augmented length is $(W_w/W_d) L$ plus L. This length is then doubled to estimate the span, L_i, of an equivalent diaphragm that has a deflected shape similar to a shear beam spanning between two shear walls.

The diaphragm DCR is calculated by Equation A1-19, using the spectral response coefficient, the sum of the diaphragm and normal wall dead load, the weight of the wall at the open front and the shear capacity of one end of the diaphragm and the required crosswall. This analysis determines the capacity of the crosswall that is required to control the displacement at the end of the diaphragm at the open front.

The required crosswall need not be right in the open front, but may be at some distance back within the building. It should also be noted that a moment frame or other system might be used in place of a crosswall in accordance with Section A111.3.5. Where the crosswall or other system is at some distance back from the open front, a second analysis is required to determine the maximum distance that is permitted. This analysis is similar to the first, except that a crosswall capacity is not included in the calculation of Equation A1-19. The maximum distance is *L*. A trial value of *L* is used in Equation A1-18, and the corresponding L_i and DCR are used to check the diaphragm. The process is repeated until an acceptable value of *L* is obtained.

SECTION A112
ANALYSIS AND DESIGN

A112.1 General. The following requirements are applicable to both the general procedure and the special procedurefor analyzing vertical elements of the lateral-force-resisting system.

❖ The requirements of this section are applicable to both the general and special procedures. The two procedures will determine different in-plane shear loading for the shear walls for buildings of equal floor area due to the difference in overall building weight and the coupling of the weight of normal walls with the shear walls.

A112.2 Existing unreinforced masonry walls.

A112.2.1 Flexural rigidity. Flexural components of deflection may be neglected in determining the rigidity of an unreinforced masonry wall.

❖ The relative rigidity of piers in a wall may be determined on the basis of net area and height as in a shear beam. Wall piers rarely have span-depth ratios that exceed that ratio where flexural stress-strain relationships exist. Use of shear deformations only for calculation of relative rigidity is an appropriate procedure.

A112.2.2 Shear walls with openings. Wall piers shall be analyzed according to the following procedure, which is diagramed in Figure A1-2.

1. For any pier,

    1.1 The pier shear capacity shall be calculated as:

    $$V_a = v_m A/1.5 \qquad \textbf{(Equation A1-20)}$$

    1.2 The pier rocking shear capacity shall be calculated as:

    $$Vr = 0.9P_D D/H \qquad \textbf{(Equation A1-21)}$$

2. The wall piers at any level are acceptable if they comply with one of the following modes of behavior:

    2.1 **Rocking controlled mode.** When the pier rocking shear capacity is less than the pier shear capacity, i.e., $V_r < V_a$ for each pier in a level, forces in the wall at that level, V_{wx}, shall be distributed to each pier in proportion to $P_D D/H$.

    For the wall at that level:

    $$0.7\,V_{wx} < \Sigma V_r \qquad \textbf{(Equation A1-22)}$$

    2.2 **Shear controlled mode.** Where the pier shear capacity is less than the pier rocking capacity,

i.e., $V_a < V_r$ in at least one pier in a level, forces in the wall at the level, V_{wx}, shall be distributed to each pier in proportion to *D/H*.

For each pier at that level:

$$V_p < V_a \qquad \textbf{(Equation A1-23)}$$

and

$$V_p < V_r \qquad \textbf{(Equation A1-24)}$$

If $V_p < V_a$ for each pier and $V_p > V_r$ for one or more piers, such piers shall be omitted from the analysis, and the procedure shall be repeated for the remaining piers, unless the wall is strengthened and reanalyzed.

3. **Masonry pier tension stress.** Unreinforced masonry wall piers need not be analyzed for tension stress.

❖ This section is for the analysis of a URM wall with windows and/or openings. The in-plane strength of the wall is the strength of the piers between the windows and doors. The pier shear capacity is its allowable shear, v_m, times the net area of the pier, A_u, and adjusted from peak shear stress to average shear stress. This adjusted shear strength is the average shear as used in the design of reinforced masonry and reinforced concrete. A second method of calculating the in-plane shear capacity is given by Equation A1-21. This rocking shear capacity assumes that the bed joint at the top and bottom of the pier has cracked and the axial load on the pier acting eccentrically on the displaced pier provides the restoring shear or rocking shear capacity. The effective lever arm of the axial loads on the ends of the pier is about 90 percent of the pier width. This restoring force is a strength limit state.

Rocking controlled mode. When rocking shear capacity of each individual pier is less than that pier's shear capacity, the pier will rotate, or rock, as a rigid block and have a stable dynamic behavior. In this case, the capacity of the wall consisting of a series of piers is the sum of the rocking shear capacities of each pier, and each pier is loaded with a portion of the total base shear. The rocking shear capacity, $P_D D/H$, is used as the relative rigidity of each pier.

Shear controlled mode. When the axial load on any wall pier is large enough to prevent bed joint cracking prior to the formation of a diagonal shear crack, a shear failure in this pier may occur at a small displacement. In this case, the load is proportioned to the piers by the relative shear rigidity. Flexural stiffness is not included in the relative rigidity calculations as the common pier shape is such that flexural deformations are those related to deep beams.

When the pier shear capacity, V_a, is less than the pier rocking capacity, V_r, in even one pier, the shear force in the wall, V_{wx}, is distributed to the piers in accordance with their relative rigidity, *D/H*. The distributed load to a pier, V_p, must be less than V_a, and if $V_p < V_a$, the shear resistance of the wall piers is adequate. If $V_p < V_r$ for each pier and $V_p > V_a$ for one or more piers, then an incompatible pier capacity exists. The piers where $V_p > V_a$ are omitted from the analysis and the procedure is repeated using only the piers where $V_p > V_a$. This procedure is shown in a flow diagram in Figure A1-2.

Masonry pier tension stress. Flexural moments and tensile stress are not calculated for unreinforced masonry piers. Flexural cracking may occur on bed joints due to out-of-plane flexure in maximum moment zones, but the pier system is stable when subjected to the dynamic loading of an earthquake.

A112.2.3 Shear walls without openings. Shear walls without openings shall be analyzed the same as for walls with openings, except that V_r shall be calculated as follows:

$$V_r = 0.9 \, (P_D + 0.5P_w) \, D/H \qquad \textbf{(Equation A1-25)}$$

❖ These walls shall be analyzed as a single pier for in-plane shear capacity. The rocking shear computation uses one-half of the wall weight as the restoring force because the product of $P_w D$ is one-half of the wall length, D.

A112.3 Plywood-sheathed shear walls. Plywood-sheathed shear walls may be used to resist lateral forces for buildings with flexible diaphragms analyzed according to provisions of Section A111. Plywood-sheathed shear walls may not be used to share lateral forces with other materials along the same line of resistance.

❖ Sheathed light-frame shear walls may be used as shear walls in buildings with flexible diaphragms in conformity with the restrictions in the building code. These sheathed walls cannot share lateral forces with other materials in the line of the sheathed light-frame walls.

A112.4 Combinations of vertical elements.

A112.4.1 Lateral-force distribution. Lateral forces shall be distributed among the vertical-resisting elements in proportion to their relative rigidities, except that moment-resisting frames shall comply with Section A112.4.2.

❖ The distribution of the lateral force should be based on realistic calculations of stiffness. The compressive modulus of unreinforced brick has been determined by testing to have values vary from 1×10^5 psi to 5×10^5 psi. Effective moduli of reinforced concrete and masonry shear walls vary from 10 to 20 percent of uncracked moduli. A special case is given in Section 112.4.2 for assignment of lateral loads, regardless of relative stiffness.

A112.4.2 Moment-resisting frames. Moment-resisting frames shall not be used with an unreinforced masonry wall in a single line of resistance unless the wall has piers that have adequate shear capacity to sustain rocking in accordance with Section A112.2.2. The frames shall be designed in accordance with the Building Code to carry 100 percent of the lateral forces tributary to that line of resistance, as determined from Equation (A1-7). The story drift ratio shall be limited to 0.0075.

❖ Use of a combination of moment frame and URM shear wall in the same line of resistance is prohibited unless the piers in the wall have a rocking shear capacity that exceeds the pier shear capacity (see Equations A1-20 and A1-21). If this condition is met, a moment frame

may be used with 100 percent of the load assigned to it. The minimum stiffness of the moment frame shall be twice that specified for other moment frames that are used to resist lateral loads (see Section A111.6.4). This section should not be interpreted to be applicable to a moment frame used in the open front of a URM building where the piers do not have significant stiffness. These masonry columns may support lintels that have a substantial load. The engineer should assess the stability of the masonry-bearing surface when the design drift of the open front occurs.

SECTION A113
DETAILED SYSTEM DESIGN REQUIREMENTS

A113.1 Wall anchorage.

A113.1.1 Anchor locations. Unreinforced masonry walls shall be anchored at the roof and floor levels as required in Section A110.2. Ceilings of plaster or similar materials, when not attached directly to roof or floor framing and where abutting masonry walls, shall either be anchored to the walls at a maximum spacing of 6 feet (1829 mm), or be removed.

❖ URM walls are anchored to all roofs and floors, and to ceilings in certain cases. If the ceiling is forced to displace with the roof diaphragm by existing wood trussing, the anchor should also be at the ceiling level. The URM wall is forced to remain vertical within the depth of the truss and the flexural crack will occur at the ceiling level. A depth of truss that triggers ceiling anchorage has not been defined. Engineering judgement should be used to determine an acceptable depth of the roof trusses not having anchorage at the ceiling level.

Ceiling systems of substantial weight act as loading on the walls. The transfer of the ceiling weight to the shear wall could be made by the diaphragm above and its shear connection, or by an independent shear connection and diaphragm at the ceiling level.

A113.1.2 Anchor requirements. Anchors shall consist of bolts installed through the wall as specified in Table A1-E, or an approved equivalent at a maximum anchor spacing of 6 feet (1829 mm). All wall anchors shall be secured to the joists to develop the required forces.

❖ When existing wall anchors have been tested to confirm the adequacy of the embedded end, the connection of the existing anchor to the wood framing shall be analyzed by rational engineering methods for its adequacy.

A113.1.3 Minimum wall anchorage. Anchorage of masonry walls to each floor or roof shall resist a minimum force determined as $0.9S_{DS}$ times the tributary weight or 200 pounds per linear foot (2920 N/m), whichever is greater, acting normal to the wall at the level of the floor or roof. Existing wall anchors, if used, must meet the requirements of this chapter or must be upgraded.

A113.1.4 Anchors at corners. At the roof and floor levels, both shear and tension anchors shall be provided within 2 feet

(610 mm) horizontally from the inside of the corners of the walls.

❖ The end pier in a system of piers has additional stiffness due to the effect of the intersecting wall acting as a flange. The placement of the first shear anchor within 2 feet (610 mm) of the end of the wall aids in transferring the shear load from the diaphragm to the corner pier.

A113.2 Diaphragm shear transfer. Bolts transmitting shear forces shall have a maximum bolt spacing of 6 feet (1829 mm) and shall have nuts installed over malleable iron or plate washers when bearing on wood, and heavy-cut washers when bearing on steel.

A113.3 Collectors. Collector elements shall be provided that are capable of transferring the seismic forces originating in other portions of the building to the element providing the resistance to those forces.

❖ The transfer of shear forces at the edge of a diaphragm to the shear resisting elements at that edge may require a collector at the edge of the diaphragm for distribution of the diaphragm shear to the elements of the shear wall. If the shear wall is the full depth of the diaphragm, no need for a collector would exist as the uniformly distributed shear load of the diaphragm would be uniformly resisted along the length of the shear wall. If the shear wall strength and stiffness were at one end of that line of resistance, the collector will accumulate load from the diaphragm and act as a tie to the shear wall element. The calculated collector force may be resisted by existing or new elements. The relative rigidity of new and existing elements should be considered in apportioning the loads to the various elements.

A pier adjacent to a corner of two intersecting shear walls is much stiffer than the usual pier analysis indicates. The effect of the flange, which is the other wall around the corner, is significant when the flange is in tension. The stiffness of this pier may cause the spandrel to have a significant collector stress. Because of this increased collector force, a minimum required spacing of the first shear bolt from the corner is specified to minimize the collector stress in the spandrel above the opening nearest to the building corner.

A113.4 Ties and continuity. Ties and continuity shall conform to the requirements of the Building Code.

A113.5 Wall bracing.

A113.5.1 General. Where a wall height-to-thickness ratio exceeds the specified limits, the wall may be laterally supported by vertical bracing members per Section A113.5.2 or by reducing the wall height by bracing per Section A113.5.3.

❖ Normal walls with excessive height-to-thickness ratios may be supported by an additional line of wall anchors that reduce the unbraced height or by vertical members that support the wall when cracked.

A113.5.2 Vertical bracing members. Vertical bracing members shall be attached to floor and roof construction for their design loads independently of required wall anchors. Horizontal spacing of vertical bracing members shall not exceed one-half of the unsupported height of the wall or 10 feet (3048 mm). Deflection of such bracing members at design loads shall not exceed one-tenth of the wall thickness.

❖ The vertical member shall be attached to the diaphragms. The reaction shall not be transferred into the URM wall and to its anchorage system. The horizontal spacing of the wall braces is limited to a distance equal to one-half of the wall height in order to provide a uniformly distributed resistance to out-of-plane displacements. This bracing does not prevent tension cracking of the URM wall. The brace should be anchored to the wall with a wall anchor at the center of the wall height as a minimum. The design loading of the wall brace is determined by the requirements of the building code, and the stiffness of the wall brace must be sufficient to meet the deflection limitatons.

A113.5.3 Intermediate wall bracing. The wall height may be reduced by bracing elements connected to the floor or roof. Horizontal spacing of the bracing elements and wall anchors shall be as required by design, but shall not exceed 6 feet (1829 mm) on center. Bracing elements shall be detailed to minimize the horizontal displacement of the wall by the vertical displacement of the floor or roof.

❖ Braces that extend between floors or roofs can reduce the unbraced height of the URM wall. A horizontal girt anchored on the face of the wall can be used to distribute the wall anchor force to braces that are spaced greater than 6 feet (1829 mm). Vertical displacement of the anchorage of the brace to the floor and roof will cause an outward displacement or bending stress in the URM wall. The deflection of the brace anchorage point caused by live loading shall be investigated and minimized.

A113.6 Parapets. Parapets and exterior wall appendages not conforming to this chapter shall be removed, or stabilized or braced to ensure that the parapets and appendages remain in their original positions.

The maximum height of an unbraced unreinforced masonry parapet above the lower of either the level of tension anchors or the roof sheathing shall not exceed the height-to-thickness ratio shown in Table A1-F. If the required parapet height exceeds this maximum height, a bracing system designed for the forces determined in accordance with the Building Code shall support the top of the parapet. Parapet corrective work must be performed in conjunction with the installation of tension roof anchors.

The minimum height of a parapet above any wall anchor shall be 12 inches (305 mm).

Exception: If a reinforced concrete beam is provided at the top of the wall, the minimum height above the wall anchor may be 6 inches (152 mm).

❖ A maximum height of unbraced parapets above the wall anchorage points is specified. This height of unreinforced wall will be dynamically stable, but may slide laterally by the diaphragm response. A minimum height of parapet above the wall anchorage is specified, as its tension capacity is dependent on the axial stress in the wall at the level of anchorage. An exception is provided for a minimum height of 6 inches (152 mm). This exception is necessary as it was specified for standard details used for parapet hazard reduction prior to the development of comprehensive requirements for earthquake hazard reduction. It is not a recommended detail, and reinforced concrete beams may be slid laterally on the top of the wall by anchorage forces. However, failure of the anchorage system has not been observed.

A113.7 Veneer.

1. Veneer shall be anchored with approved anchor ties conforming to the required design capacity specified in the Building Code and shall be placed at a maximum spacing of 24 inches (610 mm) with a maximum supported area of 4 square feet (0.372 m²).

   Exception: Existing anchor ties for attaching brick veneer to brick backing may be acceptable, provided the ties are in good condition and conform to the following minimum size and material requirements.

   Existing veneer anchor ties may be considered adequate if they are of corrugated galvanized iron strips not less than 1 inch (25.4 mm) in width, 8 inches (203 mm) in length and $^1/_{16}$ inch (1.6 mm) in thickness, or the equivalent.

2. The location and condition of existing veneer anchor ties shall be verified as follows:

   2.1. An approved testing laboratory shall verify the location and spacing of the ties and shall submit a report to the building official for approval as part of the structural analysis.

   2.2. The veneer in a selected area shall be removed to expose a representative sample of ties (not less than four) for inspection by the building official.

❖ Unbonded brick veneer and veneer poorly bonded with angled bricks or sporadic flat metal ties to the URM structural wall have been found to detach from the structural wall when shaken with 20- to 30-percent ground motion. The exception in this section was taken from the 1930 *Los Angeles Building Code* and is applicable to Seismic Zones 3 and 4.

A113.8 Nonstructural masonry walls.
Unreinforced masonry walls that carry no design vertical or lateral loads and that are not required by the design to be part of the lateral-force-resisting system shall be adequately anchored to new or existing supporting elements. The anchors and elements shall be designed for the out-of-plane forces specified in the Building Code. The height- or length-to-thickness ratio between such supporting elements for such walls shall not exceed nine.

❖ A nonstructural wall must be isolated from the structure on three sides. If the wall is not isolated from the structure, it must be considered as a shear wall and/or bearing masonry wall. The allowable height-to-thickness ratio is reduced from that allowed for a one-story bearing wall because vertical loading increases the stability of an unreinforced masonry wall. The height-to-thickness ratio may not exceed 9 or the limitations of Section A113.5.3.

A113.9 Truss and beam supports.
Where trusses and beams other than rafters or joists are supported on masonry, independent secondary columns shall be installed to support vertical loads of the roof or floor members.

Exception: Secondary supports are not required where S_{D1} is less than 0.3g. (Seismic Zones 1, 2A and 2B for the UBC).

❖ Conformance of the URM wall with the height-to-thickness limitations does not imply that these walls will be uncracked and deform as flexural beams. It is highly likely that a crack will occur at each anchorage point and in the center part of the wall where the design level of intensity of shaking occurs. These blocks of walls will rotate on these cracks but will be dynamically stable. The rotation angle between a block above and below a wall anchorage could be greater than 5 degrees. A truss or beam that has an unyielding bearing surface can load the edge of its bearing on the masonry and cause spalling of the bearing surface. The secondary support serves the same purpose as shoring that would be installed after the earthquake. A foundation is not required for the secondary support, and the secondary support loads need not be accumulative in a multistory building.

A113.10 Adjacent buildings.
Where elements of adjacent buildings do not have a separation of at least 5 inches (127 mm), the allowable height-to-thickness ratios for "all other buildings" per Table A1-B shall be used in the direction of consideration.

❖ The increases in the allowable height-to-thickness ratios of URM walls when crosswalls are present in each story were predicated on allowing the diaphragms to have dynamic displacements that cause nonlinear displacements in the crosswalls and the diaphragm or in the diaphragm alone. Figure A1-1 was plotted by nonlinear analysis using ground motions scaled to 0.4g effective peak acceleration. The acceptable demand-capacity-span boundary is equivalent to 5 inches (127 mm) of displacement of the center of the diaphragm relative to the shear walls at the end of the span. If an adjacent building restricts the diaphragm displacement, the increase in the allowable height-to-thickness ratio is not applicable.

SECTION A114
WALLS OF UNBURNED CLAY, ADOBE OR STONE MASONRY

A114.1 General. Walls of unburned clay, adobe or stone masonry construction shall conform to the following:

1. Walls of unburned clay, adobe or stone masonry shall not exceed a height- or length-to-thickness ratio specified in Table A1-G.

2. Adobe may be allowed a maximum value of 9 pounds per square inch (62.1 kPa) for shear unless higher values are justified by test.

3. Mortar for repointing may be of the same soil composition and stabilization as the brick, in lieu of cement-mortar.

TABLE A1-A—ELEMENTS REGULATED BY THIS CHAPTER

BUILDING ELEMENTS	S_{D1}			
	$\geq 0.067g < 0.133g$	$\geq 0.133g < 0.20g$	$\geq 0.20g < 0.30g$	$> 0.30g$
Parapets	X	X	X	X
Walls, anchorage	X	X	X	X
Walls, h/t ratios		X	X	X
Walls, in-plane shear		X	X	X
Diaphragms[1]			X	X
Diaphragms, shear transfer[2]		X[2]	X	X
Diaphragms, demand-capacity ratios[2]			X	X

1. Applies only to buildings designed according to the general procedures of Section A110.
2. Applies only to buildings designed according to the special procedures of Section A111.

❖ This chapter does not regulate URM elements where construction using URM is permitted by the building code. The requirements for UBC Seismic Zones 2A and 2B are separately listed, as the boundaries of those zones are 0.067 to 0.133g and 0.133 to 0.20g effective peak acceleration, respectively.

Observations of damage and analysis of URM buildings in the described seismic hazard zones provides the basis for the determination of regulated elements. Construction materials and methods used for URM buildings are very similar across the United States. The majority of URM buildings in Southern California and along the Pacific coast have been shaken with intensities correlated with UBC seismic Zones 2A and 2B. Collapse of parapets and separation of unanchored walls at the top story are observed failures. Strength and stiffness checks of existing diaphragms are required only in UBC Seismic Zones 3 and 4 and where IBC Seismic Design Category C and higher.

Diaphragms analyzed by the special procedure use Section A111.4.2 and Figure A1-1 for a determination of adequate stiffness. Diaphragms analyzed by the general procedure use the provisions of the building code.

TABLE A1-B—ALLOWABLE VALUE OF HEIGHT-TO-THICKNESS
RATIO OF UNREINFORCED MASONRY WALLS

WALL TYPES	$0.13g \leq S_{D1} < 0.25g$	$0.25g \leq S_{D1} < 0.4g$	$S_{D1} \geq 0.4g$ BUILDINGS WITH GROSSWALLS[1]	$S_{D1} > 0.4g$ ALL OTHER BUILDINGS
Walls of one-story buildings	20	16	16[2,3]	16
First-story wall of multistory building	20	18	16	15
Walls in top story of multistory building	14	14	14[2,3]	9
All other walls	20	16	16	13

1. Applies to the special procedures of Section A111 only. See Section A111.7 for other restrictions.

2. This value of height-to-thickness ratio may be used only where mortar shear tests establish a tested mortar shear strength, v_t, of not less than 100 pounds per square inch (690 kPa). This value may also be used where the tested mortar shear strength is not less than 60 pounds per square inch (414 kPa), and where a visual examination of the collar joint indicates not less than 50-percent mortar coverage.

3. Where a visual examination of the collar joint indicates not less than 50-percent mortar coverage, and the tested mortar shear strength, v_t, is greater than 30 pounds per square inch (207 kPa) but less than 60 pounds per square inch (414 kPa), the allowable height-to-thickness ratio may be determined by linear interpolation between the larger and smaller ratios in direct proportion to the tested mortar shear strength.

❖ These values are applicable to all masonry walls both solid and constructed of hollow units. Allowable values of height-to-thickness ratios were determined by dynamic testing conducted by ABK. The values in Table A1-B for the top story of a multistory building in Seismic Zones 3 and 4 correspond to the ABK research. A very substantial conservatism was introduced into all other values. This conservatism was increased by the definition of "Seismic zones" used in the building code versus the use of contours in the ABK methodology. The boundary of Seismic Zone 4 is 0.3g effective peak acceleration. The boundaries of Seismic Zones 3 and 2B are 0.2g and 0.1g effective peak acceleration, respectively.

ABK tested brick with nearly solid collar joints. These joints were not purposely filled, but the high lime mortar flowed into the collar joints. Existing walls with unfilled collar joints have had portions of the exterior wythe between the header courses fall away. This has been observed at top stories and for walls that exceed the recommended heights. Notes 2 and 3 limit increases in the height-to-thickness ratio given for buildings with crosswalls when collar joints have limited coverage and when allowable in-plane shear stress is near the minimum acceptable value.

TABLE A1-C—HORIZONTAL FORCE FACTOR, C_p

CONFIGURATION OF MATERIALS	C_p
Roofs with straight or diagonal sheathing and roofing applied directly to the sheathing, or floors with straight tongue-and-groove sheathing.	0.50
Diaphragms with double or mulitple layers of boards with edges offset, and blocked plywood systems.	0.75
Diaphragms of metal deck without topping:	
Minimal welding or mechanical attachment.	0.6
Welded or mechanically attached for seismic resistance.	0.68

❖ This table specifies the C_p factor to be used in Formula A1-13 for determination of the shear transfer force at the edge of the diaphragm. It is not the C_p force used for either a strength or relative displacement analysis of the diaphragm.

TABLE A1-D—STRENGTH VALUES FOR EXISTING MATERIALS

EXISTING MATERIALS OR CONFIGURATION OF MATERIALS[1]		STRENGTH VALUES
		× 14.594 for N/m
Horizontal diaphragms	Roofs with straight sheathing and roofing applied directly to the sheathing.	300 lbs. per ft. for seismic shear
	Roofs with diagonal sheathing and roofing applied directly to the sheathing.	750 lbs. per ft. for seismic shear
	Floors with straight tongue-and-groove sheathing.	300 lbs. per ft. for seismic shear
	Floors with straight sheathing and finished wood flooring with board edges offset or perpendicular.	1,500 lbs. per ft. for seismic shear
	Floors with diagonal sheathing and finished wood flooring.	1,800 lbs. per ft. for seismic shear
	Metal deck welded with minimal welding.[3]	1,800 lbs, per ft. for seismic shear
	Metal deck welded for seismic resistance.[4]	3,000 lbs. per ft. for seismic shear
Crosswalls[2]	Plaster on wood or metal lath.	600 lbs. per ft. for seismic shear
	Plaster on gypsum lath.	550 lbs. per ft. for seismic shear
	Gypsum wallboard, unblocked edges.	200 lbs. per ft. for seismic shear
	Gypsum wallboard, blocked edges.	400 lbs. per ft. for seismic shear
Existing footing, wood framing, structural steel, reinforcing steel	Plain concrete footings.	$f'_c = 1,500$ psi (10.34 MPa) unless otherwise shown by tests
	Douglas fir wood.	Same as D.F. No. 1
	Reinforcing steel.	$F_y = 40,000$ psi (124.1 N/mm^2) maximum
	Structural steel.	$F_y = 33,000$ psi (137.9 N/mm^2) maximum

1. Material must be sound and in good condition.

2. Shear values of these materials may be combined, except the total combined value should not exceed 900 pounds per foot (4380 N/m).

3. Minimum 22-gage steel deck with welds to supports satisfying the standards of the Steel Deck Institute.

4. Minimum 22-gage steel deck with $^3/_4 \phi$ plug welds at an average spacing not exceeding 8 inches (203 mm) and with sidelap welds appropriate for the deck span.

❖ The values for horizontal diaphragms were derived from ABK's cyclic and dynamic testing by 20-foot by 60-foot (6096 mm by 18 288 mm) diaphragms. The values given are the strength limit state capacities obtained by the experimental testing. The values for crosswalls were derived from testing conducted by the Forest Products Laboratory, the Gypsum Institute and the Southern California Hazardous Building Committee. Use of these values with the limitation of Note 2 for crosswalls is acceptable without an investigation of nailing or edge connections. The required shear connection at the edge of the diaphragm does not require renailing of the diaphragm to blocking installed between the joists and rafters or to the existing edge member of the diaphragm. The blocking installed between joists at the edge of the diaphragm should be installed tightly between the joists or rafters.

TABLE A1-E—STRENGTH VALUES OF NEW MATERIALS USED
IN CONJUNCTION WITH EXISTING CONSTRUCTION

NEW MATERIALS OR CONFIGURATION OF MATERIALS		STRENGTH VALUES
Horizontal diaphragms	Plywood sheathing applied directly over existing straight sheathing with ends of plywood sheets bearing on joists or rafters and edges of plywood located on center of individual sheathing boards.	675 lbs. per ft.
Crosswalls	Plywood sheathing applied directly over wood studs; no value should be given to plywood applied over existing plaster or wood sheathing.	1.2 times the value specified in the current Building Code.
	Drywall or plaster applied directly over wood studs.	The value specified in the current Building Code.
	Drywall or plaster applied to sheathing over existing wood studs.	50 percent of the value specified in the current Building Code.
Tension bolts[5]	Bolts extending entirely through unreinforced masonry wall secured with bearing plates on far side of a three–wythe–minimum wall with at least 30 square inches of area.[2,3]	5,400 lbs. per bolt 2,700 lbs. for two-wythe walls
Shear bolts[5]	Bolts embedded a minimum of 8 inches into unreinforced masonry walls; bolts should be centered in $2^{1}/_{2}$–inch–diameter holes with dry–pack or nonshrink grout around the circumference of the bolt.	The value for plain masonry specified for solid masonry in the current Building Code; no value larger than those given for $^{3}/_{4}$–inch bolts should be used.
Combined tension and shear bolts	Through-bolts—bolts meeting the requirements for shear and for tension bolts.[2,3]	Tension—same as for tension bolts Shear—same as for shear bolts
	Embedded bolts—bolts extending to the exterior face of the wall with a $2^{1}/_{2}$-inch round plate under the head and drilled at an angle of $22^{1}/_{2}$ degrees to the horizontal; installed as specified for shear bolts.[1,2,3]	Tension—3,600 lbs. per bolt Shear—same as for shear bolts
Infilled walls	Reinforced masonry infilled openings in existing unreinforced masonry walls; provide keys or dowels to match reinforcing.	Same as values specified for unreinforced masonry walls
Reinforced masonry[4]	Masonry piers and walls reinforced per the current Building Code.	The value specified in the current Building Code for strength design.
Reinforced concrete[4]	Concrete footings, walls and piers reinforced as specified in the current Building Code.	The value specified in the current Building Code for strength design.

For SI: 1 inch = 25.4 mm, 1 square inch = 645.16 mm², 1 pound = 4.4 N.

1. Embedded bolts to be tested as specified in Section A107.4.

2. Bolts to be $^{1}/_{2}$ inch (12.7 mm) minimum in diameter.

3. Drilling for bolts and dowels shall be done with an electric rotary drill; impact tools should not be used for drilling holes or tightening anchors and shear bolt nuts.

4. No load factors or capacity reduction factor shall be used.

5. Other bolt sizes, values and installation methods may be used, provided a testing program is conducted in accordance with UBC Standard 21–7. The useable value shall be determined by multiplying the calculated allowable value, as determined by UBC Standard 21–7, by 3.0, and the useable value shall be limited to a maximum of 1.5 times the value given in the table. Bolt spacing shall not exceed 6 feet (1829 mm) on center and shall not be less than 12 inches (305 mm) on center.

❖ The shear value of plywood applied over existing 1x boards is limited to 675 pounds per foot (306.5 kg/304.8 mm) shear. This value was determined by cyclic and dynamic testing conducted by ABK. The reason for the low value is that the nails commonly used for application of plywood do not conform to the requirements of the building code. The thickness of the 1x sheathing would limit the nail size used for plywood nailing to 4d. Substantial splitting of the old boards was also observed, especially at the closely spaced edge nailing of the perimeter of the plywood sheet. Testing of double board systems suggest that application of plywood over straight boards by use of staples of adequate length and a uniform distribution of the staples to each 1x board can make a membrane similar to the double board system. The staples should be spaced along the length of every board. Concentration of staple spacing at the perimeter of the plywood sheet is not desirable or necessary.

TABLE A1-F—MAXIMUM ALLOWABLE HEIGHT-TO-THICKNESS RATIOS FOR PARAPETS

	S_{D1}		
	$0.13g \leq S_{D1} < 0.25g$	$0.25g \leq S_{D1} < 0.4g$	$S_{D1} \geq 0.4g$
Maximum allowable height-to-thickness ratios	2.5	2.5	1.5

TABLE A1-G—MAXIMUM HEIGHT-TO-THICKNESS RATIOS FOR ADOBE OR STONE WALLS

	SEISMIC ZONE		
	2B	3	4
One-story buildings	12	10	8
Two-story buildings			
First story	14	11	9
Second story	12	10	8

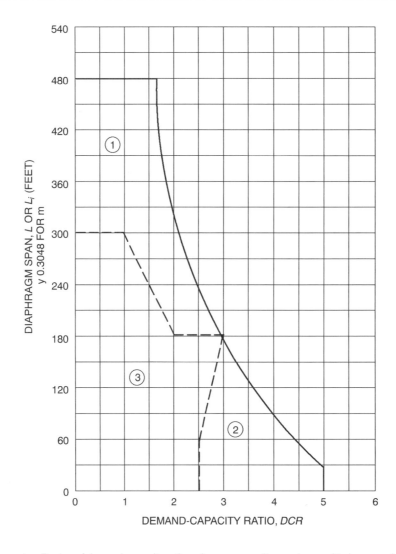

1. Region of demand-capacity ratios where crosswalls may be used to increase h/t ratios.
2. Region of demand-capacity ratios where h/t ratios of "buildings with crosswalls" may be used, whether or not crosswalls are present.
3. Region of demand-capacity ratios where h/t ratios of "all other buildings" shall be used, whether or not crosswalls are present.

Figure A1-1
ACCEPTABLE DIAPHRAGM SPAM

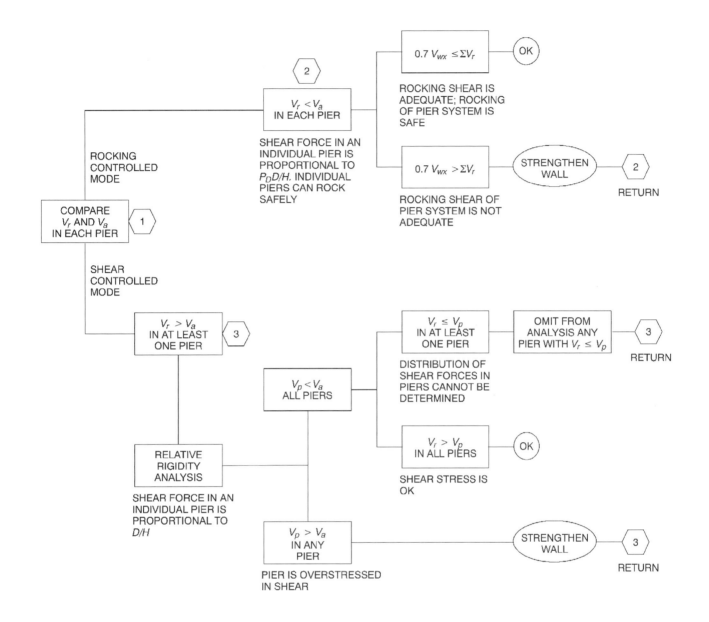

V_a = Allowable shear strength of a pier.
V_p = Shear force assigned to a pier on the basis of a relative shear rigidity analysis.
V_r = Rocking shear capacity of pier.
V_{wx} = Total shear force resisted by the wall.
SV_r = Rocking shear capacity of all piers in the wall.

Figure A1-2
ANALYSIS OF URM WALL IN-PLANE SHEAR FORCES

Bibliography

The following resource materials are referenced in this chapter or are relevant to the subject matter addressed in the this chapter.

ABK-84, *Methodology for Mitigarion of Seismic Hazards in Existing Unreinforced Masonry Buildings: The Methodology.* Topical Report 08, National Science Foundation, Contract No. NSF-C-PFR78-19200. Washington DC: Applied Science and Research Applications, 1984.

ABK-86, *Guidelines for the Evaluation of Historic Brick Masonry Buildings in Earthquake Hazard Zones.* ABK, A Joint Venture, Funded by the Department of Parks and Recreation of the State of California and the National Park Service, United States Government, January 1986.

ATC-78, *Tentative Provisions for the Development of Seismic Regulations for Buildings*, Report ATC 3-16. Redwood City, CA: Applied Technology Council.

ATC-87, *Evaluating the Seismic Resistance of Existing Buildings*, Report ATC-14. Redwood City, CA: Applied Technology Council.

BSSC-97, *NEHRP Recommended Provisions for the Development of Seismic Regulations for New Buildings, Parts I, II and maps.* Washington, DC: Building Seismic Safety Council, 1997.

GSA-76, *Earthquake Resistance of Buildings*, Vol. I-III. Washington DC: General Services Administration, 1976

IBC-00, *International Building Code.* Falls Church, VA: International Code Council, 2000.

ICBO-1988 through 1997, *Uniform Code for Building Conservation.* Whittier, CA: International Conference of Building Officials, 1988-1997.

ICBO-97, *Uniform Building Code.* Whittier, CA: International Conference of Building Officials, 1997.

ICBO-00, *Guidelines for the Seismic Retrofit of Existing Buildings.* Whittier CA: International Conference of Building Officials, 2000.

SEAOC-92, *Commentary on Appendix Chapter 1 of the Uniform Code for Building Conservation.* Los Angeles: Structural Engineers Association of California, 1992.

SSC-85, *Rehabilitation Hazardous Masonry Buildings. A Draft Model Ordinance*, Report No. SSC 85-06. Sacramento, CA: California Seismic Commission, 1985.

SSC-87, *Guidebook and Appendix,* Report No. CSSC 87-03. Sacramento, CA: California Seismic Safety Commission, 1987.

SSC-90, *Earthquake Hazard Identification and Voluntary Mitigation:* Palo Alto's City Ordinance. Report No. CSSC 90-05. Sacramento, CA: California Seismic Safety Commission, 1990.

SSC-95, *Recommended Model Ordinance for the Seismic Retrofit of Hazardous Unreinforced Masonry Bearing Wall Buildings.* Report No. CSSC 95-05. Sacramento CA: California Seismic Safety Commission, 1995.

UBC-21-8, *Uniform Building Code.* "Pointing of Unreinforced Masonry Walls." Whittier, CA: International Conference of Building Officials.

CHAPTER A2

EARTHQUAKE HAZARD REDUCTION IN EXISTING REINFORCED CONCRETE AND REINFORCED MASONRY WALL BUILDINGS WITH FLEXIBLE DIAPHRAGMS

SECTION A201
PURPOSE

The purpose of this chapter is to promote public safety and welfare by reducing the risk of death or injury that may result from the effects of earthquakes on reinforced concrete and reinforced masonry wall buildings with flexible diaphragms. Based on past earthquakes, these buildings have been categorized as being potentially hazardous and prone to significant damage, including possible collapse in a moderate to major earthquake. The provisions of this chapter are minimum standards for structural seismic resistance established primarily to reduce the risk of life loss or injury on both subject and adjacent properties. These provisions will not necessarily prevent loss of life or injury, or prevent earthquake damage to an existing building that complies with these standards.

❖ The provisions of this chapter apply to buildings designed prior to the adoption of the 1997 *Uniform Building Code* (UBC) or a code based on the 1997 *NEHRP Recommended Provisions for Seismic Regulations for New Buildings and Other Structures* (NEHRP) [such as the 2003 *International Building Code*® (IBC®)]. They address deficiencies that have been found to be direct threats to life safety during earthquakes. These deficiencies first became apparent during the 1971 San Fernando, California earthquake when tilt-up buildings collapsed or partially collapsed. Further significant damage was observed in tilt-ups and concrete block wall buildings with flexible diaphragms in the 1987 Whittier Narrows, California; 1989 Loma Prieta, California; and 1994 Northridge, California earthquakes. These deficiencies are as follows:

- A commonly used method of connecting wall panels to roofs and floors for out-of-plane loading was nailing the diaphragms to wood ledgers and bolting the wood ledgers to the walls. This method of connection allowed the wall panels to separate from the roof or floor by failure of the ledger in cross-grain bending, nail pullout from the ledger or by pulling of the nails through the edge of the wood structural panel.

- When used, wall out-of-plane anchors have often included eccentric, flexible connections that are susceptible to damage. Poor installation, including misalignment of anchors, often created additional stresses.

- Distress due to wall out-of-plane forces has occurred at the connection of the roof or floor girders to pilasters in the walls. Commonly, the anchor bolts connecting the girder to the top of the pilasters did not have adequate edge distance. Typically, this resulted in spalling of the inside face of the concrete pilasters and loss of support for the girder. The containment ties required by the building code at the top of the pilasters did not prevent this kind of damage.

- The lack of continuous ties across the full depth of the diaphragm allowed cross-grain tension failures of the framing members at joints in the plywood (wood structural panel) sheathing. This occurred in the interior of the roof or floor diaphragms. Lack of tension ties at glulam hinge locations also led to loss of support for suspended girders.

Voluntary upgrade provisions were developed in Los Angeles after the 1987 Whittier Narrows earthquake (Division 91, City of Los Angeles), and were then made mandatory in the City of Los Angeles after the 1994 Northridge earthquake. Other California jurisdictions have adopted similar provisions as either mandatory or voluntary ordinances. Where large earthquakes have previously occurred and portions of the existing wall anchorage system may have been damaged due to deficiencies in the original design, it may be appropriate to further investigate and prioritize retrofits for these structures.

This chapter adopts the concepts from the 1976 edition of the UBC, which required positive anchorage of all reinforced concrete and masonry walls to develop the out-of-plane wall loads into the floor and roof diaphragms and to prevent splitting of the ledger or tearing of the sheathing (the concept of subdiaphragms was first introduced in the 1976 edition of the UBC). However, the 1994 Northridge earthquake revealed additional deficiencies in the UBC requirements for the wall anchorage system. This chapter additionally incorporates resulting changes in design forces and detailing requirements adopted by the UBC for the design of new buildings. In addition, changes to the UBC design provisions for collector and collector connections are incorporated.

Other possible deficiencies, such as inadequate diaphragm strength, inadequate diaphragm stiffness or reinforcement at diaphragm openings, have not been reported as causing collapse, and are not addressed in this chapter because the level of risk is generally lower. Although reinforced concrete and masonry wall buildings with metal deck diaphragms are within the scope of this chapter, they are relatively rare in areas that have experienced earthquakes. Consequently, post-earthquake observations and lessons learned are very limited. It is possible that vulnerabilities associated with metal deck diaphragms will be observed in future earthquakes.

Similarly, the analysis of the walls for either in-plane or out-of-plane forces is not part of this chapter. Although walls are designed for smaller forces for out-of-plane bending, performance of walls in past earthquakes, provided that they have not been weakened by the introduction of large openings, has been good. There is a concern by some engineers that walls consisting primarily of narrow piers without proper detailing may experience substantial damage in a major earthquake due to in-plane loading. Thus, an engineer considering items beyond the scope of this chapter may want to evaluate walls consisting primarily of truck doors or large windows for in-plane shear.

SECTION A202
SCOPE

The provisions of this chapter shall apply to wall anchorage systems that resist out-of-plane forces for all buildings in Seismic Zones 2B, 3 and 4, or where Seismic Design Categories C, D, E and F are required. The date of applicability for retrofit shall be determined by the building official. Buildings designed under building codes in effect after the adoption of the 1997 edition of the *Uniform Building Code* or the adoption of the *BOCA National Building Code* or *Standard Building Code* that use the 1997 edition of the NEHRP *Recommended Provisions for Seismic Regulations*, are considered to comply with these provisions.

❖ The edition of this chapter in the 1997 *Uniform Code for Building Conservation* (UCBC) (the predecessor to the *Guidelines for the Seismic Retrofit of Existing Buildings*, titled here as Appendix A) applied to existing concrete tilt-up buildings. The scope has been expanded to include all reinforced concrete and reinforced masonry buildings with flexible diaphragms. Concrete tilt-up buildings are a subset of this type of structure. In terms of earthquake performance there is little or no difference between buildings with precast concrete walls or other heavy walls, assuming the buildings have similar diaphragm spans and wall heights. The key is the transfer of forces associated with out-of-plane earthquake forces on these walls into the flexible roof diaphragm.

The scope includes only buildings that are both located in areas of a high seismic hazard and designed under a building code in effect prior to the adoption of the seismic provisions of the 1997 UBC, 1997 NEHRP or 2000 IBC (changes based on observations in the 1994 Northridge earthquake were incorporated into both the 1997 UBC and 1997 NEHRP). Note that some jurisdictions, such as the City of Los Angeles, adopted provisions that essentially meet the requirements of the 1997 UBC prior to its adoption. Structures designed using such requirements may be considered to meet the requirements of this chapter. However, near-field effects were not considered in the interim

provisions adopted by the City of Los Angeles prior to the adoption of the 1997 UBC.

The scope of this chapter includes only elements of the wall anchorage system and collectors and their connections. The wall anchorage system is defined in the UBC as including those elements within the diaphragm required to develop wall anchorage forces including: wall anchors, struts, subdiaphragms, cross ties and continuity ties. Past performance has indicated that partial and full collapse of these structures built with deficient seismic design provisions of the past is likely at ground accelerations approaching or exceeding 0.20g (this observation is based on past earthquakes with short to moderate duration. Ground shaking of less severity but longer duration could also result in collapses). For this reason, Seismic Zones 1 and 2A have not been included in these provisions.

Seismic design forces in building codes have changed over the years, especially design forces for out-of-plane anchorage of concrete or masonry walls. The following discussion provides a history of the pertinent changes made in the UBC. The design forces listed are for the highest seismic hazard zone considered in the UBC.

Prior to 1952, the base shear forces were used for a part or portion of the building. The force for wall anchorage was 0.133W, with *W* equal to the weight of the wall tributary to the anchor. Local codes in California increased the wall anchorage force to 0.20W in 1953 for the design of anchors in concrete and masonry walls. This increase in wall design forces was adopted in the 1961 UBC.

Changes to the provisions in the 1973 UBC were motivated by observations made after the 1971 San Fernando earthquake. Changes included the increase of the wall anchor design force, which requires continuous cross-ties across the depth of the diaphragm and prohibits the use of wood ledgers in cross-grain bending and the use of toe nails or nails subject to withdrawal. Closely spaced ties were also required at the tops of pilasters to provide more resistance to anchor bolt shear failures (these provisions for closely spaced ties proved to be inadequate in subsequent earthquakes). The 1976 UBC wall anchor forces varied between 0.20W and 0.30W due to the introduction of *S* (soil profile factor) in the design force equation. The 1979 UBC modified the seismic loading on parts and portions of the building to be equal to the loading of $0.30W_p$ (regardless of soil profile), where W_p is the tributary weight of the wall component.

The format of the UBC changed substantially in 1988, but the design forces for wall anchorage did not. Provisions for designing structures with horizontal irregularities, such as reentrant corners, were introduced.

In the 1991 UBC, a provision was introduced to increase the wall anchorage design forces for the middle half of the diaphragm by 50 percent to $0.45W_p$. This modification was intended to account for the amplified response in the middle portion of the diaphragms, as observed in the 1984 Morgan Hill, 1987 Whittier Narrows and 1989 Loma Prieta earthquakes. Later, nonlinear analyses of flexible roof diaphragms determined that limiting this amplification to the middle of the diaphragm was inadequate because the flexible diaphragm more closely resembles a shear yielding beam, not a flexural beam. Yielding of the diaphragm in strong ground motions tends to create a somewhat uniform pattern of maximum response along the length of the diaphragm, rather than a clear peak in the center of the span. Second- and third-mode response of the diaphragm also increases the anchorage force adjacent to the supports. Therefore, the *1996 Accumulative Supplement to the Uniform Building Code* required that this amplification be applied to the full length of the diaphragm.

After the 1994 Northridge earthquake, wall anchor design forces were modified in the 1996 *Accumulative Supplement to the Uniform Building Code.* In Seismic Zone 4, the allowable stress design forces were increased to $0.48W_p$ throughout the length of the diaphragm for buildings of standard importance, with additional material load increases required for steel (1.7) and concrete (1.7 instead of the 1.4 typically required). The intent of these different material load increases was to make the strength of the wall anchorage system sufficient to resist amplified shaking of about 1.5g at the roof diaphragm level. Using the material load increases, the capacities for all three materials (wood, concrete and steel) are more uniform. (Although the building code is silent regarding masonry, wall anchors in masonry can be designed for the same loads as wall anchors in concrete.)

Other detailing restrictions including minimum framing member sizes (3x) in the wall anchor system, concerns for symmetry in wall anchor design and an increase in pilaster wall-anchorage design forces (due to the larger flexural stiffness of the pilaster with respect to the wall) were included in the 1996 *Accumulative Supplement to the Uniform Building Code.* Many engineers believed that the poor detailing and construction installation contributed to the observed damage as much or more than the inadequately designed wall-anchorage system itself; therefore, it was recommended that the engineer perform structural observations for wall anchor installation and testing of anchors.

The requirements in the 1997 UBC were intended to provide parity with the 1996 *Accumulative Supplement to the Uniform Building Code*, but are expressed in the context of strength design. Elements of the wall anchorage system are required to be designed with $a_p =$ 1.5 and $R_p = 3.0$. For elements at the roof level of a structure, this corresponds to $F_p = 2.0C_aI_pW_p$. Although the wall anchor design forces appear larger than those in the 1996 *UBC Supplement* (approximately $0.8W_p$ in Zone 4), it can be demonstrated that they are essentially the same by:

• Reducing the 1997 UBC forces by a factor of 1.4 to convert to allowable stress design;

• Reducing the 1997 UBC forces by a factor of 0.85 to account for the fact that the 1996 *Accumulative Supplement to the Uniform Building Code* forces have a material factor of 1.0 for wood, whereas the 1997 UBC forces have a material factor of 0.85 for wood (and similar changes for the other construction materials).

As mentioned above, the possible materials that comprise the wall anchorage system are concrete, masonry, wood and steel. The ratio of expected strength to design strength (i.e., overstrength) is not the same for each of these materials. The 1997 UBC applies different material factors to the different materials in an attempt to provide ratios of expected strength versus design strength (i.e., overstrength) for each material that are approximately equal. These material factors should be used for allowable stress and strength design. The use of "material factor" may create some confusion in that it does not have a clear meaning for allowable stress design in UBC jurisdictions. "Material load increase" is what was used in describing the factors used in the 1996 *Accumulative Supplement to the Uniform Building Code* in this chapter, but as some of the material factors in the 1997 UBC are less than one (i.e., the factor for wood is 0.85), the use of the word "increase" can also be confusing.

The 1997 UBC design forces include near-fault effects. Near-fault effects may increase wall anchor forces by as much as 50 percent for structures located within 2 km of a major fault. Modifications due to changes in soil type classification may increase or decrease wall anchor forces a relatively small amount. The wall anchorage forces in the 1997 UBC are derived using a formula that is dependent on the height of the diaphragm in the structure. Thus, in two-story structures, wall anchorage forces at the second floor are lower than corresponding forces in the 1996 *Accumulative Supplement to the Uniform Building Code.*

After the Northridge earthquake, this chapter was first included in the 1996 *Accumulative Supplement to the Uniform Building Code* as Appendix Chapter 5. The source of Appendix Chapter 5 was Division 91 of the *Los Angeles Building Code*, which was written prior to the Northridge earthquake, and consequently used smaller design forces for wall anchorage. In order to increase the forces to an acceptable level, material factors larger than those required for new buildings were required. This created some confusion that has been eliminated in this edition of the chapter by adopting the 1997 UBC material factors for the different materials.

The authors of the 1997 NEHRP, and later the authors of the 2000 IBC, took a different approach than the 1997 UBC with respect to material factors and the dependence of the wall anchor forces on the relative height of the diaphragm in the structures. Rather than provide different material factors for the different structural materials, a strength design force of $1.2W_p$ was used for geographic areas equivalent to UBC Seismic

Zone 4 for all of the materials. This force is slightly larger than the design force for steel in the 1997 UBC. This change was made partly due to the fact that the use of different material factors can become cumbersome, and because it was believed to be more appropriate to deal with inconsistencies in different material safety factors in the different material chapters. However, as a result, a significant inconsistency in design forces between the 1997 UBC and in the 2000 IBC exist. Trial designs in areas of high seismic hazard using the IBC provisions have resulted in large subdiaphragm shears and the need for substantially deeper subdiaphragms.

The 2003 IBC reduces the design force in areas of high seismic hazard to $0.8W_p$, but employs a 1.4 material factor for steel. Whereas the 1997 UBC wall anchor forces vary linearly with the height within the structure due to the h_x/h_r term in Equation 32-2 for wall anchor forces, the 2000 IBC and 2003 IBC have one equation for wall anchor forces, regardless of the height. Thus, the 2003 IBC and the 1997 UBC are reasonably consistent for wall anchorage design at the roof, with the IBC being slightly more conservative for wood components. At floor levels, the IBC is significantly more conservative.

Prior to the 1994 Northridge earthquake, the UBC required that buildings resist out-of-plane loading of 200 pounds per foot (90.8 kg/304.8 mm) regardless of the seismic zone (this requirement was introduced in the 1958 UBC, apparently for wind loading). This load was increased to 300 pounds per foot (136 kg/304.8 mm) (allowable stress loads) in the 1997 UBC for Seismic Zone 4 only. Using the 1997 UBC, this minimum load is checked against loading that is directly proportional to seismic zone coefficients. In the 2003 IBC, in Seismic Design Category B, the wall anchor force for buildings of standard importance is the larger of 10 percent of the tributary wall weight, or 40 percent of the tributary wall weight times S_{DS}, the spectral response at short periods. For higher seismic design categories (C and above), the 40-percent value is increased to 80 percent per the discussion above.

In summary, prior to the 1973 UBC, wall anchorage design was minimal and cross-grain tension and bending were not prohibited. The 1976 UBC included many of the requirements currently used for wall anchorage design today, but at much lower design force levels. The design forces increased by 50 percent in the 1979 UBC, and an additional increase of 50 percent was applied for the wall anchors in the middle of the diaphragm in the 1991 UBC. Design forces were increased again in 1994, and detailing restrictions were introduced. Therefore, structures constructed prior to adoption of the 1973 UBC represent the highest risk, with significant improvements occurring in 1976, 1979, 1991 and finally in the 1996 *Accumulative Supplement to the Uniform Building Code*. In regions that previously adopted the UBC but now adopt the IBC, the 2000 IBC represented a significant increase in design loads for wood and concrete components and a slight increase in the design loads for steel components of the wall anchor system.

The significant increases in design loads for concrete and wood were eliminated in the 2003 IBC. It may be appropriate for risk reduction programs to take the design codes used above into account when identifying or prioritizing when and which buildings should be retrofitted. Nevertheless, only those buildings designed using building codes based on the 1997 UBC, 1997 NEHRP or 2000 IBC provisions can be considered to meet the provisions of this chapter.

SECTION A203
DEFINITIONS

For the purpose of this chapter, the applicable definitions in Chapters 16, 19, 21, 22 and 23 of the 1997 *Uniform Building Code* and the following shall apply:

FLEXIBLE DIAPHRAGMS. Roofs and floors including, but not limited to, those sheathed with plywood, wood decking (1-by or 2-by) or metal decks without concrete topping slabs.

SECTION A204
SYMBOLS AND NOTATIONS

For the purpose of this chapter, the applicable symbols and notations in the Building Code shall apply.

SECTION A205
GENERAL REQUIREMENTS

A205.1 General. The seismic-resisting elements specified in this chapter shall comply with provisions of Chapter 16 of the Building Code, except as modified herein.

❖ The provisions of this chapter are intended to address deficiencies that represent direct threats to life safety. The primary concern is providing an adequate wall anchorage system. Lack of adequate collectors or collector connections at reentrant corners or internal walls can also be significant deficiencies that can lead to localized collapses. If the engineer finds that complete load paths are absent (e.g., some shear walls have no blocking for force transfer from the roof to the shear walls), the requirements of the building code for complete load paths (with reduced force levels) should be applied.

A205.2 Alterations and repairs. Alterations and repairs required to meet the provisions of this chapter shall comply with applicable structural requirements of the Building Code unless specifically modified in this chapter.

A205.3 Requirements for plans. The plans shall accurately reflect the results of the engineering investigation and design, and shall show all pertinent dimensions and sizes for plan review and construction. The following shall be provided:

1. Floor plans and roof plans shall show existing framing construction, diaphragm construction, proposed wall anchors, cross-ties and collectors. Existing nailing, an-

chors, ties and collectors shall also be shown on the plans if they are considered part of the lateral-force-resisting system.

2. At elevations where there are alterations or damage, details shall show roof and floor heights, dimensions of openings, location and extent of existing damage, and proposed repair.

3. Typical wall panel details and sections with panel thickness, height, pilasters and location of anchors shall be provided.

4. Details shall include existing and new anchors and the method of developing anchor forces into the diaphragm framing, existing and/or new cross ties, and existing and/or new or improved support of roof and floor girders at pilasters or walls.

5. The basis for design and the Building Code used for the design shall be stated on the plans.

A205.4 Structural observation. Structural observation shall be provided where required by the Building Code for all structures regulated by this chapter. The owner shall employ the engineer or architect responsible for the structural design, or another engineer or architect designated by the engineer or architect responsible for the structural design, to perform structural observations as defined in the Building Code. Observed deficiencies shall be reported in writing to the owner's representative, special inspector, contractor and the building official. The structural observer shall submit to the building official a written statement that the site visits have been made and shall identify any reported deficiencies that, to the best of the structural observer's knowledge, have not been resolved.

SECTION A206
ANALYSIS AND DESIGN

A206.1 Reinforced concrete and reinforced masonry wall anchorage. Concrete and masonry walls shall be anchored to all floors and roofs that provide lateral support for the wall. The anchorage shall provide a positive direct connection between the wall and floor or roof construction capable of resisting 75 percent of the horizontal forces specified in the Building Code.

❖ This chapter specifies that wall anchorage systems in existing buildings be evaluated using a force level of 75 percent of the design force required for new buildings. The wall anchorage system is defined in the 1997 UBC as including the elements within the diaphragm required to develop wall anchorage forces, which consist of: wall anchors, struts, subdiaphragms, cross ties and continuity ties. Given the poor performance of wall anchorage systems for buildings with flexible diaphragms in past earthquakes, and the relatively small cost associated with upgrading wall anchorage to higher force levels once some retrofit is required, it is recommended that where retrofits are performed, the engineer should consider providing capacity in excess of the 75 percent required (i.e., use the design forces in the building code).

A206.2 Special requirements for wall anchorage systems. The steel elements of the wall anchorage system shall be designed in accordance with the Building Code without the use of the 1.33 short duration allowable stress increase when using allowable stress design. A load increase of 1.4 shall be used when designing with the *Uniform Building Code* for allowable stress design. No load increase is required when using the *International Building Code*.

Wall anchors shall be provided to resist out-of-plane forces, independent of existing shear anchors.

Exception: Existing cast-in-place shear anchors may be used as wall anchors if the tie element can be readily attached to the anchors and if the engineer or architect can establish tension values for the existing anchors through the use of approved as-built plans or testing and through analysis showing that the bolts are capable of resisting the total shear load (including dead load) while being acted upon by the maximum tension force due to an earthquake. Criteria for analysis and testing shall be determined by the building official.

Expansion anchors are only allowed with special inspection and approved testing for seismic loading. Attaching the edge of plywood sheathing to steel ledgers is not considered compliant with the positive anchoring requirements of this chapter. Attaching the edge of steel decks to steel ledgers is not considered as providing the positive anchorage of this chapter unless testing and/or analysis are performed to establish shear values for the attachment perpendicular to the edge of the deck. Any installation shall be subject to special inspection.

❖ The wall anchorage forces are dependent on the span and stiffness of the diaphragm and the weight, thickness, height and flexibility of the walls. The probable dynamic forces will exceed the design force. Recent studies of strong motion records in buildings with flexible diaphragms show amplification at the roof diaphragm level of two to four times the ground acceleration (the higher amplifications typically occur at lower ground accelerations, where lower damping and no inelastic behavior occurs). Such amplifications of peak ground acceleration are expected for structures that typically have diaphragm periods corresponding with the peak of the response spectrum. Yielding of the anchor system contributes negligible energy absorption. Therefore, the capacity of the wall anchorage system should be at the expected level of earthquake loading caused by the dynamic response of the diaphragm and the walls, i.e., three times the effective ground acceleration (e.g., 1.2g at the roof diaphragm level for 0.4g ground acceleration before the near source effect and soil condition are considered).

As discussed previously, the 1997 NEHRP and the 2000 IBC do not use load factors for different materials. In the 2003 IBC and ASCE 7-02 a material factor of 1.4 is used for steel. Although it is not recommended that wall anchors yield when subjected to major earthquakes (yielding would result in large deformations that could cause damage to diaphragm nailing at the ledgers), it is still recommended that the designer provide wall anchors that will yield rather than fail in a brittle manner at reduced sections (e.g., twisted straps).

Structural steel, which has a known ductile behavior and a small coefficient of variation in material strength has the lowest ratio of yield stress to allowable stress. The material factor of 1.4 given in the 1997 UBC for strength design adjusts the strength of the steel to meet the expected anchorage loading and thus maintains consistency in the capacity of all of the materials used in the wall anchorage system. Typically only the anchor hardware is designed using this load factor. The embedment of the anchor bolt is evaluated using the concrete material factor. If the 1.33 increase for allowable stress design is used, the resulting steel member size is approximately 25 percent smaller than required by strength design. In order to provide an allowable stress design that is consistent with the strength design, the 1.33 increase should not be used.

The 1997 UBC permits use of a 0.85 material factor for wood, including nailing of the subdiaphragm and attachment of the wall anchors to the wood member. This same factor may be applied to allowable stress design even though it is not explicitly stated in the chapter. It is necessary to state clearly that the 1.4 factor for steel shall be applied to working stress, as it would be unconservative to ignore it. This is not the case with the 0.85 factor for wood.

It is also important to consider the stiffness of the steel and other materials in the anchorage. For example, the wall anchors should be substantially stiffer than the ledger nails in order to prevent them from resisting substantial load and failing in cross-grain bending. Several mechanisms for ensuring stiff wall anchors were considered in developing the provisions in the UBC, and such considerations were a motivation for increasing the minimum out-of-plane forces from 200 to 300 pounds per foot (90.8 kg to 136.2 kg per 304.8 mm) [for allowable stress design, i.e., 420 pounds per foot (190.68 kg per 304.8 m) for strength design] in Seismic Zone 4 (UBC Section 1633.2.8.1, Item 1). However, the minimum of 200 pounds per foot (90.8 kg per 304.8 mm) for allowable stress design remains unchanged in other zones (UBC 1611.4) and applies for all cases in the IBC (see Section 1604.8.2). In strength design, this minimum still applies and is equivalent to 280 pounds per foot (127.12 kg/304.8 mm). Clearly, stiff wall anchors and minimal bolt slips (e.g., no oversized holes) are goals in a good wall anchorage system.

Existing ledger bolts embedded in the concrete or masonry wall may be used as a part of the wall anchorage system if their tension values can be established by testing or through analysis. The size and spacing of these anchor bolts must be shown on the approved plans and verified by a site investigation. A strength design analysis must show that these embedded anchor bolts are capable of resisting the factored tension load in combination with the factored combination of earthquake- and gravity-induced shear loads in both the horizontal and vertical directions.

Expansion anchors are only allowed if they are installed with special inspection and if they meet the appro-priate acceptance criteria (e.g., AC 193, *Acceptance Criteria for Mechanical Expansion Anchors in Concrete Elements*). Such anchors will have an evaluation report from the ICC indicating the conditions for which the anchors are permitted to resist seismic forces.

Wall anchorages that rely on the tear-out strength (nails pulling through the edge) of the plywood (or wood structural panel sheathing) for attachment to a steel ledger are not considered as in compliance with the provisions of this chapter. This is because such connectors at the edge of the plywood sheets would be loaded with a combination of shear and tension (in-plane shear and tear-out). This means that the local demand can be much higher than the original design capacity. In addition, tear-out values of connectors adjacent to the edge of the plywood sheet are not specified in the building code.

Welding steel decking to a steel ledger angle is not considered positive anchorage, as required by this chapter, when flutes are parallel to the wall. The reasons are the same as for plywood attachment, and also because the flutes have substantial flexibility. Testing and analysis to show that the decking with flutes perpendicular to the wall could provide the positive anchorage without significant displacement normal to the wall surface, but the welds are also loaded in shear due to orthogonal loading of the diaphragm. The local demand on the welds can be much higher than the original design capacity.

A206.3 Development of anchor loads into the diaphragm. Development of anchor loads into roof and floor diaphragms shall comply with Chapter 16 of the Building Code.

Exception: If continuously tied girders are present, the maximum spacing of the continuity ties is the greater of the girder spacing or 24 feet (7315 mm).

In wood diaphragms, anchorage shall not be accomplished by use of toenails or nails subject to withdrawal. Wood ledgers, top plates or framing shall not be used in cross-grain bending or cross-grain tension. The continuous ties required in Chapter 16 of the Building Code shall be in addition to the diaphragm sheathing.

Lengths of development of anchor loads in wood diaphragms shall be based on existing field nailing of the sheathing unless existing edge nailing is positively identified on the original construction plans or at the site.

If collectors are not present at re-entrant corners, they shall be provided. New collectors shall be designed for the capacity required to develop into the diaphragm a force equal to the lesser of the rocking or shear capacity of the re-entrant wall, or the tributary shear. The capacity of the collector need not exceed the capacity of the diaphragm. A connection shall be provided from the collector to the re-entrant wall to transfer the full collector force (load). If a truss or beam other than a rafter or purlin is supported by the re-entrant wall or by a column integral with the re-entrant wall, then an independent secondary column is required to support the roof or floor members whenever rocking or shear capacity of the re-entrant wall is less than the tributary shear.

❖ This section presents the requirements of Chapter 16 of the building code for development of wall anchorage loading. Continuous ties or struts between diaphragm chords are required to distribute wall anchorage forces into the body of the diaphragm. Subdiaphragms created with subdiaphragm chords at their boundaries may be used to transmit the anchorage forces to the main cross ties.

The span-depth ratio of subdiaphragms shall not exceed 2.5. This allowable span-depth ratio is substantially less than the 4.0 value permitted in previous editions prior to the 1997 building code. This is an indirect attempt to limit subdiaphragm shear. Some engineers argue that some relaxation of this requirement is appropriate for evaluating existing buildings, as design forces have been reduced by 25 percent, and this requirement is related to the strength of the subdiaphragm. They argue that retrofitting a building that meets all of the other requirements of the chapter simply because the subdiaphragm ratio is 3 instead of 2.5 may not be necessary. However, this is a minority opinion and the requirement of retrofitting subdiaphragms with a length-to-depth ratio of more than 4.0 is a requirement of this chapter. The engineer should keep in mind that the limit on the subdiaphragm ratio of 2.5 originated because of the belief of some engineers that subdiaphragm shear demand should be limited to 300 pounds per foot (136.2 kg/304.8 mm) or less (allowable stress design). If higher subdiaphragm shears are calculated, it is recommended that the engineer increase the depth of the subdiaphragm. Girders, trusses and beams used as cross ties shall have a positive connection to the roof sheathing. This configuration of continuous orthogonal ties confines sections of the diaphragm. This confinement is needed to resist tensile forces caused by shearing stresses in the diaphragm.

The use of toe nails and nails subject to withdrawal as any part of a wall anchorage system is prohibited. Ledgers or plates on the top of walls are not allowed as a part of the wall anchorage system when the loading will result in cross-grain bending or tension.

The engineer should consider the plywood layout when selecting locations for wall anchors. Only members with edge nailing should be used as part of the wall anchorage system unless the reduced capacity for field nailing is considered (this includes locations where the plywood is staggered and the subpurlin in the first bay has edge nailing, but the subpurlin in the second bay does not). If the diaphragm nailing is unknown (no design drawings exist), the roofing should be removed locally in a number of locations to determine edge nailing. As nailing will probably vary over the length of the diaphragm, the nailing should be verified in a number of regions. If there are indications of poor construction quality, the engineer should consider removal of all roofing to verify nailing at all wall anchor locations.

At reentrant corners, the concrete or masonry return walls may have been ignored in the diaphragm analysis performed in the original design, but due to their in-plane stiffness, they function as a diaphragm support (the same is true for short interior walls or FIN walls). Deflection of the diaphragm spanning between end walls of the building may not be compatible with the deformation of the top of the short interior return walls. This deformation incompatibility results in diaphragm damage and separation of the diaphragm from the return wall. Rather than attempting to calculate a deflection compatibility approach, a simple comparison of strengths or capacities is recommended. The objective of the comparison is to find the minimum load needed to adequately tie the return wall to the diaphragm. The capacities of the diaphragm system or the return wall could be estimated as follows for a single-story structure:

- Rocking capacity of a wall: The actual rocking capacity, V_R, of the wall shall be calculated from an equilibrium equation using all the dead loads on the walls, including the weight of the wall, taken at an extreme edge of the footing. Other factors such as soil weight and contributions from continuous footings should be included, as underestimating the overturning resistance in this context is unconservative. Load factors specified in the building code that provide the maximum resistance shall be applied.

- Shear capacity of the wall: V_n shall be the nominal shear strength of the wall provided by the concrete and shear reinforcement without any strength reduction factor.

- Maximum force that can be delivered by the diaphragm:

$$V_d = (v)(L_1 + L_2)$$

where:

L_1, L_2 = Depth of the diaphragm measured on each side of the return wall.

v = Allowable diaphragm shear value from the wood chapter of the building code times a factor of 3. The 1997 NEHRP may be used for strength design values for diagonal sheathing.

No drag strut from the return wall into the diaphragm need be installed if the minimum lateral load that must be developed in the return wall does not exceed the diaphragm capacity, V_d, along the length of the return wall. It is assumed in such cases that the shear will transfer through the diaphragm even if a nonuniform shear throughout the depth of the diaphragm is required to do so.

A redundant vertical support is required beneath major framing members, such as girders or trusses supported by ends of return walls, if these return walls are limited in capacity (when either $V > V_n$ or $V > V_R$) and $V > V_d$, where V is the demand on the wall (strength design). The failure mode could be shear in the reinforced concrete or reinforced masonry wall or a dynamic response that causes rocking in the return wall. A wall that fails in in-plane shear may lose its capacity to support vertical loads. A wall that rocks on its foundation may cause damage to the bearing surface providing support for the member framing onto the return wall.

No redundant support is required for smaller members such as purlins or subpurlins. It is likely that loss of support for such minor roof-framing members will not result in significant collapse or threat to life safety.

A206.4 Anchorage at pilasters. Anchorage at pilasters shall be designed for the tributary wall-anchoring load per Section A206.1, considering the wall as a two-way slab. The edges of the two-way slab shall be considered fixed when there is continuity at pilasters and shall be considered pinned at roof and floor. The pilasters or the walls immediately adjacent to the pilasters shall be anchored directly to the roof framing such that the existing vertical anchor bolts at the top of the pilasters are bypassed without permitting tension or shear failure at the top of the pilasters.

Exception: If existing vertical anchor bolts at the top of the pilasters are used for the anchorage, additional exterior confinement shall be provided as required to resist the total anchorage force.

The minimum anchorage force at a floor or roof between the pilasters shall be that specified in Section A206.1.

❖ In the past, it has been common to anchor walls to the floor and the roof diaphragms for tributary wall loading, assuming the wall panel spans between adjacent floors or between the floor and roof. However, when pilasters are present in the walls, the stiffening effect of the pilasters on the deformed shape of the wall-pilaster system must be taken into account. Since the relative stiffness of the walls and pilasters may vary, a rational method is required to calculate the wall panel load carried by the pilaster. The reaction of a pilaster at the diaphragm level, resulting from the analysis of the wall-pilaster system, is the load applied directly to the girder through its existing anchorage to the top of the pilaster.

Three concerns must be addressed: (1) the existing bolted connection to the girder must have adequate strength for the pilaster reaction; (2) the hardware must be strong enough (including the net section); and (3) the shear strength of the existing anchors, with consideration of the edge distance of the embedded bolts in the top of the pilaster, must be adequate. If the shear strength of the existing bolts is not adequate due to their edge distance, additional exterior confinement at the top of the pilaster must be provided or the existing anchors can be bypassed with new, stiff wall anchors designed to take the entire load. The stiffness of such a bypass system must be checked to verify that it has significantly less deformation under loading than the existing system. If not, the existing system should be disconnected (e.g., remove bolts from girder).

A single-story wall panel with pilasters at each end is supported at four edges by the pilasters, the roof diaphragm and the slab-on-grade. Typically, the support at the slab-on-grade should be considered as a support without moment restraint (the panel could also be modeled with lateral restraint at both the slab-on-grade and the footing in cases where the elevations of the two are substantially different). The support condition of the panel at the pilaster can be a continuous span, a single span having moment restraint or a simple support. The moment restraint depends on whether continuity of the panel through the pilaster is shown on the drawings or whether an open joint exists at one side of the pilaster. The support at the roof level is typically considered as a support without moment restraint. Any textbook or standard for the design of slabs supported on four edges may be used to determine reactions at each edge of the panel. These references consider the aspect ratio of the panel and the rotational restraint at each panel edge. Analyses can be complicated by the presence of openings and parapets. In such cases, approximate, conservative and simplifying assumptions should be used. A logical approach regarding tributary area for the design of pilaster wall anchorage is included in the commentary of the 1999 *Bluebook* (see Figure C108.1 in the 1999 *Bluebook*).

A206.5 Symmetry. Symmetry of wall anchorage and continuity connectors about the minor axis of the framing member is required.

Exception: Eccentricity may be allowed when it can be shown that all components of forces are positively resisted. The resistance must be supported by calculations or tests.

❖ Symmetry of wall anchorage connectors that are placed on the vertical sides (i.e., would generate bending about minor axis) of the framing members is required unless the eccentricity is accounted for in the design. In the past, engineers often used allowable values in hardware manufacturer's catalogs without consideration of member capacity, and specifically the stresses caused by eccentric connections. The values in the hardware catalog typically were developed by testing the hardware on a steel jig, and thus did not consider the stresses in the wood members. The flexural stresses about the minor axis caused by unsymmetrical connectors must be combined with stresses caused by gravity loading. Eccentricity between the line of loading on the connector and the centerline of the resisting member must be resisted by providing supports normal to the line of loading or by a resisting couple. The reactions at these supports must be determined by calculations or testing. The engineer should be aware of the fact that eccentricity can arise either because the anchor design includes eccentricity or because of bolt or strap misalignment (i.e., poor installation). Placing an anchor on each vertical face of the framing member at each connection is an easy method of maintaining symmetry. An offset of anchors along the length of the framing member may be required to avoid fastener interference.

A206.6 Minimum member size. Wood members used to develop anchorage forces to the diaphragm must be at least 3-inch (76 mm) nominal members for new construction and replacement. All such members must be checked for gravity and earthquake loading as part of the wall-anchorage system.

> **Exception:** Existing 2-inch (51 mm) nominal members may be doubled and internailed to meet the strength requirement.

❖ The minimum member size of 3 inches (76 mm) nominal is only required for a new member added to the existing construction or used as replacement of an existing member. All members are to be designed for combined earthquake and gravity loading. An existing 2-inch (51 mm) nominal member may be supplemented with another 2-inch (51 mm) nominal member if the additional member is internailed to the existing subpurlin or joist for load sharing.

A206.7 Combination of anchor types. New anchors used in combination on a single framing member shall be of compatible behavior and stiffness.

❖ The sharing of the load on a single framing member between anchors of different types requires an analysis of their relative stiffness. If no data exists for calculation of relative stiffness of existing anchors, load sharing between different types of anchors is not permitted.

A206.8 Miscellaneous. Existing mezzanines relying on reinforced concrete or reinforced masonry walls for vertical and/or lateral support shall be anchored to the walls for the tributary mezzanine load. Walls depending on the mezzanine for lateral support shall be anchored per Sections A206.1, A206.2 and A206.3.

> **Exception:** Existing mezzanines that have independent lateral and vertical support need not be anchored to the walls.

Existing interior reinforced concrete, or reinforced masonry walls that extend to the floor above or to the roof diaphragm, shall be anchored for out-of-plane forces per Sections A206.1 and A206.3. Walls extending through the roof diaphragm shall be anchored for out-of-plane forces on both sides to provide diaphragm continuity. In the in-plane direction, the walls may be isolated or shall be developed into the diaphragm for a lateral force equal to the lesser of the rocking or shear capacity of the wall or the tributary shear, but need not exceed the diaphragm capacity.

❖ Existing buildings commonly have mezzanines that are used for office or light storage. Mezzanines may be dependent on the wall and pilasters for lateral support. The seismic load of mezzanines is additive to the seismic load of the walls. This combined loading should be used for calculation of wall anchorage forces at the roof diaphragm level above and below (if applicable) the mezzanine. An alternative would be to provide a lateral-load-resisting system for the mezzanine and

use the mezzanine for support of the adjacent walls. In either of these cases, the mezzanine must be anchored to the wall. If the mezzanine is isolated from the wall in accordance with the building separation requirement of the building code and has independent lateral and vertical support, no consideration of the anchorage of the mezzanine is required.

This chapter does not require that the flexural strength of the walls and/or pilasters be analyzed for lateral loads from the mezzanine. However, it is recommended that consideration be given to the existing strength of these elements before making a decision about the alternatives of providing lateral bracing below the existing mezzanine, or using the wall to transmit the seismic loading of the mezzanine to the roof and slab-on-grade levels.

This section requires that each wall that extends to the diaphragm level is provided lateral support by a wall anchorage system; if the wall is not isolated in the plane of the wall, it must be analyzed as required for a return wall in Section A206.3.

SECTION A207
MATERIALS OF CONSTRUCTION

All materials permitted by the Building Code, including their appropriate strength or allowable stresses, may be used to meet the requirements of this chapter.

Bibliography

The following resource materials are referenced in this chapter or are relevant to the subject matter addressed in this chapter.

Acceptance Criteria for Mechanical Expansion Anchors in Concrete Elements. Birmingham, AL: ICC Evaluation Service, 2004.

Accumulative Supplement to the Uniform Building Code, Uniform Mechanical Code, Uniform Code for Building Conservation, Uniform Housing Code, Uniform Fire Code. Whittier, CA: International Conference of Building Officials, 1996.

Briasco, B. "Behavior Joint Anchors Versus Wood Ledgers." *San Fernando Earthquake of 1971*, Vol. 1, Part A. Washington, DC: Department of Commerce.

City of Los Angeles Building Code. "Earthquake Hazard Reduction in Existing Tilt-up Wall Buildings." Volume 1. Division 91. Los Angeles: International Conference of Building Officials and California Building Standards Commission, 1999.

FLE 1984 Morgan Hill Earthquake. EERI Spectra Vol. 1, No.3, May 1985.

IBC-03, *International Building Code*. Falls Church: VA, 2003.

Jennings, P.C. *Engineering Features of the San Fernando Earthquake.* Pasadena, CA: California Institute of Technology, 1971.

Loma Prieta Earthquake Reconnaissance Report. EERI Spectra Vol. 6 Supplement, May 1990.

NEHRP, *Recommended Provisions for Seismic Regulations for New Buildings and Other Structures.* Part 1—Provisions. Washington, DC: Federal Emergency Management Agency, 1997.

Northridge Earthquake Reconnaissance Report, Vol. 2. EERI Spectra Vol. Supplement C to Vol. 11, January 1996.

SEAOC, *Recommended Lateral Force Requirements and Commentary.* Sacramento, CA: Structural Engineers Association of California, Seismology Committee, 1999.

SEI/ASCE 7-02, *Minimum Design Loads for Building and Other Structures.* Reston, VA: American Society of Civil Engineers, 1996.

The 1987 Whittier Earthquake. EERI Spectra Vol.4, No. 1, Feb. 1988; and Vol. 4, No. 2, May 1988.

UBC-61, *Uniform Building Code.* Pasadena, CA: International Conference of Building Officials, 1961.

UBC-73, *Uniform Building Code.* Whitter, CA: International Conference of Building Officials, 1973.

UBC-76, *Uniform Building Code.* Whitter, CA: International Conference of Building Officials, 1976.

UBC-79, *Uniform Building Code.* Whitter, CA: International Conference of Building Officials, 1979.

UBC-88, *Uniform Building Code.* Whitter, CA: International Conference of Building Officials, 1988.

UBC-91, *Uniform Building Code.* Whitter, CA: International Conference of Building Officials, 1991.

Uniform Code for Building Conservation. Whitter, CA: International Conference of Building Officials, 1997.

CHAPTER A3

PRESCRIPTIVE PROVISIONS FOR SEISMIC STRENGTHENING OF CRIPPLE WALLS AND SILL PLATE ANCHORAGE OF LIGHT, WOOD-FRAME RESIDENTIAL BUILDINGS

SECTION A301
GENERAL

❖ History has shown that light, wood-frame residential buildings with specific structural weaknesses in their original construction are susceptible to severe damage from earthquakes. The most common structural weaknesses are: (1) absence of proper connection between the exterior walls and the foundation (i.e., anchor bolts); (2) inadequate bracing of cripple walls between the foundation and first floor; and (3) discontinuous or inadequate foundations below the exterior walls (Comerio & Levin, 1982) (Steinbrugge, 1990).

Unreinforced masonry chimneys and poorly reinforced or tied reinforced masonry chimneys are also a common source of damage, and in some cases, pose a life safety threat. Chimneys are not included in the scope of this document since reduction of chimney vulnerability through pointing of mortar and bracing is not typically considered cost effective, particularly if the risks to life can be controlled by other means. For example, The Applied Technology Council (ATC) recommends adding plywood above the ceiling framing to reduce the chances of falling masonry from penetrating through the ceiling (ATC, 2002). Curtailing the occupancy and frequent use of property within the falling radius of chimneys is also an effective way of minimizing the risk of casualties. ATC recommends replacement of upper portions of damaged chimneys with light-framed construction rather than diagonal bracing. The City of Los Angeles requires the separation of chimneys from wood framing and independent bracing rather than diagonal bracing between the chimneys and wood framing because chimneys are considerably more rigid than wood framing. Otherwise, masonry-bearing wall chimneys should be replaced with light-frame construction with or without veneer (LA, 2001). Generally, the cost of replacing existing chimneys is comparable to the cost of repairing or replacing a damaged chimney after earthquakes, so in many cases it is best to forego chimney retrofits and instead take steps to reduce the exposure of occupants and neighbors and to anticipate the possibility of chimney replacement after future earthquakes.

Wood-frame buildings with continuous concrete foundations in good condition, anchorages meeting the requirements of this chapter and sheathing [plywood, oriented strand board (OSB) or diagonal sheathing] around the entire building perimeter have performed well in past earthquakes. The provisions of this chapter need not be applied to these buildings.

After the Loma Prieta, California, earthquake of 1989, the average cost to repair dwellings that suffered damage due to the above-stated weaknesses ranged from \$25,000 to \$30,000 in 1990 dollars (Gallagher, 1990) and up to \$65,000 in today's dollars (SEAOSC, 2002). Some dwellings with these deficiencies have been total losses. Experience indicates that the average cost for a licensed contractor to install sill bolts and cripple wall bracing to undamaged dwellings ranges from \$2,500 to \$5,000 in 1998 dollars (CSSC, 2002). The cost varies with accessibility and increases in dwellings with limited access due to short crawl spaces, mechanical ducts or other obstructions. If dwelling owners elect to perform the strengthening work themselves, costs can be even lower. The cost effectiveness of correcting these weaknesses becomes more favorable by considering the potential costs of emergency shelters, interim housing and lost employee productivity while damaged dwellings are being repaired (ABAG, 2000) (Comerio, 1996). Some insurance companies offer retrofit discounts that can lower earthquake insurance premiums and insured losses (CEA, 2002).

The purpose of this chapter is to provide minimum standards addressing the three structural weaknesses stated above. This chapter is intended to encourage and facilitate seismic strengthening of conventional dwellings, and to provide standardized methods for performing this work. The provisions are written in a prescriptive format to eliminate the need, in most cases, for engineering design and to encourage the direct use of the chapter by building owners. However, there is a "disturbing aspect to the wide use of these provisions. General contractors are marketing themselves as 'seismic reinforcement' specialists who offer homeowners the promise of earthquake safety without the benefit of either specialized knowledge or the direct input of a structural engineer. Most residential seismic upgrading work is being installed without the benefit of engineering expertise. The contractors' intentions are generally honorable, and some of them have the intuitive understanding and experience to design and install effective seismic systems on existing homes. In many cases, however, homeowners pay thousands of dollars for work that provides little or no improvement in safety" (Huntington, 1991). The *International Building Code*® (IBC®) and *Uniform Building Code* (UBC) allow such retrofits, provided they are no less safe than before the retrofit. Also in response to this concern, retrofit training for contractors has been periodically offered by the Federal Emergency Management Agency (FEMA) and the Association of Bay Area Governments (ABAG) (FEMA, 1995). By adopting and enforcing the provisions in this chapter, jurisdictions can further help protect the interests of homeowners and establish a standard for retrofit practice.

Note that some jurisdictions (e.g., cities of Los Angeles, San Leandro, Santa Barbara and Berkeley) have prepared drawings with approved details to further assist in retrofitting dwellings (SL, 2002). For regions outside of the western U.S., the Institute for Building and Home Safety offers an excellent retrofit guide (IBHS, 1999).

Water heaters that are not restrained from toppling in earthquakes also pose significant risks of water and fire damage that are not addressed herein (NIST, 1997). Existing premanufactured restraint systems are available at many hardware stores and have been stamped as preapproved by a government agency for limited use. They are both economical and effective for water heaters typically located adjacent to wood frame walls (DSA, 2002). However, atypical water heater bracing installations require engineered designs.

A301.1 Purpose. The provisions of this chapter are intended to promote public safety and welfare by reducing the risk of earthquake-induced damage to existing wood-frame residential buildings. The requirements contained in this chapter are prescriptive minimum standards intended to improve the seismic performance of residential buildings; however, they will not necessarily prevent earthquake damage.

This chapter sets standards for strengthening that may be approved by the building official without requiring plans or calculations prepared by an architect or engineer. The provisions of this chapter are not intended to prevent the use of any material or method of construction not prescribed herein. The building official may require that construction documents for strengthening using alternative materials or methods be prepared by an architect or engineer.

❖ In contrast to other earthquake retrofit guidelines and codes, the provisions of this chapter are not strictly designed for life safety protection. These provisions are, in fact, expected to reduce property damage, reduce the number of uninhabitable dwellings after earthquakes and avoid the increased public assistance expenditures to repair damage and provide temporary housing (ABAG, 1999).

These provisions do not guarantee that strengthened structures will not be damaged. However, it is anticipated that costly damage to the vulnerable parts of dwellings below the first floor will be greatly reduced. These requirements have not been established or calibrated using performance-based earthquake engineering, so there is no intent to state or imply particularly a performance objective or range since these requirements do not address vulnerabilities that might exist above the first floor of dwellings. "Cripple walls retrofit-

ted to these provisions are generally expected to meet a life safety performance objective" (CUREE, 2002). However, other parts of the building may exceed or perform less than this objective requires.

Analysis and design of strengthening for other structural or nonstructural components not addressed by these provisions must be performed in accordance with Section A301.3.

A301.2 Scope. The provisions of this chapter apply to light, wood-frame residential buildings that are in Seismic Design Categories D, E and F of the 2003 IBC (located in Seismic Zones 3 and 4 of the UBC), containing one or more of the structural weaknesses specified in Section A303.

Exception: The provisions of this chapter do not apply to the buildings, or elements thereof, listed below. These buildings or elements require analysis by an engineer or architect in accordance with Section A301.3 to determine appropriate strengthening.

1. Group R, Division 1 occupancies with more than four dwelling units.

2. Buildings with a lateral-force-resisting system using poles or columns embedded in the ground.

3. Cripple walls that exceed 4 feet (1219 mm) in height.

4. Buildings exceeding three stories in height and any three-story building with cripple wall studs exceeding 14 inches (356 mm) in height.

5. Buildings where the building official determines that conditions exist that are beyond the scope of the prescriptive requirements of this chapter.

The provisions of this chapter do not apply to structures, or portions thereof, constructed on a concrete slab on grade.

The details and prescriptive provisions herein are not intended to be the only acceptable strengthening methods permitted. Alternative details and methods may be used when approved by the building official. Approval of alternatives shall be based on test data showing that the method or material used is at least equivalent in terms of strength, deflection and capacity to that provided by the prescriptive methods and materials.

The provisions of this chapter may be used to strengthen historic structures, provided they are not in conflict with other related provisions and requirements that may apply.

❖ "Prescriptive" means these provisions apply to specific conditions and must be used in precisely the manner described. "Prescriptive" also means "determined in advance," without the need for case-specific analysis or design. Through the use of these provisions and the accompanying details, the dwelling owner or contractor can develop plans without the services of a design professional (civil engineer, structural engineer or architect). However, the use of other materials, proprietary systems or methods not shown by the figures and details within this chapter may require the services of a design professional.

The provisions are intended to deal with specific earthquake weaknesses; therefore, the work that is being done is structural in nature and requires submitting plans and obtaining a building permit. The weaknesses that these provisions address are listed in Section A303. It should be clearly understood that the application of these provisions is limited because the provisions are not suitable to use for strengthening hotels, motels or large multiunit apartments, since they are typically larger structures that require engineered retrofits by licensed design professionals. Also excluded are dwellings built with cripple walls with studs taller than 4 feet (1219 mm) in any location; dwellings that have columns or poles embedded in the ground as their foundation system; and buildings exceeding three stories or any three-story building with cripple wall studs exceeding 14 inches (356 mm) in height. Each of these types of buildings presents unique conditions that preclude the use of prescriptive criteria to strengthen them.

The 4-foot (1219 mm) cripple wall stud height limits the amount of overturning in braced crippled walls with lengths defined in Figure A3-10. When the height of the wall studs exceeds 4 feet (1219 mm), the owner will need to have the bracing designed by an engineer or architect.

Conventional construction provisions in Section 2308.12.4 of the IBC state that cripple walls with studs exceeding 14 inches (356 mm) in height are to be considered as first-story walls for the purpose of determining bracing. However, the bracing layout requirements in these provisions (see Figure A3-10) are provided according to the number of stories above the cripple wall.

Items 3 and 4 in the exception to Section A301.2 refer to the cripple stud height. The parameter H, represented in Figure A3-7, is greater than the stud height and is not used elsewhere in these provisions.

The building official is also allowed to exclude other residential buildings to which these provisions would otherwise apply if they have vertical or horizontal irregularities or other features not considered by these prescriptive standards. Very few dwellings are rectangular in plan. U-, T- or L-shaped plans are not necessarily problems in conventional construction with wood diaphragms. However, split-level dwellings, hillside dwellings or structures where the upper level exterior walls are horizontally offset from the line of the lower story exterior walls may be determined by the building official to be beyond the scope of these provisions. Before beginning work, it is recommended that the owner or contractor consult with the authority having jurisdiction to determine if these provisions can be applied.

Observations after past earthquakes have shown that cripple walls with substantially varying heights, such as with sloping or stepped foundations, suffer more damage than cripple walls of constant height (SEAOSC, 2002). Most of the forces are resisted by the shorter, stiffer portions of the cripple wall. For multistory buildings with more than a 2:1 height ratio between the tallest and shortest cripple wall bracing panels, it is recommended that the sheathed panel lengths be engineered along the wall line(s) in question.

The provisions of this chapter are intended for uses in high seismic regions (see Section 301.2 of the IBC). Still, for dwellings within 6.2 miles (10 km) of major active faults (Ss > 1.5) it is recommended that the panel lengths be engineered to account for potentially more severe ground motions in the design earthquake, since these provisions do not account for near-source ground motions.

Other structural weaknesses that may exist that are located above the first floor are also beyond the scope of this chapter. Situations that require analysis and design by a design professional or consultation with the building department prior to beginning any work include, but are not limited to, buildings with full-height stone veneer walls due to their added weight and dwellings built on hillsides where cripple wall heights vary substantially.

Also excluded from these provisions are buildings or portions of buildings constructed on concrete slabs-on-grade. These dwellings do not have cripple stud walls and typically would not lack bracing. These buildings may have wall anchorage deficiencies, and the provisions for wall anchorage of cripple wall buildings apply equally well to these structures. However, retrofit of these structures would require removal of wall finishes and may not be as cost effective as retrofitting buildings with crawl spaces. Also, while sliding of dwellings on slab-on-grade foundations has occasionally occurred in past earthquakes, it has not caused widespread economic and habitation losses in past earthquakes. Dwellings with stem walls (reinforced concrete or masonry foundation walls that project above the ground to the underside of the first floor framing) will experience substantial damage if the dwelling slides off the foundation. Anchorage of these structures should be considered as falling in the scope of this chapter.

A majority of dwellings constructed in California prior to 1950 were unanchored. The UBC did not begin to specify anchorage until its 1946 edition (SEAOC, 1995); however, most local governments did not uniformly adopt such model codes promptly after their publication until the late 1970s. Furthermore, some dwellings were constructed with inadequate cripple wall bracing in the 1970s and even later particularly where model codes were not enforced.

Dwellings often used horizontal wood siding or stucco as wall sheathing material. Owners of older dwellings should examine the exterior walls from within the crawl space under the first floor to determine if the sill plates are bolted to the foundation, if the bolt size and spacing comply with the building code or Table A3-A and if exterior finishes have been applied over wall sheathing materials with adequate strength, such as plywood, OSB or diagonal sheathing. Most dwellings constructed after 1950 were anchored to their foundations. However, even in high seismic regions, cripple wall bracing now considered inadequate was in common use in the 1970s and 1980s.

A301.3 Alternative design procedures. When analysis by an engineer or architect is required in accordance with Section A301.2, such analysis shall be in accordance with all requirements of the Building Code, except that the base shear may be taken as 75 percent of the horizontal forces specified in the Building Code.

❖ This section purposely omits the commonly used statement that the design must comply with all the requirements of the building code because complete code compliance is often not feasible with respect to existing buildings. For example, design professionals should not be expected to rigorously address issues such as the stiffness variations in existing flooring systems due to differences in the type or thickness of the flooring. It does, however, state that any strengthening designed by the design professional should at least be equivalent in terms of strength, deflection and capacity to that provided by the prescriptive methods. The building official is allowed to require design professionals to provide substantiating structural calculations or test data to confirm this equivalence.

This 75-percent factor applied to building code design forces accounts in a general way for differences between current design criteria and the less conservative criteria that were likely in effect when the dwelling was originally designed and built. If owners prefer, they can elect to use a greater horizontal force in order to lessen potential damage from future earthquakes.

SECTION A302
DEFINITIONS

For the purpose of this chapter, in addition to the applicable definitions in the Building Code, certain additional terms are defined as follows:

❖ Throughout this chapter there are references to "the building code." This term generally refers to the current edition of the IBC or the jurisdiction's governing code. Before using this document to determine the amount of strengthening that may be required, consult with the local building department to confirm the appropriate use of these definitions.

The definitions provided in this chapter are for terms that are not defined in the governing building codes. Other terms defined in the building code also apply to this chapter but are not repeated.

CHEMICAL ANCHOR. An assembly consisting of a threaded rod, washer, nut and chemical adhesive approved by the building official for installation in existing concrete or masonry.

COMPOSITE PANEL. A wood structural panel product composed of a combination of wood veneer and wood-based material, and bonded with waterproof adhesive.

CRIPPLE WALL. A wood-frame stud wall extending from the top of the foundation to the underside of the lowest floor framing.

EXPANSION BOLT. A single assembly approved by the building official for installation in existing concrete or masonry. For the purpose of this chapter, expansion bolts shall contain a base designed to expand when properly set, wedging the bolt in the pre-drilled hole. Assembly shall also include appropriate washer and nut.

ORIENTED STRAND BOARD (OSB). A mat-formed wood structural panel product composed of thin rectangular wood strands or wafers arranged in oriented layers and bonded with waterproof adhesive.

PERIMETER FOUNDATION. A foundation system that is located under the exterior walls of a building.

PLYWOOD. A wood structural panel product composed of sheets of wood veneer bonded together with the grain of adjacent layers oriented at right angles to one another.

SNUG-TIGHT. As tight as an individual can torque a nut on a bolt by hand, using a wrench with a 10-inch-long (254 mm) handle, and the point at which the full surface of the plate washer is contacting the wood member and slightly indenting the wood surface.

WAFERBOARD. A mat-formed wood structural panel product composed of thin rectangular wood wafers arranged in random layers and bonded with waterproof adhesive.

WOOD STRUCTURAL PANEL. A structural panel product composed primarily of wood and meeting the requirements of United States Voluntary Product Standard PS 1 and United States Voluntary Product Standard PS 2. Wood structural panels include all-veneer plywood, composite panels containing a combination of veneer and wood-based material, and mat-formed panels such as oriented strand board and waferboard.

SECTION A303
STRUCTURAL WEAKNESSES

For the purpose of this chapter, structural weaknesses shall be as specified below.

1. Sill plates or floor framing that are supported directly on the ground without an approved foundation system.

2. A perimeter foundation system that is constructed only of wood posts supported on isolated pad footings.

3. Perimeter foundation systems that are not continuous.

   Exceptions:

   1. Existing single-story exterior walls not exceeding 10 feet (3048 mm) in length, forming an extension of floor area beyond the line of an existing continuous perimeter foundation.

   2. Porches, storage rooms and similar spaces not containing fuel-burning appliances.

4. A perimeter foundation system that is constructed of unreinforced masonry or stone.

5. Sill plates that are not connected to the foundation or that are connected with less than what is required by the Building Code.

   Exception: When approved by the building official, connections of a sill plate to the foundation made with other than sill bolts may be accepted if the capacity of the connection is equivalent to that required by the Building Code.

6. Cripple walls that are not braced in accordance with the requirements of Section A304.4 and Table A3-A, or cripple walls not braced with diagonal sheathing or wood structural panels in accordance with the Building Code.

❖ This section provides criteria that allow owners and contractors to evaluate dwellings. The structural weaknesses listed in this section might not be the only weaknesses that will lead to structural damage when the building is subjected to earthquake forces. They represent conditions that can be addressed by prescriptive, nonengineered provisions that are most common and most cost effective to strengthen.

1. *Sill Plates or Floor Framing without Approved Foundation System.* Some older dwellings do not have a foundation system. Instead, the wall sill plate, and much of the floor framing, is supported directly on the ground. When subjected to earthquake-induced lateral and vertical forces, these structures can easily move because they are not anchored. Fungus, water and insect damage are also common in unapproved foundations. Foundation movement can result in a variety of structural and nonstructural damage, including broken gas and utility lines that can lead to fires. Further, these structures are highly susceptible to fungus infection and insect infestation due to inadequate wood-to-earth separation. Wood deterioration caused by this inadequacy has significantly contributed to the damage from earthquakes.

2. *Post and Pier Foundation Systems.* Many dwellings have continuous cripple walls and foundations around their perimeter and wood posts on isolated concrete pad footings (called a "post-and-pier" system) under the interior of the dwelling. Such a system is not necessarily deficient. If a post-and-pier system forms the foundation for the dwelling perimeter walls, it is considered a structural weakness because of the lack of stiff walls below the first floor. This deficient foundation system is found in some older dwellings, including Victorian era structures and buildings in areas where soil moisture is high. The posts and floor framing members of this system are usually interconnected with simple toenailed connections with no bracing between the posts. The weakness is compounded when a lack of connection occurs between the posts and the

small concrete pads that act as footings. Failures in this foundation system during earthquakes occur at the underside of the floor framing, and may lead to partial collapse of the structure.

Providing diagonal bracing members between the posts does not solve the problem. Each post would then need to be adequately connected to a foundation system. Typically, the existing footing pads are too small to make the necessary connections. Simply providing bracing between the posts only moves the point of failure from the top of the post to the bottom of the post at the pad footing. For these buildings, bracing, anchorage and provision of additional footings may be required.

Some post-and-pier-type structures may be considered as historic buildings. Care needs to be taken when performing strengthening work so the historic nature of the building is not destroyed. Many jurisdictions have adopted specific requirements for historical buildings, such as those in Chapter 10 of the code. If a dwelling utilizes this type of system and might be considered an historic building, consult with the building department before beginning any strengthening work.

3. *Noncontinuous Perimeter Foundation System.* Another deficiency is found in dwellings that do not have a continuous perimeter foundation. However, there are many variations of partial foundations, and some do not represent a significant weakness. When applying these provisions to an existing building the intent is to reduce the potential for damage to habitable portions of structures. Therefore, the standards include two exceptions to the requirement for continuous perimeter foundations.

   Exceptions:

   1. Observations after past earthquakes have shown that failures of relatively small enclosed spaces tend to be localized and may not result in the undamaged portion of the remainder of the dwelling to be uninhabitable. This exception is not intended to imply that these existing appendages should not be strengthened if they do not have a continuous foundation. Exception 1 only means that such a condition will not automatically require a new perimeter foundation. The point is that additions tend to act separately, especially if they have separate foundations, so other parts of the dwelling should not necessarily be penalized as "weak."

   2. Item 3, Exception 2 addresses nonhabitable spaces such as porches, storage rooms and other similar areas. The failure of porches is common during earth-

quakes. Such failures rarely impact the habitability of the structure, although they may temporarily limit access or egress. Storage rooms, in this exception, include those areas adjacent to the dwelling and accessible from the outside of the building only, and storage buildings or areas that are separate from the main structure. As with porches, failure of these storage areas seldom affects the habitability of the main structure. Attached rooms with water heaters or furnaces are singled out because damage to these areas could result in fires. This provision thus implies a nonstructural performance objective somewhat beyond the "risk reduction" stated in Section A301.1. Again, this exception is not intended to imply that such structures should be strengthened only if they contain fuel-burning appliances. Serious consideration should be given to strengthening any weak parts of the dwelling. There is no typical method of construction for these portions of existing buildings. Consequently, it is difficult to develop prescriptive standards to address them. Considering that this chapter is intended to reduce the potential for damage that would make dwellings uninhabitable, these small areas have been given exemptions to the general requirement for a continuous perimeter foundation.

4. *Unreinforced Masonry Perimeter Foundation.* A perimeter foundation constructed of unreinforced masonry is assumed to lack the necessary strength to resist earthquake forces. These systems are common in many older dwellings built before codes were adopted in high seismic regions and may also exist in newer dwellings where codes have not been enforced. When subjected to earthquakes, these systems are easily damaged, allowing the building to shift off foundations. Section A304.2.2 requires analysis of unreinforced masonry foundations by either an architect or an engineer.

5. *Inadequate Sill Plate Anchorage.* While sliding between an unanchored sill plate and the foundation can occur, it is actually one of the more rare sources of damage. However, it is still important that sill plate anchorage be present in order to complete the lateral force path. Bracing cripple walls without bolting the sill plate to the foundation simply moves the weak link to the interface of the sill and foundation. Compliance with either Tables A3-A and B or the building code is deemed acceptable.

Exception: Because details common in new construction are often impractical in retrofits, the provisions recognize alternative means of anchorage, including proprietary products. These proprietary products are especially useful in conditions where limited clearance prevents installation of bolts through the sill plate in the traditional fashion. Generally, cripple wall studs need to be 24 inches (610 mm) or taller to install anchor bolts with a normal drill. In some cases, drills with right-angle attachments can be used in more confined spaces.

6. *Inadequate Cripple Wall Bracing.* Past earthquakes have shown that the most common cause of major damage in dwellings is due to poorly braced cripple walls (LA, 1994).

Only wood structural panels and diagonal wood sheathing are acceptable for cripple walls in IBC Seismic Design Categories D, E and F and in all seismic zones regulated by the UBC. Gypsum board, fiberboard, particleboard, lath and plaster and gypsum sheathing boards are no longer acceptable methods for cripple wall bracing except in Seismic Design Categories A, B and C. For additional information and connection requirements for these materials, refer to Tables 2308.9.3(1) and 2308.12.4 of the IBC and referenced sections.

A common and very weak type of cripple wall can be found in older buildings constructed with horizontal, exterior wood siding. This type of siding, and its nailing, is not adequate to resist earthquake forces associated with nearby moderate or major earthquakes.

Recent earthquakes have also shown that "let-in" diagonal bracing does not adequately brace cripple walls (LA, 1994). Let-in braces are usually nominal 1 inch (25 mm) thick and placed in a notch cut into the face of the stud. Let-in braces are no longer permitted as an acceptable bracing method in the IBC for buildings located in regions where strong earthquakes are expected to occur.

Let-in braces should not be confused with diagonal wood sheathing. Diagonal wood sheathing, which is acceptable by these provisions, is composed of individual boards nominally 1 inch (25 mm) thick, laid diagonally across the face of the stud wall. These boards are laid next to one another covering the entire width and length of the wall extending from the top plate to the sill plate. If the cripple walls are covered with diagonal sheathing, the wall is adequately braced, provided the boards are nailed to each stud they cross and to the top and bottom plates. Adequate nailing consists of three 8-penny nails at each stud and the ends. If the boards or the studs are split, or if the end nails are too close to the ends of the sheathing, this system can be deficient.

The most effective cripple wall bracing system that significantly reduces the risk of damage is wood structural panel sheathing (plywood, OSB) fully nailed around the sheet perimeters to each stud and especially to the top and bottom plates.

Disturbingly, modern structures are also being found with inadequate cripple wall bracing. These dwellings have been constructed with various forms of plywood siding that was not nailed along the sill plates. Because of this lack of nailing, the cripple wall studs are free to rotate about their base until they collapse. If the dwelling has plywood sheathing as an exterior finish, check for nails spaced no more than 6 inches (152 mm) apart along all the edges of each sheet. If adequate nailing is not present, comply with the nailing requirements of this chapter.

Exterior plywood siding with vertical grooves (referred to as T1-11) can have another serious deficiency. At the edges where two adjacent panels adjoin, each panel must be nailed to the wall stud with a separate row of nails. These sheets have "lips" so that they overlap at the joints. A common, improper construction practice is providing only one row of nails through both sheets (at the overlap). This creates a weakness as the plywood thickness is only one-half of its normal thickness at the overlap, and only half the number of nails is provided. Such practice led to failures in the 1984 Morgan Hill, California, earthquake. In all cases where nailing is exposed to the elements, it is recommended that hot-dip galvanized nails be used.

A dwelling with existing Portland cement plaster (stucco) as the exterior finish might not have its cripple walls adequately braced by this material. Stucco has been a recognized bracing material for a number of years but it is only as good as the connection of the lath to the studs and plates. Many dwellings with stucco applied directly over the studs without plywood or diagonal sheathing under the stucco experienced serious damage in the 1994 Northridge earthquake (LA, 1994) (NAHB, 1994). In high seismic regions, this failure most often was due to inadequate attachment of the lath to the bottom sill plate (LA, 1994).

Through the years there have been various lath systems used for installing stucco. Stucco is normally applied in three coats [$^7/_8$ inch (22 mm) total thickness]. When subjected to high loads it can fail in diagonal tension represented by diagonal cracks. To increase the tension capacity, stucco is reinforced with wire lath. This reinforcing does not keep the stucco from cracking but helps prevent cracks from opening.

SECTION A304
STRENGTHENING REQUIREMENTS

A304.1 General.

A304.1.1 Scope. The structural weaknesses noted in Section A303 shall be strengthened in accordance with the requirements of this section. Strengthening work may include both new construction and alteration of existing construction. Except as provided herein, all strengthening work and materials shall comply with the applicable provisions of the Building Code. Alternative methods of strengthening may be used, provided such systems are designed by an engineer or architect and are approved by the building official.

❖ Use of materials, proprietary systems or methods not shown by the figures and details within this chapter requires the services of a design professional.

Where a dwelling has an unusual or irregular configuration or unusual features, the services of an engineer or architect to design a strengthening program utilizing the alternative procedures of Section A301.3 is required. The building official may require a predesign special inspection as described in Sections A304.5 and A304.5 to determine which portions of the work require the services of a design professional.

The Consortium of Universities for Research in Earthquake Engineering (CUREE) recommends a number of enhancements to these provisions that are currently under consideration by the SEAOC Existing Buildings Committee, the East Bay Chapter of the International Code Council® (ICC®) and others (CUREE, 2002).

A304.1.2 Condition of existing wood materials. All existing wood materials that will be a part of the strengthening work (sills, studs, sheathing, etc.) shall be in a sound condition and free from defects that substantially reduce the capacity of the member. Any wood material found to contain fungus infection shall be removed and replaced with new material. Any wood material found to be infested with insects or to have been infested with insects shall be strengthened or replaced with new materials to provide a net dimension of sound wood at least equal to its undamaged original dimension.

❖ Damage commonly known as "dry rot" can occur to wood framing exposed to dampness or to water leakage. Termite infestation is another cause of damage to wood members. All buildings being strengthened where damage is suspected should have a thorough inspection, but only those elements affected by retrofit need to be checked. If not repaired, dry rot and pest damage can weaken sill plates, studs and wood siding and have a substantial adverse effect on a building's response to earthquakes.

Even dwellings without dry rot or termite damage may have weaknesses due to poor construction quality. For example, insufficient nailing of plywood, OSB or diagonal sheathing will result in a structure that is unable to resist the forces imposed during earthquakes. Simply repairing the weaknesses will not be adequate if the condition of the existing wood-framing members to be utilized is in doubt. It is recommended that the exposed wood be thoroughly inspected to ensure that all wood that is part of the strengthening work is in good condition. Members that contain splits, checks (cracks) or knots affecting the ability of the member to resist earthquake forces must be strengthened or replaced.

Existing wood members showing evidence of fungus infection, commonly referred to as dry rot, or evidence of insect infestation must be removed and replaced. Fungus remains active even after the affected area is treated. Fungus infection can be found by probing the wood members with a sharp object, like a knife or awl. If the probe easily penetrates the wood, the member might have fungus infection. Sound wood will be difficult to probe. In some cases, the fungus infection will be found only inside the member because rot may affect the wood from the inside and progress outward. By the time it is noticeable on the surface as staining or softening, the wood's strength may already be significantly degraded. Thus, it is important to perform probing and visual observations prior to and during construction. In these cases, probing with a sharp tool will not always locate the infection (see Figure A304.1.2). This concealed condition may be encountered when drilling for new sill anchors. If the drill suddenly moves through the wood, a pocket of fungus infection has likely been encountered. The portion of the sill plate containing the infection will need to be cut out and replaced with a new piece of sill plate. The new sill plate must be anchored in accordance with Tables A3-A and A3-B. When replacing pieces of sill plate, pressure-treated lumber will need to be used to protect the new member from fungus infection.

Fungus growth occurs where wood is made continually or repeatedly damp by a leaking plumbing pipe or by repeated saturation and drying from an exterior wall that leaks during rains. Simply removing and replacing infected wood will not necessarily prevent the fungus infection from recurring. It is important to find the cause of the leak, repair it and allow the remaining wood to dry. Repairs will involve tracing a water stain or the actual water to a leaky pipe or fitting or other source. It is usually more difficult to trace a leak in the exterior wall covering. Hand-held moisture detectors could be used to locate moisture intrusion. Spraying the exterior of the residence with a high-pressure hose and then using the moisture detector on the inside surface of exterior walls can locate defective flashing or torn paper backing behind the stucco. A common area to check for leaks through the wall is at building corners. Evidence of the infection may be found in other members that are not involved in the strengthening work. It is recommended that other members should also be replaced and the source of the wetness eliminated by appropriate repairs.

Insect infestation, on the other hand, stops damaging the wood once the infestation has been stopped. Consequently, a member that has been significantly damaged by insects does not need to be removed if it can be strengthened. The easiest method of strength-

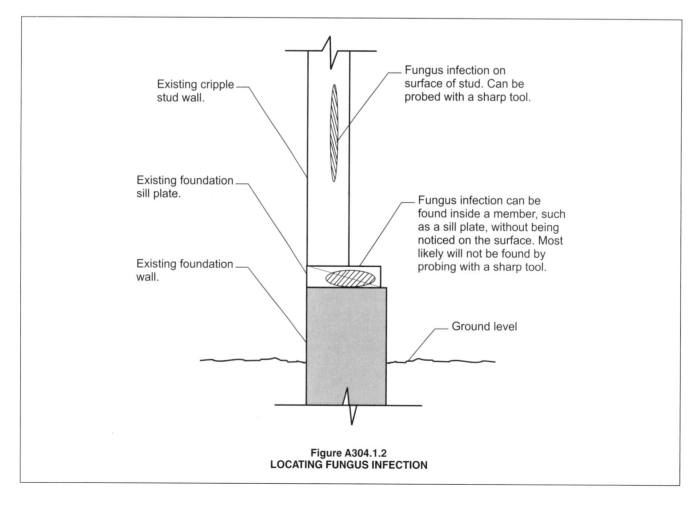

Existing cripple stud wall.

Existing foundation sill plate.

Existing foundation wall.

Fungus infection on surface of stud. Can be probed with a sharp tool.

Fungus infection can be found inside a member, such as a sill plate, without being noticed on the surface. Most likely will not be found by probing with a sharp tool.

Ground level

Figure A304.1.2
LOCATING FUNGUS INFECTION

ening is to add a new member next to the damaged member. Unfortunately, there are no clear guidelines to indicate when insect damage requires strengthening. This determination must be based on judgment gained through experience.

Prior to removing sill plates or studs for repair due to fungal or infestation damage, temporary shoring must be installed. The design of this shoring must be carefully planned by a qualified design professional with shoring design experience and installed by a contractor with shoring experience.

A304.1.3 Floor joists not parallel to foundations. Floor joists framed perpendicular or at an angle to perimeter foundations shall be restrained either by an existing nominal 2-inch-wide (51 mm) continuous rim joist or by a nominal 2-inch-wide (51 mm) full-depth blocking between alternate joists in one- and two-story buildings, and between each joist in three-story buildings. Existing blocking for multistory buildings must occur at each joist space above a braced cripple wall panel.

Existing connections at the top and bottom edges of an existing rim joist or blocking need not be verified in one-story buildings. In multistory buildings, the existing top edge connection need not be verified; however, the bottom edge connection to either the foundation sill plate or the top plate of a cripple wall shall be verified. The minimum existing bottom edge connection shall consist of 8d toenails spaced 6 inches (152 mm) apart for a continuous rim joist, or three 8d toenails per block. When this minimum bottom edge-connection is not present or cannot be verified, a supplemental connection installed as shown in Figure A3-8 shall be provided.

Where an existing continuous rim joist or the minimum existing blocking does not occur, new $^3/_4$-inch (19 mm) wood structural panel blocking installed tightly between floor joists and nailed as shown in Figure A3-8 shall be provided at the inside face of the cripple wall. In lieu of $^3/_4$-inch (19 mm) wood structural panel blocking, tight-fitting, full-depth 2-inch (51 mm) blocking may be used. New blocking may be omitted where it will interfere with vents or plumbing that penetrates the wall.

❖ For these strengthening procedures to be effective there must be a continuous horizontal load path from the exterior walls to the foundation. Where floor joists are perpendicular to a cripple wall, or frame into the cripple wall at an angle, existing rim joists (or blocking) need to be connected to either the foundation sill plate (if there are no cripple stud walls) or the top plate of the cripple wall. When reviewing existing construction, if there is a connection between the rim joist and the plate that meets the nailing requirements of this section, such a connection may be considered adequate for this link in the load path. Where these connections do not exist, new connections must be made. Rim joists will need to be toenailed with 8-penny 2$^1/_2$-inch (64 mm) long common nails, spaced 6 inches (152 mm) apart, through the joist into the plate. Blocking will need to be toenailed. Use of proprietary products for these connections might be easier than toenailing. When approved by the building official, these connec-

tions may be made by using products with current evaluation reports by an independent testing authority.

Because the forces in a single-story structure are relatively small, it is not necessary to verify these connections if the blocking or rim joists are present. In multistory buildings, the connections between the foundation and the blocking or rim joists must be verified. When these requirements are not met or cannot be verified, the provisions of this chapter apply.

In some cases, existing construction might not include a rim joist or blocking. In other cases, the members are smaller in width than a nominal 2 inches (51 mm) [1$^1/_2$ inches (38 mm)]. In these cases, either a new nominal 2-inch (51 mm) wide full-depth joist or blocking or one of the methods described in the chapter may be used to provide the load path from the floor to the sill plate or cripple wall (see Figure A3-8). In addition to providing a load path link, the rim joist or blocking provides rotational restraint for the ends of the floor joists.

A304.1.4 Floor joists parallel to foundations. Where existing floor joists are parallel to the perimeter foundations, the end joist shall be located over the foundation and, except for required ventilation openings, shall be continuous and in continuous contact with the foundation sill plate or the top plate of the cripple wall. Existing connections at the top and bottom edges of the end joist need not be verified in one-story buildings. In multistory buildings, the existing top edge connection of the end joist need not be verified; however, the bottom edge connection to either the foundation sill plate or the top plate of a cripple wall shall be verified. The minimum bottom edge connection shall be 8d toenails spaced 6 inches (152 mm) apart. If this minimum bottom edge connection is not present or cannot be verified, a supplemental connection installed as shown in Figure A3-9 shall be provided.

❖ Where floor joists are parallel to a cripple wall, the same load path concept applies as with joists perpendicular to foundations. In this condition, the end floor joist must occur over the foundation wall or cripple wall and be connected. If this member is not connected to the plate, it will need to be toenailed with 8-penny common nails spaced at 6 inches (152 mm) apart or with equivalent approved hardware. This connection need only be verified for multistory buildings, for which seismic forces are larger. If an end joist in a multistory building is not connected to the sill plate on top of the foundation, or this connection cannot be determined, the end joist may be connected to the sill plate with sheet metal angles (proprietary hardware is available). Where clearances do not permit installation of this angle, an alternative method using $^3/_4$-inch (19 mm) plywood attached to the foundation plate or cripple wall top plate and to the underside of the flooring as shown in Figure A3-9 may be used.

A304.2 Foundations.

A304.2.1 New perimeter foundations. New perimeter foundations shall be provided for structures with the structural weaknesses noted in Items 1 and 2 of Section A303. Soil investigations or geotechnical studies are not required for this work unless the building is located in a special study zone as designated by the jurisdiction or other public agency.

A304.2.2 Foundation evaluation by an engineer or architect. Partial perimeter foundations or unreinforced masonry foundations shall be evaluated by an engineer or architect for the force levels noted in Section A301.3. Test reports or other substantiating data to determine existing foundation material strengths shall be submitted for review. When approved by the building official, these foundation systems may be strengthened in accordance with the recommendations included with the evaluation in lieu of being replaced.

> **Exception:** In lieu of testing existing foundations to determine material strengths, and when approved by the building official, a new nonperimeter foundation system designed for the forces noted in Section A301.3 may be used to resist all exterior wall lateral forces.

❖ It might not be economical to replace existing partial perimeter foundations or unreinforced masonry foundations. In order to determine if existing foundation systems are adequate, an engineer or an architect should evaluate both the condition of the system as well as its ability to resist the prescribed forces. This analysis would be limited to the foundation system only. If other strengthening is to be performed, it must comply with the prescriptive provisions of this chapter.

A304.2.3 Details for new perimeter foundations. All new perimeter foundations shall be continuous and constructed according to one of the details shown in Figure A3-1 or A3-2.

> **Exceptions:**
>
> 1. When approved by the building official, the existing clearance between existing floor joists or girders and existing grade below the floor need not comply with the Building Code.
>
> • 2. When approved by the building official, and when designed by an engineer or architect, partial perimeter foundations may be used in lieu of a continuous perimeter foundation.

❖ The first three weaknesses listed in Section A303 involve buildings without complete foundation systems. These conditions will be resolved by installing a new concrete or masonry foundation system around the perimeter of the dwelling. It is the intent of these provisions that all new foundations meet the current minimum standards of the building code. Although the building code sometimes allows plain concrete foundations for one- and two-family dwellings, standard construction practice is to provide nominal horizontal reinforcing. Reinforcing is often required where the soil is expansive. These provisions, therefore, require reinforcing of all new concrete foundations with a minimum of one No. 4 reinforcing bar in the top and bottom.

The building code has specific provisions for minimum clearance under the structure for both access and ventilation. Occasionally, older construction did not provide the clearances that are required today. It is not the intent of these provisions to require current

code clearance when a new foundation must be installed. To require excavation or raising the building would be extremely difficult and costly and will not improve its resistance to earthquakes. If substantial fungal or insect infestation has occurred in the past, the owner may want to consider measures to prevent future damage. In some cases, remedial work will be required per Section A304.1.2.

An existing partial concrete foundation (weakness Type 3 in Section A303) may be replaced or it may be evaluated by a design professional to determine if it can perform in a manner equivalent to a continuous foundation. An existing unreinforced masonry or stone foundation (weakness Type 4) may be replaced with a new foundation that complies with the IBC or it may be evaluated per Section A304.2.2. Replacement might be uneconomical or aesthetically displeasing. A well-maintained unreinforced masonry foundation might be adequate to support a building for normal vertical loads, but its strength and ability to brace the building during earthquakes should be evaluated. If the existing unreinforced masonry foundation is not used to resist earthquake forces, a new foundation bracing system may be provided that is independent of the existing foundation. Examples include foundation capping and providing concrete plugs (alternative segments) in the existing foundation. Both fixes require partial removal of the existing foundation as well as shoring and/or jacking of the wood superstructure. The new system must be designed to resist all the earthquake forces from the building occurring at the foundation level. In this case, unreinforced masonry foundations do not require analysis and strengthening when an alternative foundation system is used.

Concrete foundations are typically not reinforced and commonly have cracks due to shrinkage or long-term differential settlement. Even new footings have shrinkage cracks. Common locations for these cracks are at corners and near changes in footing height or thickness. Typical shrinkage cracks in footings are straight and vertical and have uniform narrow width. Isolated cracks less than $^1/_8$ inch (3.2 mm) in width can be assumed not to significantly diminish the strength of the foundation (ATC, 2002).

A304.2.4 Required compressive strength. New concrete foundations shall have a minimum compressive strength of 2,500 pounds per square inch (17.24 MPa) at 28 days.

A304.2.5 New hollow-unit masonry foundations. New hollow-unit masonry foundations shall be solidly grouted. Mortar shall be Type M or S, and the grout and masonry units shall comply with the Building Code.

A304.2.6 Reinforcing steel. Reinforcing steel shall comply with the requirements of the Building Code.

A304.3 Foundation sill plate anchorage.

A304.3.1 Existing perimeter foundations. When the building has an existing continuous perimeter foundation, all perimeter wall sill plates shall be bolted to the foundation with chemical anchors or expansion bolts in accordance with Table A3-A.

Anchors or bolts shall be installed in accordance with Figure A3-3, with the plate washer installed between the nut and the sill plate. The nut shall be tightened to a snug-tight condition after curing is complete for chemical anchors and after expansion wedge engagement for expansion bolts. The installation of nuts on all bolts shall be subject to verification by the building official. Where existing conditions prevent anchor or bolt installation through the sill plate, this connection may be made in accordance with Figure A3-4A, A3-4B or A3-4C. The spacing of these alternate connections shall comply with the maximum spacing requirements of Table A3-A. Expansion bolts shall not be used when the installation causes surface cracking of the foundation wall at the location of the bolt.

❖ The provisions for connecting existing sill plates to existing foundations maintain the traditional code requirements for $^1/_2$-inch (12.7 mm) diameter bolts spaced a maximum of 6 feet (1829 mm) apart for one-story buildings. Two- and three-story buildings need progressively more bolts because their height and added weight result in larger forces to be resisted.

Expansion bolts and chemical anchors are acceptable for connecting to existing concrete. These connectors, due to their shorter length of embedment into the concrete, have lower capacity in concrete than anchor bolts that are cast-in-place when the foundation is poured. However, even with the required 4-inch (102 mm) embedment, these connectors will have the same capacity in the wood sill plate, which is the weakest link in this connection. Consequently, properly installed expansion bolts or chemical anchors can provide the same resistance against sliding as cast-in-place bolts.

An expansion bolt is effective when the hole is drilled the correct size and is relatively clean and when the bolt is properly tightened to set the expanding portion of the assembly in accordance with manufacturers' specifications. For expansion bolts to be fully effective, the foundation material must be able to engage the expansion portion without cracking. Where cracking indicates conditions of poor quality concrete or masonry during installation, expansion anchors may not be used. If cracks are observed during installation, installation should be stopped and a bolt should be installed at a new location at least 1 foot (305 mm) away. If the problem continues, chemical anchors or screw-type anchors should be used instead. All anchors must be installed away from the edge of the sill plate in order to be effective.

The chemical anchor is a threaded rod that uses epoxy-type adhesive to set the anchor. Chemical anchors are effective when the hole is the correct size and is completely clean. Concrete dust must be removed in accordance with manufacturers' specifications. A clean hole is more critical for chemical anchors than for expansion bolts. Chemical anchors are allowed for all types of foundations, but are required where existing concrete is in poor condition or when the installation of expansion bolts causes cracking of the concrete.

Some adhesives are viscous enough for use in horizontal holes, but others are too thin and tend to drain out of the holes before setting. To avoid this problem, consult with current ICC Evaluation Services (ICC ES^{sm}) reports on anchor systems as well as manufacturers' instructions before purchasing.

The provisions require square plate washers between the nuts and the sill plate. Because the holes that must be drilled to insert the connectors are larger than the connector diameters, the resulting holes in the sill plate are too large to provide proper bearing against the bolts. Consequently, the plate washer is installed so that the nut can be tightened sufficiently to develop the required clamping action between the sill plate and the top of the foundation wall. The use of plate washers is also intended to minimize the potential for crushing and splitting of the sill plate as the bolt is tightened. Due to the oversized hole in the wood sill, a standard round washer does not engage enough of the wood around the hole. The use of the larger plate washer will eliminate the problem of the nut recessing into the drilled hole as it is tightened. Table A3-A provides the size of the plate washers required.

The provisions call for the nut to be tightened to a "snug-tight condition" after epoxy curing is complete or after the nut has been tightened to set an expansion bolt. Tightening the nut to set the expansion bolt and tightening the nut to connect the sill plate to the foundation are separate operations. The setting requirements of expansion bolts vary according to the bolt used. The specific bolt manufacturers' procedures must be closely followed to ensure that the bolt is properly set and is capable of transmitting forces into the foundation. Because these procedures vary, the provisions only address how tight the nut should be after an expansion bolt has been properly set or the adhesive of a chemical anchor has set. If the nuts are not tight against the washer plates, there will not be sufficient clamping action between the sill plate and the foundation wall. The nut should be tightened to the point at which the full surface of the plate washer is in contact with the wood member and slightly indents the wood surface. Over-tightening beyond this snug-tight condition will cause crushing of the wood sill that will reduce the capacity of the connection. This section also gives the building official the authority to spot test the nut tightness during the required inspection.

Figures A3-4A, A3-4B and A3-4C show side plates with chemical anchors or expansion bolts into the foundation and lag bolts into the narrow face of the existing wood sill plate. Proprietary systems with ICC ES reports or other independent test reports may be used with building official approval in lieu of the connections shown if they provide an equal or greater capacity.

Figure A3-4C also shows the condition of a battered footing. This type of slanted face footing will require that the wood shim installed between the steel plate and the wood sill plate must be shaped so the steel plate will have full contact against the shim when the lag screws are tightened. Further, a beveled washer under the head of the lag screw is needed to ensure that it bears fully on the steel plate. It is recommended that the shim be nailed to the sill plate (in addition to the lag screws), but the nailing must not split the shim. Predrilling of holes may be necessary. Alternative details may be easier and faster to install and should be acceptable in principle to the building official. Discuss potential alternatives with the building department or consider hiring a licensed design professional to prepare an alternative for unique conditions.

A304.3.2 Placement of chemical anchors and expansion bolts. Chemical anchors or expansion bolts shall be placed within 12 inches (305 mm), but not less than 9 inches (229 mm), from the ends of sill plates and shall be placed in the center of the stud space closest to the required spacing. New sill plates may be installed in pieces when necessary because of existing conditions. For lengths of sill plate greater than 12 feet (3658 mm), anchors or bolts shall be spaced along the sill plate as noted in Table A3-A. For other lengths of sill plate, see Table A3-B. For lengths of sill plate less than 30 inches (762 mm), a minimum of one anchor or bolt shall be installed.

> **Exception:** Where physical obstructions such as fireplaces, plumbing or heating ducts interfere with the placement of an anchor or bolt, the anchor or bolt shall be placed as close to the obstruction as possible, but not less than 9 inches (229 mm) from the end of the plate. Center-to-center spacing of the anchors or bolts shall be reduced as necessary to provide the minimum total number of anchors required based on the full length of the wall. Center-to-center spacing shall not be less than 12 inches (305 mm).

❖ Careful attention needs to be given to the proper location and spacing of sill bolts. In order to ensure that the sills are properly connected, this section not only specifies the minimum spacing, but also limits the placement of bolts at the ends of pieces of sill plate. These provisions differ from those in the IBC in requiring the bolts to be placed no closer than 9 inches (229 mm) from the end of the sill plate. When bolts are placed closer than 9 inches (229 mm) to the end of a plate, there is a potential for that bolt to split the sill from the bolt hole to the end of the plate as the bolt is loaded from earthquake forces. When the bolt is placed more than 12 inches (305 mm) from the end of the piece, there is a tendency for the end of the plate to lift due to overturning forces on the wall (CUREE, 2002). Placing the bolt between 9 and 12 inches (229 mm and 305 mm) from the end will minimize both tendencies.

These provisions also address the fact that existing sill plates may be installed in short pieces, either where the foundation wall steps or where new pieces of a sill plate must be installed to replace sections damaged by fungus infection or insect infestation. Therefore, the provisions specify a minimum number of bolts for various lengths of sill plate.

It will not always be possible to install sill bolts at the exact spacing. There are many existing elements that can interfere with their placement, such as a fireplace, plumbing or mechanical ducts. The provisions of this chapter have taken these field situations into account and allow that where physical obstructions exist, the bolts may be omitted. However, the spacing of the remaining bolts needs to be adjusted so that the same total number of bolts is installed as though the obstruction did not exist. It is recommended that if possible, the bolts with close spacing should coincide with the sheathing locations.

A304.3.3 New perimeter foundations. Sill plates for new perimeter foundations shall be bolted as required by Table A3-A and as shown in Figure A3-1 or A3-2.

A304.4 Cripple wall bracing.

A304.4.1 General. Exterior cripple walls not exceeding 4 feet (1219 mm) in height shall use the prescriptive bracing method listed below. Cripple walls over 4 feet (1219 mm) in height require analysis by an engineer or architect in accordance with Section A301.3.

❖ When bracing a cripple wall, consideration must be given to providing adequate resistance to both the horizontal forces and the tendency for uplifting one of the ends of the wall. Any wall panel that is subject to earthquake forces has a tendency to want to lift up at one end as well as slide. This uplift can be resisted by one of two methods. In new construction a "holddown" anchor consisting of a heavy gauge metal angle is bolted to a stud and also anchored into the concrete foundation. Because this would be impractical to install in existing construction, the method used in this chapter is based on the proper proportioning of the length and height of the cripple wall bracing panels. By making the panels longer, more weight from the walls and floor above can be engaged to resist the uplift force. The basic proportion required is a minimum length of braced cripple wall panel at least two times its height (RRR, 1992). In addition, longer panels are needed as the number of stories above the wall increases. This is simply because a taller building imposes larger horizontal forces on the braced cripple wall panel.

Nonbearing walls, where floor joists run parallel to the wall, will not engage substantial weight, so holddowns at the ends of the walls may be prudent for multistory buildings. Where cripple wall bracing panels can be connected at corners of buildings and installed in combined panel lengths longer than the minimum defined in Figure A3-10, the potential for wall overturning can be reduced. In addition, continuity provided by rim joists, plates and floor framing tends to create appreciable fixity at the tops of the cripple walls that offset wall overturning.

To stay within the limits of these prescriptive methods, a maximum of 4 feet (1219 mm) for the height of the cripple wall was established to limit overturning effects. When the height of the cripple wall exceeds 4 feet (1219 mm), the dwelling owner will need to have the bracing designed by a design professional.

A304.4.1.1 Sheathing installation requirements. Wood structural panel sheathing shall not be less than $^{15}/_{32}$-inch (12 mm) thick and shall be installed in accordance with Figure A3-5 or A3-6. All individual pieces of wood structural panels shall be nailed with 8d common nails spaced 4 inches (102 mm) on center at all edges and 12 inches (305 mm) on center at each intermediate support with not less than two nails for each stud. Nails shall be driven so that their heads are flush with the surface of the sheathing and shall penetrate the supporting member a minimum of $1^1/_2$ inches (38 mm). When a nail fractures the surface, it shall be left in place and not counted as part of the required nailing. A new 8d nail shall be located within 2 inches (51 mm) of the discounted nail and be hand-driven flush with the sheathing surface. All horizontal joints must occur over nominal 2-inch-by-4-inch (51 mm by 102 mm) blocking installed with the nominal 4-inch (102 mm) dimension against the face of the plywood.

Vertical joints at adjoining pieces of wood structural panels shall be centered on existing studs such that there is a minimum $^1/_8$ inch (3.2 mm) between the panels, and such that the nails are placed a minimum of $^1/_2$ inch (12.7 mm) from the edges of the existing stud. Where such edge distances cannot be maintained because of the width of the existing stud, a new stud shall be added adjacent to the existing studs and connected in accordance with Figure A3-7.

❖ Fifteen-thirty-seconds of an inch (12 mm) thick (5 ply) plywood is prescribed as the required sheathing because of observations of ruptured $^3/_8$-inch (10 mm) thick (3 ply) plywood panels documented in the Modified Mercalli Intensity (MMI) VIII and IX intensity areas caused by the Northridge earthquake (LA, 1994).

Let-in braces have been observed to perform poorly in past earthquakes without some other form of bracing (LA, 1992). Let-in braces are no longer accepted by the 2003 IBC for Seismic Design Category D, E and F. Therefore, walls braced only with let-in braces are considered a structural weakness requiring supplemental bracing.

Even though the provisions accept existing diagonal wood sheathing (see Section A303), it is no longer cost effective for strengthening weak cripple walls. The omission of this material was not based on its ability to resist lateral forces, as it has performed well in past earthquakes and high winds (LA, 1992). Instead it was based on cost considerations and practicality, since this type of sheathing is more time consuming to install and more expensive than wood structural panels. Proprietary bracing methods may also be used when approved by the building official.

The most important component of wood structural sheathing is proper nailing. To prevent splitting of existing wood framing 8d nails are considered optimum for 2x material. If splitting of studs is observed, predrilling of holes is recommended. Predrilled holes should have a diameter of about three-fourths of the diameter of the nail. Nail guns tend to produce less splitting than

hand nailing. Minimum edge distance for nails should be maintained for plywood, the wood studs and top and sill plates to prevent splitting or premature nail failure. With this size nail, 4-inch (102 mm) spacing provides adequate capacity with the minimum bracing length permitted by Table A3-A and Figure A3-10. Further, using larger nails or closer spacing would, by comparison of capacity, require larger diameter sill bolts or closer spacing of the bolts than specified in Section A304.3.2.

When plywood is installed on the inside face of cripple walls with an exterior surface of stucco, care must be used to prevent damage to the stucco. In this situation, it is recommended that 3-inch-long (76 mm) #6 wood screws may be used instead of nails. The 3-inch (76 mm) length is needed to ensure that the shank (unthreaded) portion of the screw will have at least $^5/_8$-inch (16 mm) penetration into the studs and plates. If the threaded portion of the screws exists at the plywood-stud interface, the screws can fail in a brittle manner when the earthquake occurs.

If a nail gun is used, the operator must make sure that the nail heads do not fracture the surface of the plywood. Local variations in the density of the backing (new or existing wood-framing members) can create situations where it will be difficult to maintain consistent nail penetration. The use of a flush head attachment on a nailing gun will usually prevent overdriving. When a nail head fractures the plywood surface, the amount of force that this particular connection is capable of resisting is reduced significantly. It becomes much easier for the nail head to pull through the sheathing material. Whenever a nail head fractures the surface of the sheathing, the nail must be discounted. When a nail is discounted, it must be left in place. Removing the nail will further damage the sheathing material and could result in the rejection of the whole sheet of sheathing by the inspector (Shepherd, 1991).

When purchasing structural sheathing, one of the structural grades must be stamped on the sheets used. The correct grade of structural sheathing is important. Refer to the IBC for more information.

A304.4.2 Distribution and amount of bracing. See Table A3-A and Figure A3-10 for the distribution and amount of bracing required. Each braced panel must be at least two times the height of the cripple stud wall but not less than 48 inches (1219 mm) in length or width. Where the minimum amount of bracing prescribed in Table A3-A cannot be installed along any walls, the bracing must be designed in accordance with Section A301.3.

Exception: Where physical obstructions such as fireplaces, plumbing or heating ducts interfere with the placement of cripple wall bracing, the bracing shall then be placed as close to the obstruction as possible. The total amount of bracing required shall not be reduced because of obstructions.

❖ Table A3-A and Figure A3-10 note the location and distribution of bracing panels based on the number of stories and cripple stud height. The amount of bracing described is for each wall line.

Figure A304.4.2 shows where cripple wall bracing should be installed in conditions where the building floor plan is not a simple rectangle. While the interpretation in Figure A304.4.2 tends to penalize dwellings with reentrant corners, compared to rectangular buildings, and require more wall length in one direction

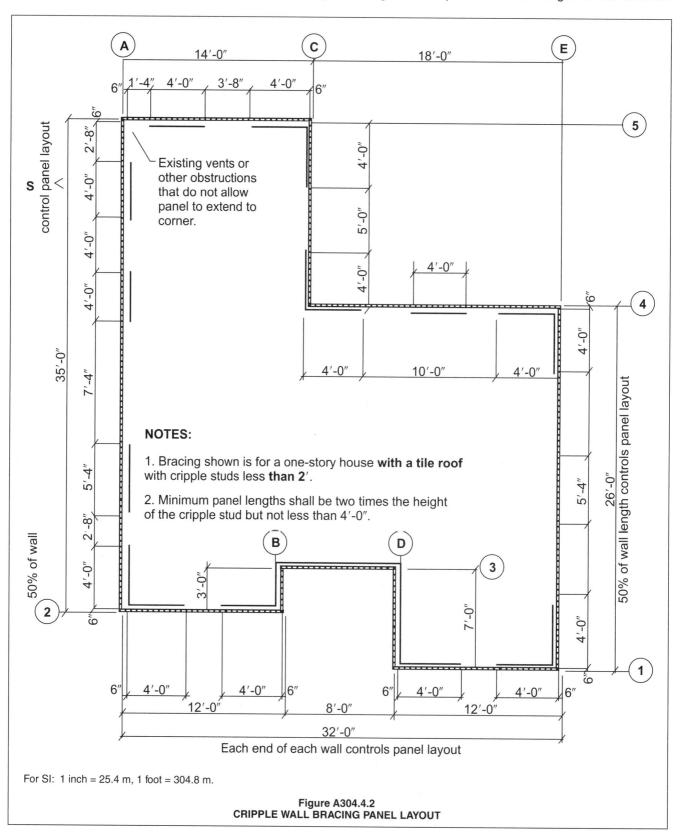

Existing vents or other obstructions that do not allow panel to extend to corner.

NOTES:

1. Bracing shown is for a one-story house **with a tile roof** with cripple studs less **than 2'**.

2. Minimum panel lengths shall be two times the height of the cripple stud but not less than 4'-0".

Each end of each wall controls panel layout

For SI: 1 inch = 25.4 m, 1 foot = 304.8 m.

Figure A304.4.2
CRIPPLE WALL BRACING PANEL LAYOUT

compared to the other, the code rules are intentionally kept simple so that homeowners and contractors can readily apply them. If owners desire to install fewer walls, they should consider hiring a design professional.

Bracing panels are required at or near each end of each wall line. Although not required, it is beneficial if the cripple wall bracing panels align with the panels above the first floor to provide a more direct load path. Thus, cripple wall bracing panels should be located under windows only when necessary. Performance can also be enhanced if all panels along a single wall line are of similar lengths rather than having one very long panel and other shorter panels. This prevents concentration of forces in one location. The existence of a small number of studs over 14 inches (356 mm) in height should not trigger this requirement.

A304.4.3 Stud space ventilation. When bracing materials are installed on the interior face of studs forming an enclosed space between the new bracing and the existing exterior finish, each braced stud space must be ventilated. Adequate ventilation and access for future inspection shall be provided by drilling one 2-inch to 3-inch-diameter (51 mm to 76 mm) round hole through the sheathing, nearly centered between each stud at the top and bottom of the cripple wall. Such holes should be spaced a minimum of 1 inch (25 mm) clear from the sill or top plates. In stud spaces containing sill bolts, the hole shall be located on the center line of the sill bolt but not closer than 1 inch (25 mm) clear from the nailing edge of the sheathing. When existing blocking occurs within the stud space, additional ventilation holes shall be placed above and below the blocking, or the existing block shall be removed and a new nominal 2-inch-by-4-inch (51 mm by 102 mm) block shall be installed with the nominal 4-inch (102 mm) dimension against the face of the plywood. For stud heights less than 18 inches (457 mm), only one ventilation hole need be provided.

❖ The most common form of cripple wall bracing will be to add sheathing to the interior face of the cripple wall from within the crawl space. When this is done, a closed space is created between each stud that does not allow natural ventilation. This can result in a buildup of moisture that will lead to fungus infection. In order to protect these concealed spaces from fungus infection, 2 inches to 3 inches diameter (51 mm to 76 mm) [3 inches (76 mm) recommended] ventilation holes must be provided (see Figure A3-7). These ventilation holes will allow the free movement of air within the stud space, thereby minimizing the risk of fungus infection. When 2x horizontal blocking is needed in the stud space to provide backing for panel joint nailing, it must be installed with the wide face oriented vertically, flush with the face of the stud on which the sheathing is being installed. This can be easily accomplished using commercially available fence rail hardware at each end to attach the block to the studs. This will eliminate blockage of ventilation inside the stud space.

Ventilation holes should be cut or drilled as close to round as possible. Hole-cutting tools are available to cut the required size hole. Square holes, or other shapes

with sharp corners or notches, can result in high concentrations of stress when the panel is loaded.

A304.4.4 Existing underfloor ventilation. Existing underfloor ventilation shall not be reduced without providing equivalent new ventilation as close to the existing ventilation as possible. Braced panels may include underfloor ventilation openings when the height of the opening, measured from the top of the foundation wall to the top of the opening, does not exceed 25 percent of the height of the cripple stud wall; however, the length of the panel shall be increased a distance equal to the length of the opening or one stud space minimum. Where an opening exceeds 25 percent of the cripple wall height, braced panels shall not be located where the opening occurs. See Figure A3-7.

Exception: For homes with a post and pier foundation system where a new continuous perimeter foundation system is being installed, new ventilation shall be provided in accordance with the Building Code.

❖ Air circulation under the floor protects the framing from fungus infection. Vents by themselves do not provide all the solutions to underfloor ventilation. It is imperative that there be cross ventilation. In order to do this, the vents must be located in opposite walls, approximately opposite each other. In many cases, heating units have been added to the dwelling and the ducts are installed within the crawl space. When this is done, the ducts block the cross ventilation and significantly reduce the efficiency of the vents. Consequently, a dwelling with vents that meet existing code might not be adequate if there are obstructions to the cross ventilation. If this condition exists, consideration should be given to providing additional vents in order to obtain the necessary cross ventilation.

A304.5 Quality control. All work shall be subject to inspection by the building official including, but not limited to:

1. Placement and installation of new chemical anchors or expansion bolts installed in existing foundations. Special inspection is not required for chemical anchors installed in existing foundations regulated by the prescriptive provisions of this chapter.

2. Installation and nailing of new cripple wall bracing.

3. Any work may be subject to special inspection when required by the building official in accordance with the Building Code.

❖ Strengthening work is only as good as the quality of the construction. In most jurisdictions the strengthening work required by this chapter will require building permits and inspections. Prior to requesting a permit, the owner or contractor should survey and determine all existing conditions, dimensions and other considerations significant to the retrofit or repair work. A plan should be prepared [11-inch by 17-inch (279 mm by 432 mm) paper with $^1/_8$-inch (3.2 mm) = 1 foot (305 mm) scale is adequate] showing the location of proposed sheathing and spacing of anchor bolts. The drawings should differentiate between new and existing components. Because of the nature of the work being performed and the materials being used, there are some additional inspections that need to be performed that are not specified in the IBC code for new construction.

1. *Placement and Installation of New Chemical Anchors or Expansion Bolts.* The building official must approve the use of expansion bolts or chemical anchors. Building officials often use evaluation reports from the ICC ES as guidance in the products they approve. These reports set allowable design values based on testing and specify requirements for construction quality control. They often call for special inspection, especially for bolts that might be subject to tension forces. (Special inspection generally involves inspection of the work while it is being performed, as opposed to when it is complete. It is performed by qualified individuals retained by the owner.)

For the purposes of Chapter 3 of the Guidelines for *Seismic Retrofit of Existing Buildings* (GSREB), special inspection is not required because the bolts in question are intended to act primarily in shear, not in tension. Even if these bolts are not set exactly as noted in the evaluation report they will still work to resist the shear forces from earthquakes. The waiving of special inspection thus represents a justifiable cost savings. While special inspection is not typically required, the building official may still require verification of proper installation per Section A304.3.1.

In lieu of special inspection, it is recommended that a post-installation torque test for expansion anchors be done together with inspection for bolt spacing, end distance and a spot check to make sure the nuts are properly tightened. Usually this inspection would be performed after the bolts were installed and before the cripple wall sheathing is placed. However, the building official may elect to perform this inspection at the same time he or she inspects the installation of the cripple wall sheathing. Vent holes in each stud space should be located and sized to allow inspectors to reach in and torque the bolts.

Both expansion bolts and chemical anchors must be approved, as stated in the definitions of Section A302. This means that they generally must have a valid evaluation report from the ICC ES or an approved equivalent independent test report. Normally chemical anchors require continuous inspection during their installation as a part of their approval for use. Continuous inspection checks that the hole is the correct depth and is sufficiently clean prior to placing the epoxy material. The purpose of all the checks is to ensure that each bolt will attain the tension strength allowed by its evaluation report.

Since sill bolts are not subject primarily to tension, however, the provisions make an exception to this rigorous special inspection requirement. However, it is recommended that a less expensive torque test of expansion anchors in lieu of tension tests is an appropriate substitute for special inspection. Torque tests must be performed on at least 25 percent of the total number of installed expansion bolts and must be done in the presence of a building inspector or a deputy inspector employed by a testing agency hired by the owner and approved by the authority having jurisdiction.

2. *Installation and Nailing of New Cripple Bracing Wall.* The final required inspection is to make sure that the connections of the wall bracing panels are installed correctly and completely. The bracing serves no purpose if the connections do not engage the proper framing members, or, in the case of wood sheathing, are overdriven.

3. *Work Subject to Special Inspection.* The building official may require special inspection where conditions on a particular job site make inspections difficult. Retrofit and strengthening work often involve unusual conditions or proprietary components that require additional levels of quality control. In order to address these problems, the building official is allowed to require that a special inspector verify work.

Most dwellings have some features that will not conform to the conditions and strengthening provisions used in this chapter. To avoid situations where those existing conditions either preclude the use of these prescriptive provisions or where other complications may occur that would make their application to a specific building difficult, the building official is encouraged to perform a predesign inspection. The cost of this inspection, however, may be in addition to the normal permit fee. Such an inspection might not be necessary if the owner provides adequate drawings supplemented by photographs to permit adequate review by the building official.

The purpose of a predesign inspection is to notify the owner or contractor of problems that may need the services of design professionals. It is not intended to be a consulting service to the owner. Typically, this inspection should focus on the following issues:

- Areas where obstructions in the crawl space along exterior walls might prevent installation of adequate lengths of bracing;
- Areas that may be questionable with respect to insect or fungal damage to wood members to be used in the strengthening;
- Foundations that may be questionable or clearly too weak to be effectively used for anchoring sill plates;
- Tests of nut snug tightness on sill bolts;
- Inadequate rim joist or blocking conditions along exterior wall lines; and
- Other concerns that the owner or contractor believes will preclude the use of the prescriptive details or methods described in this chapter.

TABLE A3-A—SILL PLATE ANCHORAGE AND CRIPPLE WALL BRACING

NUMBER OF STORIES ABOVE CRIPPLE WALLS	MINIMUM SILL PLATE CONNECTION AND MAXIMUM SPACING[1,2]	AMOUNT OF BRACING[3,4,5]	
		A Combination of Exterior Walls Finished with Portland Cement Plaster and Roofing Using Clay Tile or Concrete Tile Weighing More than 6 psf (287 N/m²)	All Other Conditions
One story	$1/_2$ inch (12.7 mm) spaced 6 feet, 0 inch (1829 mm) center-to-center with washer plate	Each end and not less than 50 percent of the wall length	Each end and not less than 40 percent of the wall length
Two stories	$1/_2$ inch (12.7 mm) spaced 4 feet, 0 inch (1219 mm) center-to-center with washer plate; or $5/_8$ inch (15.9 mm) spaced 6 feet, 0 inch (1829 mm) center-to-center with washer plate	Each end and not less than 70 percent of the wall length	Each end and not less than 50 percent of the wall length
Three stories	$5/_8$ inch (15.9 mm) spaced 4 feet, 0 inch (1219 mm) center-to-center with washer plate	100 percent of the wall length[6]	Each end and not less than 80 percent of the wall length[6]

1. Sill plate anchors shall be chemical anchors or expansion bolts in accordance with Section A304.3.1.
2. All washer plates shall be 2 inches by 2 inches by $3/_{16}$ inch (51 mm by 51 mm by 4.8 mm) minimum.
3. See Figure A3-10 for braced panel layout.
4. Braced panels at ends of walls shall be located as near to the end as possible.
5. All panels along a wall shall be nearly equal in length and shall be nearly equal in spacing along the length of the wall.
6. The minimum required underfloor ventilation openings are permitted in accordance with Section A304.4.4.

TABLE A3–B—SILL PLATE ANCHORAGE FOR VARIOUS LENGTHS OF SILL PLATE[1,2]

NUMBER OF STORIES	LENGTHS OF SILL PLATE		
	Less than 12 feet (3658 mm) to 6 feet (1829 mm)	Less than 6 feet (1829 mm) to 30 inches (762 mm)	Less than 30 inches (762 mm)[3]
One story	Three connections	Two connections	One connection
Two stories	Four connections for $1/_2$-inch (12.7 mm) anchors or bolts or Three connections for $5/_8$-inch (15.9 mm) anchors or bolts	Two connections	One connection
Three stories	Four connections	Two connections	One connection

1. Connections shall be either chemical anchors or expansion bolts.
2. See Section A304.3.2 for minimum end distances.
3. Connections shall be placed as near to the center of the length of plate as possible.

TABLE A3–C—NOT USED

A304.6 Phasing of the strengthening work. When approved by the building official, the strengthening work contained in this chapter may be completed in phases. The strengthening work in any phase shall be performed on two parallel sides of the structure at the same time.

❖ The phasing of the strengthening work can occur when approved by the building official. A new permit will be required for each phase before beginning the additional work. Phasing may benefit owners with limited budgets or scheduling conflicts with other planned alterations.

The strengthening work in any phase requires that the work be performed on at least two parallel sides and never on one side alone. This is meant to prevent rotation of the foundation anchorage, in plan, from seismic horizontal forces.

NUMBER OF STORIES	MINIMUM FOUNDATION DIMENSIONS					MINIMUM FOUNDATION REINFORCING	
	W	F	$D^{1, 2, 3}$	T	H	VERTICAL REINFORCING	
						Single-pour wall and footing	Footing poured separate from wall
1	12 inches (305 mm)	6 inches (152 mm)	12 inches (305 mm)	6 inches (152 mm)	≤ 24 inches (610 mm)	Single-pour wall and footing	Footing poured separate from wall
2	15 inches (381 mm)	7 inches (178 mm)	18 inches (457 mm)	8 inches (203 mm)	≤ 36 inches (914 mm)	#4 @ 48 inches (1219 mm) on center	#4 @ 32 inches (813 mm) on center
3	18 inches 457 mm)	8 inches (203 mm)	24 inches (610 mm)	10 inches (254 mm)	≥ 36 inches (914 mm)	#4 @ 48 inches (1219 mm) on center	#4 @ 18 inches (457 mm) on center

1. Where frost conditions occur, the minimum depth shall extend below the frost line.

2. The ground surface along the interior side of the foundation may be excavated to the elevation of the top of the footing.

3. When expansive soil is encountered, the foundation depth and reinforcement shall be as directed by the building official.

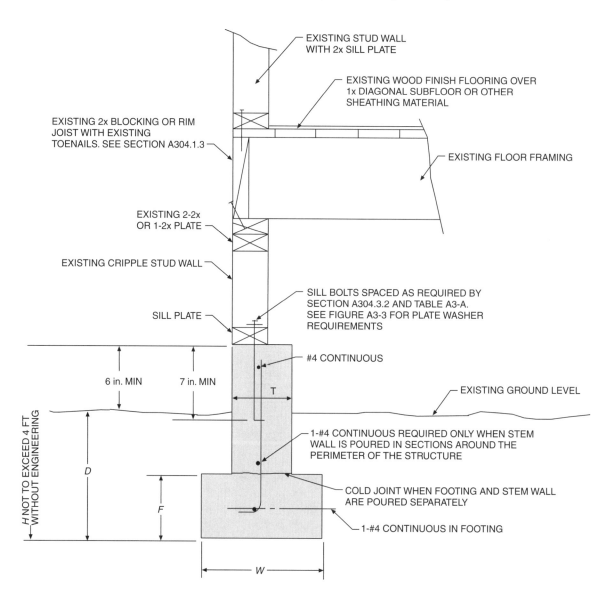

For SI: 1 inch = 25.4 mm, 1 foot = 304.8 mm.

NOTE: See Figure A3-5 or A3-6 for cripple wall bracing.

FIGURE A3-1—NEW REINFORCED CONCRETE FOUNDATION SYSTEM

NUMBER OF STORIES	MINIMUM FOUNDATION DIMENSIONS					MINIMUM FOUNDATION REINFORCING	
	W	F	$D^{1,2,3}$	T	H	VERTICAL REINFORCING	HORIZONTAL REINFORCING
1	12 inches (305 mm)	6 inches (152 mm)	12 inches (305 mm)	6 inches (152 mm)	≤ 24 inches (610 mm)	#4 @ 24 inches (610 mm) on center	#4 continuous at top of stem wall
2	15 inches (381 mm)	7 inches (178 mm)	18 inches (457 mm)	8 inches (203 mm)	≤ 24 inches (610 mm)	#4 @ 24 inches (610 mm) on center	#4 @ 16 inches (406 mm) on center
3	18 inches 457 mm)	8 inches (203 mm)	24 inches (610 mm)	10 inches (254 mm)	≥ 36 inches (914 mm)	#4 @ 24 inches (610 mm) on center	#4 @ 16 inches (406 mm) on center

1. Where frost conditions occur, the minimum depth shall extend below the frost line.

2. The ground surface along the interior side of the foundation may be excavated to the elevation of the top of the footing.

3. When expansive soil is encountered, the foundation depth and reinforcement shall be as directed by the building official.

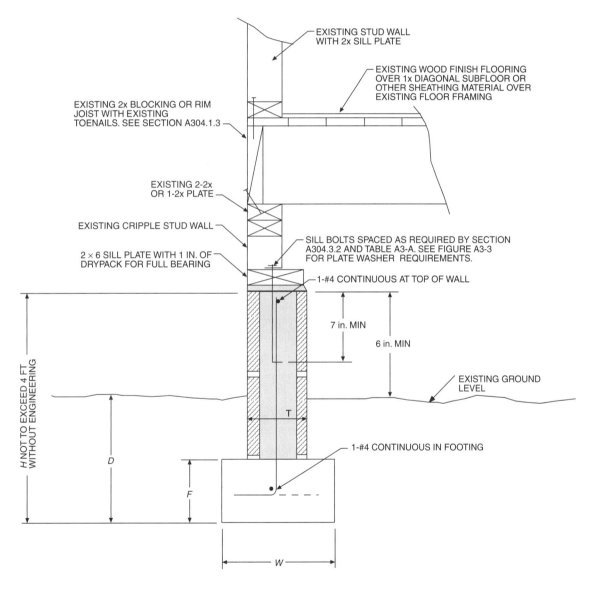

For SI: 1 inch = 25.4 mm, 1 foot = 304.8 mm.
NOTE: See Figure A3-5 or A3-6 for cripple wall bracing.

FIGURE A3-2—NEW HOLLOW-MASONRY UNIT FOUNDATION WALL

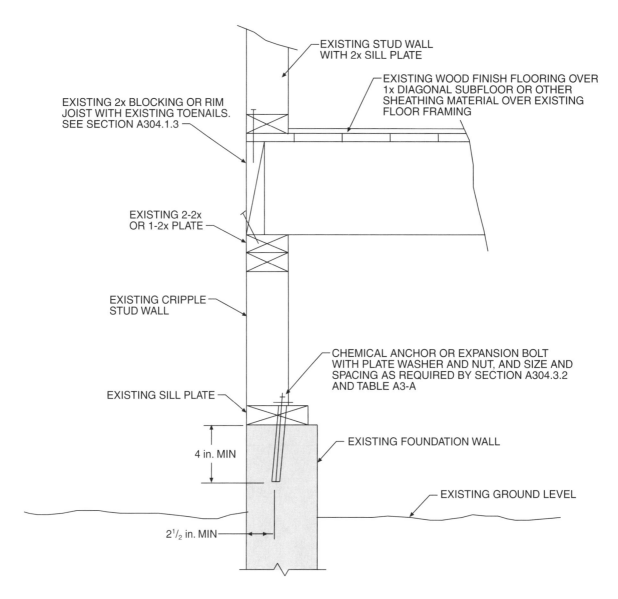

For SI: 1 inch = 25.4 mm.

NOTES:

1. Plate washers shall comply with the following:
   $\frac{1}{2}$ in. anchor or bolt—2 in. x 2 in. x $\frac{3}{16}$ in.
   $\frac{5}{8}$ in. anchor or bolt—2 in. x 2 in. x $\frac{3}{16}$ in.

2. See Figure A3-5 or A3-6 for cripple wall bracing.

FIGURE A3-3—SILL PLATE BOLTING TO EXISTING FOUNDATION

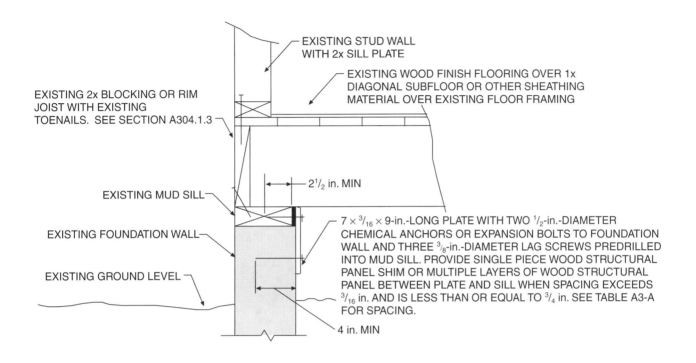

EXISTING STUD WALL WITH 2x SILL PLATE

EXISTING WOOD FINISH FLOORING OVER 1x DIAGONAL SUBFLOOR OR OTHER SHEATHING MATERIAL OVER EXISTING FLOOR FRAMING

EXISTING 2x BLOCKING OR RIM JOIST WITH EXISTING TOENAILS. SEE SECTION A304.1.3

$2^1/_2$ in. MIN

EXISTING MUD SILL

$7 \times {}^3/_{16} \times 9$-in.-LONG PLATE WITH TWO $^1/_2$-in.-DIAMETER CHEMICAL ANCHORS OR EXPANSION BOLTS TO FOUNDATION WALL AND THREE $^3/_8$-in.-DIAMETER LAG SCREWS PREDRILLED INTO MUD SILL. PROVIDE SINGLE PIECE WOOD STRUCTURAL PANEL SHIM OR MULTIPLE LAYERS OF WOOD STRUCTURAL PANEL BETWEEN PLATE AND SILL WHEN SPACING EXCEEDS $^3/_{16}$ in. AND IS LESS THAN OR EQUAL TO $^3/_4$ in. SEE TABLE A3-A FOR SPACING.

EXISTING FOUNDATION WALL

EXISTING GROUND LEVEL

4 in. MIN

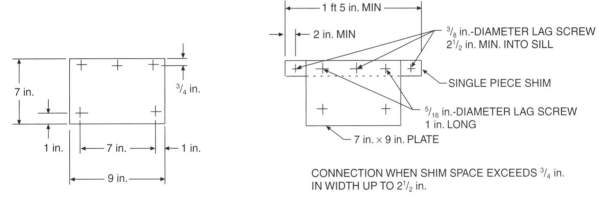

7 in.

$^3/_4$ in.

1 in.   7 in.   1 in.

9 in.

HOLE DIAMETER SHALL NOT EXCEED CONNECTOR DIAMETER BY MORE THAN $^1/_{16}$ in.

1 ft 5 in. MIN

2 in. MIN

$^3/_8$ in.-DIAMETER LAG SCREW $2^1/_2$ in. MIN. INTO SILL

SINGLE PIECE SHIM

$^5/_{16}$ in.-DIAMETER LAG SCREW 1 in. LONG

7 in. $\times$ 9 in. PLATE

CONNECTION WHEN SHIM SPACE EXCEEDS $^3/_4$ in. IN WIDTH UP TO $2^1/_2$ in.

For SI: 1 inch = 25.4 mm, 1 foot = 304.8 mm.
NOTE: If shim space exceeds $2^1/_2$ in., alternate details will be required.

FIGURE A3-4A—SILL PLATE BOLTING IN EXISTING FOUNDATION—ALTERNATE

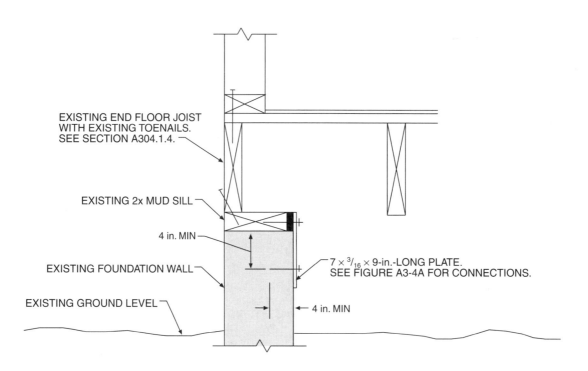

EXISTING END FLOOR JOIST WITH EXISTING TOENAILS. SEE SECTION A304.1.4.

EXISTING 2x MUD SILL

4 in. MIN

EXISTING FOUNDATION WALL

EXISTING GROUND LEVEL

$7 \times {}^{3}/_{16} \times 9$-in.-LONG PLATE. SEE FIGURE A3-4A FOR CONNECTIONS.

4 in. MIN

For SI: 1 inch = 25.4 mm.

FIGURE A3-4B—SILL PLATE BOLTING TO EXISTING FOUNDATION WITHOUT CRIPPLE WALL AND FRAMING PARALLEL TO THE FOUNDATION WALL

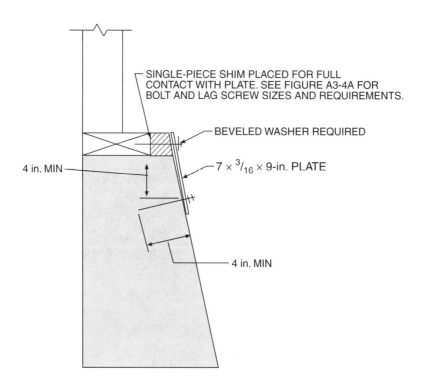

SINGLE-PIECE SHIM PLACED FOR FULL CONTACT WITH PLATE. SEE FIGURE A3-4A FOR BOLT AND LAG SCREW SIZES AND REQUIREMENTS.

BEVELED WASHER REQUIRED

$7 \times {}^{3}/_{16} \times 9$-in. PLATE

4 in. MIN

4 in. MIN

ALTERNATE CONNECTION FOR BATTERED FOOTING

For SI: 1 inch = 25.4 mm.

FIGURE A3-4C—SILL PLATE BOLTING IN EXISTING FOUNDATION—ALTERNATE

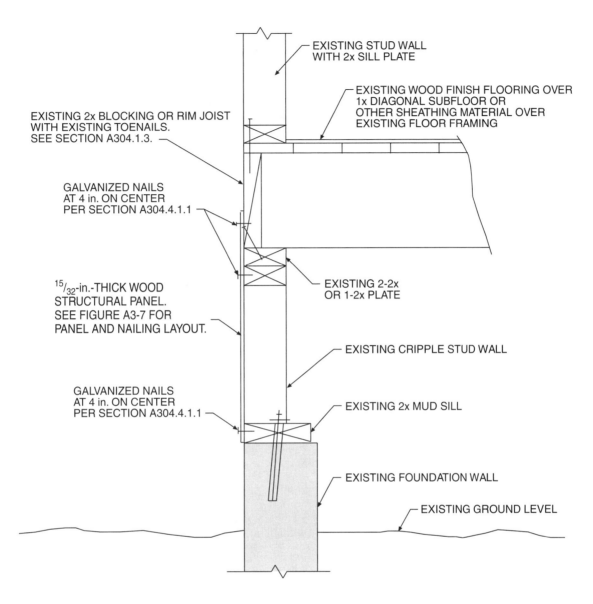

EXISTING STUD WALL
WITH 2x SILL PLATE

EXISTING WOOD FINISH FLOORING OVER
1x DIAGONAL SUBFLOOR OR
OTHER SHEATHING MATERIAL OVER
EXISTING FLOOR FRAMING

EXISTING 2x BLOCKING OR RIM JOIST
WITH EXISTING TOENAILS.
SEE SECTION A304.1.3.

GALVANIZED NAILS
AT 4 in. ON CENTER
PER SECTION A304.4.1.1

$^{15}/_{32}$-in.-THICK WOOD
STRUCTURAL PANEL.
SEE FIGURE A3-7 FOR
PANEL AND NAILING LAYOUT.

EXISTING 2-2x
OR 1-2x PLATE

EXISTING CRIPPLE STUD WALL

GALVANIZED NAILS
AT 4 in. ON CENTER
PER SECTION A304.4.1.1

EXISTING 2x MUD SILL

EXISTING FOUNDATION WALL

EXISTING GROUND LEVEL

For SI: 1 inch = 25.4 mm.
NOTE: See Figure A3-3 for sill plate bolting.

**FIGURE A3-5—CRIPPLE WALL BRACING WITH WOOD STRUCTURAL PANEL
ON EXTERIOR FACE OF CRIPPLE STUDS**

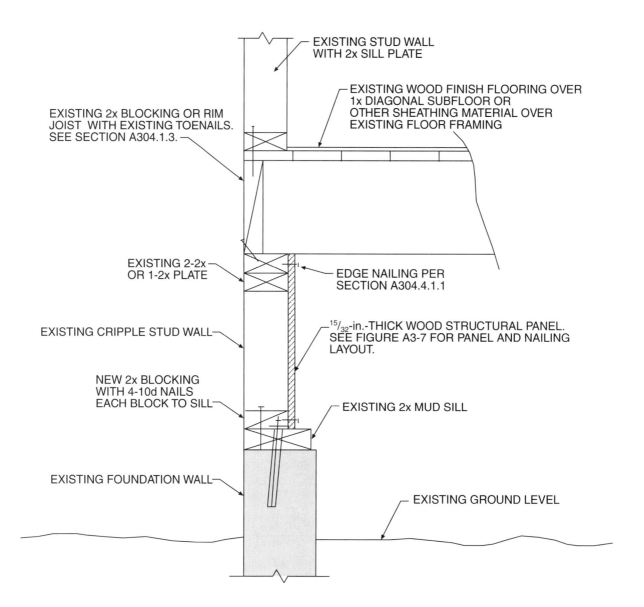

EXISTING STUD WALL
WITH 2x SILL PLATE

EXISTING WOOD FINISH FLOORING OVER
1x DIAGONAL SUBFLOOR OR
OTHER SHEATHING MATERIAL OVER
EXISTING FLOOR FRAMING

EXISTING 2x BLOCKING OR RIM
JOIST WITH EXISTING TOENAILS.
SEE SECTION A304.1.3.

EXISTING 2-2x
OR 1-2x PLATE

EDGE NAILING PER
SECTION A304.4.1.1

EXISTING CRIPPLE STUD WALL

$^{15}/_{32}$-in.-THICK WOOD STRUCTURAL PANEL.
SEE FIGURE A3-7 FOR PANEL AND NAILING
LAYOUT.

NEW 2x BLOCKING
WITH 4-10d NAILS
EACH BLOCK TO SILL

EXISTING 2x MUD SILL

EXISTING FOUNDATION WALL

EXISTING GROUND LEVEL

For SI: 1 inch = 25.4 mm.
NOTE: See Figure A3-3 for sill plate bolting.

**FIGURE A3-6—CRIPPLE WALL BRACING WITH WOOD STRUCTURAL PANEL
ON INTERIOR FACE OF CRIPPLE STUDS**

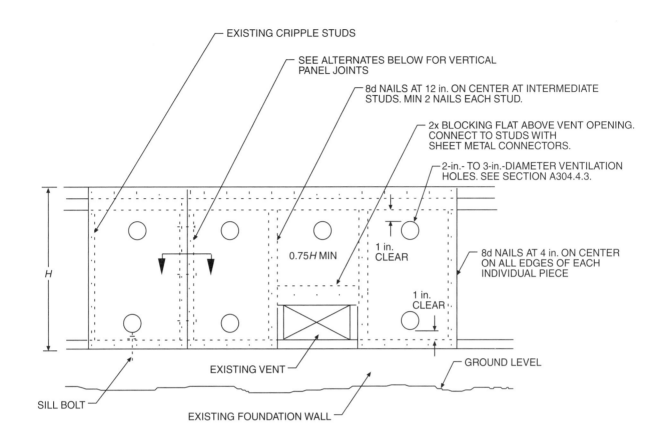

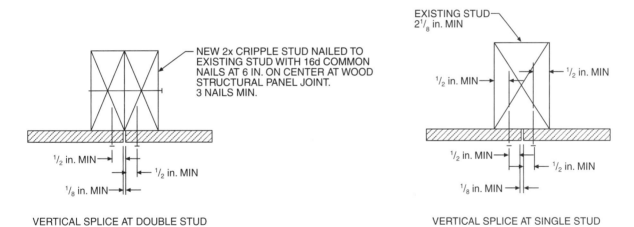

For SI: 1 inch = 25.4 mm.

FIGURE A3-7—PARTIAL CRIPPLE STUD WALL ELEVATION

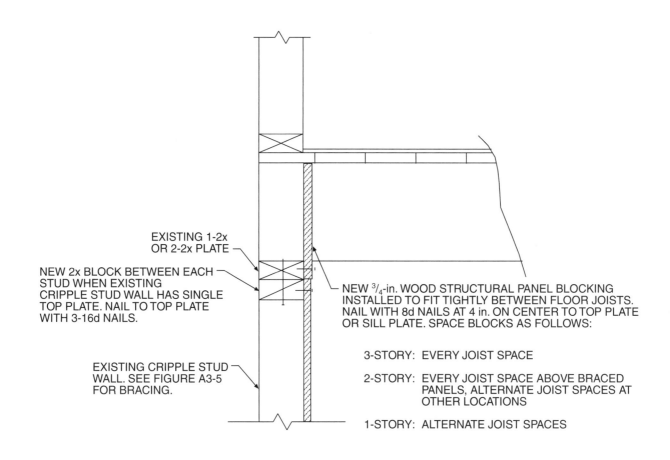

EXISTING 1-2x
OR 2-2x PLATE

NEW 2x BLOCK BETWEEN EACH
STUD WHEN EXISTING
CRIPPLE STUD WALL HAS SINGLE
TOP PLATE. NAIL TO TOP PLATE
WITH 3-16d NAILS.

EXISTING CRIPPLE STUD
WALL. SEE FIGURE A3-5
FOR BRACING.

NEW $^3/_4$-in. WOOD STRUCTURAL PANEL BLOCKING
INSTALLED TO FIT TIGHTLY BETWEEN FLOOR JOISTS.
NAIL WITH 8d NAILS AT 4 in. ON CENTER TO TOP PLATE
OR SILL PLATE. SPACE BLOCKS AS FOLLOWS:

3-STORY: EVERY JOIST SPACE

2-STORY: EVERY JOIST SPACE ABOVE BRACED
          PANELS, ALTERNATE JOIST SPACES AT
          OTHER LOCATIONS

1-STORY: ALTERNATE JOIST SPACES

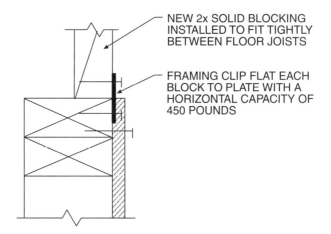

NEW 2x SOLID BLOCKING
INSTALLED TO FIT TIGHTLY
BETWEEN FLOOR JOISTS

FRAMING CLIP FLAT EACH
BLOCK TO PLATE WITH A
HORIZONTAL CAPACITY OF
450 POUNDS

For SI: 1 inch = 25.4 mm, 1 pound = 4.4 N.

**FIGURE A3-8—ALTERNATE BLOCKING WHERE RIM JOIST OR
BLOCKING HAS BEEN OMITTED**

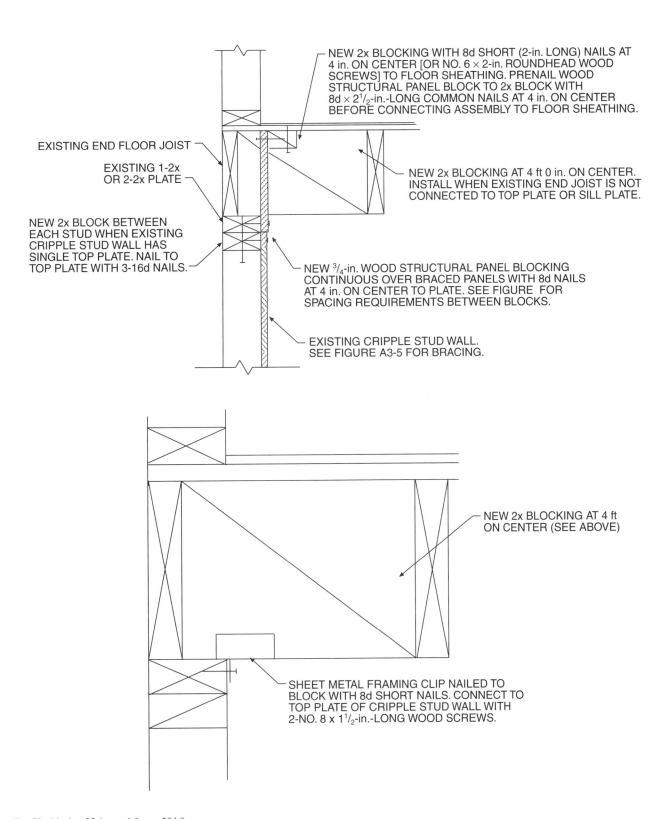

NEW 2x BLOCKING WITH 8d SHORT (2-in. LONG) NAILS AT 4 in. ON CENTER [OR NO. 6 × 2-in. ROUNDHEAD WOOD SCREWS] TO FLOOR SHEATHING. PRENAIL WOOD STRUCTURAL PANEL BLOCK TO 2x BLOCK WITH 8d × 2$\frac{1}{2}$-in.-LONG COMMON NAILS AT 4 in. ON CENTER BEFORE CONNECTING ASSEMBLY TO FLOOR SHEATHING.

EXISTING END FLOOR JOIST

EXISTING 1-2x OR 2-2x PLATE

NEW 2x BLOCK BETWEEN EACH STUD WHEN EXISTING CRIPPLE STUD WALL HAS SINGLE TOP PLATE. NAIL TO TOP PLATE WITH 3-16d NAILS.

NEW 2x BLOCKING AT 4 ft 0 in. ON CENTER. INSTALL WHEN EXISTING END JOIST IS NOT CONNECTED TO TOP PLATE OR SILL PLATE.

NEW $\frac{3}{4}$-in. WOOD STRUCTURAL PANEL BLOCKING CONTINUOUS OVER BRACED PANELS WITH 8d NAILS AT 4 in. ON CENTER TO PLATE. SEE FIGURE FOR SPACING REQUIREMENTS BETWEEN BLOCKS.

EXISTING CRIPPLE STUD WALL. SEE FIGURE A3-5 FOR BRACING.

NEW 2x BLOCKING AT 4 ft ON CENTER (SEE ABOVE)

SHEET METAL FRAMING CLIP NAILED TO BLOCK WITH 8d SHORT NAILS. CONNECT TO TOP PLATE OF CRIPPLE STUD WALL WITH 2-NO. 8 x 1$\frac{1}{2}$-in.-LONG WOOD SCREWS.

For SI: 1 inch = 25.4 mm, 1 foot = 304.8 mm.

FIGURE A3-9—CONNECTION OF CRIPPLE WALL TO FLOOR SHEATHING WHEN FLOOR FRAMING IS PARALLEL TO WALL

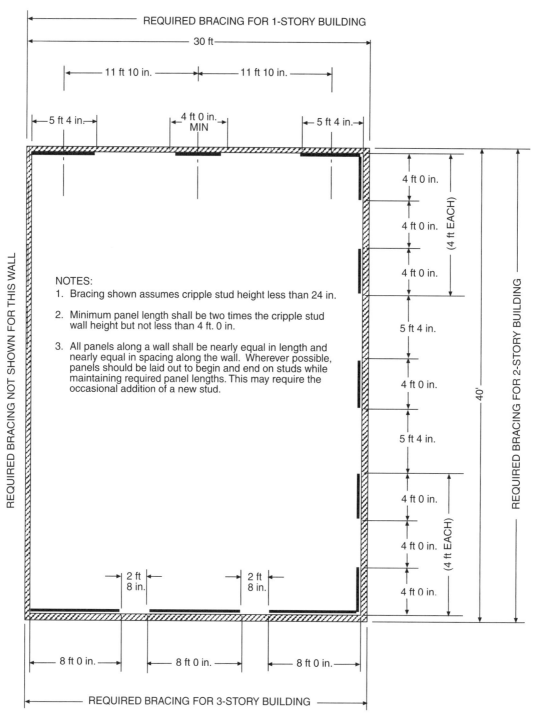

REQUIRED BRACING FOR 1-STORY BUILDING

NOTES:

1. Bracing shown assumes cripple stud height less than 24 in.

2. Minimum panel length shall be two times the cripple stud wall height but not less than 4 ft. 0 in.

3. All panels along a wall shall be nearly equal in length and nearly equal in spacing along the wall. Wherever possible, panels should be laid out to begin and end on studs while maintaining required panel lengths. This may require the occasional addition of a new stud.

REQUIRED BRACING NOT SHOWN FOR THIS WALL

REQUIRED BRACING FOR 2-STORY BUILDING

REQUIRED BRACING FOR 3-STORY BUILDING

Bracing determination:
1-story building—each end and not less than 40% of wall length.[1]
        Transverse wall—30 ft. × 0.40 = 12 ft. minimum panel length = 4 ft. 0 in.
2-story building—each end and not less than 50% of wall length.[1]
        Longitudinal wall—40 ft. × 0.50 = 20 ft. 0 in. minimum of bracing.
3-story building—each end and not less than 80% of wall length.[1]
        Transverse wall—30 ft. × 0.80 = 24 ft. 0 in. minimum of bracing.
[1]See Table A3-A for buildings with both plaster walls and roofing exceeding 6 psf (287 N/m^2).

For SI: 1 inch = 25.4 mm, 1 foot = 304.8 mm.

FIGURE A3-10—CRIPPLE WALL BRACING LAYOUT

Bibliography

The following resource materials are referenced in this chapter or are relevant to the subject matter addressed in this chapter.

ABAG 99, *Preventing the Nightmare – Designing a Model Program to Encourage Owners of Homes and Apartments to Do Earthquake Retrofits.* Association of Bay Area Governments, 1999.

ABAG 2000, *Preventing the Nightmare – Post-Earthquake Housing Issue Papers.* Association of Bay Area Governments, 2000.

ATC 2002 ATC 50, *Simplified Seismic Assessment of Detached, Single-Family, Wood-Frame Dwellings*; ATC 50-1, *Seismic Retrofit Guidelines for Detached, Single-Family Wood-Frame Dwellings;* ATC 50-2, *Safer at Home in Earthquakes a Proposed Earthquake Safety Program.* Applied Technology Council, www.atcouncil.org 2002.

CEA 2002, *Earthquake Insurance Retrofit Premium Discounts, Rates and Premiums.* Policy Information, California Earthquake Authority, http://www.earthquakeauthority.com/rates/rates_premiums.html#top, 2002.

Comerio, Mary G., John D. Landis, Cathurin J. Fispo and Juan Pablo Manzan. *Earthquake Hazards and Wood Frame Houses – What You should Know and Can Do.* Berkeley, CA: College of Environmental Design, U.C. Berkeley, 1982.

Comerio, Mary G., John D. Landis, Cathurin J. Fispo and Juan Pablo Manzan. *Residential Earthquake Recovery – Improving California's Post-Disaster Rebuilding Policies and Programs*, Berkeley, CA: California Policy Seminar, 1996.

CRERF-92, *Damaging Residential Earthquakes Since 1960.* California Residential Earthquake Recovery Fund, 1992.

CSSC-98, *Homeowner's Guide to Earthquake Safety.* California Seismic Safety Commission, California Seismic Safety Commission, www.seismic.ca.gov, 2002.

CUREE 2002, *Recommendations for Earthquake Resistance in the Design and Construction of Woodframe Buildings.* Consortium of Universities for Research in Earthquake Engineering, 2002 Final Review Draft.

DSA 2002, *Guidelines for Earthquake Bracing of Residential Water Heaters.* DSA Review and Acceptance Procedures – Manufactured Earthquake Bracing Systems for Residential Water Heaters, Division of the State Architect, Department of General Services, State of California, www.dsa.gov, 2002.

FEMA 1995, *Seismic Retrofit Training for Building Contractors and Building Inspectors - Participant Handbook.* Wiss Janney, Elstner Assoc., EMW-92-C3852, 1995.

Gallagher, Ron. *Mitigation of Principal Earthquake Hazards in Wood Frame Dwellings and Mobile Dwellings.* California Department of Insurance, June 1990.

Huntington, Craig G. "Taking the Lead on Earthquake Safety." *Civil Engineering Magazine,* June 1991.

IBHS 1999, *Is Your Home Protected from Earthquake Disaster? – A Homeowner's Guide to Earthquake Retrofit.* Institute for Building and Home Safety, www.ibhs.org, 1999.

LA 2001, *Reconstruction and Replacement of Earthquake Damaged Masonry Chimneys.* Los Angeles: Department of Building & Safety, http://www.ladbs.org/Reports_and_Publications/Information_Bulletin/information_bulletin.htm, 2001.

LA 1994, *Prescriptive Provisions for Seismic Strengthening of Light, Wood Frame, Residential Buildings.* Los Angeles: Department of Building and Safety, Joint City of Los Angeles/SEAOSC Subcommittee on Cripple Wall Buildings, 1994.

NAHB 1994, *Assessment of Damage to Residential Buildings Caused by the Northridge Earthquake.* National Association of Home Builders, http://www.nahbrc.org, 1994.

NIST 1997, *Fire Hazards and Mitigation Measures Associated with Seismic Damage of Water Heaters.* Mroz and Soong, National Institute of Standards and Technology, 1997.

RRR 1992, *Prescriptive Provisions for Voluntary Seismic Strengthening of Light, Wood-frame Residential Buildings, (with Commentary).* Residential Retrofit and Repair Committee jointly sponsored by California Office of Emergency Services, California Building Officials, American Institute of Architects – California Council, Structural Engineers Association of California, State Historical Building Safety Board, available from California Seismic Safety Commission, 1994.

SEAOC 1995, *History of Conventional Construction.* Structural Engineers Association of California, 1995.

SEAOSC 2002, *Seismic Retrofit of Existing Buildings – Retrofit of Hillside Dwellings*, Seminar Notes, Nels Roselund, *Prescriptive Provisions for Seismic Strengthening of Cripple Walls.* Ben Schmid, June 2002.

Shepherd, R. and E.O. Delos-Santos. "An Experimental Investigation of Retrofitted Cripple Walls." *Bulletin of the Seismological Society of America,* Vol 81, No. 5, pp. 2111-2126, October 1991.

SL 2002, *City of San Leandro, CA Retrofit Plan Set*, http://www.ci.san-leandro.ca.us/program.html, 2002.

Steinbrugge, Karl V. and S.T. "Ted" Algermissen. 1990, *Earthquake Losses to Single-Family Dwellings: California Experience.* US Geological Survey Bulletin 1939-A.

CHAPTER A4

EARTHQUAKE HAZARD REDUCTION
IN EXISTING WOOD-FRAME RESIDENTIAL BUILDINGS
WITH SOFT, WEAK OR OPEN-FRONT WALLS

❖ This chapter is part of the 2000 edition of the *Guidelines for Seismic Retrofit of Existing Buildings* (GSREB), published in 2001 by the International Conference of Building Officials (ICBO). It has been reprinted in its entirety as Appendix Chapter A4 in the code. For voluntary seismic retrofit, Section 707.7 specifically allows the use of GSREB provisions in lieu of the normal requirements specified in Section A407.1.1.1. This commentary is intended to explain the technical provisions of Chapter A4 of the code and their intended use.

The GSREB preface gives a brief history of the GSREB, parts of which were first published as appendices to the Uniform Code for Building Conservation (UCBC). This chapter has its origins in documents developed for the City of Los Angeles (Los Angeles, 1999) and the City of Fremont, California after the 1994 Northridge earthquake.

For convenience, this commentary uses the acronym "SWOF" to refer to buildings, wall lines, or structural conditions characterized by soft, weak, or open-front (SWOF) walls.

SECTION A401
GENERAL

A401.1 Purpose. The purpose of this chapter is to promote public welfare and safety by reducing the risk of death or injury that may result from the effects of earthquakes on existing wood-frame, multiunit residential buildings. The ground motions of past earthquakes have caused the loss of human life, personal injury and property damage in these types of buildings. This chapter creates minimum standards to strengthen the more vulnerable portions of these structures. When fully followed, these minimum standards will improve the performance of these buildings but will not necessarily prevent all earthquake-related damage.

❖ The provisions in this chapter are intended to prevent concentrations of structural damage in the vulnerable first stories of typical SWOF buildings.

These retrofit provisions are not intended to provide structural performance equivalent to that provided by new construction built in accordance with the *International Building Code*® (IBC®). Model building codes for new construction intend to safeguard against major structural failures and loss of life or, more generally, to safeguard the public health, safety and general welfare. Modern code-based designs can be expected to prevent structural collapse, limit structural and nonstructural falling hazards and provide safe egress.

In addition, due to inherent conservatism, code-based designs can also be expected to offer some measure of damage control or repairability.

To meet such a standard, an existing SWOF building would require comprehensive investigation, testing and analysis, possibly followed an by extensive structural and nonstructural retrofit. Instead, Chapter 4 of the GSREB provisions aim to "reduce the risk" with significantly less design effort, construction cost and tenant disruption. Risk reduction does not take a comprehensive approach to life safety, does not aim to protect property or function and is not equivalent to new construction under the IBC. For many owners, tenants and jurisdictions, this risk-reduction approach represents an acceptable trade-off.

Risk reduction identifies and improves the structure's more vulnerable portions, often leaving the rest of the building untouched. The SWOF condition itself is considered to be by far the most hazardous attribute of a SWOF building, and retrofit of the SWOF wall line can be expected to substantially reduce the likelihood of excessive drift or collapse. Indeed, if the SWOF condition is the building's only serious structural deficiency, proper retrofit of the SWOF condition might achieve the benefits of a full life safety retrofit. [Other examples of risk-reduction retrofit schemes include the strengthening and bolting of unbraced cripple walls in a wood-frame house without consideration of the upper stories, and the bracing of parapets in unreinforced masonry (URM) buildings without consideration of the remaining URM walls.]

As with any building or retrofit code, these provisions are formulated for certain typical conditions, and the intended performance is expected to be achieved by the great majority of the buildings to which the provisions apply. The performance of any specific building, however, might be better or worse than that intended by the provisions.

Section A401.1 refers to past earthquakes. Multi-story SWOF buildings have shown unacceptable performance, including collapse and consequent loss of life, in the 1971 San Fernando, 1978 Santa Barbara, 1989 Loma Prieta and 1994 Northridge earthquakes. For more information on their past performance, see Harris et al. (1990), Hamburger (1994), Mendes (1995) and Holmes and Somers (1996). For more information on analysis and retrofit approaches, see Vukazich (1998), LADBS (1999) and Rutherford and Chekene (2000).

A401.2 Scope. The provisions of this chapter shall apply to all existing wood-frame buildings, or portions thereof, that are used as hotels, lodging houses, congregate residences or apartment houses where:

1. The ground floor portion of the wood-frame structure contains parking or other similar open floor space that causes soft, weak or open-front wall lines as defined in this chapter, and there exists one or more levels above, or

2. The walls of any story or basement of wood construction are laterally braced with nonconforming structural materials as defined in this chapter, a soft or weak wall line exists as defined in this chapter, and there exist two or more levels above.

This chapter is applicable to Seismic Hazard Zones where S_{D1} is 0.3g or higher, or in Seismic Zones 3 and 4 of the UBC.

❖ Section A401.2 describes the construction, occupancy, deficiencies and seismic hazard levels for which this chapter is intended. In general, the provisions in this chapter were conceived to address deficiencies in a building type commonly known as a "tuckunder," so called because a groundfloor parking area is tucked under the upper stories. In California, many tuck-unders were built in the 1960s and 1970s. Structural characteristics of a typical wood-frame tuckunder of this era include:

• Perimeter walls sheathed on the exterior face with stucco (sometimes over plywood) and on the interior face with gypsum wallboard (drywall). In some cases, let-in braces were used instead of plywood, especially at upper stories.

• Interior partitions sheathed with drywall on both sides.

• No hold-down hardware.

• Floors of 2x joists with plywood sheathing, sometimes with lightweight concrete topping.

• Roofs of 2x joists with plywood sheathing.

• At the groundfloor parking or open area, steel pipe columns or wood posts supporting a glulam or large dimensional lumber (or sometimes a steel wide flange) header.

The provision refers to all existing wood-frame buildings. Design practice in the Los Angeles area changed after the 1994 Northridge earthquake, but modified building codes did not directly address the fundamental performance issues of SWOF buildings. Recent research has revealed potential deficiencies even in designs using the post-Northridge codes and may lead to more restrictive code requirements.

The provision refers to residential occupancies only. SWOF buildings can also be used for commercial or mixed occupancy. The words "or portions thereof" are intended to cover mixed-use buildings. For example, an apartment building in which the first story with the SWOF condition is occupied entirely by retail is still within the scope of this chapter. At the option of the owner, this appendix chapter provisions may also be used for SWOF buildings with only commercial occupancy.

The provision refers to SWOF wall lines "as defined in this chapter." The definitions, given in Section A402, differ somewhat from the similar definitions in the 1997 *Uniform Building Code* (UBC) and 2003 IBC.

Condition 1 is intended to capture the typical tuckunder building described above. One-story buildings are exempt because the principal risk from SWOF buildings is due to collapse of the first story under the weight of upper stories. Also exempt are nonwood-frame conditions, such as buildings in which the walls around the open parking area are concrete or URM for the full story height.

Condition 2 is intended to capture the most deficient multistory buildings without open-front wall lines. These would include, for example, a three-story building with large open areas on the groundfloor, only stucco-braced perimeter shear walls and few interior partitions to stiffen the first story (Section A402 defines "Nonconforming structural materials"). Two-story houses over stucco-braced cripple walls would not be covered by this condition, however, because single-family houses are outside the scope of the chapter (such a house might be covered by Chapter A3).

Condition 2 refers to wood basement walls. Since wood framing is almost never used for basement or retaining walls, this reference may be understood to mean the above-grade portion of a story that is only partly below grade.

Though not explicitly stated, Condition 2 is intended to apply only when the nonconforming material is part of the soft or weak wall line. Nonconforming materials are defined in Section A402.

Though not explicitly stated, a multistory building might be exempted if the portion of the building with the SWOF condition is distinct from the rest of the building and does not fall into either of the two listed categories. That is, "levels above" may be reasonably understood to mean "stories above and influenced by the SWOF condition." For example, a three-story building with a one-story parking wing might be exempt if the SWOF condition occurs only in the one-story portion of the building. Such an interpretation requires engineering judgement and might be subject to the approval of the building official.

The provision defines the threshold ground motion hazard in two ways. SD$_1$ is a seismicity parameter used by the IBC and other model codes based on ASCE 7. The S_{D1} criterion given would cover all buildings in Seismic Design Category D or E, as defined in those codes (Seismic Design Category F is for special occupancies outside the scope of this appendix chapter). Seismic zones are used to represent seismicity levels in the UBC. The intention of the provision is that only the IBC needs to be checked; it is not the intention that users must check more than one document.

While the provision is written to exempt buildings in low and moderate seismic zones, this appendix chapter may also be used for similar buildings subject to lower hazards.

SECTION A402
DEFINITIONS

Notwithstanding the applicable definitions, symbols and notations in the Building Code, the following definitions shall apply for the purposes of this chapter:

❖ In general, "the building code" means the codes, standards, regulations and interpretations in effect in a specific jurisdiction. States and local jurisdictions typically adopt, and sometimes modify, a model code. This chapter refers in some places to the 1997 UBC and 2000 IBC, which are model codes. References to "this code" or "current code" should be understood as references to "the building code."

APARTMENT HOUSE. Any building or portion thereof that contains three or more dwelling units. For the purposes of this chapter, "apartment house" includes residential condominiums.

ASPECT RATIO. The span-width ratio for horizontal diaphragms and the height-length ratio for vertical diaphragms.

CONGREGATE RESIDENCE. A congregate residence is any building or portion thereof for occupancy by other than a family that contains facilities for living, sleeping and sanitation as required by this code, and that may include facilities for eating and cooking. A congregate residence may be a shelter, convent, monastery, dormitory, fraternity or sorority house, but does not include jails, hospitals, nursing homes, hotels or lodging houses.

CRIPPLE WALL. A wood-frame stud wall extending from the top of the foundation wall to the underside of the lowest floor framing.

DWELLING UNIT. Any building or portion thereof for not more than one family that contains living facilities, including provisions for sleeping, eating, cooking and sanitation as required by this code, or congregate residence for 10 or fewer persons.

EXPANSION ANCHOR. An approved mechanical fastener placed in hardened concrete that is designed to expand in a self-drilled or pre-drilled hole of a specified size and engage the sides of the hole in one or more locations to develop shear and/or tension resistance to applied loads without grout, adhesive or drypack.

❖ "Approved" means approved by the code official. Expansion anchors used to resist earthquake effects should also have ICC Evaluation Service or equivalent evaluation for compliance with the *International Codes*®. Though the definition says "hardened concrete," expansion anchors may also be used, where approved, in grouted reinforced masonry.

An expansion anchor might also include an appropriate washer and nut as required by manufacturers' instructions. In addition, plate washers are required for some conditions (in both new construction and retrofit) to help prevent splitting of sill plates. Sill plate damage is best prevented by proper design, location and instal-

lation of hold-downs. Still, plate washers are recommended near ends of shear walls.

Installation of expansion anchors can crack weak or deteriorated concrete. Chemical, undercut or threaded screw-type anchors might be useful in these conditions. A chemical anchor may be used to resist shear and/or tension loads; it uses a structural adhesive (e.g., epoxy) to secure the metal fastener to the sides of a predrilled hole in hardened concrete or masonry. An undercut anchor, without grout or adhesive, may also be used to resist shear and/or tension; when tightened, it engages the sides and undercut surfaces of a specially predrilled hole in hardened concrete or masonry. Screw-type anchors self-thread a predrilled hole in hardened concrete and resist forces through the mechanical interlock of the anchor threads with the concrete.

GROUND FLOOR. Any floor within the wood-frame portion of a building whose elevation is immediately accessible from an adjacent grade by vehicles or pedestrians. The ground floor portion of the structure does not include any level that is completely below adjacent grades.

GUESTROOM. Any room or rooms used or intended to be used by a guest for sleeping purposes. Every 100 square feet (9.3 m²) of superficial floor area in a congregate residence shall be considered a guestroom.

HOTEL. Any building containing six or more guestrooms intended or designed to be used, rented, hired out to be occupied, or that are occupied, for sleeping purposes by guests.

LEVEL. A story, basement or underfloor space of a building with cripple walls exceeding 4 feet (1219 mm) in height.

LIFE SAFETY PERFORMANCE LEVEL. The building performance level that includes significant damage to both structural and nonstructural components during a design earthquake, though at least some margin against either partial or total structural collapse remains. Injuries may occur, but the level of risk for life-threatening injury and entrapment is low.

❖ Aside from its definition, this term is not used in the text of this chapter. As discussed in the commentary to Section A401.1, the performance objective of this chapter is risk reduction, not life safety. A risk-reduction design addresses only the most egregious deficiencies. It is not as comprehensive as a life safety design, but it might achieve the same result if the SWOF condition is the only significant deficiency in the building.

LODGING HOUSE. Any building or portion thereof containing at least one but not more than five guest rooms where rent is paid in money, goods, labor or otherwise.

MOTEL. Motel shall mean a hotel as defined in this chapter.

MULTIUNIT RESIDENTIAL BUILDINGS. Hotels, lodging houses, congregate residences and apartment houses.

NONCONFORMING STRUCTURAL MATERIALS. Wall bracing materials other than wood structural panels or diagonal sheathing.

❖ This term is used only to define the scope of this chapter in Section A401.2. By this definition, materials currently permitted for new construction by model codes such as the 1997 UBC and 2003 IBC are designated as nonconforming.

The Los Angeles and Fremont provisions for SWOF buildings define "Nonconforming structural materials" as any material that is no longer permitted for new construction or when its design value (allowable shear or aspect ratio) has been reduced since construction. Those standards are, therefore, more restrictive than the provisions given in this chapter.

Here, for purposes of identifying SWOF conditions in Section A401.2, wood structural panels and diagonal sheathing are given more credit than other bracing methods such as let-in bracing, stucco (Portland cement plaster) or straight sheathing. Though deemed "conforming," it is possible that plywood or diagonal sheathing, even in buildings from the 1970s and 1980s, will not meet current (i.e., post-Northridge) requirements for new construction (for example, in terms of material specifications, hold-downs or nailing).

OPEN-FRONT WALL LINE. An exterior wall line, without vertical elements of the lateral-force-resisting system, that requires tributary seismic forces to be resisted by diaphragm rotation or excessive cantilever beyond parallel lines of shear walls. Diaphragms that cantilever more than 25 percent of the distance between lines of lateral-force-resisting elements from which the diaphragm cantilevers shall be considered excessive. Exterior exit balconies of 6 feet (1829 mm) or less in width shall not be considered excessive cantilevers.

❖ "Without vertical elements" may be understood to mean wall lines without vertical elements sufficient to resist tributary loads in the absence of diaphragm rotation. If there are insufficient elements of the seismic-force-resisting system along a given diaphragm edge, then the wall line along that edge must meet the cantilever criteria of this definition or be considered "open."

More important than the solid wall length, though harder to calculate, is the reliance on diaphragm rotation or diaphragm cantilever. The open-front condition is largely a proxy for torsional irregularity, defined variously in 1997 UBC Table 16-M, 2000 IBC Table 1616.5.1, 2003 IBC Table 1616.5.1.1 and ASCE 7 Table 9.5.2.3.2. Torsional irregularity is often considered to be a concern only in rigid diaphragm structures. However, The Consortium of Universities for Research in Earthquake Engineering (CUREE) research has shown that torsion, or diaphragm rotation, contributes significantly to excessive drift along the open wall line, even in buildings with wood diaphragms. In lieu of difficult drift calculations for a semirigid diaphragm that would be needed to determine torsional irregularity, the provisions use the more prescriptive definition of "Open front wall line."

Figure A4-1 illustrates the 25-percent calculation, which compares the cantilever length, c, to the backspan length, b. If c/b exceeds 0.25, the cantilever is considered excessive, and Line A is considered an open-front wall line. The 25-percent value is derived from judgement for consistency with previous code provisions, not from testing, analysis or performance statistics. See Section A403.10 and its commentary for additional discussion of cantilevered diaphragms.

Exit balconies are excluded because, in typical configurations, the outside edge of the balcony does not carry substantial gravity loads from the roof or stories above.

RETROFIT. An improvement of the lateral-force-resisting system by alteration of existing structural elements or addition of new structural elements.

SOFT WALL LINE. A wall line whose lateral stiffness is less than that required by story drift limitations or deformation compatibility requirements of this chapter. In lieu of analysis, a soft wall line may be defined as a wall line in a story where the story stiffness is less than 70 percent of the story above for the direction under consideration.

❖ While the definition is for an individual wall line, the default calculation is for an entire story. According to the definition, if the story is soft in a given direction, then each wall line in that direction is considered a soft wall line. It is probably more appropriate, however, to consider individual wall lines, and to compare the stiffness of a given wall line with the corresponding wall line in the story above. The 70-percent value is consistent with similar definitions in the 1997 UBC, the 2003 IBC and ASCE 7.

For purposes of this definition, story stiffness should not include contributions from walls or partitions of "Nonconforming structural materials." Walls with hold-downs will be far stiffer than walls subject to rocking or uplift due to a lack of hold-downs. If some wall lengths are provided with hold-downs while others are not, then only those with hold-downs and those with enough gravity load to resist uplift under design level seismic forces should be counted toward the story stiffness.

Unless conservative assumptions regarding materials and details are made, stiffness calculations for partitions and other elements not shown on plans will likely require destructive investigation per Section A406.3. This can be disruptive and expensive. The relative stiffness of existing plywood can be estimated from procedures given in references such as FEMA 356.

For purposes of comparing the stiffness of adjacent stories, any reasonable material and fixity assumptions are acceptable, as long as similar assumptions are made for the two stories being compared and as long as the calculation is not given credit for undue precision. For simplicity, elastic (i.e., uncracked, nondegraded) stiffnesses may be used, and partitions and shear walls may be assumed fixed at the ground and floor levels with respect to in-plane rotation. Based on 1997 UBC Table 16-L and the exception to Section 1629.5.3, Item 2, only the lateral stiffness must be considered; that is, story stiffness may be calculated ignoring torsional effects.

STORY STRENGTH. The total strength of all seismic-resisting elements sharing the same story shear in the direction under consideration.

❖ It is preferable to calculate element capacity based on strength values. For purposes of comparing the strengths of adjacent stories, however, allowable stresses may be used.

WALL LINE. Any length of wall along a principal axis of the building used to provide resistance to lateral loads. Parallel wall lines separated by less than 4 feet (1219 mm) shall be considered one wall line for the distribution of loads.

❖ "Separated by less than 4 feet (1219 mm)" refers to out-of-plane separation; for example, where part of a wall line is recessed for architectural purposes. A given wall line might consist of several individual noncontiguous shear wall panels separated in-plane by large door or window openings or by other architectural elements.

WEAK WALL LINE. A wall line in a story where the story strength is less than 80 percent of the story above in the direction under consideration.

❖ While the definition is for an individual wall line, the calculation is for an entire story. According to the definition, if the story is weak in a given direction, then each and every wall line in that direction is considered a weak wall line. It is probably more appropriate, however, to consider individual wall lines and to compare the stiffness of a given wall line with the corresponding wall line in the story above. The 80-percent value is consistent with similar definitions in the 1997 UBC, the 2003 IBC and ASCE 7.

For the purpose of this definition, story strength is as defined above. Story strength should not include contributions from walls or partitions of nonconforming structural materials. Any reasonable material and fixity assumptions are acceptable, as long as similar assumptions are made for the two stories being compared. For simplicity, partitions and shear walls may be assumed fixed at the ground and floor levels with respect to in-plane rotation.

These assumptions might be unrealistic, especially if the shear strength is limited by rocking or uplift of piers that lack hold-downs. They are generally considered sufficient, however, for the simple purpose of determining the relative strength of a potentially critical story with respect to the story above.

Another potential shortcoming of this definition is that it does not account for different force levels in adjacent stories. In a short building, the first-story shear is substantially higher than the second-story shear. Thus, even if the two stories have the same strength, the first story will be critical. Therefore, where the lower story is just marginally strong enough to avoid classification as "weak," a comparison of demand/capacity ratios in adjacent stories is recommended. If the lower story is clearly critical and its seismic-force-resisting system does not provide ample ductility, the engineer should consider classifying the story as weak regardless of the definition.

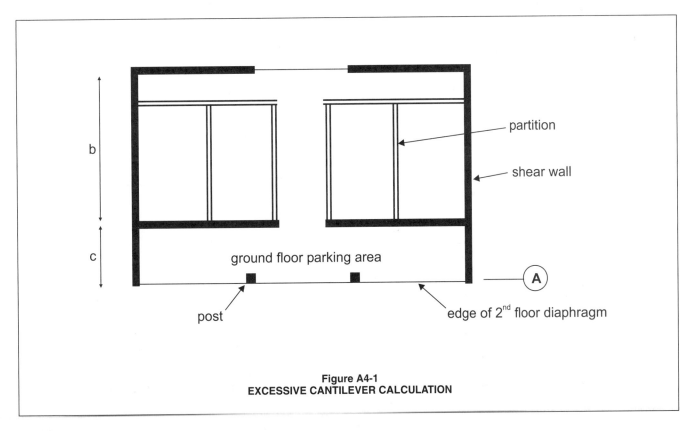

Figure A4-1
EXCESSIVE CANTILEVER CALCULATION

SECTION A403
ANALYSIS AND DESIGN

A403.1 General. Buildings within the scope of this chapter shall be analyzed, designed and constructed in conformance with the 1997 *Uniform Building Code*™ except as modified in this chapter. Prior to any analysis, an initial screening review of the buildings shall be performed as noted in Section A403.1.1. All items found to be noncompliant shall be addressed in this analysis.

No alteration of the existing lateral-force-resisting or vertical-load-carrying system shall reduce the strength or stiffness of the existing structure. When any portion of a building within the scope of this chapter is constructed on or into a slope steeper than 1 unit vertical in 3 units horizontal, the lateral-force-resisting system at and below the base level diaphragm shall be analyzed for the effects of concentrated lateral forces at the base caused by this hillside condition.

Exceptions:

1. Buildings in which all items on the applicable checklist—Tables A4-A through A4-D—are marked compliant.

2. Prescriptive measures provided in Section A405 may be used in two-story buildings of no geometrical irregularity when the roof covering of the structure is of material weighing 5 pounds per square foot (240 N/m²) or less; when the aspect ratio of the floor diaphragm meets the current code requirements; and only when deemed appropriate by the building official.

❖ The first sentence of Section A403.1 does not mean that whole buildings must be brought to conformance with the 1997 UBC or the building code. Indeed, this chapter and the building code have different performance objectives, and certain provisions in this chapter (Section A403.2, for example) specifically waive code requirements. Rather, the first sentence intends only that any modifications required by these provisions, principally new structural elements, must be made in accordance with building code requirements for new construction.

Reference to the 1997 UBC should instead be to the building code. The 1997 UBC was the model code used as a reference during development of this chapter, but it is not necessarily the building code.

Despite use of the word "shall" in the second sentence, this provision means only to suggest the sequence of work. Analysis need not be preceded by any screening process. In fact, the screening process in Section A403.1.1 is not intended to be mandatory, and items found to be noncompliant (NC) by that process do not necessarily need to be modified or remedied so that they become compliant (C). "Shall be addressed in this analysis" means that noncompliant conditions should be considered by the engineer when performing any analysis or design based on this chapter. See the commentary to Section A403.1.1 for additional discussion of the screening process.

The exceptions are to the general requirements, not exceptions to the hillside requirements in the immediately preceding sentence to the exceptions:

Exception 1: "Checklist" Tables A4-A through A4-D include items that cannot be completed without some structural analysis. Thus, Exception 1 does not really reduce the scope of work otherwise required by Section A403. In principle, however, Exception 1 offers a prescriptive alternative method: if small modifications to the building can eliminate noncompliant conditions (NC), then a full analysis and retrofit need not be performed. In other words, it is acceptable to modify the building so as to change noncompliant conditions to compliant (C) or not applicable (N/A) and thereby exempt the building from the balance of this chapter. Such an approach is advisable only when the noncompliant conditions are very few and remediable by local measures. If substantial demolition, strengthening or member replacement is required to eliminate one or more noncompliant conditions, a full analysis per Section A403 (or Section A405, if applicable) is recommended. See the commentary to Section A403.1.1 for additional discussion of Tables A4-A through A4-D.

Exception 2: Sections A405.1.2 and A405.1.3 further limit the use of the prescriptive provisions in Section A405 (see the flowchart in Figure A4-4). If the building qualifies for the prescriptive measures in Section A405, then the requirements of Section A403—including the weak story limits, drift limits, etc.—are waived. The requirements of Sections A406, A407, and A408 would still apply.

In Exception 2, "no geometrical irregularity" may be understood to mean no irregularity other than the soft and/or weak story vertical irregularities that the building is presumed to have. Plan and vertical structural irregularities are described in 1997 UBC Tables 16-L and 16-M, 2003 IBC Tables 1616.5.1.1 and 1616.5.1.2 or ASCE 7 Tables 9.5.2.3.2 and 9.5.2.3.3.

The roofing weight limit of 5 pounds per square feet (psf) (239 Pa) is intended to rule out unusually heavy structures to which the prescriptive provisions do not apply. The limit is not intended to rule out the fairly common condition of two asphalt roofings.

Limiting diaphragm aspect ratios required by Exception 2 can be found in Section 2315.1 and Table 23-II-G of the 1997 UBC or Table 2305.2.3 of the 2003 IBC. Per Table A4-D, unblocked wood structural panel diaphragms spanning less than 40 feet (12 192 mm) with aspect ratios up to 4:1 are deemed acceptable.

A403.1.1 Initial screening. Prior to any analysis, an initial screening review of the buildings shall be performed.

Each of the evaluation statements on this checklist shall be marked compliant (C), noncompliant (NC), or not applicable (N/A). Compliant statements identify issues that are acceptable according to the criteria of this chapter, while noncompliant statements identify issues that require further investigation. Certain statements may not apply to the buildings being evaluated. For noncompliant evaluation statements, the design professional may choose to conduct further investigation or comply with the prescriptive requirements of this chapter.

❖ Despite its wording, this provision means only to suggest the sequence of work. Analysis need not be preceded by any screening process. In fact, the screening process in Section A403.1.1 is not intended to be mandatory, and items found to be noncompliant (NC) by that process do not necessarily need to be modified or remedied so that they become compliant (C).

In the second paragraph, "this checklist" means Tables A4-A through A4-D, which were adapted from FEMA 310. Reference in the last sentence to "prescriptive requirements" means those in Section A405, which are allowed only in certain conditions that might or might not be related to the noncompliant evaluation statements. The words "may choose" do not necessarily mean that Section A405 may be used in lieu of Section A403 (see the flowchart in Figure A4-4).

The screening procedure, using the checklist comprising Tables A4-A through A4-D, has two purposes. First, as discussed in the commentary to Section A403.1, Exception 1, it can be used to identify buildings that are already nearly compliant and require only the correction of one or two small deficiencies. In these cases, the full analysis and retrofit design otherwise contemplated by this chapter can be avoided. Second, the checklist is recommended as a way for the designer to become familiar with the building and any likely weak links in its seismic-force-resisting system and load path without the comprehensive investigation and analysis that might be required for more than a risk reduction effort.

The presumption of the risk-reduction approach of this chapter is that the SWOF condition is by far the most egregious seismic deficiency in the building. Given that presumption, many of the lesser deficiencies listed in the checklist tables, which are geared toward life safety performance, need not be corrected to achieve the basic risk reduction. However, if certain conditions are found to be noncompliant, the designer might choose to account for them in the analytical model and detailing. For example, the risk-reduction analysis, focused on the SWOF condition, might not check the connections of walls through floors (see Table A4-B) or the adequacy of sill bolts (see Table A4-C). If those conditions are indeed noncompliant, the designer might find that they represent the next weakest links after the SWOF, and that it might be cost effective to include them in the project scope. Nevertheless, remedies for any conditions outside the scope of Section A403.2 (or Section A405, where applicable) are optional.

When Tables A4-A through A4-D are used to check Section A403.1, Exception 1, the following apply:

• If an item can not be marked C due to unknown conditions, it is to be marked NC.

• If an item is marked NC due to unknown conditions or conservative assumptions, additional inspections or testing may be used to reveal compliant conditions.

• Details and materials used to remedy any NC condition must be in conformance with the IBC.

A403.2 Scope of analysis. This chapter requires the alteration, repair, replacement or addition of structural elements and their connections to meet the strength and stiffness requirements herein. The lateral-load-path analysis shall include the resisting elements and connections from the wood diaphragm above any soft, weak or open-front wall lines to the foundation soil interface or the upper level of a Type I structure below. The top story of any building need not be analyzed. The lateral-load-path analysis for added structural elements shall also include evaluation of the allowable soil-bearing and lateral pressures in accordance with UBC Section 1805.

Exception: When an open-front, weak or soft wall line exists because of parking at the ground level of a two-level building, and the parking area is less than 20 percent of the ground floor level, then only the wall lines in the open, weak or soft directions of the enclosed parking area need comply with the provisions of this chapter.

❖ This paragraph distinguishes this chapter's risk-reduction objective from a full life safety objective. Using this chapter, the only structural elements that need consideration are those between the diaphragm above a SWOF story and the foundation soil. Stories above that level need not be checked. Further, if the wood-frame SWOF structure sits on a separate steel, concrete or masonry podium structure (a below-grade parking garage, for example), that Type I podium structure need not be checked. (Type I refers to certain fire-resistive buildings described in 1997 UBC, Section 602. The numeral should be a Roman I. Other codes may use different designations or have different requirements for Type I construction.)

Top stories are exempt for the same reason that one-story buildings were exempted in Section A401.2. These provisions focus on the specific risks posed by the collapse of SWOF stories with substantial mass above. When the SWOF story collapses, it is the crushing action of the falling upper stories that poses the greatest risk. Racking or excessive drift in a one-story SWOF building or in the top story of a SWOF building are of lesser concern.

Soil stresses need only be checked where they are affected by added structural elements such as walls, diagonal braces, frame columns, posts or other structural elements added as part of the retrofit. Where allowable soil stresses are exceeded due to these added elements, foundation elements will need to be supplemented as well.

"UBC Section 1805" refers to the 1997 UBC provision that give default allowable soil bearing and lateral pressures. The reference may be understood to refer to the corresponding requirements of the building code. The reference to default allowable stresses is not intended to prohibit the use of higher design values substantiated by testing or geotechnical investigation.

Section 1805 of the UBC gives allowable pressures suitable for use with allowable stress design procedures. The design forces prescribed in Section A403 are at a strength level, however, so appropriate adjustments may be made by increasing the allowable soil pressures for short-term loading and by reducing the prescribed design forces to allowable stress levels.

UBC Table 18-I-A, referenced by Section 1805 of the UBC, allows a 33-percent increase in allowable pressures for load combinations that include earthquake effects. By contrast, Section A108.6 in Chapter A1 allows an effective increase of 50 percent. Because each chapter of the GSREB was developed independently, it is recommended that only the 33-percent increase should be used in this chapter. Higher allowable pressures may be used when substantiated by testing or geotechnical investigation.

The exception further reduces the scope of Chapter 4 for SWOF deficiencies that are judged to be remediable by local measures. Again, the risk reduction approach acknowledges that when the most critical deficiencies are easily identified and remedied, comprehensive analysis of the entire structure should not be necessary.

The exception is intended to cover a large subclass of SWOF buildings with minimal tuckunder parking. While the provision specifically mentions "parking," it could be reasonably applied to similar unoccupied spaces used for storage or other purposes. The 20-percent value is derived from judgement, not from testing, analysis or performance statistics. Use of the term "enclosed parking area" is intended to indicate a defined area bounded in plan by walls, partitions or the edge of the building; it does not mean that the parking (or other nonoccupiable) area must be physically enclosed.

The 20-percent value is to be based on floor area. All nonoccupiable area (as opposed to just designated parking area) between the SWOF wall line and the nearest parallel wall line should be counted as the parking area. Figure A4-2 shows the type of common SWOF building for which this exception was intended. Other configurations are also eligible for the exception. For complex floor layouts, designation and calculation of the parking area might require some judgement.

With reference to Figure A4-2, Line A is assumed to be a SWOF wall line. Line B is the nearest parallel wall line. if $A_p/(L_x L_y)$ is less than 0.20, then only wall Lines A

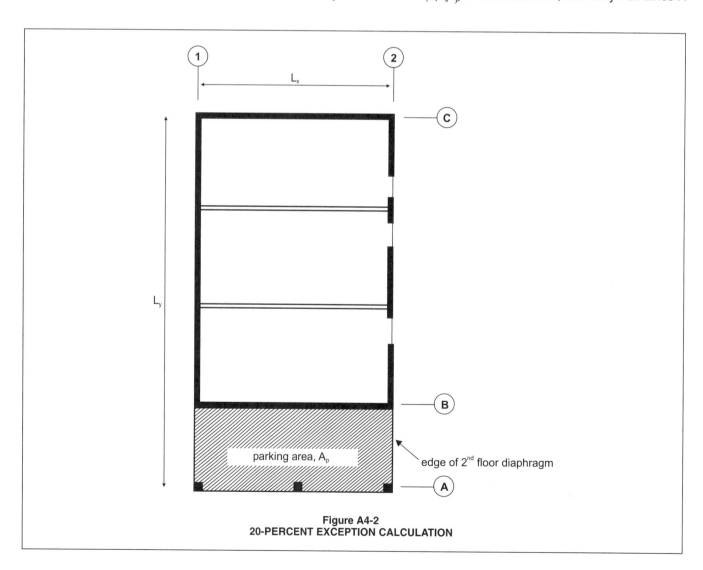

Figure A4-2
20-PERCENT EXCEPTION CALCULATION

and B need to be checked and potentially strengthened or stiffened.

The intention of this exception is only to exempt wall Lines C, 1 and 2 from drift and strength requirements. These wall lines must still be included in the structural model used to derive forces and deflections for Lines A and B. In considering Lines A and B, the load path from the second floor diaphragm to the soil-foundation interface must still be checked per Section A403.2.

This exception does not apply if the prescriptive provisions of Section A405 are used.

A403.3 Design base shear. The design base shear in a given direction shall be 75 percent of the value determined by Formulas (30-4) through (30-7) in UBC Section 1630.2.

❖ The 75-percent value represents a common allowance for existing buildings with precedent in FEMA 178. FEMA 178 specified a design base shear of 85 percent (for short period buildings) or 67 percent (for longer period buildings) of that required for new construction. At the time, the reduction was said to be related to the difference between a "mean" earthquake and a "probable" earthquake. It is not clear whether that relationship still holds for the seismicity estimates and design parameters in current codes. Nevertheless, a reduction of about 25 percent remains traditional and is consistent with Section 407.1.1.3 of the code.

"Formulas (30-4) through (30-7) in UBC Section 1630.2" refers to the 1997 UBC provision that specifies the minimum total base shear for design using the static (equivalent lateral force) procedure. The reference may be understood to refer to the corresponding requirements of the IBC. These base shear formulas lead to design forces suitable for strength (as opposed to allowable stress) design procedures.

In order to use UBC Formulas 30-4 through 30-7, one must know the seismic zone and the soil profile type. The UBC allows Soil Profile Type D to be used by default, subject to approval by the building official. Soft clays and poorly compacted deep fills are likely to need site-specific investigation (see Stewart et al., 1995, regarding the historic performance of fills). In Zone 4, one must also know the seismic source type and the distance from the site to the seismic source. In order to use the corresponding formulas in the 2003 IBC or ASCE 7, one must know the soil type (called the site class) and have access to large-scale or computerized maps of seismicity parameters. Due to small differences in the formulas for building period and base shear, the design base shear may vary depending on which building code or model code is used. These differences, and the consequent variations in base shear, are considered negligible for the purposes of this chapter. Differences in the various codes' prescribed R-values, however, might be more significant. Still, the design criteria from any recent model code should be acceptable.

The design base shear also depends on the structure's seismic-force-resisting system and its R-factor. The R-factor for existing systems should be chosen with

due consideration of the age and quality of construction, as well as the condition of the structure. R-factors assigned by the building code for new construction presume certain standards for materials, connections, detailing and quality control that might not be met by the existing structure. In those cases, a modification to the tabulated R-factor might be warranted. Section A406 gives requirements for the condition of existing materials. As for obsolete detailing (inadequate fastener spacing or member size, for example), engineering judgement, supported by appropriate research, analysis and material testing will often be needed.

With respect to existing systems comprised of light-framed walls braced by stucco or let-ins (that is, without wood structural panels), different codes might assign significantly different R-factors. The 1997 UBC would allow an R-factor of 4.5 (Table 16-N, System Type 1.1.b of the UBC), but ASCE 7 and codes based on it would allow only an R-factor of 2 (see Table 9.5.2.2 of the UBC). Provisions of the building code should apply.

A retrofitted SWOF building will frequently involve combinations of different systems. Refer to Section A403.4 and its commentary for discussion of R-factors for combined systems common to retrofitted SWOF buildings.

The building code may place limits on static procedures such as those prescribed by this chapter. In general, these limits should be heeded even for analysis of existing buildings. For example, Section 1629.8.3 of the 1997 UBC allows the static procedure only for SWOF buildings up to five stories or 65 feet (19 812 mm) tall. It also prohibits use of the static procedure for SWOF wood buildings on top of stiff podium structures because the SWOF portion is irregular. Thus, the static procedure might not be appropriate for evaluation of the existing wood building in these cases.

The simplified design procedure (1997 UBC Section 1630.2.3; ASCE 7, Section 9.5.4) is considered an acceptable conservative alternative to UBC Formulas 30-4 through 30-7. The 75-percent multiplier may be applied to the simplified base shear formula. The building code may place limits on simplified procedures, however, and these limits should be heeded even for analysis of existing buildings. For example, Section 1629.8.2 of the UBC allows the simplified procedure only for light-frame buildings up to three stories and other structure types up to two stories. Thus, the simplified procedure might not be appropriate for analysis of a three-story building retrofitted with a steel frame or masonry shear wall in the first story. In addition, while Section 1630.2.3.3 of the UBC allows distribution of design forces by story weight when the simplified base shear is used, Section A403.4 limits this to two-story buildings.

A403.4 Vertical distribution of forces. The total seismic force shall be distributed over the height of the structure based on Formula (30-15) in UBC Section 1630.5. Distribution of force by story weight shall be permitted for two-story buildings. The value of R used in the design of any story shall be less than or equal to the value of R used in the given direction for the story above.

❖ "Formula (30-15) in UBC Section 1630.5" refers to the 1997 UBC provision that prescribes the distribution of design forces over the structure height when the static (equivalent lateral force) procedure is used. The reference may be understood to refer to the corresponding requirements of the IBC. The "total seismic force" is equivalent to the "design base shear" calculated according to Section A403.3.

"Distribution of force by story weight" refers to an alternative method such as that given by 1997 UBC Formula 30-12 for use with the simplified design procedure. As discussed in the commentary to Section A403.3, simplified design base shear formulas are considered acceptable alternatives. Model codes sometimes allow the use of simplified design for buildings taller than two stories. According to Section A403.4, however, the force distribution by story weight may only be used for two-story buildings.

A retrofitted SWOF building will frequently involve combinations of different seismic-force-resisting systems. The existing wood-frame SWOF building will likely be classified as a bearing wall system of light-framed walls with either wood structural panels or stucco bracing. Structural elements commonly added for retrofit might be steel moment-resisting frames or building frame systems with wood structural panels; concrete or masonry shear walls; or steel diagonal braces. In the retrofitted building, systems might be combined in various ways: vertically (that is, with different systems in different stories); within the same story along different wall lines or in different directions; or within the same story along the same wall line. The R-factor for a given combination should be determined based on the provisions of the IBC.

Section A403.4 provides a rule for use when systems are combined vertically. This rule is essentially the same as that provided in Section 1630.4.2 of the 1997 UBC. For new construction, the purpose of this rule is to avoid concentrated inelasticity in lower stories by delaying yield in the lower story until the upper story is also near its elastic limit. For SWOF retrofits, however, it might make more sense to concentrate inelasticity in the new, properly-detailed elements so that ductility can be provided and controlled by design.

Consider, for example, a typical three-story tuckunder building with an open-front wall line along one side in the first story and stucco-braced walls in the upper stories. According to Section A403.2 only the second floor diaphragm and the first-story wall lines need to be considered for risk reduction. If a steel frame is to be installed in the open wall line, it seems entirely reasonable to design it as the building's primary source of ductility, with an R-factor of 4.5 (or even higher if special detailing is provided). After all, the brittle upper story stucco walls are not going to be touched by the retrofit, and they can not be made worse by the presence of a ductile frame below. It may seem overly

conservative to select a ductile system for retrofit and then design it for a low R-factor, in this case only 2 (according to ASCE 7). Indeed, one could argue that use of a low R-factor could result in a very strong first story that could force failure up into the second story.

Nevertheless, the provision is plain as to its intent, and until further studies are performed, the provision should be followed as written. That is, the R-factor for the design of new structural elements should be limited by the existing structure in stories above. Further, the use of a low R-factor based on upper stories does not excuse the new elements from proper detailing. That is, if a steel frame with an R-factor of 4.5 is installed, then it must be detailed as required for an R-factor of 4.5 even if its members are sized for forces based on an R-factor of 2.

A403.5 Weak story limitation. The structure shall not exceed 30 feet (9144 mm) in height or two levels if the lower level strength is less than 65 percent of the story above. Existing walls shall be strengthened as required to comply with this provision unless the weak level can resist a total lateral seismic force of Ω_0 times the design force prescribed in Section A403.4.

The story strength for each level of all other structures shall be a minimum of 80 percent of the story above.

❖ The reference to Section A403.4 may be understood to mean the force prescribed in Section A403.3 and distributed according to Section A403.4.

"Existing walls shall be strengthened" is not meant to prohibit other means of mitigating the weak story. For example, existing walls may be replaced, or new walls or other structural elements may be added.

"[T]he weak level can resist" could also mean "the weak story strength is not less than."

This provision is based on a requirement for new construction (1997 UBC Section 1629.9.1; similarly, 2000 IBC Section 1620.1.3), which allows weak stories in some buildings, subject to certain limitations. When applied to existing buildings, this provision sets the minimum strength requirements for retrofit when a weak story exists. It says that for a building up to the two-story and 30-foot (9144 mm) limits, the weak story must be strengthened to at least 65 percent of the strength of the story above—but the weak story is still allowed to be weak by the 80-percent definition, even in its retrofitted condition. For other buildings, the strength must be brought to 80 percent of the story above; that is, the weak story must be completely eliminated. In essence, this provision allows a weak story to remain in small buildings, as long as it is not excessively weak.

The overstrength exception applies to any weak story in a building of any height. That is, a weak story need only be strengthened to the lesser of: (1) Ω_0 times the story shear prescribed by Sections A403.3 and A403.4, or (2) some percentage of the strength of the story above (either 65 percent or 80 percent), depending on the height of the building.

Consider a two-story building in which the weak first story has 70 percent of the strength of the second story. Section A403.5 does not necessarily exempt this building from consideration. While its first story strength is adequate, the building still qualifies for consideration by Section A401.2 and must therefore still meet the stiffness, load path and diaphragm requirements of Section A403 (or the prescriptive provisions of Section A405, if they apply).

Whether a weak story should be allowed in even a two-story retrofitted building is a fair question. The 1997 UBC, as noted above, allows weak stories in some buildings. As a rule, retrofit provisions should not be more restrictive than the code for new construction. On the other hand, in new buildings the seismic load path is detailed to full code-level forces, the materials are controlled and the construction is inspected. Since this is not the case for retrofit, one could argue that the GSREB provision should seek to eliminate all weak stories. Indeed, Section 1620.4.1 of the 2000 IBC, Section 1620.5.1 of the 2003 IBC and Section 9.5.2.6.5.1of the ASCE 7 prohibit all weak stories for new buildings in Seismic Design Category E, which covers much of coastal California (wherever S_1 is 0.75g or greater). Until further studies are performed, the provision is accepted on the basis that requirements somewhat less stringent than the IBC may be acceptable for risk reduction in two-story buildings.

While Section A403.5 sets the minimum strength requirements for retrofit of existing weak stories, minimum stiffness requirements are given by Section A403.6. There is no similar provision that allows a soft story to remain in a two-story building.

A403.6 Story drift limitation. The calculated story drift for each retrofitted level shall not exceed the allowable deformation compatible with all vertical-load-resisting elements and 0.025 times the story height. The calculated story drift shall not be reduced by the effects of horizontal diaphragm stiffness but shall be increased when these effects produce rotation. Drift calculations shall be in accordance with UBC Section 1630.9 and 1630.10.

The effects of rotation and soil stiffness shall be included in the calculated story drift when lateral loads are resisted by vertical elements whose required depth of embedment is determined by pole formulas, such as Formulas (6-1) and (6-2)in UBC Section 1806.8.2. The range of this coefficient ofsubgrade reaction used in the deflection calculations shall be provided from an approved geotechnical engineering report or other approved methods.

❖ "UBC Section 1630.9 and 1630.10" refers to the 1997 UBC provisions for calculating interstory drift and drift limitations. The reference may be understood to refer to the corresponding requirements of the IBC. Section A403.11.2.1 makes additional requirements for drift calculation in wood shear walls.

Where the building code is based on the 1997 UBC, "the calculated story drift" means the difference between the maximum inelastic response displacement, Δ_M, at floor levels above and below the story in question. As defined in Section 1630.9.2 of the 1997 UBC, $\Delta_M = 0.7\,R\,\Delta_S$, where Δ_S is the drift calculated from a linear elastic analysis using the forces prescribed in Section A403.3. PΔ effects must be included in the analysis as required by Section A403.7 (by reference to Section 1630.1.3 of the 1997 UBC).

Where the building code is based on ASCE 7, "the calculated story drift" means the design story drift as defined in Section 9.5.5.7.1 of the ASCE, based on center of mass deflections, δ_{xe}, calculated from a linear elastic analysis using the forces prescribed in Section A403.3. PΔ effects must be included in the analysis as required by Section A403.7 (by reference to corresponding Section 9.5.5.7.2 of the ASCE 7).

Since an erratum dated March 2001, Section 1630.10.3 of the 1997 UBC allows drifts to be calculated without the base shear limits of Formula 30-6 and 30-7. Section 9.5.5.7.1 of ASCE 7 allows the first of its corresponding equations (Section 9.5.5.2.1-3) to be ignored, but not the second. In any case, this exclusion is not expected to come into play for typical SWOF buildings within the scope of this chapter.

"[E]ach retrofitted level" means each story subject to consideration according to Section A403.2. Drifts need not be checked at stories above the SWOF conditions. Once the analysis is performed, however, it is usually a trivial matter to check the drifts at upper stories. Indeed, the definition of a "Soft wall line" in Section A402 is based on calculated story drifts. According to that definition, the "70-percent rule" is acceptable as an alternative to analysis, so even if drift limits are exceeded, the story need not be classified as soft if it meets the 70-percent rule. In this case, engineering judgement should be used to determine whether flexible upper stories need to be stiffened for acceptable performance.

The limiting drift ratio of 0.025 corresponds to the limit for low- and mid-rise buildings in Section 1630.10.2 of the 1997 UBC. The 2003 IBC limit, based on Table 9.5.2.8 of ASCE 7, for most low-rise buildings is also 0.025, but only where partitions, ceilings and exterior wall systems can accommodate that deformation. Most existing SWOF buildings should not be assumed to meet that criterion, so the IBC would limit the drift ratio to 0.02. Studies by CUREE (2002) have suggested that allowable drifts should be lowered even further for open-front wood-frame buildings analyzed with the static (equivalent lateral force) procedure.

Also, ASCE 7 sets limits lower than 0.025 for masonry shear wall structures. If masonry shear walls are added as retrofit elements, these lower drift limits should be considered.

The provision does not define the "allowable deformation compatible with all vertical-load-resisting elements." Vertical-load-resisting elements in a typical wood-frame SWOF building include stud walls and partitions, sheathed or braced with plywood, stucco, plaster, gypboard or gypsum lath. If the materials are in good condition, a limit of 0.02 for these elements should be appropriate, based on Table 9.5.2.8 of ASCE 7.

Many existing SWOF buildings also have wood posts or steel pipe columns along the open side. For these, the limit of 0.025 is judged appropriate unless their condition is poor (rotted or corroded) or their connections are unable to accommodate that much deformation. SWOF buildings within the scope of this chapter would not be expected to have brittle load-bearing components (unreinforced masonry infill, for example) that would be critical in terms of deformation compatibility. If such components do exist, more stringent drift limits might be appropriate.

The provision regarding "horizontal diaphragm stiffness" requires that diaphragms be modeled as rigid relative to the walls so that the effects of actual and accidental torsion can be conservatively estimated. This is most important for evaluation of the existing SWOF wall, less so for analysis of a building retrofitted with stiff new elements along the SWOF wall line. The rigid diaphragm assumption is supported by recent testing that found the effects of a wood diaphragm on an open-front structure were sufficiently approximated by modeling the diaphragm as a rotating rigid body (CUREE, 2002). While a rigid diaphragm must be assumed for this provision, Section A403.10 also appears to require force distribution based on a flexible diaphragm assumption. This chapter thus appears to require design for the more conservative of the two possibilities.

The provision to add, but not subtract, the effects of diaphragm rotation is a conservative requirement similar to Section 9.5.5.7.1 of the ASCE 7 requirements for buildings with torsional irregularity. In those buildings, the design drift must be calculated at the most critical point along any diaphragm edge (Section 1630.9.1 of the 1997 UBC requires drift to be calculated at all "critical locations" in all structures). A test structure with obvious open-front deficiencies was tested by CUREE (2002) and performed poorly, as expected; drift calculations at the center of mass, however, would not have predicted a torsional irregularity by the letter of the code.

The "rotation" referred to in the second paragraph of this section is the rotation of an embedded pole bearing laterally against soil, either with or without restraint at the ground surface by a slab or rigid pavement. The provision is intended to address pipe columns or posts supporting gravity loads along the building's open side. Pole systems are especially flexible and vulnerable to loss of vertical capacity. This provision applies mostly to analysis of the existing structure; in the retrofitted structure, the lateral stiffness contribution of posts is generally negligible by comparison to the stiffness of new wall, frame or brace elements.

The coefficient of subgrade reaction refers to the stiffness of soil when pressed against by the embedded post. The provision refers to a range of values because an exact value is often difficult to predict with precision and because geotechnical engineers might, therefore, prefer to report a range of values.

A403.7 P Δ effects. The requirements of UBC Sections 1630.13 and 1633.2.4 shall apply except as modified herein. All structural framing elements and their connections not required by design to be part of the lateral-force-resisting system shall be designed and/or detailed to be adequate to maintain support of design dead plus live loads when subjected to the expected deformations caused by seismic forces. The stress analysis of cantilever columns shall use a buckling factor of 2.1 for the direction normal to the axis of the beam.

❖ Section 1633.2.4 of the 1997 UBC requires deformation compatibility checks for structural elements that are not part of the seismic-force-resisting system, such as posts, pipe columns, stair framing and wood-framed walls deformed out of plane. Section 1633.2.4 also requires these checks to include PΔ effects. Section 1630.1.3 gives more general requirements for consideration of PΔ effects. The provision may be understood to refer to the corresponding requirements of the IBC.

Section 9.5.5.7.2 of the ASCE 7 and Section C105.1.3 of the 1999 SEAOC Blue Book, offer a procedure for incorporating PΔ effects into the analysis. The 1999 Blue Book (Section C108.2.4 and Appendix E) also offers guidance for accommodating PΔ effects in various element types. Inelasticity is permitted when members and connections are checked for PΔ effects; the important thing is for the structure to maintain its resistance to gravity loads.

In existing SWOF buildings, the most critical elements are probably the posts along the open front, because they often have weak flexural connections and because drift is greatest along that wall line. The design check should rule out any buckling of the post and any failure of base plates or connection hardware that would compromise stability. For cantilever columns or moment frame columns, if PΔ effects are included in the analysis, then additional bending stress amplifiers in typical design equations (such as $1-f_a/F'_e$) need not be applied.

The last sentence of the provision refers to typical posts or pipe columns along the open side that are pinned at one end and, therefore, deform in single curvature: either pinned at the base and rigidly connected at the top or embedded at the base and pinned at the top. The "direction normal to the axis of the beam" generally means the direction normal to the open wall line. "Buckling factor" means the effective buckling length factor. If the column is considered pinned at both ends with respect to buckling in any direction, an effective buckling length factor of 1.0 is appropriate. Fixity assumptions depend on the nature and condition of the beam-column connection. For existing posts or pipe columns, the appropriate assumption is generally a matter of engineering judgement.

A403.8 Ties and continuity. All parts of the structure included in the scope of Section A403.2 shall be interconnected, and the connection shall be capable of resisting the seismic force created by the parts being connected. Any smaller portion of a building shall be tied to the remainder of the building with elements having a strength to resist $0.5 C_a I$ times the weight of the smaller portion. A positive connection for resisting a horizontal force acting parallel to the member shall be provided for each beam, girder or truss included in the lateral load path. This force shall not be less than $0.5 C_a I$ times the dead plus live load.

❖ This provision is based on Section 1633.2.5 of the 1997 UBC. Where the building code is based on ASCE 7 (which does not use the parameter C_a), the corresponding provision is in Section 9.5.2.6.1.1 of the ASCE.

In Section 1633.2.5 of the 1997 UBC, the 0.5 value in the last sentence of the provision was changed to 0.3 in an erratum dated January 2001. A similar reduction is appropriate for this appendix.

As noted in ASCE 7, the connection forces prescribed by this provision do not apply to the overall design of the seismic-force-resisting system. That is, they need not be added to the design forces prescribed by Sections A403.3 and A403.4 for analysis and design of the seismic-force-resisting system.

The minimum connection strengths required by this provision are prescriptive and are derived from judgement, not from testing, analysis or performance statistics.

A403.8.1 Cripple walls. Unbraced cripple walls found to be noncompliant in Table A4-C shall be analyzed and designed per Chapter 3. When a single top plate exists in the cripple wall, all end joints in the top plate shall be tied. Ties shall be connected to each end of the discontinuous top plate and shall be equal to one of the following:

1. Three-inch-by-6-inch (76 mm by 152 mm), 18-gage galvanized steel, nailed with six 8d common nails at each end.

2. One and one-fourth-inch-by-12-inch (32 mm by 305 mm), 18-gage galvanized steel, nailed with six 16d common nails at each end.

3. Two-inch-by-4-inch-by-12-inch (51 mm by 102 mm by 305 mm) wood blocking, nailed with six 16d common nails at each end.

❖ The first sentence should refer to Table A4-B, not Table A4-C. Since the use of Table A4-B is not strictly required (see commentary, Section A403.1.1), the first sentence of this provision should be understood to mean that cripple walls braced only by nonconforming materials should be evaluated and retrofitted as necessary.

The provision refers to Chapter A3, which offers prescriptive measures for wood-frame residential structures. However, Chapter A3 is intended principally for houses and explicitly does not apply to hotels and apartment houses with more than four units; these buildings require analysis. The reference to Chapter A3 might also be problematic if its prescriptive measures conflict with the engineered approach of this chapter. In general, the procedures of this chapter should be used to evaluate cripple walls as load path elements. Where analysis shows that the existing crip-

ple walls do not comply with the strength and stiffness requirements of this chapter, the provision allows that the prescriptive measures of Chapter A3, Section A304.4 and Table A3-A, may be used to specify retrofit details. If the Chapter A3 provisions do not apply, retrofit details should be designed like any other new structural element for compliance with the IBC.

The intention of this provision is that multistory wood-frame buildings with cripple walls should not be deemed compliant just because they do not have obvious SWOF conditions. For example, a large house-like building divided into more than four rental units would not be covered by Chapter A3 and might be missed by this chapter as well, if not for such a provision. Nevertheless, this provision does not apply unless the building is within the scope of the chapter defined in Section A401.2. For the case of the large house-like structure with cripple walls, the cripple story might be classified as soft or weak if it lacks the stiffening partitions of the occupied stories.

Perhaps a more common example of a SWOF building with cripple walls involves partial height masonry or concrete basement walls at the lowest level. Short wood-framed stud walls, or cripple walls, are used between the top of the basement wall and the next floor level. In this case, the cripple walls will necessarily be addressed as load path elements if the basement story has a SWOF condition.

"Ties" refers to splices. Different strap sizes are given to account for different top plate dimensions. Strap splices applied symmetrically to both sides of the top plate are preferred, though not required; symmetric splices will often be impractical if there is no other reason to remove the exterior finishes. Predrilling for nails (with a bit three-fourths the diameter of the nail) is recommended for 2x top plates where splitting due to excessive nailing is a concern.

A403.9 Collector elements. Collector elements shall be provided that can transfer the seismic forces originating in other portions of the building to the elements within the scope of Section A403.2 that provide resistance to those forces.

❖ This provision is based on Section 1633.2.6 of the 1997 UBC. "[O]ther portions" does not mean elements outside the scope of Section A403.2. That terminology remains from the UBC provision for new construction. The intent of this provision is merely to ensure adequate collectors within the load path described in Section A403.2; that is, from the diaphragm above the SWOF condition to the existing or added components of the seismic-force-resisting system.

The provision does not specify means for determining design forces for collectors. Acceptable design forces may be derived from formulas such as 33-1 in the 1997 UBC or Section 9.5.2.6.4.4 in ASCE 7, with the applied forces F_i determined from Sections A403.3 and A403.4. Provisions in Section A403.8 set a minimum on this design force.

Section 1633.2.6 of the UBC and Section 9.5.2.6.3.1 of the ASCE 7 require most collectors to be designed for special seismic loads amplified by the factor Ω_o, but they exempt light-framed wood structures. On that basis, the special seismic load combinations are not required for the wood-framed buildings within the scope of this chapter. Retrofitted SWOF buildings, however, might have steel, concrete or masonry systems that require amplified seismic forces. As written, the provision is less specific than the codes, requiring only collectors "that can transfer the seismic forces." Based on Section A403.1, which requires new elements to meet the building code, amplified forces should be used for collector design in systems other than light-framed wood.

A403.10 Horizontal diaphragms. The analysis of shear demand or capacity of an existing plywood or diagonally sheathed horizontal diaphragm need not be investigated unless the diaphragm is required to transfer lateral forces from the lateral-resisting elements above the diaphragm to other lateral-resisting elements below the diaphragm because of an offset in placement of the elements.

Wood diaphragms in structures that support floors or roofs above shall not be allowed to transmit lateral forces by rotation or cantilever except as allowed by the Building Code. However, rotational effects shall be accounted for when unsymmetric wall stiffness increases shear demands.

Exception: Diaphragms that cantilever 25 percent or less of the distance between lines of lateral-load-resisting elements from which the diaphragm cantilevers may transmit their shears by cantilever, provided that rotational effects on shear walls parallel and perpendicular to the load are taken into account.

❖ Section A403.2 explicitly requires assessment of certain floor diaphragms. The first paragraph of this provision allows an exemption. In essence, diaphragm strength must be checked only when (1) the diaphragm is of straight board sheathing (that is, not plywood or diagonal sheathing), or (2) the diaphragm is on the load path between vertical components of the seismic-force-resisting system that are offset at the floor level. In the second case, only the portions of the diaphragm that provide force transfer between the offset elements of the seismic-force-resisting system must be checked.

Since this chapter makes no requirement for in-plane stiffness of a floor diaphragm (similar to Section 1633.2.9, Item 1, of the 1997 UBC for example), this provision effectively exempts most plywood and diagonally sheathed diaphragms from any quantitative analysis.

"Wood diaphragms in structures that support floors or roofs above" is not meant to cover any diaphragms other than those already within the scope of Section A403.2.

Rigid floor diaphragms are generally considered capable of rotating as rigid bodies without substantial distortion. They are, therefore, judged capable of transmitting forces if the seismic-force-resisting system is not symmetric in plan or even if the system has elements on only three sides. By restricting transmission of forces by rotation or cantilever, this provision is saying

that wood diaphragms are inherently flexible. The existing seismic-force-resisting system must, therefore, be supplemented as needed, assuming that the diaphragm can distribute forces based on tributary area. (As noted elsewhere in this commentary, recent CUREE testing has shown that the wood diaphragm in an open-front building can behave as essentially rigid, but that the resulting rotation can impose large drifts along the open-front wall line.)

The building code is cited for exceptions. If the building code is based on the 1997 UBC, those exceptions are given in Section 2315.1. For codes based on the 2003 IBC, exceptions for rigid wood diaphragms only are in Section 2305.2.5.

For buildings within the scope of this chapter, Section 2315.1 of the 1997 UBC prohibits transmission of forces by rotation in the following cases:

- In diaphragms with straight sheathing.
- Where diaphragm cantilever exceeds the smaller of 25 feet (7620 mm) or two-thirds of the diaphragm width (with exceptions). (The width is the direction normal to the direction of the cantilever span.)
- In concrete or masonry buildings; thus, retrofit of a SWOF building with new concrete or masonry walls on three sides of open area would not be compliant.

"[R]otational effects shall be accounted for when unsymmetric wall stiffness increases shear demands" refers to torsional effects that increase the forces and drifts imposed on the seismic-force-resisting system. As required by Section A403.6, wood diaphragms must be modeled as rigid relative to the walls so that the effects of actual and accidental torsion can be conservatively estimated. In essence, the second paragraph of this provision requires a distribution of forces based on the more conservative of two analyses, one assuming a flexible diaphragm and the other assuming a rigid diaphragm.

The 25-percent value in the exception is the same as that in the definition of "Open front wall lines." It is discussed in the commentary to that definition and illustrated in Figure A4-1. The requirement to account for rotational effects on shear walls should apply as well to existing or new elements of other seismic-force-resisting systems.

A403.11 Shear walls. Shear walls shall have sufficient strength and stiffness to resist the tributary seismic loads and shall conform to the special requirements of this section.

❖ Section A403.11 addresses only wood-framed shear walls. Concrete or masonry shear walls added for retrofit are to be evaluated against the demands of Section A403.3 and detailed according to the IBC for new construction.

Use of the term "tributary" in this provision does not mean that the design forces are to be derived from tributary floor areas, nor does it mean that rigid diaphragm analysis is not required.

The term "special" in this context indicates the "particular" requirements of this section; it is not related to either special seismic-force-resisting systems or special seismic load combinations.

A403.11.1 Gypsum or cement plaster products. Gypsum or cement plaster products shall not be used to provide lateral resistance in the soft or weak story.

❖ "Gypsum" refers primarily to gypsum wallboard but also includes traditional gypsum plaster and gypsum lath (sometimes called buttonboard). "Cement plaster" refers to plaster with Portland cement, also known as stucco.

After the 1994 Northridge earthquake, some jurisdictions reduced by half the allowable shear values for gypsum wallboard and stucco and doubled the minimum wall panel lengths for conventional light-frame construction (Los Angeles, 1999). CUREE tests (2002) have shown that stucco can significantly increase the elastic stiffness and strength of wood panel shear walls, but stucco alone is still considered too brittle a material for use in vulnerable elements such as cripple walls and SWOF story walls.

"[T]he soft or weak story" means whichever story or stories triggers the retrofit according to the scope conditions in Section A401.2. The intention of the provision is that these nonconforming materials may not be counted as part of the seismic-force-resisting system even when the SWOF condition has been mitigated by retrofit. Though only soft and weak stories are mentioned, this prohibition should apply to stories with open-front wall lines as well.

A403.11.2 Wood structural panels.

A403.11.2.1 Drift limit. Wood structural panel shear walls shall meet the story drift limitation of Section A403.6. Conformance to the story drift limitation shall be determined by approved testing or calculation, or analogies drawn therefrom, and not by the use of an aspect ratio. Calculated deflection shall be determined according to UBC Standard 23-2, Section 23.223, "Calculation of Shear Wall Deflection," and 25 percent shall be added to account for inelastic action and repetitive loading. Contribution to the shear wall deflection from the anchor or tie-down slippage shall also be included. The slippage contribution shall include the vertical elongation of the connector metal components, the vertical slippage of the connectors to framing members, localized crushing of wood due to bearing loads, and shrinkage of the wood elements because of changes in moisture content as a result of aging. The total vertical slippage shall be multiplied by the shear panel aspect ratio and added to the total horizontal deflection. Individual shear panels shall be permitted to exceed the maximum aspect ratio, provided the story drift and allowable shear capacities are not exceeded.

❖ The provision's reference to UBC Standard 23-2 (which is also cited in Section 2315.1 of the 1997 UBC) may be understood to refer to corresponding requirements of the IBC. The equation in Standard 23-2 can also be found in the commentary to Section 12.4 of the 2000

NEHRP, the *APA Design/Construction Guide for Diaphragms and Shear Walls* (APA 2001), and in Section 2305.3.2 of the 2000 IBC (in the IBC, the final term of the equation is defined differently but represents the same thing as in the other versions of the equation). Its derivation and history are described in APA Form No. TT-053 (APA, 2000).

The equation defines the wall deflection as the sum of four contributions: flexural deformation of the wood, shear deformation of the wood, deformation due to nail slip and rigid body rotation due to hold-down deformation or "slippage." Nail slip data for use in the third term are given in Table A-2 of the *APA Design Construction Guidelines*. Material data from outside sources is subject to the limitations of Section A406.3.2. Additional information on wood shear wall design provisions is provided in the *APA Design Construction Guidelines* and other documents available at www.apawood.org.

The provision lists four effects that contribute to hold-down slippage, one of which is wood shrinkage. Slippage due to wood shrinkage should not be a concern in existing shear walls (unless the shrinkage occurred after an original hold-down was installed) or where wood structural panels are applied to existing framing.

A403.11.2.2 Openings. Shear walls are permitted to be designed for continuity around openings in accordance with Section 2315.1 of the UBC. Blocking and steel strapping shall be provided at corners of the openings to transfer forces from discontinuous boundary elements into adjoining panel elements. Alternatively, the perforated shear wall provisions of the IBC may be used.

❖ The provision refers to Section 2315.1 of the 1997 UBC and to the perforated shear wall provisions of the IBC. These references may be understood to refer to corresponding requirements of the building code.

The intention of provisions such as Section 2315.1 of the 1997 UBC is to ensure performance that matches the characteristics assumed for analysis and design, principally that the shear wall acts as a unit, without substantial deformations or stress concentrations at openings. The specified UBC section refers to Table 23-II-G of the 1997 UBC, which has been modified slightly by errata published in September 1997, May 1998 and January 2001: the modified table no longer allows a height-to-width ratio greater than 2:1 for wood structural panel shear walls in Seismic Zones 0, 1 and 2.

Commentary, background and design examples regarding perforated shear walls are available in Section 12.4.3 of the 2000 NEHRP commentary.

A403.11.2.3 Wood species of framing members. Allowable shear values for wood structural panels shall consider the species of the framing members. When the allowable shear values are based on Douglas fir-larch framing members, and framing members are constructed of other species of lumber, the allowable shear values shall be multiplied by the following factors: 0.82 for species with specific gravities greater than or equal to

0.42 but less than 0.49, and 0.65 for species with specific gravities less than 0.42. Redwood shall use 0.65 and hem fir shall use 0.82, unless otherwise approved.

A403.11.3 Substitution for 3-inch (76 mm) nominal width framing members. Two 2-inch (51 mm) nominal width framing members shall be permitted in lieu of any required 3-inch (76 mm) nominal width framing member when the existing and new framing members are of equal dimensions, when they are connected as required to transfer the in-plane shear between them, and when the sheathing fasteners are equally divided between them.

❖ Current design provisions for shear walls frequently require framing members of 3 inch (76 mm) nominal width. This provision recognizes that many existing buildings have only 2-inch (51 mm) nominal studs, and that it is more cost effective to double the existing member than to replace it. The requirement for members of equal dimensions is intended to ensure that the two pieces will be of roughly equivalent stiffness and will thus share loads roughly equally.

A new framing member of "wet" lumber should not be added to an existing dry member. When the wet piece shrinks, the nonuniform deformation could affect the integrity of the plywood nailing.

Connection for shear transfer should be based on calculation. Recent testing has shown that studs nailed together can experience substantial slip under seismic loads (CUREE, 2002).

Where existing members are very dry and prone to splitting, a quality assurance plan should include confirmation that existing members are not being damaged by new work. Predrilling (with a bit three-fourths of the nail diameter) is one way to reduce splitting in dry members.

The provision is based on the common condition of existing framing with nominal dimensions. If the existing framing is unsurfaced lumber, then the stud width might already be a full 2 inches (51 mm) [as opposed to 1½ inches (38 mm) for a nominal 2x member]. In that case, the existing member might be sufficient by itself.

A403.11.4 Hold-down connectors.

A403.11.4.1 Expansion anchors in tension. Expansion anchors that provide tension strength by friction resistance shall not be used to connect hold-down devices to existing concrete or masonry elements. Expansion anchors that provide tension strength by bearing (commonly referenced as "undercut" anchors) shall be permitted.

❖ Chemical (epoxy) and screw-type anchors should also be allowed for hold-downs. See the commentary under the definition of "Expansion anchor" in Section A402.

This provision prohibits friction-based expansion anchors for tension loads because of poor performance in the 1994 Northridge earthquake. Since then, approval criteria have been made more restrictive, and the design of some expansion anchors has been modified. Any anchor type with ICC ES approval for cyclic tension loads should be acceptable in this application.

A403.11.4.2 Required depth of embedment. The required depth of embedment or edge distance for the anchor used in the hold-down connector shall be provided in the concrete or masonry below any plain concrete slab unless satisfactory evidence is submitted to the building official that shows that the concrete slab and footings are of monolithic construction.

❖ "[M]onolithic construction" generally means concrete placed at the same time, with continuous reinforcing. A slab poured over the top of a previously placed footing will generally not be monotonic unless the top surface of the footing was roughened, bent reinforcing dowels connect the two pours and there is no "visqueen" vapor barrier laid between the two pours. A slab with nominal reinforcing or wire mesh provided only for crack control should be considered plain concrete.

Another reason for embedment well into the footing is that curbs and short stem walls are frequently too narrow to provide adequate edge distance for the anchor and are, therefore, unable to develop its strength.

A403.11.4.3 Required preload of bolted hold-down connectors. Bolted hold-down connectors shall be preloaded to reduce slippage of the connector. Preloading shall consist of tightening the nut on the tension anchor after the placement but before the tightening of the shear bolts in the panel flange member. The tension anchor shall be tightened until the shear bolts are in firm contact with the edge of the hole nearest the direction of the tension anchor. Hold-down connectors with self-jigging bolt standoffs shall be installed in a manner to permit preloading.

❖ Improper installation of hold-downs has been responsible for poor performance in past earthquakes (SEAOC 1999, Appendix F). Preloading is intended to ensure that the hold-down works as designed, with all of the bolts bearing simultaneously on the wood member. If existing hold-downs are exposed in the course of the work, they should be preloaded as well. Hold-down connectors with pointed self-jigging webs should have the point raised at least 1 inch (25 mm) above the sill plate to prevent the point from splitting the sill during tightening.

SECTION A404
GENERAL REQUIREMENTS
FOR PHASED CONSTRUCTION

When the building contains three or more levels, the work specified in this chapter shall be permitted to be done in the following phases. Work shall start with Phase I unless otherwise approved by the building official. When the building does not contain the conditions shown in any phase, the sequence of retrofit work shall proceed to the next phase in numerical order.

Phase 1 Work. The first phase of the retrofit work shall include the ground floor portion of the wood structure that contains parking or other similar open floor space.

Phase 2 Work. The second phase of the retrofit work shall include walls of any level of wood construction with two or more levels above that are laterally braced with nonconforming structural materials.

Phase 3 Work. The third and final phase of the retrofit work shall include the remaining portions of the building up to, but not including, the top story as specified in Section A403.2.

❖ Phased construction is sometimes beneficial to owners. This provision is intended to prioritize work related to the most hazardous conditions and to ensure that the building is not weakened by partial efforts or left in a more seismically vulnerable condition at any intermediate stage.

Phase 1 work is intended to cover the lowest wood-framed story with a SWOF condition, regardless of whether it is the groundfloor or is used for parking.

SECTION A405
PRESCRIPTIVE MEASURES
FOR WEAK STORY

A405.1 Scope. The proposed prescriptive measures provided here are intended to reduce the earthquake vulnerability of the structure and to reduce the possibility of collapse or partial collapse of the building in the event of a moderate to major earthquake.

❖ This section is not limited to weak story conditions. Any structure that meets the qualifications is eligible for these prescriptive measures.

For buildings that qualify, Section A405 is intended as an alternative to Section A403. The requirements of Sections A406, A407 and A408 still apply. The main purpose of this prescriptive alternative is to reduce the cost of engineered evaluation and design. It is expected that Section A405 could be applied by a qualified contractor without the assistance of an engineer.

Section A405 is intended for a two-story building with an uncomplicated footprint and a SWOF wall line on only one side. Figure A4-3 illustrates two such buildings in a plan. The text of Section A405 is written with tuckunder parking in mind. Parking, however, is not a prerequisite for this section; any SWOF condition and any groundfloor use may be eligible.

The provisions of this section are not merely "proposed." They are deemed acceptable for the buildings that qualify.

A405.1.1 Performance. The improved earthquake performance of the structure due to the proposed prescriptive measures varies and is greatly controlled by all of the following: proximity to the fault line; soil type; weight of roof and floor above; quality of existing walls, posts and columns, and their connections to the floor diaphragm; and the quality of construction provided to comply with the prescriptive measures. The implementation of the proposed measures is not intended to improve the earthquake performance of the building above the first story.

❖ The performance objective throughout this chapter, including Section A405, is risk reduction, as described in the commentary to Section A401.1.

The provisions of Section A405 are derived from judgement, not from testing, analysis or performance statistics. No studies or tests have been performed to confirm that these prescriptive measures will result in the same performance with the same reliability as designs engineered to the provisions of Section A403. Nevertheless, for the buildings that qualify, these prescriptive measures are expected to achieve the same risk reduction objective.

A405.1.2 Limitation. The proposed prescriptive measures rely on rotation of the second floor diaphragm to distribute the load between the side and rear wall enclosing the parking area. The owner shall provide access to ensure that the floor diaphragm is of wood structural panel or diagonal sheathing. In the absence of such a verification, a new wood structural panel diaphragm must be applied of minimum thickness of $^3/_4$ inch (19 mm) and with 10d common nails at 6 inches (152 mm) on center.

❖ See Figure A4-3 for an illustration of the side and rear wall. In the bottom part of the figure, only the portion of the side wall between Lines A and B is considered to be enclosing the parking area, and only that portion should be counted toward the required wall length under these prescriptive measures. As noted in the commentary above, the building need not include parking.

A substantial floor diaphragm above the SWOF wall line is necessary for these provisions to provide reliable performance. A diaphragm of straight board sheathing is judged inadequate. Clearly, verification of the existing diaphragm is less costly and less disruptive than installation of a new plywood diaphragm.

The provision does not identify any detail requirements for the existing wood diaphragm. Instead, the provision appears to presume that the presence of a conforming material ensures acceptable sheathing thickness, material grade, condition and nailing. No studies have been performed to determine what diaphragm strength is required for these prescriptive provisions to work reliably; in the absence of such studies, any criteria would be based on judgement alone. Still, if there is doubt as to the adequacy of the existing diaphragm, Sections 2315.3.1 and 2315.3.3 of the 1997 UBC offer criteria for new construction that may be used as conservative benchmarks.

Where a new diaphragm is required, it may be applied over the existing sheathing or to the underside of the second floor joists. Application to the underside of the floor joists will often be less disruptive. Application on top of existing nonconforming sheathing presents the additional problems of aligning the panel edges with floor framing below and providing continuity across or under the sill plates of existing partitions. In either case, appropriate collector elements must also be provided.

The provision prescribes a new or additional diaphragm of $^3/_4$ inch (19 mm) thickness with 10d nails at 6 inches (152 mm). For new diaphragms with this nailing, 3-inch (76 mm) nominal framing members are required at panel edges per Table 23-II-H of the 1997 UBC.

A405.1.3 Additional conditions. To qualify for prescriptive measures, the following additional conditions need to be satisfied:

1. Diaphragm aspect ratio = 1.5 or less.
2. Minimum length of side shear walls = 20 feet (6096 mm) with less than 10 percent openings.
3. Minimum length of rear shear walls = $^3/_4$ of rear wall length with individual walls not having more than 10 percent openings.

❖ The three listed conditions are in addition to those in Section A403.1, Exception 2.

The ratios, wall lengths and opening percentages in this provision are derived from judgement, not from testing, analysis or performance statistics. The three conditions need not be satisfied by the existing building as long as they are satisfied by the retrofit. In other words, wall elements can be added and openings modified to meet the specified criteria. Whether new or existing, only walls that comply with Section A405.2.3 should be counted for the purposes of satisfying these conditions.

The diaphragm aspect ratio provision is in error. With reference to Figure A4-3, the provision intends to require values of L/W less than or equal to 0.67, consistent with Section 2305.2.5 of the 2003 IBC.

L/W are the principal dimensions of the diaphragm bounded by the shear walls immediately adjacent to the SWOF wall line. The intention of the aspect ratio limit is to restrict the prescriptive provisions to buildings that are not prone to substantial torsion. Where the critical diaphragm is the entire second floor, as in the top part of Figure A4-3, the preferred condition has the SWOF wall line along the long side of the building so that the center of mass and the center of rigidity are close together and torsion is controlled. If the open wall line was along the short side, the three-sided structure would be subject to high torsion.

The bottom part of Figure A4-3 illustrates a potential shortcoming of this simple screening criterion. If L/W for this building were less than 0.67, it would not qualify for the prescriptive measures of Section A405. Yet this building is arguably less vulnerable to torsion than the building in the top part of the figure. In the bottom part of the figure the critical condition is the demand on the wall along Line B due to torsional response.

For the building in the bottom part of Figure A4-3, the side shear walls include only the portions between Line A and Line B. The portion between Line B and Line C is not counted toward the required 20 feet (6096 mm). Reasonable designs that make use of wall lengths not immediately adjacent to the open area (that is, lengths between Lines B and C) are possible but should be subject to engineered design. For these prescriptive measures, it is appropriate to restrict certain design choices in order to ensure reliable outcomes in the absence of engineering calculations.

The 10-percent limits on openings refer to the area of wall in the first story, not the length of wall. Only the portions of walls bounding the parking or open area should be considered.

In addition to the three listed limitations, the prescriptive measures of Section A405 should also be limited to buildings with acceptable foundations so that the engineering requirements of Section A403.2 regarding soil-bearing pressure may be waived and that the prescriptive anchor bolt and hold-down provisions may be implemented. In the absence of other criteria, minimum requirements for two-story buildings from Table 18-I-C of the 1997 UBC (or corresponding values from other model codes) should be met: the side and rear walls considered by Section A405 should have continuous concrete footings at least 6 inches (152 mm) thick and 12 inches (305 mm) wide, with bearing at least 12 inches (305 mm) below the undisturbed ground surface.

A405.2 Minimum required retrofit.

❖ For buildings that qualify, Section A405 is intended as an alternative to Section A403. The requirements of Sections A406, A407 and A408 still apply.

A405.2.1 Anchor bolt size and spacing. The anchor bolt size and spacing shall be a minimum of $^3/_4$ inch (19 mm) in diameter at 32 inches (813 mm) on center. Where existing bolts are inadequate, new steel plates bolted to the side of the foundation and nailed to the sill may be used, such as an approved connector.

❖ Anchor bolts as specified in this provision must be provided only along the length of the side and rear walls.

"[I]nadequate" refers principally to bolt size and spacing, but judgement should be applied where existing anchors indicate poor workmanship (oversized holes in sill plates, insufficient edge distance or embedment, etc.) or poor condition (corrosion, looseness, spalling, etc.).

Bolts added to comply with this provision need not be expansion anchors because they are intended principally to resist shear loads (chemical anchors are preferred) (see commentary under "Expansion anchor," Section A402).

The provision specifies $^3/_4$-inch (19.1 mm) diameter bolts. Smaller bolts are sometimes necessary for proper edge distance in narrow sill plates or concrete curbs. These smaller bolt sizes may be used if: (1) the building official approves, and (2) the designer demonstrates that the proposed bolt layout, which may combine new and existing bolts of different sizes, provides capacity equivalent to $^3/_4$-inch (19.1 mm) bolts at 32 inches (813 mm) on center. If the new bolts are substantially stiffer than the existing bolts, then the existing bolts should not be counted toward the required capacity.

If new bolts are required, they must be installed through any new blocking at the sill plate required by Section A405.2.3. Embedment must be into the footing (see commentary, Section 403.11.4.2).

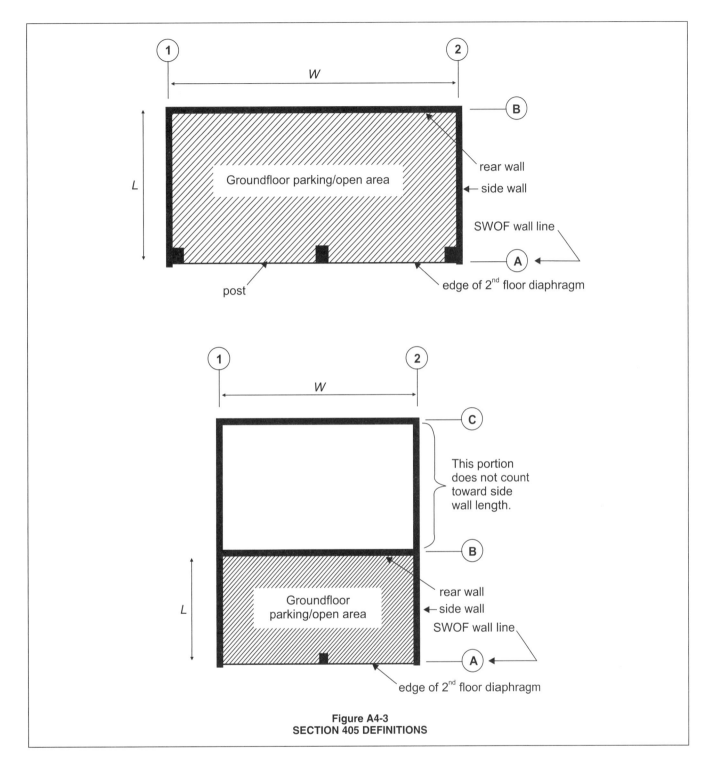

Figure A4-3
SECTION 405 DEFINITIONS

The provision allows approved steel plate connectors. These are commonly used for retrofit of cripple walls where cramped access prevents installation of new bolts vertically through the sill plate. This is less likely to be a concern in a SWOF building.

A405.2.2 Connection to floor above. Shear wall top plates shall be connected to blocking or rim joist at upper floor with a minimum of 18-gage galvanized steel angle clips 4¹/₂ inches (114 mm) long with 12-8d nails spaced no farther than 16 inches (406 mm) on center, or by equivalent shear transfer methods.

A405.2.3 Shear wall sheathing. The shear wall sheathing shall be a minimum of ¹⁵/₃₂ inch (11.9 mm) 5-Ply Structural I with 10d nails at 4 inches (102 mm) on center at edges and 12 inches (305 mm) on center at field; blocked all edges with 3 by 4 or larger. Where existing sill plates are less than 3-by thick, place flat 2-by on top of sill between studs, with flat 18-gage galvanized steel clips $4^1/_2$ inches (114 mm) long with 12-8d nails or $^3/_8$-inch-diameter (9.5 mm) lags through blocking for shear transfer to sill plate. Stagger nailing from wall sheathing between existing sill and new blocking. Anchor new blocking to foundation as specified above.

❖ The requirements for $^1/_2$-inch (12.7 mm) five-ply sheathing and 3x framing and blocking are consistent with changes made to building codes after the 1994 Northridge earthquake. For additional discussion of their background, see Harder (1994) and SEAOC (1999), Appendix F.

   The provision makes no allowance for existing $^3/_8$-inch (9.5 mm) sheathing. For these prescriptive measures, the specified sheathing is required without exception. If the existing sheathing is $^1/_2$ inch (12.7 mm), nail size and spacing must be verified; the nailing must be sufficient to develop the capacity equivalent to that required by the provision. If the existing sheathing or nailing does not meet the prescriptive requirements, the engineered approach of Section A403 may be used to take advantage of their strength contribution.

   If the required $^1/_2$-inch (12.7 mm) sheathing is applied to studs already sheathed with lesser material, care should be taken not to split the existing wall studs with new nailing. Doubled studs, predrilled holes or lag screws may be beneficial in this regard.

   Where existing framing is less than 3x nominal, studs may be doubled per Section A403.11.3.

   Sheathing applied to the inside face of existing studs should be provided with ventilation and inspection holes. Chapter 3, Section 304.4.3 of the GSREB provides some guidance.

A405.2.4 Shear wall hold-downs. Shear walls shall be provided with hold-down anchors at each end. Two hold-down anchors are required at intersecting corners. Hold-downs shall be approved connectors with a minimum $^5/_8$-inch-diameter (15.9 mm) threaded rod or other approved anchor with a minimum allowable load of 4,000 pounds (17.8 kN). Anchor embedment in concrete shall be not less than 5 inches (127 mm). Tie-rod systems shall not be less than $^5/_8$ inch (15.9 mm) in diameter unless using high strength cable. Threaded rod or high strength cable elongation shall not exceed $^5/_8$ inch (15.9 mm) using design forces.

❖ While the design provisions of Section A403 are generally waived, applicable portions of the hold-down provisions of Section A403.11.4 should be applied in addition to those given here.

   The hold-down capacity might be limited by available concrete strength or edge distance. It is acceptable to use two 2,000-pound (908 kg) hold-downs, one on each side of the post, in lieu of a single 4,000 pound (1816 kg) hold-down.

   The hold-down capacity might also be limited by the inability of a small footing and its overburden to resist uplift in the first place. For purposes of Section A405, the presumption is that the minimum footing dimensions recommended in the commentary to Sec-

tion A405.1.3 are acceptable for these prescriptive hold-down requirements.

   The minimum embedment into the footing should be the larger of 5 inches (127 mm) or the manufacturer's recommendation (see commentary, Section 403.11.4.2).

   Posts to which hold-downs are connected should be sufficient to prevent failures due to hold-down eccentricity. Posts built up from multiple 2x or 3x members are acceptable if adequately connected (see commentary, Section A403.11.3), but 4x posts are preferred.

   For the prescriptive measures of Section A405, there are no calculated design forces. Thus, the final sentence of this provision may be satisfied by ensuring that the elongation does not exceed $^5/_8$-inch (16 mm) at the allowable load of 4,000 pounds (1816 kg). This interpretation applies for hold-downs in the typical location just above the sill plate.

SECTION A406
MATERIALS OF CONSTRUCTION

A406.1 New materials. All materials approved by this code, including their appropriate allowable stresses and minimum aspect ratios, shall be permitted to meet the requirements of this chapter.

❖ "[T]his code" may be understood to mean the IBC.

A406.2 Allowable foundation and lateral pressures. Allowable foundation and lateral pressures shall be permitted to use the values from UBC Table 18-I-A. The coefficient of variation of subgrade reaction shall be established by an approved geotechnical engineering report or other approved methods when used in the deflection calculations of embedded vertical elements as required in Section A403.6.

❖ The provision's reference to Table 18-I-A of the UBC may be understood to refer to the corresponding requirements of the IBC.

   See the commentary to Section A403.2 for additional discussion of soil-bearing pressures and the commentary to Section A403.6 for additional discussion of the coefficient of subgrade reaction.

A406.3 Existing materials. All existing materials shall be in sound condition and constructed in conformance to this code before they can be used to resist the lateral loads prescribed in this chapter. The verification of existing material conditions and their conformance to these requirements shall be made by physical observation reports, material testing or record drawings as determined by the structural designer and as approved by the building official.

❖ "[T]his code" may be understood to mean the IBC.

   The provision offers no standards or criteria for condition assessment. In general, the intention of the provision is to rule out significant construction defects, deferred maintenance and unrepaired damage to structural members on the seismic load path. Based on the last sentence, the expectation of the provision is that appropriate condition assessment might require only reference to existing documents but might in some cases require destructive investigation or testing. Absent specific provisions, the appropriate inspection scope, personnel and acceptance criteria are left to the designer of record. Where the prescriptive measures of Section A405 are used, the designer of record will likely be the contractor.

Some guidance on condition assessment can be found in FEMA 356. Chapter A3, Section A304.1.2 offers provisions that might be useful for assessment of wood members with potential fungus or insect infestation.

"[C]onstructed in conformance" in the context of this provision should be understood to refer to the general quality of construction, not to the adequacy of design or detailing. The intention of the provision is to ensure that load path elements will not fail prematurely due to errors or omissions in the original construction, such as "shiner" nails, misplaced anchor bolts or hold-downs, etc. In addition, clearly archaic construction, such as an unreinforced brick foundation, is not considered to be constructed in conformance.

A406.3.1 Horizontal wood diaphragms. Existing horizontal wood diaphragms that require analysis under Section A403.10 shall be permitted to use Table A4-E for their allowable values.

❖ The design forces prescribed in Section A403 are at a strength level. When using the values in Table A4-E, the diaphragm design forces should be reduced by 1.4.

A406.3.2 Wood-structural-panel shear walls.

A406.3.2.1 Allowable nail slip values. When the required drift calculations of Section A403.11.2.1 rely on the lower slip values for common nails or surfaced dry lumber, their use in construction shall be verified by exposure. The use of box nails and unseasoned lumber may be assumed without exposure. The design value of the box nails shall be assumed to be similar to that of common nails having the same diameter. Verification of surfaced dry lumber shall be by identification conforming to UBC Section 2304.1.

❖ The reference to the 1997 UBC may be understood to refer to the corresponding requirements of the IBC.

Existing members may be assumed to be dry. The required verification need only apply to new lumber.

A406.3.2.2 Plywood panel construction. When verification of the existing plywood materials is by use of record drawings alone, the panel construction for plywood shall be assumed to be of three plies. The plywood modulus "G" shall be assumed equal to 50,000 pounds per square inch (345 MPa).

❖ Visual verification usually requires no more than a small core sample.

A406.3.3 Existing wood framing. Wood framing is permitted to use the design stresses specified in the building code under which the building was constructed or other stress criteria approved by the building official.

A406.3.4 Structural steel. All existing structural steel shall be permitted to use the allowable stresses for Grade A36. Existing pipe or tube columns shall be assumed to be of minimum wall thickness unless verified by testing or exposure.

A406.3.5 Strength of concrete. All existing concrete footings shall be permitted to use the allowable stresses for plain concrete with a compressive strength of 2,000 pounds per square inch (13.8 MPa). The strength of existing concrete with a recorded compressive strength greater than 2,000 pounds per square inch (13.8 MPa) shall be verified by testing, record drawings or department records.

❖ With reference to Section A406.3, existing unreinforced brick foundations are not considered to be constructed in conformance with applicable codes and, therefore, may not be used to resist earthquake effects in the retrofitted structure.

A406.3.6 Existing sill plate anchorage. Existing cast-in-place anchor bolts shall be permitted to use the allowable service loads for bolts with proper embedment when used for shear resistance to lateral loads.

SECTION A407
REQUIRED INFORMATION ON PLANS

A407.1 General. The plans shall show all necessary dimensions and materials for plan review and construction and shall accurately reflect the results of the engineering investigation and design. Details specific to the actual condition found shall be shown on the drawings to assure installation of all elements required for construction of the necessary complete load path. The plans shall contain a note that states that this retrofit was designed in compliance with the criteria of this chapter.

❖ The requirements of this section are in addition to any other documentation required by the building code or the code official.

Documentation requirements regarding calculated member capacities, design forces, or results of engineering analysis may be waived when the prescriptive measures of Section A405 are used.

The intent of this provision, particularly its second sentence, is to ensure that the designer has accounted for any existing conditions that might compromise the seismic load path or interfere with the intended details. An investigation should verify representative load path details. Specific conditions might require additional field investigation and detailing during the construction phase.

The note required by the last sentence of the provision should specify the 2000 edition of the GSREB. The IBC should be noted as well.

A407.2 Existing construction. The plans shall show existing diaphragm and shear wall sheathing and framing materials; fastener type and spacing; diaphragm and shear wall connections; continuity ties; and collector elements. The plans shall also show the portion of the existing materials that needs verification during construction.

❖ Details of existing construction need only be shown for the areas within the scope of the retrofit as indicated by Section A403.2.

This chapter's provisions that address verification include Sections A406.3 (general condition assessment), A405.1.2 (diaphragm type), A405.2.1 (anchor bolts), A406.3.2.1 (wood shear wall construction), A406.3.4 (steel tube thickness) and A406.3.5 (concrete strength).

A407.3 New construction.

A407.3.1 Foundation plan elements. The foundation plan shall include the size, type, location and spacing of all anchor bolts with the required depth of embedment, edge and end distance; the location and size of all columns for braced or moment frames; referenced details for the connection of braced or moment-resisting frames to their footing; and referenced sections for any grade beams and footings.

❖ Hold-down locations should be shown on the foundation plan as well.

A407.3.2 Framing plan elements. The framing plan shall include the width, location and material of shear walls; the width, location and material of frames; references on details for the column-to-beam connectors, beam-to-wall connections, and shear transfers at floor and roof diaphragms; and the required nailing and length for wall top plate splices.

A407.3.3 Shear wall schedule, notes and details. Shear walls shall have a referenced schedule on the plans that includes the correct shear wall capacity in pounds per foot (N/m); the required fastener type, length, gauge and head size; and a complete specification for the sheathing material and its thickness. The schedule shall also show the required location of 3-inch (76 mm) nominal or two 2-inch (51 mm) nominal edge members; the spacing of shear transfer elements such as framing anchors or added sill plate nails; the required hold-down with its bolt, screw or nail sizes; and the dimensions, lumber grade and species of the attached framing member.

Notes shall show required edge distance for fasteners on structural wood panels and framing members; required flush nailing at the plywood surface; limits of mechanical penetrations; and the sill plate material assumed in the design. The limits of mechanical penetrations shall also be detailed showing the maximum notching and drilled hole sizes.

❖ See the commentary to Section A407.1 regarding documentation of member capacities on plans.

The intention of this provision is to document new and modified elements. The extensive list indicates the importance of quality control for shear walls.

"[C]omplete specification" means enough information to ensure selection of the correct product by the builder and to allow inspection by the authority having jurisdiction. Reference to a material standard and a panel grade designation is generally sufficient (for example, "1997 UBC Standard 23-2, Structural I," or "United States Voluntary Product Standard PS 1-95, Structural I").

A407.3.4 General notes. General notes shall show the requirements for material testing, special inspection, structural observation and the proper installation of newly added materials.

❖ In accordance with Section A403.1, the testing and inspection requirements are the same as those for new construction. For example, see Chapter 17 of the 1997 UBC.

SECTION A408
QUALITY CONTROL

A408.1 Structural observation. All structures regulated by this chapter require structural observation. The owner shall employ the engineer or architect responsible for the structural design, or another engineer or architect designated by the engineer or architect responsible for the structural design, to perform structural observation as defined in the UBC.

❖ In accordance with Section A403.1, the structural observation requirements are generally the same as those for new construction. For example, see Chapter 17, Section 1702 of the 1997 UBC. However, retrofit projects routinely involve unanticipated conditions. The designer and builder should make allowances for such conditions when developing a project schedule and quality control plan.

BASIC STRUCTURAL CHECKLIST

TABLE A4-A—BUILDING SYSTEM

C	NC	N/A	LOAD PATH: The structure shall contain one complete load path for seismic force effects from any horizontal direction that serves to transfer the inertial forces from the mass to the foundation.
C	NC	N/A	WEAK STORY: The strength of the lateral-force-resisting system in any story shall not be less than 80 percent of the strength in an adjacent story above or below.
C	NC	N/A	SOFT STORY: The stiffness of the lateral-force-resisting system in any story shall not be less than 70 percent of the stiffness in an adjacent story above or below, or less than 80 percent of the average stiffness of the three stories above or below.
C	NC	N/A	VERTICAL DISCONTINUITIES: All vertical elements in the lateral-force-resisting systems shall be continuous to the foundation.
C	NC	N/A	DETERIORATION OF WOOD: There shall be no signs of decay, shrinkage, splitting, fire damage or sagging in any of the wood members, and none of the metal accessories shall be deteriorated, broken or loose.
C	NC	N/A	WALL ANCHORAGE: Exterior concrete or masonry walls shall be anchored for out-of-plane forces at each diaphragm level with steel anchors or straps that are developed into the diaphragm. Straps shall be minimum 7 gage.

❖ See the commentary to Section A403.1.1.

The soft story definition given here differs slightly from the one in Section A402. The definition in Section A402 should be used for checking scope requirements of Section A401.2.

TABLE A4-B—LATERAL-FORCE-RESISTING SYSTEM[1]

C	NC	N/A	REDUNDANCY: The number of lines of shear walls in each principal direction shall be greater than or equal to two.
C	NC	N/A	SHEAR STRESS CHECK: The shear stress in the shear walls shall be less than the following values: 5-Ply structural panel sheathing: 400 plf (5.8 kN/m) 3-Ply structural panel and diagonal sheathing: 200 plf (2.9 kN/m) Straight sheathing: 80 plf (1.2 kN/m)
C	NC	N/A	STUCCO (EXTERIOR PLASTER) SHEAR WALLS: Multistory buildings shall not rely on exterior stucco walls as the primary lateral-force-resisting system.
C	NC	N/A	GYPSUM WALLBOARD OR PLASTER SHEAR WALLS: Interior plaster or gypsum wallboard shall not be used as shear walls on buildings over one story in height.
C	NC	N/A	NARROW WOOD SHEAR WALLS: Narrow wood shear walls with an aspect ratio greater than 2:1 for life safety shall not be used to resist lateral forces developed in the building.
C	NC	N/A	WALLS CONNECTED THROUGH FLOORS: Shear walls shall have interconnection between stories to transfer overturning and shear forces through the floor.
C	NC	N/A	HILLSIDE SITE: For a sloping site greater than 1 vertical in 3 horizontal and with greater than one-half story above the base, the base shear in the downhill direction, including forces from the base-level diaphragm, shall be resisted through primary anchors from diaphragm struts or collectors provided in the base-level framing to the foundation.
C	NC	N/A	CRIPPLE WALLS: All cripple walls below first-floor-level shear walls shall be braced to the foundation with shear elements.
C	NC	N/A	OPENINGS: Walls with garage doors or other large openings shall be braced with plywood shear walls or shall be supported by adjacent construction through substantial positive ties.
C	NC	N/A	HOLD-DOWN ANCHORS: All walls shall have properly constructed hold-down anchors.

1. The Basic Structural Checklist shall be completed prior to completing this Supplemental Structural Checklist.

❖ The limiting values given under the column "Shear Stress Check" are appropriate for allowable stress design. Design forces based on a strength design approach (for example, the loads specified in Section A403.3) should be reduced to allowable stress levels before comparison with the limiting values provided.

The "Shear Stress Check" item includes a limiting value for straight sheathing. While this is listed for evaluation purposes, straight sheathing is not allowed when the prescriptive measures of Section A405 are used (see Section A405.1.2).

TABLE A4-C—CONNECTIONS[1]

C	NC	N/A	WOOD POSTS: There shall be a positive connection of wood posts to the foundation.
C	NC	N/A	WOOD SILLS: All wood sills shall be bolted to the foundation.
C	NC	N/A	GIRDER/COLUMN CONNECTION: There shall be a positive connection between the girder and the column support.
C	NC	N/A	WOOD SILL BOLTS: Sill bolts shall be spaced at 6 feet or less, with proper edge distance provided for wood and concrete.

For SI: 1 foot = 304.8 mm.

1. The Basic Structural Checklist shall be completed prior to completing this Supplemental Structural Checklist.

TABLE A4-D—DIAPHRAGMS[1]

C	NC	N/A	DIAPHRAGM CONTINUITY: The diaphragms shall not be composed of split-level floors. In wood buildings, the diaphragms shall not have expansion joints.
C	NC	N/A	ROOF CHORD CONTINUITY: All chord elements shall be continuous, regardless of changes in roof elevation.
C	NC	N/A	STRAIGHT SHEATHING: All straight-sheathed diaphragms shall have aspect ratios less than 2:1.
C	NC	N/A	SPANS: All wood diaphragms with spans greater than 24 feet shall consist of wood structural panels or diagonal sheathing. Wood commercial and industrial buildings may have rod-braced systems.
C	NC	N/A	UNBLOCKED DIAPHRAGMS: All unblocked wood-structural-panel diaphragms shall have horizontal spans less than 40 feet and shall have aspect ratios less than or equal to 4:1.

For SI: 1 foot = 304.8 mm.

1. The Basic Structural Checklist shall be completed prior to completing this Supplemental Structural Checklist.

TABLE A4-E—ALLOWABLE VALUES FOR EXISTING MATERIALS

EXISTING MATERIALS OR CONFIGURATIONS OF MATERIALS[1]	ALLOWABLE VALUES
	× 14.594 for N/m
1. Horizontal diaphragms[2] 1.1. Roofs with straight sheathing and roofing applied directly to the sheathing 1.2. Roofs with diagonal sheathing and roofing applied directly to the sheathing 1.3. Floors with straight tongue-and-groove sheathing 1.4. Floors with straight sheathing and finished wood flooring with board edges offset or perpendicular 1.5. Floors with diagonal sheathing and finished wood flooring	100 lbs. per ft. for seismic shear 250 lbs. per ft. for seismic shear 100 lbs. per ft. for seismic shear 500 lbs. per ft. for seismic shear 600 lbs. per ft. for seismic shear
2. Crosswalls[2, 3] 2.1. Plaster on wood or metal lath 2.2. Plaster on gypsum lath 2.3. Gypsum wallboard, unblocked edges 2.4. Gypsum wallboard, blocked edges	Per side: 200 lbs. per ft. for seismic shear 175 lbs. per ft. for seismic shear 75 lbs. per ft. for seismic shear 125 lbs. per ft. for seismic shear
3. Existing footings, wood framing, structural steel and reinforced steel 3.1. Plain concrete footings 3.2. Douglas fir wood 3.3. Reinforcing steel 3.4. Structural steel	$f_c' = 1,500$ psi (10.3 MPa) unless otherwise shown by tests[4] Allowable stress same as D.F. No. 1[4] $f_s = 18,000$ psi (124 MPa) maximum[4] $f_s = 20,000$ psi (138 MPa) maximum[4]

For SI: 1 foot = 304.8 mm.

1. Material must be sound and in good condition.
2. A one-third increase in allowable stress is not allowed.
3. Shear values of these materials may be combined, except the total combined value shall not exceed 300 pounds per foot.
4. Stresses given may be increased for combination of loads as specified in the Building Code.

❖ The values in Table A4-E are allowable values appropriate for allowable stress design. Design forces based on a strength design approach (for example, the loads specified in Section A403.3) should be reduced to allowable stress levels before comparison with the limiting values provided.

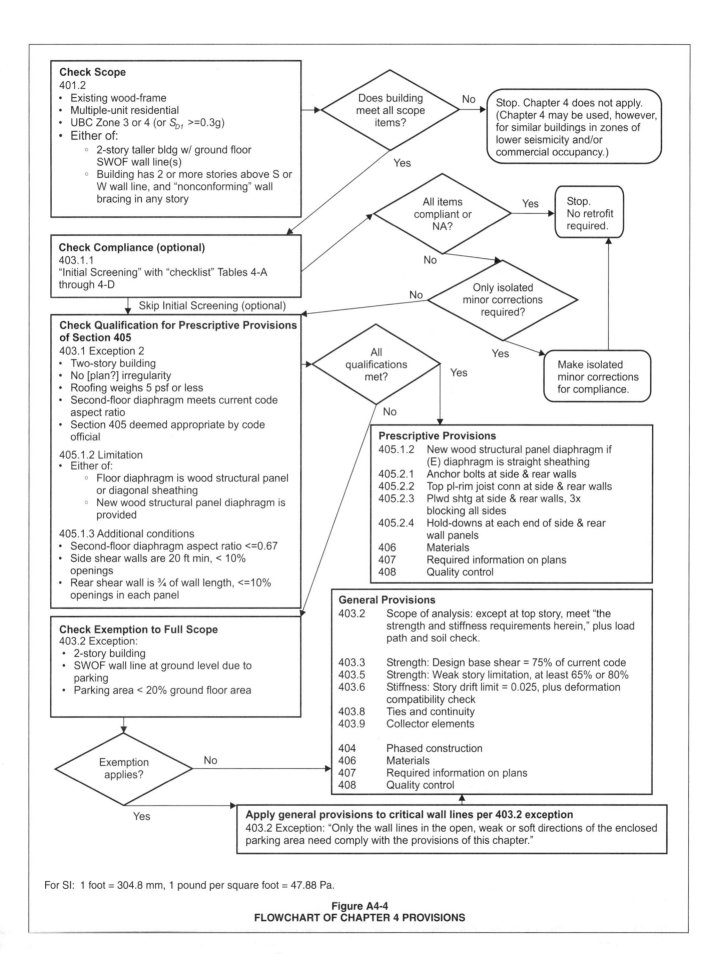

Check Scope
401.2
• Existing wood-frame
• Multiple-unit residential
• UBC Zone 3 or 4 (or S_{D1} >=0.3g)
• Either of:
  ○ 2-story taller bldg w/ ground floor SWOF wall line(s)
  ○ Building has 2 or more stories above S or W wall line, and "nonconforming" wall bracing in any story

Does building meet all scope items?

No → Stop. Chapter 4 does not apply. (Chapter 4 may be used, however, for similar buildings in zones of lower seismicity and/or commercial occupancy.)

Yes

All items compliant or NA? — Yes → Stop. No retrofit required.

No

Check Compliance (optional)
403.1.1
"Initial Screening" with "checklist" Tables 4-A through 4-D

Only isolated minor corrections required? — No

Skip Initial Screening (optional)

Yes → Make isolated minor corrections for compliance.

Check Qualification for Prescriptive Provisions of Section 405
403.1 Exception 2
• Two-story building
• No [plan?] irregularity
• Roofing weighs 5 psf or less
• Second-floor diaphragm meets current code aspect ratio
• Section 405 deemed appropriate by code official

405.1.2 Limitation
• Either of:
  ○ Floor diaphragm is wood structural panel or diagonal sheathing
  ○ New wood structural panel diaphragm is provided

405.1.3 Additional conditions
• Second-floor diaphragm aspect ratio <=0.67
• Side shear walls are 20 ft min, < 10% openings
• Rear shear wall is ¾ of wall length, <=10% openings in each panel

All qualifications met?

Yes

No

Prescriptive Provisions
405.1.2   New wood structural panel diaphragm if (E) diaphragm is straight sheathing
405.2.1   Anchor bolts at side & rear walls
405.2.2   Top pl-rim joist conn at side & rear walls
405.2.3   Plwd shtg at side & rear walls, 3x blocking all sides
405.2.4   Hold-downs at each end of side & rear wall panels
406   Materials
407   Required information on plans
408   Quality control

General Provisions
403.2   Scope of analysis: except at top story, meet "the strength and stiffness requirements herein," plus load path and soil check.

403.3   Strength: Design base shear = 75% of current code
403.5   Strength: Weak story limitation, at least 65% or 80%
403.6   Stiffness: Story drift limit = 0.025, plus deformation compatibility check
403.8   Ties and continuity
403.9   Collector elements

404   Phased construction
406   Materials
407   Required information on plans
408   Quality control

Check Exemption to Full Scope
403.2 Exception:
• 2-story building
• SWOF wall line at ground level due to parking
• Parking area < 20% ground floor area

Exemption applies? — No

Yes

Apply general provisions to critical wall lines per 403.2 exception
403.2 Exception: "Only the wall lines in the open, weak or soft directions of the enclosed parking area need comply with the provisions of this chapter."

For SI: 1 foot = 304.8 mm, 1 pound per square foot = 47.88 Pa.

Figure A4-4
FLOWCHART OF CHAPTER 4 PROVISIONS

Bibliography

The following resource materials are referenced in this chapter or are relevant to the subject matter addressed in this chapter.

APA, *Wood Structural Panel Shear Wall Deflection Formula* (Form No. TT-053). APA, Technical Services Division, August 2000. (Available from www.apawood. org.)

APA, *Diaphragms and Shear Walls: Design/construction Guide* (Form No. L350G). APA, September 2001. (Available from www.apawood.org.)

ASCE 1998, *Handbook for the Seismic Evaluation of Buildings—A Prestandard (FEMA 310)* Washington, DC: Federal Emergency Management Agency, January.

CUREE 2002, *Recommendations for Earthquake Resistance in the Design and Construction of Woodframe Buildings, Part I – Recommendations and Part II – Topical Discussions* (Final Review Draft) Richmond, CA: Consortium of Universities for Research in Earthquake Engineering, November 15.

Fremont, City of. Ordinance No. 2363.

Hamburger, R. *Lessons Learned in the Northridge Earthquake on Wood-Frame Buildings.* Seminar Papers: Northridge Earthquake, Lessons Learned, SEAONC 1994 Spring Seminar. San Francisco, CA: Structural Engineers Association of Northern California, 1994.

Harder, Robert. *City of Los Angeles Department of Building and Safety and Structural Engineers Association of Southern California Joint Task Force: Wood Frame Construction Code Recommendations.* Proceedings: 63rd Annual Convention (1994 SEAOC Convention) Structural Engineers Association of California, 1994.

Harris, S.K., Scawthorn, C., and Egan, J. *Damage in the Marina district of San Francisco in the October 17, 1989 Loma Prieta earthquake.* Tokyo: Proceedings of the 8th Japan Earthquake Engineering Symposium, 1990.

Holmes, W. and Somers, P., editors. "Northridge Earthquake of January 17, 1994". Reconnaissance Report, Volume 2. supplement C to volume 11. *Earthquake Spectra*, January 1996.

IBC-03, *International Building Code.* Country Club Hills, IL: International Code Council, 2003.

ICBO-1999, *Los Angeles Building Code*, Chapter 93. Whittier, CA: International Conference of Building Officials, 1999.

LADBS, "Multistory Wood Frame Buildings with Nonconforming Structural Materials and Tuck-under Parking Features: Summary of Wood Light-frame Building Subcommittee (Draft version 5.7.99)."

LADBS/ SCEC/SEAOSC Ground Motion Joint Task Force Committee, March, 1999.

Mendes, S. *Lessons Learned from Four Earthquake Damaged Multi-story Type V Structures.* Proceedings: 64th Annual Convention, Indian Wells, CA: 1995 SEAOC Convention, 1995.

Rutherford and Chekene. *Seismic Rehabilitation of Three Model Buildings with Tuckunder Parking: Engineering Assumptions and Cost Information.* San Jose CA: Department of Housing and Office of Emergency Services, May 2000.

Seismology Committee, SEAOC, *Recommended Lateral Force Requirements and Commentary, Seventh Edition.* Sacramento, CA: Structural Engineers Association of California, 1999.

Stewart, J., Bray, J., McMahon, D., and Kropp, A. "Seismic performance of hillside fills." *Geotechnical Special Publication,* No. 52. American Society of Civil Engineers, 1995.

UBC-97, *Uniform Building Code.* Whittier, CA: International Conference of Building Officials, 1997.

Vukazich, S. *The Apartment Owner's Guide to Earthquake Safety.* San Jose CA:, Residential Seismic Safety Program, August, 1998.

CHAPTER A5

EARTHQUAKE HAZARD REDUCTION IN EXISTING CONCRETE BUILDINGS AND CONCRETE WITH MASONRY INFILL BUILDINGS

Preface

Concrete frame buildings have been recognized as being susceptible to damage from seismic activity. The San Fernando earthquake of 1971, the Mexico City earthquake of 1985, the Whittier Narrows earthquake of 1987, the Loma Prieta earthquake of 1989, the Northridge earthquake of 1994, the Kobe earthquake of 1995, the Izmit (Turkey) earthquake of 1999 and the Gi-Gi (Taiwan) earthquake of 1999 all caused substantial damage to concrete frame buildings.

The gravity load-resisting system consisting of walls, columns and beams is generally cast-in-place concrete. Concrete frame buildings with masonry infills have residual gravity load stresses in these frames because the forms and shoring were removed before the masonry infill was constructed. Masonry infill walls are placed into the concrete frames, generally on the exterior wall line, to provide closure and fire protection. These masonry infills constrain the expected lateral deformations of the moment frame and could cause the infilled frame to respond to earthquake shaking as a "shear wall" or with significant torsional response. Infill masonry of sill height, below windows, can change the relative stiffness of the concrete columns, concentrate shear resistance in the "lower" height columns and change the expected failure mode in the columns from flexural to shear.

Code required detailing for lateral-load-resisting reinforced concrete systems changed in the 1967 *Uniform Building Code* (UBC) and were validated by damage surveys immediately after the 1971 San Fernando earthquake. These 1967 code changes were related to UBC Zones 2 and 3 in a seismic hazard mapping that has extremely little relationship to current *International Building Code*® (IBC®) hazard mapping. The detail requirements may be similar to the 1976 code but the design requirements are significantly different. Lateral-load-resisting systems such as ordinary, intermediate and special reinforced concrete frames were not considered in the 1967 code. Caution is recommended for accepting detail standards as existing in the concrete building without verification by destructive and nondestructive methods. The adoption dates given in Section A502 (1992 NBC, 1993 SBC and 1976 UBC) give reasonable confidence that special reinforced concrete moment frames will have incorporated in their construction the detailing required for structures with that definition. Assignment of a description such as "intermediate" or "ordinary" for selection of an *R*-factor should be based on the description of these definitions in the IBC.

The provisions in this chapter are derived from several sources. The Federal Emergency Management Agency (FEMA) and the Applied Technology Council (ATC) published a document (called ATC 33, later known as a Prestandard FEMA 356) that developed a methodology to analyze existing buildings. Included in this document are provisions for analysis of concrete buildings. ATC 40 is another document intended to assist engineers specifically in the analysis of concrete buildings. In response to Senate Bill 547, the City of Los Angeles took steps to develop provisions (Division 95) to be utilized for concrete frame buildings within the city's jurisdiction. The Structural Engineers Association of California (SEAOC) has had ongoing activity relating to the analysis of nonductile concrete buildings. Also, the Strong Motion Instrumentation Program (SMIP) has funded a study relating to nonductile concrete buildings with unreinforced masonry infill walls in concrete and steel frames (Kariotis et al, 1994).

The loss of lives and property in future earthquakes will be heaviest in buildings that exist today. The short-term impact of improved knowledge, codes and design practices will be limited to those relatively few recent buildings that take advantage of this improved knowledge. Thus, the dominant policy issues posed by earthquakes involve not new but existing buildings, particularly those structures that have obvious weaknesses and do not comply with the general intent and necessary requirements of current regulations. The issue before the public and the professionals is how to set standards for these noncompliant existing buildings that are consistent with both the desire for safety and the limited resources available to achieve improved safety.

It is generally accepted that the intensity of earthquakes that could reasonably be expected to occur in California would be sufficient to cause buildings with minimal seismic resistance characteristics to be seriously damaged or, perhaps, to collapse, causing serious injury or death to the occupants or passers-by.

It is reasonable, when a real hazard exists, to take steps to significantly reduce the hazard. The objective of this chapter is the reduction or mitigation of hazard to the greatest extent practicable. Application of these provisions will decrease the probability of loss of life, but loss of life cannot be prevented. Also, we should be willing to accept some major and irreparable damage as long as there is a decrease in the likelihood of loss of support for horizontal framing.

The performance goal of this chapter is lower than the performance goal set for new construction. It should be recognized that the economic costs and difficulty of strengthening existing buildings necessitates reliance on building components with seismic performance characteristics that are less than ideal.

SECTION A501
PURPOSE

The purpose of this chapter is to promote public safety and welfare by reducing the risk of death or injury that may result from the effects of earthquakes on concrete buildings and concrete frame buildings with masonry infills.

The provisions of this chapter are intended as minimum standards for structural seismic resistance, and are established primarily to reduce the risk of life loss or injury. Compliance with the provisions in this chapter will not necessarily prevent loss of life or injury or prevent earthquake damage to the rehabilitated buildings.

❖ The purpose of this chapter is to reduce the probability of earthquake damage to existing concrete buildings, with or without masonry infills designed and constructed prior to the adoption of the building codes listed in Section A502, Exception 1.

SECTION A502
SCOPE

The provisions of this chapter apply to all buildings having concrete floors and/or concrete roofs supported by reinforced concrete walls or by concrete frames and columns, and/or to concrete frames with masonry infill.

Exception: This chapter shall not apply to:

1. Buildings designed in accordance with the seismic provisions of the 1993 *BOCA National Building Code*, the 1994 *Standard Building Code*, or the 1976 *Uniform Building Code*, or later editions of these codes, unless the seismicity of the region has changed since the design of the building.

2. Concrete buildings that have a flexible diaphragm at the roof level.

3. Concrete buildings and concrete with masonry infill buildings in Seismic Zones 0 and 1, or where Seismic Design Category A is permitted.

❖ The concept of "benchmark building" used in ASCE 31, *Seismic Evaluation of Existing Buildings*, is utilized to limit the scope of this chapter. While benchmark buildings need not proceed with further evaluation, the design professional must clearly demonstrate that the building is compliant with the benchmark document. One exception to the use of benchmark buildings is if the seismicity of the region has changed since the design of the building.

Exception 2 exempts this type of building because the semiflexible or flexible roof diaphragm modifies the seismic response from that anticipated by this chapter. This exemption should not be considered as an assumption that these buildings may not be hazardous. The analyst should use a structural model that includes the appropriate stiffness of the roof diaphragm.

Exception 3 is a committee decision that the seismic hazard in this geographic region does not warrant an analysis or retrofit.

SECTION A503
DEFINITIONS

For the purposes of this chapter, the applicable definitions and notations in the Building Code and the following shall apply:

MASONRY INFILL. An unreinforced or reinforced masonry wall construction within a reinforced concrete frame.

❖ The definition of masonry infill includes partial-height, perforated and full-height masonry placed in the concrete frame. Any masonry that constrains the expected lateral deformation of the frame alone, without infill, changes the lateral stiffness of the system and the shear demand on the structure and the frame members, especially the columns. All effects of the infills must be included in the three-dimensional structural model.

SEISMIC USE GROUP III. Those buildings categorized as essential facilities or hazardous facilities, or as designated by the building official.

SECTION A504
SYMBOLS AND NOTATIONS

For the purposes of this chapter, the applicable symbols and notations in the Building Code and the following shall apply.

a_{pi} = Spectral acceleration ordinate of the trial performance point in the Acceleration-Displacement Response Spectra (ADRS) domain.

a_y = Spectral acceleration ordinate of the yield point of the capacity curve in the Acceleration-Displacement Response Spectra (ADRS) domain.

d_{pi} = Spectral displacement ordinate of the trial performance point in the Acceleration-Displacement Response Spectra (ADRS) domain.

d_y = Spectral displacement ordinate of the yield point of the capacity curve in the Acceleration-Displacement Response Spectra (ADRS) domain.

PF_1 = Participation factor for the first or primary natural vibration mode of the structure.

S_a = Spectral acceleration.

S_d = Spectral displacement.

SR_A = Modification factor for the 5-percent damped acceleration response spectra in the constant acceleration region.

SR_V = Modification factor for the 5-percent damped acceleration response spectra in the constant velocity region.

V = Total design base shear.

W = Total design seismic dead load as prescribed in the Building Code.

w_i = Portion of W that is located at or assigned to level i.

a_{LL} = Modal weight coefficient for the first or primary natural vibration mode of the structure.

D_{roof} = Roof displacement relative to the ground.

α_1 = Modal weight coefficient for the first or primary natural vibration mode of the structure.

$\varphi i, I$ = The first or primary natural vibration mode shape coordinate at floor level i in the direction of the applied seismic loading, S_a.

$\varphi roof, I$ = The first or primary natural vibration mode shape coordinate at the roof level in the direction of the applied seismic loading, S_a.

$\phi i, j$ = Displacement amplitude of floor level i in the jth natural vibration mode of the structure.

$\phi roof, j$ = Displacement amplitude of the roof level in the jth natural vibration mode of the structure.

❖ These symbols and notations are the same as given in the IBC and documents such as ASCE 31, FEMA 356, ATC 40 and textbooks.

SECTION A505
GENERAL REQUIREMENTS

A505.1 General. This chapter provides a three-tiered procedure to evaluate the need for seismic rehabilitation of existing concrete buildings and concrete buildings with masonry infills. The evaluation shall show that the existing building is in compliance with the appropriate part of the evaluation procedure as described in Sections A507, A508 and A509, or shall be modified to conform to the respective acceptance criteria. This chapter does not preclude a building from being evaluated or modified to conform to the acceptance criteria using other well-established procedures, based on rational methods of analysis in accordance with principles of mechanics and approved by the authority having jurisdiction.

Evaluation of concrete buildings with masonry infill shall be in accordance with Tier 3 analysis procedure as described in Section A509.

❖ This chapter provides a three-tiered procedure to evaluate the need for seismic rehabilitation of existing concrete buildings and concrete buildings with masonry infills. This chapter does not preclude a building from being evaluated or rehabilitated to conform to the acceptance criteria using other well-established procedures, such as ASCE 31, prestandard FEMA 356 or ATC 40. Approval of the authority having jurisdiction shall be obtained before using alternative procedures.
Concrete buildings with masonry infill should have special consideration for the seismic response of this class of buildings. It is not conservative to ignore the contribution of masonry infills to the structural stiffness. These infills may cause the building to have weak and/or soft stories or significant torsional response. Pseudo-nonlinear modeling of the infill in a Tier-3 procedure, using the testing results obtained as specified in Section A510, should include the effective stiffness of infills in the dynamic analysis.

A505.2 Properties of in-place materials. Except where specifically permitted herein, the stress-strain relationship of concrete, masonry and reinforcement shall be determined from published data or by testing. All available information, including building plans, original calculations and design criteria, site observations, testing, and records of typical materials and construction practices prevalent at the time of construction, shall be considered when determining material properties.

For Tier 3 analyses, expected material properties shall be used in lieu of nominal properties in the calculation of strength, stiffness and deformabiltity of building components.

The procedure for testing and determination of stress-strain values shall be as prescribed in Sections A505.2.1 through A505.2.5.

❖ The stress-strain relationship of existing concrete, masonry and reinforcement shall be determined from published data or by testing. Section A505.2.1 provides exceptions for Tier-1 and -2 analyses.
All available information from the sources given in this section shall be considered when determining material properties.

A505.2.1 Concrete. The compressive strength of existing concrete shall be determined by tests on cores sampled from the structure.

Exceptions:

1. For Tier 1 analysis, the compressive strength of the concrete may be determined based on the information shown on the original construction documents or based on the values shown in Table A505.1.

2. For Tier 2 analysis, the compressive strength may be determined based on the information shown on the original construction documents.

Core testing shall be performed in accordance with the following:

1. The cutting of cores shall not significantly reduce the strength of the existing structure. Cores shall not be taken in columns. Existing reinforcement shall not be cut.

2. If the construction documents do not specify a minimum compressive strength of the classes of concrete, five cores per story, with a minimum of 10 cores, shall be obtained for testing.

   Exception: If the coefficient of variation of the compressive strength does not exceed 15 percent, the number of cores per story may be reduced to two and the minimum number of tests may be reduced to five.

3. When the construction documents specify a minimum compressive strength, two cores per story per class of concrete shall be taken in the areas where that concrete was to be placed. A minimum of five cores shall be obtained for testing. If a higher strength of concrete was specified for columns than the remainder of the concrete, cores taken in the beams for verification of the specified strength of the beams shall be substituted for tests in the columns. The strength specified for columns may be used in the analyses if the specified compressive strength in the beams is verified.

TABLE A505.1—ASSUMED COMPRESSIVE STRENGTH OF STRUCTURAL CONCRETE (psi)

TIME FRAME	FOOTINGS	BEAMS	SLABS	COLUMNS	WALLS
1965 or earlier	2,000	2,000	2,000	2,000	2,000
1966-Present	3,000	3,000	3,000	3,000	3,000

For SI: 1 pound per square inch = 6.89 kPa.

4. The sampling for the concrete strength tests shall be distributed uniformly in each story. If the building has shear walls, a minimum of 50 percent of the cores shall be taken from the shear walls. Not more than 25 percent of the required cores shall be taken in floor and roof slabs. The remainder of the cores may be taken from the center of beams at mid-span. In concrete frame buildings, 75 percent of the cores shall be taken from the beams.

5. The mean value of the compressive stresses obtained from the core testing for each class of concrete shall be used in the analyses. Values of peak strain that are associated with peak compressive stress may be taken from published data for the nonlinear analyses of reinforced concrete elements.

❖ The compressive strength of cores taken shall be determined by testing for Tier-3 analyses. The engineer may use core testing for the determination of concrete compressive strength for Tier-1 and -2 analyses. Cutting of cores shall be carefully planned by the engineer to not significantly reduce the shear or flexural strength of the existing structural system. Cutting of any existing reinforcement is prohibited. The engineer will have to prepare in writing a coring program and provide an on-site representative to enforce this requirement. Cutting cores in columns is also prohibited. The total area of a column has an existing axial stress. Patching of a core hole in a column will not restore the strength of the column because the patching material is installed at near-zero stress and will be subject to shrinkage.

This section allows the compressive test results obtained from testing in areas other than columns to be assumed for the columns. If the construction documents specify a higher compressive strength for columns than that of beams and shear walls, the higher strength specified for the columns may be used in the analysis if testing shows that the strength specified for beams is verified by the cores taken from the beams.

The mean value of the compressive strengths obtained from the test program shall be used in the analyses.

A505.2.2 Solid-grouted reinforced masonry. The compressive strength of solid-grouted concrete block or brick masonry may be taken as 1,500 pounds per square inch (10.3 MPa). The strain associated with peak stress may be taken as 0.0025.

❖ The engineer may use the prism compressive strength specified in this section or may elect to cut prisms from the existing infill walls and test them in accordance with the procedures given in the IBC. The strain associated with peak stress shall be determined by prism testing or taken as 0.0025 inches (0.6 mm) per inch.

The size and number of cut prisms shall conform to the requirements of the IBC for testing of constructed masonry. Prisms should not be cut from the areas of high stress in infills. These areas are adjacent to the intersections of beams and columns and adjacent to corners of wall openings.

A505.2.3 Partially grouted masonry. A minimum of five units shall be removed from the walls and tested in conformance with UBC Standard 21-4. Compressive strength of the masonry may be determined in accordance with UBC Table 21-D, assuming Type S mortar. The strain associated with peak stress may be taken as 0.0025.

❖ The prism compressive strength of partially grouted masonry may be determined by methods given in the IBC as the "Unit Strength Method." Units are taken from the wall by removing mortar joints for unit testing. The number of units needed for testing shall conform to the IBC. The ratio of the number of unit strength method test areas to wall area should conform to that specified in the IBC Section 2105.3. The assumed strength of the grout for the unit strength method shall not be assumed as less than 2,000 pounds per square inch (psi) (13 790 kPa), or a higher compressive strength of the grout may be determined by testing grout cores. Mortar need not be analyzed for cementitious materials. Mortar may be assumed to be Type S.

A505.2.4 Unreinforced masonry. The stress-strain relationship of existing unreinforced masonry shall be determined by in-place cyclic testing. The test procedure shall conform to Section A510.

One stress-strain test per story and a minimum of five tests shall be made in the unreinforced masonry infills. The location of the tests shall be uniformly distributed throughout the building.

The average of the stress-strain values obtained from testing shall be used in the nonlinear analyses of frame infill assemblies.

❖ The procedure given in this section is intended for solid masonry, either clay or concrete. The "flat jack" method given in Section A510 cannot be used in cored brick or in hollow core unit masonry. It is recommended that the compressive strength of cored or hollow unit masonry be determined as specified in Section A505.2.3.

A505.2.5 Reinforcement. The expected yield stress of each type of new or existing reinforcement shall be taken from Table A505.2, unless the reinforcement is sampled and tested for yield stress. The axial reinforcement in columns of post-1933 buildings shall be assumed to be hard grade unless noted otherwise on the construction documents.

❖ The expected yield stress of new or existing reinforcement shall be taken from Table A505.2 unless sampled and tested for yield stress. Table A505.2 is a tabulation of expected yield stress, not a minimum yield stress. If tested yield stress values exceed those shown in Table A505.2, the mean of the test values should be used.

Underestimation of reinforcement yield stress is not a conservative assumption. The limit shear loading of a beam, column or wall is determined in accordance with Section A509.1.2.3. The probable moments should be based on expected yield moments. The concrete strain at yield rotation is determined by the sum of axial loading and yield stress of all reinforcement within the cracked concrete section. Again, understatement of probable reinforcement yield stress is nonconservative.

SECTION A506
SITE GROUND MOTION

A506.1 Site ground motion for Tier 1 analysis. The earthquake loading used for determination of demand on elements and the structure shall correspond to that required by the Building Code.

❖ The ground motion (base shear) used for a Tier-1 analysis is the same as used for the design of a new building. Seventy-five percent of the IBC base shear is used for Tier-2 and -3 analyses. The concept is that finding an existing concrete building is in full conformance with the intent of this chapter and should be based on a more restrictive standard (higher base shear), and procedures that involve more than a minimum inspection and evaluation should have a lower probability of acceptance than a more rigorous procedure.

A506.2 Site ground motion for Tier 2 analysis. The earthquake loading used for determination of demand on elements and the structure shall conform to 75 percent of that required by the Building Code.

❖ The purpose of this chapter is earthquake hazard reduction. This does not imply that existing buildings must fully conform to the standards for new construction for reduction of hazard. This reduction of the seismic loading is identical to that used in ASCE 31 for Tier 3 analyses and near identical to that used for FEMA 178 analyses.

A506.3 Site ground motion for Tier 3 analysis. The site ground motion shall be an elastic design response spectrum prepared in conformance with the Building Code but having spectral acceleration values equal to 75 percent of the code design response spectrum. The spectral acceleration values shall be increased by the occupancy importance factor when required by the Building Code.

❖ The loading criteria for a Tier-3 analysis is an elastic design response spectrum compatible with mapped spectral acceleration and velocity parameters as modified by site soil class factors. The spectral acceleration values shall be reduced as permitted for a Tier-2 analysis. The spectral acceleration values are increased by occupancy importance factors when applicable.

All Tier-3 analysis procedures are displacement based. The building is displaced relative to its base in appropriate mode shapes in its linear and nonlinear response. The acceptance criteria is a limit on material strain, sum of linear and nonlinear story drift, global strength degradation and shear strength.

SECTION A507
TIER 1 ANALYSIS PROCEDURE

A507.1 General. Structures conforming to the requirements of this section may be shown to be in conformance with this chapter by submission of a report to the building official as described in this section.

❖ This section provides a means for conducting a simplified and rapid analysis evaluation of existing concrete buildings. This section is not intended to be used for concrete buildings with masonry infill.

TABLE A505.2—ASSUMED YIELD STRESS OF EXISTING REINFORCEMENT (psi)

TYPE OF REINFORCEMENT AND ERA	ASSUMED YIELD STRESS (psi)
Pre-1940 structural and intermediate grade, plain or deformed	45,000
Pre-1940 twisted and hard grade	55,000
Post-1940 structural and intermediate grade	45,000
Post-1940 hard grade	60,000
ASTM A 615 Grade 40	50,000
ASTM A 615 Grade 60	70,000

For SI: 1 pound per square inch = 6.89 kPa.

The evaluation procedure is patterned after ASCE 31 with emphasis on Tier-1 analysis for building Type C1: Concrete Moment Frame, and Type C2: Concrete Shear Walls. Nevertheless, the analysis is based on an equivalent lateral-force procedure in lieu of the displacement method used by the referenced document.

ASCE's regions of seismicity and performance level criteria have been converted into two seismic hazard categories to closely approximate the occupancy categories of the 1997 UBC and the 2000 IBC.

A507.2 Limits. This section shall apply only to buildings for which a visual inspection can verify that their configuration is essentially regular in mass and geometry.

Exception: Buildings containing some plan or vertical irregularity may be evaluated by this section, provided rational structural calculations are performed to show that the following irregularities do not exceed the limits for a regular building classification as defined by the Building Code:

1. Weak story.
2. Soft story.
3. Geometry.
4. Vertical discontinuity.
5. Mass.
6. Torsion.

❖ In an effort to maintain the simplicity and rapid nature of the evaluation procedure, this section is limited to regular buildings.

The visual inspections required to verify the regularity of the building should include both an on-site inspection of the building and review of the record drawings. When record drawings are not available, some forms of nondestructive inspection may be required.

A507.3 Evaluation report. The engineer or architect of record shall prepare a report summarizing the analysis conducted in conformance with this section. As a minimum, the report shall include the following items:

1. Building description.
2. Site inspection summary.
3. Summary of reviewed record documents.
4. Earthquake design data used for the evaluation of the building.
5. Completed checklists.
6. Quick-check analysis calculations.
7. Summary of deficiencies.

❖ The evaluation report is intended to serve as a summary of the evaluation performed for review by the code official. It should clearly indicate one of two things: (1) only compliant items exist in the building and as such no rehabilitation measures are necessary, or (2) noncompliant items do exist in the building and the subsequent steps followed by the design professional in addressing these items. As noted in Sec-

tion A507.4, the design professional has the option of performing subsequent analyses (Tier 2 or 3) to show that the noncompliant items are acceptable or, if not acceptable, to pursue a mitigation effort to correct the deficiencies noted from the Tier-1 analysis.

When a subsequent analysis is performed, the report required under this section is intended to be used as supporting documentation for limiting the analysis to only the noncompliant items.

A507.4 Evaluation procedure. Prior to completing the required checklists, the following items shall be performed by the architect or engineer of record conducting the evaluation:

1. A site inspection shall be conducted, and any deficiencies and/or existing damage discovered shall be documented in the evaluation report per Section A507.3.
2. All available records regarding the construction, improvements and rehabilitation of the building shall besecured and reviewed.
3. Material characteristics of the building shall be determined in accordance with Section A505.2.
4. The necessary earthquake design data must be established for use in the evaluation in accordance with Section A506.1 and the Building Code.

Based on the above information, the appropriate checklist(s) shall be completed in accordance with Section A507.5. Upon completing the checklist(s), all noncompliant statements shall be summarized and included in the evaluation report.

All noncompliant items shall be mitigated by rehabilitating the structure, or shall be shown to be compliant by performing a Tier 2 or Tier 3 analysis.

❖ The design professional may seek to address noncompliant items directly within a Tier-1 analysis but it must be based on a rational analysis acceptable to the building official. Since such analysis does exist under Tier 2 or 3, the engineer or architect is encouraged to address noncompliant items accordingly.

A507.5 Evaluation checklists. Checklist selection shall be based on the Seismic Zone and occupancy or Seismic Design Category of the particular building(s) as follows:

1. Any occupancy in Seismic Zones 2A and 2B or Seismic Design Categories B and C: Basic Structural Checklist.
2. Any occupancy in Seismic Zones 3 and 4 or Seismic Design Categories D and E: Basic Structural Checklist and Supplemental Structural Checklist.

The Foundation Checklist shall be completed for all buildings.

Each evaluation statement of each required checklist shall be given one of the following marks: Compliant (C), Noncompliant (NC), or Not Applicable (N/A).

Statements that cannot be answered adequately because of a lack of information or that require further investigation beyond what is available at the time of the evaluation shall be deemed noncompliant (NC).

For buildings with a distinct lateral-force-resisting system in each principal direction, or with more than one type of lateral-force-resisting system in the same principal direction, a separate checklist evaluation shall be completed for each direction and/or system.

A507.6 Quick-check analysis procedure. Analysis under Section A507 shall be limited to quick-checks when required by the evaluation statements or the building official.

> **Exception:** Certain statements may require additional analysis not directly addressed under this section. In such conditions, the design professional shall use rational analytical methods in addressing the statement.

Buildings shall be analyzed to resist the minimum lateral forces assumed to act nonconcurrently in the direction of each principal axis of the structure in accordance with the Building Code.

Calculation of the design force level shall be in accordance with Section A506.1.

For buildings more than one story in height, the total force shall be distributed in accordance with the requirements of the Building Code.

Horizontal distribution of shear and torsional moments shall be in accordance with the Building Code. The 5-percent accidental torsion factor, the torsional amplification factor and the redundancy factor need not be considered.

❖ The lateral-force level used for performing the quick check analysis, or answering evaluation statements, is based on 100 percent of the IBC force level.

In computing the lateral force, the design professional must select an appropriate *R*-value. In making a determination, the design professional should consider the buildings confinement detailing, redundancy, toughness, overstrength, etc. As a guide, one may use the IBC in determining the appropriate *R*-value to use.

A507.6.1 Shear stress in frame columns. The average shear stress, V_{avg}, in the columns of concrete frames shall be computed in accordance with Equation (A5-1).

$$V_{avg} = \frac{n_c(V_j)}{(n_c - n_f)A_c}$$ **(Equation A5-1)**

Where:

A_c = Summation of the cross-sectional area of all columns in the story under consideration.

n_c = Total number of columns of the frames, in the direction of loading.

n_f = Total number of frames in the direction of loading.

V_j = Story shear computed in accordance with Section A507.6.

Equation (A5-1) assumes that all of the columns in the frame have similar stiffness. When the above assumption leads to an unconservative condition, the load shall be distributed in accordance with the columns' relative rigidities.

A507.6.2 Shear stress in shear walls. The average shear stress in shear walls, V_{avg}, shall be calculated in accordance with Equation (A5-2).

$$V_{avg} = V_j/A_w$$ **(Equation A5-2)**

Where:

A_w = Summation of the horizontal cross-sectional area of all shear walls in the direction of loading. Openings shall be taken into consideration when computing A_w. For masonry walls, the net area shall be used.

V_j = Story shear at level j computed in accordance with Section A507.6.

A507.6.3 Axial stress because of overturning. The axial stress of columns subjected to overturning forces, P_{ot}, shall be calculated in accordance with Equation (A5-3).

$$P_{ot} = \frac{2}{3} \times \frac{V(h_n)}{(Ln_f)}$$ **(Equation A5-3)**

Where:

h_n = Height (in feet) above the base to the roof level.

L = Total length of the frame (in feet).

n_f = Total number of frames in the direction of loading.

V = Lateral force.

SECTION A508
TIER 2 ANALYSIS PROCEDURE

A508.1 General. A Tier 2 analysis includes an analysis using the following linear methods: Static or equivalent lateral force procedures. A linear dynamic analysis may be used to determine the distribution of the base shear over the height of the structure. The analysis, as a minimum, shall address all potential deficiencies identified in Tier 1, using procedures specified in this section.

If a Tier 2 analysis identifies a nonconforming condition, such condition shall be modified to conform to the acceptance criteria. Alternatively, the design professional may choose to perform a Tier 3 analysis to verify the adequacy of the structure.

❖ The Tier-2 linear analysis procedure is a building-code type of analysis procedure, using a single "global" response modification factor, *R*, for the entire structure. This approach differs from the linear analysis procedure used in FEMA 356, which employs a "component-based" ductility related factor, *m*, in checking the acceptability of the component evaluated.

The *R*-factor for Tier-2 analyses shall be selected based on the type of existing seismic-force-resisting system, e.g., ordinary reinforced concrete moment frame, R=3; ordinary reinforced concrete shear walls, R=5; detailed plain concrete shear wall, R=3. The designer shall refer to the Chapter 16 of the IBC, or ASCE 7 for a complete list of seismic-force-resisting systems and their associated design coefficients.

A508.2 Limitations. A Tier 2 analysis procedure may be used if:

1. There is no in-plane offset in the lateral-force-resisting system.

2. There is no out-of-plane offset in the lateral-force-resisting system.

3. There is no torsional irregularity present in any story. A torsional irregularity may be deemed to exist in a story when the maximum story drift, computed including accidental torsion, at one end of the structure transverse to an axis is more than 1.2 times the average of the story drifts at the two ends of the structure.

4. There is no weak story irregularity at any floor level on any axis of the building. A weak story is one in which the story strength is less than 80 percent of that in the story above. The story strength is the total strength of all seismic-resisting elements sharing the story shear for the direction under consideration.

> **Exceptions:** Static or equivalent lateral force procedures shall not be used if:
>
> 1. The building is more than 100 feet (30 480 mm) in height.
>
> 2. The building has a vertical mass or stiffness irregularity (soft story). Mass irregularity shall be considered to exist where the effective mass of any story is more than 150 percent of the effective mass of any adjacent story. A soft story is one in which the lateral stiffness is less than 70 percent of that in the story above or less than 80 percent of the average stiffness of the three stories above.
>
> 3. The building has a vertical geometric irregularity. Vertical geometric irregularity shall be considered to exist where the horizontal dimension of the lateral-force-resisting system in any story is more than 130 percent of that in an adjacent story.
>
> 4. The building has a nonorthogonal lateral-force-resisting system.

❖ The limits for use of the linear procedures are given. Linear static procedures are most applicable to buildings that actually have sufficient strength to remain nearly elastic when subjected to the design earthquake or with regular geometries and distributions of stiffness and mass.

For buildings with plan irregularities or weak stories, the inelastic ductility demand may significantly differ from the result of linear analyses; therefore, Tier-3 analysis procedures shall be used. For buildings with vertical irregularities or significant higher mode response, linear dynamic procedures may provide a more accurate distribution of seismic demand as compared to the linear static procedure.

A508.3 Analysis procedure. A structural analysis shall be made for all structures in accordance with the requirements of the Building Code, except as modified in Section A506. The re-

sponse modification factor, R, shall be selected based on the type of seismic-force-resisting system employed.

❖ The stiffness of any component shall be calculated using the values given in Table A508.1. The building period calculated using the percentage of the gross stiffness for effective stiffness may exceed the upper limit of calculated period given in the IBC. For determining the effective stiffness of a flat plate or flat slab, the designer shall refer to the available research data. The empirical limits on elastic fundamental period given in the IBC are not appropriate for analysis of the diversity of structural systems found in existing buildings.

A508.3.1 Mathematical model. The three-dimensional mathematical model of the physical structure shall represent the spatial distribution of mass and stiffness of the structure to an extent that is adequate for the calculation of the significant features of its distribution of lateral forces. All concrete and masonry elements shall be included in the model of the physical structure.

> **Exception:** Concrete or masonry partitions that are isolated from the concrete frame members and the floor above.

Cast-in-place reinforced concrete floors with span-to-depth ratios less than three-to-one may be assumed to be rigid diaphragms. Other floors, including floors constructed of precast elements with or without a reinforced concrete topping, shall be analyzed in conformance with the Building Code to determine if they must be considered semi-rigid diaphragms. The effective in-plane stiffness of the diaphragm, including effects of cracking and discontinuity between precast elements, shall be considered. Parking structures that have ramps rather than a single floor level shall be modeled as having mass appropriately distributed on each ramp. The lateral stiffness of the ramp may be calculated as having properties based on the uncracked cross section of the slab exclusive of beams and girders.

A508.3.2 Component stiffness. Component stiffness shall be calculated based on the approximate values shown in Table A508.1.

A508.4 Design, detailing requirements and structural component load effects. The design and detailing of the components of the seismic-force-resisting system shall comply with the requirements of Section 1620 of the Building Code, unless specifically modified herein.

A508.5 Acceptance criteria. The calculated strength of a member shall not be less than the load effects on that member.

A508.5.1 Load combinations. For Load and Resistance Factor Design (Strength Design), structures and all portions thereof shall resist the most critical effects from the combinations of factored loads prescribed in the Building Code.

> **Exception:** For concrete beams and columns, the shear effect shall be determined based on the most critical load combinations prescribed in the Building Code. The shear load effect because of seismic forces shall be multiplied by a factor of C_d, or $0.7 R$ for use with the UBC, but combined shear load effect needs not be greater than V_e, as calculated in ac-

cordance with Equation (A5-4). M_{pr1} and M_{pr2} are the end moments, assumed to be in the same direction (clock-wise or counter clock-wise), based on steel tensile stress being equal to $1.25 f_y$ where f_y is the specified yield strength.

$$V_e = \frac{M_{pr1} + M_{pr2}}{L} \pm \frac{W_g}{2} \qquad \textbf{(Equation A5-4)}$$

Where:

W_g = Total gravity loads on the beam.

A508.5.2 Determination of the strength of members. The strength of a member shall be determined by multiplying the nominal strength of the member by a strength reduction factor, ϕ. The nominal strength of the member shall be determined in accordance with the Building Code.

SECTION A509
TIER 3 ANALYSIS PROCEDURE

A509.1 General. A Tier 3 evaluation shall be performed using the procedure prescribed in Section A509.2. Alternatively, the procedures prescribed in Sections A509.3 and A509.4 may also be used where specifically permitted herein.

❖ The Tier 3 analysis procedures include a pseudo-nonlinear dynamic procedure, a capacity spectrum analysis procedure based on ATC and the nonlinear static procedure based on FEMA 356. Review of these documents is recommended before using these analysis procedures. Revisions and commentary on these documents is suggested by FEMA 440.

This chapter imposes more restrictions on use of ATC 40 and FEMA 356 when plan irregularities exist in the three-dimensional model of the existing structure than that specified in those documents. Near collapse and partial collapse of nonductile concrete buildings has been determined to be related in many cases to torsional response to orthogonal earthquake shaking. The pseudo-nonlinear dynamic has no limitation on plan irregularities. The procedure requires concurrent orthogonal response spectrum analysis and degradation of system and element stiffness in each iterative analysis. However, the engineer should consider introduction of new elements that will reduce the torsional irregularities into the existing structural model prior to beginning the analysis. This will increase the probability that the pseudo-nonlinear analysis will have an acceptable closure. The same option is available for use of the nonlinear static procedure and the capacity spectrum method. The plan irregularity may be reduced to the limits specified in this chapter prior to beginning the procedure.

The first procedure described in this section is called a pseudo-nonlinear dynamic analysis because it is an iterative procedure that modifies the stiffness matrix for each subsequent iteration. The effective stiffness used in the iterative procedure is the element or system stiffness taken as a linear secant stiffness from a nonlinear force-displacement plot for that element or system.

The pseudo-nonlinear dynamic procedure requires two tools, a linear elastic analysis program such as SAP 2000 that estimates story displacements on orthogonal axes for each modal response spectra analysis and a nonlinear analysis program such as FEM / I or a similar nonlinear analysis program that was written specifically for reinforced concrete or masonry (Kunnath, et al, 1992). The nonlinear analysis program should be capable of mimicking the behavior of experimental test specimens. Programs capable of replicating experimental behavior typically include post-yield

TABLE A508.1—COMPONENT STIFFNESS

COMPONENT	FLEXURAL RIGIDITY	SHEAR RIGIDITY[1]	AXIAL RIGITY
Beam, nonprestressed	$0.3 - 0.5\ E_c I_g$	$0.4\ E_c A_w$	$E_c A_g$
T- or L-shape beams, nonprestressed	$0.25 - 0.45\ I_g$	$0.4\ E_c A_w$	$E_c A_g$
Beam, prestressed	$10.\ E_c I_g$	$0.4\ E_c A_w$	$E_c A_g$
Column in compression ($P > 0.5\ f'_c A_g$)	$0.7 - 0.9\ E_c I_g$	$0.4\ E_c A_w$	$E_c A_g$
Column in compression ($P \leq 0.5\ f'_c A_g$)	$0.5 - 0.7\ E_c I_g$	$0.4\ E_c A_w$	$E_c A_g$
Column in tension	$0.3 - 0.5\ E_c I_g$	$0.4\ E_c A_w$	$E_s A_s$
Walls	To be determined based on rational procedures	$0.4\ E_c A_w$	$E_c A_g$
Flat slab, nonprestressed	To be determined based on rational procedures		
Flat slab, prestressed	To be determined based on rational procedures		

1. For shear stiffness, the quantity $0.4\ E_c$ has been used to represent the shear modulus, G.

strain hardening of the reinforcement, modeling of tension-stiffening during crack development in the concrete and utilizing the compressive stress-strain relationship to track strength degradation. These capabilities are needed for the nonlinear static procedure and the capacity spectrum method to replicate the nonlinear behavior of reinforced concrete.

The nonlinear static procedure uses the approximation that the sum of the linear and nonlinear displacement of a structure is equivalent to the displacement of an identical structure using its yield stiffness unchanged by pseudo-stresses in excess of yield stress. The structure is displaced by a load distributed over the height of the structure to a target displacement. This target displacement is based on the spectrum and the elastic period (stiffness) of the structure. The structure is modeled in a nonlinear analysis program and displaced by the applied loading to the specified target displacement. It is recommended that the distribution of the load on the height of the structure be verified by a response spectrum analysis at significant changes in the displaced shape as determined by the nonlinear analysis. The analyst should determine an effective yield point in the nonlinear analysis to verify the preliminary yield stiffness used for determining the target displacement. This chapter requires that the structure be displaced to 150 percent of the basic target displacement specified by FEMA 356. The element strain acceptance is verified at the structural displacement specified in this chapter.

The capacity spectrum method is described in Chapter 8 of ATC 40. The brief description given in this commentary is given only for assistance to the analyst in choosing an analysis program suitable for his or her analytical capabilities. The capacity spectrum method utilizes a nonlinear push-over such as required by the nonlinear static procedure to determine a force (spectral acceleration)-spectral displacement relationship. The top displacement of the structure is related to the displacement of a single-degree-of-freedom (SDOF) substitute structure at the height of its equivalent mass. This allows the "capacity" to be plotted on the acceleration displacement response spectrum. The capacity spectrum method allows consideration of the equivalent viscous damping of the SDOF spectrum. Viscous damping is used in preparation of spectra for convenience in response-spectra analysis. However, it is generally agreed that hysteretic damping due to nonlinear behavior is the predominate damping. A recommended reference for conversion of hysteretic damping to viscous damping is *Dynamics of Structures* by Anil K. Chopra. The analysis tools needed for the capacity spectrum method are the same as needed for the alternative Tier-3 procedures.

A509.1.1 Mathematical model. The three-dimensional mathematical model of the physical structure shall represent the spatial distribution of mass and stiffness of the structure to an extent that is adequate for the calculation of the significant features of its dynamic response. All concrete and masonry elements shall be included in the model of the physical structure.

Exception: Concrete or masonry partitions that are isolated from the concrete frame members and the floor above.

Cast-in-place reinforced concrete floors with span-to-depth ratios less than three-to-one may be assumed to be rigid diaphragms. Other floors, including floors constructed of precast elements with or without a reinforced concrete topping, shall be analyzed in conformance with the Building Code to determine if they must be considered semi-rigid diaphragms. The effective in-plane stiffness of the diaphragm, including effects of cracking and discontinuity between precast elements, shall be considered. Parking structures that have ramps rather than a single floor level shall be modeled as having mass appropriately distributed on each ramp. The lateral stiffness of the ramp may be calculated as having properties based on the uncracked cross section of the slab exclusive of beams and girders.

❖ The mathematical model must represent the existing building as discovered in the survey. All building elements that extend the full story height in any story or restrict the curvature of any element must be included in the model.

Cast-in-place reinforced floor diaphragms that have a span-depth ratio of less than that specified in the document may be assumed to be rigid diaphragms in the model. Floor diaphragms constructed of precast elements, either with or without cast-in-place reinforced concrete topping, shall also be analyzed to determine if they must be considered as semirigid diaphragms. Semirigid diaphragms must be represented in the mathematical model with a cracked (effective) stiffness. The probable elastic response of the diaphragm to the seismic loading should be used to estimate the effective stiffness.

Parking structures that have diaphragms that ramp between story heights shall have their tributary and self-weight distributed along their span length. The span of the diaphragm may be taken as the distance between the level floors that are at the story levels used in the mathematical model.

A509.1.2 Acceptance criteria.

❖ This section has three separate limits for acceptance. They are maximum values of compressive strain, story drift and shear strength. The lesser of the limits shall be used for acceptance. The compressive strain in elements or systems is determined for story drift as specified in Section A509.1.2.2. The limits for acceptance of compressive strain are specified in Section A509.1.2.2, Items 1 and 2.

A509.1.2.1 Compressive strain determination. Compressive strain in columns, shear walls and infills may be determined by the nonlinear analysis or a procedure that assumes plane sections remain plane.

Compressive strain shall be determined for combined flexure and axial loading. The seismic flexural moments and axial load shall be taken from the response spectrum analysis for frame or shear wall buildings, and from the substructure model for infill frames. The combination of critical effects for analysis of compressive strain shall be those given in the Building

Code for Strength Design or Load and Factor Resistance Design.

❖ Compressive strain in reinforced concrete elements may be calculated from nonlinear analyses that include warping of originally plane sections or from simplified analyses that assume plane sections remain plane, such as BIAX (Wallace and Moehle, 1989).

The critical strains shall be determined from the unreduced element loading calculated by the dynamic response of the mathematical model. The loading in the mathematical model shall be the load combinations of the IBC for that described as strength design or load-resistance factor design.

A509.1.2.2 Story drift limitation. Story drift is the displacement of one level relative to the level above or below, calculated by the response spectrum analysis using the appropriate effective stiffness. The story drift is limited to displacement that causes any of the following effects:

1. Compressive strain of 0.003 in the frame confining infill or in a shear wall.

2. Compressive strain of 0.004 in a reinforced concrete column, unless the engineer can show by published experimental research that the existing confinement reinforcement justifies higher values of strain.

3. Peak strain in masonry infills as determined by experimental data or by physical testing as prescribed in Section A510.

4. Displacement that was calculated by the nonlinear analysis as to when strength degradation of any element began.

   Exception: Item 4 may be taken as the displacement that causes a strength degradation in that line of resistance equal to 10 percent of the sum of the strength of the elements in that line of resistance.

❖ This section has four separate limits on story drift. The story drift that causes the lesser of the four limits shall be used for acceptance. The limitations are as follows.

A story drift that causes a compressive strain of 0.003 inch (0.8 mm) per inch is a limit for shear walls and concrete frames that confine masonry infills. These strains are considered acceptable for unconfined concrete. The acceptable compressive strain for concrete beams that is typical for frame buildings should be the same as that specified for beams that confine infills. The maximum strain for semiconfined (shell) concrete, which is the concrete outside the confinement reinforcement, is limited to 0.004 inch (.10 mm) per inch. The peak strain in masonry infills is limited to the strain shown acceptable by published experimental data or by the prescribed physical testing of the infill masonry.

The relative displacement at any story level is limited to the strength level displacement in that story. The strength level displacement is defined as the displacement when strength degradation begins in the nonlinear force-displacement relationship of a line of resistance in any story level. Strength degradation is defined as a point in the nonlinear force-displacement re-

lationship when a tangent to the plotted relationship is negative. It is recommended that the nonlinear displacement is increased beyond where the first point of negative slope is detected to verify that the drop in resistance is not an anomaly in the calculations. This limitation is modified by the exception that permits the beginning of strength degradation to be defined as a reduction of strength on a line of resistance, which is comprised of several elements, of 10 percent of the peak strength.

A509.1.2.3 Shear strength limitation. The required in-plane shear strength of all columns, piers and shear walls shall be the shear associated with the moments induced at the ends of columns or piers and at the base of shear walls by the story displacements. No strength reduction factors shall be used in the determination of strength.

❖ The shear strength of columns, piers and shear walls must exceed that calculated from the flexural strength of these elements. No strength reduction factors shall be used in the determination of flexural strengths. The nonlinear force-displacement relationship of an infilled frame panel is determined by an analysis that must include a failure mode. This failure mode can be either compressive strain or shear failure.

A509.2 Pseudo-nonlinear dynamic analysis procedure. Structures shall be analyzed for seismic forces acting concurrently on the orthogonal axes of the structure. The effects of the loading on two orthogonal axes shall be combined by SRSS methods. The analysis shall include all torsional effects. Accidental torsional effects need not be considered.

❖ This procedure requires a three-dimensional mathematical model of the building. This three-dimensional model is analyzed for concurrent seismic excitation on orthogonal axes of the building. The demand effects for the concurrent loading are combined by square root of the sum of the squares (SRSS) methods. This three-dimensional dynamic analysis procedure combines concurrent translational and torsional effects and does not require that torsional effects be magnified by accidental torsion.

A509.2.1 Determination of the effective stiffness.

❖ This section provides two methods for determination of the effective stiffness of reinforced concrete elements and of infilled frames. The pseudo-nonlinear dynamic analysis procedure uses the current effective stiffness of elements of the three-dimensional mathematical model to predict the displacements in the next iteration of a pseudo-nonlinear analysis. The current effective stiffness is a secant stiffness estimated from the force-displacement relationship obtained from a nonlinear analysis of each element of a story on a line of resistance. The iterative procedure concludes when the effective stiffness used is appropriate for the displacement calculated by the last dynamic analysis.

A509.2.1.1 General. The effective stiffness of concrete and masonry elements or systems shall be calculated as the secant stiffness of the element or system with due consideration of the effects of tensile cracking and compression strain. The secant stiffness shall be taken from the force-displacement relationship of the element or system. The secant stiffness shall be measured as the slope from the origin to the intersection of the force-displacement relationship at the assumed displacement. The force-displacement relationship shall be determined by a nonlinear analysis. The force-displacement analysis shall include the calculation of the displacement at which strength degradation begins.

Exception: The initial effective moment of inertia of beams and columns in shear wall or infilled frame buildings may be estimated using Table A508.1. The ratio of effective moment of inertia used for the beams and for the columns shall be verified by Equations (A5-5), (A5-6) and (A5-7). The estimates shall be revised if the ratio used exceeds the ratio calculated by more than 20 percent.

$$I_e = \left(\frac{M_{cr}}{M_a}\right)^3 I_g + \left[1 - \left(\frac{M_{cr}}{M_a}\right)^3\right] I_{cr} \quad \textbf{(Equation A5-5)}$$

Where:

$$M_{cr} = \frac{f_r I_g}{y_t} \quad \textbf{(Equation A5-6)}$$

and

$$f_r = 7.5\sqrt{f_c'} \quad \textbf{(Equation A5-7)}$$

❖ The effective stiffness of concrete and masonry for infills, which are covered in Section A509.2.1.2, is determined by a nonlinear analysis with consideration of the effects of tensile cracking of concrete and masonry and of the reduction of the chord compressive modulus of elasticity due to strain. The force-displacement relationship of elements shall include a determination of the relative displacement at the strength limit state. A force-displacement relationship is determined for the system or element. Judgement for using a primary mode shape or a multimode shape shall be used. The calculated story forces will provide guidance for the distribution of loading used in the nonlinear analysis. An exception to the requirement for a nonlinear analysis is provided for beams and columns in shear wall and infilled frame buildings. It is assumed that for such buildings, the shear walls or infilled frames will provide the principal resistance to seismic loading. The initial estimates used in the analyses must be revised if the effective stiffness calculated from the moments differs from the effective stiffness determined by the final analysis and varies from the initial estimates by more than 20 percent.

A509.2.1.2 Effective stiffness of infills. The effective stiffness of an infill shall be determined from a nonlinear analysis of the infill and the confining frame. The effect of the infill on the stiffness of the system shall be determined by differentiating the force-displacement relationship of the frame-infill system from the frame-only system.

❖ The effective stiffness of infills is determined by nonlinear analyses of representative infill panels. Representative panels are based on height-length ratios and location of openings (or no openings) within the panel. The confining frame without any infill is also analyzed. The stiffening effect of the infill is the difference between the force-displacement relationships of the frame with infill and without infill.

A509.2.1.3 Model of infill. The mathematical model of an infilled frame structure shall include the stiffness effects of the infill as a pair of diagonals in the bays of the frame. The diagonals shall be considered as having concrete properties and only axial loads. Their lines of action shall intersect the beam-column joints. The secant stiffness of the force-displacement relationship, calculated as prescribed in Section A509.2.1.2, shall be used to determine the effective area of the diagonals. The effective stiffness of the frame shall be determined as specified in Section A509.2.1.1. Other procedures that provide the same effective stiffness for the combination of infill and frame may be used when approved by the building official.

❖ The stiffening effects of the infill are modeled as a pair of isotropic diagonals. This simplification is for symmetry in the model for fully reversing cyclical loading. Experimental testing of infilled reinforced concrete frames shows that using only a portion of the infill as a brace within the frames may provide a strength comparable to the experimental model but will significantly overestimate the displacement for each increment of loading.

A509.2.2 Description of analysis procedures. The pseudo-nonlinear dynamic analysis is an iterative response spectrum analysis procedure using effective stiffness as the stiffness of the structural components. The response spectrum analysis shall use the peak dynamic response of all modes having a significant contribution to total structural response. Peak modal responses are calculated using the ordinates of the appropriate response spectrum curve that corresponds to the modal periods. Maximum modal contributions are combined in a statistical manner to obtain an approximate total structural response.

The effective stiffnesses shall be determined by an iterative method. The mathematical model using assumed effective stiffnesses shall be used to calculate dynamic displacements. The effective stiffness of all concrete and masonry elements shall be modified to represent the secant stiffness obtained from the nonlinear force-displacement analysis of the element or system at the calculated displacement. A re-analysis of the mathematical model shall be made using the adjusted effective stiffness of existing and supplemental elements and systems until closure of the iterative process is obtained. A difference of 10 percent from the effective stiffness used and that recalculated may be assumed to constitute closure of the iterative process.

❖ Each iterative step of the pseudo-nonlinear dynamic analysis is identical to a conventional elastic response spectrum analysis, except the effective stiffness of elements are used in lieu of an elastic stiffness. The initial estimate of an effective stiffness could be the yield stiffness for elements that are likely to have an in-plane capacity in excess of probable demand. The iterative procedure is deemed closed when the stiffness used in the current analysis does not deviate by more than 10 percent from that determined from the force-displacement relationship. This closure criterion is for all elements of the structural model.

A509.2.2.1 Number of modes. At least 90 percent of the participating mass of the structure is included in the calculation of response for each principal horizontal direction.

❖ This subsection is common to that required by current codes.

A509.2.2.2 Combining modes. The peak displacements for each mode shall be combined by recognized methods. Modal interaction effects of three-dimensional models shall be considered when combining modal maxima.

❖ This subsection is common to that required by current codes.

A509.3 Capacity spectrum analysis procedure.

A509.3.1 General. This section presents an alternative procedure for a nonlinear static analysis for verification of acceptable performance by comparing the available capacity to the earthquake demand.

Where inelastic torsional response is a dominant feature of overall response, the engineer shall use either a retrofit that reduces the torsional response or an alternative analysis procedure. Inelastic torsional response may be deemed to exist if torsional irregularity as defined in Section A508.2 is present in any story.

The behavior of foundation components and the effects of soil-structure interaction shall be modeled or shown to be insignificant to building response.

❖ The capacity spectrum analysis procedure is contained in ATC 40.

Static inelastic methods and dynamic elastic methods are not able to adequately represent the full effect of torsional response. Response amplitudes associated with inelastic torsion may be much larger than those indicated by these approaches. For structures influenced by inelastic torsion, it often is more appropriate to use simple models or procedures to identify approximately the effect of the irregularity on torsional response, and to apply this effect independently to either a two- or three-dimensional static inelastic analysis of the building. Available research may provide insight into the required analysis process (Goel and Chopra 1991; Sedarat and Bertero 1990; Otani and Li 1984). Where inelastic torsional response is expected to be a dominant feature of the overall response, it usually is preferable to engineer a retrofit strategy that reduces the torsional response, rather than try to engineer an analysis procedure to represent inelastic torsion (see commentary, Section 508.2).

Soil structure interaction refers to response modification because of interaction effects that could include a reduction or increase in the roof displacement and modeling of the foundation soil-superstructure system.

Soil flexibility results in a period elongation and an increase in damping. In the context of inelastic static analysis as described in this methodology, the main relevant impacts of soil-structure interaction are to provide additional flexibility at the base level that may relieve inelastic deformation demands in the super structure. Because the net effect is not readily assessed before carrying out the detailed analysis, it is recommended that foundation flexibility be routinely included in the analysis model.

A509.3.2 Modeling of building components.

A509.3.2.1 Component initial stiffness. Component initial stiffness shall be represented by a secant value defined by the effective yield point of the component. The effective initial stiffness shall be calculated using principles of mechanics, with due consideration of the effects of tensile cracking and compression strain.

Exception: Component effective initial stiffness may be calculated using the approximate values shown in Table A508.1.

❖ The stiffness values provided in Table A508.1 represent values expected for typical proportions and reinforcement ratios. Some adjustment up or down, depending on the actual proportions and reinforcement ratios, is acceptable.

A509.3.2.2 Component strength. The strength of building components shall be calculated using the procedures outlined in the appropriate section of the Building Code.

Exception: Component properties may be calculated using the principles of mechanics as verified by experimental results.

❖ The component strength is the effective yield strength as calculated by the procedures outlined in the IBC using the expected compressive strength and yield stress given in Tables A505.1 and A505.2. The exception allows the use of experimentally verified data for computation of component strength.

A509.3.2.3 Component deformability. The deformability of building components shall be obtained from nonlinear load-deformation relationships that are appropriate for the component being considered. The nonlinear load-deformation relationship shall include information on the plastic deformation capacity at which lateral strength degrades, the plastic deformation capacity at which gravity-load resistance degrades, and the residual strength of the component after strength degradation.

The nonlinear load-deformation relationships of building components shall be determined from nonlinear analyses based on the principles of mechanics, experimental data or established values published in technical literature, as approved by the building official.

❖ The deformability (ductility) of building components shall be determined from nonlinear load-deformation relationships. These load-deformation relationships shall include the strength limit state and the rate of strength degradation in the post-strength limit state.

A509.3.3 Description of analysis procedures.

A509.3.3.1 Determination of the capacity curve. The structure's capacity shall be represented by a capacity curve, which is a plot of the building's base shear versus roof displacement. The capacity curve shall be determined by performing a series of sequential analyses with increasing lateral load, using a mathematical model that accounts for reduced resistance of yielding components. The analysis should include the effect of gravity loads on the building's response to lateral loads.

Lateral forces shall be applied to the structure in proportion to the product of mass and fundamental mode shape.

Exceptions:

1. For buildings with weak stories, the vertical distribution of lateral forces shall be modified to reflect the changed fundamental mode shape after yielding of the weak story.

2. For buildings over 100 feet (30 480 mm) in height or buildings with irregularities that cause significant participation from modes of vibration other than the fundamental mode, the vertical distribution of lateral forces shall reflect the contribution of higher modes.

❖ The capacity curve is generally constructed to represent the first mode response of the structure based on the assumption that the fundamental mode of vibration is the predominant response of the structure. This is generally valid for buildings with fundamental periods of vibration up to about one second. For more flexible buildings with a fundamental period greater than one second, the engineer should consider addressing higher mode effects in the analysis. See Section A509.4.5 for a description of how to include higher mode effects in pushover analyses.

A509.3.3.2 Conversion of the capacity curve to thecapacity spectrum. The capacity curve calculated in Section A509.3.3.1 shall be converted to the capacity spectrum, which is a representation of the capacity curve in the Acceleration-Displacement Response Spectra (ADRS) format. Each point on the capacity curve shall be converted using Equations (A5-8) and (A5-9).

$$a = \frac{V/W}{\alpha_1} \qquad \text{(Equation A5-8)}$$

$$d = \frac{\Delta_{roof}}{PF_1 \phi_{roof,1}} \qquad \text{(Equation A5-9)}$$

Where:

$$PF_1 = \frac{\sum_{i=1}^{N} W_i \phi_{i1}/g}{\sum_{i=1}^{N} W_i \phi^2_{i1}/g} \qquad \text{(Equation A5-10)}$$

$$\alpha_1 = \frac{\left[\sum_{i=1}^{N} W_i \phi_{i1}/g\right]^2}{\left[\sum_{i=1}^{N} W_i/g\right]\left[\sum_{i=1}^{N} W_i \phi^2_{i1}/g\right]} \qquad \text{(Equation A5-11)}$$

❖ In order to utilize the capacity spectrum analysis procedure, both the demand response spectra and the capacity curve need to be plotted in the spectral acceleration versus spectral displacement domain. Spectra plotted in this format are known as acceleration-displacement response spectra (ADRS).

A509.3.3.3 Bilinear representation of the capacity spectrum. A bilinear representation of the capacity spectrum curve obtained in Section A509.3.3.2 shall be used in estimating the appropriate reduction of spectral demand. The first segment of the bilinear representation of the capacity spectrum shall be a line from the origin at the initial stiffness of the building using the component initial stiffness specified in Table A508.1. The second segment of the bilinear representation of the capacity spectrum shall be a line back from the trial performance point, a_{pi}, d_{pi}, at a slope that results in the area under the bilinear representation being approximately equal to the area under the actual capacity spectrum curve. The intersection of the twosegments of the bilinear representation of the capacity spectrum shall determine the yield point a_y, d_y.

❖ This subsection describes the procedure for conversion of the actual capacity spectrum curve to a bilinear representation.

A509.3.3.4 Development of the demand spectrum. The demand spectrum is a plot of the spectral acceleration and spectral displacement of the demand earthquake ground motion in the Acceleration-Displacement Response Spectra (ADRS) format. The 5-percent damped acceleration response spectra in Section A506 shall be modified for use in the capacity spectrum analysis procedure as follows:

1. In the constant acceleration region, the 5-percent damped acceleration spectra shall be multiplied by:

$$SR_A = 1.51 - 0.32 \ln\left[\frac{21(a_y d_{pi} - d_y a_{pi})}{a_{pi} d_{pi}} + 5\right] \geq 0.56$$

$$\text{(Equation A5-12)}$$

2. In the constant velocity region, the 5-percent damped acceleration spectra shall be multiplied by:

$$SR_v = 1.40 - 0.25 \ln\left[\frac{21(a_y d_{pi} - d_y a_{pi})}{a_{pi} d_{pi}} + 5\right] \geq 0.67$$

$$\text{(Equation A5-13)}$$

3. The spectral displacement ordinate, S_d, for a corresponding spectral acceleration, S_a, shall be determined from:

$$S_d = S_a \left(\frac{T}{4\pi}\right)^2 g \qquad \text{(Equation A5-14)}$$

❖ In the ATC 40 document, the equations for SR_A and SR_V involve a factor κ (kappa). This κ-factor depends on the structural behavior of the building, which in turn depends on the quality of the seismic-resisting system and the duration of ground shaking. There are three categories of behavior for quantification of the κ-factor in the ATC 40 document. One category of behavior represents stable, reasonably full hysteretic loops, the second category of behavior represents a moderate reduction of the hysteretic loop and the third category represents poor hysteretic behavior.

For the development of the equations for SR_A and SR_V for this code section, an assumption has been made regarding the type of behavior that the existing structure will experience. It has been assumed that the existing buildings will experience poor hysteretic behavior, and the corresponding κ-factor has been incorporated into the SR_A and SR_V equations.

A509.3.3.5 Calculation of the performance point. The performance point shall represent the maximum roof displacement expected for the demand earthquake ground motion. When the displacement of intersection of the capacity spectrum defined in Section A509.3.3.2 and the demand spectrum defined in Section A509.3.3.4 is within 5 percent of the displacement of the trial performance point, a_{pi}, d_{pi}, used in Section A509.3.3.3, the trial performance point shall be considered the performance point. If the intersection of the capacity spectrum and the demand spectrum is not within the acceptable tolerance of 5 percent, a new trial performance point shall be selected and the analysis shall be repeated.

❖ This subsection describes the procedure for calculation of the performance point (expected nonlinear displacement of the roof level of the building). It also provides the criteria for acceptance.

A509.3.4 Response limits. The inter-story drift between floors of the building and the corresponding strains in building components shall be checked at the performance point to verify acceptability under the demand earthquake ground motion. Performance shall be considered acceptable if building response parameters do not exceed the limitations outlined in Section A509.1.2.

A509.4 Displacement coefficient analysis procedure.

A509.4.1 General. This section presents a procedure for generalized nonlinear static analysis for verification of acceptable performance by comparing the available capacity to the earthquake demand.

Where inelastic torsional response is a dominant feature of overall response, the engineer shall use either a retrofit that reduces the torsional response or an alternative analysis procedure. Inelastic torsional response may be deemed to exist if there is torsional irregularity as defined in Section A508.2 present in any story.

The mathematical model of the building shall be determined in accordance with Section A509.1. The general procedure for execution of the displacement coefficient analysis shall be determined in accordance with Section A509.4.5.

Results of the displacement coefficient analysis procedure shall be checked using the applicable acceptance criteria specified in Section A509.1.2.

For three-dimensional analyses, the static lateral forces shall be imposed on the three-dimensional mathematical model corresponding to the mass distribution at each story level. Effects of accidental torsion shall be considered.

For two-dimensional analyses, the mathematical model describing the framing along each axis of the building shall be developed. The effects of horizontal torsion shall be considered by increasing the target displacement (see Section A509.4.2) by a displacement multiplier, η. The displacement multiplier is the ratio of the maximum displacement at any point on any floor diaphragm (including torsional effects for actual torsion and accidental torsion) to the average displacement on that diaphragm.

The behavior of foundation components and effects of soil-structure interaction shall be modeled or shown to be insignificant to building response.

A509.4.2 Target displacement (δ_t). The target displacement of the control node (typically the center of mass of the building's roof) shall be determined using the following equation:

$$\delta_t = C_0 C_1 C_2 S_a \frac{T_e^2}{4\pi^2} g$$

Where:

C_0 = Modification factor to relate spectral displacement to expected building roof displacement. Value of C_0 can be estimated using any one of the following:

    1. The first modal participation factor at the level of the control node.

    2. The modal participation factor at the level of the control node computed using a shape vector corresponding to the deflected shape of the building at the target displacement.

    3. The appropriate value from Table A509.4.2.

TABLE A509.4.2—VALUES OF MODIFICATION FACTOR, C_0

NUMBER OF STORIES	C_0
1	1.0
2	1.2
3	1.3
5	1.4
10+	1.5

Linear interpolation shall be used to calculate intermediate values.

C_1 = Modification factor to relate expected maximum inelastic displacements to displacements for linear elastic response. C_1 shall not be taken as less than 1.0.

    = 1.0 for $T_e \geq T_0$

    = $[1.0 + (R - 1)T_0/T_e]/R$ for $T_e < T_0$

Where:

$$R = \text{Strength ratio} = \frac{S_a}{V_y / W} \frac{1}{C_0}$$

V_y = Yield strength calculated using the results of static pushover analysis where the nonlinear base-shear roof-displacement curve of the building is characterized by a bilinear relation (see Section A509.4.5).

T_0 = Characteristic period of the response spectrum, defined as the period associated with the transition from the constant acceleration segment of the spectrum to the constant velocity segment of the spectrum.

C_2 = Modification factor to represent the effect of hysteresis shape on maximum displacement response.

= 1.3. where $T > T_0$

= 1.1. where $T \geq T_0$

> **Exception:** Where the stiffness of the structural component in a lateral-force-resisting system, which resists no less than 30 percent of the story shear, does not deteriorate at the target displacement level, C_2 may be assumed to be equal to 1.0.

S_a = Response spectral acceleration at the effective fundamental period and damping ratio of the building, g, in the direction under consideration.

T_e = Effective fundamental period of the building in the direction under consideration, per Section A509.4.5.

❖ Target displacement is an estimate of the likely displacement of the structure under the design earthquake. It uses data obtained from statistical studies on bilinear and trilinear, nonstrength degrading, SDOF systems with viscous damping equal to 5 percent of critical. The target displacement of the multidegree-of-freedom (MDOF) system is computed by modifying the SDOF displacement by using a number of modification factors.

Coefficient C_0 accounts for the difference between the roof displacement of a MDOF building and the displacement of the equivalent SDOF system. Using only the first mode shape and elastic behavior, coefficient C_0 is equal to the first mode participation factor at the roof level. The actual shape vector may take a different form, especially since it is intended to simulate the time-varying deflection profile of the building responding elastically to ground motion. Based on past studies, the use of a shape vector corresponding to the deflected shape at the target displacement level is more appropriate. This shape will likely be different from the elastic first-mode shape. The tabulated values given in FEMA 356 (which are based on the number of stories) are based on a straight-line vector with equal masses at each floor level. These may be approximate, especially if there is a great variation in the floor masses over the height of the building.

Coefficient C_1 accounts for the difference between peak displacement amplitude for nonlinear response as compared with the linear response in structures with relatively stable and full hysteretic loops. The values of the coefficient are based on analytical and ex-

perimental investigations of seismic response of yielding structures. The quantity R in the code represents the ratio of the required elastic strength to the yielding strength of the structure. Some recent studies suggest that maximum elastic and inelastic displacement amplitudes may differ considerably if either the strength ratio R is large or if the building is located in the near field of a causative fault. If the value of R exceeds 5, it is recommended that a displacement larger than the elastic displacement be used as the basis for calculating the target displacement.

Coefficient C_2 represents the effect of the hysteretic shape on the maximum displacement response. If the hysteretic loops exhibit significant pinching or stiffness degradation, the energy absorption and dissipation capacities decrease, resulting in larger displacements. This effect is known to be important for short-period, low-strength structures with very pinched hysteretic loops. Since pinching is a manifestation of the structural damage, the smaller the degree of nonlinear response, then the smaller the degree of pinching. Thus, the values of coefficient C_2 are smaller for systems with periods higher than T_0 and smaller levels of damage.

A509.4.3 Lateral load patterns. Two different vertical distributions of loads shall be used. The first load pattern, termed as the uniform pattern, shall be based on lateral forces proportional to the mass at each story level. The second pattern, called the modal pattern, shall be selected from one of the following:

1. A lateral load pattern represented by C_{vx}, if more than 75 percent of mass participates in the fundamental mode in the direction under consideration. C_{vx} is given by the following expression:

$$C = \frac{W_x h^k}{\sum_{i1}^{=n} w h_i}$$

Where:

w_i = Portion of the total building weight, W, located on or assigned to floor level i.

h_i = Height in feet from base to floor level i.

w_x = Portion of the total building weight, W, located on or assigned to floor level x.

h_x = Height in feet from base to floor level x.

k = 1.0 for $T_e \leq 0.5$ sec.

= 2.0 for $T_e \leq 2.5$ sec.

Linear interpolation shall be used to estimate k for intermediate values of T_e.

2. A lateral load pattern proportional to the story inertia forces consistent with the story shear distribution computed by combination of modal responses using response spectrum analysis of the building, including a sufficient number of modes to capture 90 percent of the total seismic mass and the appropriate ground motion spectrum.

❖ The distribution of lateral inertia forces varies continuously during earthquake response. The extremes of the distribution depend upon various factors, like severity of earthquake shaking (or degree of nonlinear response), frequency characteristics of the building and the ground motion, etc. The distribution of inertia forces determines relative magnitudes of shears, moments, deformations, etc. The loading profile that is critical for one design quantity may be different from the one that is critical for another design quantity. To account for these issues, FEMA 356 requires that at least two different lateral load patterns be considered. It is assumed that with the two load patterns recommended in FEMA 356, the range of design actions occurring during actual seismic response will be approximately bound.

The uniform load pattern is recommended because it emphasizes demands in lower stories over demand in upper stories and magnifies the relative importance of story shear forces compared to overturning moments. The other load patterns, which are based on C_{vx} or modal patterns, are recommended so as to give credit to at least the elastic higher mode effects.

A509.4.4 Period determination. The effective fundamental period, T_e, in the direction under consideration, shall bedetermined using the force-displacement relation of the nonlinear static pushover analysis. The nonlinear relation between the base shear and target displacement of the control node shall be replaced by a bilinear relation to estimate the effective lateral stiffness, K_e, and the yield strength, V_y, of the building. The effective lateral stiffness shall be taken as the secant stiffness calculated at a base shear force equal to 60 percent of the yield strength. The effective fundamental period, T_e, shall then be calculated as:

$$T_e = T_i \sqrt{\frac{K_i}{K_e}}$$

Where:

T_i = Elastic fundamental period in the direction under consideration calculated by elastic dynamic analysis.

K_i = Elastic lateral stiffness of the building in the direction under consideration.

K_e = Effective lateral stiffness of the building in the direction under consideration.

❖ The fundamental period of the structure changes as the structure deforms into the inelastic range. The elastic response spectra provide only an approximation of response once the structure has entered the nonlinear range, regardless of the reference period used initially. Thus, in order to simplify the analysis process, a reference period corresponding to 60 percent of the yield strength is recommended. Determination of this period requires that the structure first be loaded laterally to large deformation levels and the overall load-deformation relation be then graphically examined.

It is important to note that it is not appropriate to use empirical code equations for the determination of period because these provide low estimates of fundamental period that result in higher spectral design

forces. The code equations are appropriate for the linear analysis procedures because higher spectral forces would result in higher force and deformation demands. On the contrary, it is more conservative to use a high estimate of fundamental period for nonlinear procedures as this will result in a higher target displacement.

A509.4.5 General execution procedure for the displacement coefficient analysis procedure. The general procedure for the execution of the displacement coefficient analysis procedure shall be as follows:

1. An elastic structural model shall be created that includes all components (existing and new) contributing significantly to the weight, strength, stiffness or stability of the structure, and whose behavior is important in satisfying the intended seismic performance.

2. The structural model shall be loaded with gravity loads before application of the lateral loads.

3. The mathematical model shall be subjected to in-cremental lateral loads using one of the lateral loadpatterns described in Section A509.4.3. At least twodifferent load patterns shall be used in each principal direction.

4. The intensity of the lateral load shall be monotonically increased until the weakest component reaches a deformation at which there is a significant change in its stiffness. The stiffness properties of this "yielded" component shall be modified to reflect the post-yield behavior, and the modified structure shall be subjected to an increase in lateral loads (for load control) or displacements (for displacement control) using the same lateral load pattern.

5. The previous step shall be repeated as more components reach their yield strengths. At each stage, the internal forces and deformations (both elastic and plastic) of all components shall be computed.

6. The forces and deformations from all previous loading stages shall be accumulated to obtain the total force and deformations of all components at all stages.

7. The loading process shall be continued until unacceptable performance is detected or until a roof displacement is obtained that is larger than the maximum displacement expected in the design earthquake at the control node.

8. A plot of the control node displacement versus base shear at various stages shall be created. This plot is indicative of the nonlinear response of the structure, and changes in the slope of this load-displacement curve are indicative of the yielding of various components.

9. The load-displacement curve obtained in Item 8 shall be used to compute the effective period of the structure, which would then be used to estimate the target displacement (Section A509.4.2).

10. Once the target displacement has been determined, the accumulated forces and deformations at this displacement shall be used to evaluate the performance of various components.

11. If either the force-demands in the nonductile components or deformation-demands in the ductile components exceed the permissible values, then the component shall be deemed to violate the performance criterion, indicating that rehabilitation be performed for such elements.

The relation between base shear force and lateral displacement of the control node shall be established for control node displacements ranging between zero and 150 percent of the target displacement, δ_t.

❖ The procedure for pushover analysis presented in the following considers the effect of higher modes. For details, refer to the paper "Adaptive Spectra-Based Pushover Procedure for Seismic Evaluation of Structures" by Gupta and Kunnath. The main differences between the procedure presented in Section A509.4 and the procedure recommended in the following section are two-fold. First, the recommended procedure uses a site-specific spectrum to define the loading characteristics, and second, the applied load pattern is changed continuously depending on the instantaneous dynamic properties of the system. In the following procedure, the spectral estimates become the basis for determining the incremental lateral forces to be applied in the pushover analysis itself. Also, the load pattern in the recommended method can consider as many modes as deemed important during the course of the analysis. The basic steps involved in executing the pushover analysis using the recommended procedure are as follows:

1. Create a mathematical model of the structure.

2. Specify the nonlinear force-deformation relations for various elements in the structure. In the simplest form, this entails specifying the initial stiffness, the yield moment and the post-yield stiffness of the element. Alternatively, it is possible to define a more detailed force-deformation envelope that includes cracking, yielding and P-delta softening. An available section analysis program, such as BIAX, can be used to generate the expected section behavior.

3. Compute the damped elastic response spectrum for the site-specific ground motion to be used for evaluation. This is required to obtain the modal spectral accelerations (elastic force demands) at various steps. Suitable damping constants, depending upon the structural material and type, should be used. For example, 5 percent of critical damping is a reasonable value for RC structures.

4. Perform an eigenvalue analysis of the structural model at the current stiffness state (for the first step this will be the initial stiffness) of the structure to compute periods and eigenvalues of the system. Using the story weights (masses) and the computed eigenvalues, determine the modal participation factors as given by the following expression:

$$\Gamma_j \frac{1}{g}\sum_{i=1}^{i=N} W_i \phi_{ij} \qquad \text{(Equation 1)}$$

where:

Γ_j = Modal participation factor for j^{th} mode.

ϕ_{ij} = Mass normalized mode shape value at i^{th} level and for j^{th} mode.

W_i = Weight of ith story.

g = Acceleration due to gravity.

N = Number of stories.

Note that Equation 1 has been normalized such that $\Sigma W\Phi^2 = 1$.

5. Compute the story forces at each story level for each of the n modes to be included in the analysis using the following relationship:

$$F_{ij} = \Gamma_j\phi_{ij}W_iS_a(j) \qquad \text{(Equation 2)}$$

where:

F_{ij} = Lateral story force at i^{th} level for j^{th} mode $(1 \leq j \leq n)$.

$S_a(j)$ = Spectral acceleration corresponding to j^{th} mode.

6. Compute modal base shears (V_j) and combine them using SRSS to compute building base shear (V) as shown below:

$$Vj = \sum_{i=1}^{i=N} F_{ij} \qquad \text{(Equation 3)}$$

$$V = \sqrt{\sum_{j=1}^{N} V_j^2} \qquad \text{(Equation 4)}$$

7. The story forces computed in Step 5 are uniformly scaled using the scaling factor S_n indicated below:

$$\overline{V}_j = S_n V_j \qquad \text{(Equation 5a)}$$

where:

$$S_n = \frac{V_B}{N_S V} \qquad \text{(Equation 5b)}$$

where:

V_B is the base shear estimate for the entire structure and N_S is the number of uniform steps over which the base shear is to be applied.

NOTE: The process of applying lateral forces to the structural model commences at this step in the first iteration. The lateral force is applied incrementally in small steps to avoid excessive overshooting of element yield forces. Initial increments during the elastic phase of the response can be considerably larger than the final increments in the post-yield phase. In the procedure described here, equal increments are assumed for simplicity. For example, assume that the base shear estimate is 40 percent of W (W

being the seismic weight of the building) and the number of steps in which the total force will be applied is 100. If the building base shear computed in Step 6 above is 25 percent of W, then all the story forces would be scaled by a factor of 0.016 (=0.40/0.25/100). Scaled story forces would then result in a building base shear of 0.004W, which is equal to the base shear to be applied in one increment.

8. Perform a static analysis of the structure using the scaled incremental story forces computed in the previous step corresponding to each mode independently. This means that for modes other than the fundamental mode, the structure will be pushed and pulled simultaneously.

9. Compute element forces, displacements, story drifts, member rotations, etc., by SRSS combination of the respective modal quantities for this step and add these to the same from the previous step.

10. At the end of every step, compare the accumulated member forces with their respective yield values. If any member has yielded, recompute the member and global stiffness matrices and return to Step 4.

11. Repeat the process until either the maximum base shear has been reached or the global drift exceeds the specified limit.

It is clear from the above description that the applied load pattern keeps varying continuously based on the instantaneous dynamic characteristics of the structure. The computation of the story forces is, at any step, identical to traditional response spectrum analysis. Whenever one or more elements yield, a new structure is created by changing the stiffness of the yielded element(s) and the response spectral analysis is repeated. Since one ground motion results in one pushover curve, a suite of ground motions will produce a family of curves, which can be used to generate mean response parameters.

A509.4.6 Acceptance criteria. The inter-story drift between floors of the building and the corresponding strains in building components shall be checked at 150 percent of the target displacement, δ_t, to verify acceptability under the demand earthquake ground motion. Performance shall be considered acceptable if building response parameters do not exceed the limitations outlined in Section A509.1.2.

Exception: Where the effective stiffness, K_e, and the yield strength, V_y, of the building can be determined through rational analysis, the acceptance criteria may be determined based on 100 percent of the target displacement, δ_t.

❖ The compressive strains in the components of the structural system caused by the interstory drift associated with a structural system displacement of 150 percent of the target displacement shall be equal or less than that specified in Section A509.1.2.2, Items 1

through 3. The tangent to the load-displacement curve at 150 percent of the target displacement shall not have a negative slope except as permitted by the exception to Section A509.1.2.2, Item 4.

An exception to the requirement that 150 percent of the target displacement be used for acceptance is given for a structure when rational analysis can determine an effective stiffness and a yield strength for the components of the structure at the target displacement and these calculated stiffness and strengths correspond to that predicted by the structural model at the target displacement. A deviation of less than 20 percent between the independently calculated values and the system values should be considered confirmation of the system performance at the target displacement.

SECTION A510
DETERMINATION OF THE STRESS-STRAIN RELATIONSHIP OF EXISTING UNREINFORCED MASONRY

A510.1 Scope. This section covers procedures for determining the expected compressive modulus, peak strain and peak compressive stress of unreinforced brick masonry used for infills in frame buildings.

❖ The unreinforced masonry used in existing infill buildings may have stress-strain characteristics significantly different from those used in design codes for new buildings. Earthquake hazard reduction guidelines use expected material values for stress-strain relationships. These expected material values also should include apparent yield and peak strengths. The analyses of existing buildings anticipate nonlinear response to probable earthquake shaking. A nonlinear analysis of an element, such as a confined panel of unreinforced masonry, requires that a simulation of cyclic compressive loading be used to determine the degradation, if any, of the compressive modulus of the masonry.

A510.2 General procedure. The outer wythe of multiple wythe brick masonry shall be tested by inserting two flat jacks into the mortar joints of the outer wythe. The prism height (the vertical distance between the flat jacks) shall be five bricks high. The test location shall have adequate overburden and/or vertical confinement to resist the flat jack forces.

❖ This section is applicable to solid masonry units. Grouted masonry, reinforced or unreinforced, is rarely found in infilled panels and its testing is described in Section A505.2.2. Flat jacks have not been successfully used in partially grouted or ungrouted cored brick or hollow unit masonry. The flat jacks will expand into the voids damaging the flat jacks and not providing any useful data on the material characteristics of the masonry. If the clay tile is laid in the traditional manner for bearing walls, with cores horizontal, the pressure from the flat jacks will likely fracture the thin shell of the unit in flexure and damage the flat jack. Data obtained prior to the failure of the shell will not be useful because the loaded surface does not represent the net area of the clay tile unit. The net area is the area of the mortared end of the unit if laid with the cores vertical, or the mortared cross-webs if laid with the cores horizontal. Use of bearing plates to span cores or cells of masonry units and provide uniform confinement of the surface of the flat jack generally will require a space greater than the height of the existing mortar joint. Attempts to provide a space that is adequate for a flat jack and bearing plates above and below the flat jack failed to provide useful stress-strain data. The roughness of the masonry bearing surfaces concealed the zero point for a secant that represents the stress-strain relationship of the masonry with cores or open cells. This type of masonry should be tested to determine its stress-strain relationship by displacement-controlled loading of a five-unit height prism.

The location of the test and the available overburden or vertical confinement above the flat jack should be determined by the engineer.

A510.3 Preparation for the test. Remove a mortar joint at the top and bottom of the test prism by saw-cutting or drilling and grinding to a smooth surface. The cuts for inserting the flat jacks shall not have a deviation from parallel of more than $^3/_8$ inch (9.53 mm). The deviation from parallel shall be measured at the ends of the flat jacks. The width of the saw cut shall not exceed the width of the mortar joint. The length of the saw cut on the face of the wall may exceed the length of the flat jacks by not more than twice the thickness of the outer wythe plus 1 inch (25.4 mm).

❖ The smoothness of the cut surface and a consistent width of the cut made for insertion of the flat jack and the steel shims is critical. Use of partial length shims in a cut without parallel edges will result in misleading data. Use of a masonry saw resting on a guide attached to the masonry wall is the most successful method of cutting the slot.

The restriction on the parallelism of the cuts given in this section is for the two cuts made for isolating the prism from the wall.

A510.4 Required equipment. The flat jacks shall be rectangular or with semi-circular ends to mimic the radius of the saw blade used to cut the slot for the flat jack. The length of the flat jack shall be 18 inches (457 mm) maximum and 16 inches (406 mm) minimum. This length shall be measured on the longest edge of a flat jack with semi-circular ends. The maximum width of the flat jack shall not exceed the average width of the wythe of brick that is loaded. The minimum width of a flat jack shall be $3^1/_2$ inches (89 mm) measured out-to-out of the flat jack. The flat jack shall have a minimum of two ports to allow air in the flat jack to be replaced by hydraulic fluid. The unused port shall be sealed after all of the air is forced out of the flat jack. The thickness of the flat jack shall not exceed three-quarters of the minimum height of the mortar joint. It is recommended that the height of the flat jack be about one-half of the width of the slot cut for installation of the flat jack. The remain-

ing space can be filled with steel shim plates having plan dimensions equal to the flat jack.

❖ Flat jacks are commonly manufactured-to-order for the project but useable jacks may be in the inventory of testing agencies. The life of a flat jack is relatively short. Failure (leakage) is common due to fracture of the weld joining the upper and lower surfaces of the flat jack. The life of the flat jack is extended by using multiple steel shims of varying thickness to completely fill the cut made for the flat jack.

A510.5 Data acquisition equipment. The strain in the tested prism shall be recorded by gauges or similar recording equipment having a minimum range of 0.0001 inch (0.0025 mm). The compressive strain shall be measured on the surface of the prism and shall have a gauge length, measured vertically on the face of the prism, of 10 inches (254 mm) minimum. The gauge points shall be fixed to the wall by drilled-in anchors or by anchors set in epoxy or similar material. The support for the data-recording apparatus shall be isolated from the wall by a minimum of $^1/_{16}$ inch (1.5 mm), so that the gauge length used in the calculation of strain can be taken as the measured length between the anchors of the equipment supports. The gauging equipment shall be as close to the face of the prism as possible, to minimize the probability of erroneous strain measurements caused by bulging of the prism outward from its original plane.

The compressive strain data shall be measured at a minimum of two points on the vertical face of the prism. These points shall be the one-third points of the length of the flat jacks plus or minus $^1/_2$ inch (12.7 mm). As an alternative, the strain may be measured at three points on the face of the prism.

These points shall be spaced at one-quarter of the flat jack length plus or minus $^1/_2$ inch (12.7 mm).

A horizontal gauge at midheight of the prism may be used to record Poisson strain, but this recording data should be considered secondary in importance to the vertical gauges, and the horizontal gauge's placement shall not interfere with placing the vertical gauging as close as possible to the face of the prism.

❖ A critical step in obtaining useful stress-strain data is the mounting of strain gauges at the prescribed distance from the face of the wall [$^1/_{16}$ inch (1.6 mm)]. The five-unit-high prism will bulge from the Poisson effect as the strain approaches peak values. Bulging of the prism may cause a relative rotation of the extensions of the gauge points on the wall. This relative rotation could alter the measured value from the strain in the prism.

A510.6 Loading and recording data. The loading shall be applied by hydraulic pumps that add hydraulic fluid to the flat jacks in a controlled method. The application of load shall be incremental and held constant while strains are recorded. The increasing loading for each cycle of loading shall be divided into a minimum of four equal load increments. The strain shall be recorded at each load step. The decrease in loading shall be divided into a minimum of two equal unloading increments. Strain shall be recorded on the decreasing load steps. The hydraulic pressure shall be reduced to zero, and the permanent strain caused by this cycle of loading shall be recorded. This procedure shall be used for each cycle of loading.

The load applied in each cycle of loading shall be determined by estimating the peak compressive stress of the existing brick masonry. The hydraulic pressure needed to cause this peak compressive stress in the prism shall be calculated by assuming that the area of the loaded prism is equal to the area of the flat jack. A maximum of one-third of this pressure, rounded to the nearest 25 pounds per square inch (172 kPa), shall be applied in the specified increments to the peak pressure prescribed for the first cycle of loading. After recording the strain data, this pressure shall be reduced in a controlled manner, for each of the specified increments for unloading and for recording data. The maximum jack pressure on the subsequent cycles shall be one-half, two-thirds, five-sixths and estimated peak pressure. If the estimated peak compressive stress is less than the existing peak compressive stress, the cyclic loading and unloading shall continue using increments of increasing pressure equal to those used prior to the application of estimated peak pressure.

All strain data shall be recorded to 0.0001 inch (0.0025 mm). Jack pressure shall be recorded in increments of 25 pounds per square inch (172 kPa) pressure.

❖ The flow of hydraulic fluid to the flat jack must be controlled as in a cyclic test using displacement control. A hand-operated pump is adequate and recommended. Observation of the pressure gauge is used for determination of the specified incremental loading. Flow of hydraulic fluid to the flat jack must be stopped for recording of strain data.

It is very important to obtain incremental unloading and reloading stress-strain data. Recording permanent set at the end of each loading cycle is also very important. Permanent compressive set will cause the hysteretic behavior of the compressive strut within the infill panel to become pinched and degrade the secant stiffness of the infilled panel.

A510.7 Quality control. The flat jack shall be calibrated before use by placing the flat jack between bearing plates of 2-inch (51 mm) minimum thickness in a calibrated testing machine. A calibration curve to convert hydraulic pressure in the flat jack to total load shall be prepared and included in the report of the test results. Flat jacks shall be recalibrated after three uses.

The hydraulic pressure in the flat jacks shall be determined by a calibrated dial indicator having a subdivision of 25 pounds per square inch (172 kPa) or less. The operator of the hydraulic pump shall use this dial indicator to control the required increments of hydraulic pressure in loading and unloading.

❖ This section describes the requirements for calibration of the flat jacks and the maximum intervals of pressure indication.

A510.8 Interpretation of the data. The data obtained from the testing required by Section A505.2.4 shall be averaged both in the expected peak compressive stress and the corresponding peak strain. The envelope of the averaged stress-strain relation-

ship of all tests shall be used for the material model of the masonry in the infilled frame. If two strain measurements have been made on the surface of the prism, these strain measurements shall be averaged for determination of the stress-strain relationship for the test. If three strain measurements have been made on the surface of the prism, the data recorded by the center gauge shall be given a weight of two for preparing the average stress-strain relationship for the test.

❖ The total quantity of stress-strain data obtained shall be averaged for calculation of expected peak compressive stress and strain. An averaged shape of the envelope of the stress-strain relationship should be prepared. A nonlinear push-over analysis of the infilled frame requires that this envelope be used for stiffness degradation of the infill element.

Experimental testing by flat jacks has shown that use of three strain measurement devices on the face of the prism gives data that should not have equal weighting. The loading effect of the flat jack extends beyond the length of the flat jack. This effect is minimized by using strain gauges at one-third length points on the prism.

TABLE A5-A—BASIC STRUCTURAL CHECKLIST

This basic structural checklist shall be completed when required by Section A507.5 prior to completing the corresponding supplemental structural checklist. Each of the evaluation statements on this checklist shall be marked compliant (C), noncompliant (NC), or not applicable (N/A) for a Tier I evaluation. Compliant statements identify issues that are acceptable according to the criteria of this chapter and the Building Code, while noncompliant statements identify issues that require further investigation. Certain statements may not apply to the buildings being evaluated. Noncompliant items shall be mitigated by rehabilitating the structure, or shall be shown to be compliant by performing a Tier 2 or Tier 3 analysis.

BUILDING SYSTEM
General

C	NC	N/A	LOAD PATH: The structure shall contain one complete load path for seismic force effects from any horizontal direction that serves to transfer the inertial forces from the mass to the foundation.
C	NC	N/A	ADJACENT BUILDINGS: An adjacent building shall not be located next to the structure being evaluated closer than 4 percent of the height.
C	NC	N/A	MEZZANINES: Interior mezzanine levels shall be braced independently of the main structure, or shall be anchored to the lateral-force-resisting elements of the main structure.

Configuration

C	NC	N/A	WEAK STORY: The strength of the lateral-force-resisting system in any story shall not be less than 80 percent of the strength in an adjacent story above or below.
C	NC	N/A	SOFT STORY: The stiffness of the lateral-force-resisting system in any story shall not be less than 70 percent of the stiffness in an adjacent story above or below, or less than 80 percent of the average stiffness of the three stories above or below.
C	NC	N/A	GEOMETRY: There shall be no changes in horizontal dimension of the lateral-force-resisting system of more than 30 percent in a story relative to adjacent stories, excluding one-story penthouses.
C	NC	N/A	VERTICAL DISCONTINUITIES: All vertical elements in the lateral-force-resisting system shall be continuous to the foundation.
C	NC	N/A	MASS: There shall be no change in effective mass of more than 50 percent from one story to the next.
C	NC	N/A	TORSION: The distance between the story center of mass and the story center of regidity shall be less than 20 percent of the building width in either plan dimension.

Condition of Materials

C	NC	N/A	DETERIORATION OF CONCRETE: There shall be no visible deterioration of concrete or reinforcing steel in any of the vertical- or lateral-force-resisting elements.
C	NC	N/A	POST-TENSIONING ANCHORS: There shall be no evidence of corrosion or spalling in the vicinity of post-tensioning or end fittings. Coil anchors shall not have been used.
C	NC	N/A	MASONRY UNITS: There shall be no visible deterioration of masonry units.
C	NC	N/A	MASONRY JOINTS: The mortar shall not be easily scraped away from the joints by hand with a metal tool, and there shall be no areas of eroded mortar.
C	NC	N/A	CONCRETE WALL CRACKS: All existing diagonal cracks in wall elements shall be less than $1/8$ inch for Seismic Use Groups other than Group III and less than $1/16$ inch for Seismic Use Group III; shall not be concentrated in one location; and shall not form an X pattern.
C	NC	N/A	REINFORCED MASONRY WALL CRACKS: All existing diagonal cracks in wall elements shall be less than $1/8$ inch for Seismic Use Groups other than Group III and less than $1/16$ inch for Seismic Use Group III; shall not be concentrated in one location; and shall not form an X pattern.
C	NC	N/A	CRACKS IN BOUNDARY COLUMS: There shall be no existing diagonal cracks wider $1/8$ inch for Seismic Use Groups other than Group III and wider than $1/16$ inch for Seismic Use Group III in concrete colums that encase masonry infills.

LATERAL-FORCE-RESISTING SYSTEM
Moment Frames
General

C	NC	N/A	REDUNDANCY: The number of lines of moment frames in each principal direction shall be greater than or equal to two for Seismic Use Groups I, II and III. The number of bays of moment frames in each line shall be greater than or equal to two for Seismic Use Groups other than Group III, and three for Seismic Use Group III.

Moment Frames with Infill Walls

C	NC	N/A	INTERFERING WALLS: All infill walls placed in moment frames shall be isolated from structural elements.

(Continued)

TABLE A5-A—BASIC STRUCTURAL CHECKLIST—(Continued)

			Concrete Moment Frames
C	NC	N/A	SHEAR STRESS CHECK: The sear stress in the concrete columns, calculated using the quick-check procedure of Section A507.6.1, shall be less than 100 psi or $2\sqrt{f_c'}$.
C	NC	N/A	AXIAL STESS CHECK: The axial stress because of gravity loads in columns subjected to overturning forces shall be less than $0.10 \times f_c'$. Alternatively, the axial stresses because of overturning forces alone, calculated using the quick-check procedure of Section A507.6.3, shall be less than $0.30 \times f_c'$.
			Frames Not Part of the Lateral-Force-Resisting System
C	NC	N/A	COMPLETE FRAMES: Steel or concrete frames classified as nonlateral-force-resisting components shall form a complete vertical-load-carrying system.
			Shear Walls **General**
C	NC	N/A	REDUNDANCY: The number of lines of shear walls in each principal direction shall be greater than or equal to two.
			Concrete Shear Walls
C	NC	N/A	SHEAR STRESS CHECK: The shear stress in the concrete shear walls, calculated using the quick-check procedure of Section A507.6.2, shall be less than 100 psi or $2\sqrt{f_c'}$.
C	NC	N/A	REINFORCING STEEL: The ratio of reinforcing steel area to gross concrete area shall be greater than 0.0015 in vertical direction and greater than 0.0025 in the horizontal direction. The spacing of reinforcing steel shall be equal to or less than 18 inches.
			CONNECTIONS **Shear Transfer**
C	NC	N/A	TRANSFER TO SHEAR WALLS: The diaphragm shall be reinforced and connected for transfer of loads to the shear walls for Seismic Use Groups I and II, and the connections shall be able to develop the shear strength of the walls for Seismic Use Group III.
			Vertical Components
C	NC	N/A	CONCRETE COLUMNS: All concrete columns shall be doweled into the foundation for Seismic Use Groups I and II, and the dowels shall be able to develop the tensile capacity of the column for Seismic Use Group III.
C	NC	N/A	WALL REINFORCING: Walls shall be doweled into the foundation for Seismic Use Groups I and II, and the dowels shall be able to develop the strength of the walls for Seismic Use Group III.

For SI: 1 inch = 25.4 mm.

TABLE A5-B—SUPPLEMENTAL STRUCTURAL CHECKLIST

This supplemental structural checklist shall be completed when required by Section A507.5. The basic structural checklist shall be completed prior to completing this supplemental structural checklist.

			LATERAL-FORCE-RESISTING SYSTEM **Moment Frames** **Concrete Moment Frames**
C	NC	N/A	FLAT SLAB FRAMES: The lateral-force-resisting system shall not be a frame consisting of columns and a flat slab/plate without beams.
C	NC	N/A	PRESTRESSED FRAME ELEMENTS: The lateral-load-resisting frames shall not include any prestressed or post-tensioned elements.
C	NC	N/A	SHORT CAPTIVE COLUMNS: There shall be no columns at a level with height-depth ratios less than 50 percent of the nominal height-depth ratio of the typical columns at that level for Seismic Use Groups other than Group III, and less than 75 percent for Seismic Use Group III.
C	NC	N/A	NO SHEAR FAILURES: The shear capacity of frame members shall be able to develop the moment capacity at the top and bottom of the columns.
C	NC	N/A	STRONG COLUMN/WEAK BEAM: The sum of the moment capacity of the columns shall be 20-percent greater than that of the beams at frame joints.
C	NC	N/A	BEAM BARS: At least two longitudinal top and two longitudinal bottom bars shall extend continuously throughout the length of each frame beam. At least 25 percent of the longitudinal bars provided at the joints for either positive or negative moment shall be continuous throughout the length of the members.
C	NC	N/A	COLUMN-BAR SPLICES: All column bar lap-splice lengths shall be greater than 35 d_b for Seismic Use Groups other than Group III and greater than 50 d_b for Seismic Use Group III, and shall be enclosed by ties spaced at or less than 8 d_b for all Seismic Use Groups.
C	NC	N/A	BEAM-BAR SPLICES: The lap splices for longitudinal beam reinforcing shall not be located within $l_b/4$ of the joints and shall not be located within the vicinity of potential plastic hinge locations.
C	NC	N/A	COLUMN-TIE SPACING: Frame columns shall have ties spaced at or less than $d/4$ throughout their length and at or less than 8 d_b at all potential plastic hinge locations.
C	NC	N/A	STIRRUP SPACING: All beams shall have stirrups spaced at or less than $d/2$ throughout their length. At potential plastic hinge locations, stirrups shall be spaced at or less than the minimum of 8 d_b or $d/4$.
C	NC	N/A	JOINT REINFORCING: Beam-column joints shall have ties spaced at or less than 8 d_b.
C	NC	N/A	JOINT ECCENTRICITY: For Seismic Use Group III, there shall be no eccentricities larger than 20 percent of the smallest column plan dimension between girder and column centerlines.
C	NC	N/A	STIRRUP AND TIE HOOKS: For Seismic Use Group III, the beam stirrups and column ties shall be anchored into the member cores with hooks of 135° or more.
			Frames Not Part of the Lateral-Force-Resisting System
C	NC	N/A	DEFORMATION COMPATIBILITY: Nonlateral-force-resisting components shall have the shear capacity to develop the flexural strength of the elements for Seismic Use Groups other than Group III and shall have ductile detailing for Seismic Use Group III.
C	NC	N/A	FLAT SLABS: Flat slabs/plates classified as nonlateral-force-resisting components shall have continuous bottom steel through the column joints for Seismic Use Groups other than Group III. Flat slabs/plates shall not be permitted for Seismic Use Group III.
			Shear Walls **Concrete Shear Walls**
C	NC	N/A	COUPLING BEAMS: The stirrups in all coupling beams over means of egress shall be spaced at or less than $d/2$ and shall be anchored into the core with hooks of 135° or more. In addition, the beams shall have the capacity in shear to develop the uplift capacity of the adjacent wall for Seismic Use Group III.
C	NC	N/A	OVERTURNING: For Seismic Use Group III, all shear walls shall have aspect ratios less than 4:1. Wall piers need not be considered.
C	NC	N/A	CONFINEMENT REINFORCING: For shear walls in Seismic Use Group III with aspect ratios greater than 2.0, boundary elements shall be confined with spirals or ties with spacing less than 8 d_b.
C	NC	N/A	REINFORCING AT OPENINGS: For Seismic Use Group III, there shall be added trim reinforcement around all wall openings.
C	NC	N/A	WALL THICKNESS: For Seismic Use Group III, thickness of bearing walls shall not be less than $^1/_{25}$ the minimum unsupported height or length, or less than 4 inches.

(Continued)

TABLE A5-B—SUPPLEMENTAL STRUCTURAL CHECKLIST—(Continued)

			DIAPHRAGMS General
C	NC	N/A	DIAPHRAGM CONTINUITY: The diaphragms shall not be composed of split-level floors.
C	NC	N/A	DIAPHRAGM OPENINGS ADJACENT TO SHEAR WALLS: Diaphragm openings immediately adjacent to the shear walls shall be less than 25 percent of the wall length for Seismic Use Groups I and II, and less than 15 percent of the wall length for Seismic Use Group III.
C	NC	N/A	PLAN IRREGULARITIES: There shall be tensile capacity to develop the strength of the diaphragm at re-entrant corners or other locations of plan irregularities. This statement shall apply to Seismic Use Group III only.
C	NC	N/A	DIAPHRAGM REINFORCEMENT AT OPENINGS: There shall be reinforcement around all diaphragm openings larger than 50 percent of the building width in either major plan dimension. This statement shall apply to Seismic Use Group III only.
			Other Diaphragms
C	NC	N/A	OTHER DIAPHRAGMS: The diaphragm shall not consist of a system other than those described in Section A502.
			CONNECTIONS Vertical Components
C	NC	N/A	LATERAL LOAD AT PILE CAPS: Pile caps shall have top reinforcement, and piles shall be anchored to the pile caps for Seismic Use Groups I and II. The pile cap reinforcement and pile anchorage shall be able to develop the tensile capacity of the piles for Seismic Use Group III.

For SI: 1 inch = 25.4 mm.

TABLE A5-C—GEOLOGIC SITE HAZARD AND FOUNDATION CHECKLIST

This geologic site hazard and foundation checklist shall be completed when required by Section A507.5.

Each of the evaluation statements on this checklist shall be marked compliant (C), noncompliant (NC), or not applicable (N/A) for a Tier I evaluation. Compliant statements identify issues that are acceptable according to the criteria of this chapter and the Building Code, while noncompliant statements identify issues that require further investigation. Certain statements may not apply to the buildings being evaluated. Noncompliant items shall be mitigated by rehabilitating the structure, or shall be shown to be compliant by performing a Tier 2 or Tier 3 analysis.

			Geologic Site Hazards
C	NC	N/A	LIQUEFACTION: Liquefaction-susceptible, saturated, loose granular soils that could jeopardize the building's seismic performance shall not exist in the foundation soils at depths within 50 feet under the building for Seismic Use Groups I, II, and III.
C	NC	N/A	SLOPE FAILURE: The building site shall either be sufficiently remote from potential earthquake-induced slope failures or rockfalls to be unaffected by such failures, or shall be capable of accommodating any predicted movements without failure.
C	NC	N/A	SURFACE FAULT RUPTURE: Surface fault rupture and surface displacement at the building site are not anticipated.
			Condition Foundations
			The following statement shall be completed for all Tier I building evaluations.
C	NC	N/A	FOUNDATION PERFORMANCE: There shall be no evidence of excessive foundation movement such as settlement or heave that would affect the integrity or strength of the structure.
			The following statement shall be completed for buildings in regions of high or moderate seismicity being evaluated to Seismic Use Group III:
C	NC	N/A	DETERIORATION: There shall not be evidence that foundation elements have deteriorated because of corrosion, sulfate attack, material breakdown or other reasons in a manner that would affect the integrity or strength of the structure.
			Capacity of Foundations
			The following statement shall be completed for all Tier I building evaluations.
C	NC	N/A	POLE FOUNDATIONS: Pole foundations shall have a minimum embedment depth of 4 feet for Seismic Use Groups I, II, and III.
			The following statements shall be completed for buildings in regions of high seismicity and for buildings in regions of moderate seismicity evaluated to Seismic Use Group III:
C	NC	N/A	OVERTURNING: The ratio of the effective horizontal dimension at the foundation level of the lateral-force-resisting system to the building height (base/height) shall be greater than $0.6S_a$.
C	NC	N/A	TIES BETWEEN FOUNDATION ELEMENTS: The foundation shall have ties adequate to resist seismic forces where footings, piles and piers are not restrained by beams, slabs or soils classified as Class A, B or C.
C	NC	N/A	DEEP FOUNDATIONS: Piles and piers shall be capable of transferring the lateral forces between the structure and the soil. This statement shall apply to Seismic Use Group III only.
C	NC	N/A	SLOPING SITES: The grade difference from one side of the building to another shall not exceed one-half the story height at the location of embedment. This statement shall apply to Seismic Use Group III Performance Level only.

For SI: 1 foot = 308.8 mm.

❖ Tables A5-A, A5-B and A5-C were patterned after ASCE 31, with emphasis on Tier-1 analysis for building Type C1:Concrete Moment Frame, and Type C2: Concrete Shear Walls. The analyses needed for completion of the Tier-1 review are based on an equivalent lateral force procedure. The commentary for the similar sections in ASCE 31 is useful for information on the development of a quick-pass document for existing buildings.

Bibliography

The following resource materials are referenced in this chapter or are relevant to the subject matter addressed in this chapter.

ATC-40, *Recommended Methodology for Seismic Evaluation and Retrofit of Existing Concrete Buildings*. Report No. SSC 96-01. ATC, Aug. 1996.

Chopra, Anil K. *Dynamics of Structures*. Prentiss-Hall, 1995.

Ewing et al. *FEM/I: A Finite Element Computer Program for the Nonlinear Static Analysis of Reinforced Masonry Building Components*. Boulder, CO: The Masonry Society, Dec. 1987 (revised June 1990).

FEMA 356, *Prestandard and Commentary for the Seismic Rehabilitation of Buildings*. Washington, DC: Federal Emergency Management Agency, Nov. 2000.

FEMA 440, *Improvement of Nonlinear Static Seismic Analysis Procedures*, Preprint Edition. Washington, DC: Federal Emergency Management Agency, Feb. 2005.

Ghassan Al- Chaar. *Evaluating Strength and Stiffness of Unreinforced Masonry Infill Structures*, ERDC/CERL TR-02-1. Champaign, IL: US Corp of Engineers, Engineering Research and Development Center (EDRC), 2002.

"Gi-Gi (Taiwan) Earthquake, 1999." *EERI Spectra*. Supplement to Vol.17, April 2001 and Vol. 17, No. 4, Nov. 2001.

Goel and Chopra. "Inelastic Seismic Response of One-story Asymmetric-plan Systems: Effects of System Parameters and Yielding." *Earthquake Engineering and Structural Dynamics*. Vol. 20, No. 3, Mar. 1991.

Gupta and Kunnath. "Adaptive Spectra-Based Pushover Procedure for Seismic Evaluation of Structures." *Earthquake Spectra*. Vol. 16, No.2. Oakland, CA: EERI.

"Izmut (Turkey) Earthquake, 1999." *EERI Spectra.* Supplement A to Vol. 16, Dec. 2000.

Kariotis et al. *Simulation of the Recorded Response of Unreinforced Masonry (URM) Infill Buildings*. CSMIP/94-05 Oct. 1994.

"Kobe Earthquake, 1995." *EERI Spectra*.

Kunnath et al. IDARC, Technical Report NCEER-92-0022, 1992.

"Loma Prieta Earthquake, 1989." *EERI Spectra*. Vol. 6 Supplement, May 1990.

Maheny et al. *The Capacity Spectrum Method for Evaluating Structural Response During the Loma Prieta Earthquake*. Memphis, TN: National Earthquake Conference, 1993.

"Mexico City Earthquake, 1985." *EERI Spectra*. Vol. 4, No. 3, Aug. 1988.

Otani and Li. *Kurayoshi City East Building Damaged by 1983 Tattori Earthquake*. Cornell University: Proceedings, 2nd Workshop Seismic Performance of Existing Buildings, July 1989.

"San Fernando Earthquake, 1971." *EERI Spectra*.

Sedarat and Bertero. *Effects of Torsion on the Nonlinear Inelastic Seismic Response of Multi-story Structure*. Cerrito, CA: Fourth US National Conference on Earthquake Engineering, Proceedings Vol. 2, EERI, 1990.

Wallace and Moehle. BIAX- Report No. UCB/SEMM-89/12. UC Berkeley, 1989.

"Whittier Earthquake, 1987." *EERI Spectra*. Vol. 4. No. 1, Feb. 1988 and Vol. 4 No. 2, May 1988.

APPENDIX A

REFERENCED STANDARDS

ASTM

American Society for Testing and Materials
100 Bar Harbor Drive
West Conshohocken, PA 19428-2959

Standard reference number	Title	Referenced in code section number
C496—96	Standard Test Method for Splitting Tensile Strength of Cylindrical Concrete Specimens	A104, A106.3.3.2
E519—00e1	Standard Test Method for Diagonal Tension (Shear) in Masonry Assemblages	A104, A106.3.3.2

DOC

U.S. Department of Commerce
National Institute of Standards and Technology
100 Bureau Drive Stop 3460
Gaithersburg, MD 20899

Standard reference number	Title	Referenced in code section number
PS-1—95	Construction and Industrial Plywood	A302
PS-2—92	Performance Standard for Wood-based Structural-Use Panels	A302

FEMA

Federal Emergency Management Agency
Federal Center Plaza
500 C Street S.W.
Washington, DC 20472

Standard reference number	Title	Referenced in code section number
FEMA 302	1997 NEHRP Recommended Provisions for Seismic Regulations for New Buildings and Other Structures	A103, A108.1, A202

ICC

International Code Council
5203 Leesburg Pike, Suite 708
Falls Church, VA 22041

Standard reference number	Title	Referenced in code section number
IBC—00	International Building Code®	A102.2, A103, A104, A108.2, A206.2, A301.2, A403.11.2.2, B102.2.1
UBC—97	Uniform Building Code®	A102.2, A103, A104, A108.2, A202, A203, A206.2, A301.2, A401.2, A403.1, A403.2, A403.3, A403.4, A405.2.3, A403.7, A403.11.2.2, A406.2, A406.3.2.1, A408.1, A502
UBC—Standard 21-4	Hollow and Solid Load-bearing Concrete Masonry Units	A106.2, A505.2.3
UBC—Standard 21-6	In-Place Masonry Shear Tests	A104
UBC—Standard 21-7	Tests of Anchors in Unreinforced Masonry	A105.3, A107.3, A107.4, Table A1-E
UBC—Standard 21-8	Pointing of Unreinforced Masonry Walls	A103, A106.3.3.9
UBC—Standard 23-2	Construction and Industrial Plywood	A403.11.2.1

APPENDIX B

SUPPLEMENTARY ACCESSIBILITY REQUIREMENTS FOR EXISTING BUILDINGS AND FACILITIES

The provisions contained in this appendix are not mandatory unless specifically referenced in the adopting ordinance.

General Comments

As stated in Section 101.7, the provisions of the appendices do not apply unless specifically adopted.

Chapter 11 of the *International Building Code®* (IBC®) contains provisions that set forth requirements for accessibility to buildings and their associated sites and facilities for people with physical disabilities. Sections 406, 506, 606, 706, 806, 905, 1004, 1005 and 1308 in the code address accessibility provisions and alternatives permitted in existing buildings. Appendix B was added to address accessibility in construction for items that are not typically enforceable through the traditional building code enforcement process.

Section B101 deals with historical facilities that are registered through the National Historic Preservation Act or a similar state or local law. If such a facility undergoes an alteration or change of occupancy, it must comply with accessibility regulations as specified in Section 1004 and 1005. If the required alterations would adversely affect some historic aspect of the structure, alternatives are available. Before these alternatives can be utilized, the owner must go through a process to evaluate the requirements and implications of each. Section B101.3 describes the process if the building is registered through the National Historic Preservation Act. Section B101.4 describes the process if the building is registered through

a state or local registration law. Section B101.5 provides alternatives for displays in historical buildings.

Section B102 lists criteria specific to fixed transportation facilities and stations, such as train stations.

Section B103 enumerates the standards referenced in this appendix.

Purpose

Appendix B includes scoping requirements found in the new *Americans with Disabilities Act Accessibility Guidelines* (ADAAG) that are not in the accessibility provisions in the IBC or the code. Items in this appendix deal with topics not typically addressed in building codes. For example, Section B101 specifies the process to follow regarding alterations or change of occupancy in historically registered buildings.

This appendix is being included in the code for the convenience of building owners and designers who are required to comply with the the new ADAAG, and for the benefit of jurisdictions that may wish to go beyond the traditional boundary of the code and formally adopt these additional ADAAG requirements. The requirements within the code indicate what is required and enforceable by the code official. The appendix includes items that are outside the scope of code enforcement, but must be complied with in addition to the code requirements to satisfy ADAAG regulations.

SECTION 101
QUALIFIED HISTORICAL BUILDINGS AND FACILITIES

B101.1 General. Qualified historic buildings and facilities shall comply with Sections B101.2 through B101.5.

❖ Special considerations are given to registered historic buildings (see commentary, Sections 1004 and 1005.15).

B101.2 Qualified historic buildings and facilities. These procedures shall apply to buildings and facilities designated as historic structures that undergo alterations or a change of occupancy.

❖ This section provides a process to evaluate historical registered buildings in the event where accessibility requirements may adversely affect the historical significance of the building.

B101.3 Qualified historic buildings and facilities subject to Section 106 of the National Historic Preservation Act. Where an alteration or change of occupancy is undertaken to a

qualified historic building or facility that is subject to Section 106 of the National Historic Preservation Act, the federal agency with jurisdiction over the undertaking shall follow the Section 106 process. Where the State Historic Preservation Officer or Advisory Council on Historic Preservation determines that compliance with the requirements for accessible routes, ramps, entrances, or toilet facilities would threaten or destroy the historic significance of the building or facility, the alternative requirements of Section 1005 for that element are permitted.

❖ If a facility is subject to Section 106 of the National Historic Preservation Act, this section outlines the procedure to follow with the federal agency involved to allow the facility to utilize the alternatives offered in Section 1004 of the code.

B101.4 Qualified historic buildings and facilities not subject to Section 106 of the National Historic Preservation Act. Where an alteration or change of occupancy is undertaken to a qualified historic building or facility that is not subject to Section 106 of the National Historic Preservation Act, and the entity undertaking the alterations believes that compliance with the requirements for accessible routes, ramps, entrances, or toilet facilities would threaten or destroy the historic significance of the building or facility, the entity shall consult with the State Historic Preservation Officer. Where the State Historic Preservation Officer determines that compliance with the accessibility requirements for accessible routes, ramps, entrances, or toilet facilities would threaten or destroy the historical significance of the building or facility, the alternative requirements of Section 1005 for that element are permitted.

❖ If a facility is not subject to Section 106 of the National Historic Preservation Act, this section outlines the procedure to follow with the state historic preservation officer to allow the facility to utilize the alternatives offered in Section 1004 of the code.

B101.4.1 Consultation with interested persons. Interested persons shall be invited to participate in the consultation process, including state or local accessibility officials, individuals with disabilities, and organizations representing individuals with disabilities.

❖ Some type of public forum or meeting must be held to allow for the input of interested parties that may wish to utilize the facilities offered in the historic building.

B101.4.2 Certified local government historic preservation programs. Where the State Historic Preservation Officer has delegated the consultation responsibility for purposes of this section to a local government historic preservation program that has been certified in accordance with Section 101 of the National Historic Preservation Act of 1966 [(16 U.S.C. 470a(c)] and implementing regulations (36 CFR 61.5), the responsibility shall be permitted to be carried out by the appropriate local government body or official.

❖ Where the state has certified a local group in accordance with the National Historic Preservation Act to oversee activity within an historic district or area, that group can take on the role of the state historic preservation officer.

B101.5 Displays. In qualified historic buildings and facilities where alternative requirements of Section 1005 are permitted, displays and written information shall be located where they can be seen by a seated person. Exhibits and signs displayed horizontally shall be 44 inches (1120 mm) maximum above the floor.

❖ Displays, exhibits and signs in historic buildings must be located so that a person in a wheelchair can look at them.

SECTION 102
FIXED TRANSPORTATION FACILITIES AND STATIONS

B102.1 General. Existing fixed transportation facilities and stations shall comply with Section B102.2.

❖ Fixed transportation facilities and stations, such as train or light rail stations, provide access to public transportation and are considered an essential ser-

vice; therefore, accessibility in these facilities is of special concern. Requirements for new construction can be found in Appendix E of the IBC.

B102.2 Existing facilities—key stations. Rapid rail, light rail, commuter rail, intercity rail, high-speed rail and other fixed guide-way systems, altered stations, and intercity rail and key stations, as defined under criteria established by the Department of Transportation in Subpart C of 49 CFR Part 37, shall comply with Sections B102.2.1 through B102.2.3.

❖ Existing guideway system facilities are required to meet a lesser degree of accessibility specified in Sections B102.2.1 through B102.2.3.

B102.2.1 Accessible route. At least one accessible route from an accessible entrance to those areas necessary for use of the transportation system shall be provided. The accessible route shall include the features specified in Appendix E109.2 of the *International Building Code*, except that escalators shall comply with *International Building Code* Section 3005.2.2. Where technical unfeasibility in existing stations requires the accessible route to lead from the public way to a paid area of the transit system, an accessible fare collection machine complying with *International Building Code* Appendix E109.2.3 shall be provided along such accessible route.

❖ At least one accessible route is required throughout passenger areas in an existing facility. Existing escalators in subway stations are not required to meet the criteria for a 32 inch (813 mm) minimum clear width in Section 3005.2.2. If it is technically infeasible to provide an accessible route from the entrance to a ticket sales area, some type of fare-taking machinery may be provided along the accessible route to the train platform.

B102.2.2 Platform and vehicle floor coordination. Station platforms shall be positioned to coordinate with vehicles in accordance with applicable provisions of 36 CFR Part 1192. Low-level platforms shall be 8 inches (250 mm) minimum above top of rail.

Exception: Where vehicles are boarded from sidewalks or street-level, low-level platforms shall be permitted to be less than 8 inches (250 mm).

❖ In existing facilities, the vertical separation between the platform and the cars is slightly less restrictive than in new construction. The gap between the platform and the cars is only measured at accessible cars. Retrofitted cars may have an even larger variance. When train cars cannot meet the vertical separation and gap requirements, alternatives such as lifts and ramps may be utilized.

B102.2.3 Direct connections. New direct connections to commercial, retail, or residential facilities shall, to the maximum extent feasible, have an accessible route complying with Section 506.2 from the point of connection to boarding platforms and transportation system elements used by the public. Any elements provided to facilitate future direct connections shall be on an accessible route connecting boarding platforms and transportation system elements used by the public.

❖ New buildings that provide a direct connection to the existing transit systems must also have an accessible route to the transit system. The exceptions in Section 506.2 are applicable.

SECTION B103
DWELLING UNITS AND SLEEPING UNITS

B103.1 Communication features. Where dwelling units and sleeping units are altered or added, the requirements of Section E104.3 of the *International Building Code* shall apply only to the units being altered or added until the number of units with accessible communication features complies with the minimum number required for new construction.

SECTION B104
REFERENCED STANDARDS

Y3.H626 2P National Historic Preservation J101.2, 43/933 Act of 1966, as amended J101.3, 3rd Edition, Washington, DC: J101.3.2 US Government Printing Office, 1993.

2003 *International Building Code.*

49 CFR Part 37.43 (c), Alteration of Transportation Facilities by Public Entities, Department of Transportation, 400 7th Street SW, Room 8102, Washington, DC 20590-0001.

RESOURCE A

GUIDELINE ON FIRE RATINGS OF ARCHAIC MATERIALS AND ASSEMBLIES

Introduction

The *International Existing Building Code* (IEBC) is a comprehensive code with the goal of addressing all aspects of work taking place in existing buildings and providing user friendly methods and tools for regulation and improvement of such buildings. This resource document is included within the cover of the IEBC with that goal in mind and as a step towards accomplishing that goal.

In the process of repair and alteration of existing buildings, based on the nature and the extent of the work, the IEBC might require certain upgrades in the fire resistance rating of building elements, at which time it becomes critical for the designers and the code officials to be able to determine the fire resistance rating of the existing building elements as part of the overall evaluation for the assessment of the need for improvements. This resource document provides a guideline for such an evaluation for fire resistance rating of archaic materials that is not typically found in the modern model building codes.

Resource A is only a guideline and is not intended to be a document for specific adoption as it is not written in the format or language of ICC's *International Codes* and is not subject to the code development process

PURPOSE

The *Guideline on Fire Ratings of Archaic Materials and Assemblies* focuses upon the fire-related performance of archaic construction. "Archaic" encompasses construction typical of an earlier time, generally prior to 1950. "Fire-related performance" includes fire resistance, flame spread, smoke production, and degree of combustibility.

The purpose of this guideline is to update the information which was available at the time of original construction, for use by architects, engineers, and code officials when evaluating the fire safety of a rehabilitation project. In addition, information relevant to the evaluation of general classes of materials and types of construction is presented for those cases when documentation of the fire performance of a particular archaic material or assembly cannot be found.

It has been assumed that the building materials and their fastening, joining, and incorporation into the building structure are sound mechanically. Therefore, some determination must be made that the original manufacture, the original construction practice, and the rigors of aging and use have not weakened the building. This assessment can often be difficult because process and quality control was not good in many industries, and variations among locally available raw materials and manufacturing techniques often resulted in a product which varied widely in its strength and durability. The properties of iron and steel, for example, varied widely, depending on the mill and the process used.

There is nothing inherently inferior about archaic materials or construction techniques. The pressures that promote fundamental change are most often economic or technological—matters not necessarily related to concerns for safety. The high cost of labor made wood lath and plaster uneconomical. The high cost of land and the congestion of the cities provided the impetus for high-rise construction. Improved technology made it possible. The difficulty with archaic materials is not a question of suitability, but familiarity.

Code requirements for the fire performance of key building elements (e.g., walls, floor/ceiling assemblies, doors, shaft enclosures) are stated in performance terms: hours of fire resistance. It matters not whether these elements were built in 1908 or 1980, only that they provide the required degree of fire resistance. The level of performance will be defined by the local community, primarily through the enactment of a building or rehabilitation code. This guideline is only a tool to help evaluate the various building elements, regardless of what the level of performance is required to be.

The problem with archaic materials is simply that documentation of their fire performance is not readily available. The application of engineering judgment is more difficult because building officials may not be familiar with the materials or construction method involved. As a result, either a full-scale fire test is required or the archaic construction in question removed and replaced. Both alternatives are time consuming and wasteful.

This guideline and the accompanying Appendix are designed to help fill this information void. By providing the necessary documentation, there will be a firm basis for the continued acceptance of archaic materials and assemblies.

1
FIRE–RELATED PERFORMANCE OF ARCHAIC MATERIALS AND ASSEMBLIES

1.1
FIRE PERFORMANCE MEASURES

This guideline does not specify the level of performance required for the various building components. These requirements are controlled by the building occupancy and use and are set forth in the local building or rehabilitation code.

The fire resistance of a given building element is established by subjecting a sample of the assembly to a "standard" fire test which follows a "standard" time-temperature curve. This test method has changed little since the 1920's. The test results tabulated in the Appendix have been adjusted to reflect current test methods.

The current model building codes cite other fire-related properties not always tested for in earlier years: flame spread, smoke production, and degree of combustibility. However, they can generally be assumed to fall within well defined values because the principal combustible component of archaic materials is cellulose. Smoke production is more important today because of the increased use of plastics. However, the early flame spread tests, developed in the early 1940's, also included a test for smoke production.

"Plastics," one of the most important classes of contemporary materials, were not found in the review of archaic materials. If plastics are to be used in a rehabilitated building, they should be evaluated by contemporary standards. Information and documentation of their fire-related properties and performance is widely available.

Flame spread, smoke production and degree of combustibility are discussed in detail below. Test results for eight common species of lumber, published in an Underwriter's Laboratories' report (104), are noted in the following table:

TUNNEL TEST RESULTS FOR EIGHT SPECIES OF LUMBER

SPECIES OF LUMBER	FLAME SPREAD	FUEL CONTRIBUTED	SMOKE DEVELOPED
Western White Pine	75	50-60	50
Northern White Pine	120-215	120-140	60-65
Ponderosa Pine	80-215	120-135	100-110
Yellow Pine	180-190	130-145	275-305
Red Gum	140-155	125-175	40-60
Yellow Birch	105-110	100-105	45-65
Douglas Fir	65-100	50-80	10-100
Western Hemlock	60-75	40-65	40-120

Flame Spread

The flame spread of interior finishes is most often measured by the ASTM E-84 "tunnel test." This test measures how far and how fast the flames spread across the surface of the test sample. The resulting flame spread rating (FSR) is expressed as a number on a continuous scale where cement-asbestos board is 0 and red oak is 100. (Materials with a flame spread greater than red oak have an FSR greater than 100.) The scale is divided into distinct groups or classes. The most commonly used flame spread classifications are: Class I or A*, with a 0-25 FSR; Class II or B, with a 26-75 FSR; and Class III or C, with a 76-200 FSR. The *NFPA Life Safety Code* also has a Class D (201-500 FSR) and Class E (over 500 FSR) interior finish.

These classifications are typically used in modern building codes to restrict the rate of fire spread. Only the first three classifications are normally permitted, though not all classes of materials can be used in all places throughout a building. For example, the interior finish of building materials used in exits or in corridors leading to exits is more strictly regulated than materials used within private dwelling units.

In general, inorganic archaic materials (e.g., bricks or tile) can be expected to be in Class I. Materials of whole wood are mostly Class II. Whole wood is defined as wood used in the same form as sawn from the tree. This is in contrast to the contemporary reconstituted wood products such as plywood, fiberboard, hardboard, or particle board. If the organic archaic material is not whole wood, the flame spread classification could be well over 200 and thus would be particularly unsuited for use in exits and other critical locations in a building. Some plywoods and various wood fiberboards have flame spreads over 200. Although they can be treated with fire retardants to reduce their flame spread, it would be advisable to assume that all such products have a flame spread over 200 unless there is information to the contrary.

Smoke Production

The evaluation of smoke density is part of the ASTM E-84 tunnel test. For the eight species of lumber shown in the table above, the highest levels are 275-305 for Yellow Pine, but most of the others are less smoky than red oak which has an index of 100. The advent of plastics caused substantial increases in the smoke density values measured by the tunnel test. The ensuing limitation of the smoke production for wall and ceiling materials by the model building codes has been a reaction to the introduction of plastic materials. In general, cellulosic materials fall in the 50-300 range of smoke density which is below the general limitation of 450 adopted by many codes.

Degree of Combustibility

The model building codes tend to define "noncombustibility" on the basis of having passed ASTM E-136 or if the material is totally inorganic. The acceptance of gypsum wallboard as noncombustible is based on limiting paper thickness to not over $1/8$ inch and a 0-50 flame spread rating by ASTM E-84. At times there were provisions to define a Class I or A material (0-25 FSR) as noncombustible, but this is not currently recognized by most model building codes.

If there is any doubt whether or not an archaic material is noncombustible, it would be appropriate to send out samples for evaluation. If an archaic material is determined to be noncombustible according to ASTM E-136, it can be expected that it will not contribute fuel to the fire.

1.2
COMBUSTIBLE CONSTRUCTION TYPES

One of the earliest forms of timber construction used exterior load-bearing masonry walls with columns and/or wooden walls supporting wooden beams and floors in the interior of the building. This form of construction, often called "mill" or "heavy timber" construction, has approximately 1 hour fire resistance. The exterior walls will generally contain the fire within the building.

With the development of dimensional lumber, there was a switch from heavy timber to "balloon frame" construction. The balloon frame uses load-bearing exterior wooden walls which have long timbers often extending from foundation to roof. When longer lumber became scarce, another form of construction, "platform" framing, replaced the balloon framing. The difference between the two systems is significant because platform framing is automatically fire-blocked at every floor while balloon framing commonly has concealed spaces that extend unblocked from basement to attic. The architect, engineer, and code official must be alert to the details of construction and the ease with which fire can spread in concealed spaces.

2
BUILDING EVALUATION

A given rehabilitation project will most likely go through several stages. The preliminary evaluation process involves the designer in surveying the prospective building. The fire resistance of existing building materials and construction systems is identified; potential problems are noted for closer study. The final evaluation phase includes: developing design solutions to upgrade the fire resistance of building elements, if necessary; preparing working drawings and specifications; and the securing of the necessary code approvals.

2.1
PRELIMINARY EVALUATION

A preliminary evaluation should begin with a building survey to determine the existing materials, the general arrangement of the structure and the use of the occupied spaces, and the details of construction. The designer needs to know "what is there" before a decision can be reached about what to keep and what to remove during the rehabilitation process. This preliminary evaluation should be as detailed as necessary to make initial plans. The fire-related properties need to be determined from the applicable building or rehabilitation code, and the materials and assemblies existing in the building then need to be evaluated for these properties. Two work sheets are shown below to facilitate the preliminary evaluation.

Two possible sources of information helpful in the preliminary evaluation are the original building plans and the building code in effect at the time of original construction. Plans may be on file with the local building department or in the offices of the original designers (e.g., architect, engineer) or their successors. If plans are available, the investigator should verify that the building was actually constructed as called for in the plans, as well as incorporate any later alterations or

---

* Some codes are Roman numerals, others use letters

changes to the building. Earlier editions of the local building code should be on file with the building official. The code in effect at the time of construction will contain fire performance criteria. While this is no guarantee that the required performance was actually provided, it does give the investigator some guidance as to the level of performance which may be expected. Under some code administration and enforcement systems, the code in effect at the time of construction also defines the level of performance that must be provided at the time of rehabilitation.

Figure 1 illustrates one method for organizing preliminary field notes. Space is provided for the materials, dimensions, and condition of the principal building elements. Each floor of the structure should be visited and the appropriate information obtained. In practice, there will often be identical materials and construction on every floor, but the exception may be of vital importance. A schematic diagram should be prepared of each floor showing the layout of exits and hallways and indicating where each element described in the field notes fits into the structure as a whole. The exact arrangement of interior walls within apartments is of secondary importance from a fire safety point of view and need not be shown on the drawings unless these walls are required by code to have a fire resistance rating.

The location of stairways and elevators should be clearly marked on the drawings. All exterior means of escape (e.g., fire escapes) should be identified.*

The following notes explain the entries in Figure 1.

Exterior Bearing Walls: Many old buildings utilize heavily constructed walls to support the floor/ceiling assemblies at the exterior of the building. There may be columns and/or interior bearing walls within the structure, but the exterior walls are an important factor in assessing the fire safety of a building.

The field investigator should note how the floor/ceiling assemblies are supported at the exterior of the building. If columns are incorporated in the exterior walls, the walls may be considered non-bearing.

Interior Bearing Walls: It may be difficult to determine whether or not an interior wall is load bearing, but the field investigator should attempt to make this determination. At a later stage of the rehabilitation process, this question will need to be determined exactly. Therefore, the field notes should be as accurate as possible.

Exterior Non-Bearing Walls: The fire resistance of the exterior walls is important for two reasons. These walls (both bearing and non-bearing) are depended upon to: a) contain a fire within the building of origin; or b) keep an exterior fire *outside* the building. It is therefore important to indicate on the drawings where any openings are located as well as the materials and construction of all doors or shutters. The drawings should indicate the presence of wired glass, its thickness and framing, and identify the materials used for windows and door frames. The protection of openings adjacent to exterior means of escape (e.g., exterior stairs, fire escapes) is particularly important. The ground floor drawing should locate the building on the property and indicate the precise distances to adjacent buildings.

Interior Non-Bearing Walls (Partitions): A partition is a "wall that extends from floor to ceiling and subdivides space within any story of a building." (48) Figure 1 has two categories (A & B) for Interior Non-Bearing Walls (Partitions) which can be used for different walls, such as hallway walls as compared to inter-apartment walls. Under some circumstances there may be only one type of wall construction; in others, three or more types of wall construction may occur.

---

* Problems providing adequate exiting are discussed at length in the *Egress Guideline for Residential Rehabilitation.*

FIGURE 1   PRELIMINARY EVALUATION FIELD NOTES

Building Element		Materials	Thickness	Condition	Notes
Exterior Bearing Walls					
Interior Bearing Walls					
Exterior Non-Bearing Walls					
Interior Non-Bearing Walls or Partitions:	A				
	B				
Strucural Frame: Columns					
Beams					
Other					
Floor/Ceiling Structural System Spanning					
Roofs					
Doors (including frame and hardware): a) Enclosed vertical exitway					
b) Enclosed horizontal exitway					
c) Other					

The field investigator should be alert for differences in function as well as in materials and construction details. In general, the details within apartments are not as important as the major exit paths and stairwells. The preliminary field investigation should attempt to determine the thickness of all walls. A term introduced below called "thickness design" will depend on an accurate ($\pm$ $^1/_4$ inch) determination. Even though this initial field survey is called "preliminary," the data generated should be as accurate and complete as possible.

The field investigator should note the exact location from which observations are recorded. For instance, if a hole is found through a stairwell wall which allows a cataloguing of the construction details, the field investigation notes should reflect the location of the "find." At the preliminary stage it is not necessary to core every wall; the interior details of construction can usually be determined at some location.

Structural Frame: There may or may not be a complete skeletal frame, but usually there are columns, beams, trusses, or other like elements. The dimensions and spacing of the structural elements should be measured and indicated on the drawings. For instance, if there are ten inch square columns located on a thirty foot square grid throughout the building, this should be noted. The structural material and cover or protective materials should be identified wherever possible. The thickness of the cover materials should be determined to an accuracy of $\pm$ $^1/_4$ inch. As discussed above, the preliminary field survey usually relies on accidental openings in the cover materials rather than a systematic coring technique.

Floor/Ceiling Structural Systems: The span between supports should be measured. If possible, a sketch of the cross-section of the system should be made. If there is no location where accidental damage has opened the floor/ceiling construction to visual inspection, it is necessary to make such an opening. An evaluation of the fire resistance of a floor/ceiling assembly requires detailed knowledge of the materials and their arrangement. Special attention should be paid to the cover on structural steel elements and the condition of suspended ceilings and similar membranes.

Roofs: The preliminary field survey of the roof system is initially concerned with water-tightness. However, once it is apparent that the roof is sound for ordinary use and can be retained in the rehabilitated building, it becomes necessary to evaluate the fire performance. The field investigator must measure the thickness and identify the types of materials which have been used. Be aware that there may be several layers of roof materials.

<u>Doors:</u> Doors to stairways and hallways represent some of the most important fire elements to be considered within a building. The uses of the spaces separated largely controls the level of fire performance necessary. Walls and doors enclosing stairs or elevator shafts would normally require a higher level of performance than between a the bedroom and bath. The various uses are differentiated in Figure 1.

Careful measurements of the thickness of door panels must be made, and the type of core material within each door must be determined. It should be noted whether doors have self-closing devices; the general operation of the doors should be checked. The latch should engage and the door should fit tightly in the frame. The hinges should be in good condition. If glass is used in the doors, it should be identified as either plain glass or wired glass mounted in either a wood or steel frame.

<u>Materials:</u> The field investigator should be able to identify ordinary building materials. In situations where an unfamiliar material is found, a sample should be obtained. This sample should measure at least 10 cubic inches so that an ASTM E-136 fire test can be conducted to determine if it is combustible.

<u>Thickness:</u> The thickness of all materials should be measured accurately since, under certain circumstances, the level of fire resistance is very sensitive to the material thickness.

<u>Condition:</u> The method of attaching the various layers and facings to one another or to the supporting structural element should be noted under the appropriate building element. The "secureness" of the attachmnent and the general condition of the layers and facings should be noted here.

<u>Notes:</u> The "Notes" column can be used for many purposes, but it might be a good idea to make specific references to other field notes or drawings.

After the building survey is completed, the data collected must be analyzed. A suggested work sheet for organizing this information is given below as Figure 2.

The required fire resistance and flame spread for each building element are normally established by the local building or rehabilitation code. The fire performance of the existing materials and assemblies should then be estimated, using one of the techniques described below. If the fire performance of the existing building element(s) is equal to or greater than that required, the materials and assemblies may remain. If the fire performance is less than required, then corrective measures must be taken.

The most common methods of upgrading the level of protection are to either remove and replace the existing building element(s) or to repair and upgrade the existing materials and assemblies. Other fire protection measures, such as automatic sprinklers or detection and alarm systems, also could be considered, though they are beyond the scope of this guideline. If the upgraded protection is still less than that required or deemed to be acceptable, additional corrective measures must be taken. This process must continue until an acceptable level of performance is obtained.

FIGURE 1   PRELIMINARY EVALUATION WORKSHEET

Building Element		Required Fire Resistance	Required Flame Spread	Estimated Fire Resistance	Estimated Flame Spread	Method of Upgrading	Estimated Upgraded Protection	Notes
Exterior Bearing Walls								
Interior Bearing Walls								
Exterior Non-Bearing Walls								
Interior Non-Bearing Walls or Partitions:	A							
	B							
Structural Frame: Columns								
Beams								
Other								
Floor/Ceiling Structural System Spanning								
Roofs								
Doors (including frame and hardware): a) Enclosed vertical exitway								
b) Enclosed horizontal exitway								
c) Other								

2.2
FIRE RESISTANCE OF EXISTING BUILDING ELEMENTS

The fire resistance of the existing building elements can be estimated from the tables and histograms contained in the Appendix. The Appendix is organized first by type of building element: walls, columns, floor/ceiling assemblies, beams, and doors. Within each building element, the tables are organized by type of construction (e.g., masonry, metal, wood frame), and then further divided by minimum dimensions or thickness of the building element.

A histogram precedes every table that has 10 or more entries. The X-axis measures fire resistance in hours; the Y-axis shows the number of entries in that table having a given level of fire resistance. The histograms also contain the location of each entry within that table for easy cross-referencing.

The histograms, because they are keyed to the tables, can speed the preliminary investigation. For example, Table 1.3.2, *Wood Frame Walls 4" to Less Than 6" Thick*, contains 96 entries. Rather than study each table entry, the histogram shows that every wall assembly listed in that table has a fire resistance of less than 2 hours. If the building code required the wall to

have 2 hours fire resistance, the designer, with a minimum of effort, is made aware of a problem that requires closer study.

Suppose the code had only required a wall of 1 hour fire resistance. The histogram shows far fewer complying elements (19) than noncomplying ones (77). If the existing assembly is not one of the 19 complying entries, there is a strong possibility the existing assembly is deficient. The histograms can also be used in the converse situation. If the existing assembly is not one of the smaller number of entries with a lower than required fire resistance, there is a strong possibility the existing assembly will be acceptable.

At some point, the existing building component or assembly must be located within the tables. Otherwise, the fire resistance must be determined through one of the other techniques presented in the guideline. Locating the building component in the Appendix Tables not only guarantees the accuracy of the fire resistance rating, but also provides a source of documentation for the building official.

2.3
EFFECTS OF PENETRATIONS IN FIRE RESISTANT ASSEMBLIES

There are often many features in existing walls or floor/ceiling assemblies which were not included in the original certification or fire testing. The most common examples are pipes and utility wires passed through holes poked through an assembly. During the life of the building, many penetrations are added, and by the time a building is ready for rehabilitation it is not sufficient to just consider the fire resistance of the assembly as originally constructed. It is necessary to consider all penetrations and their relative impact upon fire performance. For instance, the fire resistance of the corridor wall may be less important than the effect of plain glass doors or transoms. In fact, doors are the most important single class of penetrations.

A fully developed fire generates substantial quantities of heat and excess gaseous fuel capable of penetrating any holes which might be present in the walls or ceiling of the fire compartment. In general, this leads to a severe degradation of the fire resistance of those building elements and to a greater potential for fire spread. This is particularly applicable to penetrations located high in a compartment where the positive pressure of the fire can force the unburned gases through the penetration.

Penetrations in a floor/ceiling assembly will generally completely negate the barrier qualities of the assembly and will lead to rapid spread of fire to the space above. It will not be a problem, however, if the penetrations are filled with noncombustible materials strongly fastened to the structure. The upper half of walls are similar to the floor/ceiling assembly in that a positive pressure can reasonably be expected in the top of the room, and this will push hot and/or burning gases through the penetration unless it is completely sealed.

Building codes require doors installed in fire resistive walls to resist the passage of fire for a specified period of time. If the door to a fully involved room is not closed, a large plume of fire will typically escape through the doorway, preventing anyone from using the space outside the door while allowing the fire to spread. This is why door closers are so important. Glass in doors and transoms can be expected to rapidly shatter unless constructed of listed or approved wire glass in a steel frame. As with other building elements, penetrations or non-rated portions of doors and transoms must be upgraded or otherwise protected.

Table 5.1 in Section V of the Appendix contains 41 entries of doors mounted in sound tightfitting frames. Part 3.4 below outlines one procedure for evaluating and possibly upgrading existing doors.

3
FINAL EVALUATION AND DESIGN SOLUTION

The final evaluation begins after the rehabilitation project has reached the final design stage and the choices made to keep certain archaic materials and assemblies in the rehabilitated building. The final evaluation process is essentially a more refined and detailed version of the preliminary evaluation. The specific fire resistance and flame spread requirements are determined for the project. This may involve local building and fire officials reviewing the preliminary evaluation as depicted in Figures 1 and 2 and the field drawings and notes. When necessary, provisions must be made to upgrade existing building elements to provide the required level of fire performance.

There are several approaches to design solutions that can make possible the continued use of archaic materials and assemblies in the rehabilitated structure. The simplest case occurs when the materials and assembly in question are found within the Appendix Tables and the fire performance properties satisfy code requirements. Other approaches must be used, though, if the assembly cannot be found within the Appendix or the fire performance needs to be upgraded. These approaches have been grouped into two classes: experimental and theoretical.

3.1
THE EXPERIMENTAL APPROACH

If a material or assembly found in a building is not listed in the Appendix Tables, there are several other ways to evaluate fire performance. One approach is to conduct the appropriate fire test(s) and thereby determine the fire-related properties directly. There are a number of laboratories in the United States which routinely conduct the various fire tests. A current list can be obtained by writing the Center for Fire Research, National Bureau of Standards, Washington, D.C. 20234.

The contract with any of these testing laboratories should require their observation of specimen preparation as well as the testing of the specimen. A complete description of where and how the specimen was obtained from the building, the transportation of the specimen, and its preparation for testing should be noted in detail so that the building official can be satisfied that the fire test is representative of the actual use.

The test report should describe the fire test procedure and the response of the material or assembly. The laboratory usually submits a cover letter with the report to describe the provisions of the fire test that were satisfied by the material or assembly under investigation. A building official will generally require this cover letter, but will also read the report to confirm that the material or assembly complies with the code requirements. Local code officials should be involved in all phases of the testing process.

The experimental approach can be costly and time consuming because specimens must be taken from the building and transported to the testing laboratory. When a load bearing assembly has continuous reinforcement, the test specimen must be removed from the building, transported, and tested in one piece. However, when the fire performance cannot be determined by other means, there may be no alternative to a full-scale test.

A "non-standard" small-scale test can be used in special cases. Sample sizes need only be 10-25 square feet, while full-scale tests require test samples of either 100 or 180 square feet in size. This small-scale test is best suited for testing non-load bearing assemblies against thermal transmission only.

3.2
THE THEORETICAL APPROACH

There will be instances when materials and assemblies in a building undergoing rehabilitation cannot be found in the Appendix Tables. Even where test results are available for more or less similar construction, the proper classification may not be immediately apparent. Variations in dimensions, loading conditions, materials, or workmanship may markedly affect the performance of the individual building elements, and the extent of such a possible effect cannot be evaluated from the tables.

Theoretical methods being developed offer an alternative to the full-scale fire tests discussed above. For example, Section 4302(b) of the 1979 edition of the *Uniform Building Code* specifically allows an engineering design for fire resistance in lieu of conducting full-scale tests. These techniques draw upon computer simulation and mathematical modeling, thermodynamics, heat-flow analysis, and materials science to predict the fire performance of building materials and assemblies.

One theoretical method, known as the "Ten Rules of Fire Endurance Ratings," was published by T. Z. Harmathy in the May, 1965 edition of *Fire Technology*. (35) Harmathy's Rules provide a foundation for extending the data within the Appendix Tables to analyze or upgrade current as well as archaic building materials or assemblies.

HARMATHY'S TEN RULES

Rule 1: The "thermal" * fire endurance of a construction consisting of a number of parallel layers is greater than the sum of the "thermal" fire endurances characteristic of the individual layers when exposed separately to fire.

The minimum performance of an untested assembly can be estimated if the fire endurance of the individual components is known. Though the exact rating of the assembly cannot be stated, the endurance of the assembly is greater than the sum of the endurance of the components.

When a building assembly or component is found to be deficient, the fire endurance can be upgraded by providing a protective membrane. This membrane could be a new layer of brick, plaster, or drywall. The fire endurance of this membrane is called the "finish rating." Appendix Tables 1.5.1 and 1.5.2 contain the finish ratings for the most commonly employed materials. (See also the notes to Rule 2).

The test criteria for the finish rating is the same as for the thermal fire endurance of the total assembly: average temperature increases of 250°F above ambient or 325°F above ambient at any one place with the membrane being exposed to the fire. The temperature is measured at the interface of the assembly and the protective membrane.

Rule 2: The fire endurance of a construction does not decrease with the addition of further layers.

Harmathy notes that this rule is a consequence of the previous rule. Its validity follows from the fact that the additional layers increase both the resistance to heat flow and the heat capacity of the construction. This, in turn, reduces the rate of temperature rise at the unexposed surface.

This rule is not just restricted to "thermal" performance but affects the other fire test criteria: direct flame passage, cotton waste ignition, and load bearing performance. This means that certain restrictions must be imposed on the materials to be added and on the loading conditions. One restriction is that a new layer, if applied to the exposed surface, must not produce additional thermal stresses in the construction, i.e., its thermal expansion characteristics must be similar to those of the adjacent layer. Each new layer must also be capable of contributing enough additional strength to the assembly to sustain the added dead load. If this requirement is not fulfilled, the allowable live load must be reduced by an amount equal to the weight of the new layer. Because of these limitations, this rule should not be applied without careful consideration.

Particular care must be taken if the material added is a good thermal insulator. Properly located, the added insulation could improve the "thermal" performance of the assembly. Improperly located, the insulation could block necessary thermal transmission through the assembly, thereby subjecting the structural elements to greater temperatures for longer periods of time, and could cause premature structural failure of the supporting members.

Rule 3: The fire endurance of constructions containing continuous air gaps or cavities is greater than the fire endurance of similar constructions of the same weight, but containing no air gaps or cavities.

By providing for voids in a construction, additional resistances are produced in the path of heat flow. Numerical heat flow analyses indicate that a 10 to 15 percent increase in fire endurance can be achieved by creating an air gap at the midplane of a brick wall. Since the gross volume is also increased by the presence of voids, the air gaps and cavities have a beneficial effect on stability as well. However, constructions containing combustible materials within an air gap may be regarded as exceptions to this rule because of the possible development of burning in the gap.

There are numerous examples of this rule in the tables. For instance:

Table 1.1.4; Item W-8-M-82: Cored concrete masonry, nominal 8 inch thick wall with one unit in wall thickness and with 62% minimum of solid material in each unit, load bearing (80 PSI). Fire endurance: 2¹/₂ hours.

Table 1.1.5; Item W-10-M-11: Cored concrete mansonry, nominal 10 inch thick wall with two units in wall thickness and a 2 inch air space, load bearing (80 PSI). The units are essentially the same as item W-8-M-82. Fire endurance: 3¹/₂ hours.

These walls show 1 hour greater fire endurance by the addition of the 2 inch air space.

Rule 4: The farther an air gap or cavity is located from the exposed surface, the more beneficial is its effect on the fire endurance.

---

* The "thermal" fire endurance is the time at which the average temperature on the unexposed side of a construction exceeds its initial value by 250° when the other side is exposed to the "standard" fire specified by ASTM Test Method E-19.

Radiation dominates the heat transfer across an air gap or cavity, and it is markedly higher where the temperature is higher. The air gap or cavity is thus a poor insulator if it is located in a region which attains high temperatures during fire exposure.

Some of the clay tile designs take advantage of these factors. The double cell design, for instance, ensures that there is a cavity near the unexposed face. Some floor/ceiling assemblies have air gaps or cavities near the top surface and these enhance their thermal performance.

Rule 5: The fire endurance of a construction cannot be increased by increasing the thickness of a completely enclosed air layer.

Harmathy notes that there is evidence that if the thickness of the air layer is larger than about $^1/_2$ inch, the heat transfer through the air layer depends only on the temperature of the bounding surfaces, and is practically independent of the distance between them. This rule is not applicable if the air layer is not completely enclosed, i.e., if there is a possibility of fresh air entering the gap at an appreciable rate.

Rule 6: Layers of materials of low thermal conductivity are better utilized on that side of the construction on which fire is more likely to happen.

As in Rule 4, the reason lies in the heat transfer process, though the conductivity of the solid is much less dependent on the ambient temperature of the materials. The low thermal conductor creates a substantial temperature differential to be established across its thickness under transient heat flow conditions. This rule may not be applicable to materials undergoing physico-chemical changes accompanied by significant heat absorption or heat evolution.

Rule 7: The fire endurance of asymmetrical constructions depends on the direction of heat flow.

This rule is a consequence of Rules 4 and 6 as well as other factors. This rule is useful in determining the relative protection of corridors and stairwells from the surrounding spaces. In addition, there are often situations where a fire is more likely, or potentially more severe, from one side or the other.

Rule 8: The presence of moisture, if it does not result in explosive spalling, increases the fire endurance.

The flow of heat into an assembly is greatly hindered by the release and evaporation of the moisture found within cementitious materials such as gypsum, portland cement, or magnesium oxychloride. Harmathy has shown that the gain in fire endurance may be as high as 8 percent for each percent (by volume) of moisture in the construction. It is the moisture chemically bound within the construction material at the time of manufacture or processing that leads to increased fire endurance. There is no direct relationship between the relative humidity of the air in the pores of the material and the increase in fire endurance.

Under certain conditions there may be explosive spalling of low permeability cementitious materials such as dense concrete. In general, one can assume that extremely old concrete has developed enough minor cracking that this factor should not be significant.

Rule 9: Load-supporting elements, such as beams, girders and joists, yield higher fire endurances when subjected to fire endurance tests as parts of floor, roof, or ceiling assemblies than they would when tested separately.

One of the fire endurance test criteria is the ability of a load-supporting element to carry its design load. The element will be deemed to have failed when the load can no longer be supported.

Failure usually results for two reasons. Some materials, particularly steel and other metals, lose much of their structural strength at elevated temperatures. Physical deflection of the supporting element, due to decreased strength or thermal expansion, causes a redistribution of the load forces and stresses throughout the element. Structural failure often results because the supporting element is not designed to carry the redistributed load.

Roof, floor, and ceiling assemblies have primary (e.g., beams) and secondary (e.g., floor joists) structural members. Since the primary load-supporting elements span the largest distances, their deflection becomes significant at a stage when the strength of the secondary members (including the roof or floor surface) is hardly affected by the heat. As the secondary members follow the deflection of the primary load-supporting element, an increasingly larger portion of the load is transferred to the secondary members.

When load-supporting elements are tested separately, the imposed load is constant and equal to the design load throughout the test. By definition, no distribution of the load is possible because the element is being tested by itself. Without any other structural members to which the load could be transferred, the individual elements cannot yield a higher fire endurance than they do when tested as parts of a floor, roof or ceiling assembly.

Rule 10: The load-supporting elements (beams, girders, joists, etc.) of a floor, roof, or ceiling assembly can be replaced by such other load-supporting elements which, when tested separately, yielded fire endurances not less than that of the assembly.

This rule depends on Rule 9 for its validity. A beam or girder, if capable of yielding a certain performance when tested separately, will yield an equally good or better performance when it forms a part of a floor, roof, or ceiling assembly. It must be emphasized that the supporting element of one assembly must not be replaced by the supporting element of another assembly if the performance of this latter element is not known from a separate (beam) test. Because of the load-reducing effect of the secondary elements that results from a test performed on an assembly, the performance of the supporting element alone cannot be evaluated by simple arithmetic. This rule also indicates the advantage of performing separate fire tests on primary load-supporting elements.

ILLUSTRATION OF HARMATHY'S RULES

Harmathy provided one schematic figure which illustrated his Rules.* It should be useful as a quick reference to assist in applying his Rules.

---

* Reproduced from the May 1065 *Fire Technology* (Vol. 1, No. 2). Copyright National Fire Protection Association, Boston. Reproduced by permission.

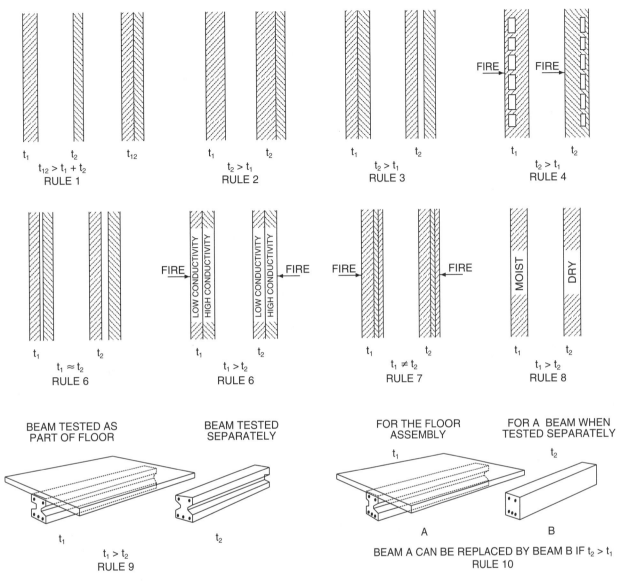

Diagrammatic illustration of ten rules.
t = fire endurance

EXAMPLE APPLICATION OF HARMATHY'S RULES

The following examples, based in whole or in part upon those presented in Harmathy's paper (35), show how the Rules can be applied to practical cases.

Example 1

Problem

A contractor would like to keep a partition which consists of a $3^3/_4$ inch thick layer of red clay brick, a $1^1/_4$ inch thick layer of plywood, and a $^3/_8$ inch thick layer of gypsum wallboard, at a location where 2 hour fire endurance is required. Is this assembly capable of providing a 2 hour protection?

Solution

(1) This partition does not appear in the Appendix Tables.

(2) Bricks of this thickness yield fire endurances of approximately 75 minutes (Table 1.1.2, Item W-4-M-2).

(3) The $1^1/_4$ inch thick plywood has a finish rating of 30 minutes.

(4) The $^3/_8$ inch gypsum wallboard has a finish rating of 10 minutes.

(5) Using the recommended values from the tables and applying Rule 1, the fire endurance (FI) of the assembly is larger than the sum of the individual layers, or

$$FI > 75 + 30 + 10 = 115 \text{ minutes}$$

Discussion

This example illustrates how the Appendix Tables can be utilized to determine the fire resistance of assemblies not explicitly listed.

Example 2

Problem

(1) A number of buildings to be rehabilitated have the same type of roof slab which is supported with different structural elements.

(2) The designer and contractor would like to determine whether or not this roof slab is capable of yielding a 2 hour fire endurance. According to a rigorous interpretation of ASTM E-119, however, only the roof assembly, including the roof slab as well as the cover and the supporting elements, can be subjected to a fire test. Therefore, a fire endurance classification cannot be issued for the slabs separately.

(3) The designer and contractor believe this slab will yield a 2 hour fire endurance even without the cover, and any beam of at least 2 hour fire endurance will provide satisfactory support. Is it possible to obtain a classification for the slab separately?

Solution

(1) The answer to the question is yes.

(2) According to Rule 10 it is not contrary to common sense to test and classify roofs and supporting elements separately. Furthermore, according to Rule 2, if the roof slabs actually yield a 2 hour fire endurance, the endurance of an assembly, including the slabs, cannot be less than 2 hours.

(3) The recommended procedure would be to review the tables to see if the slab appears as part of any tested roof or floor/ceiling assembly. The supporting system can be regarded as separate from the slab specimen, and the fire endurance of the assembly listed in the table is at least the fire endurance of the slab. There would have to be an adjustment for the weight of the roof cover in the allowable load if the test specimen did not contain a cover.

(4) The supporting structure or element would have to have at least a 2 hour fire endurance when tested separately.

Discussion

If the tables did not include tests on assemblies which contained the slab, one procedure would be to assemble the roof slabs on any convenient supporting system (not regarded as part of the specimen) and to subject them to a load which, besides the usually required superimposed load, includes some allowances for the weight of the cover.

Example 3

Problem

A steel-joisted floor and ceiling assembly is known to have yielded a fire endurance of 1 hour and 35 minutes. At a certain location, a 2 hour endurance is required. What is the most economical way of increasing the fire endurance by at least 25 minutes?

Solution

(1) The most effective technique would be to increase the ceiling plaster thickness. Existing coats of paint would have to be removed and the surface properly prepared before the new plaster could be applied. Other materials (e.g., gypsum wallboard) could also be considered.

(2) There may be other techniques based on other principles, but an examination of the drawings would be necessary.

Discussion

(1) The additional plaster has at least three effects:

  a) The layer of plaster is increased and thus there is a gain of fire endurance (Rule 1).

  b) There is a gain due to shifting the air gap farther from the exposed surface (Rule 4).

  c) There is more moisture in the path of heat flow to the structural elements (Rules 7 and 8).

(2) The increase in fire endurance would be at least as large as that of the finish rating for the added thickness of plaster. The combined effects in (1) above would further increase this by a factor of 2 or more, depending upon the geometry of the assembly.

Example 4

Problem

The fire endurance of item W-l0-M-l in Table 1.1.5 is 4 hours. This wall consists of two $3^3/_4$ inch thick layers of structural tiles separated by a 2 inch air gap and $^3/_4$ inch portland cement plaster or stucco on both sides. If the actual wall in the building is identical to item W-10-M-1 except that it has a 4 inch air gap, can the fire endurance be estimated at 5 hours?

Solution

The answer to the question is no for the reasons contained in Rule 5.

Example 5

Problem

In order to increase the insulating value of its precast roof slabs, a company has decided to use two layers of different concretes. The lower layer of the slabs, where the strength of the concrete is immaterial (all the tensile load is carried by the steel reinforcement), would be made with a concrete of low strength but good insulating value. The upper layer, where the concrete is supposed to carry the compressive load, would remain the original high strength, high thermal conductivity concrete. How will the fire endurance of the slabs be affected by the change?

Solution

The effect on the thermal fire endurance is beneficial:

(1) The total resistance to heat flow of the new slabs has been increased due to the replacement of a layer of high thermal conductivity by one of low conductivity.

(2) The layer of low conductivity is on the side more likely to be exposed to fire, where it is more effectively utilized according to Rule 6. The layer of low thermal conductivity also provides better protection for the steel reinforcement, thereby extending the time before reaching the temperature at which the creep of steel becomes significant.

3.3 "THICKNESS DESIGN" STRATEGY

The "thickness design" strategy is based upon Harmathy's Rules 1 and 2. This design approach can be used when the construction materials have been identified and measured, but the specific assembly cannot be located within the tables. The tables should be surveyed again for thinner walls of like material and construction detail that have yielded the desired or greater fire endurance. If such an assembly can be found, then the thicker walls in the building have more than enough fire resistance. The thickness of the walls thus becomes the principal concern.

This approach can also be used for floor/ceiling assemblies, except that the thickness of the cover* and the slab become the central concern. The fire resistance of the untested assembly will be at least the fire resistance of an assembly listed in the table having a similar design but with less cover and/or thinner slabs. For other structural elements (e.g., beams and columns), the element listed in the table must also be of a similar design but with less cover thickness.

3.4
EVALUATION OF DOORS

A separate section on doors has been included because the process for evaluation presented below differs from those suggested previously for other building elements. The impact of unprotected openings or penetrations in fire resistant assemblies has been detailed in Part 2.3 above. It is sufficient to note here that openings left unprotected will likely lead to failure of the barrier under actual fire conditions.

For other types of building elements (e.g., beams, columns), the Appendix Tables can be used to establish a minimum level of fire performance. The benefit to rehabilitation is that the need for a full-scale fire test is then eliminated. For doors, however, this cannot be done. The data contained in Appendix Table 5.1, Resistance of Doors to Fire Exposure, can only provide guidance as to whether a successful fire test is even feasible.

For example, a door required to have 1 hour fire resistance is noted in the tables as providing only 5 minutes. The likelihood of achieving the required 1 hour, even if the door is upgraded, is remote. The ultimate need for replacement of the doors is reasonably clear, and the expense and time needed for testing can be saved. However, if the performance documented in the table is near or in excess of what is being required, then a fire test should be conducted. The test documentation can then be used as evidence of compliance with the required level of performance.

The table entries cannot be used as the sole proof of performance of the door in question because there are too many unknown variables which could measurably affect fire performance. The wood may have dried over the years; coats of flammable varnish could have been added. Minor deviations in the internal construction of a door can result in significant differences in performance. Methods of securing inserts in panel doors can vary. The major non-destructive method of analysis, an x-ray, often cannot provide the necessary detail. It is for these, and similar reasons, that a fire test is still felt to be necessary.

It is often possible to upgrade the fire performance of an existing door. Sometimes, "as is" and modified doors are evaluated in a single series of tests when failure of the unmodified door is expected. Because doors upgraded after an initial failure must be tested again, there is a potential savings of time and money.

The most common problems encountered are plain glass, panel inserts of insufficient thickness, and improper fit of a door in its frame. The latter problem can be significant because a fire can develop a substantial positive pressure, and the fire will work its way through otherwise innocent-looking gaps between door and frame.

One approach to solving these problems is as follows. The plain glass is replaced with approved or listed wire glass in a steel frame. The panel inserts can be upgraded by adding an additional layer of material. Gypsum wallboard is often used for this purpose. Intumescent paint applied to the edges of the door and frame will expand when exposed to fire, forming an effective seal around the edges. This seal, coupled with the generally even thermal expansion of a wood door in a wood frame, can prevent the passage of flames and other fire gases. Figure 3 below illustrates these solutions.

Because the interior construction of a door cannot be determined by a visual inspection, there is no absolute guarantee that the remaining doors are identical to the one(s) removed from the building and tested. But the same is true for doors constructed today, and reason and judgment must be applied. Doors that appear identical upon visual inspection can be weighed. If the weights are reasonably close, the doors can be assumed to be identical and therefore provide the same level of fire performance. Another approach is to fire test more than one door or to dismantle doors selected at random to see if they had been constructed in the same manner. Original building plans showing door details or other records showing that doors were purchased at one time or obtained from a single supplier can also be evidence of similar construction.

More often though, it is what is visible to the eye that is most significant. The investigator should carefully check the condition and fit of the door and frame, and for frames out of plumb or separating from the wall. Door closers, latches, and hinges must be examined to see that they function properly and are tightly secured. If these are in order and the door and frame have passed a full-scale test, there can be a reasonable basis for allowing the existing doors to remain.

4
SUMMARY

This section summarizes the various approaches and design solutions discussed in the preceding sections of the guideline. The term "structural system" includes: frames, beams, columns, and other structural elements. "Cover" is a protective layer(s) of materials or membrane which slows the flow of heat to the structural elements. It cannot be stressed too strongly that the fire endurance of actual building elements can be greatly reduced or totally negated by removing part of the cover to allow pipes, ducts, or conduits to pass through the element. This must be repaired in the rehabilitation process.

The following approaches shall be considered equivalent.

---

* Cover: the protective layer or membrane of material which slows the flow of heat to the structural elements.

FIGURE 3

MODIFICATION DETAILS

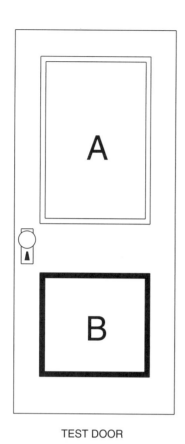

TEST DOOR

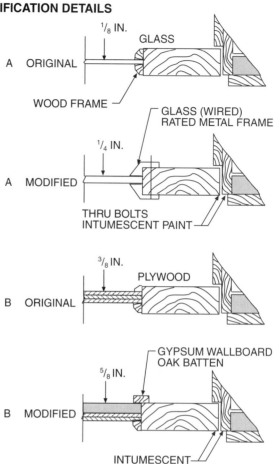

4.1 The fire resistance of a building element can be established from the Appendix Tables. This is subject to the following limitations:

- The building element in the rehabilitated building shall be constructed of the same materials with the same nominal dimensions as stated in the tables.
- All penetrations in the building element or its cover for services such as electricity, plumbing, and HVAC shall be packed with noncombustible cementitious materials and so fixed that the packing material will not fall out when it loses its water of hydration.

The effects of age and wear and tear shall be repaired so that the building element is sound and the original thickness of all components, particularly covers and floor slabs, is maintained.

This approach essentially follows the approach taken by model building codes. The assembly must appear in a table either published in or accepted by the code for a given fire resistance rating to be recognized and accepted.

4.2 The fire resistance of a building element which does not explicitly appear in the Appendix Tables can be established if one or more elements of same design but different dimensions have been listed in the tables. For walls, the existing element must be thicker than the one listed. For floor/ceiling assemblies, the assembly listed in the table must have the same or less cover and the same or thinner slab constructed of the same material as the actual floor/ceiling assembly. For other structural elements, the element listed in the table must be of a similar design but with less cover thickness. The fire resistance in all instances shall be the fire resistance recommended in the table. This is subject to the following limitations:

- The actual element in the rehabilitated building shall be constructed of the same materials as listed in the table. Only the following dimensions may vary from those specified: for walls, the overall thickness must exceed that specified in the table; for floor/ceiling assemblies, the thickness of the cover and the slab must be greater than, or equal to, that specified in the table; for other structural elements, the thickness of the cover must be greater than that specified in the table.
- All penetrations in the building element or its cover for services such as electricity, plumbing, or HVAC shall be packed with noncombustible cementitious materials and so fixed that the packing material will not fall out when it loses its water of hydration.
- The effects of age and wear and tear shall be repaired so that the building element is sound and the original thickness of all components, particularly covers and floor slabs, is maintained.

This approach is an application of the "thickness design" concept presented in Part 3.3 of the guideline. There should be many instances when a thicker building element was utilized than the one listed in the Appendix Tables. This guideline recognizes the inherent superiority of a thicker design. Note: "thickness design" for floor/ceiling assemblies and structural elements refers to cover and slab thickness rather than total thickness.

The "thickness design" concept is essentially a special case of Harmathy's Rules (specifically Rules 1 and 2). It should be recognized that the only source of data is the Appendix Tables. If other data are used, it must be in connection with the approach below.

4.3 The fire resistance of building elements can be established by applying Harmathy's Ten Rules of Fire Resistance Ratings as set forth in Part 3.2 of the guideline. This is subject to the following limitations:

- The data from the tables can be utilized subject to the limitations in 4.2 above.
- Test reports from recognized journals or published papers can be used to support data utilized in applying Harmathy's Rules.
- Calculations utilizing recognized and well established computational techniques can be used in applying Harmathy's Rules. These include, but are not limited to, analysis of heat flow, mechanical properties, deflections, and load bearing capacity.

APPENDIX

Introduction

The fire resistance tables that follow are a part of Resource A and provide a tabular form of assigning fire resistance ratings to various archaic building elements and assemblies.

These tables for archaic materials and assemblies do for archaic materials what Tables 720.1(1), 720.1(2), and 720.1(3) of the *International Building Code* do for more modern building elements and assemblies. The fire resistance tables of Resource A should be used as described in the "Purpose and Procedure" that follows the table of contents for these tables.

RESOURCE A TABLE OF CONTENTS

PURPOSE AND PROCEDURE

The tables and histograms which follow are to be used only within the analytical framework detailed in the main body of this guideline.

Histograms precede any table with 10 or more entries. The use and interpretation of these histograms is explained in Part 2 of the guideline. The tables are in a format similar to that found in the model building codes. The following example, taken from an entry in Table 1.1.2, best explains the table format.

1. Item Code: The item code consists of a four place series in the general form w-x-y-z in which each member of the series denotes the following:

   w = Type of building element (e.g., W=Walls; F=Floors, etc.)

   x = The building element thickness rounded down to the nearest one inch increment (e.g., $4^5/_8$" is rounded off to 4")

   y = The general type of material from which the building element is constructed (e.g., M=Masonry; W=Wood, etc.)

   z = The item number of the particular building element in a given table

   The item code shown in the example W-4-M-50 denotes the following:

   W = Wall, as the building element

   4 = Wall thickness in the range of 4" to less than 5"

   M = Masonry construction

   50 = The 50th entry in Table 1.1.2

2. The specific name or heading of this column identifies the dimensions which, if varied, has the greatest impact on fire resistance. The critical dimension for walls, the example here, is thickness. It is different for other building elements (e.g., depth for beams; membrane thickness for some floor/ceiling assemblies). The table entry is the named dimension of the building element measured at the time of actual testing to within $\pm$ $^1/_8$ inch tolerance. The thickness tabulated includes facings where facings are a part of the wall construction.

3. Construction Details: The construction details provide a brief description of the manner in which the building element was constructed.

4. Performance: This heading is subdivided into two columns. The column labeled "Load" will either list the load that the building element was subjected to during the fire test or it will contain a note number which will list the load and any other significant details. If the building element was not subjected to a load during the test, this column will contain "n/a," which means "not applicable."

   The second column under performance is labeled "Time" and denotes the actual fire endurance time observed in the fire test.

5. Reference Number: This heading is subdivided into three columns: Pre-BMS-92; BMS-92; and Post-BMS-92. The table entry under this column is the number in the Bibliography of the original source reference for the test data.

6. Notes: Notes are provided at the end of each table to allow a more detailed explanation of certain aspects of the test. In certain tables the notes given to this column have also been listed under the "Construction Details" and/or "Load" columns.

7. Rec Hours: This column lists the recommended fire endurance rating, in hours, of a building element. In some cases, the recommended fire endurance will be less than that listed under the "Time" column. In no case is the "Rec Hours" greater than given in the "Time" column.

ITEM CODE	THICKNESS	CONSTRUCTION DETAILS	PERFORMANCE		REFERENCE NUMBER			NOTES	REC. HOURS
			LOAD	TIME	PRE-BMS-92	BMS-92	POST-BMS-92		
W-4-M-5	$4^5/_8$"	Core: structural clay tile, See notes 12, 16, 21; Facings on unexposed side only, see not 18	n/a	25 min.		1		3, 4, 24	$^1/_3$

SECTION I-WALLS
FIGURE 1.1.1—WALLS—MASONRY
0" TO LESS THAN 4" THICK

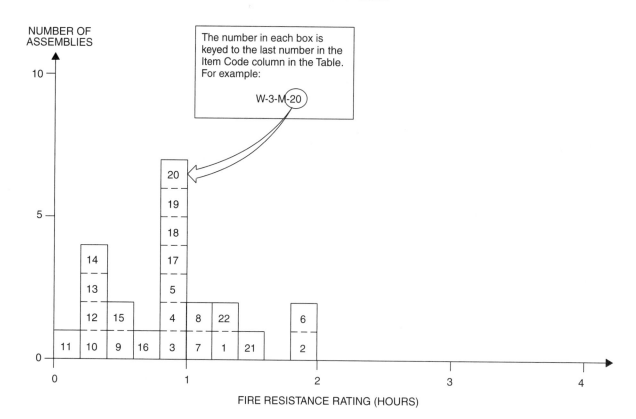

TABLE 1.1.1—MASONRY WALLS
0" TO LESS THAN 4" THICK

| ITEM CODE | THICKNESS | CONSTRUCTION DETAILS | PERFORMANCE | | REFERENCE NUMBER | | | NOTES | REC. HOURS |
			LOAD	TIME	PRE-BMS-92	BMS-92	POST-BMS-92		
W-2-M-1	$2^{1}/_{4}$"	Solid partition; $^{3}/_{4}$" gypsum plank-10' × 1'6"; $^{3}/_{4}$ plus gypsum plaster each side.	N/A	1 hr. 22 min.			7	1	$1^{1}/_{4}$
W-3-M-2	3"	Concrete block (18" × 9" × 3") of fuel ash, portland cement and plasticizer; cement/sand mortar.	N/A	2 hrs.			7	2, 3	2
W-2-M-3	2"	Solid gypsum block wall; No facings	N/A	1 hr.		1		4	1
W-3-M-4	3"	Solid gypsum blocks, laid in 1:3 sanded gypsum mortar.	N/A	1 hr.		1		4	1
W-3-M-5	3"	Magnesium oxysulfate wood fiber blocks; 2" thick, laid in portland cement-lime mortar; Facings: $^{1}/_{2}$" of 1:3 sanded gypsum plaster on both sides.	N/A	1 hr.		1		4	1
W-3-M-6	3"	Magnesium oxysulfate bound wood fiber blocks; 3" thick; laid in portland cement-lime mortar; Facings: $^{1}/_{2}$" of 1:3 sanded gypsum plaster on both sides.	N/A	2 hrs.		1		4	2

(Continued)

TABLE 1.1.1—MASONRY WALLS
0" TO LESS THAN 4" THICK—(Continued)

ITEM CODE	THICKNESS	CONSTRUCTION DETAILS	PERFORMANCE		REFERENCE NUMBER			NOTES	REC. HOURS
			LOAD	TIME	PRE-BMS-92	BMS-92	POST-BMS-92		
W-3-M-7	3"	Clay tile; Ohio fire clay; single cell thick; Face plaster: $^5/_8$" (both sides) 1:3 sanded gypsum; Design "E," Construction "A."	N/A	1 hr. 6 min.	0.		2	5, 6, 7, 11, 12, 39	1
W-3-M-8	3"	Clay tile; Illinois surface clay; single cell thick; Face plaster: $^5/_8$" (both sides) 1:3 sanded gypsum; Design "A," Construction "E."	N/A	1 hr. 1 min			2	5, 8, 9, 11, 12, 39	1
W-3-M-9	3"	Clay tile; Illinois surface clay; single cell thick; No face plaster; Design "A," Construction "C."	N/A	25 min.			2	5, 10, 11, 12, 39	$^1/_3$
W-3-M-10	$3^7/_8$"	8" × $4^7/_8$" glass blocks; weight 4 lbs. each; portland cement-lime mortar; horizontal mortar joints reinforced with metal lath.	N/A	15 min.		1		4	$^1/_4$
W-3-M-11	3"	Core: structural clay tile; see Notes 14, 18, 23; No facings.	N/A	10 min.		1		5, 11, 26	$^1/_6$
W-3-M-12	3"	Core: structural clay tile; see Notes 14, 19, 23; No facings.	N/A	20 min.		1		5, 11, 26	$^1/_3$
W-3-M-13	$3^5/_8$"	Core: structural clay tile; see Notes 14, 18, 23; Facings: unexposed side; see Note 20.	N/A	20 min.		1		5, 11, 26	$^1/_3$
W-3-M-14	$3^5/_8$"	Core: structural clay tile; see Notes 14, 19, 23; Facings: unexposed side only; see Note 20.	N/A	20 min.		1		5, 11, 26	$^1/_3$
W-3-M-15	$3^5/_8$"	Core: clay structural tile; see Notes 14, 18, 23; Facings: side exposed to fire; see Note 20.	N/A	30 min.		1		5, 11, 26	$^1/_2$
W-3-M-16	$3^5/_8$"	Core: clay structural tile; see Notes 14, 19, 23; Facings: side exposed to fire; see Note 20.	N/A	45 min.		1		5, 11, 26	$^3/_4$
W-2-M-17	2"	2" thick solid gypsum blocks; see Note 27.	N/A	1 hr.		1		27	1
W-3-M-18	3"	Core: 3" thick gypsum blocks 70% solid; see Note 2; No facings.	N/A	1 hr.		1		27	1
W-3-M-19	3"	Core: hollow concrete units; see Notes 29, 35, 36, 38; No facings.	N/A	1 hr.		1		27	1
W-3-M-20	3"	Core: hollow concrete units; see Notes 28, 35, 36, 37, 38; No facings.	N/A	1 hr.		1			1
W-3-M-21	$3^1/_2$"	Core: hollow concrete units; see Notes 28, 35, 36, 37, 38; Facings: one side; see Note 37.	N/A	$1^1/_2$ hrs.		1			$1^1/_2$

(Continued)

TABLE 1.1.1—MASONRY WALLS
0" TO LESS THAN 4" THICK—(Continued)

ITEM CODE	THICKNESS	CONSTRUCTION DETAILS	PERFORMANCE		REFERENCE NUMBER			NOTES	REC. HOURS
			LOAD	TIME	PRE-BMS-92	BMS-92	POST-BMS-92		
W-3-M-22	$3^1/_2$"	Core: hollow concrete units; see Notes 29, 35, 36, 38; Facings: one side, see Note 37.	N/A	$1^1/_4$ hrs.		1			$1^1/_4$

Notes:

1. Failure mode - flame thru.
2. Passed 2 hour fire test (Grade "C" fire res. - British).
3. Passed hose stream test.
4. Tested at NBS under ASA Spec. No. A2-1934. As nonload bearing partitions.
5. Tested at NBS under ASA Spec. No. 42-1934 (ASTM C 19-33) except that hose stream testing where carried was run on test specimens exposed for full test duration, not for a reduced period as is contemporarily done.
6. Failure by thermal criteria - maximum temperature rise 325°F.
7. Hose stream failure.
8. Hose stream - pass.
9. Specimen removed prior to any failure occurring.
10. Failure mode - collapse.
11. For clay tile walls, unless the source or density of the clay can be positively identified or determined, it is suggested that the lowest hourly rating for the fire endurance of a clay tile partition of that thickness be followed. Identified sources of clay showing longer fire endurance can lead to longer time recommendations.
12. See appendix for construction and design details for clay tile walls.
13. Load: 80 psi for gross wall area.
14. One cell in wall thickness.
15. Two cells in wall thickness.
16. Double shells plus one cell in wall thickness.
17. One cell in wall thickness, cells filled with broken tile, crushed stone, slag cinders or sand mixed with mortar.
18. Dense hard-burned clay or shale tile.
19. Medium-burned clay tile.
20. Not less than $5/_8$ inch thickness of 1:3 sanded gypsum plaster.
21. Units of not less than 30 percent solid material.
22. Units of not less than 40 percent solid material.
23. Units of not less than 50 percent solid material.
24. Units of not less than 45 percent solid material.
25. Units of not less than 60 percent solid material.
26. All tiles laid in portland cement-lime mortar.
27. Blocks laid in 1:3 sanded gypsum mortar voids in blocks not to exceed 30 percent.
28. Units of expanded slag or pumice aggregate.
29. Units of crushed limestone, blast furnace, slag, cinders and expanded clay or shale.
30. Units of calcareous sand and gravel. Coarse aggregate, 60 percent or more calcite and dolomite.
31. Units of siliceous sand and gravel. Ninety percent or more quartz, chert or flint.
32. Unit at least 49 percent solid.
33. Unit at least 62 percent solid.
34. Unit at least 65 percent solid.
35. Unit at least 73 percent solid.
36. Ratings based on one unit and one cell in wall thickness.
37. Minimum of $1/_2$ inch - 1:3 sanded gypsum plaster.
38. Nonload bearing.
39. See Clay Tile Partition Design Construction drawings, below.

(Continued)

TABLE 1.1.1—MASONRY WALLS
0" TO LESS THAN 4" THICK—(Continued)

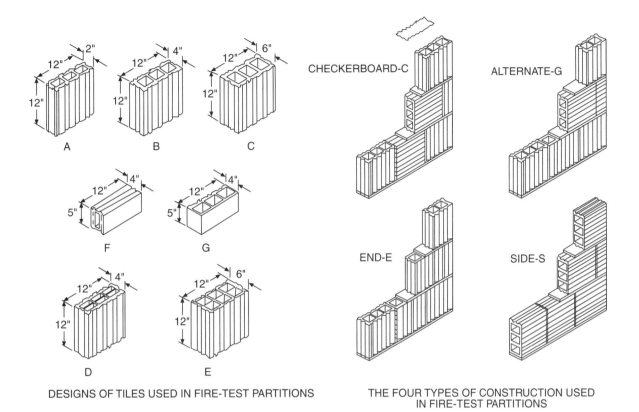

DESIGNS OF TILES USED IN FIRE-TEST PARTITIONS

THE FOUR TYPES OF CONSTRUCTION USED
IN FIRE-TEST PARTITIONS

FIGURE 1.1.2—WALLS—MASONRY
4" TO LESS THAN 6" THICK

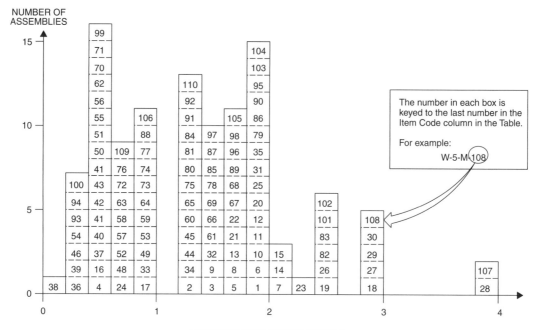

TABLE 1.1.2—MASONRY WALLS
4" TO LESS THAN 6" THICK

ITEM CODE	THICKNESS	CONSTRUCTION DETAILS	PERFORMANCE		REFERENCE NUMBER			NOTES	REC. HOURS
			LOAD	TIME	PRE-BMS-92	BMS-92	POST-BMS-92		
W-4-M-1	4"	Solid 3" thick, gypsum blocks laid in 1:3 sanded gypsum mortar; Facings: $^1/_2$" of 1:3 sanded gypsum plaster (both sides).	N/A	2 hrs.		1		1	2
W-4-M-2	4"	Solid clay or shale brick.	N/A	1 hr. 15 min		1		1, 2	$1^1/_4$
W-4-M-3	4"	Concrete; No facings.	N/A	1 hr. 30 min.		1		1	$1^1/_2$
W-4-M-4	4"	Clay tile; Illinois surface clay; single cell thick; No face plaster; Design "B," Construction "C."	N/A	25 min.			2	3-7, 36	$^1/_3$
W-4-M-5	4"	Solid sand-lime brick.	N/A	1 hr. 45 min.		1		1	$1^3/_4$
W-4-M-6	4"	Solid wall; 3" thick block; $^1/_2$" plaster each side; $17^3/_4$" × $8^3/_4$" × 4" "Breeze Blocks"; portland cement/sand mortar.	N/A	1 hr.52 min.			7	2	$1^3/_4$
W-4-M-7	4"	Concrete (4020 psi); Reinforcement: vertical $^3/_8$"; horizontal $^1/_4$"; 6" × 6" grid.	N/A	2 hrs. 10 min.			7	2	2
W-4-M-8	4"	Concrete wall (4340 psi crush); reinforcement $^1/_4$" diameter rebar on 8" centers (vertical and horizontal).	N/A	1 hr. 40 min.			7	2	$1^2/_3$

(Continued)

TABLE 1.1.2—MASONRY WALLS
4" TO LESS THAN 6" THICK—(Continued)

ITEM CODE	THICKNESS	CONSTRUCTION DETAILS	PERFORMANCE		REFERENCE NUMBER			NOTES	REC. HOURS
			LOAD	TIME	PRE-BMS-92	BMS-92	POST-BMS-92		
W-4-M-9	$4^3/_{16}$"	$4^3/_{16}$" $\times$ $2^5/_8$" cellular fletton brick (1873 psi) with $^1/_2$" sand mortar; bricks are U-shaped yielding hollow cover (approx. 2" $\times$ 4") in final cross-section configuration.	N/A	1 hr. 25 min.			7	2	$1^3/_4$
W-4-M-10	$4^1/_4$"	$4^1/_4$" $\times$ $2^1/_2$" fletton (1831 psi) brick in $^1/_2$" sand mortar.	N/A	1 hr. 53 min			7	2	$1^3/_4$
W-4-M-11	$4^1/_4$"	$4^1/_4$" $\times$ $2^1/_2$" London stock (683 psi) brick; $^1/_2$" grout.	N/A	1 hr. 52 min.			7	2	$1^3/_4$
W-4-M-12	$4^1/_2$"	$4^1/_4$" $\times$ $2^1/_2$" Leicester red, wire-cut brick (4465 psi) in $^1/_2$" sand mortar.	N/A	1 hr. 56 min.			7	6	$1^3/_4$
W-4-M-13	$4^1/_4$"	$4^1/_4$" $\times$ $2^1/_2$" stairfoot brick (7527 psi) $^1/_2$" sand mortar.	N/A	1 hr. 37 min.			7	2	$1^1/_2$
W-4-M-14	$4^1/_4$"	$4^1/_4$" $\times$ $2^1/_2$" sand-lime brick (2603 psi) $^1/_2$" sand mortar.	N/A	2 hrs. 6 min.			7	2	2
W-4-M-15	$4^1/_4$"	$4^1/_4$" $\times$ $2^1/_2$" concrete brick (2527 psi) $^1/_2$" sand mortar.	N/A	2 hrs. 10 min.			7	2	2
W-4-M-16	$4^1/_2$"	4" thick clay tile; Ohio fire clay; single cell thick; No plaster exposed face; $^1/_2$" 1:2 gypsum back face; Design "F," Construction "S."	N/A	31 min.			2	3-6, 36	$^1/_2$
W-4-M-17	$4^1/_2$"	4" thick clay tile; Ohio fire clay; single cell thick; Plaster exposed face; $^1/_2$" 1:2 sanded gypsum; Back Face: none; Construction "S," Design "F."	80 psi	50 min.			2	3-5, 8, 36	$^3/_4$
W-4-M-18	$4^1/_2$"	Core: solid sand-lime brick; $^1/_2$" sanded gypsum plaster facings on both sides.	80 psi	3 hrs.		1		1, 11	3
W-4-M-19	$4^1/_2$"	Core: solid sand-lime brick; $^1/_2$" sanded gypsum plaster facings on both sides.	80 psi	2 hrs. 30 min.		1		1, 11	$2^1/_2$
W-4-M-20	$4^1/_2$"	Core: concrete brick $^1/_2$" of 1:3 sanded gypsum plaster facings on both sides.	80 psi	2 hrs.		1		1, 11	2
W-4-M-21	$4^1/_2$"	Core: solid clay or shale brick; $^1/_2$" thick, 1:3 sanded gypsum plaster facings on fire sides.	80 psi	1 hr. 45 min.		1		1, 2, 11	$1^3/_4$
W-4-M-22	$4^3/_4$"	4" thick clay tile; Ohio fire clay; single cell thick; cells filled with cement and broken tile concrete; Plaster on exposed face; none on unexposed face; $^3/_4$" 1:3 sanded gypsum; Design "G," Construction "E."	N/A	1 hr. 48 min.			2	2, 3-5, 9, 36	$1^3/_4$
W-4-M-23	$4^3/_4$"	4" thick clay tile; Ohio fire clay; single cell thick; cells filled with cement and broken tile concrete; No plaster exposed faced; $^3/_4$" neat gypsum plaster on unexposed face; Design "G," Construction "E."	N/A	2 hrs. 14 min.			2	2, 3-5, 9, 36	2

(Continued)

TABLE 1.1.2—MASONRY WALLS
4" TO LESS THAN 6" THICK—(Continued)

ITEM CODE	THICKNESS	CONSTRUCTION DETAILS	PERFORMANCE		REFERENCE NUMBER			NOTES	REC. HOURS
			LOAD	TIME	PRE-BMS-92	BMS-92	POST-BMS-92		
W-5-M-24	5"	3" × 13" air space; 1" thick metal reinforced concrete facings on both sides; faces connected with wood splines.	2,250 lbs./ft.	45 min.		1		1	$^3/_4$
W-5-M-25	5"	Core: 3" thick void filled with "nondulated" mineral wool weighing 10 lbs./ft.^{3}; 1" thick metal reinforced concrete facings on both sides.	2,250 lbs./ft.	2 hrs.		1		1	2
W-5-M-26	5"	Core: solid clay or shale brick; $^1/_2$" thick, 1:3 sanded gypsum plaster facings on both sides.	40 psi	2 hrs. 30 min.		1		1, 2, 11	$2^1/_2$
W-5-M-27	5"	Core: solid 4" thick gypsum blocks, laid in 1:3 sanded gypsum mortar; $^1/_2$" of 1:3 sanded gypsum plaster facings on both sides.	N/A	3 hrs.		1		1	3
W-5-M-28	5"	Core: 4" thick hollow gypsum blocks with 30% voids; blocks laid in 1:3 sanded gypsum mortar; No facings.	N/A	4 hrs.		1		1	4
W-5-M-29	5"	Core: concrete brick; $^1/_2$" of 1:3 sanded gypsum plaster facings on both sides.	160 psi	3 hrs.		1		1	3
W-5-M-30	$5^1/_4$"	4" thick clay tile; Illinois surface clay; double cell thick; Plaster: $^5/_8$" sanded gypsum 1:3 both faces; Design "D," Construction "S."	N/A	2 hrs. 53 min.			2	2-5, 9, 36	$2^3/_4$
W-5-M-31	$5^1/_4$"	4" thick clay tile; New Jersey fire clay; double cell thick; Plaster: $^5/_8$" sanded gypsum 1:3 both faces; Design "D," Construction "S."	N/A	1 hr. 52 min.			2	2-5, 9, 36	$1^3/_4$
W-5-M-32	$5^1/_4$"	4" thick clay tile; New Jersey fire clay; single cell thick; Plaster: $^5/_8$" sanded gypsm 1:3 both faces; Design "D," Construction "S."	N/A	1 hr. 34 min.	2		2	2-5, 9, 36	$1^1/_2$
W-5-M-33	$5^1/_4$"	4" thick clay tile; New Jersey fire clay; single cell thick; Face plaster: $^5/_8$" both sides; 1:3 sanded gypsum; Design "B," Construction "S."	N/A	50 min.			2	3-5, 8, 36	$^3/_4$
W-5-M-34	$5^1/_4$"	4" thick clay tile; Ohio fire clay; single cell thick; Face plaster: $^5/_8$" both sides; 1:3 sanded gypsum; Design "B," Construction "A."	N/A	1 hr. 19 min.			2	2-5, 9, 36	$1^1/_4$
W-5-M-35	$5^1/_4$"	4" thick clay tile; Illinois surface clay; single cell thick; Face plaster: $^5/_8$" both sides; 1:3 sanded gypsum; Design "B," Construction "S."	N/A	1 hr. 59 min.			2	2-5, 10 36	$1^3/_4$
W-5-M-36	4"	Core: structural clay tile; see Notes 12, 16, 21; No facings.	N/A	15 min.		1		3, 4, 24	$^1/_4$

(Continued)

TABLE 1.1.2—MASONRY WALLS
4" TO LESS THAN 6" THICK—(Continued)

ITEM CODE	THICKNESS	CONSTRUCTION DETAILS	PERFORMANCE		REFERENCE NUMBER			NOTES	REC. HOURS
			LOAD	TIME	PRE-BMS-92	BMS-92	POST-BMS-92		
W-4-M-37	4"	Core: structural clay tile; see Notes 12, 17, 21; No facings.	N/A	25 min.		1		3, 4, 24	$^1/_3$
W-4-M-38	4"	Core: structural clay tile; see Notes 12, 16, 20; No facings.	N/A	10 min.		1		3, 4, 24	$^1/_6$
W-4-M-39	4"	Core: structural clay tile; see Notes 12, 17, 20; No facings.	N/A	20 min.		1		3, 4, 24	$^1/_3$
W-4-M-40	4"	Core: structural clay tile; see Notes 13, 16, 23; No facings.	N/A	30 min.		1		3, 4, 24	$^1/_2$
W-4-M-41	4"	Core: structural clay tile; see Notes 13, 17, 23; No facings.	N/A	35 min.		1		3, 4, 24	$^1/_2$
W-4-M-42	4"	Core: structural clay tile; see Notes 13, 16, 21; No facings.	N/A	25 min.		1		3, 4, 24	$^1/_3$
W-4-M-43	4"	Core: structural clay tile; see Notes 13, 17, 21; No facings.	N/A	30 min.		1		3, 4, 24	$^1/_2$
W-4-M-44	4"	Core: structural clay tile; see Notes 15, 16, 20; No facings	N/A	1 hr. 15 min.		1		3, 4, 24	$1^1/_4$
W-4-M-45	4"	Core: structural clay tile; see Notes 15, 17, 20; No facings.	N/A	1 hr. 15 min.	2	1		3, 4, 24	$1^1/_4$
W-4-M-46	4"	Core: structural clay tile; see Notes 14, 16, 22; No facings.	N/A	20 min.		1		3, 4, 24	$^1/_3$
W-4-M-47	4"	Core: structural clay tile; see Notes 14, 17, 22; No facings.	N/A	25 min.		1		3, 4, 24	$^1/_3$
W-4-M-48	$4^1/_4$"	Core: structural clay tile; see Notes 12, 16, 21; Facings: both sides; see Note 18.	N/A	45 min.		1		3, 4, 24	$^3/_4$
W-4-M-49	$4^1/_4$"	Core: structural clay tile; see Notes 12, 17, 21; Facings: both sides; see Note 18.	N/A	1 hr.		1		3, 4, 24	1
W-4-M-50	$4^5/_8$"	Core: structural clay tile; see Notes 12, 16, 21; Facings: unexposed side only; see Note 18.	N/A	25 min.		1		3, 4, 24	$^1/_3$
W-4-M-51	$4^5/_8$"	Core: structural clay tile; see Notes 12, 17, 21; Facings: unexposed side only; see Note 18.	N/A	30 min.		1		3, 4, 24	$^1/_2$
W-4-M-52	$4^5/_8$"	Core: structural clay tile; see Notes 12, 16, 21; Facings: unexposed side only; see Note 18.	N/A	45 min.		1		3, 4, 24	$^3/_4$
W-4-M-53	$4^5/_8$"	Core: structural clay tile; see Notes 12, 17, 21; Facings: fire side only; see Note 18.	N/A	1 hr.		1		3, 4, 24	1
W-4-M-54	$4^5/_8$"	Core: structural clay tile; see Notes 12, 16, 20; Facings: unexposed side; see Note 18.	N/A	20 min.		1		3, 4, 24	$^1/_3$
W-4-M-55	$4^5/_8$"	Core: structural clay tile; see Notes 12, 17, 20; Facings: exposed side; see Note 18.	N/A	25 min.		1		3, 4, 24	$^1/_3$
W-4-M-56	$4^5/_8$"	Core: structural clay tile; see Notes 12, 16, 20; Facings: fire side only; see Note 18.	N/A	30 min.		1		3, 4, 24	$^1/_2$
W-4-M-57	$4^5/_8$"	Core: structural clay tile; see Notes 12, 17, 20; Facings: fire side only; see Note 18.	N/A	45 min.		1		3, 4, 24	$^3/_4$

(Continued)

TABLE 1.1.2—MASONRY WALLS
4" TO LESS THAN 6" THICK—(Continued)

ITEM CODE	THICKNESS	CONSTRUCTION DETAILS	PERFORMANCE		REFERENCE NUMBER			NOTES	REC. HOURS
			LOAD	TIME	PRE-BMS-92	BMS-92	POST-BMS-92		
W-4-M-58	$4^5/_8$"	Core: structural clay tile; see Notes 13, 16, 23; Facings: unexposed side only; see Note 18.	N/A	40 min.		1		3, 4, 24	$^2/_3$
W-4-M-59	$4^5/_8$"	Core: structural clay tile; see Notes 13, 17, 23; Facings: unexposed side only; see Note 18.	N/A	1 hr.		1		3, 4, 24	1
W-4-M-60	$4^5/_8$"	Core: structural clay tile; see Notes 13, 16, 23; Facings: fire side only; see Note 18.	N/A	1 hr. 15 min.		1		3, 4, 24	$1^1/_4$
W-4-M-61	$4^5/_8$"	Core: structural clay tile; see Notes 13, 17, 23; Facings: fire side only; see Note 18.	N/A	1 hr. 30 min.		1		3, 4, 24	$1^1/_2$
W-4-M-62	$4^5/_8$"	Core: structural clay tile; see Notes 13, 16, 21; Facings: unexposed side only; see Note 18.	N/A	35 min.		1		3, 4, 24	$^1/_2$
W-4-M-63	$4^5/_8$"	Core: structural clay tile; see Notes 13, 17, 21; Facings: unexposed face only; see Note 18.	N/A	45 min.		1		3, 4, 24	$^3/_4$
W-4-M-64	$4^5/_8$"	Core: structural clay tile; see Notes 13, 16, 23; Facings: exposed face only; see Note 18.	N/A	1 hr.		1		3, 4, 24	1
W-4-M-65	$4^5/_8$"	Core: structural clay tile; see Notes 13, 17, 21; Facings: exposed side only; see Note 18.	N/A	1 hr. 15 min.		1		3, 4, 24	$1^1/_4$
W-4-M-66	$4^5/_8$"	Core: structural clay tile; see Notes 15, 17, 20; Facings: unexposed side only; see Note 18	N/A	1 hr. 30 min.		1		3, 4, 24	$1^1/_2$
W-4-M-67	$4^5/_8$"	Core: structural clay tile; see Notes 15, 16, 20; Facings: exposed side only; see Note 18.	N/A	1 hr. 45 min.		1		3, 4, 24	$1^3/_4$
W-4-M-68	$4^5/_8$"	Core: structural clay tile; see Notes 15, 17, 20; Facings: exposed side only; see Note 18.	N/A	1 hr. 45 min.		1		3, 4, 24	$1^3/_4$
W-4-M-69	$4^5/_8$"	Core: structural clay tile; see Notes 15, 16, 20; Facings: unexposed side only; see Note 18.	N/A	1 hr. 30 min.		1		3, 4, 24	$1^3/_4$
W-4-M-70	$4^5/_8$"	Core: structural clay tile; see Notes 14, 16, 22; Facings: unexposed side only; see Note 18.	N/A	30 min.		1		3, 4, 24	$^1/_2$
W-4-M-71	$4^5/_8$"	Core: structural clay tile; see Notes 14, 17, 22; Facings: exposed side only; see Note 18.	N/A	35 min.		1		3, 4, 24	$^1/_2$
W-4-M-72	$4^5/_8$"	Core: structural clay tile; see Notes 14, 16, 22; Facings: fire side of wall only; see Note 18.	N/A	45 min.		1		3, 4, 24	$^3/_4$
W-4-M-73	$4^5/_8$"	Core: structural clay tile; see Notes 14, 17, 22; Facings: fire side of wall only; see Note 18.	N/A	1 hr.		1		3, 4, 24	1
W-4-M-74	$5^1/_4$"	Core: structural clay tile; see Notes 12, 16, 21; Facings: both sides; see Note 18.	N/A	1 hr.		1		3, 4, 24	1

(Continued)

TABLE 1.1.2—MASONRY WALLS
4" TO LESS THAN 6" THICK—(Continued)

ITEM CODE	THICKNESS	CONSTRUCTION DETAILS	PERFORMANCE		REFERENCE NUMBER			NOTES	REC. HOURS
			LOAD	TIME	PRE-BMS-92	BMS-92	POST-BMS-92		
W-5-M-75	$5^1/_4$"	Core: structural clay tile; see Notes 12, 17, 21; Facings: both sides; see Note 18	N/A	1 hr. 15 min.		1		3, 4, 24	$1^1/_4$
W-5-M-76	$5^1/_4$"	Core: structural clay tile; see Notes 12, 16, 20; Facings: both sides; see Note 18.	N/A	45 min.		1		3, 4, 24	$^3/_4$
W-5-M-77	$5^1/_4$"	Core: structural clay tile; see Notes 12, 17, 20; Facings: both sides; see Note 18.	N/A	1 hr.		1		3, 4, 24	1
W-5-M-78	$5^1/_4$"	Core: structural clay tile; see Notes 13, 16, 23; Facings: both sides of wall; see Note 18.	N/A	1 hr. 30 min.		1		3, 4, 24	$1^1/_2$
W-5-M-79	$5^1/_4$"	Core: structural clay tile; see Notes 13, 17, 23; Facings: both sides of wall; see Note 18.	N/A	2 hrs.		1		3, 4, 24	2
W-5-M-80	$5^1/_4$"	Core: structural clay tile; see Notes 13, 16, 21; Facings: both sides of wall; see Note 18.	N/A	1 hr. 15 min.		1		3, 4, 24	$1^1/_4$
W-5-M-81	$5^1/_4$"	Core: structural clay tile; see Notes 13, 16, 21; Facings: both sides of wall; see Note 18.	N/A	1 hr. 30 min.		1		3, 4, 24	$1^1/_2$
W-5-M-82	$5^1/_4$"	Core: structural clay tile; see Notes 15, 16, 20; Facings: both sides; see Note 18.	N/A	2 hrs. 30 min.		1		3, 4, 24	$2^1/_2$
W-5-M-83	$5^1/_4$"	Core: structural clay tile; see Notes 15, 17, 20; Facings: both sides; see Note 18.	N/A	2 hrs. 30 min.		1		3, 4, 24	$2^1/_2$
W-5-M-84	$5^1/_4$"	Core: structural clay tile; see Notes 14, 16, 22; Facings: both sides of wall; see Note 18.	N/A	1 hr. 15 min.		1		3, 4, 24	$1^1/_4$
W-5-M-85	$5^1/_4$"	Core: structural clay tile; see Notes 14, 17, 22; Facings: both sides of wall; see Note 18.	N/A	1 hr. 30 min.		1		3, 4, 24	$1^1/_2$
W-4-M-86	4"	Core: 3" thick gypsum blocks 70% solid; see Note 26; Facings: both sides; see Note 25.	N/A	2 hrs.		1			2
W-4-M-87	4"	Core: hollow concrete units; see Notes 27, 34, 35; No facings.	N/A	1 hr. 30 min.		1			$1^1/_2$
W-4-M-88	4"	Core: hollow concrete units; see Notes 28, 33, 35; No facings.	N/A	1 hr.		1			1
W-4-M-89	4"	Core: hollow concrete units; see Notes 28, 34, 35; Facings: both sides; see Note 25.	N/A	1 hr. 45 min.		1			$1^3/_4$
W-4-M-90	4"	Core: hollow concrete units; see Notes 27, 34, 35; Facings: both sides; see Note 25.	N/A	2 hrs.		1			2
W-4-M-91	4"	Core: hollow concrete units; see Notes 27, 32, 35; No facings.	N/A	1 hr. 15 min.		1			$1^1/_4$
W-4-M-92	4"	Core: hollow concrete units; see Notes 28, 34, 35; No facings.	N/A	1 hr. 15 min.		1			$1^1/_4$
W-4-M-93	4"	Core: hollow concrete units; see Notes 29, 32, 35; No facings.	N/A	20 min.		1			$^1/_3$

(Continued)

TABLE 1.1.2—MASONRY WALLS
4" TO LESS THAN 6" THICK—(Continued)

ITEM CODE	THICKNESS	CONSTRUCTION DETAILS	PERFORMANCE		REFERENCE NUMBER			NOTES	REC. HOURS
			LOAD	TIME	PRE-BMS-92	BMS-92	POST-BMS-92		
W-4-M-94	4"	Core: hollow concrete units; see Notes 30, 34, 35; No facings.	N/A	15 min.		1			$^1/_4$
W-4-M-95	$4^1/_2$"	Core: hollow concrete units; see Notes 27, 34, 35; Facings: one side only; see Note 25.	N/A	2 hrs.		1			2
W-4-M-96	$4^1/_2$"	Core: hollow concrete units; see Notes 27, 32, 35; Facings: one side only; see Note 25.	N/A	1 hr. 45 min.		1			$1^3/_4$
W-4-M-97	$4^1/_2$"	Core: hollow concrete units; see Notes 28, 33, 35; Facings: one side; see Note 25.	N/A	1 hr. 30 min.		1			$1^1/_2$
W-4-M-98	$4^1/_2$"	Core: hollow concrete units; see Notes 28, 34, 35; Facings: one side only; see Note 25.	N/A	1 hr. 45 min.		1			$1^3/_4$
W-4-M-99	$4^1/_2$"	Core: hollow concrete units; see Notes 29, 32, 35; Facings: one side; see Note 25.	N/A	30 min.		1			$^1/_2$
W-4-M-100	$4^1/_2$"	Core: hollow concrete units; see Notes 30, 34, 35; Facings: one side; see Note 25.	N/A	20 min.		1			$^1/_3$
W-5-M-101	5"	Core: hollow concrete units; see Notes 27, 34, 35; Facings: both sides; see Note 25.	N/A	2 hrs. 30 min.		1			$2^1/_2$
W-5-M-102	5"	Core: hollow concrete units; see Notes 27, 32, 35; Facings: both sides; see Note 25.	N/A	2 hrs. 30 min.		1			$2^1/_2$
W-5-M-103	5"	Core: hollow concrete units; see Notes 28, 33, 35; Facings: both sides; see Note 25.	N/A	2 hrs.		1			2
W-5-M-104	5"	Core: hollow concrete units; see Notes 28, 31, 35; Facings: both sides; see Note 25.	N/A	2 hrs.		1			2
W-5-M-105	5"	Core: hollow concrete units; see Notes 29, 32, 35; Facings: both sides; see Note 25.	N/A	1 hr. 45 min.		1			$1^3/_4$
W-5-M-106	5"	Core: hollow concrete units; see Notes 30, 34, 35; Facings: both sides; see Note 25.	N/A	1 hr.		1			1
W-5-M-107	5"	Core: 5" thick solid gypsum blocks; see Note 26; No facings.	N/A	4 hrs.		1			4
W-5-M-108	5"	Core: 4" thick hollow gypsum blocks; see Note 26; Facings: both sides; see Note 25.	N/A	3 hrs.		1			3
W-5-M-109	4"	Concrete with 4" × 4" No. 6 welded wire mesh at wall center.	100 psi	45 min.			43	2	$^3/_4$

(Continued)

TABLE 1.1.2—MASONRY WALLS
4" TO LESS THAN 6" THICK—(Continued)

ITEM CODE	THICKNESS	CONSTRUCTION DETAILS	PERFORMANCE		REFERENCE NUMBER			NOTES	REC. HOURS
			LOAD	TIME	PRE-BMS-92	BMS-92	POST-BMS-92		
W-4-M-110	4"	Concrete with 4" × 4" No. 6 welded wire mesh at wall center.	N/A	1 hr. 15 min.			43	2	$1^1/_4$

Notes:

1. Tested as NBS under ASA Spec. No. A 2-1934.
2. Failure mode - maximum temperature rise.
3. Treated at NBS under ASA Spec. No. 42-1934 (ASTM C 19-53) except that hose stream testing where carried out was run on test specimens exposed for full test duration, not for or reduced period as is contemporarily done.
4. For clay tile walls, unless the source the clay can be positively identified, it is suggested that the most pessimistic hour rating for the fire endurance of a clay tile partition of that thickness to be followed. Identified sources of clay showing longer fire endurance can lead to longer time recommendations.
5. See appendix for construction and design details for clay tile walls.
6. Failure mode - flame thru or crack formation showing flames.
7. Hole formed at 25 minutes; partition collapsed at 42 minutes or removal from furnace.
8. Failure mode - collapse.
9. Hose stream pass.
10. Hose stream hole formed in specimen.
11. Load: 80 psi for gross wall cross sectional area.
12. One cell in wall thickness.
13. Two cells in wall thickness.
14. Double cells plus one cell in wall thickness.
15. One cell in wall thickness, cells filled with broken tile, crushed stone, slag, cinders or sand mixed with mortar.
16. Dense hard-burned clay or shale tile.
17. Medium-burned clay tile.
18. Not less than $5/_8$ inch thickness of 1:3 sanded gypsum plaster.
19. Units of not less than 30 percent solid material.
20. Units of not less than 40 percent solid material.
21. Units of not less than 50 percent solid material.
22. Units of not less than 45 percent solid material.
23. Units of not less than 60 percent solid material.
24. All tiles laid in portland cement-lime mortar.
25. Minimum $1/_2$ inch - 1:3 sanded gypsum plaster.
26. Laid in 1:3 sanded gypsum mortar. Voids in hollow units not to exceed 30 percent.
27. Units of expanded slag or pumice aggregate.
28. Units of crushed limestone, blast furnace slag, cinders and expanded clay or shale.
29. Units of calcareous sand and gravel. Coarse aggregate, 60 percent or more calcite and dolomite.
30. Units of siliceous sand and gravel. Ninety percent or more quartz, chert or flint.
31. Unit at least 49 percent solid.
32. Unit at least 62 percent solid.
33. Unit at least 65 percent solid.
34. Unit at least 73 percent solid.
35. Ratings based on one unit and one cell in wall thickness.
36. See Clay Tile Partition Design Construction drawings, below.

(Continued)

TABLE 1.1.2—MASONRY WALLS
4" TO LESS THAN 6" THICK—(Continued)

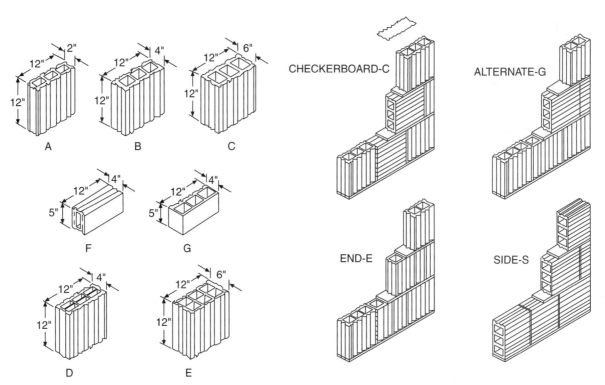

DESIGNS OF TILES USED IN FIRE-TEST PARTITIONS

THE FOUR TYPES OF CONSTRUCTION USED
IN FIRE-TEST PARTITIONS

FIGURE 1.1.3—WALLS—MASONRY
6" TO LESS THAN 8" THICK

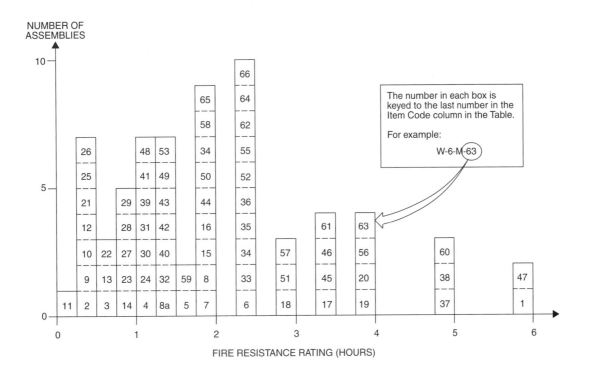

TABLE 1.1.3—MASONRY WALLS
6" TO LESS THAN 8" THICK

| ITEM CODE | THICKNESS | CONSTRUCTION DETAILS | PERFORMANCE | | REFERENCE NUMBER | | | NOTES | REC. HOURS |
			LOAD	TIME	PRE-BMS-92	BMS-92	POST-BMS-92		
W-6-M-1	6"	Core: 5" thick, solid gypsum blocks laid in 1:3 sanded gypsum mortar; $^1/_2$" of 1:3 sanded gypsum plaster facings on both sides.	N/A	6 hrs.	1				6
W-6-M-2	6"	6" clay tile; Ohio fire clay; single cell thick; No plaster; Design "C," Construction "A."	N/A	17 min.			2	1, 3, 4, 6, 55	$^1/_4$
W-6-M-3	6"	6" clay tile; Illinois surface clay; double cell thick; No plaster; Design "E," Construction "C."	N/A	45 min.			2	1-4, 7, 55	$^3/_4$
W-6-M-4	6"	6" clay tile; New Jersey fire clay; double cell thick; No plaster; Design "E," Construction "S."	N/A	1 hr. 1 min.			2	1-4, 8, 55	1
W-7-M-5	$7^1/_4$"	6" clay tile; Illinois surface clay; double cell thick; Plaster: $^5/_8$" - 1:3 sanded gypsum both faces; Design "E," Construction "A."	N/A	1 hr. 41 min.			2	1-4, 55	$1^2/_3$
W-7-M-6	$7^1/_4$"	6" clay tile; New Jersey fire clay; double cell thick; Plaster: $^5/_8$" - 1:3 sanded gypsum both faces; Design "E," Construction "S."	N/A	2 hrs. 23 min.			2	1-4, 9, 55	$2^1/_3$
W-7-M-7	$7^1/_4$"	6" clay tile; Ohio fire clay; single cell thick; Plaster: $^5/_8$" sanded gypsum; 1:3 both faces; Design "C," Construction "A."	N/A	1 hr. 54 min.			2	1-4, 9, 55	$2^3/_4$

(Continued)

TABLE 1.1.3—MASONRY WALLS
6" TO LESS THAN 8" THICK—(Continued)

ITEM CODE	THICKNESS	CONSTRUCTION DETAILS	PERFORMANCE		REFERENCE NUMBER			NOTES	REC. HOURS
			LOAD	TIME	PRE-BMS-92	BMS-92	POST-BMS-92		
W-7-M-8	$7^1/_4$"	6" clay tile; Illinois surface clay; single cell thick; Plaster: $^5/_8$" sanded gypsum 1:3 both faces; Design "C," Construction "S."	N/A	2 hrs.			2	1, 3, 4, 9, 10, 55	2
W-7-M-8a	$7^1/_4$"	6" clay tile; Illinois surface clay; single cell thick; Plaster: $^5/_8$" sanded gypsum 1:3 both faces; Design "C," Construction "E."	N/A	1 hr. 23 min			2	1-4, 9, 10, 55	$1^3/_4$
W-6-M-9	6"	Core: structural clay tile; see Notes 12, 16, 20; No facings.	N/A	20 min.		1		3, 5, 24	$^1/_3$
W-6-M-10	6"	Core: structural clay tile; see Notes 12, 17, 20; No facings.	N/A	25 min.		1		3, 5, 24	$^1/_3$
W-6-M-11	6"	Core: structural clay tile; see Notes 12, 16, 19; No facings.	N/A	15 min.		1		3, 5, 24	$^1/_4$
W-6-M-12	6"	Core: structural clay tile; see Notes 12, 17, 19; No facings.	N/A	20 min.		1		3, 5, 24	$^1/_3$
W-6-M-13	6"	Core: structural clay tile; see Notes 13, 16, 22; No facings.	N/A	45 min.		1		3, 5, 24	$^3/_4$
W-6-M-14	6"	Core: structural clay tile; see Notes 13, 17, 22; No facings.	N/A	1 hr.		1		3, 5, 24	1
W-6-M-15	6"	Core: structural clay tile; see Notes 15, 17, 19; No facings.	N/A	2 hrs.		1		3, 5, 24	2
W-6-M-16	6"	Core: structural clay tile; see Notes 15, 16, 19; No facings.	N/A	2 hrs.		1		3, 5, 24	2
W-6-M-17	6"	Cored concrete masonry; see Notes 12, 34, 36, 38, 41; No facings.	80 psi	3 hrs. 30 min.		1		5, 25	$3^1/_2$
W-6-M-18	6"	Cored concrete masonry; see Notes 12, 33, 36, 38, 41; No facings.	80 psi	3 hrs.		1		5, 25	3
W-6-M-19	$6^1/_2$"	Cored concrete masonry; see Notes 12, 34, 36, 38, 41; Facings: side 1; see Note 35.	80 psi	4 hrs.		1		5, 25	4
W-6-M-20	$6^1/_2$"	Cored concrete masonry; see Notes 12, 33, 36, 38, 41; Facings: side 1; see Note 35.	80 psi	4 hrs.		1		5, 25	4
W-6-M-21	$6^5/_8$"	Core: structural clay tile; see Notes 12, 16, 20; Facings: unexposed face only; see Note 18.	N/A	30 min.		1		3, 5, 24	$^1/_2$
W-6-M-22	$6^5/_8$"	Core: structural clay tile; see Notes 12, 17, 20; Facings: unexposed face only; see Note 18.	N/A	40 min.		1		3, 5, 24	$^2/_3$
W-6-M-23	$6^5/_8$"	Core: structural clay tile; see Notes 12, 16, 20; Facings: exposed face only; see Note 18.	N/A	1 hr.		1		3, 5, 24	1
W-6-M-24	$6^5/_8$"	Core: structural clay tile; see Notes 12, 17, 20; Facings: exposed face only; see Note 18.	N/A	1 hr. 5 min.		1		3, 5, 24	1
W-6-M-25	$6^5/_8$"	Core: structural clay tile; see Notes 12, 16, 19; Facings: unexposed side only; see Note 18.	N/A	25 min.		1		3, 5, 24	$^1/_3$

(Continued)

TABLE 1.1.3—MASONRY WALLS
6" TO LESS THAN 8" THICK—(Continued)

ITEM CODE	THICKNESS	CONSTRUCTION DETAILS	PERFORMANCE		REFERENCE NUMBER			NOTES	REC. HOURS
			LOAD	TIME	PRE-BMS-92	BMS-92	POST-BMS-92		
W-6-M-26	$6^5/_8$"	Core: structural clay tile; see Notes 12, 7, 19; Facings: unexposed face only; see Note 18.	N/A	30 min.		1		3, 5, 24	$^1/_2$
W-6-M-27	$6^5/_8$"	Core: structural clay tile; see Notes 12, 16, 19; Facings: exposed side only; see Note 18.	N/A	1 hr.		1		3, 5, 24	1
W-6-M-28	$6^5/_8$"	Core: structural clay tile; see Notes 12, 17, 19; Facings: fire side only; see Note 18.	N/A	1 hr.		1		3, 5, 24	1
W-6-M-29	$6^5/_8$"	Core: structural clay tile; see Notes 13, 16, 22; Facings: unexposed side only; see Note 18.	N/A	1 hr.		1		3, 5, 24	1
W-6-M-30	$6^5/_8$"	Core: structural clay tile; see Notes 13, 17, 22; Facings: unexposed side only; see Note 18.	N/A	1 hr. 15 min.		1		3, 5, 24	$1^1/_4$
W-6-M-31	$6^5/_8$"	Core: structural clay tile; see Notes 13, 16, 22; Facings: fire side only; see Note 18.	N/A	1 hr. 15 min.		1		3, 5, 24	$1^1/_4$
W-6-M-32	$6^5/_8$"	Core: structural clay tile; see Notes 13, 17, 22; Facings: fire side only; see Note 18.	N/A	1 hr. 30 min.		1		3, 5, 24	$1^1/_2$
W-6-M-33	$6^5/_8$"	Core: structural clay tile; see Notes 15, 16, 19; Facings: unexposed side only; see Note 18.	N/A	2 hrs. 30 min.		1		3, 5, 24	$2^1/_2$
W-6-M-34	$6^5/_8$"	Core: structural clay tile; see Notes 15, 17, 19; Facings: unexposed side only; see Note 18.	N/A	2 hrs. 30 min.		1		3, 5, 24	$2^1/_2$
W-6-M-35	$6^5/_8$"	Core: structural clay tile; see Notes 15, 16, 19; Facings: fire side only; see Note 18.	N/A	2 hrs. 30 min.		1		3, 5, 24	$2^1/_2$
W-6-M-36	$6^5/_8$"	Core: structural clay tile; see Notes 15, 17, 19; Facings: fire side only; see Note 18.	N/A	2 hrs. 30 min.		1		3, 5, 24	$2^1/_2$
W-6-M-37	7"	Cored concrete masonry; see Notes 12, 34, 36, 38, 41; see Note 35 for facings on both sides.	80 psi	5 hrs.		1		5, 25	5
W-6-M-38	7"	Cored concrete masonry; see Notes 12, 33, 36, 38, 41; see Note 35 for facings.	80 psi	5 hrs.		1		5, 25	5
W-6-M-39	$7^1/_4$"	Core: structural clay tile; see Notes 12, 16, 20; Facings: both sides; see Note 18.	N/A	1 hr. 15 min.		1		3, 5, 24	$1^1/_4$
W-6-M-40	$7^1/_4$"	Core: structural clay tile; see Notes 12, 17, 20; Facings: both sides; see Note 18.	N/A	1 hr. 30 min.		1		3, 5, 24	$1^1/_2$
W-6-M-41	$7^1/_4$"	Core: structural clay tile; see Notes 12, 16, 19; Facings: both sides; see Note 18.	N/A	1 hr. 15 min.		1		3, 5, 24	$1^1/_4$
W-6-M-42	$7^1/_4$"	Core: structural clay tile; see Notes 12, 17, 19; Facings: both sides; see Note 18.	N/A	1 hr. 30 min.		1		3, 5, 24	$1^1/_2$

(Continued)

TABLE 1.1.3—MASONRY WALLS
6" TO LESS THAN 8" THICK—(Continued)

ITEM CODE	THICKNESS	CONSTRUCTION DETAILS	PERFORMANCE		REFERENCE NUMBER			NOTES	REC. HOURS
			LOAD	TIME	PRE-BMS-92	BMS-92	POST-BMS-92		
W-7-M-43	7¼"	Core: structural clay tile; see Notes 13, 16, 22; Facings: both sides of wall; see Note 18.	N/A	1 hr. 30 min.		1		3, 5, 24	1½
W-7-M-44	7¼"	Core: structural clay tile; see Notes 13, 17, 22; Facings: both sides of wall; see Note 18.	N/A	2 hrs.		1		3, 5, 24	1½
W-7-M-45	7¼"	Core: structural clay tile; see Notes 15, 16, 19; Facings: both sides; see Note 18.	N/A	3 hrs. 30 min.		1		3, 5, 24	3½
W-7-M-46	7¼"	Core: structural clay tile; see Notes 15, 17, 19; Facings: both sides; see Note 18.	N/A	3 hrs. 30 min.		1		3, 5, 24	3½
W-6-M-47	6"	Core: 5" thick solid gypsum blocks; see Note 45; Facings: both sides; see Note 45.	N/A	6 hrs.		1			6
W-6-M-48	6"	Core: hollow concrete units; see Notes 47, 50, 54; No facings.	N/A	1 hr. 15 min.		1			1¼
W-6-M-49	6"	Core: hollow concrete units; see Notes 46, 50, 54; No facings.	N/A	1 hr. 30 min.		1			1½
W-6-M-50	6"	Core: hollow concrete units; see Notes 46, 41, 54; No facings.	N/A	2 hrs.		1			2
W-6-M-51	6"	Core: hollow concrete units; see Notes 46, 53, 54; No facings.	N/A	3 hrs.		1			3
W-6-M-52	6"	Core: hollow concrete units; see Notes 47, 53, 54; No facings.	N/A	2 hrs. 30 min.		1			2½
W-6-M-53	6"	Core: hollow concrete units; see Notes 47, 51, 54; No facings.	N/A	1 hr. 30 min.		1			1½
W-6-M-54	6½"	Core: hollow concrete units; see Notes 46, 50, 54; Facings: one side only; see Note 35.	N/A	2 hrs.		1			2
W-6-M-55	6½"	Core: hollow concrete units; see Notes 4, 51, 54; Facings: one side; see Note 35.	N/A	2 hrs. 30 min.		1			2½
W-6-M-56	6½"	Core: hollow concrete units; see Notes 46, 53, 54; Facings: one side; see Note 35.	N/A	4 hrs.		1			4
W-6-M-57	6½"	Core: hollow concrete units; see Notes 47, 53, 54; Facings: one side; see Note 35.	N/A	3 hrs.		1			3
W-6-M-58	6½"	Core: hollow concrete units; see Notes 47, 51, 54; Facings: one side; see Note 35.	N/A	2 hrs.		1			2
W-6-M-59	6½"	Core: hollow concrete units; see Notes 47, 50, 54; Facings: one side; see Note 35.	N/A	1 hr. 45 min.		1			1¾
W-7-M-60	7"	Core: hollow concrete units; see Notes 46, 53, 54; Facings: both sides; see Note 35.	N/A	5 hrs.		1			5
W-7-M-61	7"	Core: hollow concrete units; see Notes 46, 51, 54; Facings: both sides; see Note 35.	N/A	3 hrs. 30 min.		1			3½

(Continued)

TABLE 1.1.3—MASONRY WALLS
6" TO LESS THAN 8" THICK—(Continued)

ITEM CODE	THICKNESS	CONSTRUCTION DETAILS	PERFORMANCE		REFERENCE NUMBER			NOTES	REC. HOURS
			LOAD	TIME	PRE-BMS-92	BMS-92	POST-BMS-92		
W-7-M-62	7"	Core: hollow concrete units; see Notes 46, 50, 54; Facings: both sides; see Note 35.	N/A	2 hrs. 30 min.		1			$2^{1}/_{2}$
W-7-M-63	7"	Core: hollow concrete units; see Notes 47, 53, 54; Facings: both sides; see Note 35.	N/A	4 hrs.		1			4
W-7-M-64	7"	Core: hollow concrete units; see Notes 47, 51, 54; Facings: both sides; see Note 35.	N/A	2 hrs. 30 min.		1			$2^{1}/_{2}$
W-7-M-65	7"	Core: hollow concrete units; see Notes 47, 50, 54; Facings: both sides; see Note 35.	N/A	2 hrs.		1			2
W-6-M-66	6"	Concrete wall with 4" × 4" No. 6 wire fabric (welded) near wall center for reinforcement.	N/A	2 hrs. 30 min.			43	2	$2^{1}/_{2}$

Notes:
1. Tested at NBS under ASA Spec. No. 43-1934 (ASTM C 19-53) except that hose stream testing where carried out was run on test specimens exposed for full test duration, not for a reduced period as is contemporarily done.
2. Failure by thermal criteria - maximum temperature rise.
3. For clay tile walls, unless the source or density of the clay can be positively identified or determined, it is suggested that the lowest hourly rating for the fire endurance of a clay tile partition of that thickness be followed. Identified sources of clay showing longer fire endurance can lead to longer time recommendations.
4. See Note 55 for construction and design details for clay tile walls.
5. Tested at NBS under ASA Spec. No. A2-1934.
6. Failure mode - collapse.
7. Collapsed on removal from furnace at 1 hour 9 minutes.
8. Hose stream - failed.
9. Hose stream - passed.
10. No end point met in test.
11. Wall collapsed at 1 hour 28 minutes.
12. One cell in wall thickness.
13. Two cells in wall thickness.
14. Double shells plus one cell in wall thickness.
15. One cell in wall thickness, cells filled with broken tile, crushed stone, slag, cinders or sand mixed with mortar.
16. Dense hard-burned clay or shale tile.
17. Medium-burned clay tile.
18. Not less than $^{5}/_{8}$ inch thickness of 1:3 sanded gypsum plaster.
19. Units of not less than 30 percent solid material.
20. Units of not less than 40 percent solid material.
21. Units of not less than 50 percent solid material.
22. Units of not less than 45 percent solid material.
23. Units of not less than 60 percent solid material.
24. All tiles laid in portland cement-lime mortar.
25. Load: 80 psi for gross cross sectional area of wall.
26. Three cells in wall thickness.
27. Minimum percent of solid material in concrete units = 52.
28. Minimum percent of solid material in concrete units = 54.
29. Minimum percent of solid material in concrete units = 55.
30. Minimum percent of solid material in concrete units = 57.
31. Minimum percent of solid material in concrete units = 62.
32. Minimum percent of solid material in concrete units = 65.
33. Minimum percent of solid material in concrete units = 70.
34. Minimum percent of solid material in concrete units = 76.
35. Not less than $^{1}/_{2}$ inch of 1:3 sanded gypsum plaster.
36. Noncombustible or no members framed into wall.
37. Combustible members framed into wall.
38. One unit in wall thickness.
39. Two units in wall thickness.
40. Three units in wall thickness.
41. Concrete units made with expanded slag or pumice aggregates.
42. Concrete units made with expanded burned clay or shale, crushed limestone, air cooled slag or cinders.
43. Concrete units made with calcareous sand and gravel. Coarse aggregate, 60 percent or more calcite and dolomite.
44. Concrete units made with siliceous sand and gravel. Ninety percent or more quartz, chert or flint.
45. Laid in 1:3 sanded gypsum mortar.

(Continued)

TABLE 1.1.3—MASONRY WALLS
6" TO LESS THAN 8" THICK—(Continued)

46. Units of expanded slag or pumice aggregate.
47. Units of crushed limestone, blast furnace, slag, cinder and expanded clay or shale.
48. Units of calcareous sand and gravel. Coarse aggregate, 60 percent or more calcite and dolomite.
49. Units of siliceous sand and gravel. Ninety percent or more quartz, chert or flint.
50. Unit minimum 49 percent solid.
51. Unit minimum 62 percent solid.
52. Unit minimum 65 percent solid.
53. Unit minimum 73 percent solid.
54. Ratings based on one unit and one cell in wall section.
55. See Clay Tile Partition Design Construction drawings, below.

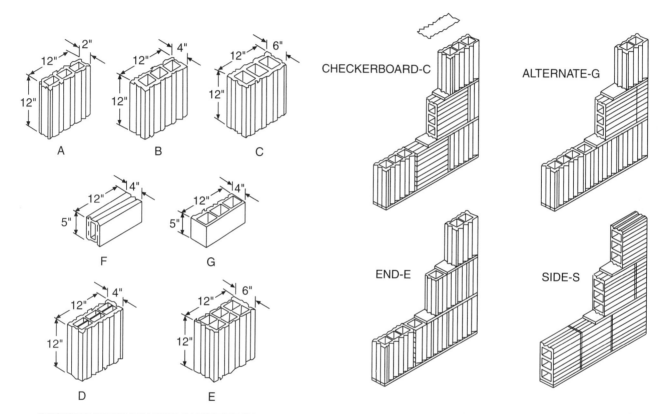

DESIGNS OF TILES USED IN FIRE-TEST PARTITIONS

THE FOUR TYPES OF CONSTRUCTION USED
IN FIRE-TEST PARTITIONS

FIGURE 1.1.4—WALLS—MASONRY
8" TO LESS THAN 10" THICK

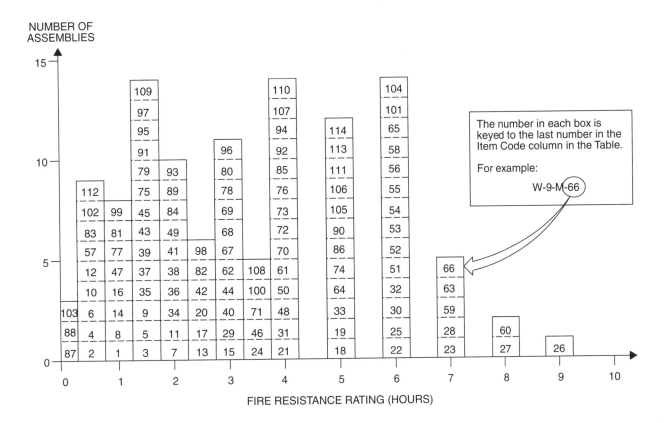

TABLE 1.1.4—MASONRY WALLS
8" TO LESS THAN 10" THICK

ITEM CODE	THICKNESS	CONSTRUCTION DETAILS	PERFORMANCE		REFERENCE NUMBER			NOTES	REC. HOURS
			LOAD	TIME	PRE-BMS-92	BMS-92	POST-BMS-92		
W-8-M-1	8"	Core: clay or shale structural tile; Units in wall thickness: 1; Cells in wall thickness: 2; Minimum % solids in units: 40.	80 psi	1 hr. 15 min.		1		1, 20	$1^1/_4$
W-8-M-2	8"	Core: clay or shale structural tile; Units in wall thickness: 1; Cells in wall thickness: 2; Minimum % solids in units: 40; No facings; Result for wall with combustible members framed into interior.	80 psi	45 min.		1		1, 20	$^3/_4$
W-8-M-3	8"	Core: clay or shale structural tile; Units in wall thickness: 1; Cells in wall thickness: 2; Minimum % solids in units: 43.	80 psi	1 hr. 30 min.		1		1, 20	$1^1/_2$
W-8-M-4	8"	Core: clay or shale structural tile; Units in wall thickness: 1; Cells in wall thickness: 2; Minimum % solids in units: 43; No facings; Combustible members framed into wall.	80 psi	45 min.		1		1, 20	$^3/_4$
W-8-M-5	8"	Core: clay or shale structural tile; No facings.	See Notes	1 hr. 30 min.		1		1, 2, 5, 10, 18, 20, 21	$1^1/_2$

(Continued)

TABLE 1.1.4—MASONRY WALLS
8" TO LESS THAN 10" THICK—(Continued)

ITEM CODE	THICKNESS	CONSTRUCTION DETAILS	PERFORMANCE		REFERENCE NUMBER			NOTES	REC. HOURS
			LOAD	TIME	PRE-BMS-92	BMS-92	POST-BMS-92		
W-8-M-6	8"	Core: clay or shale structural tile; No facings.	See Notes	45 min.		1		1, 2, 5, 10, 19, 20, 21	$^3/_4$
W-8-M-7	8"	Core: clay or shale structural tile; No facings	See Notes	2 hrs.		1		1, 2, 5, 13, 18, 20, 21	2
W-8-M-8	8"	Core: clay or shale structural tile; No facings.	See Notes	1 hr. 45 min.		1		1, 2, 5, 13, 19, 20, 21	$1^1/_4$
W-8-M-9	8"	Core: clay or shale structural tile; No facings.	See Notes	1 hr. 15 min.		1		1, 2, 6, 9, 18, 20, 21	$1^3/_4$
W-8-M-10	8"	Core: clay or shale structural tile; No facings.	See Notes	45 min.		1		1, 2, 6, 9, 19, 20, 21	$^3/_4$
W-8-M-11	8"	Core: clay or shale structural tile; No facings.	See Notes	2 hrs.		1		1, 2, 6, 10, 18, 20, 21	2
W-8-M-12	8"	Core: clay or shale structural tile; No facings.	See Notes	45 min.		1		1, 2, 6, 10, 19, 20, 21	$^3/_4$
W-8-M-13	8"	Core: clay or shale structural tile; No facings.	See Notes	2 hrs. 30 min.		1		1, 3, 6, 12, 18, 20, 21	$2^1/_2$
W-8-M-14	8"	Core: clay or shale structural tile; No facings.	See Notes	1 hr.		1		1, 2, 6, 12, 19, 20, 21	1
W-8-M-15	8"	Core: clay or shale structural tile; No facings.	See Notes	3 hrs.		1		1, 2, 6, 16, 18, 20, 21	3
W-8-M-16	8"	Core: clay or shale structural tile; No facings.	See Notes	1 hr. 15 min.		1		1, 2, 6, 16, 19, 20, 21	$1^1/_4$
W-8-M-17	8"	Cored clay or shale brick; Units in wall thickness: 1; Cells in wall thickness: 1; Minimum % solids: 70; No facings.	See Notes	2 hrs. 30 min.		1		1, 44	$2^1/_2$
W-8-M-18	8"	Cored clay or shale brick; Units in wall thickness: 2; Cells in wall thickness: 2; Minimum % solids: 87; No facings.	See Notes	5 hrs.		1		1, 45	5
W-8-M-19	8"	Core: solid clay or shale brick; No facings.	See Notes	5 hrs.		1		1, 22, 45	5
W-8-M-20	8"	Core: hollow rolok of clay or shale.	See Notes	2 hrs. 30 min.		1		1, 22, 45	$2^1/_2$
W-8-M-21	8"	Core: hollow rolok bak of clay or shale; No facings.	See Notes	4 hrs.		1		1, 45	4
W-8-M-22	8"	Core: concrete brick; No facings.	See Notes	6 hrs.		1		1, 45	6
W-8-M-23	8"	Core: sand-lime brick; No facings.	See Notes	7 hrs.		1		1, 45	7

(Continued)

TABLE 1.1.4—MASONRY WALLS
8" TO LESS THAN 10" THICK—(Continued)

ITEM CODE	THICKNESS	CONSTRUCTION DETAILS	PERFORMANCE		REFERENCE NUMBER			NOTES	REC. HOURS
			LOAD	TIME	PRE-BMS-92	BMS-92	POST-BMS-92		
W-8-M-24	8"	Core: 4", 40% solid clay or shale structural tile; 1 side 4" brick facing.	See Notes	3 hrs. 30 min.		1		1, 20	$3^1/_2$
W-8-M-25	8"	Concrete wall (3220 psi); Reinforcing vertical rods 1" from each face and 1" diameter; horizontal rods $^5/_8$" diameter.	22,200 lbs./ft.	6 hrs.			7		6
W-8-M-26	8"	Core: sand-line brick; $^1/_2$" of 1:3 sanded gypsum plaster facings on one side.	See Notes	9 hrs.		1		1, 45	9
W-8-M-27	$8^1/_2$"	Core: sand-line brick; $^1/_2$" of 1:3 sanded gypsum plaster facings on one side.	See Notes	8 hrs.		1		1, 45	8
W-8-M-28	$8^1/_2$"	Core: concrete; $^1/_2$" of 1:3 sanded gypsum plaster facings on one side.	See Notes	7 hrs.		1		1, 45	7
W-8-M-29	$8^1/_2$"	Core: hollow rolok of clay or shale; $^1/_2$" of 1:3 sanded gypsum plaster facings on one side.	See Notes	3 hrs.		1		1, 45	3
W-8-M-30	$8^1/_2$"	Core: solid clay or shale brick $^1/_2$" thick, 1:3 sanded gypsum plaster facings on one side.	See Notes	6 hrs.		1		1, 22, 45	6
W-8-M-31	$8^1/_2$"	Core: cored clay or shale brick; Units in wall thickness: 1; Cells in wall thickness: 1; Minimum % solids: 70; $^1/_2$" of 1:3 sanded gypsum plaster facings on both sides.	See Notes	4 hrs.		1		1, 44	4
W-8-M-32	$8^1/_2$"	Core: cored clay or shale brick; Units in wall thickness: 2; Cells in wall thickness: 2; Minimum % solids: 87; $^1/_2$" of 1:3 sanded gypsum plaster facings on one side.	See Notes	6 hrs.		1		1, 45	6
W-8-M-33	$8^1/_2$"	Core: hollow Rolok Bak of clay or shale; $^1/_2$" of 1:3 sanded gypsum plaster facings on one side.	See Notes	5 hrs.		1		1, 45	5
W-8-M-34	$8^5/_8$"	Core: clay or shale structural tile; Units in wall thickness: 1; Cells in wall thickness: 2; Minimum % solids in units: 40; $^5/_8$" of 1:3 sanded gypsum plaster facings on one side.	See Notes	2 hrs.		1		1, 20 21	2
W-8-M-35	$8^5/_8$"	Core: clay or shale structural tile; Units in wall thickness: 1; Cells in wall thickness: 2; Minimum % solids in units: 40; Exposed face: $^5/_8$" of 1:3 sanded gypsum plaster.	See Notes	1 hr. 30 min.		1		1, 20, 21	$1^1/_2$

(Continued)

TABLE 1.1.4—MASONRY WALLS
8" TO LESS THAN 10" THICK—(Continued)

ITEM CODE	THICKNESS	CONSTRUCTION DETAILS	PERFORMANCE		REFERENCE NUMBER			NOTES	REC. HOURS
			LOAD	TIME	PRE-BMS-92	BMS-92	POST-BMS-92		
W-8-M-36	8⁵/₈"	Core: clay or shale structural tile; Units in wall thickness: 1; Cells in wall thickness: 2; Minimum % solids in units: 43; ⁵/₈" of 1:3 sanded gypsum plaster facings on one side.	See Notes	2 hrs.				1, 20, 21	2
W-8-M-37	8⁵/₈"	Core: clay or shale structural tile; Units in wall thickness: 1; Cells in wall thickness: 2; Minimum % solids in units: 43; ⁵/₈" of 1:3 sanded gypsum plaster of the exposed face only.	See Notes	1 hr. 30 min.		1		1, 20, 21	1¹/₂
W-8-M-38	8⁵/₈"	Core: clay or shale structural tile; Facings: side 1; see Note 17.	See Notes	2 hrs.		1		1, 2, 5, 10, 18, 20, 21	2
W-8-M-39	8⁵/₈"	Core: clay or shale structural tile; Facings: exposed side only; see Note 17.	See Notes	1 hr. 30 min.		1		1, 2, 5, 10, 19, 20, 21	1¹/₂
W-8-M-40	8⁵/₈"	Core: clay or shale structural tile; Facings: exposed side only; see Note 17.	See Notes	3 hrs.		1		1, 2, 5, 13, 18, 20, 21	3
W-8-M-41	8⁵/₈"	Core: clay or shale structural tile; Facings: exposed side only; see Note 17.	See Notes	2 hrs.		1		1, 2, 5, 13, 19, 20, 21	2
W-8-M-42	8⁵/₈"	Core: clay or shale structural tile; Facings: side 1; see Note 17.	See Notes	2 hrs. 30 min.		1		1, 2, 9, 18, 20, 21	2¹/₂
W-8-M-43	8⁵/₈"	Core: clay or shale structural tile; Facings: exposed side only; see Note 17.	See Notes	1 hr. 30 min.		1		1, 2, 6, 9, 19, 20, 21	1¹/₂
W-8-M-44	8⁵/₈"	Core: clay or shale structural tile; Facings: side 1, see Note 17; side 2, none.	See Notes	3 hrs.		1		1, 2, 10, 18, 20, 21	3
W-8-M-45	8⁵/₈"	Core: clay or shale structural tile; Facings: fire side only; see Note 17.	See Notes	1 hr. 30 min.		1		1, 2, 6, 10, 19, 20, 21	1¹/₂
W-8-M-46	8⁵/₈"	Core: clay or shale structural tile; Facings: side 1, see Note 17; side 2, none.	See Notes	3 hrs. 30 min.		1		1, 2, 6, 12, 18, 20, 21	3¹/₂
W-8-M-47	8⁵/₈"	Core: clay or shale structural tile; Facings: exposed side only; see Note 17.	See Notes	1 hr. 45 min.		1		1, 2, 6, 12, 19, 20, 21	1³/₄
W-8-M-48	8⁵/₈"	Core: clay or shale structural tile; Facings: side 1, see Note 17; side 2, none.	See Notes	4 hrs.		1		1, 2, 6, 16, 18, 20, 21	4
W-8-M-49	8⁵/₈"	Core: clay or shale structural tile; Facings: fire side only; see Note 17.	See Notes	2 hrs.		1		1, 2, 6, 16, 19, 20, 21	2

(Continued)

TABLE 1.1.4—MASONRY WALLS
8" TO LESS THAN 10" THICK—(Continued)

ITEM CODE	THICKNESS	CONSTRUCTION DETAILS	PERFORMANCE		REFERENCE NUMBER			NOTES	REC. HOURS
			LOAD	TIME	PRE-BMS-92	BMS-92	POST-BMS-92		
W-8-M-50	$8^5/_8$"	Core: 4", 40% solid clay or shale clay structural tile; 4" brick plus $^5/_8$" of 1:3 sanded gypsum plaster facings on one side.	See Notes	4 hrs.		1		1, 20	4
W-8-M-51	$8^3/_4$"	$8^3/_4$" × $2^1/_2$" and 4" × $2^1/_2$" cellular fletton (1873 psi) single and triple cell hollow brick set in $^1/_2$" sand mortar in alternate courses.	3.6 tons/ft.	6 hrs.			7	23, 29	6
W-8-M-52	$8^3/_4$"	$8^3/_4$" thick cement brick (2527 psi) with P.C. and sand mortar.	3.6 tons/ft.	6 hrs.			7	23, 24	6
W-8-M-53	$8^3/_4$"	$8^3/_4$" × $2^1/_2$" fletton brick (1831 psi) in $^1/_2$" sand mortar.	3.6 tons/ft.	6 hrs.			7	23, 24	6
W-8-M-54	$8^3/_4$"	$8^3/_4$" × $2^1/_2$" London stock brick (683 psi) in $^1/_2$" P.C. - sand mortar.	7.2 tons/ft.	6 hrs.			7	23, 24	6
W-9-M-55	9"	9" × $2^1/_2$" Leicester red wire-cut brick (4465 psi) in $^1/_2$" P.C. - sand mortar.	6.0 tons/ft.	6 hrs.			7	23, 24	6
W-9-M-56	9"	9" × 3" sand-lime brick (2603 psi) in $^1/_2$" P.C. - sand mortar.	3.6 tons/ft.	6 hrs.			7	23, 24	6
W-9-M-57	9"	2 layers $2^7/_8$" fletton brick (1910 psi) with $3^1/_4$" air space; Cement and sand mortar.	1.5 tons/ft.	32 min.			7	23, 25	$^1/_3$
W-9-M-58	9"	9" × 3" stairfoot brick (7527 psi) in $^1/_2$" sand-cement mortar.	7.2 tons/ft.	6 hrs.			7	23, 24	6
W-9-M-59	9"	Core: solid clay or shale brick; $^1/_2$" thick; 1:3 sanded gypsum plaster facings on both sides.	See Notes	7 hrs.		1		1, 22, 45	7
W-9-M-60	9"	Core: concrete brick; $^1/_2$" of 1:3 sanded gypsum plaster facings on both sides.	See Notes	8 hrs.		1		1, 45	8
W-9-M-61	9"	Core: hollow Rolok of clay or shale; $^1/_2$" of 1:3 sanded gypsum plaster facings on both sides.	See Notes	4 hrs.		1		1, 45	4
W-9-M-62	9"	Cored clay or shale brick; Units in wall thickness: 1; Cells in wall thickness: 1; Minimum % solids: 70; $^1/_2$" of 1:3 sanded gypsum plaster facings on one side.	See Notes	3 hrs.		1		1, 44	3
W-9-M-63	9"	Cored clay or shale brick; Units in wall thickness: 2; Cells in wall thickness: 2; Minimum % solids: 87; $^1/_2$" of 1:3 sanded gypsum plaster facings on both sides.	See Notes	7 hrs.		1		1, 45	7
W-9-M-64	9-10"	Core: cavity wall of clay or shale brick; No facings.	See Notes	5 hrs.		1		1, 45	5
W-9-M-65	9-10"	Core: cavity construction of clay or shale brick; $^1/_2$" of 1:3 sanded gypsum plaster facings on one side.	See Notes	6 hrs.		1		1, 45	6

(Continued)

TABLE 1.1.4—MASONRY WALLS
8" TO LESS THAN 10" THICK—(Continued)

ITEM CODE	THICKNESS	CONSTRUCTION DETAILS	PERFORMANCE		REFERENCE NUMBER			NOTES	REC. HOURS
			LOAD	TIME	PRE-BMS-92	BMS-92	POST-BMS-92		
W-9-M-66	9-10"	Core: cavity construction of clay or shale brick; $1/2$" of 1:3 sanded gypsum plaster facings on both sides.	See Notes	7 hrs.		1		1, 45	7
W-9-M-67	$9^1/_4$"	Core: clay or shale structural tile; Units in wall thickness: 1; Cells in wall thickness: 2; Minimum % solids in units: 40; $5/_8$" of 1:3 sanded gypsum plaster facings on both sides.	See Notes	3 hrs.		1		1, 20, 21	3
W-9-M-68	$9^1/_4$"	Core: clay or shale structural tile; Units in wall thickness: 1; Cells in wall thickness: 2; Minimum % solids in units: 43; $5/_8$" of 1:3 sanded gypsum plaster facings on both sides.	See Notes	3 hrs.		1		1, 20, 21	3
W-9-M-69	$9^1/_4$"	Core: clay or shale structural tile; Facings: sides 1 and 2; see Note 17.	See Notes	3 hrs.		1		1, 2, 5, 10, 18, 20, 21	3
W-9-M-70	$9^1/_4$"	Core: clay or shale structural tile; Facings: sides 1 and 2; see Note 17.	See Notes	4 hrs.		1		1, 2, 5, 13, 18, 20, 21	4
W-9-M-71	$9^1/_4$"	Core: clay or shale structural tile; Facings: sides 1 and 2; see Note 17.	See Notes	3 hrs. 30 min.		1		1, 2, 6, 9, 18, 20, 21	$3^1/_2$
W-9-M-72	$9^1/_4$"	Core: clay or shale structural tile; Facings: sides 1 and 2; see Note 17.	See Notes	4 hrs.		1		1, 2, 6, 10, 18, 20, 21	4
W-9-M-73	$9^1/_4$"	Core: clay or shale structural tile; Facings: sides 1 and 2; see Note 17.	See Notes	4 hrs.		1		1, 2, 6, 12, 18, 20, 21	4
W-9-M-74	$9^1/_4$"	Core: clay or shale structural tile; Facings: sides 1 and 2; see Note 17.	See Notes	5 hrs.		1		1, 2, 6 16, 18, 20, 21	5
W-9-M-75	8"	Cored concrete masonry; see Notes 2, 19, 26, 34, 40; No facings.	80 psi	1 hr. 30 min.		1		1, 20	$1^1/_2$
W-8-M-76	8"	Cored concrete masonry; see Notes 2, 18, 26, 34, 40; No facings	80 psi	4 hrs.		1		1, 20	4
W-8-M-77	8"	Cored concrete masonry; see Notes 2, 19, 26, 31, 40; No facings.	80 psi	1 hr. 15 min.		1		1, 20	$1^1/_4$
W-8-M-78	8"	Cored concrete masonry; see Notes 2, 18, 26, 31, 40; No facings.	80 psi	3 hrs.		1		1, 20	3
W-8-M-79	8"	Cored concrete masonry; see Notes 2, 19, 26, 36, 42; No facings.	80 psi	1 hr. 30 min.		1		1, 20	$1^1/_2$
W-8-M-80	8"	Cored concrete masonry; see Notes 2, 18, 26, 36, 41; No facings.	80 psi	3 hrs.		1		1, 20	3

(Continued)

TABLE 1.1.4—MASONRY WALLS
8" TO LESS THAN 10" THICK—(Continued)

ITEM CODE	THICKNESS	CONSTRUCTION DETAILS	PERFORMANCE		REFERENCE NUMBER			NOTES	REC. HOURS
			LOAD	TIME	PRE-BMS-92	BMS-92	POST-BMS-92		
W-8-M-81	8"	Cored concrete masonry; see Notes 2, 19, 26, 34, 41; No facings.	80 psi	1 hr.		1		1, 20	1
W-8-M-82	8"	Cored concrete masonry; see Notes 2, 18, 26, 34, 41; No facings.	80 psi	2 hrs. 30 min.		1		1, 20	$2^1/_2$
W-8-M-83	8"	Cored concrete masonry; see Notes 2, 19, 26, 29, 41; No facings.	80 psi	45 min.		1		1, 20	$^3/_4$
W-8-M-84	8"	Cored concrete masonry; see Notes 2, 18, 26, 29, 41; No facings.	80 psi	2 hrs.		1		1, 20	2
W-8-M-85	$8^1/_2$"	Cored concrete masonry; see Notes 3, 18, 26, 34, 41; Facings: $2^1/_4$" brick.	80 psi	4 hrs.		1		1, 20	4
W-8-M-86	8"	Cored concrete masonry; see Notes 3, 18, 26, 34, 41; Facings: $3^3/_4$" brick face.	80 psi	5 hrs.		1		1, 20	5
W-8-M-87	8"	Cored concrete masonry; see Notes 2, 19, 26, 30, 43; No facings.	80 psi	12 min.		1		1, 20	$^1/_5$
W-8-M-88	8"	Cored concrete masonry; see Notes 2, 18, 26, 30, 43; No facings.	80 psi	12 min.		1		1, 20	$^1/_5$
W-8-M-89	$8^1/_2$"	Cored concrete masonry; see Notes 2, 19, 26, 34, 40; Facings: fire side only; see Note 38.	80 psi	2 hrs.		1		1, 20	2
W-8-M-90	$8^1/_2$"	Cored concrete masonry; see Notes 2, 18, 26, 34, 40; Facings: side 1; see Note 38.	80 psi	5 hrs.		1		1, 20	5
W-8-M-91	$8^1/_2$"	Cored concrete masonry; see Notes 2, 19, 26, 31, 40; Facings: fire side only; see Note 38.	80 psi	1 hr. 45 min.		1		1, 20	$1^3/_4$
W-8-M-92	$8^1/_2$"	Cored concrete masonry; see Notes 2, 18, 26, 31, 40; Facings: one side; see Note 38.	80 psi	4 hrs.		1		1, 20	4
W-8-M-93	$8^1/_2$"	Cored concrete masonry; see Notes 2, 19, 26, 36, 41; Facings: fire side only; see Note 38.	80 psi	2 hrs.		1		1, 20	2
W-8-M-94	$8^1/_2$"	Cored concrete masonry; see Notes 2, 18, 26, 36, 41; Facings: fire side only; see Note 38.	80 psi	4 hrs.		1		1, 20	4
W-8-M-95	$8^1/_2$"	Cored concrete masonry; see Notes 2, 19, 26, 34, 41; Facings: fire side only; see Note 38.	80 psi	1 hr. 30 min.		1		1, 20	$1^1/_2$
W-8-M-96	$8^1/_2$"	Cored concrete masonry; see Notes 2, 18, 26, 34, 41; Facings: one side; see Note 38.	80 psi	3 hrs.		1		1, 20	3
W-8-M-97	$8^1/_2$"	Cored concrete masonry; see Notes 2, 19, 26, 29, 41; Facings: fire side only; see Note 38.	80 psi	1 hr. 30 min.		1		1, 20	$1^1/_2$

(Continued)

TABLE 1.1.4—MASONRY WALLS
8" TO LESS THAN 10" THICK—(Continued)

ITEM CODE	THICKNESS	CONSTRUCTION DETAILS	PERFORMANCE		REFERENCE NUMBER			NOTES	REC. HOURS
			LOAD	TIME	PRE-BMS-92	BMS-92	POST-BMS-92		
W-8-M-98	$8^1/_2$"	Cored concrete masonry; see Notes 2, 18, 26, 29, 41; Facings: one side; see Note 38.	80 psi	2 hrs. 30 min.		1		1, 20	$2^1/_2$
W-8-M-99	$8^1/_2$"	Cored concrete masonry; see Notes 3, 19, 23, 27, 41; No facings.	80 psi	1 hr. 15 min.		1		1, 20	$1^1/_4$
W-8-M-100	$8^1/_2$"	Cored concrete masonry; see Notes 3, 18, 23, 27, 41; No facings.	80 psi	3 hrs. 30 min.		1		1, 20	$3^1/_2$
W-8-M-101	$8^1/_2$"	Cored concrete masonry; see Notes 3, 18, 26, 34, 41; Facings: $3^3/_4$" brick face; one side only; see Note 38.	80 psi	6 hrs.		1		1, 20	6
W-8-M-102	$8^1/_2$"	Cored concrete masonry; see Notes 2, 19, 26, 30, 43; Facings: fire side only; see Note 38.	80 psi	30 min.		1		1, 20	$1/_2$
W-8-M-103	$8^1/_2$"	Cored concrete masonry; see Notes 2, 18, 26, 30, 43; Facings: one side only; see Note 38.	80 psi	12 min.		1		1, 20	$1/_5$
W-8-M-104	9"	Cored concrete masonry; see Notes 2, 18, 26, 34, 40; Facings: both sides; see Note 38.	80 psi	6 hrs.		1		1, 20	6
W-8-M-105	9"	Cored concrete masonry; see Notes 2, 18, 26, 31, 40; Facings: both sides; see Note 38.	80 psi	5 hrs.		1		1, 20	5
W-8-M-106	9"	Cored concrete masonry; see Notes 2, 18, 26, 36, 41; Facings: both sides of wall; see Note 38.	80 psi	5 hrs.		1		1, 20	5
W-8-M-107	9"	Cored concrete masonry; see Notes 2, 18, 26, 34, 41; Facings: both sides; see Note 38.	80 psi	4 hrs.		1		1, 20	4
W-8-M-108	9"	Cored concrete masonry; see Notes 2, 18, 26, 29, 41; Facings: both sides; see Note 38.	80 psi	3 hrs. 30 min.		1		1, 20	$3^1/_2$
W-8-M-109	9"	Cored concrete masonry; see Notes 3, 19, 23, 27, 40; Facings: fire side only; see Note 38.	80 psi	1 hr. 45 min.		1		1, 20	$1^3/_4$
W-8-M-110	9"	Cored concrete masonry; see Notes 3, 18, 23, 27, 41; Facings: one side only; see Note 38.	80 psi	4 hrs.		1		1, 20	4
W-8-M-111	9"	Cored concrete masonry; see Notes 3, 18, 26, 34, 41; $2^1/_4$" brick face on one side only; see Note 38.	80 psi	5 hrs.		1		1, 20	5
W-8-M-112	9"	Cored concrete masonry; see Notes 2, 18, 26, 30, 43; Facings: both sides; see Note 38.	80 psi	30 min.		1		1, 20	$1/_2$

(Continued)

TABLE 1.1.4—MASONRY WALLS
8" TO LESS THAN 10" THICK—(Continued)

ITEM CODE	THICKNESS	CONSTRUCTION DETAILS	PERFORMANCE		REFERENCE NUMBER			NOTES	REC. HOURS
			LOAD	TIME	PRE-BMS-92	BMS-92	POST-BMS-92		
W-9-M-113	9$\frac{1}{2}$"	Cored concrete masonry; see Notes 3, 18, 23, 27, 41; Facings: both sides; see Note 38.	80 psi	5 hrs.		1		1, 20	5
W-8-M-114	8"		200 psi	5 hrs.			43	22	5

Notes:

1. Tested at NBS under ASA Spec. No. 43-1934 (ASTM C 19-53).
2. One unit in wall thickness.
3. Two units in wall thickness.
4. Two or three units in wall thickness.
5. Two cells in wall thickness.
6. Three or four cells in wall thickness.
7. Four or five cells in wall thickness.
8. Five or six cells in wall thickness.
9. Minimum percent of solid materials in units = 40%.
10. Minimum percent of solid materials in units = 43%.
11. Minimum percent of solid materials in units = 46%.
12. Minimum percent of solid materials in units = 48%.
13. Minimum percent of solid materials in units = 49%.
14. Minimum percent of solid materials in units = 45%.
15. Minimum percent of solid materials in units = 51%.
16. Minimum percent of solid materials in units = 53%.
17. Not less than $\frac{5}{8}$ inch thickness of 1:3 sanded gypsum plaster.
18. Noncombustible or no members framed into wall.
19. Combustible members framed into wall.
20. Load: 80 psi for gross cross-sectional area of wall.
21. Portland cement-lime mortar.
22. Failure mode thermal.
23. British test.
24. Passed all criteria.
25. Failed by sudden collapse with no preceding signs of impending failure.
26. One cell in wall thickness.
27. Two cells in wall thickness.
28. Three cells in wall thickness.
29. Minimum percent of solid material in concrete units = 52.
30. Minimum percent of solid material in concrete units = 54.
31. Minimum percent of solid material in concrete units = 55.
32. Minimum percent of solid material in concrete units = 57.
33. Minimum percent of solid material in concrete units = 60.
34. Minimum percent of solid material in concrete units = 62.
35. Minimum percent of solid material in concrete units = 65.
36. Minimum percent of solid material in concrete units = 70.
37. Minimum percent of solid material in concrete units = 76.
38. Not less than $\frac{1}{2}$ inch of 1:3 sanded gypsum plaster.
39. Three units in wall thickness.
40. Concrete units made with expanded slag or pumice aggregates.
41. Concrete units made with expanded burned clay or shale, crushed limestone, air cooled slag or cinders.
42. Concrete units made with calcareous sand and gravel. Coarse aggregate, 60 percent or more calcite and dolomite.
43. Concrete units made with siliceous sand and gravel. Ninety percent or more quartz, chert and dolomite.
44. Load: 120 psi for gross cross-sectional area of wall.
45. Load: 160 psi for gross cross-sectional area of wall.

FIGURE 1.1.5—WALLS—MASONRY
10" TO LESS THAN 12" THICK

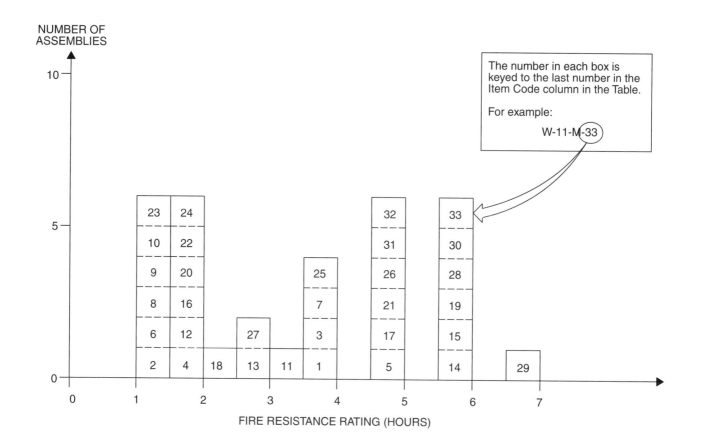

TABLE 1.1.5—WALLS—MASONRY
10" TO LESS THAN 12" THICK

| ITEM CODE | THICKNESS | CONSTRUCTION DETAILS | PERFORMANCE | | REFERENCE NUMBER | | | NOTES | REC. HOURS |
			LOAD	TIME	PRE-BMS-92	BMS-92	POST-BMS-92		
W-10-M-1	10"	Core: two 3³/₄", 40% solid clay or shale structural tiles with 2" air space between; Facings: ³/₄" portland cement plaster on stucco on both sides.	80 psi	4 hrs.		1		1, 20	4
W-10-M-2	10"	Core: cored concrete masonry, 2" air cavity; see Notes 3, 19, 27, 34, 40; No facings.	80 psi	1 hr. 30 min.		1		1, 20	1¹/₂
W-10-M-3	10"	Cored concrete masonry; see Notes 3, 18, 27, 34, 40; No facings.	80 psi	4 hrs.		1		1, 20	4
W-10-M-4	10"	Cored concrete masonry; see Notes 2, 19, 26, 34, 40; No facings.	80 psi	2 hrs.		1		1, 20	2
W-10-M-5	10"	Cored concrete masonry; see Notes 2, 18, 26, 33, 40; No facings.	80 psi	5 hrs.		1		1, 20	5
W-10-M-6	10"	Cored concrete masonry; see Notes 2, 19, 26, 33, 41; No facings.	80 psi	1 hr. 30 min.		1		1, 20	1¹/₂
W-10-M-7	10"	Cored concrete masonry; see Notes 2, 18, 26, 33, 41; No facings.	80 psi	4 hrs.		1		1, 20	4

(Continued)

TABLE 1.1.5—MASONRY WALLS
10" TO LESS THAN 12" THICK—(Continued)

ITEM CODE	THICKNESS	CONSTRUCTION DETAILS	PERFORMANCE		REFERENCE NUMBER			NOTES	REC. HOURS
			LOAD	TIME	PRE-BMS-92	BMS-92	POST-BMS-92		
W-10-M-8	10"	Cored concrete masonry (cavity type 2" air space); see Notes 3, 19, 27, 34, 42; No facings.	80 psi	1 hr. 15 min.		1		1, 20	$1^1/_4$
W-10-M-9	10"	Cored concrete masonry (cavity type 2" air space); see Notes 3, 18, 27, 34, 42; No facings.	80 psi	1 hr. 15 min.		1		1, 20	$1^1/_4$
W-10-M-10	10"	Cored concrete masonry (cavity type 2" air space); see Notes 3, 19, 27, 34, 41; No facings.	80 psi	1 hr. 15 min.		1		1, 20	$1^1/_4$
W-10-M-11	10"	Cored concrete masonry (cavity type 2" air space); see Notes 3, 18, 27, 34, 41; No facings.	80 psi	3 hrs. 30 min.		1		1, 20	$3^1/_2$
W-10-M-12	10"	9" thick concrete block ($11^3/_4$" × 9" × $4^1/_4$") with two 2" thick voids included; $3/_8$" P.C. plaster $1/_8$" neat gypsum.	N/A	1 hr. 53 min.			7	23, 44	$1^3/_4$
W-10-M-13	10"	Holly clay tile block wall - $8^1/_2$" block with two 3" voids in each $8^1/_2$" section; $3/_4$" gypsum plaster - each face.	N/A	2 hrs. 42 min.			7	23, 25	$2^1/_2$
W-10-M-14	10"	Two layers $4^1/_4$" brick with $1^1/_2$" air space; No ties sand cement mortar. (Fletton brick - 1910 psi).	N/A	6 hrs.			7	23, 24	6
W-10-M-15	10"	Two layers $4^1/_4$" thick Fletton brick (1910 psi); $1^1/_2$" air space; Ties: 18" o.c. vertical; 3' o.c. horizontal.	N/A	6 hrs.			7	23, 24	6
W-10-M-16	$10^1/_2$"	Cored concrete masonry; 2" air cavity; see Notes 3, 19, 27, 34, 40; Facings: fire side only; see Note 38.	80 psi	2 hrs.		1		1, 20	2
W-10-M-17	$10^1/_2$"	Cored concrete masonry; see Notes 3, 18, 27, 34, 40; Facings: side 1 only; see Note 38.	80 psi	5 hrs.		1		1, 20	5
W-10-M-18	$10^1/_2$"	Cored concrete masonry; see Notes 2, 19, 26, 33, 40; Facings: fire side only; see Note 38.	80 psi	2 hrs. 30 min.		1		1, 20	$2^1/_2$
W-10-M-19	$10^1/_2$"	Cored concrete masonry; see Notes 2, 18, 26, 33, 40; Facings: one side; see Note 38.	80 psi	6 hrs.		1		1, 20	6
W-10-M-20	$10^1/_2$"	Cored concrete masonry; see Notes 2, 19, 26, 33, 41; Facings: fire side of wall only; see Note 38.	80 psi	2 hrs.		1		1, 20	2
W-10-M-21	$10^1/_2$"	Cored concrete masonry; see Notes 2, 18, 26, 33, 41; Facings: one side only; see Note 38.	80 psi	5 hrs.		1		1, 20	5
W-10-M-22	$10^1/_2$"	Cored concrete masonry (cavity type 2" air space); see Notes 3,19, 27, 34, 42; Facings: fire side only; see Note 38.	80 psi	1 hr. 45 min.		1		1, 20	$1^3/_4$

(Continued)

TABLE 1.1.5—MASONRY WALLS
10" TO LESS THAN 12" THICK—(Continued)

ITEM CODE	THICKNESS	CONSTRUCTION DETAILS	PERFORMANCE		REFERENCE NUMBER			NOTES	REC. HOURS
			LOAD	TIME	PRE-BMS-92	BMS-92	POST-BMS-92		
W-10-M-23	$10^1/_2$"	Cored concrete masonry (cavity type 2" air space); see Notes 3, 18, 27, 34, 42; Facings: one side only; see Note 38.	80 psi	1 hr. 15 min.		1		1, 20	$1^1/_4$
W-10-M-24	$10^1/_2$"	Cored concrete masonry (cavity type 2" air space); see Notes 3, 19, 27, 34, 41; Facings: fire side only; see Note 38.	80 psi	2 hrs.		1		1, 20	2
W-10-M-25	$10^1/_2$"	Cored concrete masonry (cavity type 2" air space); see Notes 3, 18, 27, 34, 41; Facings: one side only; see Note 38.	80 psi	4 hrs.		1		1, 20	4
W-10-M-26	$10^5/_8$"	Core: 8", 40% solid tile plus 2" furring tile; $^5/_8$" sanded gypsum plaster between tile types; Facings: both sides $^3/_4$" portland cement plaster or stucco.	80 psi	5 hrs.		1		1, 20	5
W-10-M-27	$10^5/_8$"	Core: 8", 40% solid tile plus 2" furring tile; $^5/_8$" sanded gypsum plaster between tile types; Facings: one side $^3/_4$" portland cement plaster or stucco.	80 psi	3 hrs. 30 min.		1		1, 20	$3^1/_2$
W-11-M-28	11"	Cored concrete masonry; see Notes 3, 18, 27, 34, 40; Facings: both sides; see Note 38.	80 psi	6 hrs.		1		1, 20	6
W-11-M-29	11"	Cored concrete masonry; see Notes 2, 18, 26, 33, 40; Facings: both sides; see Note 38.	80 psi	7 hrs.		1		1, 20	7
W-11-M-30	11"	Cored concrete masonry; see Notes 2, 18, 26, 33, 41; Facings: both sides of wall; see Note 38.	80 psi	6 hrs.		1		1, 20	6
W-11-M-31	11"	Cored concrete masonry (cavity type 2" air space); see Notes 3, 18, 27, 34, 42; Facings: both sides; see Note 38.	80 psi	5 hrs.		1		1, 20	5
W-11-M-32	11"	Cored concrete masonry (cavity type 2" air space); see Notes 3, 18, 27, 34, 41; Facings: both sides; see Note 38.	80 psi	5 hrs.		1		1, 20	5

(Continued)

TABLE 1.1.5—MASONRY WALLS
10" TO LESS THAN 12" THICK—(Continued)

ITEM CODE	THICKNESS	CONSTRUCTION DETAILS	PERFORMANCE		REFERENCE NUMBER			NOTES	REC. HOURS
			LOAD	TIME	PRE-BMS-92	BMS-92	POST-BMS-92		
W-11-M-33	11"	Two layers brick (4$^{1}/_{2}$" Fletton, 2,428 psi) 2" air space; galvanized ties; 18" o.c. - horizontal; 3' o.c. - vertical.	3 tons/ft.	6 hrs.			7	23, 24	6

Notes:
1. Tested at NBS - ASA Spec. No. A2-1934.
2. One unit in wall thickness.
3. Two units in wall thickness.
4. Two or three units in wall thickness.
5. Two cells in wall thickness.
6. Three or four cells in wall thickness.
7. Four or five cells in wall thickness.
8. Five or six cells in wall thickness.
9. Minimum percent of solid materials in units = 40%.
10. Minimum percent of solid materials in units = 43%.
11. Minimum percent of solid materials in units = 46%.
12. Minimum percent of solid materials in units = 48%.
13. Minimum percent of solid materials in units = 49%.
14. Minimum percent of solid materials in units = 45%.
15. Minimum percent of solid materials in units = 51%.
16. Minimum percent of solid materials in units = 53%.
17. Not less than $^{5}/_{8}$ inch thickness of 1:3 sanded gypsum plaster.
18. Noncombustible or no members framed into wall.
19. Combustible members framed into wall.
20. Load: 80 psi for gross cross sectional area of wall.
21. Portland cement-lime mortar.
22. Failure mode - thermal.
23. British test.
24. Passed all criteria.
25. Failed by sudden collapse with no preceding signs of impending failure.
26. One cell in wall thickness.
27. Two cells in wall thickness.
28. Three cells in wall thickness.
29. Minimum percent of solid material in concrete units = 52%.
30. Minimum percent of solid material in concrete units = 54%.
31. Minimum percent of solid material in concrete units = 55%.
32. Minimum percent of solid material in concrete units = 57%.
33. Minimum percent of solid material in concrete units = 60%.
34. Minimum percent of solid material in concrete units = 62%.
35. Minimum percent of solid material in concrete units = 65%.
36. Minimum percent of solid material in concrete units = 70%.
37. Minimum percent of solid material in concrete units = 76%.
38. Not less than $^{1}/_{2}$" of 1:3 sanded gypsum plaster.
39. Three units in wall thickness.
40. Concrete units made with expanded slag or pumice aggregates.
41. Concrete units made with expanded burned clay or shale, crushed limestone, air cooled slag or cinders.
42. Concrete units made with calcareous sand and gravel. Coarse aggregate, 60 percent or more calcite and dolomite.

FIGURE 1.1.6—WALLS—MASONRY
12" TO LESS THAN 14" THICK

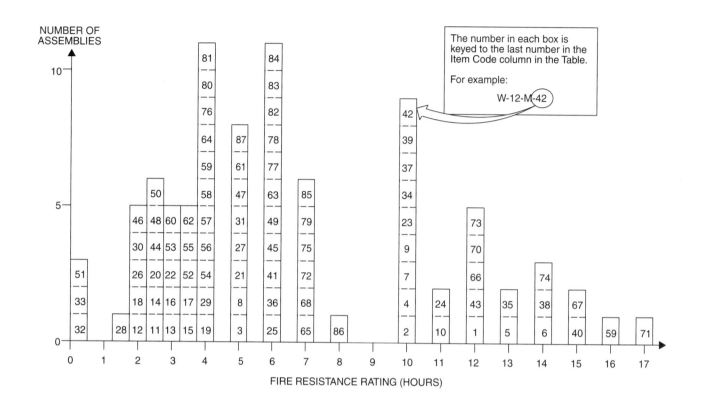

TABLE 1.1.6—WALLS—MASONRY
12" TO LESS THAN 14" THICK

ITEM CODE	THICKNESS	CONSTRUCTION DETAILS	PERFORMANCE		REFERENCE NUMBER			NOTES	REC. HOURS
			LOAD	TIME	PRE-BMS-92	BMS-92	POST-BMS-92		
W-12-M-1	12"	Core: solid clay or shale brick; No facings.	N/A	12 hrs.		1		1	12
W-12-M-2	12"	Core: solid clay or shale brick; No facings.	160 psi	10 hrs.		1		1, 44	10
W-12-M-3	12"	Core: hollow Rolok of clay or shale; No facings.	160 psi	5 hrs.		1		1, 44	5
W-12-M-4	12"	Core: hollow Rolok Bak of clay or shale; No facings.	160 psi	10 hrs.		1		1, 44	10
W-12-M-5	12"	Core: concrete brick; No facings.	160 psi	13 hrs.		1		1, 44	13
W-12-M-6	12"	Core: sand-lime brick; No facings.	N/A	14 hrs.		1		1	14
W-12-M-7	12"	Core: sand-lime brick; No facings.	160 psi	10 hrs.		1		1, 44	10
W-12-M-8	12"	Cored clay or shale brick; Units in wall thickness: 1; Cells in wall thickness: 2; Minimum % solids: 70; No facings.	120 psi	5 hrs.		1		1, 45	5
W-12-M-9	12"	Cored clay or shale brick; Units in wall thickness: 3; Cells in wall thickness: 3; Minimum % solids: 87; No facings.	160 psi	10 hrs.		1		1, 44	10

(Continued)

TABLE 1.1.6—WALLS—MASONRY
12" TO LESS THAN 14" THICK—(Continued)

ITEM CODE	THICKNESS	CONSTRUCTION DETAILS	PERFORMANCE		REFERENCE NUMBER			NOTES	REC. HOURS
			LOAD	TIME	PRE-BMS-92	BMS-92	POST-BMS-92		
W-12-M-10	12"	Cored clay or shale brick; Units in wall thickness: 3; Cells in wall thickness: 3; Minimum % solids: 87; No facings.	N/A	11 hrs.		1		1	11
W-12-M-11	12"	Core: clay or shale structural tile; see Notes 2, 6, 9, 18; No facings.	80 psi	2 hrs.		1		1, 20	$2^1/_2$
W-12-M-12	12"	Core: clay or shale structural tile; see Notes 2, 4, 9, 19; No facings.	80 psi	2 hrs.		1		1, 20	2
W-12-M-13	12"	Core: clay or shale structural tile; see Notes 2, 6, 14, 19; No facings.	80 psi	3 hrs.		1		1, 20	3
W-12-M-14	12"	Core: clay or shale structural tile; see Notes 2, 6,14, 18; No facings.	80 psi	2 hrs. 30 min.		1		1, 20	$2^1/_2$
W-12-M-15	12"	Core: clay or shale structural tile; see Notes 2, 4, 13, 18; No facings.	80 psi	3 hrs. 30 min.		1		1, 20	$3^1/_2$
W-12-M-16	12"	Core: clay or shale structural tile; see Notes 2, 4, 13, 19; No facings.	80 psi	3 hrs.		1		1, 20	3
W-12-M-17	12"	Core: clay or shale structural tile; see Notes 3, 6, 9, 18; No facings.	80 psi	3 hrs. 30 min.		1		1, 20	$3^1/_2$
W-12-M-18	12"	Core: clay or shale structural tile; see Notes 3, 6, 9, 19; No facings.	80 psi	2 hrs.		1		1, 20	2
W-12-M-19	12"	Core: clay or shale structural tile; see Notes 3, 6, 14, 18; No facings.	80 psi	4 hrs.		1		1, 20	4
W-12-M-20	12"	Core: clay or shale structural tile; see Notes 3, 6, 14, 19; No facings.	80 psi	2 hrs. 30 min.		1		1, 20	$2^1/_2$
W-12-M-21	12"	Core: clay or shale structural tile; see Notes 3, 6, 16, 18; No facings.	80 psi	5 hrs.		1		1, 20	5
W-12-M-22	12"	Core: clay or shale structural tile; see Notes 3, 6, 16, 19; No facings.	80 psi	3 hrs.		1		1, 20	3
W-12-M-23	12"	Core: 8", 70% solid clay or shale structural tile; 4" brick facings on one side.	80 psi	10 hrs.		1		1, 20	10
W-12-M-24	12"	Core: 8", 70% solid clay or shale structural tile; 4" brick facings on one side.	N/A	11 hrs.		1		1	11
W-12-M-25	12"	Core: 8", 40% solid clay or shale structural tile; 4" brick facings on one side.	80 psi	6 hrs.		1		1, 20	6
W-12-M-26	12"	Cored concrete masonry; see Notes 1, 9, 15, 16, 20; No facings.	80 psi	2 hrs.		1		1, 20	2
W-12-M-27	12"	Cored concrete masonry; see Notes 2, 18, 26, 34, 41; No facings.	80 psi	5 hrs.		1		1, 20	5
W-12-M-28	12"	Cored concrete masonry; see Notes 2, 19, 26, 31, 41; No facings.	80 psi	1 hr. 30 min.		1		1, 20	$1^1/_2$
W-12-M-29	12"	Cored concrete masonry; see Notes 2, 18, 26, 31, 41; No facings.	80 psi	4 hrs.		1		1, 20	4

(Continued)

TABLE 1.1.6—WALLS—MASONRY
12" TO LESS THAN 14" THICK—(Continued)

ITEM CODE	THICKNESS	CONSTRUCTION DETAILS	PERFORMANCE		REFERENCE NUMBER			NOTES	REC. HOURS
			LOAD	TIME	PRE-BMS-92	BMS-92	POST-BMS-92		
W-12-M-30	12"	Cored concrete masonry; see Notes 3, 19, 27, 31, 43; No facings.	80 psi	2 hrs.		1		1, 20	2
W-12-M-31	12"	Cored concrete masonry; see Notes 3, 18, 27, 31, 43; No facings.	80 psi	5 hrs.		1		1, 20	5
W-12-M-32	12"	Cored concrete masonry; see Notes 2, 19, 26, 32, 43; No facings.	80 psi	25 min.		1		1, 20	$^1/_3$
W-12-M-33	12"	Cored concrete masonry; see Notes 2, 18, 26, 32, 43; No facings.	80 psi	25 min.		1		1, 20	$^1/_3$
W-12-M-34	12$^1/_2$"	Core: solid clay or shale brick; $^1/_2$" of 1:3 sanded gypsum plaster facings on one side.	160 psi	10 hrs.		1		1, 44	10
W-12-M-35	12$^1/_2$"	Core: solid clay or shale brick; $^1/_2$" of 1:3 sanded gypsum plaster facings on one side.	N/A	13 hrs.		1		1	13
W-12-M-36	12$^1/_2$"	Core: hollow Rolok of clay or shale; $^1/_2$" of 1:3 sanded gypsum plaster facings on one side.	160 psi	6 hrs.		1		1, 44	6
W-12-M-37	12$^1/_2$"	Core: hollow Rolok Bak of clay or shale; $^1/_2$" of 1:3 sanded gypsum plaster facings on one side.	160 psi	10 hrs.		1		1, 44	10
W-12-M-38	12$^1/_2$"	Core: concrete; $^1/_2$" of 1:3 sanded gypsum plaster facings on one side.	160 psi	14 hrs.		1		1, 44	14
W-12-M-39	12$^1/_2$"	Core: sand-lime brick; $^1/_2$" of 1:3 sanded gypsum plaster facings on one side.	160 psi	10 hrs.		1		1, 44	10
W-12-M-40	12$^1/_2$"	Core: sand-lime brick; $^1/_2$" of 1:3 sanded gypsum plaster facings on one side.	N/A	15 hrs.		1		1	15
W-12-M-41	12$^1/_2$"	Cored clay or shale brick; Units in wall thickness: 1; Cells in wall thickness: 2; Minimum % solids: 70; $^1/_2$" of 1:3 sanded gypsum plaster facings on one side.	120 psi	6 hrs.		1		1, 45	6
W-12-M-42	12$^1/_2$"	Cored clay or shale brick; Units in wall thickness: 3; Cells in wall thickness: 3; Minimum % solids: 87; $^1/_2$" of 1:3 sanded gypsum plaster facings on one side.	160 psi	10 hrs.		1		1, 44	10
W-12-M-43	12$^1/_2$"	Cored clay or shale brick; Units in wall thickness: 3; Cells in wall thickness: 3; Minimum % solids: 87; $^1/_2$" of 1:3 sanded gypsum plaster facings on one side.	N/A	12 hrs.		1		1	12

(Continued)

TABLE 1.1.6—WALLS—MASONRY
12" TO LESS THAN 14" THICK—(Continued)

ITEM CODE	THICKNESS	CONSTRUCTION DETAILS	PERFORMANCE		REFERENCE NUMBER			NOTES	REC. HOURS
			LOAD	TIME	PRE-BMS-92	BMS-92	POST-BMS-92		
W-12-M-44	12$\frac{1}{2}$"	Cored concrete masonry; see Notes 2, 19, 26, 34, 41; Facings: fire side only; see Note 38.	80 psi	2 hrs. 30 min.		1		1, 20	2$\frac{1}{2}$
W-12-M-45	12$\frac{1}{2}$"	Cored concrete masonry; see Notes 2, 18, 26, 34, 39, 41; Facings: one side only; see Note 38.	80 psi	6 hrs.		1		1, 20	6
W-12-M-46	12$\frac{1}{2}$"	Cored concrete masonry; see Notes 2, 19, 26, 31, 41; Facings: fire side only; see Note 38.	80 psi	2 hrs.		1		1, 20	2
W-12-M-47	12$\frac{1}{2}$"	Cored concrete masonry; see Notes 2, 18, 26, 31, 41; Facings: one side of wall only; see Note 38.	80 psi	5 hrs.		1		1, 20	5
W-12-M-48	12$\frac{1}{2}$"	Cored concrete masonry; see Notes 3, 19, 27, 31, 43; Facings: fire side only; see Note 38.	80 psi	2 hrs. 30 min.		1		1, 20	2$\frac{1}{2}$
W-12-M-49	12$\frac{1}{2}$"	Cored concrete masonry; see Notes 3, 18, 27, 31, 43; Facings: one side only; see Note 38.	80 psi	6 hrs.		1		1, 20	6
W-12-M-50	12$\frac{1}{2}$"	Cored concrete masonry; see Notes 2, 19, 26, 32, 43; Facings: fire side only; see Note 38.	80 psi	2 hrs. 30 min.		1		1, 20	2$\frac{1}{2}$
W-12-M-51	12$\frac{1}{2}$"	Cored concrete masonry; see Notes 2, 18, 26, 32, 43; Facings: one side only; see Note 38.	80 psi	25 min.		1		1, 20	$\frac{1}{3}$
W-12-M-52	12$\frac{5}{8}$"	Clay or shale structural tile; see Notes 2, 6, 9, 18; Facings: side 1, see Note 17; side 2, none.	80 psi	3 hrs. 30 min.		1		1, 20	3$\frac{1}{2}$
W-12-M-53	12$\frac{5}{8}$"	Clay or shale structural tile; see Notes 2, 6, 9, 19; Facings: fire side only; see Note 17.	80 psi	3 hrs.		1		1, 20	3
W-12-M-54	12$\frac{5}{8}$"	Clay or shale structural tile; see Notes 2, 6, 14, 19; Facings: side 1, see Note 17; side 2, none.	80 psi	4 hrs.		1		1, 20	4
W-12-M-55	12$\frac{5}{8}$"	Clay or shale structural tile; see Notes 2, 6, 14, 18; Facings: exposed side only; see note 17.	80 psi	3 hrs. 30 min.		1		1, 20	3$\frac{1}{2}$
W-12-M-56	12$\frac{5}{8}$"	Clay or shale structural tile; see Notes 2, 4, 13, 18; Facings: side 1, see Note 17; side 2, none.	80 psi	4 hrs.		1		1, 20	4
W-12-M-57	12$\frac{5}{8}$"	Clay or shale structural tile; see Notes 1, 4, 13, 19; Facings: fire side only; see Note 17.	80 psi	4 hrs.		1		1, 20	4
W-12-M-58	12$\frac{5}{8}$"	Clay or shale structural tile; see Notes 3, 6, 9, 18; Facings: side 1, see Note 17; side 2, none.	80 psi	4 hrs.		1		1, 20	4
W-12-M-59	12$\frac{5}{8}$"	Clay or shale structural tile; see Notes 3, 6, 9, 19; Facings: fire side only; see Note 17.	80 psi	3 hrs.		1		1, 20	3

(Continued)

TABLE 1.1.6—WALLS—MASONRY
12" TO LESS THAN 14" THICK—(Continued)

ITEM CODE	THICKNESS	CONSTRUCTION DETAILS	PERFORMANCE		REFERENCE NUMBER			NOTES	REC. HOURS
			LOAD	TIME	PRE-BMS-92	BMS-92	POST-BMS-92		
W-12-M-60	$12^5/_8$"	Clay or shale structural tile; see Notes 3, 6, 14, 18; Facings: side 1, see Note 17; side 2, none.	80 psi	5 hrs.		1		1, 20	5
W-12-M-61	$12^5/_8$"	Clay or shale structural tile; see Notes 3, 6, 14, 19; Facings: fire side only; see Note 17.	80 psi	3 hrs. 30 min.		1		1, 20	$3^1/_2$
W-12-M-62	$12^5/_8$"	Clay or shale structural tile; see Notes 3, 6, 16, 18; Facings: side 1, see Note 17; side 2, none.	80 psi	6 hrs.		1		1, 20	6
W-12-M-63	$12^5/_8$"	Clay or shale structural tile; see Notes 3, 6, 16, 19; Facings: fire side only; see Note 17.	80 psi	4 hrs.		1		1, 20	4
W-12-M-64	$12^5/_8$"	Core: 8", 40% solid clay or shale structural tile; Facings: 4"brick plus $^5/_8$" of 1:3 sanded gypsum plaster on one side.	80 psi	7 hrs.		1		1, 20	7
W-13-M-65	13"	Core: solid clay or shale brick; $^1/_2$" of 1:3 sanded gypsum plaster facings on both sides.	160 psi	12 hrs.		1		1, 44	12
W-13-M-66	13"	Core: solid clay or shale brick; $^1/_2$" of 1:3 sanded gypsum plaster facings on both sides.	N/A	15 hrs.		1		1, 20	15
W-13-M-67	13"	Core: solid clay or shale brick; $^1/_2$" of 1:3 sanded gypsum plaster facings on both sides.	N/A	15 hrs.		1		1	15
W-13-M-68	13"	Core: hollow Rolok of clay or shale; $^1/_2$" of 1:3 sanded gypsum plaster facings on both sides.	80 psi	7 hrs.		1		1, 20	7
W-13-M-69	13"	Core: concrete brick; $^1/_2$" of 1:3 sanded gypsum plaster facings on both sides.	160 psi	16 hrs.		1		1, 44	16
W-13-M-70	13"	Core: sand-lime brick; $^1/_2$" of 1:3 sanded gypsum plaster facings on both sides.	160 psi	12 hrs.		1		1, 44	12
W-13-M-71	13"	Core: sand-lime brick; $^1/_2$" of 1:3 sanded gypsum plaster facings on both sides.	N/A	17 hrs.		1		1	17
W-13-M-72	13"	Cored clay or shale brick; Units in wall thickness: 1; Cells in wall thickness: 2; Minimum % solids: 70; $^1/_2$" of 1:3 sanded gypsum plaster facings on both sides.	120 psi	7 hrs.		1		1, 45	7
W-13-M-73	13"	Cored clay or shale brick; Units in wall thickness: 3; Cells in wall thickness: 3; Minimum % solids: 87; $^1/_2$" of 1:3 sanded gypsum plaster facings on both sides.	160 psi	12 hrs.		1		1, 44	12
W-13-M-74	13"	Cored clay or shale brick; Units in wall thickness: 3; Cells in wall thickness: 2; Minimum % solids: 87; $^1/_2$" of 1:3 sanded gypsum plaster facings on both sides.	N/A	14 hrs.		1		1	14

(Continued)

TABLE 1.1.6—WALLS—MASONRY
12" TO LESS THAN 14" THICK—(Continued)

ITEM CODE	THICKNESS	CONSTRUCTION DETAILS	PERFORMANCE		REFERENCE NUMBER			NOTES	REC. HOURS
			LOAD	TIME	PRE-BMS-92	BMS-92	POST-BMS-92		
W-13-M-75	13"	Cored concrete masonry; see Notes 18, 23, 28, 39, 41; No facings.	80 psi	7 hrs.		1		1, 20	7
W-13-M-76	13"	Cored concrete masonry; see Notes 19, 23, 28, 39, 41; No facings.	80 psi	4 hrs.		1		1, 20	4
W-13-M-77	13"	Cored concrete masonry; see Notes 3, 18, 27, 31, 43; Facings: both sides; see Note 38.	80 psi	6 hrs.		1		1, 20	6
W-13-M-78	13"	Cored concrete masonry; see Notes 2, 18, 26, 31, 41; Facings: both sides; see Note 38.	80 psi	6 hrs.		1		1, 20	6
W-13-M-79	13"	Cored concrete masonry; see Notes 2, 18, 26, 34, 41; Facings: both sides of wall; see Note 38.	80 psi	7 hrs.		1		1, 20	7
W-13-M-80	13$\frac{1}{4}$"	Core: clay or shale structural tile; see Notes 2, 6, 9, 18; Facings: both sides; see Note 17.	80 psi	4 hrs.		1		1, 20	4
W-13-M-81	13$\frac{1}{4}$"	Core: clay or shale structural tile; see Notes 2, 6, 14, 19; Facings: both sides; see Note 17.	80 psi	4 hrs.		1		1, 20	4
W-13-M-82	13$\frac{1}{4}$"	Core: clay or shale structural tile; see Notes 2, 4, 13, 18; Facings: both sides; see Note 17.	80 psi	6 hrs.		1		1, 20	6
W-13-M-83	13$\frac{1}{4}$"	Core: clay or shale structural tile; see Notes 3, 6, 9, 18; Facings: both sides; see Note 17.	80 psi	6 hrs.		1		1, 20	6
W-13-M-84	13$\frac{1}{4}$"	Core: clay or shale structural tile; see Notes 3, 6, 14, 18; Facings: both sides; see Note 17.	80 psi	6 hrs.		1		1, 20	6
W-13-M-85	13$\frac{1}{4}$"	Core: clay or shale structural tile; see Notes 3, 6, 16, 18; Facings: both sides; see Note 17.	80 psi	7 hrs.		1		1, 20	7
W-13-M-86	13$\frac{1}{2}$"	Cored concrete masonry; see Notes 18, 23, 28, 39, 41; Facings: one side only; see Note 38.	80 psi	8 hrs.		1		1, 20	8

(Continued)

TABLE 1.1.6—WALLS—MASONRY
12" TO LESS THAN 14" THICK—(Continued)

ITEM CODE	THICKNESS	CONSTRUCTION DETAILS	PERFORMANCE		REFERENCE NUMBER			NOTES	REC. HOURS
			LOAD	TIME	PRE-BMS-92	BMS-92	POST-BMS-92		
W-13-M-87	13$^1/_2$"	Cored concrete masonry; see Notes 19, 23, 28, 39, 41; Facings: fire side only; see Note 38.	80 psi	5 hrs.		1		1, 20	5

Notes:
1. Tested at NBS - ASA Spec. No. A2-1934.
2. One unit in wall thickness.
3. Two units in wall thickness.
4. Two or three units in wall thickness.
5. Two cells in wall thickness.
6. Three or four cells in wall thickness.
7. Four or five cells in wall thickness.
8. Five or six cells in wall thickness.
9. Minimum percent of solid materials in units = 40%.
10. Minimum percent of solid materials in units = 43%.
11. Minimum percent of solid materials in units = 46%.
12. Minimum percent of solid materials in units = 48%.
13. Minimum percent of solid materials in units = 49%.
14. Minimum percent of solid materials in units = 45%.
15. Minimum percent of solid materials in units = 51%.
16. Minimum percent of solid materials in units = 53%.
17. Not less than $^5/_8$ inch thickness of 1:3 sanded gypsum plaster.
18. Noncombustible or no members framed into wall.
19. Combustible members framed into wall.
20. Load: 80 psi for gross area.
21. Portland cement-lime mortar.
22. Failure mode - thermal.
23. British test.
24. Passed all criteria.
25. Failed by sudden collapse with no preceding signs of impending failure.
26. One cell in wall thickness.
27. Two cells in wall thickness.
28. Three cells in wall thickness.
29. Minimum percent of solid material in concrete units = 52%.
30. Minimum percent of solid material in concrete units = 54%.
31. Minimum percent of solid material in concrete units = 55%.
32. Minimum percent of solid material in concrete units = 57%.
33. Minimum percent of solid material in concrete units = 60%.
34. Minimum percent of solid material in concrete units = 62%.
35. Minimum percent of solid material in concrete units = 65%.
36. Minimum percent of solid material in concrete units = 70%.
37. Minimum percent of solid material in concrete units = 76%.
38. Not less than $^1/_2$" of 1:3 sanded gypsum plaster.
39. Three units in wall thickness.
40. Concrete units made with expanded slag or pumice aggregates.
41. Concrete units made with expanded burned clay or shale, crushed limestone, air cooled slag or cinders.
42. Concrete units made with calcareous sand and gravel. Coarse aggregate, 60 percent or more calcite and dolomite.
43. Concrete units made with siliceous sand and gravel. Ninety percent or more quartz, chert or flint.
44. Load: 160 psi of gross wall cross sectional area.
45. Load: 120 psi of gross wall cross sectional area.

FIGURE 1.1.7—WALLS—MASONRY
14" OR MORE THICK

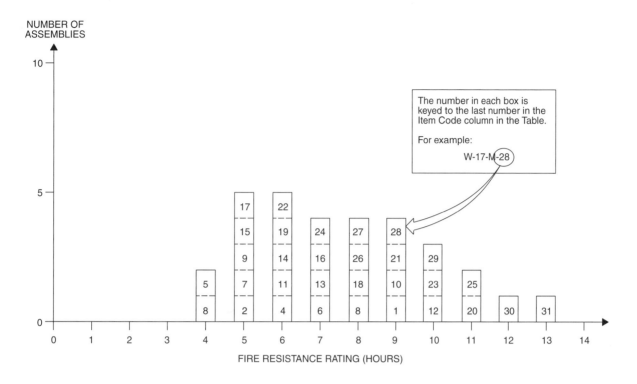

TABLE 1.1.7—WALLS—MASONRY
14" OR MORE THICK

ITEM CODE	THICKNESS	CONSTRUCTION DETAILS	PERFORMANCE		REFERENCE NUMBER			NOTES	REC. HOURS
			LOAD	TIME	PRE-BMS-92	BMS-92	POST-BMS-92		
W-14-M-1	14"	Core: cored masonry; see Notes 18, 28, 33, 39, 41; Facings: both sides; see Note 38.	80 psi	9 hrs.		1		1, 20	9
W-16-M-2	16"	Core: clay or shale structural tile; see Notes 4, 7, 9, 19; No facings.	80 psi	5 hrs.		1		1, 20	5
W-16-M-3	16"	Core: clay or shale structural tile; see Notes 4, 7, 9, 19; No facings.	80 psi	4 hrs.		1		1, 20	4
W-16-M-4	16"	Core: clay or shale structural tile; see Notes 4, 7, 10, 18; No facings.	80 psi	6 hrs.		1		1, 20	6
W-16-M-5	16"	Core: clay or shale structural tile; see Notes 4, 7, 10, 19; No facings.	80 psi	4 hrs.		1		1, 20	4
W-16-M-6	16"	Core: clay or shale structural tile; see Notes 4, 7, 11, 18; No facings.	80 psi	7 hrs.		1		1, 20	7
W-16-M-7	16"	Core: clay or shale structural tile; see Notes 4, 7, 11, 19; No facings.	80 psi	5 hrs.		1		1, 20	5
W-16-M-8	16"	Core: clay or shale structural tile; see Notes 4, 8, 13, 18; No facings.	80 psi	8 hrs.		1		1, 20	8
W-16-M-9	16"	Core: clay or shale structural tile; see Notes 4, 8, 13, 19; No facings.	80 psi	5 hrs.		1		1, 20	5
W-16-M-10	16"	Core: clay or shale structural tile; see Notes 4, 8, 15, 18; No facings.	80 psi	9 hrs.		1		1, 20	9
W-16-M-11	16"	Core: clay or shale structural tile; see Notes 3, 7, 14, 18; No facings.	80 psi	6 hrs.		1		1, 20	6

(Continued)

TABLE 1.1.7—WALLS—MASONRY
14" OR MORE THICK—(Continued)

ITEM CODE	THICKNESS	CONSTRUCTION DETAILS	PERFORMANCE		REFERENCE NUMBER			NOTES	REC. HOURS
			LOAD	TIME	PRE-BMS-92	BMS-92	POST-BMS-92		
W-16-M-12	16"	Core: clay or shale structural tile; see Notes 4, 8, 16, 18; No facings.	80 psi	10 hrs.		1		1, 20	10
W-16-M-13	16"	Core: clay or shale structural tile; see Notes 4, 6, 16, 19; No facings.	80 psi	7 hrs.		1		1, 20	7
W-16-M-14	$16^5/_8$"	Core: clay or shale structural tile; see Notes 4, 7, 9, 18; Facings: side 1, see Note 17; side 2, none.	80 psi	6 hrs.		1		1, 20	6
W-16-M-15	$16^5/_8$"	Core: clay or shale structural tile; see Notes 4, 7, 9, 19; Facings: fire side only; see Note 17.	80 psi	5 hrs.		1		1, 20	5
W-16-M-16	$16^5/_8$"	Core: clay or shale structural tile; see Notes 4, 7, 10, 18; Facings: side 1, see Note 17; side 2, none.	80 psi	7 hrs.		1		1, 20	7
W-16-M-17	$16^5/_8$"	Core: clay or shale structural tile; see Notes 4, 7, 10, 19; Facings: fire side only; see Note 17.	80 psi	5 hrs.		1		1, 20	5
W-16-M-18	$16^5/_8$"	Core: clay or shale structural tile; see Notes 4, 7, 11, 18; Facings: side 1, see Note 17; side 2, none.	80 psi	5 hrs.		1		1, 20	8
W-16-M-19	$16^5/_8$"	Core: clay or shale structural tile; see Notes 4, 7, 11, 19; Facings: fire side only; see Note 17.	80 psi	6 hrs.		1		1, 20	6
W-16-M-20	$16^5/_8$"	Core: clay or shale structural tile; see Notes 4, 8, 13, 18; Facings: sides 1 and 2; see Note 17.	80 psi	11 hrs.		1		1, 20	11
W-16-M-21	$16^5/_8$"	Core: clay or shale structural tile; see Notes 4, 8, 13 18; Facings: side 1, see Note 17; side 2, none.	80 psi	9 hrs.		1		1, 20	9
W-16-M-22	$16^5/_8$"	Core: clay or shale structural tile; see Notes 4, 8, 13, 19; Facings: fire side only; see Note 17.	80 psi	6 hrs.		1		1, 20	6
W-16-M-23	$16^5/_8$"	Core: clay or shale structural tile; see Notes 4, 8, 15, 18; Facings: side 1, see Note 17; side 2, none.	80 psi	10 hrs.		1		1, 20	10
W-16-M-24	$16^5/_8$"	Core: clay or shale structural tile; see Notes 4, 8, 15, 19; Facings: fire side only; see Note 17.	80 psi	7 hrs.		1		1, 20	7
W-16-M-25	$16^5/_8$"	Core: clay or shale structural tile; see Notes 4, 6, 16, 18; Facings: side 1, see Note 17; side 2, none.	80 psi	11 hrs.		1		1, 20	11
W-16-M-26	$16^5/_8$"	Core: clay or shale structural tile; see Notes 4, 6, 16, 19; Facings: fire side only; see Note 17.	80 psi	8 hrs.		1		1, 20	8
W-17-M-27	$17^1/_4$"	Core: clay or shale structural tile; see Notes 4, 7, 9, 18; Facings: sides 1 and 2; see Note 17.	80 psi	8 hrs.		1		1, 20	8
W-17-M-28	$17^1/_4$"	Core: clay or shale structural tile; see Notes 4, 7, 10, 18; Facings: sides 1 and 2; see Note 17.	80 psi	9 hrs.		1		1, 20	9

(Continued)

TABLE 1.1.7—MASONRY WALLS
14" OR MORE THICK—(Continued)

ITEM CODE	THICKNESS	CONSTRUCTION DETAILS	PERFORMANCE		REFERENCE NUMBER			NOTES	REC. HOURS
			LOAD	TIME	PRE-BMS-92	BMS-92	POST-BMS-92		
W-17-M-29	17¹/₄"	Core: clay or shale structural tile; see Notes 4, 7, 11, 18; Facings: sides 1 and 2; see Note 17.	80 psi	10 hrs.		1		1, 20	10
W-17-M-30	17¹/₄"	Core: clay or shale structural tile; see Notes 4, 8, 15, 18; Facings: sides 1 and 2; see Note 17.	80 psi	12 hrs.		1		1, 20	12
W-17-M-31	17¹/₄"	Core: clay or shale structural tile; see Notes 4, 6, 16, 18; Facings: sides 1 and 2; see Note 17.	80 psi	13 hrs.		1		1, 20	13

Notes:
1. Tested at NBS - ASA Spec. No. A2-1934.
2. One unit in wall thickness.
3. Two units in wall thickness.
4. Two or three units in wall thickness.
5. Two cells in wall thickness.
6. Three or four cells in wall thickness.
7. Four or five cells in wall thickness.
8. Five or six cells in wall thickness.
9. Minimum percent of solid materials in units = 40%.
10. Minimum percent of solid materials in units = 43%.
11. Minimum percent of solid materials in units = 46%.
12. Minimum percent of solid materials in units = 48%.
13. Minimum percent of solid materials in units = 49%.
14. Minimum percent of solid materials in units = 45%.
15. Minimum percent of solid materials in units = 51%.
16. Minimum percent of solid materials in units = 53%.
17. Not less than ⁵/₈ inch thickness of 1:3 sanded gypsum plaster.
18. Noncombustible or no members framed into wall.
19. Combustible members framed into wall.
20. Load: 80 psi for gross area.
21. Portland cement-lime mortar.
22. Failure mode - thermal.
23. British test.
24. Passed all criteria.
25. Failed by sudden collapse with no preceding signs of impending failure.
26. One cell in wall thickness.
27. Two cells in wall thickness.
28. Three cells in wall thickness.
29. Minimum percent of solid material in concrete units = 52%.
30. Minimum percent of solid material in concrete units = 54%.
31. Minimum percent of solid material in concrete units = 55%.
32. Minimum percent of solid material in concrete units = 57%.
33. Minimum percent of solid material in concrete units = 60%.
34. Minimum percent of solid material in concrete units = 62%.
35. Minimum percent of solid material in concrete units = 65%.
36. Minimum percent of solid material in concrete units = 70%.
37. Minimum percent of solid material in concrete units = 76%.
38. Not less than ¹/₂" of 1:3 sanded gypsum plaster.
39. Three units in wall thickness.
40. Concrete units made with expanded slag or pumice aggregates.
41. Concrete units made with expanded burned clay or shale, crushed limestone, air cooled slag or cinders.
42. Concrete units made with calcareous sand and gravel. Coarse aggregate, 60 percent or more calcite and dolomite.
43. Concrete units made with siliceous sand and gravel. Ninety percent or more quartz, chert or flint.

FIGURE 1.2.1—WALLS—METAL FRAME
0" TO LESS THAN 4" THICK

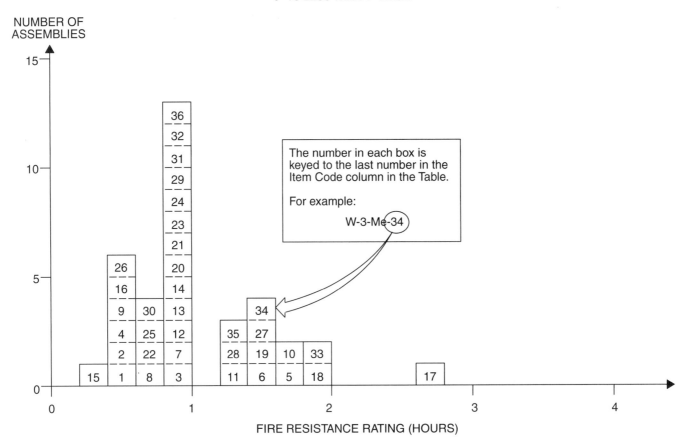

TABLE 1.2.1—WALLS—METAL FRAME
0" TO LESS THAN 4" THICK

ITEM CODE	THICKNESS	CONSTRUCTION DETAILS	PERFORMANCE		REFERENCE NUMBER			NOTES	REC. HOURS
			LOAD	TIME	PRE-BMS-92	BMS-92	POST-BMS-92		
W-3-Me-1	3"	Core: steel channels having three rows of 4" × $^1/_8$" staggered slots in web; core filled with heat expanded vermiculite weighing 1.5 lbs./ft.^{2} of wall area; Facings: sides 1 and 2, 18 gage steel, spot welded to core.	N/A	25 min.		1			$^1/_3$
W-3-Me-2	3"	Core: steel channels having three rows of 4" × $^1/_8$" staggered slots in web; core filled with heat expanded vermiculite weighing 2 lbs./ft.^{2} of wall area; Facings: sides 1 and 2, 18 gage steel, spot welded to core.	N/A	30 min.		1			$^1/_2$
W-3-Me-3	$2^1/_2$"	Solid partition: $^3/_8$" tension rods (vertical) 3' o.c. with metal lath; Scratch coat: cement/sand/lime plaster; Float coats: cement/sand/lime plaster; Finish coats: neat gypsum plaster.	N/A	1 hr.			7	1	1
W-2-Me-4	2"	Solid wall: steel channel per Note 1; 2" thickness of 1:2; 1:3 portland cement on metal lath.	N/A	30 min.		1			$^1/_2$

(Continued)

TABLE 1.2.1—WALLS—METAL FRAME
0" TO LESS THAN 4" THICK

ITEM CODE	THICKNESS	CONSTRUCTION DETAILS	PERFORMANCE		REFERENCE NUMBER			NOTES	REC. HOURS
			LOAD	TIME	PRE-BMS-92	BMS-92	POST-BMS-92		
W-2-Me-5	2"	Solid wall: steel channel per Note 1; 2" thickness of neat gypsum plaster on metal lath.	N/A	1 hr. 45 min.		1			$1^3/_4$
W-2-Me-6	2"	Solid wall: steel channel per Note 1; 2" thickness of 1:1$^1/_2$; 1:1$^1/_2$ gypsum plaster on metal lath.	N/A	1 hr. 30 min.		1			$1^1/_2$
W-2-Me-7	2"	Solid wall: steel channel per Note 2; 2" thickness of 1:1; 1:1 gypsum plaster on metal lath.	N/A	1 hr.		1			1
W-2-Me-8	2"	Solid wall: steel channel per Note 1; 2" thickness of 1:2; 1:2 gypsum plaster on metal lath.	N/A	45 min.		1			$^3/_4$
W-2-Me-9	2$^1/_4$"	Solid wall: steel channel per Note 2; 2$^1/_4$" thickness of 1:2; 1:3 portland cement on metal lath.	N/A	30 min.		1			$^1/_2$
W-2-Me-10	2$^1/_4$"	Solid wall: steel channel per Note 2; 2$^1/_4$" thickness of neat gypsum plaster on metal lath.	N/A	2 hrs.		1			2
W-2-Me-11	2$^1/_4$"	Solid wall: steel channel per Note 2; 2$^1/_4$" thickness of 1:$^1/_2$; 1:$^1/_2$ gypsum plaster on metal lath.	N/A	1 hr. 45 min.		1			$1^3/_4$
W-2-Me-12	2$^1/_4$"	Solid wall: steel channel per Note 2; 2$^1/_4$" thickness of 1:1; 1:1 gypsum plaster on metal lath.	N/A	1 hr. 15 min.		1			$1^1/_4$
W-2-Me-13	2$^1/_4$"	Solid wall: steel channel per Note 2; 2$^1/_4$" thickness of 1:2; 1:2 gypsum plaster on metal lath.	N/A	1 hr.		1			1
W-2-Me-14	2$^1/_2$"	Solid wall: steel channel per Note 1; 2$^1/_2$" thickness of 4.5:1:7; 4.5:1:7 portland cement, sawdust and sand sprayed on wire mesh; see Note 3.	N/A	1 hr.		1			1
W-2-Me-15	2$^1/_2$"	Solid wall: steel channel per Note 2; 2$^1/_2$" thickness of 1:4; 1:4 portland cement sprayed on wire mesh; see Note 3.	N/A	20 min.		1			$^1/_3$
W-2-Me-16	2$^1/_2$"	Solid wall: steel channel per Note 2; 2$^1/_2$" thickness of 1:2; 1:3 portland cement on metal lath.	N/A	30 min.		1			$^1/_2$
W-2-Me-17	2$^1/_2$"	Solid wall: steel channel per Note 2; 2$^1/_2$" thickness of neat gypsum plaster on metal lath.	N/A	2 hrs. 30 min.		1			$2^1/_2$
W-2-Me-18	2$^1/_2$"	Solid wall: steel channel per Note 2; 2$^1/_2$" thickness of 1:$^1/_2$; 1:$^1/_2$ gypsum plaster on metal lath.	N/A	2 hrs.		1			2
W-2-Me-19	2$^1/_2$"	Solid wall: steel channel per Note 2; 2$^1/_2$" thickness of 1:1; 1:1 gypsum plaster on metal lath.	N/A	1 hr. 30 min.		1			$1^1/_2$
W-2-Me-20	2$^1/_2$"	Solid wall: steel channel per Note 2; 2$^1/_2$" thickness of 1:2; 1:2 gypsum plaster on metal lath.	N/A	1 hr.		1			1

(Continued)

TABLE 1.2.1—WALLS—METAL FRAME
0" TO LESS THAN 4" THICK—(Continued)

ITEM CODE	THICKNESS	CONSTRUCTION DETAILS	PERFORMANCE		REFERENCE NUMBER			NOTES	REC. HOURS
			LOAD	TIME	PRE-BMS-92	BMS-92	POST-BMS-92		
W-2-Me-21	$2^1/_2$"	Solid wall: steel channel per Note 2; $2^1/_2$" thickness of 1:2; 1:3 gypsum plaster on metal lath.	N/A	1 hr.		1			1
W-3-Me-22	3"	Core: steel channel per Note 2; 1:2; 1:2 gypsum plaster on $3/_4$" soft asbestos lath; plaster thickness 2".	N/A	45 min.		1			$3/_4$
W-3-Me-23	$3^1/_2$"	Solid wall: steel channel per Note 2; $2^1/_2$" thickness of 1:2; 1:2 gypsum plaster on $3/_4$" asbestos lath.	N/A	1 hr.		1			1
W-3-Me-24	$3^1/_2$"	Solid wall: steel channel per Note 2; lath over and $1:2^1/_2$; $1:2^1/_2$ gypsum plaster on 1" magnesium oxysulfate wood fiberboard; plaster thickness $2^1/_2$".	N/A	1 hr.		1			1
W-3-Me-25	$3^1/_2$"	Core: steel studs; see Note 4; Facings: $3/_4$" thickness of $1:^1/_{30}:2$; $1:^1/_{30}:3$ portland cement and asbestos fiber plaster.	N/A	45 min.		1			$3/_4$
W-3-Me-26	$3^1/_2$"	Core: steel studs; see Note 4; Facings: both sides $3/_4$" thickness of 1:2; 1:3 portland cement.	N/A	30 min.		1			$1/_2$
W-3-Me-27	$3^1/_2$"	Core: steel studs; see Note 4; Facings: both sides $3/_4$" thickness of neat gypsum plaster.	N/A	1 hr. 30 min.		1			$1^1/_2$
W-3-Me-28	$3^1/_2$"	Core: steel studs; see Note 4; Facings: both sides $3/_4$" thickness of $1:^1/_2$; $1:^1/_2$ gypsum plaster.	N/A	1 hr. 15 min.		1			$1^1/_4$
W-3-Me-29	$3^1/_2$"	Core: steel studs; see Note 4; Facings: both sides $3/_4$" thickness of 1:2; 1:2 gypsum plaster.	N/A	1 hr.		1			1
W-3-Me-30	$3^1/_2$"	Core: steel studs; see Note 4; Facings: both sides $3/_4$" thickness of 1:2; 1:3 gypsum plaster.	N/A	45 min.		1			$3/_4$
W-3-Me-31	$3^3/_4$"	Core: steel studs; see Note 4; Facings: both sides $7/_8$" thickness of $1:^1/_{30}:2$; $1:^1/_{30}:3$ portland cement and asbestos fiber plaster.	N/A	1 hr.		1			1
W-3-Me-32	$3^3/_4$"	Core: steel studs; see Note 4; Facings: both sides $7/_8$" thickness of 1:2; 1:3 portland cement.	N/A	45 min.		1			$3/_4$
W-3-Me-33	$3^3/_4$"	Core: steel studs; see Note 4; Facings: both sides $7/_8$" thickness of neat gypsum plaster.	N/A	2 hrs.		1			2
W-3-Me-34	$3^3/_4$"	Core: steel studs; see Note 4; Facings: both sides $7/_8$" thickness of $1:^1/_2$; $1:^1/_2$ gypsum plaster.	N/A	1 hr. 30 min.		1			$1^1/_2$
W-3-Me-35	$3^3/_4$"	Core: steel studs; see Note 4; Facings: both sides $7/_8$" thickness of 1:2; 1:2 gypsum plaster.	N/A	1 hr. 15 min.		1			$1^1/_4$

(Continued)

TABLE 1.2.1—WALLS—METAL FRAME
0" TO LESS THAN 4" THICK—(Continued)

ITEM CODE	THICKNESS	CONSTRUCTION DETAILS	PERFORMANCE		REFERENCE NUMBER			NOTES	REC. HOURS
			LOAD	TIME	PRE-BMS-92	BMS-92	POST-BMS-92		
W-3-Me-36	$3^3/_4$"	Core: steel; see Note 4; Facings: $^7/_8$" thickness of 1:2; 1:3 gypsum plaster on both sides.	N/A	1 hr.		1			1

Notes:
1. Failure mode - local temperature rise - back face.
2. Three-fourths inch or 1 inch channel framing - hot-rolled or strip-steel channels.
3. Reinforcement is 4-inch square mesh of No. 6 wire welded at intersections (no channels).
4. Ratings are for any usual type of nonload-bearing metal framing providing 2 inches (or more) air space.

General Note:
The construction details of the wall assemblies are as complete as the source documentation will permit. Data on the method of attachment of facings and the gauge of steel studs was provided when known. The cross-sectional area of the steel stud can be computed, thereby permitting a reasoned estimate of actual loading conditions. For load-bearing assemblies, the maximum allowable stress for the steel studs has been provided in the table "Notes." More often, it is the thermal properties of the facing materials, rather than the specific gauge of the steel, that will determine the degree of fire resistance. This is particularly true for non-bearing wall assemblies.

FIGURE 1.2.2—WALLS—METAL FRAME
4" TO LESS THAN 6" THICK

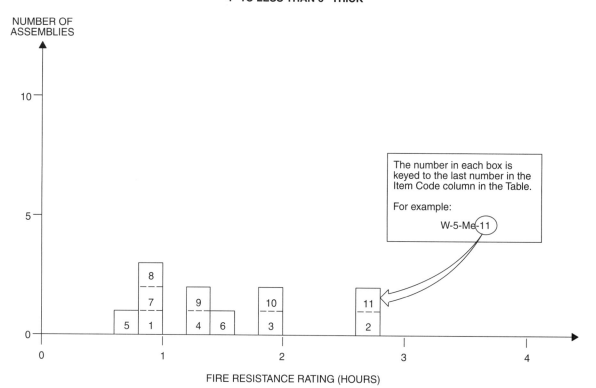

FIRE RESISTANCE RATING (HOURS)

TABLE 1.2.2—WALLS—METAL FRAME
4" TO LESS THAN 6" THICK

| ITEM CODE | THICKNESS | CONSTRUCTION DETAILS | PERFORMANCE | | REFERENCE NUMBER | | | NOTES | REC. HOURS |
			LOAD	TIME	PRE-BMS-92	BMS-92	POST-BMS-92		
W-5-Me-1	$5^1/_2$"	3" cavity with 16 ga. channel studs ($3^1/_2$" o.c.) of $^1/_2$" × $^1/_2$" channel and 3" spacer; Metal lath on ribs with plaster (three coats) $^3/_4$" over face of lath; Plaster (each side): scratch coat, cement/lime/sand with hair; float coat, cement/lime/sand; finish coat, neat gypsum.	N/A	1 hr. 11 min.			7	1	1
W-4-Me-2	4"	Core: steel studs; see Note 2; Facings: both sides 1" thickness of neat gypsum plaster.	N/A	2 hrs. 30 min.		1			$2^1/_2$
W-4-Me-3	4"	Core: steel studs; see Note 2; Facings: both sides 1" thickness of 1:$^1/_2$; 1:$^1/_2$ gypsum plaster.	N/A	2 hrs.		1			2
W-4-Me-4	4"	Core: steel; see Note 2; Facings: both sides 1" thickness of 1:2; 1:3 gypsum plaster.	N/A	1 hr. 15 min.		1			$1^1/_4$
W-4-Me-5	$4^1/_2$"	Core: lightweight steel studs 3" in depth; Facings: both sides $^3/_4$" thick sanded gypsum plaster, 1:2 scratch coat, 1:3 brown coat applied on metal lath.	See Note 4	45 min.		1		5	$^3/_4$
W-4-Me-6	$4^1/_2$"	Core: lightweight steel studs 3" in depth; Facings: both sides $^3/_4$" thick neat gypsum plaster on metal lath.	See Note 4	1 hr. 30 min.		1		5	$1^1/_2$

(Continued)

TABLE 1.2.2—WALLS—METAL FRAME
4" TO LESS THAN 6" THICK—(Continued)

ITEM CODE	THICKNESS	CONSTRUCTION DETAILS	PERFORMANCE		REFERENCE NUMBER			NOTES	REC. HOURS
			LOAD	TIME	PRE-BMS-92	BMS-92	POST-BMS-92		
W-4-Me-7	$4^1/_2$"	Core: lightweight steel studs 3" in depth; Facings: both sides $^3/_4$" thick sanded gypsum plaster, 1:2 scratch and brown coats applied on metal lath.	See Note 4	1 hr.		1		5	1
W-4-Me-8	$4^3/_4$"	Core: lightweight steel studs 3" in depth; Facings: both sides $^7/_8$" thick sanded gypsum plaster, 1:2 scratch coat, 1:3 brown coat, applied on metal lath.	See Note 4	1 hr.		1		5	1
W-4-Me-9	$4^3/_4$"	Core: lightweight steel studs 3" in depth; Facings: both sides $^7/_8$" thick sanded gypsum plaster, 1:2 scratch and 1:3 brown coats applied on metal lath.	See Note 4	1 hr. 15 min.		1		5	$1^1/_4$
W-5-Me-10	5"	Core: lightweight steel studs 3" in depth; Facings: both sides 1" thick neat gypsum plaster on metal lath.	See Note 4	2 hrs.		1		5	2
W-5-Me-11	5"	Core: lightweight steel studs 3" in depth; Facings: both sides 1" thick neat gypsum plaster on metal lath.	See Note 4	2 hrs. 30 min.		1		5, 6	$2^1/_2$

Notes:
1. Failure mode - local back face temperature rise.
2. Ratings are for any usual type of non-bearing metal framing providing a minimum 2 inches air space.
3. Facing materials secured to lightweight steel studs not less than 3 inches deep.
4. Rating based on loading to develop a maximum stress of 7270 psi for net area of each stud.
5. Spacing of steel studs must be sufficient to develop adequate rigidity in the metal-lath or gypsum-plaster base.
6. As per Note 4 but load/stud not to exceed 5120 psi.

General Note:
The construction details of the wall assemblies are as complete as the source documentation will permit. Data on the method of attachment of facings and the gauge of steel studs was provided when known. The cross sectional area of the steel stud can be computed, thereby permitting a reasoned estimate of actual loading conditions. For load-bearing assemblies, the maximum allowable stress for the steel studs has been provided in the table "Notes." More often, it is the thermal properties of the facing materials, rather than the specific gauge of the steel, that will determine the degree of fire resistance. This is particularly true for non-bearing wall assemblies.

**TABLE 1.2.3—WALLS—METAL FRAME
6" TO LESS THAN 8" THICK**

ITEM CODE	THICKNESS	CONSTRUCTION DETAILS	PERFORMANCE		REFERENCE NUMBER			NOTES	REC. HOURS
			LOAD	TIME	PRE-BMS-92	BMS-92	POST-BMS-92		
W-6-Me-1	$6^5/_8$"	On one side of 1" magnesium oxysulfate wood fiberboard sheathing attached to steel studs (see Notes 1 and 2), 1" air space, $3^3/_4$" brick secured with metal ties to steel frame every fifth course; Inside facing of $^7/_8$" 1:2 sanded gypsum plaster on metal lath secured directly to studs; Plaster side exposed to fire.	See Note 2	1 hr. 45 min.		1		1	$1^3/_4$
W-6-Me-2	$6^5/_8$"	On one side of 1" magnesium oxysulfate wood fiberboard sheathing attached to steel studs (see Notes 1 and 2), 1" air space, $3^3/_4$" brick secured with metal ties to steel frame every fifth course; Inside facing of $^7/_8$" 1:2 sanded gypsum plaster on metal lath secured directly to studs; Brick face exposed to fire.	See Note 2	4 hrs.		1		1	4
W-6-Me-3	$6^5/_8$"	On one side of 1" magnesium oxysulfate wood fiberboard sheathing attached to steel studs (see Notes 1 and 2), 1" air space, $3^3/_4$" brick secured with metal ties to steel frame every fifth course; Inside facing of $^7/_8$" vermiculite plaster on metal lath secured directly to studs; Plaster side exposed to fire.	See Note 2	2 hrs.		1		1	2

Notes:

1. Lightweight steel studs (minimum 3 inches deep) used. Stud spacing dependent on loading, but in each case, spacing is to be such that adequate rigidity is provided to the metal lath plaster base.
2. Load is such that stress developed in studs is not greater than 5120 psi calculated from net stud area.

General Note:

The construction details of the wall assemblies are as complete as the source documentation will permit. Data on the method of attachment of facings and the gauge of steel studs was provided when known. The cross sectional area of the steel stud can be computed, thereby permitting a reasoned estimate of actual loading conditions. For load-bearing assemblies, the maximum allowable stress for the steel studs has been provided in the table "Notes." More often, it is the thermal properties of the facing materials, rather than the specific gauge of the steel, that will determine the degree of fire resistance. This is particularly true for non-bearing wall assemblies.

**TABLE 1.2.4—WALLS—METAL FRAME
8" TO LESS THAN 10" THICK**

ITEM CODE	THICKNESS	CONSTRUCTION DETAILS	PERFORMANCE		REFERENCE NUMBER			NOTES	REC. HOURS
			LOAD	TIME	PRE-BMS-92	BMS-92	POST-BMS-92		
W-9-Me-1	$9^{1}/_{16}$"	On one side of $^{1}/_{2}$" wood fiberboard sheathing next to studs, $^{3}/_{4}$" air space formed with $^{3}/_{4}$" × $1^{5}/_{8}$" wood strips placed over the fiberboard and secured to the studs, paper backed wire lath nailed to strips $3^{3}/_{4}$" brick veneer held in place by filling a $^{3}/_{4}$" space between the brick and paper backed lath with mortar; Inside facing of $^{3}/_{4}$" neat gypsum plaster on metal lath attached to $^{5}/_{16}$" plywood strips secured to edges of steel studs; Rated as combustible because of the sheathing; See Notes 1 and 2; Plaster exposed.	See Note 2	1 hr. 45 min.		1		1	$1^{3}/_{4}$
W-9-Me-2	$9^{1}/_{16}$"	Same as above with brick exposed.	See Note 2	4 hrs.		1		1	4
W-8-Me-3	$8^{1}/_{2}$"	On one side of paper backed wire lath attached to studs and $3^{3}/_{4}$" brick veneer held in place by filling a 1" space between the brick and lath with mortar; Inside facing of 1" paper-enclosed mineral wool blanket weighing .6 lb./ft.^{2} attached to studs, metal lath or paper backed wire lath laid over the blanket and attached to the studs, $^{3}/_{4}$" sanded gypsum plaster 1:2 for the scratch coat and 1:3 for the brown coat; See Notes 1 and 2; Plaster face exposed.	See Note 2	4 hrs.		1		1	4
W-8-Me-4	$8^{1}/_{2}$"	Same as above with brick exposed.	See Note 2	5 hrs.		1		1	5

Notes:
1. Lightweight steel studs ≥ 3 inches in depth. Stud spacing dependent on loading, but in any case, the spacing is to be such that adequate rigidity is provided to the metal-lath plaster base.
2. Load is such that stress developed in studs is ≤ 5120 psi calculated from the net area of the stud.

General Note:
The construction details of the wall assemblies are as complete as the source documentation will permit. Data on the method of attachment of facings and the gauge of steel studs was provided when known. The cross sectional area of the steel stud can be computed, thereby permitting a reasoned estimate of actual loading conditions. For load-bearing assemblies, the maximum allowable stress for the steel studs has been provided in the table "Notes." More often, it is the thermal properties of the facing materials, rather than the specific gauge of the steel, that will determine the degree of fire resistance. This is particularly true for non-bearing wall assemblies.

TABLE 1.3.1—WOOD FRAME WALLS
0" TO LESS THAN 4" THICK

ITEM CODE	THICKNESS	CONSTRUCTION DETAILS	PERFORMANCE		REFERENCE NUMBER			NOTES	REC. HOURS
			LOAD	TIME	PRE-BMS-92	BMS-92	POST-BMS-92		
W-3-W-1	$3^3/_4$"	Solid wall: $2^1/_4$" wood-wool slab core; $^3/_4$" gypsum plaster each side.	N/A	2 hrs.			7	1, 6	2
W-3-W-2	$3^7/_8$"	2 × 4 stud wall; $^3/_{16}$" thick cement asbestos board on both sides of wall.	360 psi net area	10 min.		1		2-5	$^1/_6$
W-3-W-3	$3^7/_8$"	Same as W-3-W-2 but stud cavities filled with 1 lb./ft.^{2} mineral wool batts.	360 psi net area	40 min.		1		2-5	$^2/_3$

Notes:
1. Achieved "Grade C" fire resistance (British).
2. Nominal 2 × 4 wood studs of No. 1 common or better lumber set edgewise, 2 × 4 plates at top and bottom and blocking at mid height of wall.
3. All horizontal joints in facing material backed by 2 × 4 blocking in wall.
4. Load: 360 psi of net stud cross sectional area.
5. Facings secured with 6d casing nails. Nail holes predrilled and 0.02 inch to 0.03 inch smaller than nail diameter.
6. The wood-wool core is a pressed excelsior slab which possesses insulating properties similar to cellulosic insulation.

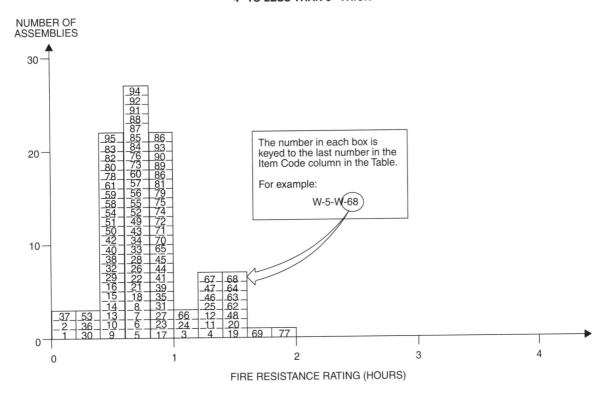

FIGURE 1.3.2—WOOD FRAME WALLS
4" TO LESS THAN 6" THICK

TABLE 1.3.2—WOOD FRAME WALLS
4" TO LESS THAN 6" THICK

ITEM CODE	THICKNESS	CONSTRUCTION DETAILS	PERFORMANCE		REFERENCE NUMBER			NOTES	REC. HOURS
			LOAD	TIME	PRE-BMS-92	BMS-92	POST-BMS-92		
W-4-W-1	4"	2" × 4" stud wall; $^{3}/_{16}$" CAB; no insulation; Design A.	35 min.	10 min.			4	1-10	$^{1}/_{6}$
W-4-W-2	$4^{1}/_{8}$"	2" × 4" stud wall; $^{3}/_{16}$" CAB; no insulation; Design A.	38 min.	9 min.			4	1-10	$^{1}/_{6}$
W-4-W-3	$4^{3}/_{4}$"	2" × 4" stud wall; $^{3}/_{16}$" CAB and $^{3}/_{8}$" gypsum board face (both sides); Design B.	62 min.	64 min.			4	1-10	1
W-5-W-4	5"	2" × 4" stud wall; $^{3}/_{16}$" CAB and $^{1}/_{2}$" gypsum board (both sides); Design B.	79 min.	Greater than 90 min.			4	1-10	1
W-4-W-5	$4^{3}/_{4}$"	2" × 4" stud wall; $^{3}/_{16}$" CAB and $^{3}/_{8}$" gypsum board (both sides); Design B.	45 min.	45 min.			4	1-12	—
W-5-W-6	5"	2" × 4" stud wall; $^{3}/_{16}$" CAB and $^{1}/_{2}$" gypsum board face (both sides); Design B.	45 min.	45 min.			4	1-10, 12, 13	—
W-4-W-7	4"	2" × 4" stud wall; $^{3}/_{16}$" CAB face; $3^{1}/_{2}$" mineral wool insulation; Design C.	40 min.	42 min.			4	1-10	$^{2}/_{3}$
W-4-W-8	4"	2" × 4" stud wall; $^{3}/_{16}$" CAB face; $3^{1}/_{2}$" mineral wool insulation; Design C.	46 min.	46 min.			4	1-10, 43	$^{2}/_{3}$
W-4-W-9	4"	2" × 4" stud wall; $^{3}/_{16}$" CAB face; $3^{1}/_{2}$" mineral wool insulation; Design C.	30 min.	30 min.			4	1-10, 12, 14	—

(Continued)

TABLE 1.3.2—WOOD FRAME WALLS
4" TO LESS THAN 6" THICK—(Continued)

ITEM CODE	THICKNESS	CONSTRUCTION DETAILS	PERFORMANCE		REFERENCE NUMBER			NOTES	REC. HOURS
			LOAD	TIME	PRE-BMS-92	BMS-92	POST-BMS-92		
W-4-W-10	$4^1/_8$"	2" × 4" stud wall; $^3/_{16}$" CAB face; $3^1/_2$" mineral wool insulation; Design C.	—	30 min.			4	1-8, 12, 14	—
W-4-W-11	$4^3/_4$"	2" × 4" stud wall; $^3/_{16}$" CAB face; $^3/_8$" gypsum strips over studs; $5^1/_2$" mineral wool insulation; Design D.	79 min.	79 min.			4	1-10	1
W-4-W-12	$4^3/_4$"	2" × 4" stud wall; $^3/_{16}$" CAB face; $^3/_8$" gypsum strips at stud edges; $7^1/_2$" mineral wool insulation; Design D.	82 min.	82 min.			4	1-10	1
W-4-W-13	$4^3/_4$"	2" × 4" stud wall; $^3/_{16}$" CAB face; $^3/_8$" gypsum board strips over studs; $5^1/_2$" mineral wool insulation; Design D.	30 min.	30 min.			4	1-12	—
W-4-W-14	$4^3/_4$"	2" × 4" stud wall; $^3/_{16}$" CAB face; $^3/_8$" gypsum board strips over studs; 7" mineral wool insulation; Design D.	30 min.	30 min.			4	1-12	—
W-5-W-15	$5^1/_2$"	2" × 4" stud wall; Exposed face: CAB shingles over 1" × 6"; Unexposed face: $^1/_8$" CAB sheet; $^7/_{16}$" fiberboard (wood); Design E.	34 min.	—			4	1-10	$^1/_2$
W-5-W-16	$5^1/_2$"	2" × 4" stud wall; Exposed face: $^1/_8$" CAB sheet; $^7/_{16}$" fiberboard; Unexposed face: CAB shingles over 1" × 6"; Design E.	32 min.	33 min.			4	1-10	$^1/_2$
W-5-W-17	$5^1/_2$"	2" × 4" stud wall; Exposed face: CAB shingles over 1" × 6"; Unexposed face: $^1/_8$" CAB sheet; gypsum at stud edges; $3^1/_2$" mineral wood insulation; Design F.	51 min.	—			4	1-10	$^3/_4$
W-5-W-18	$5^1/_2$"	2" × 4" stud wall; Exposed face: $^1/_8$" CAB sheet; gypsum board at stud edges; Unexposed face: CAB shingles over 1" × 6"; $3^1/_2$" mineral wool insulation; Design F.	42 min.	—			4	1-10	$^2/_3$
W-5-W-19	$5^5/_8$"	2" × 4" stud wall; Exposed face: CAB shingles over 1" × 6"; Unexposed face: $^1/_8$" CAB sheet; gypsum board at stud edges; $5^1/_2$" mineral wool insulation; Design G.	74 min.	85 min.			4	1-10	1
W-5-W-20	$5^5/_8$"	2" × 4" stud wall; Exposed face: $^1/_8$" CAB sheet; gypsum board at $^3/_{16}$" stud edges; $^7/_{16}$" fiberboard; Unexposed face: CAB shingles over 1" × 6"; $5^1/_2$" mineral wool insulation; Design G.	79 min.	85 min.			4	1-10	$1^1/_4$
W-5-W-21	$5^5/_8$"	2" × 4" stud wall; Exposed face: CAB shingles 1" × 6" sheathing; Unexposed face: CAB sheet; gypsum board at stud edges; $5^1/_2$" mineral wool insulation; Design G.	38 min.	38 min.			4	1-10, 12, 14	—

(Continued)

TABLE 1.3.2—WOOD FRAME WALLS
4" TO LESS THAN 6" THICK—(Continued)

ITEM CODE	THICKNESS	CONSTRUCTION DETAILS	PERFORMANCE		REFERENCE NUMBER			NOTES	REC. HOURS
			LOAD	TIME	PRE-BMS-92	BMS-92	POST-BMS-92		
W-5-W-22	$5^5/_8$"	2" × 4" stud wall; Exposed face: CAB sheet; gypsum board at stud edges; Unexposed face: CAB shingles 1" × 6" sheathing; $5^1/_2$" mineral wool insulation; Design G.	38 min.	38 min.			4	1-12	—
W-6-W-23	6"	2" × 4" stud wall; 16" o.c.; $^1/_2$" gypsum board each side; $^1/_2$" gypsum plaster each side.	N/A	60 min.			7	15	1
W-6-W-24	6"	2" × 4" stud wall; 16" o.c.; $^1/_2$" gypsum board each side; $^1/_2$" gypsum plaster each side.	N/A	68 min.			7	16	1
W-6-W-25	$6^7/_8$"	2" × 4" stud wall; 18" o.c.; $^3/_4$" gypsum plank each side; $^3/_{16}$" gypsum plaster each side.	N/A	80 min.			7	15	$1^1/_3$
W-5-W-26	$5^1/_8$"	2" × 4" stud wall; 16" o.c.; $^3/_8$" gypsum board each side; $^3/_{16}$" gypsum plaster each side.	N/A	37 min.			7	15	$^1/_2$
W-5-W-27	$5^3/_4$"	2" × 4" stud wall; 16" o.c.; $^3/_8$" gypsum lath each side; $^1/_2$" gypsum plaster each side.	N/A	52 min.			7	15	$^3/_4$
W-5-W-28	5"	2" × 4" stud wall; 16" o.c.; $^1/_2$" gypsum board each side.	N/A	37 min.			7	16	$^1/_2$
W-5-W-29	5"	2" × 4" stud wall; $^1/_2$" fiberboard both sides 14% M.C. with F.R. paint at 35 gm./ft.^{2}.	N/A	28 min.			7	15	$^1/_3$
W-4-W-30	$4^3/_4$"	2" × 4" stud wall; Fire side: $^1/_2$" (wood) fiberboard; Back side: $^1/_4$" CAB; 16" o.c.	N/A	17 min.			7	15, 16	$^1/_4$
W-5-W-31	$5^1/_8$"	2" × 4" stud wall; 16" o.c.; $^1/_2$" fiberboard insulation with $^1/_{32}$" asbestos (both sides of each board).	N/A	50 min.			7	16	$^3/_4$
W-4-W-32	$4^1/_4$"	2" × 4" stud wall; $^3/_8$"thick gypsum wallboard on both faces; insulated cavities.	See Note 23	25 min.		1		17, 18, 23	$^1/_3$
W-4-W-33	$4^1/_2$"	2" × 4" stud wall; $^1/_2$" thick gypsum wallboard on both faces.	See Note 17	40 min.		1		17, 23	$^1/_3$
W-4-W-34	$4^1/_2$"	2" × 4" stud wall; $^1/_2$" thick gypsum wallboard on both faces; insulated cavities.	See Note 17	45 min.		1		17, 18, 23	$^3/_4$
W-4-W-35	$4^1/_2$"	2" × 4" stud wall; $^1/_2$" thick gypsum wallboard on both faces; insulated cavities.	N/A	1 hr.		1		17, 18, 24	1
W-4-W-36	$4^1/_2$"	2" × 4" stud wall; $^1/_2$" thick, 1.1 lbs./ft.^{2} wood fiberboard sheathing on both faces.	See Note 23	15 min.		1		17, 23	$^1/_4$
W-4-W-37	$4^1/_2$"	2" × 4" stud wall; $^1/_2$" thick, 0.7 lb./ft.^{2} wood fiberboard sheathing on both faces.	See Note 23	10 min.		1		17, 23	$^1/_6$
W-4-W-38	$4^1/_2$"	2" × 4" stud wall; $^1/_2$" thick, flameproofed 1.6 lbs./ft.^{2} wood fiberboard sheathing on both faces.	See Note 23	30 min.		1		17, 23	$^1/_2$

(Continued)

TABLE 1.3.2—WOOD FRAME WALLS
4" TO LESS THAN 6" THICK—(Continued)

ITEM CODE	THICKNESS	CONSTRUCTION DETAILS	PERFORMANCE		REFERENCE NUMBER			NOTES	REC. HOURS
			LOAD	TIME	PRE-BMS-92	BMS-92	POST-BMS-92		
W-4-W-39	$4^1/_2$"	2" × 4" stud wall; $^1/_2$" thick gypsum wallboard on both faces; insulated cavities.	See Note 23	1 hr.		1		17, 18, 23	1
W-4-W-40	$4^1/_2$"	2" × 4" stud wall; $^1/_2$" thick, 1:2; 1:3 gypsum plaster on wood lath on both faces.	See Note 23	30 min.		1		17, 21, 23	$^1/_2$
W-4-W-41	$4^1/_2$"	2" × 4" stud wall; $^1/_2$", 1:2; 1:3 gypsum plaster on wood lath on both faces; insulated cavities.	See Note 23	1 hr.		1		17, 18, 21, 24	1
W-4-W-42	$4^1/_2$"	2" × 4" stud wall; $^1/_2$", 1:5; 1:7.5 lime plaster on wood lath on both wall faces.	See Note 23	30 min.		1		17, 21, 23	$^1/_2$
W-4-W-43	$4^1/_2$"	2" × 4" stud wall; $^1/_2$" thick 1:5; 1:7.5 lime plaster on wood lath on both faces; insulated cavities.	See Note 23	45 min.		1		17, 18, 21, 23	$^3/_4$
W-4-W-44	$4^5/_8$"	2" × 4" stud wall; $^3/_{16}$" thick cement-asbestos over $^3/_8$" thick gypsum board on both faces.	See Note 23	1 hr.		1		23, 25, 26, 27	1
W-4-W-45	$4^5/_8$"	2" × 4" stud wall; studs faced with 4" wide strips of $^3/_8$" thick gypsum board; $^3/_{16}$" thick gypsum cement-asbestos board on both faces; insulated cavities.	See Note 23	1 hr.		1		23, 25, 27, 28	1
W-4-W-46	$4^5/_8$"	Same as W-4-W-45 but nonload bearing.	N/A	1 hr. 15 min.		1		24, 28	$1^1/_4$
W-4-W-47	$4^7/_8$"	2" × 4" stud wall; $^3/_{16}$" thick cement-asbestos board over $^1/_2$" thick gypsum sheathing on both faces.	See Note 23	1 hr. 15 min.		1		23, 25, 26, 27	$1^1/_4$
W-4-W-48	$4^7/_8$"	Same as W-4-W-47 but nonload bearing.	N/A	1 hr. 30 min.		1		24, 27	$1^1/_2$
W-5-W-49	5"	2" × 4" stud wall; Exterior face: $^3/_4$" wood sheathing; asbestos felt 14 lbs./100 ft.^{2} and $^5/_{32}$" cement-asbestos shingles; Interior face: 4" wide strips of $^3/_8$" gypsum board over studs; wall faced with $^3/_{16}$" thick cement-asbestos board.	See Note 23	40 min.		1		18, 23, 25, 26, 29	$^2/_3$
W-5-W-50	5"	2" × 4" stud wall; Exterior face: as per W-5-W-49; Interior face: $^9/_{16}$" composite board consisting of $^7/_{16}$" thick wood fiberboard faced with $^1/_8$" thick cement-asbestos board; Exterior side exposed to fire.	See Note 23	30 min.		1		23, 25, 26, 30	$^1/_2$
W-5-W-51	5"	Same as W-5-W-50 but interior side exposed to fire.	See Note 23	30 min.		1		23, 25, 26	$^1/_2$
W-5-W-52	5"	Same as W-5-W-49 but exterior side exposed to fire.	See Note 23	45 min.		1		18, 23, 25, 26	$^3/_4$
W-5-W-53	5"	2" × 4" stud wall; $^3/_4$" thick T&G wood boards on both sides.	See Note 23	20 min.		1		17, 23	$^1/_3$

(Continued)

TABLE 1.3.2—WOOD FRAME WALLS
4" TO LESS THAN 6" THICK—(Continued)

ITEM CODE	THICKNESS	CONSTRUCTION DETAILS	PERFORMANCE		REFERENCE NUMBER			NOTES	REC. HOURS
			LOAD	TIME	PRE-BMS-92	BMS-92	POST-BMS-92		
W-5-W-54	5"	Same as W-5-W-53 but with insulated cavities.	See Note 23	35 min.		1		17, 18, 23	$^1/_2$
W-5-W-55	5"	2" × 4" stud wall; $^3/_4$" thick T&G wood boards on both sides with 30 lbs./100 ft.^{2} asbestos; paper, between studs and boards.	See Note 23	45 min.		1		17, 23	$^3/_4$
W-5-W-56	5"	2" × 4" stud wall; $^1/_2$" thick, 1:2; 1:3 gypsum plaster on metal lath on both sides of wall.	See Note 23	45 min.		1		17, 21, 34	$^3/_4$
W-5-W-57	5"	2" × 4" stud wall; $^3/_4$" thick 2:1:8; 2:1:12 lime and Keene's cement plaster over metal lath on both sides of wall.	See Note 23	45 min.		1		17, 21, 23	$^1/_2$
W-5-W-58	5"	2" × 4" stud wall; $^3/_4$" thick 2:1:8; 2:1:10 lime portland cement plaster over metal lath on both sides of wall.	See Note 23	30 min.		1		17, 21, 23	$^1/_2$
W-5-W-59	5"	2" × 4" stud wall; $^3/_4$" thick 1:5; 1:7.5 lime plaster on metal lath on both sides of wall.	See Note 23	30 min.		1		17, 21, 23	$^1/_2$
W-5-W-60	5"	2" × 4" stud wall; $^3/_4$" thick 1:$^1/_{30}$:2; 1:$^1/_{30}$:3 portland cement, asbestos fiber plaster on metal lath on both sides of wall.	See Note 23	45 min.		1		17, 21, 23	$^3/_4$
W-5-W-61	5"	2" × 4" stud wall; $^3/_4$" thick 1:2; 1:3 portland cement plaster on metal lath on both sides of wall.	See Note 23	30 min.		1		17, 21, 23	$^1/_2$
W-5-W-62	5"	2" × 4" stud wall; $^3/_4$" thick neat gypsum plaster on metal lath on both sides of wall.	N/A	1 hr. 30 min.		1		17, 22, 24	$1^1/_2$
W-5-W-63	5"	2" × 4" stud wall; $^3/_4$" thick neat gypsum plaster on metal lath on both sides of wall.	See Note 23	1 hr. 30 min.		1		17, 21, 23	$1^1/_2$
W-5-W-64	5"	2" × 4" stud wall; $^3/_4$" thick 1:2; 1:2 gypsum plaster on metal lath on both sides of wall; insulated cavities.	See Note 23	1 hr. 30 min.		1		17, 18, 21, 23	$1^1/_2$
W-5-W-65	5"	2" × 4" stud wall; same as W-5-W-64 but cavities not insulated.	See Note 23	1 hr.		1		17, 21, 23	1
W-5-W-66	5"	2" × 4" stud wall; $^3/_4$" thick 1:2; 1:3 gypsum plaster on metal lath on both sides of wall; insulated cavities.	See Note 23	1 hr. 15 min.		1		17, 18, 21, 23	$1^1/_4$
W-5-W-67	$5^1/_{16}$"	Same as W-5-W-49 except cavity insulation of 1.75 lbs./ft.^{2} mineral wool bats; rating applies when either wall side exposed to fire.	See Note 23	1 hr. 15 min.		1		23, 26, 25	$1^1/_4$
W-5-W-68	$5^1/_4$"	2" × 4" stud wall; $^7/_8$" thick 1:2; 1:3 gypsum plaster on metal lath on both sides of wall; insulated cavities.	See Note 23	1 hr. 30 min.		1		17, 18, 21, 23	$1^1/_2$

(Continued)

TABLE 1.3.2—WOOD FRAME WALLS
4" TO LESS THAN 6" THICK—(Continued)

ITEM CODE	THICKNESS	CONSTRUCTION DETAILS	PERFORMANCE		REFERENCE NUMBER			NOTES	REC. HOURS
			LOAD	TIME	PRE-BMS-92	BMS-92	POST-BMS-92		
W-5-W-69	$5^1/_4$"	2" × 4" stud wall; $^7/_8$" thick neat gypsum plaster applied on metal lath on both sides of wall.	N/A	1 hr. 45 min.		1		17, 22, 24	$1^3/_4$
W-5-W-70	$5^1/_4$"	2" × 4" stud wall; $^1/_2$" thick neat gypsum plaster on $^3/_8$" plain gypsum lath on both sides of wall.	See Note 23	1 hr.		1		17, 22, 23	1
W-5-W-71	$5^1/_4$"	2" × 4" stud wall; $^1/_2$" thick of 1:2; 1:2 gypsum plaster on $^3/_8$" thick plain gypsum lath with $1^3/_4$" x $1^3/_4$" metal lath pads nailed 8" o.c. vertically and 16" o.c. horizontally on both sides of wall.	See Note 23	1 hr.		1		17, 21, 23	1
W-5-W-72	$5^1/_4$"	2" × 4" stud wall; $^1/_2$" thick of 1:2; 1:2 gypsum plaster on $^3/_8$" perforated gypsum lath, one $^3/_4$" diameter hole or larger per 16" square of lath surface, on both sides of wall.	See Note 23	1 hr.		1		17, 21, 23	1
W-5-W-73	$5^1/_4$"	2" × 4" stud wall; $^1/_2$" thick of 1:2; 1:2 gypsum plaster on $^3/_8$" gypsum lath (plain, indented or perforated) on both sides of wall.	See Note 23	45 min.		1		17, 21, 23	$^3/_4$
W-5-W-74	$5^1/_4$"	2" × 4" stud wall; $^7/_8$" thick of 1:2; 1:3 gypsum plaster over metal lath on both sides of wall.	See Note 23	1 hr.		1		17, 21, 23	1
W-5-W-75	$5^1/_4$"	2" × 4" stud wall; $^7/_8$" thick of $1:^1/_{30}:2$; $1:^1/_{30}:3$ portland cement, asbestos plaster applied over metal lath on both sides of wall.	See Note 23	1 hr.		1		17, 21, 23	1
W-5-W-76	$5^1/_4$"	2" × 4" stud wall; $^7/_8$" thick of 1:2; 1:3 portland cement plaster over metal lath on both sides of wall.	See Note 23	45 min.		1		17, 21, 23	$^3/_4$
W-5-W-77	$5^1/_2$"	2" × 4" stud wall; 1" thick neat gypsum plaster over metal lath on both sides of wall; nonload bearing.	N/A	2 hrs.		1		17, 22, 24	2
W-5-W-78	$5^1/_2$"	2" × 4" stud wall; $^1/_2$" thick of 1:2; 1:2 gypsum plaster on $^1/_2$" thick, 0.7 lb./ft.^{2} wood fiberboard on both sides of wall.	See Note 23	35 min.		1		17, 21, 23	$^1/_2$
W-4-W-79	$4^3/_4$"	2" × 4" wood stud wall; $^1/_2$" thick of 1:2; 1:2 gypsum plaster over wood lath on both sides of wall; mineral wool insulation.	N/A	1 hr.			43	21, 31, 35, 38	1
W-4-W-80	$4^3/_4$"	Same as W-4-W-79 but uninsulated.	N/A	35 min.			43	21, 31, 35	$^1/_2$
W-4-W-81	$4^3/_4$"	2" × 4" wood stud wall; $^1/_2$" thick of 3:1:8; 3:1:12 lime, Keene's cement, sand plaster over wood lath on both sides of wall; mineral wool insulation.	N/A	1 hr.			43	21, 31, 35, 40	1

(Continued)

TABLE 1.3.2—WOOD FRAME WALLS
4" TO LESS THAN 6" THICK—(Continued)

ITEM CODE	THICKNESS	CONSTRUCTION DETAILS	PERFORMANCE		REFERENCE NUMBER			NOTES	REC. HOURS
			LOAD	TIME	PRE-BMS-92	BMS-92	POST-BMS-92		
W-4-W-82	$4^3/_4$"	$2" \times 4"$ wood stud wall; $1/_2$" thick of $1:6^1/_4$; $1:6^1/_4$ lime Keene's cement plaster over wood lath on both sides of wall; mineral wool insulation.	N/A	30 min.			43	21, 31, 35, 40	$1/_2$
W-4-W-83	$4^3/_4$"	$2" \times 4"$ wood stud wall; $1/_2$" thick of $1:5$; $1:7.5$ lime plaster over wood lath on both sides of wall.	N/A	30 min.			43	21, 31, 35	$1/_2$
W-5-W-84	$5^1/_8$"	$2" \times 4"$ wood stud wall; $^{11}/_{16}$" thick of $1:5$; $1:7.5$ lime plaster over wood lath on both sides of wall; mineral wool insulation.	N/A	45 min.			43	21, 31, 35, 39	$1/_2$
W-5-W-85	$5^1/_4$"	$2" \times 4"$ wood stud wall; $^3/_4$" thick of $1:5$; $1:7$ lime plaster over wood lath on both sides of wall; mineral wool insulation.	N/A	40 min.			43	21, 31, 35, 40	$2/_3$
W-5-W-86	$5^1/_4$"	$2" \times 4"$ wood stud wall; $1/_2$" thick of $2:1:12$ lime, Keene's cement and sand scratch coat; $1/_2$" thick $2:1:18$ lime, Keene's cement and sand brown coat over wood lath on both sides of wall; mineral wool insulation.	N/A	1 hr.			43	21, 31, 35, 40	1
W-5-W-87	$5^1/_4$"	$2" \times 4"$ wood stud wall; $1/_2$" thick of $1:2$; $1:2$ gypsum plaster over $^3/_8$" plaster board on both sides of wall.	N/A	45 min.			43	21, 31	$^3/_4$
W-5-W-88	$5^1/_4$"	$2" \times 4"$ wood stud wall; $1/_2$" thick of $1:2$; $1:2$ gypsum plaster over $^3/_8$" gypsum lath on both sides of wall.	N/A	45 min.			43	21, 31	$^3/_4$
W-5-W-89	$5^1/_4$"	$2" \times 4"$ wood stud wall; $1/_2$" thick of $1:2$; $1:2$ gypsum plaster over $^3/_8$" gypsum lath on both sides of wall.	N/A	1 hr.			43	21, 31, 33	1
W-5-W-90	$5^1/_4$"	$2" \times 4"$ wood stud wall; $1/_2$" thick neat plaster over $^3/_8$" thick gypsum lath on both sides of wall.	N/A	1 hr.			43	21, 22, 31	1
W-5-W-91	$5^1/_4$"	$2" \times 4"$ wood stud wall; $1/_2$" thick of $1:2$; $1:2$ gypsum plaster over $^3/_8$" thick indented gypsum lath on both sides of wall.	N/A	45 min.			43	21, 31	$^3/_4$
W-5-W-92	$5^1/_4$"	$2" \times 4"$ wood stud wall; $1/_2$" thick of $1:2$; $1:2$ gypsum plaster over $^3/_8$" thick perforated gypsum lath on both sides of wall.	N/A	45 min.			43	21, 31, 34	$^3/_4$
W-5-W-93	$5^1/_4$"	$2" \times 4"$ wood stud wall; $1/_2$" thick of $1:2$; $1:2$ gypsum plaster over $^3/_8$" perforated gypsum lath on both sides of wall.	N/A	1 hr.			43	21, 31	1
W-5-W-94	$5^1/_4$"	$2" \times 4"$ wood stud wall; $1/_2$" thick of $1:2$; $1:2$ gypsum plaster over $^3/_8$" thick perforated gypsum lath on both sides of wall.	N/A	45 min.			43	21, 31, 34	$^3/_4$

(Continued)

TABLE 1.3.2—WOOD FRAME WALLS
4" TO LESS THAN 6" THICK—(Continued)

| ITEM CODE | THICKNESS | CONSTRUCTION DETAILS | PERFORMANCE | | REFERENCE NUMBER | | | NOTES | REC. HOURS |
			LOAD	TIME	PRE-BMS-92	BMS-92	POST-BMS-92		
W-5-W-95	$5^1/_2$"	2" × 4" wood stud wall; $^1/_2$" thick of 1:2; 1:2 gypsum plaster over $^1/_2$" thick wood fiberboard plaster base on both sides of wall.	N/A	35 min.			43	21, 31, 36	$^1/_2$
W-5-W-96	$5^3/_4$"	2" × 4" wood stud wall; $^1/_2$" thick of 1:2; 1:2 gypsum plaster over $^7/_8$" thick flameproofed wood fiberboard on both sides of wall.	N/A	1 hr.			43	21, 31, 37	1

Notes:
1. All specimens 8 feet or 8 feet 8 inches by 10 feet, 4 inches, i.e. one-half of furnace size. See Note 42 for design cross section.
2. Specimens tested in tandem (two per exposure).
3. Test per ASA No. A2-1934 except where unloaded. Also, panels were of "half" size of furnace opening. Time value signifies a thermal failure time.
4. Two-inch by 4-inch studs: 16 inches on center.; where 10 feet 4 inches, blocking at 2-foot 4-inch height.
5. Facing 4 feet by 8 feet, cement-asbestos board sheets, $^3/_{16}$ inch thick.
6. Sheathing (diagonal): $^{25}/_{22}$ inch by $5^1/_2$ inch, 1 inch by 6 inches pine.
7. Facing shingles: 24 inches by 12 inches by $^5/_{32}$ inch where used.
8. Asbestos felt: asphalt sat between sheathing and shingles.
9. Load: 30,500 pounds or 360 psi/stud where load was tested.
10. Walls were tested beyond achievement of first test end point. A load-bearing time in excess of performance time indicates that although thermal criteria were exceeded, load-bearing ability continued.
11. Wall was rated for one hour combustible use in original source.
12. Hose steam test specimen. See table entry of similar design above for recommended rating.
13. Rated one and one-fourth hour load bearing. Rated one and one-half hour nonload bearing.
14. Failed hose stream.
15. Test terminated due to flame penetration.
16. Test terminated - local back face temperature rise.
17. Nominal 2-inch by 4-inch wood studs of No. 1 common or better lumber set edgewise. Two-inch by four-inch plates at top and bottom and blocking ad mid height of wall.
18. Cavity insulation consists of rock wool bats 1.0 lb./ft.[2] of filled cavity area.
19. Cavity insulation consists of glass wool bats 0.6 lb./ft.[2] of filled cavity area.
20. Cavity insulation consists of blown-in forck wool 2.0 lbs./ft.[2] of filled cavity area
21. Mix proportions for plastered walls as follows: first ratio indicates scratch coat mix, weight of dry plaster: dry sand; second ratio indicates brown coat mix.
22. "Neat" plaster is taken to mean unsanded wood-fiber gypsum plaster.
23. Load: 360 psi of net stud cross sectional area.
24. Rated as nonload bearing.
25. Nominal 2-inch by 4-inch studs per Note 17, spaced at 16 inches on center.
26. Horizontal joints in facing material supported by 2-inch by 4-inch blocking within wall.
27. Facings secured with 6d casing nails. Nail holes predrilled and were 0.02 to 0.03 inch smaller than nail diameter.
28. Cavity insulation consists of mineral wool bats weighing 2 lbs./ft.[2] of filled cavity area.
29. Interior wall face exposed to fire.
30. Exterior wall faced exposed to fire.
31. Nominal 2-inch by 4-inch studs of yellow pine or Douglas-fir spaced 16 inches on center in a single row.
32. Studs as in Note 31 except double row, with studs in rows staggered.
33. Six roofing nails with metal-lath pads around heats to each 16-inch by 48-inch lath.
34. Areas of holes less than $2^3/_4$ percent of area of lath.
35. Wood laths were nailed with either 3d or 4d nails, one nail to each bearing, and the end joining broken every seventh course.
36. One-half-inch thick fiberboard plaster base nailed with 3d or 4d common wire nails spaced 4 to 6 inches on center.
37. Seven-eighths-inch thick fiberboard plaster base nailed with 5d common wire nails spaced 4 to 6 inches on center.
38. Mineral wood bats 1.05 to 1.25 lbs./ft.[2] with waterproofed-paper backing.
39. Blown-in mineral wool insulation, 2.2 lbs./ft.[2].
40. Mineral wool bats, 1.4 lbs./ft.[2] with waterproofed-paper backing.
41. Mineral wood bats, 0.9 lb./ft.[2].
42. See wall design diagram, below.

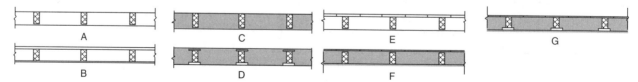

43. Duplicate specimen of W-4-W-7, tested simultaneously with W-4-W-7 in 18-foot test furnace.

TABLE 1.3.3—WOOD FRAME WALLS
6" TO LESS THAN 8" THICK

ITEM CODE	THICKNESS	CONSTRUCTION DETAILS	PERFORMANCE		REFERENCE NUMBER			NOTES	REC. HOURS
			LOAD	TIME	PRE-BMS-92	BMS-92	POST-BMS-92		
W-6-W-1	$6^1/_4$"	2×4 stud wall; $^1/_2$" thick, 1:2; 1:2 gypsum plaster on $^7/_8$" flameproofed wood fiberboard weighing 2.8 lbs./ft.^{2} on both sides of wall.	See Note 3	1 hr.		1		1-3	1
W-6-W-2	$6^1/_2$"	2×4 stud wall; $^1/_2$" thick, 1:3; 1:3 gypsum plaster on 1" thick magnesium oxysulfate wood fiberboard on both sides of wall.	See Note 3	45 min.		1		1-3	$^3/_4$
W-7-W-3	$7^1/_4$"	Double row of 2×4 studs, $^1/_2$" thick of 1:2; 1:2 gypsum plaster applied over $^3/_8$" thick perforated gypsum lath on both sides of wall; mineral wool insulation.	N/A	1 hr.			43	2, 4, 5	1
W-7-W-4	$7^1/_2$"	Double row of 2×4 studs, $^5/_8$" thick of 1:2; 1:2 gypsum plaster applied over $^3/_8$" thick perforated gypsum lath over laid with 2" $\times$ 2", 16 gage wire fabric, on both sides of wall.	N/A	1 hr. 15 min.			43	2, 4	$1^1/_4$

Notes:
1. Nominal 2-inch by 4-inch wood studs of No. 1 common or better lumber set edgewise. Two-inch by 4-inch plates at top and bottom and blocking at mid height of wall.
2. Mix proportions for plastered walls as follows: first ratio indicates scratch coat mix, weight of dry plaster:dry sand; second ratio indicates brown coat mix.
3. Load: 360 psi of net stud cross sectional area.
4. Nominal 2-inch by 4-inch studs of yellow pine of Douglas-fir spaced 16 inches in a double row, with studs in rows staggered.
5. Mineral wool bats, 0.19 lb./ft.^{2}.

TABLE 1.4.1—WALLS—MISCELLANEOUS MATERIALS
0" TO LESS THAN 4" THICK

ITEM CODE	THICKNESS	CONSTRUCTION DETAILS	PERFORMANCE		REFERENCE NUMBER			NOTES	REC. HOURS
			LOAD	TIME	PRE-BMS-92	BMS-92	POST-BMS-92		
W-3-Mi-1	$3^7/_8$"	Glass brick wall: (bricks $5^3/_4$" $\times$ $5^3/_4$" $\times$ $3^7/_8$") $^1/_4$" mortar bed, cement/lime/sand; mounted in brick (9") wall with mastic and $^1/_2$" asbestos rope.	N/A	1 hr.			7	1, 2	1
W-3-Mi-2	3"	Core: 2" magnesium oxysulfate wood-fiber blocks; laid in portland cement-lime mortar; Facings: on both sides; see Note 3.	N/A	1 hr.		1		3	1
W-3-Mi-3	$3^7/_8$"	Core: 8" $\times$ $4^7/_8$" glass blocks $3^7/_8$" thick weighing 4 lbs. each; laid in portland cement-lime mortar; horizontal mortar joints reinforced with metal lath.	N/A	15 min.		1			$^1/_4$

Notes:
1. No failure reached at 1 hour.
2. These glass blocks are assumed to be solid based on other test data available for similar but hollow units which show significantly reduced fire endurance.
3. Minimum of $^1/_2$ inch of 1:3 sanded gypsum plaster required to develop this rating.

TABLE 1.4.2—WALLS—MISCELLANEOUS MATERIALS
4" TO LESS THAN 6" THICK

ITEM CODE	THICKNESS	CONSTRUCTION DETAILS	PERFORMANCE		REFERENCE NUMBER			NOTES	REC. HOURS
			LOAD	TIME	PRE-BMS-92	BMS-92	POST-BMS-92		
W-4-Mi-1	4"	Core: 3" magnesium oxysulfate wood-fiber blocks; laid in portland cement mortar; Facings: both sides; see Note 1.	N/A	2 hrs.		1			2

Notes:

1. One-half inch sanded gypsum plaster. Voids in hollow blocks to be not more than 30 percent.

2003 INTERNATIONAL EXISTING BUILDING CODE® COMMENTARY

FIGURE 1.5.1—FINISH RATINGS—INORGANIC MATERIALS

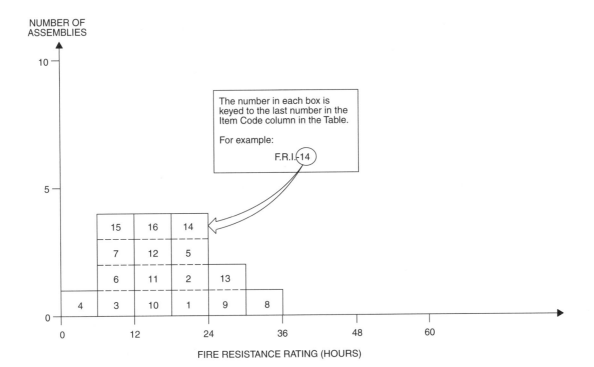

TABLE 1.5.1—FINISH RATINGS—INORGANIC MATERIALS

ITEM CODE	THICKNESS	CONSTRUCTION DETAILS	PERFORMANCE FINISH RATING	REFERENCE NUMBER PRE-BMS-92	REFERENCE NUMBER BMS-92	REFERENCE NUMBER POST-BMS-92	NOTES	REC. F.R. (MIN.)
F.R.-I-1	$9/16$"	$3/8$" gypsum wallboard faced with $3/16$" cement-asbestos board.	20 minutes		1		1, 2	15
F.R.-I-2	$11/16$"	$1/2$" gypsum sheathing faced with $3/16$" cement-asbestos board.	20 minutes		1		1, 2	20
F.R.-I-3	$3/16$"	$3/16$" cement-asbestos board over uninsulated cavity.	10 minutes		1		1, 2	5
F.R.-I-4	$3/16$"	$3/16$" cement-asbestos board over insulated cavities.	5 minutes		1		1, 2	5
F.R.-I-5	$3/4$"	$3/4$" thick 1:2; 1:3 gypsum plaster over paper backed metal lath.	20 minutes		1		1, 2, 3	20
F.R.-I-6	$3/4$"	$3/4$" thick portland cement plaster on metal lath.	10 minutes		1		1, 2	10
F.R.-I-7	$3/4$"	$3/4$" thick 1:5; 1:7.5 lime plaster on metal lath.	10 minutes		1		1, 2	10
F.R.-I-8	1"	1" thick neat gypsum plaster on metal lath.	35 minutes		1		1, 2, 4	35
F.R.-I-9	$3/4$"	$3/4$" thick neat gypsum plaster on metal lath.	30 minutes		1		1, 2, 4	30
F.R.-I-10	$3/4$"	$3/4$" thick 1:2; 1:2 gypsum plaster on metal lath.	15 minutes		1		1, 2, 3	15
F.R.-I-11	$1/2$"	Same as F.R.-1-7, except $1/2$" thick on wood lath.	15 minutes		1		1, 2, 3	15
F.R.-I-12	$1/2$"	$1/2$" thick 1:2; 1:3 gypsum plaster on wood lath.	15 minutes		1		1, 2, 3	15

(Continued)

TABLE 1.5.1—FINISH RATINGS—INORGANIC MATERIALS—(Continued)

ITEM CODE	THICKNESS	CONSTRUCTION DETAILS	PERFORMANCE FINISH RATING	REFERENCE NUMBER PRE-BMS-92	BMS-92	POST-BMS-92	NOTES	REC. F.R. (MIN.)
F.R.-I-13	$^7/_8$"	$^1/_2$" thick 1:2; 1:2 gypsum plaster on $^3/_8$" perforated gypsum lath.	20 minutes		1		1, 2, 3	30
F.R.-I-14	$^7/_8$"	$^1/_2$" thick 1:2; 1:2 gypsum plaster on $^3/_8$" thick plain or indented gypsum plaster.	20 minutes		1		1, 2, 3	20
F.R.-I-15	$^3/_8$"	$^3/_8$" gypsum wallboard.	10 minutes		1		1, 2	10
F.R.-I-16	$^1/_2$"	$^1/_2$" gypsum wallboard.	5 minutes		1		1, 2	15

Notes:
1. The finish rating is the time required to obtain an average temperature rise of 250°F., or a single point rise of 325°F., at the interface between the material being rated and the substrate being protected.
2. Tested in accordance with the Standard Specifications for Fire Tests of Building Construction and Materials, ASA No. A2-1932.
3. Mix proportions for plasters as follows: first ratio, dry weight of plaster: dry weight of sand for scratch coat; second ratio, plaster: sand for brown coat.
4. Neat plaster means unsanded wood-fiber gypsum plaster.

General Note:
The finish rating of modern building materials can be found in the current literature.

TABLE 1.5.2—FINISH RATINGS—ORGANIC MATERIALS

ITEM CODE	THICKNESS	CONSTRUCTION DETAILS	PERFORMANCE FINISH RATING	REFERENCE NUMBER PRE-BMS-92	BMS-92	POST-BMS-92	NOTES	REC. F.R. (MIN.)
F.R.-O-1	$^9/_{16}$"	$^7/_{16}$" wood fiberboard faced with $^1/_8$" cement-asbestos board.	15 minutes		1		1, 2	15
F.R.-O-2	$^{29}/_{32}$"	$^3/_4$" wood sheathing, asbestos felt weighing 14 lbs./100 ft.^{2} and $^5/_{32}$" cement-asbestos shingles.	20 minutes		1		1, 2	20
F.R.-O-3	$1^1/_2$"	1" thick magnesium oxysulfate wood fiberboard faced with 1:3; 1:3 gypsum plaster, $1/_2$" thick.	20 minutes		1		1, 2, 3	20
F.R.-O-4	$^1/_2$"	$^1/_2$" thick wood fiberboard.	5 minutes		1		1, 2	5
F.R.-O-5	$^1/_2$"	$^1/_2$" thick flameproofed wood fiberboard.	10 minutes		1		1, 2	10
F.R.-O-6	1"	$^1/_2$" thick wood fiberboard faced with $^1/_2$" thick 1:2; 1:2 gypsum plaster.	15 minutes		1		1, 2, 3	30
F.R.-O-7	$1^3/_8$"	$^7/_8$" thick flameproofed wood fiberboard faced with $^1/_2$" thick 1:2; 1:2 gypsum plaster.	30 minutes		1		1, 2, 3	30
F.R.-O-8	$1^1/_4$"	$1^1/_4$" thick plywood.	30 minutes			35		30

Notes:
1. The finish rating is the time required to obtain an average temperature rise of 250°F., or a single point rise of 325°F., at the interface between the material being rated and he substrate being protected.
2. Tested in accordance with the Standard Specifications for Fire Tests of Building Construction and Materials, ASA No. A2-1932.
3. Plaster ratios as follows: first ratio is for scratch coat, weight of dry plaster: weight of dry sand; second ratio is for the brown coat.

General Note:
The finish rating of thinner materials, particularly thinner woods, have not been listed because the possible effects of shrinkage, warpage and aging cannot be predicted.

SECTION II—COLUMNS

TABLE 2.1.1—REINFORCED CONCRETE COLUMNS
MINIMUM DIMENSION 0" TO LESS THAN 6"

ITEM CODE	MINIMUM DIMENSION	CONSTRUCTION DETAILS	PERFORMANCE		REFERENCE NUMBER			NOTES	REC. HOURS
			LOAD	TIME	PRE-BMS-92	BMS-92	POST-BMS-92		
C-6-RC-1	6"	6" × 6" square columns; gravel aggregate concrete (4030 psi); Reinforcement: vertical, four $^7/_8$" rebars; horizontal, $^5/_{16}$" ties at 6" pitch; Cover: 1".	34.7 tons	62 min.			7	1, 2	1
C-6-RC-2	6"	6" × 6" square columns; gravel aggregate concrete (4200 psi); Reinforcement: vertical, four $^1/_2$" rebars; horizontal, $^5/_{16}$" ties at 6" pitch; Cover: 1".	21 tons	69 min.			7	1, 2	1

Notes:
1. Collapse.
2. British Test.

FIGURE 2.1.2—REINFORCED CONCRETE COLUMNS
MINIMUM DIMENSION 10" TO LESS THAN 12"

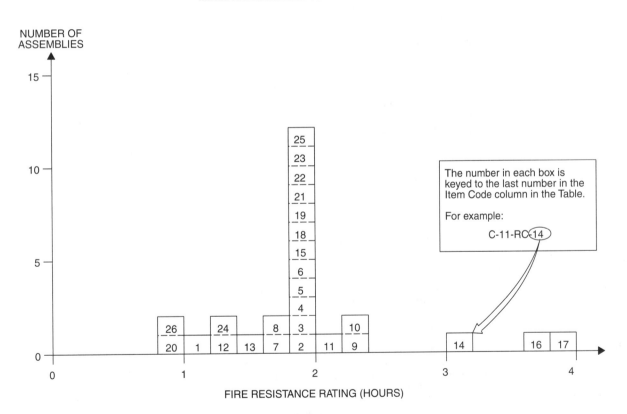

**TABLE 2.1.2—REINFORCED CONCRETE COLUMNS
MINIMUM DIMENSION 10" TO LESS THAN 12"**

ITEM CODE	MINIMUM DIMENSION	CONSTRUCTION DETAILS	PERFORMANCE		REFERENCE NUMBER			NOTES	REC. HOURS
			LOAD	TIME	PRE-BMS-92	BMS-92	POST-BMS-92		
C-10-RC-1	10"	10" square columns; aggregate concrete (4260 psi); Reinforcement: vertical, four $1^1/_4$" rebars; horizontal, $^3/_8$" ties at 6 pitch; Cover: $1^1/_4$".	92.2 tons	1 hr. 2 min.			7	1	1
C-10-RC-2	10"	10" square columns; aggregate concrete (2325 psi); Reinforcement: vertical, four $^1/_2$" rebars; horizontal, $^5/_{16}$" ties at 6" pitch; Cover: 1".	46.7 tons	1 hr. 52 min.			7	1	$1^3/_4$
C-10-RC-3	10"	10" square columns; aggregate concrete (5370 psi); Reinforcement: vertical, four $^1/_2$" rebars; horizontal, $^5/_{16}$" ties at 6" pitch; Cover: 1".	46.5 tons	2 hrs.			7	2, 3, 11	2
C-10-RC-4	10"	10" square columns; aggregate concrete (5206 psi); Reinforcement: vertical, four $^1/_2$" rebars; horizontal, $^5/_{16}$" ties at 6" pitch; Cover: 1".	46.5 tons	2 hrs.			7	2, 7	2
C-10-RC-5	10"	10" square columns; aggregate concrete (5674 psi); Reinforcement: vertical, four $^1/_2$" rebars; horizontal, $^5/_{16}$" ties at 6" pitch; Cover: 1".	46.7 tons	2 hrs.			7	1	2
C-10-RC-6	10"	10" square columns; aggregate concrete (5150 psi); Reinforcement: vertical, four $1^1/_2$" rebars; horizontal, $^5/_{16}$" ties at 6" pitch; Cover: 1".	66 tons	1 hr. 43 min.			7	1	$1^3/_4$
C-10-RC-7	10"	10" square columns; aggregate concrete (5580 psi); Reinforcement: vertical, four $^1/_2$" rebars; horizontal, $^5/_{16}$" ties at 6" pitch; Cover: $1^1/_8$".	62.5 tons	1 hr. 38 min.			7	1	$1^1/_2$
C-10-RC-8	10"	10" square columns; aggregate concrete (4080 psi); Reinforcement: vertical, four $1^1/_8$" rebars; horizontal, $^5/_{16}$" ties at 6" pitch; Cover: $1^1/_8$".	72.8 tons	1 hr. 48 min.			7	1	$1^3/_4$
C-10-RC-9	10"	10" square columns; aggregate concrete (2510 psi); Reinforcement: vertical, four $^1/_2$" rebars; horizontal, $^5/_{16}$" ties at 6" pitch; Cover: 1".	51 tons	2 hrs. 16 min.			7	1	$2^1/_4$
C-10-RC-10	10"	10" square columns; aggregate concrete (2170 psi); Reinforcement: vertical, four $^1/_2$" rebars; horizontal, $^5/_{16}$" ties at 6" pitch; Cover: 1".	45 tons	2 hrs. 14 min.			7	12	$2^1/_4$
C-10-RC-11	10"	10" square columns; gravel aggregate concrete (4015 psi); Reinforcement: vertical, four $^1/_2$" rebars; horizontal, $^5/_{16}$" ties at 6" pitch; Cover: $1^1/_8$".	46.5 tons	2 hrs. 6 min.			7	1	2

(Continued)

TABLE 2.1.2—REINFORCED CONCRETE COLUMNS
MINIMUM DIMENSION 10" TO LESS THAN 12"—(Continued)

ITEM CODE	MINIMUM DIMENSION	CONSTRUCTION DETAILS	PERFORMANCE		REFERENCE NUMBER			NOTES	REC. HOURS
			LOAD	TIME	PRE-BMS-92	BMS-92	POST-BMS-92		
C-11-RC-12	11"	11" square columns; gravel aggregate concrete (4150 psi); Reinforcement: vertical, four $1^1/_4$" rebars; horizontal, $^3/_8$" ties at $7^1/_2$" pitch; Cover: $1^1/_2$".	61 tons	1 hr. 23 min.			7	1	$1^1/_4$
C-11-RC-13	11"	11" square columns; gravel aggregate concrete (4380 psi); Reinforcement: vertical, four $1^1/_4$" rebars; horizontal, $^3/_8$" ties at $7^1/_2$" pitch; Cover: $1^1/_2$".	61 tons	1 hr. 26 min.			7	1	$1^1/_4$
C-11-RC-14	11"	11" square columns; gravel aggregate concrete (4140 psi); Reinforcement: vertical, four $1^1/_4$" rebars; horizontal, $^3/_8$" ties at $7^1/_2$" pitch; steel mesh around reinforcement; Cover: $1^1/_2$".	61 tons	3 hrs. 9 min.			7	1	3
C-11-RC-15	11"	11" square columns; slag aggregate concrete (3690 psi); Reinforcement: vertical, four $1^1/_4$" rebars; horizontal, $^3/_8$" ties at $7^1/_2$" pitch; Cover: $1^1/_2$".	91 tons	2 hrs.			7	2, 3, 4, 5	2
C-11-RC-16	11"	11" square columns; limestone aggregate concrete (5230 psi); Reinforcement: vertical, four $1^1/_4$" rebars; horizontal, $^3/_8$" ties at $7^1/_2$" pitch; Cover: $1^1/_2$".	91.5 tons	3 hrs. 41 min.			7	1	$3^1/_2$
C-11-RC-17	11"	11" square columns; limestone aggregate concrete (5530 psi); Reinforcement: vertical, four $1^1/_4$" rebars; horizontal, $^3/_8$" ties at $7^1/_2$" pitch; Cover: $1^1/_2$".	91.5 tons	3 hrs. 47 min.			7	1	$3^1/_2$
C-11-RC-18	11"	11" square columns; limestone aggregate concrete (5280 psi); Reinforcement: vertical, four $1^1/_4$" rebars; horizontal, $^3/_8$" ties at $7^1/_2$" pitch; Cover: $1^1/_2$".	91.5 tons	2 hrs.			7	2, 3, 4, 6	2
C-11-RC-19	11"	11" square columns; limestone aggregate concrete (4180 psi); Reinforcement: vertical, four $^5/_8$" rebars; horizontal, $^3/_8$" ties at 7″ pitch; Cover: $1^1/_2$".	71.4 tons	2 hrs.			7	2, 7	2
C-11-RC-20	11"	11" square columns; gravel concrete (4530 psi); Reinforcement: vertical, four $^5/_8$" rebars; horizontal, $^3/_8$" ties at 7" pitch; Cover: $1^1/_2$" with $^1/_2$" plaster.	58.8 tons	2 hrs.			7	2, 3, 9	$1^1/_4$
C-11-RC-21	11"	11" square columns; gravel concrete (3520 psi); Reinforcement: vertical, four $^5/_8$" rebars; horizontal, $^3/_8$" ties at 7" pitch; Cover: $1^1/_2$".	Variable	1 hr. 24 min.			7	1, 8	2

(Continued)

**TABLE 2.1.2—REINFORCED CONCRETE COLUMNS
MINIMUM DIMENSION 10" TO LESS THAN 12"—(Continued)**

ITEM CODE	MINIMUM DIMENSION	CONSTRUCTION DETAILS	PERFORMANCE		REFERENCE NUMBER			NOTES	REC. HOURS
			LOAD	TIME	PRE-BMS-92	BMS-92	POST-BMS-92		
C-11-RC-22	11"	11" square columns; aggregate concrete (3710 psi); Reinforcement: vertical, four $^5/_8$" rebars; horizontal, $^3/_8$" ties at 7" pitch; Cover: $1^1/_2$".	58.8 tons	2 hrs.			7	2, 3, 10	2
C-11-RC-23	11"	11" square columns; aggregate concrete (3190 psi); Reinforcement: vertical, four $^5/_8$" rebars; horizontal, $^3/_8$" ties at 7" pitch; Cover: $1^1/_2$".	58.8 tons	2 hrs.			7	2, 3, 10	2
C-11-RC-24	11"	11" square columns; aggregate concrete (4860 psi); Reinforcement: vertical, four $^5/_8$" rebars; horizontal, $^3/_8$" ties at 7" pitch; Cover: $1^1/_2$".	86.1 tons	1 hr. 20 min.			7	1	$1^1/_3$
C-11-RC-25	11"	11" square columns; aggregate concrete (4850 psi); Reinforcement: vertical, four $^5/_8$" rebars; horizontal, $^3/_8$" ties at 7" pitch; Cover: $1^1/_2$".	58.8 tons	1 hr. 59 min.			7	1	$1^3/_4$
C-11-RC-26	11"	11" square columns; aggregate concrete (3834 psi); Reinforcement: vertical, four $^5/_8$" rebars; horizontal, $^5/_{16}$" ties at $4^1/_2$" pitch; Cover: $1^1/_2$".	71.4 tons	53 min.			7	1	$^3/_4$

1. Failure mode - collapse.
2. Passed 2 hour fire exposure.
3. Passed hose stream test.
4. Reloaded effectively after 48 hours but collapsed at load in excess of original test load.
5. Failing load was 150 tons.
6. Failing load was 112 tons.
7. Failed during hose stream test.
8. Range of load 58.8 tons (initial) to 92 tons (92 minutes) to 60 tons (80 minutes).
9. Collapsed at 44 tons in reload after 96 hours.
10. Withstood reload after 72 hours.
11. Collapsed on reload after 48 hours.

TABLE 2.1.3—REINFORCED CONCRETE COLUMNS
MINIMUM DIMENSION 12" TO LESS THAN 14"

ITEM CODE	MINIMUM DIMENSION	CONSTRUCTION DETAILS	PERFORMANCE		REFERENCE NUMBER			NOTES	REC. HOURS
			LOAD	TIME	PRE-BMS-92	BMS-92	POST-BMS-92		
C-12-RC-1	12"	12" square columns; gravel aggregate concrete (2647 psi); Reinforcement: vertical, four $^5/_8$" rebars; horizontal, $^5/_{16}$" ties at $4^1/_2$" pitch; Cover: 2".	78.2 tons	38 min.		1	7	1	$^1/_2$
C-12-RC-2	12"	Reinforced columns with $1^1/_2$" concrete outside of reinforced steel; Gross diameter or side of column: 12"; Group I, Column A.	—	6 hrs.		1		2, 3	6
C-12-RC-3	12"	Description as per C-12-RC-2; Group I, Column B.	—	4 hrs.		1		2, 3	4
C-12-RC-4	12"	Description as per C-12-RC-2; Group II, Column A.	—	4 hrs.		1		2, 3	4
C-12-RC-5	12"	Description as per C-12-RC-2; Group II, Column B.	—	2 hrs. 30 min.		1		2, 3	$2^1/_2$
C-12-RC-6	12"	Description as per C-12-RC-2; Group III, Column A.	—	3 hrs.		1		2, 3	3
C-12-RC-7	12"	Description as per C-12-RC-2; Group III, Column B.	—	2 hrs.		1		2, 3	2
C-12-RC-8	12"	Description as per C-12-RC-2; Group IV, Column A.	—	2 hrs.		1		2, 3	2
C-12-RC-9	12"	Description as per C-12-RC-2; Group IV, Column B.	—	1 hr. 30 min.		1		2, 3	$1^1/_2$

Notes:

1. Failure mode - unspecified structural.
2. Group I: includes concrete having calcareous aggregate containing a combined total of not more than 10 percent of quartz, chert and flint for the coarse aggregate.

   Group II: includes concrete having trap-rock aggregate applied without metal ties and also concrete having cinder, sandstone or granite aggregate, if held in place with wire mesh or expanded metal having not larger than 4-inch mesh, weighing not less than 1.7 lbs./yd.^{2}, placed not more than 1 inch from the surface of the concrete.

   Group III: includes concrete having cinder, sandstone or granite aggregate tied with No. 5 gage steel wire, wound spirally over the column section on a pitch of 8 inches, or equivalent ties, and concrete having siliceous aggregates containing a combined total of 60 percent or more of quartz, chert and flint, if held in place with wire mesh or expanded metal having not larger than 4-inch mesh, weighing not less than 1.7 lbs./yd.^{2}, placed not more than 1 inch from the surface of the concrete.

   Group IV: includes concrete having siliceous aggregates containing a combined total of 60 percent or more of quartz, chert and flint, and tied with No. 5 gage steel wire wound spirally over the column section on a pitch of 8 inches, or equivalent ties.
3. Groupings of aggregates and ties are the same as for structural steel columns protected solidly with concrete, the ties to be placed over the vertical reinforcing bars and the mesh where required, to be placed within 1 inch from the surface of the column.

   Column A: working loads are assumed as carried by the area of the column inside of the lines circumscribing the reinforcing steel.

   Column B: working loads are assumed as carried by the gross area of the column.

TABLE 2.1.4—REINFORCED CONCRETE COLUMNS
MINIMUM DIMENSION 14" TO LESS THAN 16"

ITEM CODE	MINIMUM DIMENSION	CONSTRUCTION DETAILS	PERFORMANCE		REFERENCE NUMBER			NOTES	REC. HOURS
			LOAD	TIME	PRE-BMS-92	BMS-92	POST-BMS-92		
C-14-RC-1	14"	14″ square columns; gravel aggregate concrete (4295 psi); Reinforcement: vertical four $^3/_4$″ rebars; horizontal: $^1/_4$″ ties at 9″ pitch; Cover: $1^1/_2$″.	86 tons	1 hr. 22 min.			7	1	$1^1/_4$
C-14-RC-2	14"	Reinforced concrete columns with $1^1/_2$″ concrete outside reinforcing steel; Gross diameter or side of column: 12″; Group I, Column A.	—	7 hrs.		1		2, 3	7
C-14-RC-3	14"	Description as per C-14-RC-2; Group II, Column B.	—	5 hrs.		1		2, 3	5
C-14-RC-4	14"	Description as per C-14-RC-2; Group III, Column A.	—	5 hrs.		1		2, 3	5
C-14-RC-5	14"	Description as per C-14-RC-2; Group IV, Column B.	—	3 hrs. 30 min.		1		2, 3	$3^1/_2$
C-14-RC-6	14"	Description as per C-14-RC-2; Group III, Column A.	—	4 hrs.		1		2, 3	4
C-14-RC-7	14"	Description as per C-14-RC-2; Group III, Column B.	—	2 hrs. 30 min.		1		2, 3	$2^1/_2$
C-14-RC-8	14"	Description as per C-14-RC-2; Group IV, Column A.	—	2 hrs. 30 min.		1		2, 3	$2^1/_2$
C-14-RC-9	14"	Description as per C-14-RC-2; Group IV, Column B.	—	1 hr. 30 min.		1		2, 3	$1^1/_2$

Notes:
1. Failure mode - main rebars buckled between links at various points.
2. Group I: includes concrete having calcareous aggregate containing a combined total of not more than 10 percent of quartz, chert and flint for the coarse aggregate.
   Group II: includes concrete having trap-rock aggregate applied without metal ties and also concrete having cinder, sandstone or granite aggregate, if held in place with wire mesh or expanded metal having not larger than 4-inch mesh, weighing not less than 1.7 lbs./yd.^{2}, placed not more than 1 inch from the surface of the concrete.
   Group III: includes concrete having cinder, sandstone or granite aggregate tied with No. 5 gage steel wire, wound spirally over the column section on a pitch of 8 inches, or equivalent ties, and concrete having siliceous aggregates containing a combined total of 60 percent or more of quartz, chert and flint, if held in place with wire mesh or expanded metal having not larger than 4-inch mesh, weighing not less than 1.7 lbs./yd.^{2}, placed not more than 1 inch from the surface of the concrete.
   Group IV: includes concrete having siliceous aggregates containing a combined total of 60 percent or more of quartz, chert and flint, and tied with No. 5 gage steel wire wound spirally over the column section on a pitch of 8 inches, or equivalent ties.
3. Groupings of aggregates and ties are the same as for structural steel columns protected solidly with concrete, the ties to be placed over the vertical reinforcing bars and the mesh where required, to be placed within 1 inch from the surface of the column.
   Column A: working loads are assumed as carried by the area of the column inside of the lines circumscribing the reinforcing steel.
   Column B: working loads are assumed as carried by the gross area of the column.

FIGURE 2.1.5—REINFORCED CONCRETE COLUMNS
MINIMUM DIMENSION 16" TO LESS THAN 18"

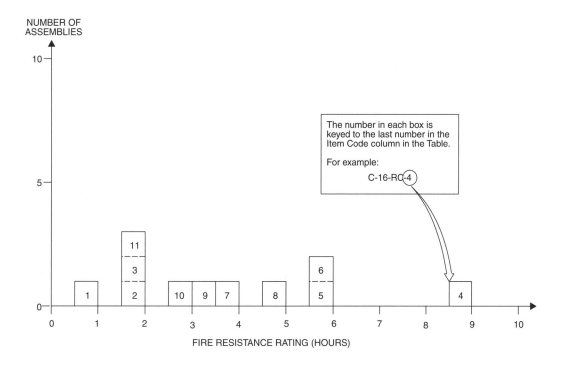

TABLE 2.1.5—REINFORCED CONCRETE COLUMNS
MINIMUM DIMENSION 16" TO LESS THAN 18"

ITEM CODE	MINIMUM DIMENSION	CONSTRUCTION DETAILS	PERFORMANCE		REFERENCE NUMBER			NOTES	REC. HOURS
			LOAD	TIME	PRE-BMS-92	BMS-92	POST-BMS-92		
C-16-RC-1	16"	16" square columns; gravel aggregate concrete (4550 psi); Reinforcement: vertical, eight $1^3/_8$" rebars; horizontal, $^5/_{16}$" ties at 6" pitch $1^3/_8$" below column surface and $^5/_{16}$" ties at 6" pitch linking center rebars of each face forming a smaller square in column cross section.	237 tons	1 hr			7	1, 2, 3	1
C-16-RC-2	16"	16" square columns; gravel aggregate concrete (3360 psi); Reinforcement: vertical, eight $1^3/_8$" rebars; horizontal, $^5/_{16}$" ties at 6" pitch; Cover: $1^3/_8$".	210 tons	2 hrs.			7	2, 4, 5, 6	2
C-16-RC-3	16"	16" square columns; gravel aggregate concrete (3980 psi); Reinforcement: vertical, four $^7/_8$" rebars; horizontal, $^3/_8$" ties at 6" pitch; Cover: 1".	123.5 tons	2 hrs.			7	2, 4, 7	2
C-16-RC-4	16"	Reinforced concrete columns with $1^1/_2$" concrete outside reinforcing steel; Gross diameter or side of column: 16"; Group I, Column A.	—	9 hrs.		1		8, 9	9
C-16-RC-5	16"	Description as per C-16-RC-4; Group I, Column B.	—	6 hrs.		1		8, 9	6

(Continued)

**TABLE 2.1.5—REINFORCED CONCRETE COLUMNS
MINIMUM DIMENSION 16" TO LESS THAN 18"—(Continued)**

ITEM CODE	MINIMUM DIMENSION	CONSTRUCTION DETAILS	PERFORMANCE		REFERENCE NUMBER			NOTES	REC. HOURS
			LOAD	TIME	PRE-BMS-92	BMS-92	POST-BMS-92		
C-16-RC-6	16"	Description as per C-16-RC-4; Group II, Column A.	—	6 hrs.		1		8, 9	6
C-16-RC-7	16"	Description as per C-16-RC-4; Group II, Column B.	—	4 hrs.		1		8, 9	4
C-16-RC-8	16"	Description as per C-16-RC-4; Group III, Column A.	—	5 hrs.		1		8, 9	5
C-16-RC-9	16"	Description as per C-16-RC-4; Group III, Column B.	—	3 hrs. 30 min.		1		8, 9	$3^1/_2$
C-16-RC-10	16"	Description as per C-16-RC-4; Group IV, Column A.	—	3 hrs.		1		8, 9	3
C-16-RC-11	16"	Description as per C-16-RC-4; Group IV, Column B.	—	2 hrs.		1		8, 9	2

Notes:

1. Column passed 1 hour fire test.
2. Column passed hose stream test.
3. No reload specified.
4. Column passed 2 hour fire test.
5. Column reloaded successfully after 24 hours.
6. Reinforcing details same as C-16-RC-1.
7. Column passed reload after 72 hours.
8. Group I: includes concrete having calcareous aggregate containing a combined total of not more than 10 percent of quartz, chert and flint for the coarse aggregate.

   Group II: includes concrete having trap-rock aggregate applied without metal ties and also concrete having cinder, sandstone or granite aggregate, if held in place with wire mesh or expanded metal having not larger than 4-inch mesh, weighing not less than 1.7 lbs./yd.[2], placed not more than 1 inch from the surface of the concrete.

   Group III: includes concrete having cinder, sandstone or granite aggregate tied with No. 5 gage steel wire, wound spirally over the column section on a pitch of 8 inches, or equivalent ties, and concrete having siliceous aggregates containing a combined total of 60 percent or more of quartz, chert and flint, if held in place with wire mesh or expanded metal having not larger than 4-inch mesh, weighing not less than 1.7 lbs./yd.[2], placed not more than 1 inch from the surface of the concrete.

   Group IV: includes concrete having siliceous aggregates containing a combined total of 60 percent or more of quartz, chert and flint, and tied with No. 5 gage steel wire wound spirally over the column section on a pitch of 8 inches, or equivalent ties.
9. Groupings of aggregates and ties are the same as for structural steel columns protected solidly with concrete, the ties to be placed over the vertical reinforcing bars and the mesh where required, to be placed within 1 inch from the surface of the column.

   Column A: working loads are assumed as carried by the area of the column inside of the lines circumscribing the reinforcing steel.

   Column B: working loads are assumed as carried by the gross area of the column.

TABLE 2.1.6—REINFORCED CONCRETE COLUMNS
MINIMUM DIMENSION 18" TO LESS THAN 20"

ITEM CODE	MINIMUM DIMENSION	CONSTRUCTION DETAILS	PERFORMANCE		REFERENCE NUMBER			NOTES	REC. HOURS
			LOAD	TIME	PRE-BMS-92	BMS-92	POST-BMS-92		
C-18-RC-1	18"	Reinforced concrete columns with 1¹/₂" concrete outside reinforced steel; Gross diameter or side of column: 18"; Group I, Column A.	—	11 hrs.		1		1, 2	11
C-18-RC-2	18"	Description as per C-18-RC-1; Group I, Column B.	—	8 hrs.		1		1, 2	8
C-18-RC-3	18"	Description as per C-18-RC-1; Group II, Column A.	—	7 hrs.		1		1, 2	7
C-18-RC-4	18"	Description as per C-18-RC-1; Group II, Column B.	—	5 hrs.		1		1, 2	5
C-18-RC-5	18"	Description as per C-18-RC-1; Group III, Column A.	—	6 hrs.		1		1, 2	6
C-18-RC-6	18"	Description as per C-18-RC-1; Group III, Column B.	—	4 hrs.		1		1, 2	4
C-18-RC-7	18"	Description as per C-18-RC-1; Group IV, Column A.	—	3 hrs. 30 min.		1		1, 2	3¹/₂
C-18-RC-8	18"	Description as per C-18-RC-1; Group IV, Column B.	—	2 hrs. 30 min.		1		1, 2	2¹/₂

Notes:

1. Group I: includes concrete having calcareous aggregate containing a combined total of not more than 10 percent of quartz, chert and flint for the coarse aggregate.

   Group II: includes concrete having trap-rock aggregate applied without metal ties and also concrete having cinder, sandstone or granite aggregate, if held in place with wire mesh or expanded metal having not larger than 4-inch mesh, weighing not less than 1.7 lbs./yd.[2], placed not more than 1 inch from the surface of the concrete.

   Group III: includes concrete having cinder, sandstone or granite aggregate tied with No. 5 gage steel wire, wound spirally over the column section on a pitch of 8 inches, or equivalent ties, and concrete having siliceous aggregates containing a combined total of 60 percent or more of quartz, chert and flint, if held in place with wire mesh or expanded metal having not larger than 4-inch mesh, weighing not less than 1.7 lbs./yd.[2], placed not more than 1 inch from the surface of the concrete.

   Group IV: includes concrete having siliceous aggregates containing a combined total of 60 percent or more of quartz, chert and flint and, tied with No. 5 gage steel wire wound spirally over the column section on a pitch of 8 inches, or equivalent ties.

2. Groupings of aggregates and ties are the same as for structural steel columns protected solidly with concrete, the ties to be placed over the vertical reinforcing bars and the mesh where required, to be placed within 1 inch from the surface of the column.

   Column A: working loads are assumed as carried by the area of the column inside of the lines circumscribing the reinforcing steel.

   Column B: working loads are assumed as carried by the gross area of the column.

**FIGURE 2.1.7—REINFORCED CONCRETE COLUMNS
MINIMUM DIMENSION 20" TO LESS THAN 22"**

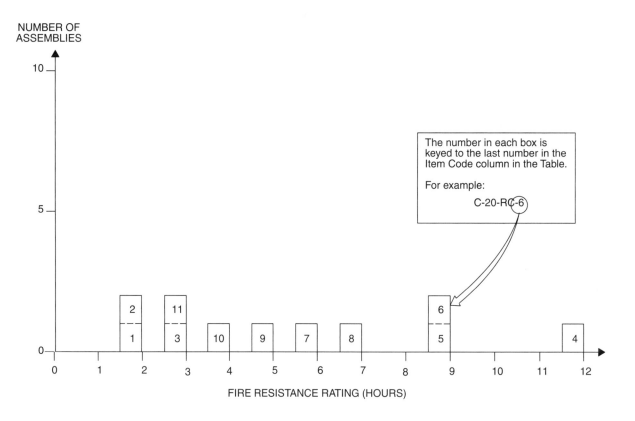

**TABLE 2.1.7—REINFORCED CONCRETE COLUMNS
MINIMUM DIMENSION 20" TO LESS THAN 22"**

ITEM CODE	MINIMUM DIMENSION	CONSTRUCTION DETAILS	PERFORMANCE		REFERENCE NUMBER			NOTES	REC. HOURS
			LOAD	TIME	PRE-BMS-92	BMS-92	POST-BMS-92		
C-20-RC-1	20"	20" square columns; gravel aggregate concrete (6690 psi); Reinforcement: vertical, four $1^3/_4$" rebars; horizontal, $^3/_8$" wire at 6" pitch; Cover $1^3/_4$".	367 tons	2 hrs.			7	1, 2, 3	2
C-20-RC-2	20"	20" square columns; gravel aggregate concrete (4330 psi); Reinforcement: vertical, four $1^3/_4$" rebars; horizontal, $^3/_8$" ties at 6" pitch; Cover $1^3/_4$".	327 tons	2 hrs.			7	1, 2, 4	2
C-20-RC-3	$20^1/_4$"	20" square columns; gravel aggregate concrete (4230 psi); Reinforcement: vertical, four $1^1/_8$" rebars; horizontal, $^3/_8$" wire at 5" pitch; Cover $1^1/_8$".	199 tons	2 hrs. 56 min.			7	5	$2^3/_4$
C-20-RC-4	20"	Reinforced concrete columns with $1^1/_2$" concrete outside of reinforcing steel; Gross diameter or side of column: 20"; Group I, Column A.	—	12 hrs.		1		6, 7	12
C-20-RC-5	20"	Description as per C-20-RC-4; Group I, Column B.	—	9 hrs.		1		6, 7	9
C-20-RC-6	20"	Description as per C-20-RC-4; Group II, Column A.	—	9 hrs.		1		6, 7	9

(Continued)

TABLE 2.1.7—REINFORCED CONCRETE COLUMNS
MINIMUM DIMENSION 20" TO LESS THAN 22"—(Continued)

ITEM CODE	MINIMUM DIMENSION	CONSTRUCTION DETAILS	PERFORMANCE		REFERENCE NUMBER			NOTES	REC. HOURS
			LOAD	TIME	PRE-BMS-92	BMS-92	POST-BMS-92		
C-20-RC-7	20"	Description as per C-20-RC-4; Group II, Column B.	—	6 hrs		1		6, 7	6
C-20-RC-8	20"	Description as per C-20-RC-4; Group III, Column A.	—	7 hrs.		1		6, 7	7
C-20-RC-9	20"	Description as per C-20-RC-4; Group III, Column B.	—	5 hrs.		1		6, 7	5
C-20-RC-10	20"	Description as per C-20-RC-4; Group IV, Column A.	—	4 hrs.		1		6, 7	4
C-20-RC-11	20"	Description as per C-20-RC-4; Group IV, Column B.	—	3 hrs.		1		6, 7	3

Notes:
1. Passed 2 hour fire test.
2. Passed hose stream test.
3. Failed during reload at 300 tons.
4. Passed reload after 72 hours.
5. Failure mode - collapse.
6. Group I: includes concrete having calcareous aggregate containing a combined total of not more than 10 percent of quartz, chert and flint for the coarse aggregate.

   Group II: includes concrete having trap-rock aggregate applied without metal ties and also concrete having cinder, sandstone or granite aggregate, if held in place with wire mesh or expanded metal having not larger than 4-inch mesh, weighing not less than 1.7 lbs./yd.^{2}, placed not more than 1 inch from the surface of the concrete.

   Group III: includes concrete having cinder, sandstone or granite aggregate tied with No. 5 gage steel wire, wound spirally over the column section on a pitch of 8 inches, or equivalent ties, and concrete having siliceous aggregates containing a combined total of 60 percent or more of quartz, chert and flint, if held in place with wire mesh or expanded metal having not larger than 4-inch mesh, weighing not less than 1.7 lbs./yd.^{2}, placed not more than 1 inch from the surface of the concrete.

   Group IV: includes concrete having siliceous aggregates containing a combined total of 60 percent or more of quartz, chert and flint, and tied with No. 5 gage steel wire wound spirally over the column section on a pitch of 8 inches, or equivalent ties.
7. Groupings of aggregates and ties are the same as for structural steel columns protected solidly with concrete, the ties to be placed over the vertical reinforcing bars and the mesh where required, to be placed within 1 inch from the surface of the column.

   Column A: working loads are assumed as carried by the area of the column inside of the lines circumscribing the reinforcing steel.

   Column B: working loads are assumed as carried by the gross area of the column.

TABLE 2.1.8—HEXAGONAL REINFORCED CONCRETE COLUMNS
MINIMUM DIMENSION 12" TO LESS THAN 14"

ITEM CODE	MINIMUM DIMENSION	CONSTRUCTION DETAILS	PERFORMANCE		REFERENCE NUMBER			NOTES	REC. HOURS
			LOAD	TIME	PRE-BMS-92	BMS-92	POST-BMS-92		
C-12-HRC-1	12"	12" hexagonal columns; gravel aggregate concrete (4420 psi); Reinforcement: vertical, eight $\frac{1}{2}$" rebars; horizontal, $\frac{5}{16}$" helical winding at $1\frac{1}{2}$" pitch; Cover: $\frac{1}{2}$".	88 tons	58 min.			7	1	$\frac{3}{4}$
C-12-HRC-2	12"	12" hexagonal columns; gravel aggregate concrete (3460 psi); Reinforcement: vertical, eight $\frac{1}{2}$" rebars; horizontal, $\frac{5}{16}$" helical winding at $1\frac{1}{2}$" pitch; Cover: $\frac{1}{2}$".	78.7 tons	1 hr.			7	2	1

Notes:
1. Failure mode - collapse.
2. Test stopped at 1 hour.

TABLE 2.1.9—HEXAGONAL REINFORCED CONCRETE COLUMNS
MINIMUM DIMENSION 14" TO LESS THAN 16"

ITEM CODE	MINIMUM DIMENSION	CONSTRUCTION DETAILS	PERFORMANCE		REFERENCE NUMBER			NOTES	REC. HOURS
			LOAD	TIME	PRE-BMS-92	BMS-92	POST-BMS-92		
C-14-HRC-1	14"	14" hexagonal columns; gravel aggregate concrete (4970 psi); Reinforcement: vertical, eight $^1/_2$" rebars; horizontal, $^5/_{16}$" helical winding on 2" pitch; Cover: $^1/_2$".	90 tons	2 hrs.			7	1, 2, 3	2

Notes:
1. Withstood 2 hour fire test.
2. Withstood hose stream test.
3. Withstood reload after 48 hours.

TABLE 2.1.10—HEXAGONAL REINFORCED CONCRETE COLUMNS
DIAMETER — 16" TO LESS THAN 18"

ITEM CODE	MINIMUM DIMENSION	CONSTRUCTION DETAILS	PERFORMANCE		REFERENCE NUMBER			NOTES	REC. HOURS
			LOAD	TIME	PRE-BMS-92	BMS-92	POST-BMS-92		
C-16-HRC-1	16"	16" hexagonal columns; gravel concrete (6320 psi); Reinforcement: vertical, eight $^5/_8$" rebars; horizontal, $^5/_{16}$" helical winding on $^3/_4$" pitch; Cover: $^1/_2$".	140 tons	1 hr. 55 min.			7	1	$1^3/_4$
C-16-HRC-2	16"	16" hexagonal columns; gravel aggregate concrete (5580 psi); Reinforcement: vertical, eight $^5/_8$" rebars; horizontal, $^5/_{16}$" helical winding on $1^3/_4$" pitch; Cover: $^1/_2$"	124 tons	2 hrs.			7	2	2

Notes:
1. Failure mode - collapse.
2. Failed on furnace removal.

TABLE 2.1.11—HEXAGONAL REINFORCED CONCRETE COLUMNS
DIAMETER — 20" TO LESS THAN 22"

ITEM CODE	MINIMUM DIMENSION	CONSTRUCTION DETAILS	PERFORMANCE		REFERENCE NUMBER			NOTES	REC. HOURS
			LOAD	TIME	PRE-BMS-92	BMS-92	POST-BMS-92		
C-20-HRC-1	20"	20" hexagonal columns; gravel concrete (6080 psi); Reinforcement: vertical, $^3/_4$" rebars; horizontal, $^5/_6$" helical winding on $1^3/_4$" pitch; Cover: $^1/_2$".	211 tons	2 hrs.			7	1	2
C-20-HRC-2	20"	20" hexagonal columns; gravel concrete (5080 psi); Reinforcement: vertical, $^3/_4$" rebars; horizontal, $^5/_{16}$" wire on $1^3/_4$" pitch; Cover: $^1/_2$".	184 tons	2 hrs. 15 min.			7	2, 3, 4	$2^1/_4$

Notes:
1. Column collapsed on furnace removal.
2. Passed $2^1/_4$ hour fire test.
3. Passed hose stream test.
4. Withstood reload after 48 hours.

TABLE 2.2—ROUND CAST IRON COLUMNS

ITEM CODE	MINIMUM DIMENSION	CONSTRUCTION DETAILS	PERFORMANCE		REFERENCE NUMBER			NOTES	REC. HOURS
			LOAD	TIME	PRE-BMS-92	BMS-92	POST-BMS-92		
C-7-CI-1	7" O.D.	Column: .6" minimum metal thickness; unprotected.	—	30 min.		1			$^1/_2$
C-7-CI-2	7" O.D.	Column: .6" minimum metal thickness concrete filled, outside unprotected.	—	45 min.		1			$^3/_4$
C-11-CI-3	11" O.D.	Column: .6" minimum metal thickness; Protection: $1^1/_2$" portland cement plaster on high ribbed metal lath, $^1/_2$" broken air space.	—	3 hrs.		1			3
C-11-CI-4	11" O.D.	Column: .6" minimum metal thickness; Protection: 2" concrete other than siliceous aggregate.	—	2 hrs. 30 min.		1			$2^1/_2$
C-12-CI-5	12.5" O.D.	Column: 7" O.D. .6" minimum metal thickness; Protection: 2" porous hollow tile, $^3/_4$" mortar between tile and column, outside wire ties.	—	3 hrs.		1			3
C-7-CI-6	7.6" O.D.	Column: 7" I.D., $^3/_{10}$" minimum metal thickness, concrete filled unprotected.	—	30 min.		1			$^1/_2$
C-8-CI-7	8.6" O.D.	Column: 8" I.D., $^3/_{10}$" minimum metal thickness; concrete filled reinforced with four $3^1/_2$" $\times$ $^3/_8$" angles, in fill; unprotected outside.	—	1 hr.		1			1

FIGURE 2.3—STEEL COLUMNS—GYPSUM ENCASEMENTS

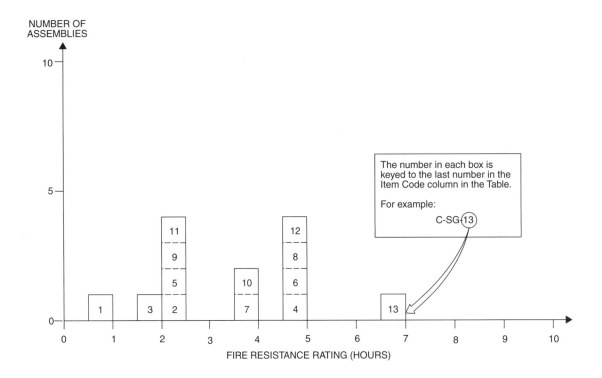

TABLE 2.3—STEEL COLUMNS—GYPSUM ENCASEMENTS

ITEM CODE	MINIMUM AREA OF SOLID MATERIAL	CONSTRUCTION DETAILS	PERFORMANCE		REFERENCE NUMBER			NOTES	REC. HOURS
			LOAD	TIME	PRE-BMS-92	BMS-92	POST-BMS-92		
C-SG-1	—	Steel protected with $^3/_4$" 1:3 sanded gypsum or 1" 1:2$^1/_2$ portland cement plaster on wire or lath; one layer.	—	1 hr.		1			1
C-SG-2	—	Same as C-SG-1; two layers.	—	2 hrs. 30 min.		1			2$^1/_2$
C-SG-3	130 in.^{2}	2" solid blocks with wire mesh in horizontal joints; 1" mortar on flange; reentrant space filled with block and mortar.	—	2 hrs.		1			2
C-SG-4	150 in.^{2}	Same as C-130-SG-3 with $^1/_2$" sanded gypsum plaster.	—	5 hrs.		1			5
C-SG-5	130 in.^{2}	2" solid blocks with wire mesh in horizontal joints; 1" mortar on flange; reentrant space filled with gypsum concrete.	—	2 hrs. 30 min.		1			2$^1/_2$
C-SG-6	150 in.^{2}	Same as C-130-SG-5 with $^1/_2$" sanded gypsum plaster.	—	5 hrs.		1			5
C-SG-7	300 in.^{2}	4" solid blocks with wire mesh in horizontal joints; 1" mortar on flange; reentrant space filled with block and mortar.	—	4 hrs.		1			4
C-SG-8	300 in.^{2}	Same as C-300-SG-7 with reentrant space filled with gypsum concrete.	—	5 hrs.		1			5

(Continued)

TABLE 2.3—STEEL COLUMNS—GYPSUM ENCASEMENTS—(Continued)

ITEM CODE	MINIMUM AREA OF SOLID MATERIAL	CONSTRUCTION DETAILS	PERFORMANCE		REFERENCE NUMBER			NOTES	REC. HOURS
			LOAD	TIME	PRE-BMS-92	BMS-92	POST-BMS-92		
C-SG-9	85 in.^{2}	2" solid blocks with cramps at horizontal joints; mortar on flange only at horizontal joints; reentrant space not filled.	—	2 hrs. 30 min.		1			$2^1/_2$
C-SG-10	105 in.^{2}	Same as C-85-SG-9 with $^1/_2$" sanded gypsum plaster.	—	4 hrs.		1			4
C-SG-11	95 in.^{2}	3" hollow blocks with cramps at horizontal joints; mortar on flange only at horizontal joints; reentrant space not filled.	—	2 hrs. 30 min.		1			$2^1/_2$
C-SG-12	120 in.^{2}	Same as C-95-SG-11 with $^1/_2$" sanded gypsum plaster.	—	5 hrs.		1			5
C-SG-13	130 in.^{2}	2" neat fibered gypsum reentrant space filled poured solid and reinforced with 4" × 4" wire mesh$^1/_2$" sanded gypsum plaster.	—	7 hrs.		1			7

TABLE 2.4—TIMBER COLUMNS MINIMUM DIMENSION

ITEM CODE	MINIMUM DIMENSION	CONSTRUCTION DETAILS	PERFORMANCE		REFERENCE NUMBER			NOTES	REC. HOURS
			LOAD	TIME	PRE-BMS-92	BMS-92	POST-BMS-92		
C-11-TC-1	11"	With unprotected steel plate cap.	—	30 min.		1		1, 2	$^1/_2$
C-11-TC-2	11"	With unprotected cast iron cap and pintle.	—	45 min.		1		1, 2	$^3/_4$
C-11-TC-3	11"	With concrete or protected steel or cast iron cap.	—	1 hr. 15 min.		1		1, 2	$1^1/_4$
C-11-TC-4	11"	With $^3/_8$" gypsum wallboard over column and over cast iron or steel cap.	—	1 hr. 15 min.		1		1, 2	$1^1/_4$
C-11-TC-5	11"	With 1" portland cement plaster on wire lath over column and over cast iron or steel cap; $^3/_4$" air space.	—	2 hrs.		1		1, 2	2

Notes:
1. Minimum area: 120 square inches.
2. Type of wood: long leaf pine or Douglas fir.

TABLE 2.5.1.1—STEEL COLUMNS—CONCRETE ENCASEMENTS MINIMUM DIMENSION LESS THAN 6"

ITEM CODE	MINIMUM DIMENSION	CONSTRUCTION DETAILS	PERFORMANCE		REFERENCE NUMBER			NOTES	REC. HOURS
			LOAD	TIME	PRE-BMS-92	BMS-92	POST-BMS-92		
C-5-SC-1	5"	5" × 6" outer dimensions; 4" × 3" × 10 lbs. "H" beam; Protection: gravel concrete (4900 psi) 6" × 4" - 13 SWG mesh.	12 tons	1 hr. 29 min.			7	1	$1^1/_4$

Notes:
1. Failure mode - collapse.

TABLE 2.5.1.2—STEEL COLUMNS—CONCRETE ENCASEMENTS
6" TO LESS THAN 8" THICK

ITEM CODE	MINIMUM DIMENSION	CONSTRUCTION DETAILS	PERFORMANCE		REFERENCE NUMBER			NOTES	REC. HOURS
			LOAD	TIME	PRE-BMS-92	BMS-92	POST-BMS-92		
C-7-SC-1	7"	7" × 8" column; 4" × 3" × 10 lbs. "H" beam; Protection: brick filled concrete (6220 psi); 6" × 4" mesh - 13 SWG; 1" below column surface.	12 tons	2 hrs. 46 min.			7	1	3
C-7-SC-2	7"	7" × 8" column; 4" × 3" × 10 lbs. "H" beam; Protection: gravel concrete (5140 psi); 6" × 4" 13 SWG mesh 1" below surface.	12 tons	3 hrs. 1 min.			7	1	$2^3/_4$
C-7-SC-3	7"	7" × 8" column; 4" × 3" × 10 lbs. "H" beam; Protection: concrete (4540 psi); 6" × 4" - 13 SWG mesh; 1" below column surface.	12 tons	3 hrs. 9 min.			7	1	3
C-7-SC-4	7"	7" × 8" column; 4" × 3" × 10 lbs. "H" beam; Protection: gravel concrete (5520 psi); 4" × 4" mesh; 16 SWG.	12 tons	2 hrs. 50 min.			7	1	$2^3/_4$

Notes:

1. Failure mode - collapse.

FIGURE 2.5.1.3—STEEL COLUMNS—CONCRETE ENCASEMENTS
MINIMUM DIMENSION 8" TO LESS THAN 10"

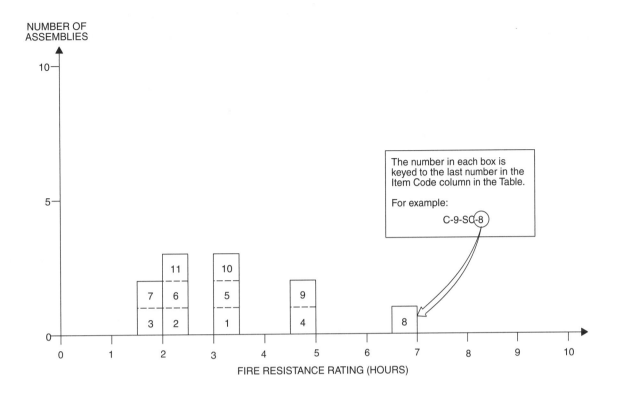

TABLE 2.5.1.3—STEEL COLUMNS—CONCRETE ENCASEMENTS
MINIMUM DIMENSION 8" TO LESS THAN 10"

ITEM CODE	MINIMUM DIMENSION	CONSTRUCTION DETAILS	PERFORMANCE		REFERENCE NUMBER			NOTES	REC. HOURS
			LOAD	TIME	PRE-BMS-92	BMS-92	POST-BMS-92		
C-8-SC-1	$8^1/_2$"	$8^1/_2$" × 10" column; 6" × $4^1/_2$" × 20 lbs. "H" beam; Protection: gravel concrete (5140 psi); 6" × 4" - 13 SWG mesh.	39 tons	3 hrs. 8 min.			7	1	3
C-8-SC-2	8"	8" × 10" column; 8" × 6" × 35 lbs. "I" beam; Protection: gravel concrete (4240 psi); 6" × 4" - 13 SWG mesh; $^1/_2$" cover.	90 tons	2 hrs. 1 min.			7	1	2
C-8-SC-3	8"	8" × 10" concrete encased column; 8" × 6" × 35 lbs. "H" beam; protection: aggregate concrete (3750 psi); 4" mesh - 16 SWG reinforcing $^1/_2$" below column surface.	90 tons	1 hr. 58 min.			7	1	$1^3/_4$
C-8-SC-4	8"	6" × 6" steel column; 2" outside protection; Group I.	—	5 hrs.		1		2	5
C-8-SC-5	8"	6" × 6" steel column; 2" outside protection; Group II.	—	3 hrs. 30 min.		1		2	$3^1/_2$
C-8-SC-6	8"	6" × 6" steel column; 2" outside protection; Group III.	—	2 hrs. 30 min.		1		2	$2^1/_2$
C-8-SC-7	8"	6" × 6" steel column; 2" outside protection; Group IV.	—	1 hr. 45 min.		1		2	$1^3/_4$

(Continued)

TABLE 2.5.1.3—STEEL COLUMNS—CONCRETE ENCASEMENTS
MINIMUM DIMENSION 8" TO LESS THAN 10"—(Continued)

ITEM CODE	MINIMUM DIMENSION	CONSTRUCTION DETAILS	PERFORMANCE		REFERENCE NUMBER			NOTES	REC. HOURS
			LOAD	TIME	PRE-BMS-92	BMS-92	POST-BMS-92		
C-9-SC-8	9"	6" × 6" steel column; 3" outside protection; Group I.	—	7 hrs.		1		2	7
C-9-SC-9	9"	6" × 6" steel column; 3" outside protection; Group II.	—	5 hrs.		1		2	5
C-9-SC-10	9"	6" × 6" steel column; 3" outside protection; Group III.	—	3 hrs. 30 min.		1		2	$3^1/_2$
C-9-SC-11	9"	6" × 6" steel column; 3" outside protection; Group IV.	—	2 hrs. 30 min.		1		2	$2^1/_2$

Notes:

1. Failure mode - collapse.

2. Group I: includes concrete having calcareous aggregate containing a combined total of not more than 10 percent of quartz, chert and flint for the coarse aggregate.

   Group II: includes concrete having trap-rock aggregate applied without metal ties and also concrete having cinder, sandstone or granite aggregate, if held in place with wire mesh or expanded metal having not larger than 4-inch mesh, weighing not less than 1.7 lbs./yd.^{2}, placed not more than 1 inch from the surface of the concrete.

   Group III: includes concrete having cinder, sandstone or granite aggregate tied with No. 5 gage steel wire, wound spirally over the column section on a pitch of 8 inches, or equivalent ties, and concrete having siliceous aggregates containing a combined total of 60 percent or more of quartz, chert and flint, if held in place with wire mesh or expanded metal having not larger than 4-inch mesh, weighing not less than 1.7 lbs./yd.^{2}, placed not more than1 inch from the surface of the concrete.

   Group IV: includes concrete having siliceous aggregates containing a combined total of 60 percent or more of quartz, chert and flint, and tied with No. 5 gage steel wire wound spirally over the column section on a pitch of 8 inches, or equivalent ties.

**FIGURE 2.5.1.4—STEEL COLUMNS—CONCRETE ENCASEMENTS
MINIMUM DIMENSION 10" TO LESS THAN 12"**

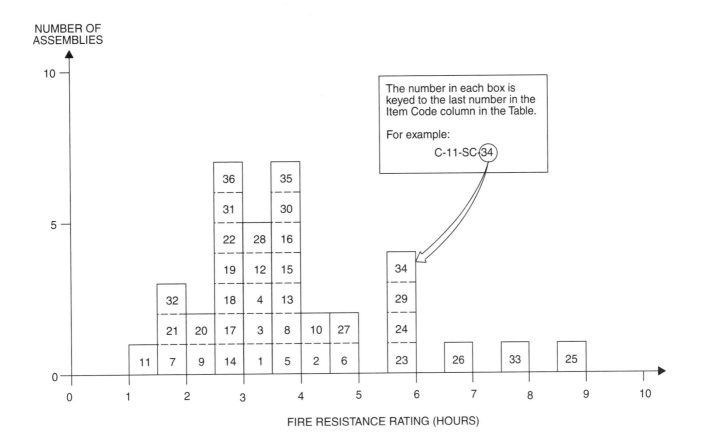

**TABLE 2.5.1.4—STEEL COLUMNS—CONCRETE ENCASEMENTS
MINIMUM DIMENSION 10" TO LESS THAN 12"**

ITEM CODE	MINIMUM DIMENSION	CONSTRUCTION DETAILS	PERFORMANCE		REFERENCE NUMBER			NOTES	REC. HOURS
			LOAD	TIME	PRE-BMS-92	BMS-92	POST-BMS-92		
C-10-SC-1	10"	10" × 12" concrete encased steel column; 8" × 6" × 35 lbs. "H" beam; Protection: gravel aggregate concrete (3640 psi); Mesh 6" × 4" 13 SWG, 1" below column surface.	90 tons	7 hrs.			7	1,2	3
C-10-SC-2	10"	10" × 16" column; 8" × 6" × 35 lbs. "H" beam; Protection: clay brick concrete (3630 psi); 6" × 4" mesh; 13 SWG, 1" below column surface.	90 tons	5 hrs.			7	2	4
C-10-SC-3	10"	10" × 12" column; 8" × 6" × 35 lbs. "H" beam; Protection: crushed stone and sand concrete (3930 psi); 6" × 4" - 13 SWG mesh; 1" below column surface.	90 tons	3 hrs. 30 min.			7	2	$3^1/_4$
C-10-SC-4	10"	10" × 12" column; 8" × 6" × 35 lbs. "H" beam; Protection: crushed basalt and sand concrete (4350 psi); 6" × 4" - 13 SWG mesh; 1" below column surface.	90 tons	2 hrs. 30 min.			7	2	$3^1/_3$

(Continued)

TABLE 2.5.1.4—STEEL COLUMNS—CONCRETE ENCASEMENTS
MINIMUM DIMENSION 10" TO LESS THAN 12"—(Continued)

ITEM CODE	MINIMUM DIMENSION	CONSTRUCTION DETAILS	PERFORMANCE		REFERENCE NUMBER			NOTES	REC. HOURS
			LOAD	TIME	PRE-BMS-92	BMS-92	POST-BMS-92		
C-10-SC-5	10"	10" × 12" column; 8" × 6" × 35 lbs. "H" beam; Protection: gravel aggregate concrete (5570 psi); 6" × 4" mesh; 13 SWG.	90 tons	3 hrs. 39 min.			7	2	$3^1/_2$
C-10-SC-6	10"	10" × 16" column; 8" × 6" × 35 lbs. "I" beam; Protection: gravel concrete (4950 psi); mesh; 6" × 4" 13 SWG 1" below column surface.	90 tons	4 hrs. 32 min.			7	2	$4^1/_2$
C-10-SC-7	10"	10" × 12" concrete encased steel column; 8" × 6" × 35 lbs. "H" beam; Protection: aggregate concrete (1370 psi); 6" × 4" mesh; 13 SWG reinforcing 1" below column surface.	90 tons	2 hrs.			7	3, 4	2
C-10-SC-8	10"	10" × 12" concrete encased steel column; 8" × 6" × 35 lbs. "H" column; Protection: aggregate concrete (4000 psi); 13 SWG iron wire loosely around column at 6" pitch about 2" beneath column surface.	86 tons	3 hrs. 36 min.			7	2	$3^1/_2$
C-10-SC-9	10"	10" × 12" concrete encased steel column; 8" × 6" × 35 lbs. "H" beam; Protection: aggregate concrete (3290 psi); 2" cover minimum.	86 tons	2 hrs. 8 min.			7	2	2
C-10-SC-10	10"	10" × 14" concrete encased steel column; 8" × 6" × 35 lbs. "H" column; Protection: crushed brick filled concrete (5310 psi); 6" × 4" mesh; 13 SWG reinforcement 1" below column surface.	90 tons	4 hrs. 28 min.			7	2	$4^1/_3$
C-10-SC-11	10"	10" × 14" concrete encased column; 8" × 6" 35 lbs. "H" beam; Protection: aggregate concrete (342 psi); 6" × 4" mesh; 13 SWG reinforcement 1" below surface.	90 tons	1 hr. 2 min.			7	2	1
C-10-SC-12	10"	10" × 12" concrete encased steel column; 8" × 6" × 35 lbs. "H" beam; Protection: aggregate concrete (4480 psi); four $3/_8$" vertical bars at "H" beam edges with $3/_{16}$" spacers at beam surface at 3' pitch and $3/_{16}$" binders at 10" pitch; 2" concrete cover.	90 tons	3 hrs. 2 min.			7	2	3

(Continued)

TABLE 2.5.1.4—STEEL COLUMNS—CONCRETE ENCASEMENTS
MINIMUM DIMENSION 10" TO LESS THAN 12"—(Continued)

ITEM CODE	MINIMUM DIMENSION	CONSTRUCTION DETAILS	PERFORMANCE		REFERENCE NUMBER			NOTES	REC. HOURS
			LOAD	TIME	PRE-BMS-92	BMS-92	POST-BMS-92		
C-10-SC-13	10"	10" × 12" concrete encased steel column; 8" × 6" × 35 lbs. "H" beam; Protection: aggregate concrete (5070 psi); 6" × 4" mesh; 13 SWG reinforcing at 6" beam sides wrapped and held by wire ties across (open) 8" beam face; reinforcements wrapped in 6" × 4" mesh; 13 SWG throughout; $^1/_2$" cover to column surface.	90 tons	3 hrs. 59 min.			7	2	$3^3/_4$
C-10-SC-14	10"	10" × 12" concrete encased steel column; 8" × 6" × 35 lbs. "H" beam; Protection: aggregate concrete (4410 psi); 6" × 4" mesh; 13 SWG reinforcement $1^1/_4$" below column surface; $^1/_2$" limestone cement plaster with $^3/_8$" gypsum plaster finish.	90 tons	2 hrs. 50 min.			7	2	$2^3/_4$
C-10-SC-15	10"	10" × 12" concrete encased steel column; 8" × 6" × 35 lbs. "H" beam; Protection: crushed clay brick filled concrete (4260 psi); 6" × 4" mesh; 13 SWG reinforcing 1" below column surface.	90 tons	3 hrs. 54 min.			7	2	$3^3/_4$
C-10-SC-16	10"	10" × 12" concrete encased steel column; 8" × 6" × 35 lbs. "H" beam; Protection: limestone aggregate concrete (4350 psi); 6" × 4" mesh; 13 SWG reinforcing 1" below column surface.	90 tons	3 hrs. 54 min.			7	2	$3^3/_4$
C-10-SC-17	10"	10" × 12" concrete encased steel column; 8" × 6" × 35 lbs. "H" beam; Protection: limestone aggregate concrete (5300 psi); 6" × 4"; 13 SWG wire mesh 1" below column surface.	90 tons	3 hrs.			7	4, 5	3
C-10-SC-18	10"	10" × 12" concrete encased steel column; 8" × 6" × 35 lbs. "H" beam; Protection: limestone aggregate concrete (4800 psi) with 6" × 4"; 13 SWG mesh reinforcement 1" below surface.	90 tons	3 hrs.			7	4, 5	3
C-10-SC-19	10"	10" × 14" concrete encased steel column; 12" × 8" × 65 lbs. "H" beam; Protection: aggregate concrete (3900 psi); 4" mesh; 16 SWG reinforcing $^1/_2$" below column surface.	118 tons	2 hrs. 42 min.			7	2	2
C-10-SC-20	10"	10" × 14" concrete encased steel column; 12" × 8" × 65 lbs. "H" beam; Protection: aggregate concrete (4930 psi); 4" mesh; 16 SWG reinforcing $^1/_2$" below column surface.	177 tons	2 hrs. 8 min.			7	2	2

(Continued)

**TABLE 2.5.1.4—STEEL COLUMNS—CONCRETE ENCASEMENTS
MINIMUM DIMENSION 10" TO LESS THAN 12"—(Continued)**

ITEM CODE	MINIMUM DIMENSION	CONSTRUCTION DETAILS	PERFORMANCE		REFERENCE NUMBER			NOTES	REC. HOURS
			LOAD	TIME	PRE-BMS-92	BMS-92	POST-BMS-92		
C-10-SC-21	$10^3/_8$"	$10^3/_8$" × $12^3/_8$" concrete encased steel column; 8" × 6" × 35 lbs. "H" beam; Protection: aggregate concrete (835 psi) with 6" × 4" mesh; 13 SWG reinforcing $1^3/_{16}$" below column surface; $^3/_{16}$" gypsum plaster finish.	90 tons	2 hrs.			7	3, 4	2
C-11-SC-22	11"	11" × 13" concrete encased steel column; 8" × 6" × 35 lbs. "H" beam; Protection: "open texture" brick filled concrete (890 psi) with 6" × 4" mesh; 13 SWG reinforcing $1^1/_2$" below column surface; $^3/_8$" lime cement plaster; $^1/_8$" gypsum plaster finish.	90 tons	3 hrs.			7	6, 7	3
C-11-SC-23	11"	11" × 12" column; 4" × 3" × 10 lbs. "H" beam; gravel concrete (4550 psi); 6" × 4" - 13 SWG mesh reinforcing; 1" below column surface.	12 tons	6 hrs.			7	7, 8	6
C-11-SC-24	11"	11" × 12" column; 4" × 3" × 10 lbs. "H" beam; Protection: gravel aggregate concrete (3830 psi); with 4" × 4" mesh; 16 SWG, 1" below column surface.	16 tons	5 hrs. 32 min.			7	2	$5^1/_2$
C-10-SC-25	10"	6" × 6" steel column with 4" outside protection; Group I.	—	9 hrs.		1		9	9
C-10-SC-26	10"	Description as per C-SC-25; Group II.	—	7 hrs.		1		9	7
C-10-SC-27	10"	Description as per C-10-SC-25; Group III.	—	5 hrs.		1		9	5
C-10-SC-28	10"	Description as per C-10-SC-25; Group IV.	—	3 hrs. 30 min.		1		9	$3^1/_2$
C-10-SC-29	10"	8" × 8" steel column with 2" outside protection; Group I.	—	6 hrs.		1		9	6
C-10-SC-30	10"	Description as per C-10-SC-29; Group II.	—	4 hrs.		1		9	4
C-10-SC-31	10"	Description as per C-10-SC-29; Group III.	—	3 hrs.		1		9	3
C-10-SC-32	10"	Description as per C-10-SC-29; Group IV.	—	2 hrs.		1		9	2
C-11-SC-33	11"	8" × 8" steel column with 3" outside protection; Group I.	—	8 hrs.		1		9	8
C-11-SC-34	11"	Description as per C-10-SC-33; Group II.	—	6 hrs.		1		9	6
C-11-SC-35	11"	Description as per C-10-SC-33; Group III.	—	4 hrs.		1		9	4

(Continued)

TABLE 2.5.1.4—STEEL COLUMNS—CONCRETE ENCASEMENTS
MINIMUM DIMENSION 10" TO LESS THAN 12"—(Continued)

ITEM CODE	MINIMUM DIMENSION	CONSTRUCTION DETAILS	PERFORMANCE		REFERENCE NUMBER			NOTES	REC. HOURS
			LOAD	TIME	PRE-BMS-92	BMS-92	POST-BMS-92		
C-11-SC-36	11"	Description as per C-10-SC-33; Group IV.	—	3 hrs.		1		9	3

Notes:

1. Tested under total restraint load to prevent expansion - minimum load 90 tons.
2. Failure mode - collapse.
3. Passed 2 hour fire test (Grade "C," British).
4. Passed hose stream test.
5. Column tested and passed 3 hour grade fire resistance (British).
6. Column passed 3 hour fire test.
7. Column collapsed during hose stream testing.
8. Column passed 6 hour fire test.
9. Group I: includes concrete having calcareous aggregate containing a combined total of not more than 10 percent of quartz, chert and flint for the coarse aggregate.

   Group II: includes concrete having trap-rock aggregate applied without metal ties and also concrete having cinder, sandstone or granite aggregate, if held in place with wire mesh or expanded metal having not larger than 4-inch mesh, weighing not less than 1.7 lbs./yd.^{2}, placed not more than 1 inch from the surface of the concrete.

   Group III: includes concrete having cinder, sandstone or granite aggregate tied with No. 5 gage steel wire, wound spirally over the column section on a pitch of 8 inches, or equivalent ties, and concrete having siliceous aggregates containing a combined total of 60 percent or more of quartz, chert and flint, if held in place with wire mesh or expanded metal having not larger than 4-inch mesh, weighing not less than 1.7 lbs./yd.^{2}, placed not more than 1 inch from the surface of the concrete.

   Group IV: includes concrete having siliceous aggregates containing a combined total of 60 percent or more of quartz, chert and flint, and tied with No. 5 gage steel wire wound spirally over the column section on a pitch of 8 inches, or equivalent ties.

FIGURE 2.5.1.5—STEEL COLUMNS—CONCRETE ENCASEMENTS
MINIMUM DIMENSION 12" TO LESS THAN 14"

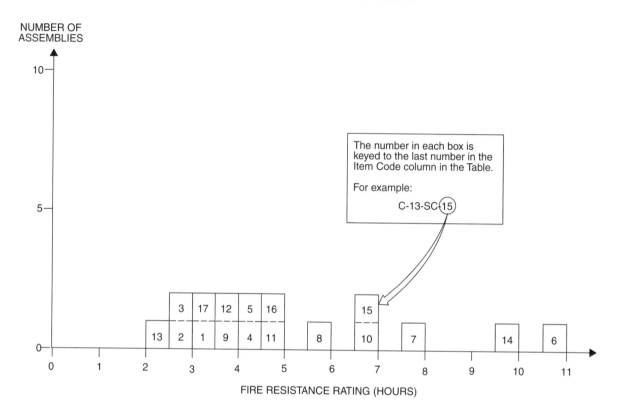

TABLE 2.5.1.5—STEEL COLUMNS—CONCRETE ENCASEMENTS
MINIMUM DIMENSION 12" TO LESS THAN 14"

ITEM CODE	MINIMUM DIMENSION	CONSTRUCTION DETAILS	PERFORMANCE		REFERENCE NUMBER			NOTES	REC. HOURS
			LOAD	TIME	PRE-BMS-92	BMS-92	POST-BMS-92		
C-12-SC-1	12"	12" × 14" concrete encased steel column; 8" × 6" × 35 lbs. "H" beam; Protection: aggregate concrete (4150 psi) with 4" mesh; 16 SWG reinforcing 1" below column surface.	120 tons	3 hrs. 24 min.			7	1	$3^1/_3$
C-12-SC-2	12"	12" × 16" concrete encased column; 8" × 6" × 35 lbs. "H" beam; Protection: aggregate concrete (4300 psi) with 4" mesh; 16 SWG reinforcing 1" below column surface.	90 tons	2 hrs. 52 min.			7	1	$2^3/_4$
C-12-SC-3	12"	12" × 16" concrete encased steel column; 12" × 8" × 65 lbs. "H" column; Protection: gravel aggregate concrete (3550 psi) with 4" mesh; 16 SWG reinforcement 1" below column surface.	177 tons	2 hrs. 31 min.			7	1	$2^1/_2$
C-12-SC-4	12"	12" × 16" concrete encased column; 12" × 8" × 65 lbs. "H" beam; Protection: aggregate concrete (3450 psi) with 4" mesh; 16 SWG reinforcement 1" below column surface.	118 tons	4 hrs. 4 min.			7	1	4

(Continued)

**TABLE 2.5.1.5—STEEL COLUMNS—CONCRETE ENCASEMENTS
MINIMUM DIMENSION 12" TO LESS THAN 14"—(Continued)**

ITEM CODE	MINIMUM DIMENSION	CONSTRUCTION DETAILS	PERFORMANCE		REFERENCE NUMBER			NOTES	REC. HOURS
			LOAD	TIME	PRE-BMS-92	BMS-92	POST-BMS-92		
C-12-SC-5	12^1/$_2$"	12^1/$_2$" × 14" column; 6" × 4^1/$_2$" × 20 lbs. "H" beam; Protection: gravel aggregate concrete (3750 psi) with 4" × 4" mesh; 16 SWG reinforcing 1" below column surface.	52 tons	4 hrs. 29 min.			7	1	4^1/$_3$
C-12-SC-6	12"	8" × 8" steel column; 2" outside protection; Group I.	—	11 hrs.			1	2	11
C-12-SC-7	12"	Description as per C-12-SC-6; Group II.	—	8 hrs.		1		2	8
C-12-SC-8	12"	Description as per C-12-SC-6; Group III.	—	6 hrs.		1		2	6
C-12-SC-9	12"	Description as per C-12-SC-6; Group IV.	—	4 hrs.		1		2	4
C-12-SC-10	12"	10" × 10" steel column; 2" outside protection; Group I.	—	7 hrs.		1		2	7
C-12-SC-11	12"	Description as per C-12-SC-10; Group II.	—	5 hrs.		1		2	5
C-12-SC-12	12"	Description as per C-12-SC-10; Group III.	—	4 hrs.		1		2	4
C-12-SC-13	12"	Description as per C-12-SC-10; Group IV.	—	2 hrs. 30 min.		1		2	2^1/$_2$
C-13-SC-14	13"	10" × 10" steel column; 3" outside protection; Group I.	—	10 hrs.		1		2	10
C-13-SC-15	13"	Description as per C-12-SC-14; Group II.	—	7 hrs.		1		2	7
C-13-SC-16	13"	Description as per C-12-SC-14; Group III.	—	5 hrs.		1		2	5
C-13-SC-17	13"	Description as per C-12-SC-14; Group IV.	—	3 hrs. 30 min.		1		2	3^1/$_2$

Notes:

1. Failure mode - collapse.
2. Group I: includes concrete having calcareous aggregate containing a combined total of not more than 10 percent of quartz, chert and flint for the coarse aggregate.

   Group II: includes concrete having trap-rock aggregate applied without metal ties and also concrete having cinder, sandstone or granite aggregate, if held in place with wire mesh or expanded metal having not larger than 4-inch mesh, weighing not less than 1.7 lbs./yd.^{2}, placed not more than 1 inch from the surface of the concrete.

   Group III: includes concrete having cinder, sandstone or granite aggregate tied with No. 5 gage steel wire, wound spirally over the column section on a pitch of 8 inches, or equivalent ties, and concrete having siliceous aggregates containing a combined total of 60 percent or more of quartz, chert and flint, if held in place with wire mesh or expanded metal having not larger than 4-inch mesh, weighing not less than 1.7 lbs./yd.^{2}, placed not more than 1 inch from the surface of the concrete.

   Group IV: includes concrete having siliceous aggregates containing a combined total of 60 percent or more of quartz, chert and flint, and tied with No. 5 gage steel wire wound spirally over the column section on a pitch of 8 inches, or equivalent ties.

**FIGURE 2.5.1.6—STEEL COLUMNS—CONCRETE ENCASEMENTS
MINIMUM DIMENSION 14" TO LESS THAN 16"**

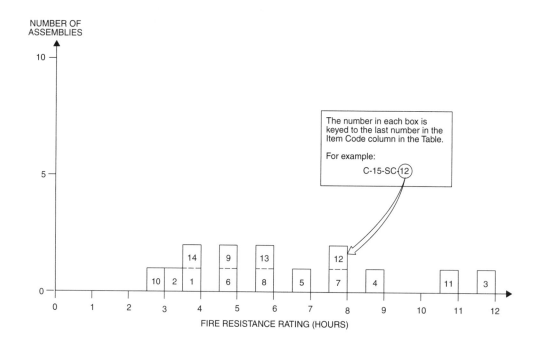

**TABLE 2.5.1.6—STEEL COLUMNS—CONCRETE ENCASEMENTS
MINIMUM DIMENSION 14" TO LESS THAN 16"**

ITEM CODE	MINIMUM DIMENSION	CONSTRUCTION DETAILS	PERFORMANCE		REFERENCE NUMBER			NOTES	REC. HOURS
			LOAD	TIME	PRE-BMS-92	BMS-92	POST-BMS-92		
C-14-SC-1	14"	24" × 16" concrete encased steel column; 8" × 6" × 35 lbs. "H" column; Protection: aggregate concrete (4240 psi); 4" mesh - 16 SWG reinforcing 1" below column surface.	90 tons	3 hrs. 40 min.			7	1	3
C-14-SC-2	14"	14" × 18" concrete encased steel column; 12" × 8" × 65 lbs. "H" beam; Protection: gravel aggregate concrete (4000 psi) with 4" - 16 SWG wire mesh reinforcement 1" below column surface.	177 tons	3 hrs. 20 min.			7	1	3
C-14-SC-3	14"	10" × 10" steel column; 4" outside protection; Group I.	—	12 hrs.		1		2	12
C-14-SC-4	14"	Description as per C-14-SC-3; Group II.	—	9 hrs.		1		2	9
C-14-SC-5	14"	Description as per C-14-SC-3; Group III.	—	7 hrs.		1		2	7
C-14-SC-6	14"	Description as per C-14-SC-3; Group IV.	—	5 hrs.		1		2	5
C-14-SC-7	14"	12" × 12" steel column; 2" outside protection; Group I.	—	8 hrs.		1		2	8
C-14-SC-8	14"	Description as per C-14-SC-7; Group II.	—	6 hrs.		1		2	6

(Continued)

**TABLE 2.5.1.6—STEEL COLUMNS—CONCRETE ENCASEMENTS
MINIMUM DIMENSION 14" TO LESS THAN 16"—(Continued)**

ITEM CODE	MINIMUM DIMENSION	CONSTRUCTION DETAILS	PERFORMANCE		REFERENCE NUMBER			NOTES	REC. HOURS
			LOAD	TIME	PRE-BMS-92	BMS-92	POST-BMS-92		
C-14-SC-9	14"	Description as per C-14-SC-7; Group III.	—	5 hrs.		1		2	5
C-14-SC-10	14"	Description as per C-14-SC-7; Group IV	—	3 hrs.		1		2	3
C-15-SC-11	15"	12" × 12" steel column; 3" outside protection; Group I.	—	11 hrs.		1		2	11
C-15-SC-12	15"	Description as per C-15-SC-11; Group II.	—	8 hrs.		1		2	8
C-15-SC-13	15"	Description as per C-15-SC-11; Group III.	—	6 hrs.		1		2	6
C-15-SC-14	15"	Description as per C-15-SC-11; Group IV.	—	4 hrs.		1		2	4

Notes:
1. Collapse.
2. Group I: includes concrete having calcareous aggregate containing a combined total of not more than 10 percent of quartz, chert and flint for the coarse aggregate.

Group II: includes concrete having trap-rock aggregate applied without metal ties and also concrete having cinder, sandstone or granite aggregate, if held in place with wire mesh or expanded metal having not larger than 4-inch mesh, weighing not less than 1.7 lbs./yd.², placed not more than 1 inch from the surface of the concrete.

Group III: includes concrete having cinder, sandstone or granite aggregate tied with No. 5 gage steel wire, wound spirally over the column section on a pitch of 8 inches, or equivalent ties, and concrete having siliceous aggregates containing a combined total of 60 percent or more of quartz, chert and flint, if held in place with wire mesh or expanded metal having not larger than 4-inch mesh, weighing not less than 1.7 lbs./yd.², placed not more than 1 inch from the surface of the concrete.

Group IV: includes concrete having siliceous aggregates containing a combined total of 60 percent or more of quartz, chert and flint, and tied with No. 5 gage steel wire wound spirally over the column section on a pitch of 8 inches, or equivalent ties.

**TABLE 2.5.1.7—STEEL COLUMNS—CONCRETE ENCASEMENTS
MINIMUM DIMENSION 16" TO LESS THAN 18"**

ITEM CODE	MINIMUM DIMENSION	CONSTRUCTION DETAILS	PERFORMANCE		REFERENCE NUMBER			NOTES	REC. HOURS
			LOAD	TIME	PRE-BMS-92	BMS-92	POST-BMS-92		
C-16-SC-13	16"	12" × 12" steel column; 4" outside protection; Group I.	—	14 hrs.		1		1	14
C-16-SC-2	16"	Description as per C-16-SC-1; Group II.	—	10 hrs.		1		1	10
C-16-SC-3	16"	Description as per C-16-SC-1; Group III.	—	8 hrs.		1		1	8
C-16-SC-4	16"	Description as per C-16-SC-1; Group IV.	—	5 hrs.		1		1	5

Notes:
1. Group I: includes concrete having calcareous aggregate containing a combined total of not more than 10 percent of quartz, chert and flint for the coarse aggregate.

Group II: includes concrete having trap-rock aggregate applied without metal ties and also concrete having cinder, sandstone or granite aggregate, if held in place with wire mesh or expanded metal having not larger than 4-inch mesh, weighing not less than 1.7 lbs./yd.², placed not more than 1 inch from the surface of the concrete.

Group III: includes concrete having cinder, sandstone or granite aggregate tied with No. 5 gage steel wire, wound spirally over the column section on a pitch of 8 inches, or equivalent ties, and concrete having siliceous aggregates containing a combined total of 60 percent or more of quartz, chert and flint, if held in place with wire mesh or expanded metal having not larger than 4-inch mesh, weighing not less than 1.7 lbs./yd.², placed not more than 1 inch from the surface of the concrete.

Group IV: includes concrete having siliceous aggregates containing a combined total of 60 percent or more of quartz, chert and flint, and tied with No. 5 gage steel wire wound spirally over the column section on a pitch of 8 inches, or equivalent ties.

TABLE 2.5.2.1—STEEL COLUMNS—BRICK AND BLOCK ENCASEMENTS
MINIMUM DIMENSION 10" TO LESS THAN 12"

ITEM CODE	MINIMUM DIMENSION	CONSTRUCTION DETAILS	PERFORMANCE		REFERENCE NUMBER			NOTES	REC. HOURS
			LOAD	TIME	PRE-BMS-92	BMS-92	POST-BMS-92		
C-10-SB-1	$10^1/_2$"	$10^1/_2$" × 13" brick encased steel columns; 8" × 6" × 35 lbs. "H" beam; Protection. Fill of broken brick and mortar; 2" brick on edge; joints broken in alternate courses; cement-sand grout; 13 SWG wire reinforcement in every third horizontal joint.	90 tons	3 hrs. 6 min.			7	1	3
C-10-SB-2	$10^1/_2$"	$10^1/_2$" × 13" brick encased steel columns; 8" × 6" × 35 lbs. "H" beam; Protection: 2" brick; joints broken in alternate courses; cement-sand grout; 13 SWG iron wire reinforcement in alternate horizontal joints.	90 tons	2 hrs.			7	2, 3, 4	2
C-10-SB-3	10"	10" × 12" block encased columns; 8" × 6" × 35 lbs. "H" beam; Protection: 2" foamed slag concrete blocks; 13 SWG wire at each horizontal joint; mortar at each joint.	90 tons	2 hrs.			7	5	2
C-10-SB-4	$10^1/_2$"	$10^1/_2$" × 12" block encased steel columns; 8" × 6" × 35 lbs. "H" beam; Protection: gravel aggregate concrete fill (unconsolidated) 2" thick hollow clay tiles with mortar at edges.	86 tons	56 min.			7	1	$^3/_4$
C-10-SB-5	$10^1/_2$"	$10^1/_2$" × 12" block encased steel columns; 8" × 6" × 35 lbs. "H" beam; Protection: 2" hollow clay tiles with mortar at edges.	86 tons	22 min.			7	1	$^1/_4$

Notes:
1. Failure mode - collapse.
2. Passed 2 hour fire test (Grade "C" - British).
3. Passed hose stream test.
4. Passed reload test.
5. Passed 2 hour fire exposure but collapsed immediately following hose stream test.

TABLE 2.5.2.2—STEEL COLUMNS—BRICK AND BLOCK ENCASEMENTS
MINIMUM DIMENSION 12" TO LESS THAN 14"

ITEM CODE	MINIMUM DIMENSION	CONSTRUCTION DETAILS	PERFORMANCE		REFERENCE NUMBER			NOTES	REC. HOURS
			LOAD	TIME	PRE-BMS-92	BMS-92	POST-BMS-92		
C-12-SB-1	12"	12" × 15" brick encased steel columns; 8" × 6" × 35 lbs. "H" beam; Protection: $2^5/_8$" thick brick; joints broken in alternate courses; cement-sand grout; fill of broken brick and mortar.	90 tons	1 hr. 49 min.			7	1	$1^3/_4$

Notes:
1. Failure mode – collapse.

TABLE 2.5.2.3—STEEL COLUMNS—BRICK AND BLOCK ENCASEMENTS
MINIMUM DIMENSION 14" TO LESS THAN 16"

ITEM CODE	MINIMUM DIMENSION	CONSTRUCTION DETAILS	PERFORMANCE		REFERENCE NUMBER			NOTES	REC. HOURS
			LOAD	TIME	PRE-BMS-92	BMS-92	POST-BMS-92		
C-15-SB-1	15"	15" × 17" brick encased steel columns; 8" × 6" × 35 lbs. "H" beam; Protection: 4$\frac{1}{2}$" thick brick; joints broken in alternate courses; cement-sand grout; fill of broken brick and mortar.	45 tons	6 hrs.			7	1	6
C-15-SB-2	15"	15" × 17" brick encased steel columns; 8" × 6" × 35 lbs. "H" beam; Protection. Fill of broken brick and mortar; 4$\frac{1}{2}$" brick; joints broken in alternate courses; cement-sand grout.	86 tons	6 hrs.			7	2, 3, 4	6
C-15-SB-3	15"	15" × 18" brick encased steel columns; 8" × 6" × 35 lbs. "H" beam; Protection: 4$\frac{1}{2}$" brick work; joints alternating; cement-sand grout.	90 tons	4 hrs.			7	5, 6	4
C-15-SB-4	14"	14" × 16" block encased steel columns; 8" × 6" × 35 lbs. "H" beam; Protection: 4" thick foam slag concrete blocks; 13 SWG wire reinforcement in each horizontal joint; mortar in joints.	90 tons	5 hrs. 52 min.			7	7	4$\frac{3}{4}$

Notes:
1. Only a nominal load was applied to specimen.
2. Passed 6 hour fire test (Grade "A" - British).
3. Passed (6 minute) hose stream test.
4. Reload not specified.
5. Passed 4 hour fire exposure.
6. Failed by collapse between first and second minute of hose stream exposure.
7. Mode of failure - collapse.

TABLE 2.5.3.1—STEEL COLUMNS—PLASTER ENCASEMENTS
MINIMUM DIMENSION 6" TO LESS THAN 8"

ITEM CODE	MINIMUM DIMENSION	CONSTRUCTION DETAILS	PERFORMANCE		REFERENCE NUMBER			NOTES	REC. HOURS
			LOAD	TIME	PRE-BMS-92	BMS-92	POST-BMS-92		
C-7-SP-1	7$\frac{1}{2}$"	7$\frac{1}{2}$ × 9$\frac{1}{2}$" plaster protected steel columns; 8" × 6" × 35 lbs. "H" beam; Protection: 24 SWG wire metal lath; 1$\frac{1}{4}$" lime plaster.	90 tons	57 min.			7	1	$\frac{3}{4}$
C-7-SP-2	7$\frac{7}{8}$"	7$\frac{7}{8}$" × 10" plaster protected steel columns; 8" × 6" × 35 lbs. "H" beam; Protection: $\frac{3}{8}$" gypsum bal wire wound with 16 SWG wire helically wound at 4" pitch; 1/2"gypsum plaster.	90 tons	1 hr. 13 min.			7	1	1
C-7-SP-3	7$\frac{1}{4}$"	7$\frac{1}{4}$" × 9$\frac{3}{8}$" plaster protected steel columns; 8" × 6" × 35 lbs. "H" beam; Protection: $\frac{3}{8}$" gypsum board; wire helically wound 16 SWG at 4" pitch; $\frac{1}{4}$" gypsum plaster finish.	90 tons	1 hr. 14 min.			7	1	1

Notes:
1. Failure mode – collapse.

**TABLE 2.5.3.2—STEEL COLUMNS—PLASTER ENCASEMENTS
MINIMUM DIMENSION 8" TO LESS THAN 10"**

ITEM CODE	MINIMUM DIMENSION	CONSTRUCTION DETAILS	PERFORMANCE		REFERENCE NUMBER			NOTES	REC. HOURS
			LOAD	TIME	PRE-BMS-92	BMS-92	POST-BMS-92		
C-8-SP-1	8"	8" × 10" plaster protected steel columns; 8" × 6" × 35 lbs. "H" beam; Protection: 24 SWG wire lath; 1" gypsum plaster.	86 tons	1 hr. 23 min.			7	1	$1^1/_4$
C-8-SP-2	$8^1/_2$"	$8^1/_2$" × $10^1/_2$" plaster protected steel columns; 8" × 6" × 35 lbs. "H" beam; Protection: 24 SWG metal lath wrap; $1^1/_4$" gypsum plaster.	90 tons	1 hr. 36 min.			7	1	$1^1/_2$
C-9-SP-3	9"	9" × 11" plaster protected steel columns; 8" × 6" × 35 lbs. "H" beam; Protection: 24 SWG metal lath wrap; $1/_8$" M.S. ties at 12" pitch wire netting $1^1/_2$" × 22 SWG between first and second plaster coats; $1^1/_2$" gypsum plaster.	90 tons	1 hr. 33 min.			7	1	$1^1/_2$
C-8-SP-4	$8^3/_4$"	$8^3/_4$" × $10^3/_4$" plaster protected steel columns; 8" × 6" × 35 lbs. "H" beam; Protection: $3/_4$" gypsum board; wire wound spirally (#16 SWG) at $1^1/_2$" pitch; $1/_2$" gypsum plaster.	90 tons	2 hrs.			7	2, 3, 4	2

Notes:
1. Failure mode - collapse.
2. Passed 2 hour fire exposure test (Grade "C" - British).
3. Passed hose stream test.
4. Passed reload test.

**TABLE 2.5.4.1—STEEL COLUMNS—MISCELLANEOUS ENCASEMENTS
MINIMUM DIMENSION 6" TO LESS THAN 8"**

ITEM CODE	MINIMUM DIMENSION	CONSTRUCTION DETAILS	PERFORMANCE		REFERENCE NUMBER			NOTES	REC. HOURS
			LOAD	TIME	PRE-BMS-92	BMS-92	POST-BMS-92		
C-7-SM-1	$7^5/_8$"	$7^5/_8$" × $9^1/_2$" (asbestos plaster) protected steel columns; 8" × 6" × 35 lbs. "H" beam; Protection: 20 gage $1/_2$" metal lath; $9/_{16}$" asbestos plaster (minimum).	90 tons	1 hr. 52 min.			7	1	$1^3/_4$

Notes:
1. Failure mode - collapse.

**TABLE 2.5.4.2—STEEL COLUMNS—MISCELLANEOUS ENCASEMENTS
MINIMUM DIMENSION 8" TO LESS THAN 10"**

ITEM CODE	MINIMUM DIMENSION	CONSTRUCTION DETAILS	PERFORMANCE		REFERENCE NUMBER			NOTES	REC. HOURS
			LOAD	TIME	PRE-BMS-92	BMS-92	POST-BMS-92		
C-9-SM-1	$9^5/_8$"	$9^5/_8$" × $11^3/_8$" asbestos slab and cement plaster protected columns; 8" × 6" × 35 lbs. "H" beam; Protection: 1" asbestos slab; wire wound; $5/_8$" plaster.	90 tons	2 hrs.			7	1, 2	2

Notes:
1. Passed 2 hour fire exposure test.
2. Collapsed during hose stream test.

**TABLE 2.5.4.3—STEEL COLUMNS—MISCELLANEOUS ENCASEMENTS
MINIMUM DIMENSION 10" TO LESS THAN 12"**

ITEM CODE	MINIMUM DIMENSION	CONSTRUCTION DETAILS	PERFORMANCE		REFERENCE NUMBER			NOTES	REC. HOURS
			LOAD	TIME	PRE-BMS-92	BMS-92	POST-BMS-92		
C-11-SM-1	11^1/$_2$"	11^1/$_2$" × 13^1/$_2$" wood wool and plaster protected steel columns; 8" × 6" × 35 lbs. "H" beam; Protection: wood-wool-cement paste as fill and to 2" cover over beam; ^{3}/$_4$" gypsum plaster finish.	90 tons	2 hrs.			7	1, 2, 3	2
C-10-SM-1	10"	10" × 12" asbestos protected steel columns; 8" × 6" × 35 lbs. "H" beam; Protection: sprayed on asbestos paste to 2" cover over column.	90 tons	4 hrs.			7	2, 3, 4	4

Notes:

1. Passed 2 hour fire exposure (Grade "C" - British).
2. Passed hose stream test.
3. Passed reload test.
4. Passed 4 hour fire exposure test.

**TABLE 2.5.4.4—STEEL COLUMNS—MISCELLANEOUS ENCASEMENTS
MINIMUM DIMENSION 12" TO LESS THAN 14"**

ITEM CODE	MINIMUM DIMENSION	CONSTRUCTION DETAILS	PERFORMANCE		REFERENCE NUMBER			NOTES	REC. HOURS
			LOAD	TIME	PRE-BMS-92	BMS-92	POST-BMS-92		
C-12-SM-1	12"	12" × 14^1/$_4$" cement and asbestos protected columns; 8" × 6" × 35 lbs. "H" beam; Protection: fill of asbestos packing pieces 1" thick 1' 3" o.c.; cover of 2" molded asbestos inner layer; 1" molded asbestos outer layer; held in position by 16 SWG nichrome wire ties; wash of refractory cement on outer surface.	86 tons	4 hrs. 43 min.			7	1, 2, 3	4^2/$_3$

Notes:

1. Passed 4 hour fire exposure (Grade "B" - British).
2. Passed hose stream test.
3. Passed reload test.

SECTION III—FLOOR/CEILING ASSEMBLIES

FIGURE 3.1—FLOOR/CEILING ASSEMBLIES—REINFORCED CONCRETE

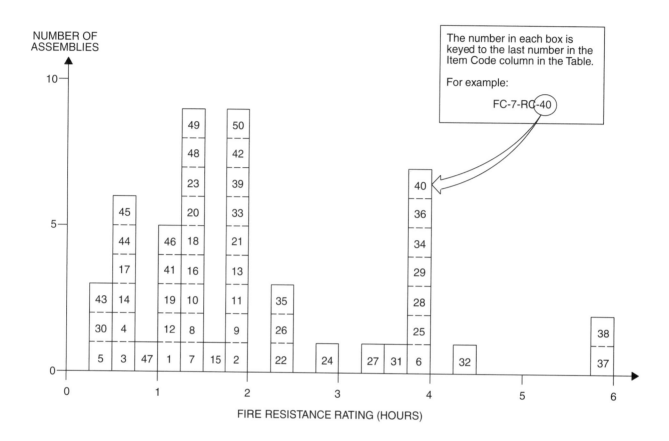

TABLE 3.1—FLOOR/CEILING ASSEMBLIES—REINFORCED CONCRETE

| ITEM CODE | ASSEMBLY THICKNESS | CONSTRUCTION DETAILS | PERFORMANCE | | REFERENCE NUMBER | | | NOTES | REC. HOURS |
			LOAD	TIME	PRE-BMS-92	BMS-92	POST-BMS-92		
F/C-3-RC-1	$3^3/_4$"	$3^3/_4$" thick floor; $3^1/_4$" (5475 psi) concrete deck; $^1/_2$" plaster under deck; $^3/_8$" main reinforcement bars at $5^1/_2$" pitch with $^7/_8$" concrete cover; $^3/_8$" main reinforcement bars at $4^1/_2$" pitch perpendicular with $^1/_2$" concrete cover; 13'1" span restrained.	195 psf	24 min.			7	1, 2	1
F/C-3-RC-2	$3^1/_4$"	$3^1/_4$" deep (3540 psi) concrete deck; $^3/_8$" main reinforcement bars at $5^1/_2$" pitch with $^7/_8$" cover; $^3/_8$" main reinforcement bars at $4^1/_2$" pitch perpendicular with $^1/_2$" cover; 13'1" span restrained.	195 psf	2 hrs.			7	1, 3, 4	$1^3/_4$
F/C-3-RC-3	$3^1/_4$"	$3^1/_4$" deep (4175 psi) concrete deck; $^3/_8$" main reinforcement bars at $5^1/_2$" pitch with $^7/_8$" cover; $^3/_8$" main reinforcement bars at $4^1/_2$" pitch perpendicular with $^1/_2$" cover; 13'1" span restrained.	195 psf	31 min.			7	1, 5	$^1/_2$

(Continued)

TABLE 3.1—FLOOR/CEILING ASSEMBLIES—REINFORCED CONCRETE—(Continued)

ITEM CODE	ASSEMBLY THICKNESS	CONSTRUCTION DETAILS	PERFORMANCE		REFERENCE NUMBER			NOTES	REC. HOURS
			LOAD	TIME	PRE-BMS-92	BMS-92	POST-BMS-92		
F/C-3-RC-4	$3^1/_4$"	$3^1/_4$" deep (4355 psi) concrete deck; $3/_8$" main reinforcement bars at $5^1/_2$" pitch with $7/_8$" cover; $3/_8$" main reinforcement bars at $4^1/_2$" pitch perpendicular with $1/_2$" cover; 13'1" span restrained.	195 psf	41 min.			7	1, 5, 6	$1/_2$
F/C-3-RC-5	$3^1/_4$"	$3^1/_4$" thick (3800 psi) concrete deck; $3/_8$" main reinforcement bars at $5^1/_2$" pitch with $7/_8$" cover; $3/_8$" main reinforcement bars at $4^1/_2$" pitch perpendicular with $1/_2$" cover; 13'1" span restrained.	195 psf	1 hr. 5 min.			7	1, 5	$1/_4$
F/C-4-RC-6	$4^1/_4$"	$4^1/_4$" thick; $3^1/_4$" (4000 psi) concrete deck; 1" sprayed asbestos lower surface; $3/_8$" main reinforcement bars at $5^7/_8$" pitch with $7/_8$" concrete cover; $3/_8$" main reinforcement bars at $4^1/_2$" pitch perpendicular with $1/_2$" concrete cover; 13'1" span restrained.	195 psf	4 hrs.			7	1, 7	4
F/C-4-RC-7	4"	4" (5025 psi) concrete deck; $1/_4$" reinforcement bars at $7^1/_2$" pitch with $3/_4$" cover; $3/_8$" main reinforcement bars at $3^3/_4$" pitch perpendicular with $1/_2$" cover; 13'1" span restrained.	140 psf	1 hr. 16 min.			7	1, 2	$1^1/_4$
F/C-4-RC-8	4"	4" thick (4905 psi) deck; $1/_4$" reinforcement bars at $7^1/_2$" pitch with $7/_8$" cover; $3/_8$" main reinforcement bars at $3^3/_4$" pitch perpendicular with $1/_2$" cover; 13'1" span restrained.	100 psf	1 hr. 23 min.			7	1, 2	$1^1/_3$
F/C-4-RC-9	4"	4" deep (4370 psi); $1/_4$" reinforcement bars at 6" pitch with $3/_4$" cover; $1/_4$" main reinforcement bars at 4" pitch perpendicular with $1/_2$" cover; 13'1" span restrained.	150 psf	2 hrs.			7	1, 3	2
F/C-4-RC-10	4"	4" thick (5140 psi) deck; $1/_4$" reinforcement bars at $7^1/_2$" pitch with $7/_8$" cover; $3/_8$" main reinforcement bars at $3^3/_4$" pitch perpendicular with $1/_2$" cover; 13'1" span restrained.	140 psf	1 hr. 16 min.			7	1, 5	$1^1/_4$
F/C-4-RC-11	4"	4" thick (4000 psi) concrete deck; 3" × $1^1/_2$" × 4 lbs. R.S.J.; 2'6" C.R.S.; flush with top surface; 4" × 6" x 13 SWG mesh reinforcement 1" from bottom of slab; 6'6" span restrained.	150 psf	2 hrs.			7	1, 3	2
F/C-4-RC-12	4"	4" deep (2380 psi) concrete deck; 3" × $1^1/_2$" × 4 lbs. R.S.J.; 2'6" C.R.S.; flush with top surface; 4" × 6" x 13 SWG mesh reinforcement 1" from bottom surface; 6'6" span restrained.	150 psf	1 hr. 3 min.			7	1, 2	1

(Continued)

TABLE 3.1—FLOOR/CEILING ASSEMBLIES—REINFORCED CONCRETE—(Continued)

ITEM CODE	ASSEMBLY THICKNESS	CONSTRUCTION DETAILS	PERFORMANCE		REFERENCE NUMBER			NOTES	REC. HOURS
			LOAD	TIME	PRE-BMS-92	BMS-92	POST-BMS-92		
F/C-4-RC-13	$4^1/_2$"	$4^1/_2$" thick (5200 psi) deck; $^1/_4$" reinforcement bars at $7^1/_4$" pitch with $^7/_8$" cover; $^3/_8$" main reinforcement bars at $3^3/_4$" pitch perpendicular with $^1/_2$" cover; 13'1" span restrained.	140 psf	2 hrs.			7	1, 3	2
F/C-4-RC-14	$4^1/_2$"	$4^1/_2$" deep (2525 psi) concrete deck; $^1/_4$" reinforcement bars at $7^1/_2$" pitch with $^7/_8$" cover; $^3/_8$" main reinforcement bars at $3^3/_8$" pitch perpendicular with $^1/_2$" cover; 13'1" span restrained.	150 psf	42 min.			7	1, 5	$^2/_3$
F/C-4-RC-15	$4^1/_2$"	$4^1/_2$" deep (4830 psi) concrete deck; $1^1/_2$" × No. 15 gauge wire mesh; $^3/_8$" reinforcement bars at 15" pitch with 1" cover; $^1/_2$" main reinforcement bars at 6" pitch perpendicular with $^1/_2$" cover; 12' span simply supported.	75 psf	1 hr. 32 min.			7	1, 8	$1^1/_2$
F/C-4-RC-16	$4^1/_2$"	$4^1/_2$" deep (4595 psi) concrete deck; $^1/_4$" reinforcement bars at $7^1/_2$" pitch with $^7/_8$" cover; $^3/_8$" main reinforcement bars at $3^1/_2$" pitch perpendicular with $^1/_2$" cover; 12' span simply supported.	75 psf	1 hr. 20 min.			7	1, 8	$1^1/_3$
F/C-4-RC-17	$4^1/_2$"	$4^1/_2$" deep (3625 psi) concrete deck; $^1/_4$" reinforcement bars at $7^1/_2$" pitch with $^7/_8$" cover; $^3/_8$" main reinforcement bars at $3^1/_2$" pitch perpendicular with $^1/_2$" cover; 12' span simply supported.	75 psf	35 min.			7	1, 8	$^1/_2$
F/C-4-RC-18	$4^1/_2$"	$4^1/_2$" deep (4410 psi) concrete deck; $^1/_4$" reinforcement bars at $7^1/_2$" pitch with $^7/_8$" cover; $^3/_8$" main reinforcement bars at $3^1/_2$" pitch perpendicular with $^1/_2$" cover; 12' span simply supported.	85 psf	1 hr. 27 min.			7	1, 8	$1^1/_3$
F/C-4-RC-19	$4^1/_2$"	$4^1/_2$" deep (4850 psi) deck; $^3/_8$" reinforcement bars at 15" pitch with 1" cover; $^1/_2$" main reinforcement bars at 6" pitch perpendicular with $^1/_2$" cover; 12' span simply supported.	75 psf	2 hrs. 15 min.			7	1, 9	$1^1/_4$
F/C-4-RC-20	$4^1/_2$"	$4^1/_2$" deep (3610 psi) deck; $^1/_4$" reinforcement bars at $7^1/_2$" pitch with $^7/_8$" cover; $^3/_8$" main reinforcement bars at $3^1/_2$" pitch perpendicular with $^1/_2$" cover; 12' span simply supported.	75 psf	1 hr. 22 min.			7	1, 8	$1^1/_3$
F/C-5-RC-21	5"	5" deep; $4^1/_2$" (5830 psi) concrete deck; $^1/_2$" plaster finish bottom of slab; $^1/_4$" reinforcement bars at $7^1/_2$" pitch with $^7/_8$" cover; $^3/_8$" main reinforcement bars at $3^1/_2$" pitch perpendicular with $^1/_2$" cover; 12' span simply supported.	69 psf	2 hrs.			7	1, 3	2

(Continued)

TABLE 3.1—FLOOR/CEILING ASSEMBLIES—REINFORCED CONCRETE—(Continued)

ITEM CODE	ASSEMBLY THICKNESS	CONSTRUCTION DETAILS	PERFORMANCE		REFERENCE NUMBER			NOTES	REC. HOURS
			LOAD	TIME	PRE-BMS-92	BMS-92	POST-BMS-92		
F/C-5-RC-22	5"	$4^1/_2$" (5290 psi) concrete deck; $^1/_2$" plaster finish bottom of slab; $^1/_4$" reinforcement bars at $7^1/_2$" pitch with $^7/_8$" cover; $^3/_8$" main reinforcement bars at $3^1/_2$" pitch perpendicular with $^1/_2$" cover; 12' span simply supported.	No load	2 hrs. 28 min.			7	1, 10, 11	$2^1/_4$
F/C-5-RC-23	5"	5" (3020 psi) concrete deck; 3" × $1^1/_2$" × 4 lbs. R.S.J.; 2' C.R.S. with 1" cover on bottom and top flanges; 8' span restrained.	172 psf	1 hr. 24 min.			7	1, 2, 12	$1^1/_2$
F/C-5-RC-24	$5^1/_2$"	5" (5180 psi) concrete deck; $^1/_2$" retarded plaster underneath slab; $^1/_4$" reinforcement bars at $7^1/_2$" pitch with $1^3/_8$" cover; $^3/_8$" main reinforcement bars at $3^1/_2$" pitch perpendicular with 1" cover; 12' span simply supported.	60 psf	2 hrs. 48 min.			7	1, 10	$2^3/_4$
F/C-6-RC-25	6"	6" deep (4800 psi) concrete deck; $^1/_4$" reinforcement bars at $7^1/_2$" pitch with $^7/_8$" cover; $^3/_8$" main reinforcement bars at $3^1/_2$" pitch perpendicular with $^7/_8$" cover; 13'1" span restrained.	195 psf	4 hrs.			7	1, 7	4
F/C-6-RC-26	6	6" (4650 psi) concrete deck; $^1/_4$" reinforcement bars at $7^1/_2$" pitch with $^7/_8$" cover; $^3/_8$" main reinforcement bars at $3^1/_2$" pitch perpendicular with $^1/_2$" cover; 13'1" span restrained.	195 psf	2 hrs. 23 min.			7	1, 2	$2^1/_4$
F/C-6-RC-27	6"	6" deep (6050 psi) concrete deck; $^1/_4$" reinforcement bars at $7^1/_2$" pitch $^7/_8$" cover; $^3/_8$" reinforcement bars at $3^1/_2$" pitch perpendicular with $^1/_2$" cover; 13'1" span restrained.	195 psf	3 hrs. 30 min.			7	1, 10	$3^1/_2$
F/C-6-RC-28	6"	6" deep (5180 psi) concrete deck; $^1/_4$" reinforcement bars at 8" pitch $^3/_4$" cover; $^1/_4$" reinforcement bars at $5^1/_2$" pitch perpendicular with $^1/_2$" cover; 13'1" span restrained.	150 psf	4 hrs.			7	1, 7	4
F/C-6-RC-29	6"	6" thick (4180 psi) concrete deck; 4" × 3" × 10 lbs. R.S.J.; 2'6" C.R.S. with 1" cover on both top and bottom flanges; 13'1" span restrained.	160 psf	3 hrs. 48 min.			7	1, 10	$3^3/_4$
F/C-6-RC-30	6"	6" thick (3720 psi) concrete deck; 4" × 3" × 10 lbs. R.S.J.; 2'6" C.R.S. with 1" cover on both top and bottom flanges; 12' span simply supported.	115 psf	29 min.			7	1, 5, 13	$^1/_4$
F/C-6-RC-31	6"	6" deep (3450 psi) concrete deck; 4" × $1^3/_4$" × 5 lbs. R.S.J.; 2'6" C.R.S. with 1" cover on both top and bottom flanges; 12' span simply supported.	25 psf	3 hrs. 35 min.			7	1, 2	$3^1/_2$

(Continued)

TABLE 3.1—FLOOR/CEILING ASSEMBLIES—REINFORCED CONCRETE—(Continued)

ITEM CODE	ASSEMBLY THICKNESS	CONSTRUCTION DETAILS	PERFORMANCE		REFERENCE NUMBER			NOTES	REC. HOURS
			LOAD	TIME	PRE-BMS-92	BMS-92	POST-BMS-92		
F/C-6-RC-32	6"	6" deep (4460 psi) concrete deck; 4" × 1³/₄" × 5 lbs. R.S.J.; 2' C.R.S.; with 1" cover on both top and bottom flanges; 12' span simply supported.	60 psf	4 hrs. 30 min.			7	1, 10	4¹/₂
F/C-6-RC-33	6"	6" deep (4360 psi) concrete deck; 4" × 1³/₄" × 5 lbs. R.S.J.; 2' C.R.S.; with 1" cover on both top and bottom flanges; 13'1" span restrained.	60 psf	2 hrs.			7	1, 3	2
F/C-6-RC-34	6¹/₄"	6¹/₄" thick; 4³/₄" (5120 psi) concrete core; 1" T&G board flooring; ¹/₂" plaster undercoat; 4" × 3" × 10 lbs. R.S.J.; 3' C.R.S. flush with top surface concrete; 12' span simply supported; 2" × 1'3" clinker concrete insert.	100 psf	4 hrs.			7	1, 7	4
F/C-6-RC-35	6¹/₄"	4³/₄" (3600 psi) concrete core; 1" T&G board flooring; ¹/₂" plaster undercoat; 4" × 3" × 10 lbs. R.S.J.; 3' C.R.S.; flush with top surface concrete; 12' span simply supported; 2" × 1'3" clinker concrete insert.	100 psf	2 hrs. 30 min.			7	1, 5	2¹/₂
F/C-6-RC-36	6¹/₄"	4³/₄" (2800 psi) concrete core; 1" T&G board flooring; ¹/₂" plaster undercoat; 4" × 3" × 10 lbs. R.S.J.; 3' C.R.S.; flush with top surface concrete; 12" span simply supported; 2" × 1'3" clinker concrete insert.	80 psf	4 hrs.			7	1, 7	4
F/C-7-RC-37	7"	(3640 psi) concrete deck; ¹/₄" reinforcement bars at 6" pitch with 1¹/₂" cover; ¹/₄" reinforcement bars at 5" pitch perpendicular with 1¹/₂" cover; 13'1" span restrained.	169 psf	6 hrs.			7	1, 14	6
F/C-7-RC-38	7"	(4060 psi) concrete deck; 4" × 3" × 10 lbs. R.S.J.; 2'6" C.R.S. with 1¹/₂" cover on both top and bottom flanges; 4" × 6" × 13 SWG mesh reinforcement 1¹/₂" from bottom of slab; 13'1" span restrained.	175 psf	6 hrs.			7	1, 14	6
F/C-7-RC-39	7¹/₄"	5³/₄" (4010 psi) concrete core; 1" T&G board flooring; ¹/₂" plaster undercoat; 4" × 3" × 10 lbs. R.S.J.; 2'6" C.R.S.; 1" down from top surface of concrete; 12' simply supported span; 2" × 1'3" clinker concrete insert.	95 psf	2 hrs.			7	1, 3	2
F/C-7-RC-40	7¹/₄"	5³/₄" (3220 psi) concrete core; 1" T&G flooring; ¹/₂" plaster undercoat; 4" × 3" × 10 lbs. R.S.J.; 2'6" C.R.S.; 1" down from top surface of concrete; 12' simply supported span; 2" × 1'3" clinker concrete insert.	95 psf	4 hrs.			7	1, 7	4

(Continued)

TABLE 3.1—FLOOR/CEILING ASSEMBLIES—REINFORCED CONCRETE—(Continued)

ITEM CODE	ASSEMBLY THICKNESS	CONSTRUCTION DETAILS	PERFORMANCE		REFERENCE NUMBER			NOTES	REC. HOURS
			LOAD	TIME	PRE-BMS-92	BMS-92	POST-BMS-92		
F/C-7-RC-41	10" (2$\frac{1}{4}$" Slab)	Ribbed floor, see Note 15 for details; slab 2$\frac{1}{2}$" deep (3020 psi); $\frac{1}{4}$" reinforcement bars at 6" pitch with $\frac{3}{4}$" cover; beams 7$\frac{1}{2}$" deep × 5" wide; 24" C.R.S.; $\frac{5}{8}$" reinforcement bars two rows $\frac{1}{2}$" vertically apart with 1" cover; 13'1" span restricted.	195 psf	1 hr. 4 min.			7	1, 2, 15	1
F/C-5-RC-42	5$\frac{1}{2}$"	Composite ribbed concrete slab assembly; see Note 17 for details.	See Note 16	2 hrs.			43	16, 17	2
F/C-3-RC-43	3"	2500 psi concrete; $\frac{5}{8}$" cover; fully restrained at test.	See Note 16	30 min.			43	16	$\frac{1}{2}$
F/C-3-RC-44	3"	2000 psi concrete; $\frac{5}{8}$" cover; free or partial restraint at test.	See Note 16	45 min.			43	16	$\frac{3}{4}$
F/C-4-RC-45	4"	2500 psi concrete; $\frac{5}{8}$" cover; fully restrained at test.	See Note 16	40 min.			43	16	$\frac{2}{3}$
F/C-4-RC-46	4"	2000 psi concrete; $\frac{3}{4}$" cover; free or partial restraint at test.	See Note 16	1 hr. 15 min.			43	16	1$\frac{1}{4}$
F/C-5-RC-47	5"	2500 psi concrete; $\frac{3}{4}$" cover; fully restrained at test.	See Note 16	1 hr.			43	16	1
F/C-5-RC-48	5"	2000 psi concrete; $\frac{3}{4}$" cover; free or partial restraint at test.	See Note 16	1 hr. 30 min.			43	16	1$\frac{1}{2}$
F/C-6-RC-49	6"	2500 psi concrete; 1" cover; fully restrained at test.	See Note 16	1 hr. 30 min.			43	16	1$\frac{1}{2}$
F/C-6-RC-50	6"	2000 psi concrete; 1" cover; free or partial restraint at test.	See Note 16	2 hrs.			43	16	2

Notes:
1. British test.
2. Failure mode - local back face temperature rise.
3. Tested for Grade "C" (2 hour) fire resistance.
4. Collapse imminent following hose stream.
5. Failure mode - flame thru.
6. Void formed with explosive force and report.
7. Achieved Grade "B" (4 hour) fire resistance (British).
8. Failure mode - collapse.
9. Test was run to 2 hours, but specimen was partially supported by the furnace at 1$\frac{1}{4}$ hours.
10. Failure mode - average back face temperature.
11. Recommended endurance for nonload bearing performance only.
12. Floor maintained load bearing ability to 2 hours at which point test was terminated.
13. Test was run to 3 hours at which time failure mode 2 (above) was reached in spite of crack formation at 29 minutes.
14. Tested for Grade "A" (6 hour) fire resistance.
15.

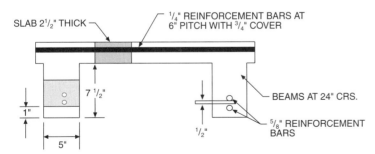

16. Load unspecified.
17. Total assembly thickness 5$\frac{1}{2}$ inches. Three-inch thick blocks of molded excelsior bonded with portland cement used as inserts with 2$\frac{1}{2}$-inch cover (concrete) above blocks and $\frac{3}{4}$-inch gypsum plaster below. Nine-inch wide ribs containing reinforcing steel of unspecified size interrupted 20-inch wide segments of slab composite (i.e., plaster, excelsior blocks, concrete cover).

FIGURE 3.2—FLOOR/CEILING ASSEMBLIES—STEEL STRUCTURAL ELEMENTS

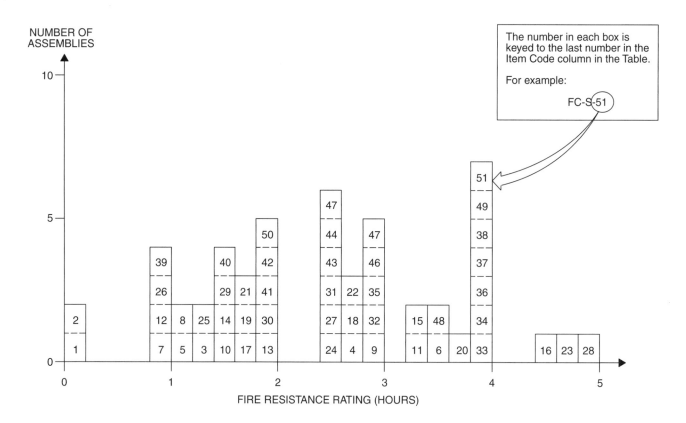

TABLE 3.2—FLOOR/CEILING ASSEMBLIES—STEEL STRUCTURAL ELEMENTS

ITEM CODE	MEMBRANE THICKNESS	CONSTRUCTION DETAILS	PERFORMANCE		REFERENCE NUMBER			NOTES	REC. HOURS
			LOAD	TIME	PRE-BMS-92	BMS-92	POST-BMS-92		
F/C-S-1	0"	- 10' × 13'6"; S.J. 103 - 24" o.c.; Deck: 2" concrete; Membrane: none.	145 psf	7 min.			3	1, 2, 3, 8	0
F/C-S-2	0"	- 10' × 13'6"; S.J. 103 - 24" o.c.; Deck: 2" concrete; Membrane: none	145 psf	7 min.			3	1, 2, 3, 8	0
F/C-S-3	$^1/_2$"	- 10' × 13'6"; S.J. 103 - 24" o.c.; Deck: 2" concrete 1:2:4; Membrane: furring 12" o.c.; Clips A, B, G; No extra reinforcement; $^1/_2$" plaster - 1.5:2.5.	145 psf	1 hr. 15 min.			3	2, 3, 8	$1^1/_4$
F/C-S-4	$^1/_2$"	- 10' × 13'6"; S.J. 103 - 24" o.c.; Deck: 2" concrete 1:2:4; Membrane: furring 16" o.c.; Clips D, E, F, G; Diagonal wire reinforcement; $^1/_2$" plaster - 1.5:2.5.	145 psf	2 hrs. 46 min.			3	3, 8	$2^3/_4$
F/C-S-5	$^1/_2$"	- 10' × 13'6"; S.J. 103 - 24" o.c.; Deck: 2" concrete 1:2:4; Membrane: furring 16" o.c.; Clips A, B, G; No extra reinforcement; $^1/_2$" plaster - 1.5:2.5.	145 psf	1 hr. 4 min.			3	2, 3, 8	1

(Continued)

TABLE 3.2—FLOOR/CEILING ASSEMBLIES—STEEL STRUCTURAL ELEMENTS—(Continued)

ITEM CODE	MEMBRANE THICKNESS	CONSTRUCTION DETAILS	PERFORMANCE		REFERENCE NUMBER			NOTES	REC. HOURS
			LOAD	TIME	PRE-BMS-92	BMS-92	POST-BMS-92		
F/C-S-6	$1/2$"	10' × 13'6"; S.J. 103 - 24" o.c.; Deck: 2" concrete 1:2:4; Membrane: furring 16" o.c.; Clips D, E, F, G; Hexagonal mesh reinforcement; $1/2$" plaster.	145 psf	3 hrs. 28 min.			3	2, 3, 8	$2^1/_3$
F/C-S-7	$1/2$"	10' × 13'6"; S.J. 103 - 24" o.c.; Deck: 4 lbs. rib lath; 6" × 6" - 10 × 10 ga. reinforcement; 2" deck gravel concrete; Membrane: furring 16" o.c.; Clips C, E; Reinforcement: none; $1/2$" plaster - 1.5:2.5 mill mix.	N/A	55 min.			3	5, 8	$3/_4$
F/C-S-8	$1/2$"	Spec. 9' × 4'4"; S.J. 103 bar joists - 18" o.c.; Deck: 4 lbs. rib lath base; 6" × 6" - 10 × 10 ga. reinforcement; 2" deck 1:2:4 gravel concrete; Membrane: furring, $3/_4$" C.R.S., 16" o.c.; Clips C, E; Reinforcement: none; $1/2$" plaster - 1.5:2.5 mill mix.	300 psf	1 hr. 10 min.			3	2, 3, 8	1
F/C-S-9	$5/_8$"	10' × 13'6"; S.J. 103 - 24" o.c.; Deck: 2" concrete 1:2:4; Membrane: furring 12" o.c.; Clips A, B, G; Extra "A" clips reinforcement; $5/_8$" plaster - 1.5:2; 1.5:3.	145 psf	3 hrs.			3	6, 8	3
F/C-S-10	$5/_8$"	18' × 13'6"; Joists, S.J. 103 - 24" o.c.; Deck: 4 lbs. rib lath; 6" × 6" - 10 × 10 ga. reinforcement; 2" deck 1:2:3.5 gravel concrete; Membrane: furring, spacing 16" o.c.; Clips C, E; Reinforcement: none; $5/_8$" plaster - 1.5:2.5 mill mix.	145 psf	1 hr. 25 min.			3	2, 3, 8	$1^1/_3$
F/C-S-11	$5/_8$"	10' × 13'6"; S.J. 103 - 24" o.c.; Deck: 2" concrete 1:2:4; Membrane: furring 12" o.c.; Clips D, E, F, G; Diagonal wire reinforcement; $5/_8$" plaster - 1.5:2; 0.5:3.	145 psf	3 hrs. 15 min.			3	2, 4, 8	$3^1/_4$
F/C-S-12	$5/_8$"	10' × 13'6"; Joists, S.J. 103 - 24" o.c.; Deck: 3.4 lbs. rib lath; 6" × 6" - 10 × 10 ga. reinforcement; 2" deck 1:2:4 gravel concrete; Membrane: furring 16" o.c.; Clips D, E, F, G; Reinforcement: none; $5/_8$" plaster - 1.5:2.5.	145 psf	1 hr.			3	7, 8	1
F/C-S-13	$3/_4$"	Spec. 9' × 4'4"; S.J. 103 - 18" o.c.; Deck: 4 lbs. rib lath; 6" × 6" - 10 × 10 ga. reinforcement; 2" deck 1:2:4 gravel concrete; Membrane: furring, $3/_4$" C.R.S., 16" o.c.; Clips C, E; Reinforcement: none; $3/_4$" plaster - 1.5:2.5 mill mix.	300 psf	1 hr. 56 min.			3	3, 8	$1^3/_4$

(Continued)

TABLE 3.2—FLOOR/CEILING ASSEMBLIES—STEEL STRUCTURAL ELEMENTS—(Continued)

ITEM CODE	MEMBRANE THICKNESS	CONSTRUCTION DETAILS	PERFORMANCE		REFERENCE NUMBER			NOTES	REC. HOURS
			LOAD	TIME	PRE-BMS-92	BMS-92	POST-BMS-92		
F/C-S-14	$^7/_8$"	Floor finish: 1" concrete; plate cont. weld; 4" - 7.7 lbs. "I" beams; Ceiling: $^1/_4$" rods 12" o.c.; $^7/_8$" gypsum sand plaster.	105 psf	1 hr. 35 min.			6	2, 4, 9, 10	$1^1/_2$
F/C-S-15	1"	Floor finish: $1^1/_2$" L.W. concrete; $^1/_2$" limestone cement; plate cont. weld; 5" - 10 lbs. "I" beams; Ceiling: $^1/_4$" rods 12" o.c. tack welded to beams metal lath; 1" P. C. plaster.	165 psf	3 hrs. 20 min.			6	4, 9, 11	
F/C-S-16	1"	10' × 13'6"; S.J. 103 - 24" o.c.; Deck: 2" concrete 1:2:4; Membrane: furring 12" o.c.; Clips D, E, F, G; Hexagonal mesh reinforcement; 1" thick plaster - 1.5:2; 1.5:3.	145 psf	4 hrs. 26 min.			3	2, 4, 8	$4^1/_3$
F/C-S-17	1"	10' × 13'6"; Joists - S.J. 103 - 24" o.c.; Deck: 3.4 lbs. rib lath; 6" × 6" - 10 × 10 ga. reinforcement; 2" deck 1:2:4 gravel concrete; Membrane: furring 16" o.c.; Clips D, E, F, G; 1" plaster.	145 psf	1 hr. 42 min.			3	2, 4, 8	$1^2/_3$
F/C-S-18	$1^1/_8$"	10' × 13'6"; S. J. 103 - 24" o.c.; Deck: 2" concrete 1:2:4; Membrane: furring 12" o.c.; Clips C, E, F, G; Diagonal wire reinforcement; $1^1/_8$" plaster.	145 psf	2 hrs. 44 min.			3	2, 4, 8	$2^2/_3$
F/C-S-19	$1^1/_8$"	10' × 13'6"; Joists - S.J. 103 - 24" o.c.; Deck: $1^1/_2$" gypsum concrete over; $^1/_2$" gypsum board; Membrane: furring 12" o.c.; Clips D, E, F, G; $1^1/_8$" plaster - 1.5:2; 1.5:3.	145 psf	1 hr. 40 min.			3	2, 3, 8	$1^2/_3$
F/C-S-20	$1^1/_8$"	$2^1/_2$" cinder concrete; $^1/_2$" topping; plate 6" welds 12" o.c.; 5" - 18.9 lbs. "H" center; 5" - 10 lbs. "I" ends; 1" channels 18" o.c.; $1^1/_8$" gypsum sand plaster.	150 psf	3 hrs 43 min.			6	2, 4, 9, 11	$3^2/_3$
F/C-S-21	$1^1/_4$"	10' × 13'6"; Joists S.J. 103 24" o.c.; Deck: $1^1/_2$" gypsum concrete over; $^1/_2$" gypsum board base; Membrane: furring 12" o.c.; Clips D, E, F, G; $1^1/_4$" plaster - 1.5:2; 1.5:3.	145 psf	1 hr. 48 min.			3	2, 3, 8	$1^7/_3$
F/C-S-22	$1^1/_4$"	Floor finish: $1^1/_2$" limestone concrete; $^1/_2$" sand cement topping; plate to beams $3^1/_2$"; 12" o.c. welded; 5" - 10 lbs. "I" beams; 1" channels 18" o.c.; $1^1/_4$" wood fiber gypsum sand plaster on metal lath.	292 psf	2 hrs. 45 min.			6	2, 4, 9, 10	$2^3/_4$
F/C-S-23	$1^1/_2$"	$2^1/_2$" L.W. (gas exp.) concrete; Deck: $^1/_2$" topping; plate $6^1/_4$" welds 12" o.c.; Beams: 5" - 18.9 lbs. "H" center; 5" - 10 lbs. "I" ends; Membrane: 1" channels 18" o.c.; $1^1/_2$" gypsum sand plaster.	150 psf	4 hrs. 42 min.			6	2, 4, 9	$4^2/_3$

(Continued)

TABLE 3.2—FLOOR/CEILING ASSEMBLIES—STEEL STRUCTURAL ELEMENTS—(Continued)

ITEM CODE	MEMBRANE THICKNESS	CONSTRUCTION DETAILS	PERFORMANCE		REFERENCE NUMBER			NOTES	REC. HOURS
			LOAD	TIME	PRE-BMS-92	BMS-92	POST-BMS-92		
F/C-S-24	$1^1/_2$"	Floor finish: $1^1/_2$" limestone concrete; $^1/_2$" cement topping; plate $3^1/_2$" - 12" o.c. welded; 5" - 10 lbs. "I" beams; Ceiling: 1" channels 18" o.c.; $1^1/_2$" gypsum plaster.	292 psf	2 hrs. 34 min.			6	2, 4, 9, 10	$2^1/_2$
F/C-S-25	$1^1/_2$"	Floor finish: $1^1/_2$" gravel concrete on exp. metal; plate cont. weld; 4" - 7.7 lbs. "I" beams; Ceiling: $^1/_4$" rods 12" o.c. welded to beams; $1^1/_2$" fiber gypsum sand plaster.	70 psf	1 hr. 24 min.			6	2, 4, 9, 10	$1^1/_3$
F/C-S-26	$2^1/_2$"	Floor finish: bare plate; $6^1/_4$" welding - 12" o.c.; 5" - 18.9 lbs. "H" girders (inner); 5" - 10 lbs "I" girders (two outer); 1" channels 18" o.c.; 2" reinforced gypsum tile; $^1/_2$" gypsum sand plaster.	122 psf	1 hr.			6	7, 9, 11	1
F/C-S-27	$2^1/_2$"	Floor finish: 2" gravel concrete; plate to beams $3^1/_2$" - 12" o.c. welded; 4" - 7.7 lbs. "I" beams; 2" gypsum ceiling tiles; $^1/_2$" 1:3 gypsum sand plaster.	105 psf	2 hrs. 31 min.			6	2, 4, 9, 10	$2^1/_2$
F/C-S-28	$2^1/_2$"	Floor finish: $1^1/_2$" gravel concrete; $^1/_2$" gypsum asphalt; plate continuous weld; 4" - 7.7 lbs. "I" beams; 12" - 31.8 lbs. "I" beams - girder at 5' from one end; 1" channels 18" o.c.; 2" reinforcement gypsum tile; $^1/_2$" 1:3 gypsum sand plaster.	200 psf	4 hrs. 55 min.			6	2, 4, 9, 11	$4^2/_3$
F/C-S-29	$^3/_4$"	Floor: 2" reinforced concrete or 2" precast reinforced gypsum tile; Ceiling: $^3/_4$" portland cement-sand plaster 1:2 for scratch coat and 1:3 for brown coat with 15 lbs. hydrated lime and 3 lbs. of short asbestos fiber bag per cement or $^3/_4$" sanded gypsum plaster 1:2 for scratch coat and 1:3 for brown coat.	See Note 12	1 hr. 30 min.		1		12, 13, 14	$1^1/_2$
F/C-S-30	$^3/_4$"	Floor: $2^1/_4$" reinforced concrete or 2" reinforced gypsum tile; the latter with $^1/_4$" mortar finish; Ceiling: $^3/_4$" sanded gypsum plaster; 1:2 for scratch coat and 1:3 for brown coat.	See Note 12	2 hrs.		1		12, 13, 14	2
F/C-S-31	$^3/_4$"	Floor: $2^1/_2$" reinforced concrete or 2" reinforced gypsum tile; the latter with $^1/_4$" mortar finish; Ceiling: 1" neat gypsum plaster or $^3/_4$" gypsum-vermiculite plaster, ratio of gypsum to fine vermiculite 2:1 to 3:1.	See Note 12	2 hrs. 30 min.		1		12, 13, 14	$2^1/_2$

(Continued)

TABLE 3.2—FLOOR/CEILING ASSEMBLIES—STEEL STRUCTURAL ELEMENTS—(Continued)

ITEM CODE	MEMBRANE THICKNESS	CONSTRUCTION DETAILS	PERFORMANCE		REFERENCE NUMBER			NOTES	REC. HOURS
			LOAD	TIME	PRE-BMS-92	BMS-92	POST-BMS-92		
F/C-S-32	$^3/_4$"	Floor: $2^1/_2$" reinforced concrete or 2" reinforced gypsum tile; the latter with $^1/_2$" mortar finish; Ceiling: 1" neat gypsum plaster or $^3/_4$" gypsum-vermiculite plaster, ratio of gypsum to fine vermiculite 2:1 to 3:1.	See Note 12	3 hrs.		1		12, 13, 14	3
F/C-S-33	1"	Floor: $2^1/_2$" reinforced concrete or 2" reinforced gypsum slabs; the latter with $^1/_2$" mortar finish; Ceiling: 1" gypsum-vermiculite plaster applied on metal lath and ratio 2:1 to 3:1 gypsum to vermiculite by weight.	See Note 12	4 hrs.		1		12, 13, 14	4
F/C-S-34	$2^1/_2$"	Floor: 2" reinforced concrete or 2" precast reinforced portland cement concrete or gypsum slabs; precast slabs to be finished with $^1/_4$" mortar top coat; Ceiling: 2" precast reinforced gypsum tile, anchored into beams with metal ties or clips and covered with $^1/_2$" 1:3 sanded gypsum plaster.	See Note 12	4 hrs.		1		12, 13, 14	4
F/C-S-35	1"	Floor: 1:3:6 portland cement, sand and gravel concrete applied directly to the top of steel units and $1^1/_2$" thick at top of cells, plus $^1/_2$" $1:2^1/_2$" cement-sand finish, total thickness at top of cells, 2"; Ceiling: 1" neat gypsum plaster, back of lath 2" or more from underside of cellular steel.	See Note 15	3 hrs.		1		15, 16, 17, 18	3
F/C-S-36	1"	Floor: same as F/C-S-35; Ceiling: 1" gypsum-vermiculite plaster (ratio of gypsum to vermiculite 2:1 to 3:1), the back of lath 2" or more from under-side of cellular steel.	See Note 15	4 hrs.		1		15, 16, 17, 18	4
F/C-S-37	1"	Floor: same as F/C-S-35; Ceiling: 1" neat gypsum plaster; back of lath 9" or more from underside of cellular steel.	See Note 15	4 hrs.		1		15, 16, 17, 18	4
F/C-S-38	1"	Floor: same as F/C-S-35; Ceiling: 1" gypsum-vermiculite plaster (ratio of gypsum to vermiculite 2:1 to 3:1), the back of lath being 9" or more from underside of cellular steel.	See Note 15	5 hrs.		1		15, 16, 17, 18	5
F/C-S-39	$^3/_4$"	Floor: asbestos paper 14 lbs./100 ft.^{2} cemented to steel deck with waterproof linoleum cement, wood screeds and $^7/_8$" wood floor; Ceiling: $^3/_4$" sanded gypsum plaster 1:2 for scratch coat and 1:3 for brown coat.	See Note 19	1 hr.		1		19, 20, 21, 22	1

(Continued)

TABLE 3.2—FLOOR/CEILING ASSEMBLIES—STEEL STRUCTURAL ELEMENTS—(Continued)

ITEM CODE	MEMBRANE THICKNESS	CONSTRUCTION DETAILS	PERFORMANCE		REFERENCE NUMBER			NOTES	REC. HOURS
			LOAD	TIME	PRE-BMS-92	BMS-92	POST-BMS-92		
F/C-S-40	$^3/_4$"	Floor: $1^1/_2$", 1:2:4 portland cement concrete; Ceiling: $^3/_4$" sanded gypsum plaster 1:2 for scratch coat and 1:3 for brown coat.	See Note 19	1 hr. 30 min.		1		19, 20, 21, 22	$1^1/_2$
F/C-S-41	$^3/_4$"	Floor: 2", 1:2:4 portland cement concrete; Ceiling: $^3/_4$" sanded gypsum plaster, 1:2 for scratch coat and 1:3 for brown coat.	See Note 19	2 hrs.		1		19, 20, 21, 22	2
F/C-S-42	1"	Floor: 2", 1:2:4 portland cement concrete; Ceiling: 1" portland cement-sand plaster with 10 lbs. of hydrated lime for @ bag of cement 1:2 for scratch coat and $1:2^1/_2$" for brown coat.	See Note 19	2 hrs.		1		19, 20, 21, 22	2
F/C-S-43	$1^1/_2$"	Floor: 2", 1:2:4 portland cement concrete; Ceiling: $1^1/_2$", 1:2 sanded gypsum plaster on ribbed metal lath.	See Note 19	2 hrs. 30 min.		1		19, 20, 21, 22	$2^1/_2$
F/C-S-44	$1^1/_8$"	Floor: 2", 1:2:4 portland cement concrete; Ceiling: $1^1/_8$", 1:1 sanded gypsum plaster.	See Note 19	2 hrs. 30 min.		1		19, 20, 21, 22	$2^1/_2$
F/C-S-45	1"	Floor: $2^1/_2$", 1:2:4 portland cement concrete; Ceiling: 1", 1:2 sanded gypsum plaster.	See Note 19	2 hrs. 30 min.		1		19, 20, 21, 22	$2^1/_2$
F/C-S-46	$^3/_4$"	Floor: $2^1/_2$", 1:2:4 portland cement concrete; Ceiling: 1" neat gypsum plaster or $^3/_4$" gypsum-vermiculite plaster, ratio of gypsum to vermiculite 2:1 to 3:1.	See Note 19	3 hrs.		1		19, 20, 21, 22	3
F/C-S-47	$1^1/_8$"	Floor: $2^1/_2$", 1:2:4 portland cement, sand and cinder concrete plus $^1/_2$", $1:2^1/_2$" cement-sand finish; total thickness 3"; Ceiling: $1^1/_8$", 1:1 sanded gypsum plaster.	See Note 19	3 hrs.		1		19, 20, 21, 22	3
F/C-S-48	$1^1/_8$"	Floor: $2^1/_2$", gas expanded portland cement-sand concrete plus $^1/_2$", 1:2.5 cement-sand finish; total thickness 3"; Ceiling: $1^1/_8$", 1:1 sanded gypsum plaster.	See Note 19	3 hrs. 30 min.		1		19, 20, 21, 22	$3^1/_2$
F/C-S-49	1"	Floor: $2^1/_2$", 1:2:4 portland cement concrete; Ceiling: 1" gypsum-vermiculite plaster; ratio of gypsum to vermiculite 2:1 to 3:1.	See Note 19	4 hrs.		1		19, 20, 21, 22	4
F/C-S-50	$2^1/_2$"	Floor: 2", 1:2:4 portland cement concrete; Ceiling: 2" interlocking gypsum tile supported on upper face of lower flanges of beams, $^1/_2$" 1:3 sanded gypsum plaster.	See Note 19	2 hrs.		1		19, 20, 21, 22	2

(Continued)

TABLE 3.2—FLOOR/CEILING ASSEMBLIES—STEEL STRUCTURAL ELEMENTS—(Continued)

ITEM CODE	MEMBRANE THICKNESS	CONSTRUCTION DETAILS	PERFORMANCE		REFERENCE NUMBER			NOTES	REC. HOURS
			LOAD	TIME	PRE-BMS-92	BMS-92	POST-BMS-92		
F/C-S-51	$2^1/_2$"	Floor: 2", 1:2:4 portland cement concrete; Ceiling: 2" precast metal reinforced gypsum tile, $^1/_2$" 1:3 sanded gypsum plaster (tile clipped to channels which are clipped to lower flanges of beams).	See Note 19	4 hrs.		1		19, 20, 21, 22	4

Notes:

1. No protective membrane over structural steel.

2. Performance time indicates first endpoint reached only several tests were continued to points where other failures occurred.

3. Load failure.

4. Thermal failure.

5. This is an estimated time to load bearing failure. The same joist and deck specimen ws used for a later test with different membrane protection.

6. Test stopped at 3 hours to reuse specimen; no endpoint reached.

7. Test stopped at 1 hour to reuse specimen; no endpoint reached.

8. All plaster used = gypsum.

9. Specimen size - 18 feet by $13^1/_2$ inches. Floor deck - base material - $^1/_4$-inch by 18-foot steel plate welded to "I" beams.

10. "I" beams - 24 inches o.c.

11. "I" beams - 48 inches o.c.

12. Apply to open web joists, pressed steel joists or rolled steel beams, which are not stressed beyond 18,000 lbs./in.^{2} in flexure for open-web pressed or light rolled joists, and 20,000 lbs./in.^{2} for American standard or heavier rolled beams.

13. Ratio of weight of portland cement to fine and coarse aggregates combined for floor slabs shall be not less than $1:6^1/_2$.

14. Plaster for ceiling shall be applied on metal lath which shall be tied to supports to give the equivalent of single No. 18 gage steel wires 5 inches o.c.

15. Load: maximum fiber stress in steel not to exceed 16,000 psi.

16. Prefabricated units 2 feet wide with length equal to the span, composed of two pieces of No. 18 gage formed steel welded together to give four longitudinal cells.

17. Depth not less than 3 inches and distance between cells no less than 2 inches.

18. Ceiling: metal lath tied to furring channels secured to runner channels hung from cellular steel.

19. Load: rolled steel supporting beams and steel plate base shall not be stressed beyond 20,000 psi in flexure.
    Formed steel (with wide upper flange) construction shall not be stressed beyond 16,000 psi.

20. Some type of expanded metal or woven wire shall be embedded to prevent cracking in concrete flooring.

21. Ceiling plaster shall be metal lath wired to rods or channels which are clipped or welded to steel construction. Lath shall be no smaller than 18 gage steel wire and not more than 7 inches o.c.

22. The securing rods or channels shall be at least as effective as single $^3/_{16}$-inch rods with 1-inch of their length bent over the lower flanges of beams with the rods or channels tied to this clip with 14 gage iron wire.

FIGURE 3.3—FLOOR/CEILING ASSEMBLIES—WOOD JOIST

FIRE RESISTANCE RATING (HOURS)

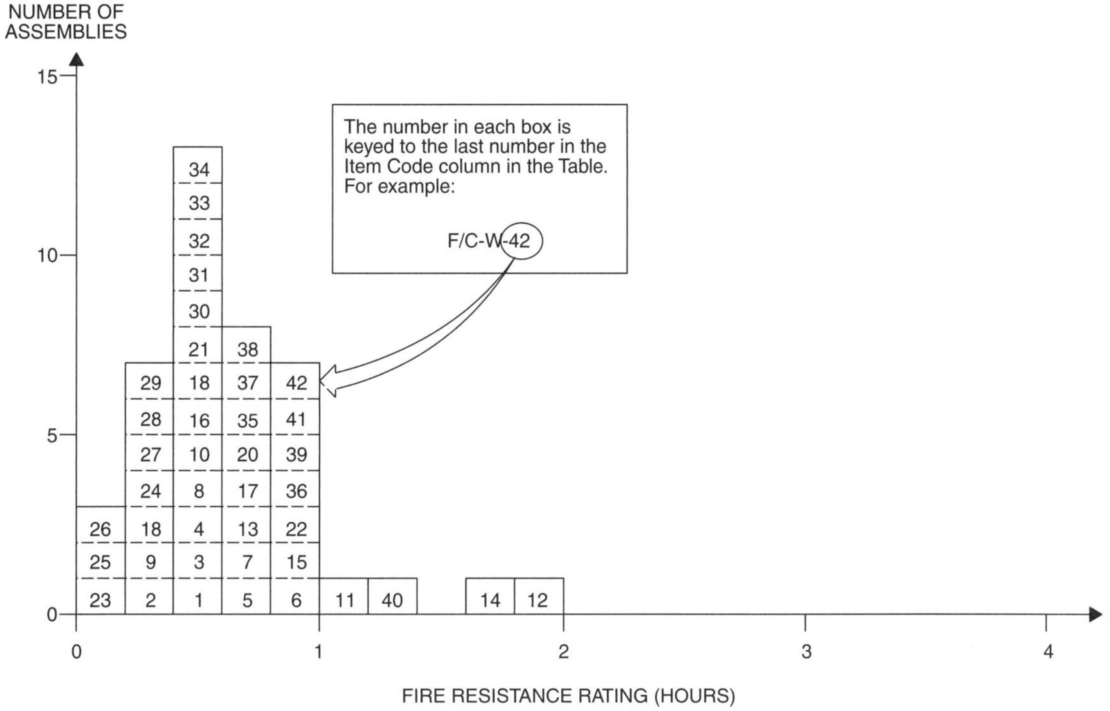

TABLE 3.3—FLOOR/CEILING ASSEMBLIES—WOOD JOIST

ITEM CODE	MEMBRANE THICKNESS	CONSTRUCTION DETAILS	PERFORMANCE		REFERENCE NUMBER			NOTES	REC. HOURS
			LOAD	TIME	PRE-BMS-92	BMS-92	POST-BMS-92		
F/C-W-1	$3/8$"	12' clear span - 2" × 9" wood joists; 18" o.c.; Deck: 1" T&G; Filler: 3" of ashes on $1/2$" boards nailed to joist sides 2" from bottom; 2" air space; Membrane: $3/8$" gypsum board.	60 psf	36 min.			7	1, 2	$1/2$
F/C-W-2	$1/2$"	12' clear span - 2" × 7" joists; 15" o.c.; Deck: 1" nominal lumber; Membrane: $1/2$" fiber board.	60 psf	22 min.			7	1, 2, 3	$1/4$
F/C-W-3	$1/2$"	12' clear span - 2" × 7" wood joists; 16" o.c.; 2" × $1 1/2$" bridging at center; Deck: 1″ T&G; Membrane: $1/2$" fiber board; 2 coats "distemper" paint.	30 psf	28 min.			7	1, 3, 15	$1/3$
F/C-W-4	$3/16$"	12' clear span - 2" × 7" wood joists; 16" o.c.; 2" × $1 1/2$" bridging at center span; Deck: 1" nominal lumber; Membrane: $1/2$" fiber board under $3/16$" gypsum plaster.	30 psf	32 min.			7	1, 2	$1/2$
F/C-W-5	$5/8$"	As per previous F/C-W-4 except membrane is $5/8$" lime plaster.	70 psf	48 min.			7	1, 2	$3/4$
F/C-W-6	$5/8$"	As per previous F/C-W-5 except membrane is $5/8$" gypsum plaster on 22 gage $3/8$" metal lath.	70 psf	49 min.			7	1, 2	$3/4$

(Continued)

TABLE 3.3—FLOOR/CEILING ASSEMBLIES—WOOD JOIST—(Continued)

ITEM CODE	MEMBRANE THICKNESS	CONSTRUCTION DETAILS	PERFORMANCE		REFERENCE NUMBER			NOTES	REC. HOURS
			LOAD	TIME	PRE-BMS-92	BMS-92	POST-BMS-92		
F/C-W-7	$^1/_2$"	As per previous F/C-W-6 except membrane is $^1/_2$" fiber board under $^1/_2$" gypsum plaster.	60 psf	43 min.			7	1, 2, 3	$^2/_3$
F/C-W-8	$^1/_2$"	As per previous F/C-W-7 except membrane is $^1/_2$" gypsum board.	60 psf	33 min.			7	1, 2, 3	$^1/_2$
F/C-W-9	$^9/_{16}$"	12' clear span - 2" × 7" wood joists; 15" o.c.; 2" × 1$^1/_2$" bridging at center; Deck: 1" nominal lumber; Membrane: $^3/_8$" gypsum board; $^3/_{16}$" gypsum plaster.	60 psf	24 min.			7	1, 2, 3	$^1/_3$
F/C-W-10	$^5/_8$"	As per F/C-W-9 except membrane is $^5/_8$" gypsum plaster on wood lath.	60 psf	27 min.			7	1, 2, 3	$^1/_3$
F/C-W-11	$^7/_8$"	12' clear span - 2" × 9" wood joists; 15" o.c.; 2" × 1$^1/_2$" bridging at center span; Deck: 1" T&G; Membrane: original ceiling joists have $^3/_8$" plaster on wood lath; 4" metal hangers attached below joists creating 15" chases filled with mineral wool and closed with $^7/_8$" plaster (gypsum) on $^3/_8$" S.W.M. metal lath to form new ceiling surface.	75 psf	1 hr. 10 min.			7	1, 2	1
F/C-W-12	$^7/_8$"	12' clear span - 2" × 9" wood joists; 15" o.c.; 2" × 1$^1/_2$" bridging at center; Deck: 1" T&G; Membrane: 3" mineral wood below joists; 3" hangers to channel below joists; $^7/_8$" gypsum plaster on metal lath attached to channels.	75 psf	2 hrs.			7	1, 4	2
F/C-W-13	$^7/_8$"	12' clear span - 2" × 9" wood joists; 16" o.c.; 2" × 1$^1/_2$" bridging at center span; Deck: 1" T&G on 1" bottoms on $^3/_4$" glass wool strips on $^3/_4$" gypsum board nailed to joists; Membrane: $^3/_4$" glass wool strips on joists; $^3/_8$" perforated gyspum lath; $^1/_2$" gypsum plaster.	60 psf	41 min.			7	1, 3	$^2/_3$
F/C-W-14	$^7/_8$"	12' clear span - 2" × 9" wood joists; 15" o.c.; Deck: 1" T&G; Membrane: 3" foam concrete in cavity on $^1/_2$" boards nailed to joists; wood lath nailed to 1" × 1$^1/_4$" straps 14 o.c. across joists; $^7/_8$" gypsum plaster.	60 psf	1 hr. 40 min.			7	1, 5	1$^2/_3$
F/C-W-15	$^7/_8$"	12' clear span - 2" × 9" wood joists; 18" o.c.; Deck: 1" T&G; Membrane: 2" foam concrete on $^1/_2$" boards nailed to joist sides 2" from joist bottom; 2" air space; 1" × 1$^1/_4$" wood straps 14" o.c. across joists; $^7/_8$" lime plaster on wood lath.	60 psf	53 min.			7	1, 2	$^3/_4$

(Continued)

2003 INTERNATIONAL EXISTING BUILDING CODE® COMMENTARY

TABLE 3.3—FLOOR/CEILING ASSEMBLIES—WOOD JOIST—(Continued)

ITEM CODE	MEMBRANE THICKNESS	CONSTRUCTION DETAILS	PERFORMANCE		REFERENCE NUMBER			NOTES	REC. HOURS
			LOAD	TIME	PRE-BMS-92	BMS-92	POST-BMS-92		
F/C-W-16	$7/8$"	12' clear span - 2" × 9" wood joists; Deck: 1" T&G; Membrane: 3" ashes on $1/2$" boards nailed to joist sides 2" from joist bottom; 2" air space; 1" × $1 1/4$" wood straps 14" o.c.; $7/8$" gypsum plaster on wood lath.	60 psf	28 min.			7	1, 2	$1/3$
F/C-W-17	$7/8$"	As per previous F/C-W-16 but with lime plaster mix.	60 psf	41 min.			7	1, 2	$2/3$
F/C-W-18	$7/8$"	12' clear span - 2" × 9" wood joists; 18" o.c.; 2" × $1 1/2$" bridging at center; Deck: 1" T&G; Membrane: $7/8$" gypsum plster on wood lath.	60 psf	36 min.			7	1, 2	$1/2$
F/C-W-19	$7/8$"	As per previous F/C-W-18 except with lime plaster membrane and deck is 1" nominal boards (plain edge).	60 psf	19 min.			7	1, 2	$1/4$
F/C-W-20	$7/8$"	As per F/C-W-19, except deck is 1" T&G boards.	60 psf	43 min.			7	1, 2	$2/3$
F/C-W-21	1	12' clear span - 2" × 9" wood joists; 16" o.c.; 2" × $1 1/2$" bridging at center; Deck: 1" T&G; Membrane: $3/8$" gypsum base board; $5/8$" gypsum plaster.	70 psf	29 min.			7	1, 2	$1/3$
F/C-W-22	$1 1/8$"	12' clear span - 2" × 9" wood joists; 16" o.c.; 2" × 2" wood bridging at center; Deck: 1" T&G; Membrane: hangers, channel with $3/8$" gypsum baseboard affixed under $3/4$" gypsum plaster.	60 psf	1 hr.			7	1, 2, 3	1
F/C-W-23	$3/8$"	Deck: 1" nominal lumber; Joists: 2" × 7"; 15" o.c.; Membrane: $3/8$" plasterboard with plaster skim coat.	60 psf	$11 1/2$ min.			12	2, 6	$1/6$
F/C-W-24	$1/2$"	Deck: 1" T&G lumber; Joists: 2" × 9"; 16" o.c.; Membrane: $1/2$" plasterboard.	60 psf	18 min.			12	2, 7	$1/4$
F/C-W-25	$1/2$"	Deck: 1" T&G lumber; Joists: 2" × 7"; 16" o.c.; Membrane: $1/2$" fiber insulation board.	30 psf	8 min.			12	2, 8	$2/15$
F/C-W-26	$1/2$"	Deck: 1" nominal lumber; Joists: 2" × 7"; 15" o.c.; Membrane: $1/2$" fiber insulation board.	60 psf	8 min.			12	2, 9	$2/15$
F/C-W-27	$5/8$"	Deck: 1" nominal lumber; Joists: 2" × 7"; 15" o.c.; Membrane: $5/8$" gypsum plaster on wood lath.	60 psf	17 min.			12	2, 10	$1/4$
F/C-W-28	$5/8$"	Deck: 1" T&G lumber; Joists: 2" × 9"; 16" o.c.; Membrane: $1/2$" fiber insulation board; $1/2$" plaster.	60 psf	20 min.			12	2, 11	$1/3$
F/C-W-29	No Membrane	Exposed wood joists.	See Note 13	15 min.		1		1, 12, 13, 14	$1/4$

(Continued)

TABLE 3.3—FLOOR/CEILING ASSEMBLIES—WOOD JOIST—(Continued)

ITEM CODE	MEMBRANE THICKNESS	CONSTRUCTION DETAILS	PERFORMANCE		REFERENCE NUMBER			NOTES	REC. HOURS
			LOAD	TIME	PRE-BMS-92	BMS-92	POST-BMS-92		
F/C-W-30	$^3/_8$"	Gypsum wallboard: $^3/_8$" or $^1/_2$" with $1^1/_2$" No. 15 gage nails with $^3/_{16}$" heads spaced 6" centers with asbestos paper applied with paperhangers paste and finished with casein paint.	See Note 13	25 min.		1		1, 12, 13, 14	$^1/_2$
F/C-W-31	$^1/_2$"	Gypsum wallboard: $^1/_2$" with $1^3/_4$" No. 12 gage nails with $^1/_2$" heads, 6" o.c., and finished with casein paint.	See Note 13	25 min.		1		1, 12, 13, 14	$^1/_2$
F/C-W-32	$^1/_2$"	Gypsum wallboard: $^1/_2$" with $1^1/_2$" No. 12 gage nails with $^1/_2$" heads, 18" o.c., with asbestos paper applied with paperhangers paste and secured with $1^1/_2$" No. 15 gage nails with $^3/_{16}$" heads and finished with casein paint; combined nail spacing 6" o.c.	See Note 13	30 min.		1		1, 12, 13, 14	$^1/_2$
F/C-W-33	$^3/_8$"	Gypsum wallboard: two layers $^3/_8$" secured with $1^1/_2$" No. 15 gage nails with $^3/_8$" heads, 6" o.c.	See Note 13	30 min.		1		1, 12, 13, 14	$^1/_2$
F/C-W-34	$^1/_2$"	Perforated gypsum lath: $^3/_8$", plastered with $1^1/_8$" No. 13 gage nails with $^5/_{16}$" heads, 4" o.c.; $^1/_2$" sanded gypsum plaster.	See Note 13	30 min.		1		1, 12, 13, 14	$^1/_2$
F/C-W-35	$^1/_2$"	Same as F/C-W-34, except with $1^1/_8$" No. 13 gage nails with $^3/_8$" heads, 4" o.c.	See Note 13	45 min.		1		1, 12, 13, 14	$^3/_4$
F/C-W-36	$^1/_2$"	Perforated gypsum lath: $^3/_8$", nailed with $1^1/_8$" No. 13 gage nails with $^3/_8$" heads, 4" o.c.; joints covered with 3" strips of metal lath with $1^3/_4$" No. 12 nails with $^1/_2$" heads, 5" o.c.; $^1/_2$" sanded gypsum plaster.	See Note 13	1 hr.		1		1, 12, 13, 14	1
F/C-W-37	$^1/_2$"	Gypsum lath: $^3/_8$" and lower layer of $^3/_8$" perforated gypsum lath nailed with $1^3/_4$" No. 13 nails with $^5/_{16}$" heads, 4" o.c.; $^1/_2$" sanded gypsum plaster or $^1/_2$" portland cement plaster.	See Note 13	45 min.		1		1, 12, 13, 14	$^3/_4$
F/C-W-38	$^3/_4$"	Metal lath: nailed with $1^1/_4$" No. 11 nails with $^3/_8$" heads or 6d common driven 1" and bent over, 6" o.c.; $^3/_4$" sanded gypsum plaster.	See Note 13	45 min.		1		1, 12, 13, 14	$^3/_4$
F/C-W-39	$^3/_4$"	Same as F/C-W-38, except nailed with $1^1/_2$" No. 11 barbed roof nails with $^7/_{16}$" heads, 6" o.c.	See Note 13	1 hr.		1		1, 12, 13, 14	1

(Continued)

TABLE 3.3—FLOOR/CEILING ASSEMBLIES—WOOD JOIST—(Continued)

ITEM CODE	MEMBRANE THICKNESS	CONSTRUCTION DETAILS	PERFORMANCE		REFERENCE NUMBER			NOTES	REC. HOURS
			LOAD	TIME	PRE-BMS-92	BMS-92	POST-BMS-92		
F/C-W-40	$^3/_4$"	Same as F/C-W-38, except with lath nailed to joists with additional supports for lath 27" o.c.; attached to alternate joists and consisting of two nails driven $1^1/_4$", 2" above bottom on opposite sides of the joists, one loop of No. 18 wire slipped over each nail; the ends twisted together below lath.	See Note 13	1 hr. 15 min.		1 1		1, 12, 13, 14	$1^1/_4$
F/C-W-41	$^3/_4$"	Metal lath: nailed with $1^1/_2$" No. 11 barbed roof nails with $^7/_{16}$" heads, 6 o.c., with $^3/_4$" portland cement plaster for scratch coat and 1:3 for brown coat, 3 lbs. of asbestos fiber and 15 lbs. of hydrated lime/94 lbs. bag of cement.	See Note 13	1 hr.		1		1, 12, 13, 14	1
F/C-W-42	$^3/_4$"	Metal lath: nailed with 8d, No. $11^1/_2$ gage barbed box nails, $2^1/_2$" driven, $1^1/_4$" on slant and bent over, 6" o.c.; $^3/_4$" sanded gypsum plaster, 1:2 for scratch coat and 1:3 for below coat.	See Note 13	1 hr.		1		1, 12, 13, 14	1

Notes:

1. Thickness indicates thickness of first membrane protection on ceiling surface.
2. Failure mode - flame thru.
3. Failure mode - collapse.
4. No endpoint reached at termination of test.
5. Failure imminent - test terminated.
6. Joist failure - 11.5 minutes; flame thru - 13.0 minutes.; collapse - 24 minutes.
7. Joist failure - 17 minutes; flame thru - 18 minutes.; collapse - 33 minutes.
8. Joist failure - 18 minutes; flame thru - 8 minutes.; collapse - 30 minutes.
9. Joist failure - 12 minutes; flame thru - 8 minutes.; collapse - 22 minutes.
10. Joist failure - 11 minutes; flame thru - 17 minutes.; collapse - 27 minutes.
11. Joist failure - 17 minutes; flame thru - 20 minutes.; collapse - 43 minutes.
12. Joists: 2-inch by 10-inch southern pine or Douglas fir; No. 1 common or better. Subfloor: $^3/_4$-inch wood sheating diaphragm of asbestos paper, and finish of tongue-and-groove wood flooring.
13. Loadings: not more than 1,000 psi maximum fiber stress in joists.
14. Perforations in gypsum lath are to be not less than $^3/_4$-inch diameter with one perforation for not more than 16/in.^{2} diameter.
15. "Distemper" is a British term for a water-based paint such as white wash or calcimine.

FIGURE 3.4—FLOOR/CEILING ASSEMBLIES—HOLLOW CLAY TILE WITH REINFORCED CONCRETE

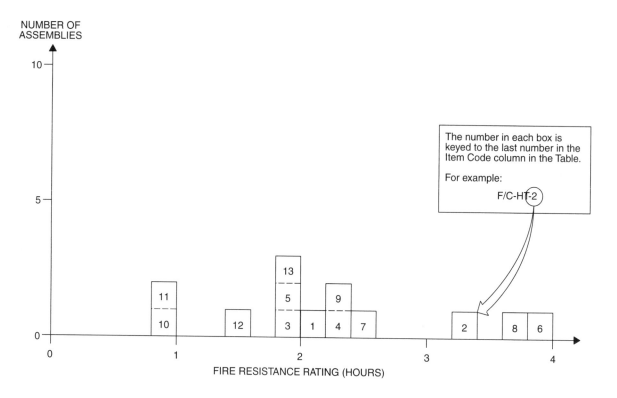

TABLE 3.4—FLOOR/CEILING ASSEMBLIES—HOLLOW CLAY TILE WITH REINFORCED CONCRETE

| ITEM CODE | ASSEMBLY THICKNESS | CONSTRUCTION DETAILS | PERFORMANCE | | REFERENCE NUMBER | | | NOTES | REC. HOURS |
			LOAD	TIME	PRE-BMS-92	BMS-92	POST-BMS-92		
F/C-HT-1	6"	Cover: $1^1/_2$" concrete (6080 psi); three cell hollow clay tiles, 12" × 12" × 4"; $3^1/_4$" concrete between tiles including two $^1/_2$" rebars with $^3/_4$" concrete cover; $^1/_2$" plaster cover, lower.	75 psf	2 hrs. 7 min.			7	1, 2, 3	2
F/C-HT-2	6"	Cover: $1^1/_2$" concrete (5840 psi); three cell hollow clay tiles, 12" × 12" × 4"; $3^1/_4$" concrete between tiles including two $^1/_2$" rebars each with $^1/_2$" concrete cover and $^5/_8$" filler tiles between hollow tiles; $^1/_2$" plaster cover, lower.	61 psf	3 hrs. 23 min.			7	3, 4, 6	$3^1/_3$
F/C-HT-3	6"	Cover: $1^1/_2$" concrete (6280 psi); three cell hollow clay tiles, 12" × 12" × 4"; $3^1/_4$" concrete between tiles including two $^1/_2$" rebars with $^1/_2$" cover; $^1/_2$" plaster cover, lower.	122 psf	2 hrs.			7	1, 3, 5, 8	2
F/C-HT-4	6"	Cover: $1^1/_2$" concrete (6280 psi); three cell hollow clay tiles, 12" × 12" × 4"; $3^1/_4$" concrete between tiles including two $^1/_2$" rebars with $^3/_4$" cover; $^1/_2$" plaster cover, lower.	115 psf	2 hrs. 23 min.			7	1, 3, 7	$2^1/_3$

(Continued)

TABLE 3.4—FLOOR/CEILING ASSEMBLIES—HOLLOW CLAY TILE WITH REINFORCED CONCRETE

ITEM CODE	ASSEMBLY THICKNESS	CONSTRUCTION DETAILS	PERFORMANCE		REFERENCE NUMBER			NOTES	REC. HOURS
			LOAD	TIME	PRE-BMS-92	BMS-92	POST-BMS-92		
F/C-HT-5	6"	Cover: $1^1/_2$" concrete (6470 psi); three cell hollow clay tiles, 12" × 12" × 4"; $3^1/_4$" concrete between tiles including two $^1/_2$" rebars with $^1/_2$" cover; $^1/_2$" plaster cover, lower.	122 psf	2 hrs.			7	1, 3, 5, 8	2
F/C-HT-6	8"	Floor cover: $1^1/_2$" gravel cement (4300 psi); three cell, 12" × 12" × 6"; $3^1/_2$" space between tiles including two $^1/_2$" rebars with 1" cover from concrete bottom; $^1/_2$" plaster cover, lower.	165 psf	4 hrs.			7	1, 3, 9, 10	4
F/C-HT-7	9" (nom.)	Deck: $^7/_8$" T&G on 2" × $1^1/_2$" bottoms (18" o.c.) $1^1/_2$" concrete cover (4600 psi); three cell hollow clay tiles, 12" × 12" × 4"; 3" concrete between tiles including one $^3/_4$" rebar $^3/_4$" from tile bottom; $^3/_4$" plaster cover.	95 psf	2 hrs. 26 min.			7	4, 11, 12, 13	$2^1/_3$
F/C-HT-8	9" (nom.)	Deck: $^7/_8$" T&G on 2" × $1^1/_2$" bottoms (18" o.c.) $1^1/_2$" concrete cover (3850 psi); three cell hollow clay tiles, 12" × 12" × 4"; 3" concrete between tiles including one $^3/_4$" rebar $^3/_4$" from tile bottoms; $^1/_2$" plaster cover.	95 psf	3 hrs. 28 min.			7	4, 11, 12, 13	
F/C-HT-9	9" (nom.)	Deck: $^7/_8$" T&G on 2" × $1^1/_2$" bottoms (18" o.c.) $1^1/_2$" concrete cover (4200 psi); three cell hollow clay tiles, 12" × 12" × 4"; 3" concrete between tiles including one $^3/_4$" rebar $^3/_4$" from tile bottoms; $^1/_2$" plaster cover.	95 psf	2 hrs. 14 min.			7	3, 5, 8, 11	
F/C-HT-10	$5^1/_2$"	Fire clay tile (4" thick); $1^1/_2$" concrete cover; for general details, see Note 15.	See Note 14	1 hr.			43	15	1
F/C-HT-11	8"	Fire clay tile (6" thick); 2" cover.	See Note 14	1 hr.			43	15	1
F/C-HT-12	$5^1/_2$"	Fire clay tile (4" thick); $1^1/_2$" cover; $^5/_8$" gypsum plaster, lower.	See Note 14	1 hr. 30 min.			43	15	$1^1/_2$
F/C-HT-13	8"	Fire clay tile (6" thick); 2" cover; $^5/_8$" gypsum plaster, lower.	See Note 14	2 hrs.			43	15	$1^1/_2$

Notes:

1. A generalized cross section of this floor type follows:

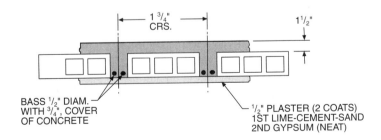

2. Failure mode - structural.
3. Plaster: base coat - lime-cement-sand; top coat - gypsum (neat).

(Continued)

TABLE 3.4—FLOOR/CEILING ASSEMBLIES—HOLLOW CLAY TILE WITH REINFORCED CONCRETE—(Continued)

4. Failure mode - collapse.
5. Test stopped before any endpoints were reached.
6. A generalized cross section of this floor type follows:

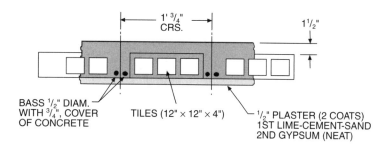

7. Failure mode - thermal - back face temperature rise.
8. Passed hose stream test.
9. Failed hose stream test.
10. Test stopped at 4 hours before any endpoints were reached.

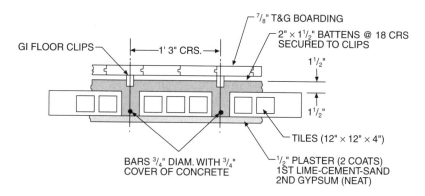

11. A generalized cross section of this floor type follows:
12. Plaster: base coat - retarded hemihydrate gypsum-sand; second coat - neat gypsum.
13. Concrete in Item 7 is P.C. based but with crushed brick aggregates while in Item 8 river sand and river gravels are used with the P.C.
14. Load - unspecified.
15. The 12-inch by 12-inch fire-clay tiles were laid end to end in rows spaced $2^1/_2$ inches or 4 inches apart. The reinforcing steel was placed between these rows and the concrete cast around them and over the tile to form the structural floor.

SECTION IV—BEAMS

TABLE 4.1.1—REINFORCED CONCRETE BEAMS
DEPTH 10" TO LESS THAN 12"

| ITEM CODE | DEPTH | CONSTRUCTION DETAILS | PERFORMANCE | | REFERENCE NUMBER | | | NOTES | REC. HOURS |
			LOAD	TIME	PRE-BMS-92	BMS-92	POST-BMS-92		
B-11-RC-1	11"	24" wide × 11" deep reinforced concrete "T" beam (3290 psi); Details: see Note 5 figure.	8.8 tons	4 hrs. 2 min.			7	1, 2, 14	4
B-10-RC-2	10"	24" wide × 10" deep reinforced concrete "T" beam (4370 psi); Details: see Note 6 figure.	8.8 tons	1 hr. 53 min.			7	1, 3	$1^3/_4$
B-10-RC-3	$10^1/_2$"	24" wide × $10^1/_2$" deep reinforced concrete "T" beam (4450 psi); Details: see Note 7 figure.	8.8 tons	2 hrs. 40 min.			7	1, 3	$2^2/_3$
B-11-RC-4	11"	24" wide × 11" deep reinforced concrete "T" beam (2400 psi); Details: see Note 8 figure.	8.8 tons	3 hrs. 32 min.			7	1, 3, 14	$3^1/_2$
B-11-RC-5	11"	24" wide × 11" deep reinforced concrete "T" beam (4250 psi); Details: see Note 9 figure.	8.8 tons	3 hrs. 3 min.			7	1, 3, 14	3
B-11-RC-6	11"	Concrete flange: 4" deep × 2' wide (4895 psi) concrete; Concrete beam: 7" deep × $6^1/_2$" wide beam; "I" beam reinforcement; 10" × $4^1/_2$" × 25 lbs. R.S.J.; 1" cover on flanges; Flange reinforcement: $3/_8$" diameter bars at 6" pitch parallel to "T"; $1/_4$" diameter bars perpendicular to "T"; Beam reinforcement: 4" × 6" wire mesh No. 13 SWG; Span: 11' restrained; Details: see Note 10 figure.	10 tons	6 hrs.			7	1, 4	6
B-11-RC-7	11"	Concrete flange: 6" deep × $1'6^1/_2$" wide (3525 psi) concrete; Concrete beam: 5" deep × 8" wide precast concrete blocks $8^3/_4$" long; "I" beam reinforcement; 7" × 4" × 16 lbs. R.S.J.; 2" cover on bottom; $1^1/_2$" cover on top; Flange reinforcement: two rows $1/_2$" diameter rods parallel to "T"; Beam reinforcement: $1/_8$" wire mesh perpendicular to 1"; Span: 1'3" simply supported; Details: see Note 11 figure.	3.9 tons	4 hrs.			7	1, 2	4
B-11-RC-8	11"	Concrete flange: 4" deep × 2' wide (3525 psi) concrete; Concrete beam 7" deep × $4^1/_2$" wide; (scaled from drawing); "I" beam reinforcement; 10" × $4^1/_2$" × 25 lbs. R.S.J.; no concrete cover on bottom; Flange reinforcement: $3/_8$" diameter bars at 6 pitch parallel to "T"; $1/_4$" diameter bars perpendicular to "T"; Span: 11' restricted.	10 tons	4 hrs.			7	1, 2, 12	4

(Continued)

**TABLE 4.1.1—REINFORCED CONCRETE BEAMS
DEPTH 10" TO LESS THAN 12"—(Continued)**

ITEM CODE	DEPTH	CONSTRUCTION DETAILS	PERFORMANCE		REFERENCE NUMBER			NOTES	REC. HOURS
			LOAD	TIME	PRE-BMS-92	BMS-92	POST-BMS-92		
B-11-RC-9	11¹/₂"	24" wide × 11¹/₂" deep reinforced concrete "T" beam (4390 psi); Details: see Note 12 figure.	8.8 tons	3 hrs. 24 min.			7	1, 3	3¹/₃

Notes:

1. Load concentrated at mid span.

2. Achieved 4 hour performance (Class "B," British).

3. Failure mode – collapse.

4. Achieved 6 hour performance (Class "A," British).

5.

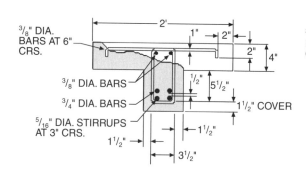

6.

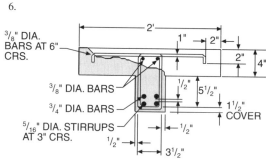

7.

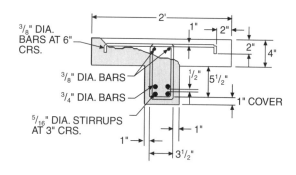

8.

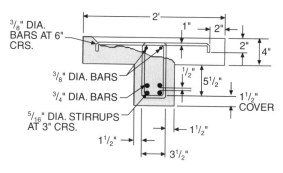

9.

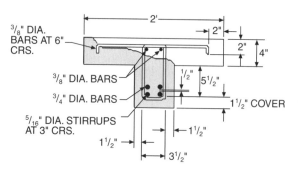

10.

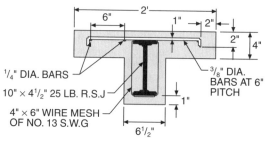

(Continued)

TABLE 4.1.1—REINFORCED CONCRETE BEAMS
DEPTH 10" TO LESS THAN 12"—(Continued)

11.

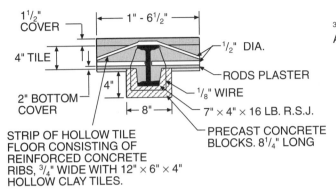

STRIP OF HOLLOW TILE
FLOOR CONSISTING OF
REINFORCED CONCRETE
RIBS, $3/4$" WIDE WITH 12" × 6" × 4"
HOLLOW CLAY TILES.

SPAN AND END CONDITIONS:-10'-3" (CLEAR).
SIMPLY SUPPORTED.

12.

13.

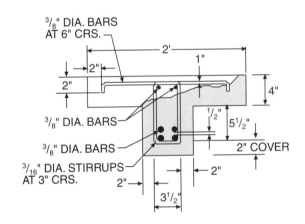

14. The different performances achieved by B-11-RC-1, B-11-RC-4 and B-11-RC-5 are attributable to differences in concrete aggregate compositions reported in the source document but unreported in this table. This demonstrates the significance of material composition in addition to other details.

TABLE 4.1.2—REINFORCED CONCRETE BEAMS
DEPTH 12" TO LESS THAN 14"

ITEM CODE	DEPTH	CONSTRUCTION DETAILS	PERFORMANCE		REFERENCE NUMBER			NOTES	REC. HOURS
			LOAD	TIME	PRE-BMS-92	BMS-92	POST-BMS-92		
B-12-RC-1	12"	12" × 8" section; 4160 psi aggregate concrete; Reinforcement: 4-$\frac{7}{8}$" rebars at corners; 1" below each surface; $\frac{1}{4}$" stirrups 10" o.c.	5.5 tons	2 hrs.			7	1	2
B-12-RC-2	12"	Concrete flange: 4" deep × 2' wide (3045 psi) concrete at 35 days; Concrete beam: 8" deep; "I" beam reinforcement: 10" × 4$\frac{1}{2}$" × 25 lbs. R.S.J.; 1" cover on flanges; Flange reinforcement: $\frac{3}{8}$" diameter bars at 6" pitch parallel to "T"; $\frac{1}{4}$" diameter bars perpendicular to "T"; Beam reinforcement: 4" × 6" wire mesh No. 13 SWG; Span: 10'3" simply supported.	10 tons	4 hrs.			7	2, 3, 5	4
B-13-RC-3	13"	Concrete flange: 4" deep × 2' wide (3825 psi) concrete at 46 days; Concrete beam: 9" deep × 8$\frac{1}{2}$" wide; (scaled from drawing); "I" beam reinforcement: 10" × 4$\frac{1}{2}$" × 25 lbs. R.S.J.; 3" cover on bottom flange; 1" cover on top flange; Flange reinforcement: $\frac{3}{8}$" diameter bars at 6" pitch parallel to "T"; $\frac{1}{4}$" diameter bars perpendicular to "T"; Beam reinforcement: 4" × 6" wire mesh No. 13 SWG; Span: 11' restrained.	10 tons	6 hrs.			7	2, 3, 6, 8, 9	4
B-12-RC-4	12"	Concrete flange: 4" deep × 2' wide (3720 psi) concrete at 42 days; Concrete beam: 8" deep × 8$\frac{1}{2}$" wide; (scaled from drawing); "I" beam reinforcement: 10" × 4$\frac{1}{2}$" × 25 lbs. R.S.J.; 2" cover bottom flange; 1" cover top flange; Flange reinforcement: $\frac{3}{8}$" diameter bars at 6" pitch parallel to "T"; $\frac{1}{4}$" diameter bars perpendicular to "T"; Beam reinforcement: 4" × 6" wire mesh No. 13 SWG; Span: 11' restrained.	10 tons	6 hrs.			7	2, 3, 4, 7, 8, 9 / 1	4

Notes:
1. Qualified for 2 hour use. (Grade "C," British) Test included hose stream and reload at 48 hours.
2. Load concentrated at mid span.
3. British test.
4. British test - qualified for 6 hour use (Grade "A").

(Continued)

**TABLE 4.1.2—REINFORCED CONCRETE BEAMS
DEPTH 12" TO LESS THAN 14"—(Continued)**

5.

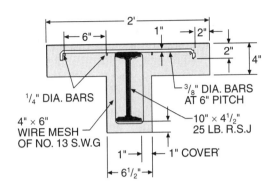

6.

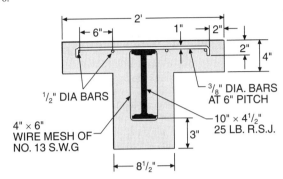

7.

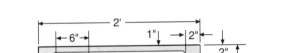
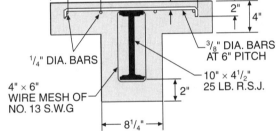

8. See Table 4.1.3, Note 5.

9. Hourly rating based upon B-12-RC-2 above.

TABLE 4.1.3—REINFORCED CONCRETE BEAMS
DEPTH 14" TO LESS THAN 16"

ITEM CODE	DEPTH	CONSTRUCTION DETAILS	PERFORMANCE		REFERENCE NUMBER			NOTES	REC. HOURS
			LOAD	TIME	PRE-BMS-92	BMS-92	POST-BMS-92		
B-15-RC-1	15"	Concrete flange: 4" deep × 2' wide (3290 psi) concrete; Concrete beam: 10" deep × 8¹/₂" wide; "I" beam reinforcement: 10" × 4¹/₂" × 25 lbs. R.S.J.; 4" cover on bottom flange; 1" cover on top flange; Flange reinforcement: ³/₈" diameter bars at 6" pitch parallel to "T"; ¹/₄" diameter bars perpendicular to "T"; Beam reinforcement: 4" × 6" wire mesh No. 13 SWG; Span: 11' restrained.	10 tons	6 hrs.			7	1, 2, 3 5, 6	4
B-15-RC-2	15"	Concrete flange: 4" deep × 2' wide (4820 psi) concrete; Concrete beam: 10" deep × 8¹/₂" wide; "I" beam reinforcement: 10" × 4¹/₂" × 25 lbs. R.S.J.; 1" cover over wire mesh on bottom flange; 1" cover on top flange; Flange reinforcement: ³/₈" diameter bars at 6" pitch parallel to "T"; ¹/₄" diameter bars perpendicular to "T"; Beam reinforcement: 4" × 6" wire mesh No. 13 SWG; Span: 11' restrained.	10 tons	6 hrs.			7	1, 2, 4, 5, 6	4

Notes:

1. Load concentrated at mid span.
2. Achieved 6 hour fire rating (Grade "A," British).
3.
4.

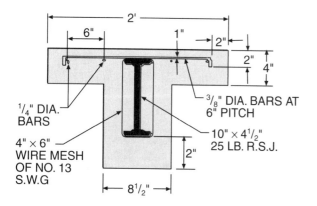

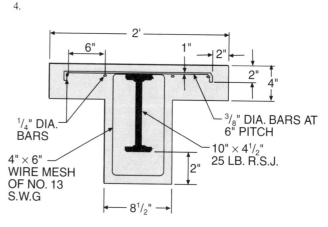

5. Section 43.147 of the 1979 edition of the *Uniform Building Code Standards* provides:

   "A restrained condition in fire tests, as used in this standard, is one in which expansion at the supports of a load-carrying element resulting from the effects of the fire is resisted by forces external to the element. An unrestrained condition is one in which the load-carrying element is free to expand and rotate at its support.

   "(R)estraint in buildings is defined as follows: Floor and roof assemblies and individual beams in buildings shall be considered restrained when the surrounding or supporting structure is capable of resisting the thermal expansion throughout the range of anticipated elevated temperatures. Construction not complying . . . is assumed to be free to rotate and expand and shall be considered as unrestrained.

   "Restraint may be provided by the lateral stiffness of supports for floor and roof assemblies and intermediate beams forming part of the assembly. In order to develop restraint, connections must adequately transfer thermal thrusts to such supports. The rigidity of adjoining panels or structures shall be considered in assessing the capability of a structure to resist therm expansion."

   Because it is difficult to determine whether an existing building's structural system is capable of providing the required restraint, the lower hourly ratings of a similar but unrestrained assembly have been recommended.

6. Hourly rating based upon Table 4.2.1, Item B-12-RC-2.

TABLE 4.2.1—REINFORCED CONCRETE BEAMS—UNPROTECTED
DEPTH 10" TO LESS THAN 12"

ITEM CODE	DEPTH	CONSTRUCTION DETAILS	PERFORMANCE		REFERENCE NUMBER			NOTES	REC. HOURS
			LOAD	TIME	PRE-BMS-92	BMS-92	POST-BMS-92		
B-SU-1	10"	10" × 4$\frac{1}{2}$" × 25 lbs. "I" beam.	10 tons	39 min.			7	1	$\frac{1}{3}$

Notes:

1. Concentrated at mid span.

TABLE 4.2.2—STEEL BEAMS—CONCRETE PROTECTION
DEPTH 10" TO LESS THAN 12"

ITEM CODE	DEPTH	CONSTRUCTION DETAILS	PERFORMANCE		REFERENCE NUMBER			NOTES	REC. HOURS
			LOAD	TIME	PRE-BMS-92	BMS-92	POST-BMS-92		
B-SU-1	10"	10" × 8" rectangle; aggregate concrete (4170 psi) with 1" top cover and 2" bottom cover; No. 13 SWG iron wire loosely wrapped at approximately 6" pitch about 7" × 4" × 16 lbs. "I" beam.	3.9 tons	3 hrs. 46 min.			7	1, 2, 3	3$\frac{3}{4}$
B-SU-1	10"	10" × 8" rectangle; aggregate concrete (3630 psi) with 1" top cover and 2" bottom cover; No. 13 SWG iron wire loosely wrapped at approximately 6" pitch about 7" × 4" × 16 lbs. "I" beam.	5.5 tons	5 hrs. 26 min.			7	1, 4, 5, 6, 7	3$\frac{3}{4}$

Notes:

1. Load concentrated at mid span.
2. Specimen 10-foot 3-inch clear span simply supported.
3. Passed Grade "C" fire resistance (British) including hose stream and reload.
4. Specimen 11-foot clear span - restrained.
5. Passed Grade "B" fire resistance (British) including hose stream and reload.
6. See Table 4.1.3, Note 5.
7. Hourly rating based upon B-SC-1 above.

SECTION V—DOORS

FIGURE 5.1—RESISTANCE OF DOORS TO FIRE EXPOSURE

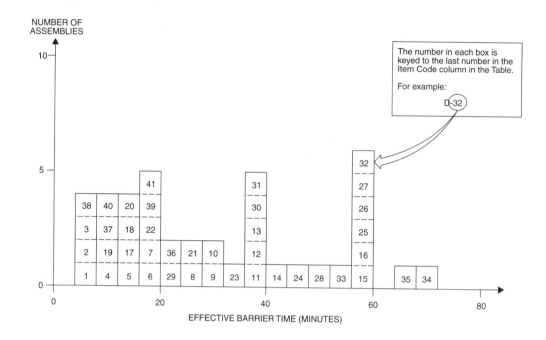

TABLE 5.1—RESISTANCE OF DOORS TO FIRE EXPOSURE

ITEM CODE	DOOR MINIMUM THICKNESS	CONSTRUCTION DETAILS	PERFORMANCE		REFERENCE NUMBER			NOTES	REC. (MIN.)
			EFFECTIVE BARRIER	EDGE FLAMING	PRE-BMS-92	BMS-92	POST-BMS-92		
D-1	$^3/_8$"	Panel door; pine perimeter ($1^3/_8$"); painted (enamel).	5 min. 10 sec.	N/A			90	1, 2	5
D-2	$^3/_8$"	As above, with two coats U.L. listed intumescent coating.	5 min. 30 sec.	5 min.			90	1, 2, 7	5
D-3	$^3/_8$"	As D-1, with standard primer and flat interior paint.	5 min. 55 sec.	N/A			90	1, 3, 4	5
D-4	$2^5/_8$"	As D-1, with panels covered each side with $^1/_2$" plywood; edge grouted with sawdust filled plaster; door faced with $^1/_8$" hardboard each side; paint see (5).	11 min. 15 sec.	3 min. 45 sec.			90	1, 2, 5, 7	10
D-5	$^3/_8$"	As D-1, except surface protected with glass fiber reinforced intumescent fire retardant coating.	16 min.	N/A			90	1, 3, 4, 7	15
D-6	$1^5/_8$"	Door detail: As D-4, except with $^1/_8$" cement asbestos board facings with aluminum foil; door edges protected by sheet metal.	17 min.	10 min. 15 sec.			90	1, 3, 4	15
D-7	$1^5/_8$"	Door detail with $^1/_8$" hardboard cover each side as facings; glass fiber reinforced intumescent coating applied.	20 min.	N/A			90	1, 3, 4, 7	20
D-8	$1^5/_8$"	Door detail same as D-4; paint was glass reinforced epoxy intumescent.	26 min.	24 min. 45 sec.			90	1, 3, 4, 6, 7	25

(Continued)

TABLE 5.1—RESISTANCE OF DOORS TO FIRE EXPOSURE—(Continued)

ITEM CODE	DOOR MINIMUM THICKNESS	CONSTRUCTION DETAILS	PERFORMANCE		REFERENCE NUMBER			NOTES	REC. (MIN.)
			EFFECTIVE BARRIER	EDGE FLAMING	PRE-BMS-92	BMS-92	POST-BMS-92		
D-9	$1^5/_8$"	Door detail same as D-4 with facings of $^1/_8$" cement asbestos board.	29 min.	3 min. 15 sec.			90	1, 2	5
D-10	$1^5/_8$"	As per D-9.	31 min. 30 sec.	7 min. 20 sec.			90	1, 3, 4	6
D-11	$1^5/_8$"	As per D-7; painted with epoxy intumescent coating including glass fiber roving.	36 min. 25 sec.	N/A			90	1, 3, 4	35
D-12	$1^5/_8$"	As per D-4 with intumescent fire retardant paint.	37 min. 30 sec.	24 min. 40 sec.			90	1, 3, 4	30
D-13	$1^1/_2$" (nom.)	As per D-4, except with 24 ga. galvanized sheet metal facings.	39 min.	39 min.			90	1, 3, 4	39
D-14	$1^5/_8$"	As per D-9.	41 min. 30 sec.	17 min. 20 sec.			90	1, 3, 4, 6	20
D-15	—	Class C steel fire door.	60 min.	58 min.			90	7, 8	60
D-16	—	Class B steel fire door.	60 min.	57 min.			90	7, 8	60
D-17	$1^3/_4$"	Solid core flush door; core staves laminated to facings but not each other; Birch plywood facings $^1/_2$" rebate in door frame for door; $^3/_{32}$" clearance between door and wood frame.	15 min.	13 min.			37	11	13
D-18	$1^3/_4$"	As per D-17.	14 min.	13 min.			37	11	13
D-19	$1^3/_4$"	Door same as D-17, except with 16 ga. steel; $^3/_{32}$" door frame clearance.	12 min.	—			37	9, 11	10
D-20	$1^3/_4$"	As per D-19.	16 min.	—			37	10, 11	10
D-21	$1^3/_4$"	Doors as per D-17; intumescent paint applied to top and side edges.	26 min.	—			37	11	25
D-22	$1^3/_4$"	Door as per D-17, except with $^1/_2$" × $^1/_8$" steel strip set into edges of door at top and side facing stops; matching strip on stop.	18 min.	6 min.			37	11	18
D-23	$1^3/_4$"	Solid oak door.	36 min.	22 min.			15	13	25
D-24	$1^7/_8$"	Solid oak door.	35 min.	35 min.			15	13	35
D-25	$1^7/_8$"	Solid teak door.	58 min.	34 min.			15	13	35
D-26	$1^7/_8$"	Solid (pitch) pine door.	57 min.	36 min.			15	13	35
D-27	$1^7/_8$"	Solid deal (pine) door.	57 min.	30 min.			15	13	30
D-28	$1^7/_8$"	Solid mahogany door.	49 min.	40 min.			15	13	45
D-29	$1^7/_8$"	Solid poplar door.	24 min.	3 min.			15	13, 14	5
D-30	$1^7/_8$"	Solid oak door.	40 min.	33 min.			15	13	35
D-31	$1^7/_8$"	Solid walnut door.	40 min.	15 min.			15	13	20
D-32	$2^5/_8$"	Solid Quebec pine.	60 min.	60 min.			15	13	60
D-33	$2^5/_8$"	Solid pine door.	55 min.	39 min.			15	13	40
D-34	$2^5/_8$"	Solid oak door.	69 min.	60 min.			15	13	60
D-35	$2^5/_8$"	Solid teak door.	65 min.	17 min.			15	13	60
D-36	$1^1/_2$"	Solid softwood door.	23 min.	8.5 min.			15	13	10
D-37	$^3/_4$"	Panel door.	8 min.	7.5 min.			15	13	5
D-38	$^5/_{16}$"	Panel door.	5 min.	5 min.			15	13	5

(Continued)

TABLE 5.1—RESISTANCE OF DOORS TO FIRE EXPOSURE—(Continued)

ITEM CODE	DOOR MINIMUM THICKNESS	CONSTRUCTION DETAILS	PERFORMANCE		REFERENCE NUMBER			NOTES	REC. (MIN.)
			EFFECTIVE BARRIER	EDGE FLAMING	PRE-BMS-92	BMS-92	POST-BMS-92		
D-39	$3/4$"	Panel door, fire retardant treated.	$17^1/_2$ min.	3 min.			15	13	8
D-40	$3/4$"	Panel door, fire retardant treated.	$8^1/_2$ min.	$8^1/_2$ min.			15	13	8
D-41	$3/4$"	Panel door, fire retardant treated.	$16^3/_4$ min.	$11^1/_2$ min.			15	13	8

Notes:

1. All door frames were of standard lumber construction.
2. Wood door stop protected by asbestos millboard.
3. Wood door stop protected by sheet metal.
4. Door frame protected with sheet metal and weather strip.
5. Surface painted with intumescent coating.
6. Door edge sheet metal protected.
7. Door edge intumescent paint protected.
8. Formal steel frame and door stop.
9. Door opened into furnace at 12 feet.
10. Similar door opened into furnace at 12 feet.
11. The doors reported in these tests represent the type contemporaries used as 20-minute solid-core wood doors. The test results demonstrate the necessity of having wall anchored metal frames, minimum cleaners possible between door, frame and stops. They also indicate the utility of long throw latches and the possible use of intumescent paints to seal doors to frames in event of a fire.
12. Minimum working clearance and good latch closure are absolute necessities for effective containment for all such working door assemblies.
13. Based on British tests.
14. Failure at door - frame interface.

Bibliography

1. Central Housing Committee on Research, Design, and Construction; Subcommittee on Fire Resistance Classifications, "Fire-Resistance Classifications of Building Constructions," Building Materials and Structures, Report BMS 92, National Bureau of Standards, Washington, Oct. 1942. (Available from NTIS No. COM-73-10974)

2. Foster, H. D., Pinkston, E. R., and Ingberg, S. H., "Fire Resistance of Structural Clay Tile Partitions," Building Materials and Structures, Report BMS 113, National Bureau of Standards, Washington, Oct. 1948.

3. Ryan, J. V., and Bender, E.W., "Fire Endurance of Open-Web Steel-Joist Floors with Concrete Slabs and Gypsum Ceilings," Building Materials and Structures, Report BMS 141, National Bureau of Standards, Washington, Aug. 1954.

4. Mitchell, N. D., "Fire Tests of Wood-Framed Walls and Partitions with Asbestos-Cement Facings," Building Materials and Structures, Report BMS 123, National Bureau of Standards, Washington, May 1951.

5. Robinson, H. E., Cosgrove, L. A., and Powell, F. J., "Thermal Resistance of Airspace and Fibrous Insulations Bounded by Reflective Surfaces," Building Materials and Structures, Report BMS 151, National Bureau of Standards, Washington, Nov. 1957.

6. Shoub, H., and Ingberg, S. H., "Fire Resistance of Steel Deck Floor Assemblies," Building Science Series, 11, National Bureau of Standards, Washington, Dec. 1967.

7. Davey, N., and Ashton, L. A., "Investigations on Building Fires, Part V: Fire Tests of Structural Elements," National Building Studies, Research Paper, No. 12, Dept. of Scientific and Industrial Research (Building Research Station), London, 1953.

8. National Board of Fire Underwriters, Fire Resistance Ratings of Beam, Girder, and Truss Protections, Ceiling Constructions, Column Protections, Floor and Ceiling Constructions, Roof Constructions, Walls and Partitions, New York, April 1959.

9. Mitchell, N.D., Bender, E.D., and Ryan, J.V., "Fire Resistance of Shutters for Moving-Stairway Openings," Building Materials and Structures, Report BMS 129, National Bureau of Standards, Washington, March 1952.

10. National Board of Fire Underwriters, National Building Code; an Ordinance Providing for Fire Limits, and Regulations Governing the Construction, Alteration, Equipment, or Removal of Buildings or Structures, New York, 1949.

11. Department of Scientific and Industrial Research and of the Fire Offices' Committee, Joint Committee of the Building Research Board, "Fire Gradings of Buildings, Part I: General Principles and Structural Precautions," Post-War Building Studies, No. 20, Ministry of Works, London, 1946.

12. Lawson, D. I.,Webster, C. T., and Ashton, L. A., "Fire Endurance of Timber Beams and Floors," National Building Studies, Bulletin, No. 13, Dept. of Scientific and Industrial Research and Fire Offices' Committee (Joint Fire Research Organization), London, 1951.

13. Parker, T. W., Nurse, R. W., and Bessey, G. E., "Investigations on Building Fires. Part I: The Estimation of the Maximum Temperature Attained in Building Fires from Examination of the Debris, and Part II: The Visible Change in Concrete or Mortar Exposed to High Temperatures," National Building Studies, Technical Paper, No. 4, Dept. of Scientific and Industrial Research (Building Research Station), London, 1950.

14. Bevan, R. C., and Webster, C. T., "Investigations on Building Fires, Part III: Radiation from Building Fires," National Building Studies, Technical Paper, No. 5, Dept. of Scientific and Industrial Research (Building Research Station), London, 1950.

15. Webster, D. J., and Ashton, L. A., "Investigations on Building Fires, Part IV: Fire Resistance of Timber Doors," National Building Studies, Technical Paper, No. 6, Dept. of Scientific and Industrial Research (Building Research Station), London, 1951.

16. Kidder, F. E., Architects' and Builders' Handbook: Data for Architects, Structural Engineers, Contractors, and Draughtsmen, comp. by a Staff of Specialists and H. Parker, editor-in-chief, 18th ed., enl., J. Wiley, New York, 1936.

17. Parker, H., Gay, C. M., and MacGuire, J. W., Materials and Methods of Architectural Construction, 3rd ed., J. Wiley, New York, 1958.

18. Diets, A. G. H., Dwelling House Construction, The MIT Press, Cambridge, 1971.

19. Crosby, E. U., and Fiske, H. A., Handbook of Fire Protection, 5th ed., The Insurance Field Company, Louisville, Ky., 1914.

20. Crosby, E. U., Fiske, H. A., and Forster, H.W., Handbook of Fire Protection, 8th ed., R. S. Moulton, general editor, National Fire Protection Association, Boston, 1936.

21. Kidder, F. E., Building Construction and Superintendence, rev. and enl., by T. Nolan, W. T. Comstock, New York, 1909-1913, 2 vols.

22. National Fire Protection Association, Committee on Fire-Resistive Construction, The Baltimore Conflagration, 2nd ed., Chicago, 1904.

23. Przetak, L., Standard Details for Fire-Resistive Building Construction, McGraw-Hill Book Co., New York, 1977.

24. Hird, D., and Fischl, C. F., "Fire Hazard of Internal Linings," National Building Studies, Special Report, No. 22, Dept. of Scientific and Industrial Research and Fire Offices' Committee (Joint Fire Research Organization), London, 1954.

25. Menzel, C. A., Tests of the Fire-Resistance and Strength of Walls Concrete Masonry Units, Portland Cement Association, Chicago, 1934.

26. Hamilton, S. B., "A Short History of the Structural Fire Protection of Buildings Particularly in England," National Building Studies, Special Report, No. 27, Dept. of Scientific and Industrial Research (Building Research Station), London, 1958.

27. Sachs, E. O., and Marsland, E., "The Fire Resistance of Doors and Shutters being Tabulated Results of Fire Tests Conducted by the Committee," Journal of the British Fire Prevention Committee, No. VII, London, 1912.

28. Egan, M. D., Concepts in Building Firesafety, J. Wiley, New York, 1978.

29. Sachs, E. O., and Marsland, E., "The Fire Resistance of Floors Being Tabulated Results of Fire Tests Conducted by the Committee," Journal of the British Fire Prevention Committee, No. VI, London, 1911.

30. Sachs, E. O., and Marsland, E., "The Fire Resistance of Partitions Being Tabulated Results of Fire Tests Conducted by the Committee," Journal of the British Fire Prevention Committee, No. IX, London, 1914.

31. Ryan, J. V., and Bender, E. W., "Fire Tests of Precast Cellular Concrete Floors and Roofs," National Bureau of Standards Monograph, 45, Washington, April 1962.

32. Kingberg, S. H., and Foster, H. D., "Fire Resistance of Hollow Load-Bearing Wall Tile," National Bureau of Standards Research Paper, No. 37, (Reprint from NBS Journal of Research, Vol. 2) Washington, 1929.

33. Hull, W. A., and Ingberg, S. H., "Fire Resistance of Concrete Columns," Technologic Papers of the Bureau of Standards, No. 272, Vol. 18, Washington, 1925, pp. 635-708.

34. National Board of Fire Underwriters, Fire Resistance Ratings of Less than One Hour, New York, Aug. 1956.

35. Harmathy, T. Z., "Ten Rules of Fire Endurance Rating," Fire Technology, Vol. 1, May 1965, pp. 93-102.

36. Son, B. C., "Fire Endurance Test on a Steel Tubular Column Protected with Gypsum Board," National Bureau of Standards, NBSIR, 73-165, Washington, 1973.

37. Galbreath, M., "Fire Tests of Wood Door Assemblies," Fire Study, No. 36, Div. of Building Research, National Research Council Canada, Ottawa, May 1975.

38. Morris, W. A., "An Investigation into the Fire Resistance of Timber Doors," Fire Research Note, No. 855, Fire Research Station, Boreham Wood, Jan. 1971.

39. Hall, G. S., "Fire Resistance Tests of Laminated Timber Beams," Timber Association Research Report, WR/RR/1, High Sycombe, July 1968.

40. Goalwin, D. S., "Fire Resistance of Concrete Floors," Building Materials and Structures, Report BMS 134, National Bureau of Standards, Washington, Dec. 1952.

41. Mitchell, N. D., and Ryan, J. V., "Fire Tests of Steel Columns Encased with Gypsum Lath and Plaster," Building Materials and Structures, Report BMS 135, National Bureau of Standards, Washington, April 1953.

42. Ingberg, S. H., "Fire Tests of Brick Walls," Building Materials and Structures, Report BMS 143, National Bureau of Standards, Washington, Nov. 1954.

43. National Bureau of Standards, "Fire Resistance and Sound-Insulation Ratings for Walls, Partitions, and Floors," Technical Report on Building Materials, 44, Washington, 195X.

44. Malhotra, H. L., "Fire Resistance of Brick and Block Walls," Fire Note, No. 6, Ministry of Technology and Fire Offices' Committee Joint Fire Research Organization, London, HMSO, 1966.

45. Mitchell, N. D., "Fire Tests of Steel Columns Protected with Siliceous Aggregate Concrete," Building Materials and Structures, Report BMS 124, National Bureau of Standards, Washington, May 1951.

46. Freitag, J. K., Fire Prevention and Fire Protection as Applied to Building Construction; a Handbook of Theory and Practice, 2nd ed., J. Wiley, New York, 1921.

47. Ingberg, S. H., and Mitchell, N. D., "Fire Tests of Wood and Metal-Framed Partition," Building Materials and Structures, Report BMS 71, National Bureau of Standards, Washington, 1941.

48. Central Housing Committee on Research, Design, and Construction, Subcommittee on Definitions, "A Glossary of Housing Terms," Building Materials and Structures, Report BMS 91, National Bureau of Standards, Washington, Sept. 1942.

49. Crosby, E. U., Fiske, H. A., and Forster, H.W., Handbook of Fire Protection, 7th ed., D. Van Nostrand Co., New York 1924.

50. Bird, E. L., and Docking, S. J., Fire in Buildings, A. & C. Black, London, 1949.

51. American Institute of Steel Construction, Fire Resistant Construction in Modern Steel-Framed Buildings, New York, 1959.

52. Central Dockyard Laboratory, "Fire Retardant Paint Tests - a Critical Review," CDL Technical Memorandum, No. P87/73, H. M. Naval Base, Portsmouth, Dec. 1973.

53. Malhotra, H. L., "Fire Resistance of Structural Concrete Beams," Fire Research Note, No. 741, Fire Research Station, Borehamwood, May 1969.

54. Abrams, M. S., and Gustaferro, A. H., "Fire Tests of Poke-Thru Assemblies," Research and Development Bulletin, 1481-1, Portland Cement Association, Skokie, 1971.

55. Bullen, M. L., "A Note on the Relationship Between Scale Fire Experiments and Standard Test Results," Building Research Establishment Note, N51/75, Borehamwood, May 1975.

56. The America Fore Group of Insurance Companies, Research Department, Some Characteristic Fires in Fire Resistive Buildings, Selected from twenty years record in the files of the N.F.P.A. "Quarterly," New York, c. 1933.

57. Spiegelhalter, F., "Guide to Design of Cavity Barriers and Fire Stops," Current Paper, CP 7/77, Building Research Establishment, Borehamwood, Feb. 1977.

58. Wardle, T. M. "Notes on the Fire Resistance of Heavy Timber Construction," Information Series, No. 53, New Zealand Forest Service, Wellington, 1966.

59. Fisher, R. W., and Smart, P. M. T., "Results of Fire Resistance Tests on Elements of Building Construction," Building Research Establishment Report, G R6, London, HMSO, 1975.

60. Serex, E. R., "Fire Resistance of Alta Bates Gypsum Block Non-Load Bearing Wall," Report to Alta Bates Community Hospital, Structural Research Laboratory Report, ES-7000, University of Calif., Berkeley, 1969.

61. Thomas, F. G., and Webster, C. T., "Investigations on Building Fires, Part VI: The Fire Resistance of Reinforced Concrete Columns," National Building Studies, Research Paper, No. 18, Dept. of Scientific and Industrial Research (Building Research Station), London, HMSO, 1953.

62. Building Research Establishment, "Timber Fire Doors," Digest, 220, Borehamwood, Nov. 1978.

63. Massachusetts State Building Code; Recommended Provisions, Article 22: Repairs, Alterations, Additions, and Change of Use of Existing Buildings, Boston, Oct. 23, 1978.

64. Freitag, J. K., Architectural Engineering; with Especial Reference to High Building Construction, Including Many Examples of Prominent Office Buildings, 2nd ed., rewritten, J. Wiley, New York, 1906.

65. Architectural Record, Sweet's Indexed Catalogue of Building Construction for the Year 1906, New York, 1906.

66. Dept. of Commerce, Building Code Committee, "Recommended Minimum Requirements for Fire Resistance in Buildings," Building and Housing, No. 14, National Bureau of Standards, Washington, 1931.

67. British Standards Institution, "Fire Tests on Building Materials and Structures," British Standards, 476, Pt. 1, London, 1953.

68. Lönberg-Holm, K., "Glass," The Architectural Record, Oct. 1930, pp. 345-357.

69. Structural Clay Products Institute, "Fire Resistance," Technical Notes on Brick and Tile Construction, 16 rev., Washington, 1964.

70. Ramsey, C. G., and Sleeper, H. R., Architectural Graphic Standards for Architects, Engineers, Decorators, Builders, and Draftsmen, 3rd ed., J. Wiley, New York, 1941.

71. Underwriters' Laboratories, Fire Protection Equipment List, Chicago, Jan. 1957.

72. Underwriters' Laboratories, Fire Resistance Directory; with Hourly Ratings for Beams, Columns, Floors, Roofs, Walls, and Partitions, Chicago, Jan. 1977.

73. Mitchell, N. D., "Fire Tests of Gunite Slabs and Partitions," Building Materials and Structures, Report BMS 131, National Bureau of Standards, Washington, May 1952.

74. Woolson, I. H., and Miller, R. P., "Fire Tests of Floors in the United States," Proceedings International Association for Testing Materials, VIth Congress, New York, 1912, Section C, pp. 36-41.

75. Underwriters' Laboratories, "An Investigation of the Effects of Fire Exposure Upon Hollow Concrete Building Units, Conducted for American Concrete Institute, Concrete Products Association, Portland Cement Association, Joint Submittors," Retardant Report, No. 1555, Chicago, May 1924.

76. Dept. of Scientific & Industrial Research and of the Fire Offices' Committee, Joint Committee of the Building Research Board, "Fire Gradings of Buildings. Part IV: Chimneys and Flues," Post-War Building Studies, No. 29, London, HMSO, 1952.

77. National Research Council of Canada. Associate Committee on the National Building Code, Fire Performance Ratings, Suppl. No. 2 to the National Building Code of Canada, Ottawa, 1965.

78. Associated Factory Mutual Fire Insurance Companies, The National Board of Fire Underwriters, and the Bureau of Standards, Fire Tests of Building Columns; an Experimental Investigation of the Resistance of Columns, Loaded and Exposed to Fire or to Fire and Water, with Record of Characteristic Effects, Jointly Conducted at Underwriters. Laboratories, Chicago, 1917-19.

79. Malhotra, H. L., "Effect of Age on the Fire Resistance of Reinforced Concrete Columns," Fire Research Memorandum, No. 1, Fire Research Station, Borehamwood, April 1970.

80. Bond, H., ed., Research on Fire; a Description of the Facilities, Personnel and Management of Agencies Engaged in Research on Fire, a Staff Report, National Fire Protection Association, Boston, 1957.

81. California State Historical Building Code, Draft, 1978.

82. Fisher, F. L., et al., "A Study of Potential Flashover Fires in Wheeler Hall and the Results from a Full Scale Fire Test of a Modified Wheeler Hall Door Assembly," Fire Research Laboratory Report, UCX 77-3; UCX-2480, University of Calif., Dept. of Civil Eng., Berkeley, 1977.

83. Freitag, J. K., The Fireproofing of Steel Buildings, 1st ed., J. Wiley, New York, 1906.

84. Gross, D., "Field Burnout Tests of Apartment Dwellings Units," Building Science Series, 10, National Bureau of Standards, Washington, 1967.

85. Dunlap, M. E., and Cartwright, F. P., "Standard Fire Tests for Combustible Building Materials," Proceedings of the American Society for Testing Materials, vol. 27, Philadelphia, 1927, pp. 534-546.

86. Menzel, C. A., "Tests of the Fire Resistance and Stability of Walls of Concrete Masonry Units," Proceedings of the American Society for Testing Materials, vol. 31, Philadelphia, 1931, pp. 607-660.

87. Steiner, A. J., "Method of Fire-Hazard Classification of Building Materials," Bulletin of the American Society for Testing and Materials, March 1943, Philadelphia, 1943, pp. 19-22.

88. Heselden, A. J. M., Smith, P. G., and Theobald, C. R., "Fires in a Large Compartment Containing Structural Steelwork; Detailed Measurements of Fire Behavior," Fire Research Note, No. 646, Fire Research Station, Borehamwood, Dec. 1966.

89. Ministry of Technology and Fire Offices' Committee Joint Fire Research Organization, "Fire and Structural Use of Timber in Buildings; Proceedings of the Symposium Held at the Fire Research Station, Borehamwood, Herts on 25th October, 1967," Symposium, No. 3, London, HMSO, 1970.

90. Shoub, H., and Gross, D., "Doors as Barriers to Fire and Smoke," Building Science Series, 3, National Bureau of Standards, Washington, 1966.

91. Ingberg, S. H., "The Fire Resistance of Gypsum Partitions," Proceedings of the American Society for Testing and Materials, vol. 25, Philadelphia, 1925, pp. 299-314.

92. Ingberg, S.H., "Influence of Mineral Composition of Aggregates on Fire Resistance of Concrete," Proceedings of the American Society for Testing and Materials, vol. 29, Philadelphia, 1929, pp. 824-829.

93. Ingberg, S. H., "The Fire Resistive Properties of Gypsum," Proceedings of the American Society for Testing andMaterials, vol. 23, Philadelphia, 1923, pp. 254-256.

94. Gottschalk, F.W., "Some Factors in the Interpretation of Small-Scale Tests for Fire-RetardantWood," Bulletin of the American Society for Testing and Materials, October 1945, pp. 40-43.

95. Ministry of Technology and Fire Offices' Committee Joint Fire Research Organization, "Behaviour of Structural Steel in Fire; Proceedings of the Symposium Held at the Fire Research Station Borehamwood, Herts on 24th January, 1967," Symposium, No. 2, London, HMSO, 1968.

96. Gustaferro, A. H., and Martin, L. D., Design for Fire Resistance of Pre-cast Concrete, prep. for the Prestressed Concrete Institute Fire Committee, 1st ed., Chicago, PCI, 1977.

97. "The Fire Endurance of Concrete; a Special Issue," Concrete Construction, vol. 18, no. 8, Aug. 1974, pp. 345-440.

98. The British Constructional Steelwork Association, "Modern Fire Protection for Structural Steelwork," Publication, No. FPl, London, 1961.

99. Underwriters' Laboratories, "Fire Hazard Classification of Building Materials," Bulletin, No. 32, Sept. 1944, Chicago, 1959.

100. Central Housing Committee on Research, Design, and Construction, Subcommittee on Building Codes, "Recommended Building Code Requirements for New Dwelling Construction with Special Reference to War Housing; Report," Building Materials and Structures, Report BMS 88, National Bureau of Standards, Washington, Sept. 1942.

101. De Coppet Bergh, D., Safe Building Construction; a Treatise Giving in Simplest Forms Possible Practical and Theoretical Rules and Formulae Used in Construction of Buildings and General Instruction, new ed., thoroughly rev. Macmillan Co., New York, 1908.

102. Cyclopedia of Fire Prevention and Insurance; a General Reference Work on Fire and Fire Losses, Fireproof Construction, Building Inspection..., prep. by architects, engineers, underwriters and practical insurance men. American School of Correspondence, Chicago, 1912.

103. Setchkin, N. P., and Ingberg, S. H., "Test Criterion for an Incombustible Material," Proceedings of the American Society for Testing Materials, vol. 45, Philadelphia, 1945, pp. 866-877.

104. Underwriters' Laboratories, "Report on Fire Hazard Classification of Various Species of Lumber," Retardant, 3365, Chicago, 1952.

105. Steingiser, S., "A Philosophy of Fire Testing," Journal of Fire & Flammability, vol. 3, July 1972, pp. 238-253.

106. Yuill, C. H., Bauerschlag, W. H., and Smith, H. M., "An Evaluation of the Comparative Performance of 2.4.1 Plywood and Two-Inch Lumber Roof Decking under Equivalent Fire Exposure," Fire Protection Section, Final Report, Project No. 717A-3-211, Southwest Research Institute, Dept. of Structural Research, San Antonio, Dec. 1962.

107. Ashton, L. A., and Smart, P.M. T., Sponsored Fire-Resistance Tests on Structural Elements, London, Dept. of Scientific and Industrial Research and Fire Offices. Committee, London, 1960.

108. Butcher, E. G., Chitty, T. B., and Ashton, L. A., "The Temperature Attained by Steel in Building Fires," Fire Research Technical Paper, No. 15, Ministry of Technology and Fire Offices. Committee, Joint Fire Research Organization, London, HMSO, 1966.

109. Dept. of the Environment and Fire Offices' Committee, Joint Fire Research Organization, "Fire-Resistance Requirements for Buildings - a New Approach; Proceedings of the Symposium Held at the Connaught Rooms, London, 28 September 1971," Symposium, No. 5, London, HMSO, 1973.

110. Langdon Thomas, G. J., "Roofs and Fire," Fire Note, No. 3, Dept. of Scientific and Industrial Research and Fire Offices' Committee, Joint Fire Research Organization, London, HMSO, 1963.

111. National Fire Protection Association and the National Board of Fire Underwriters, Report on Fire the Edison Phonograph Works, Thomas A. Edison, Inc., West Orange, N.J., December 9, 1914, Boston, 1915.

112. Thompson, J. P., Fire Resistance of Reinforced Concrete Floors, Portland Cement Association, Chicago, 1963.

113. Forest Products Laboratory, "Fire Resistance Tests of Plywood Covered Wall Panels," Information reviewed and reaffirmed, Forest Service Report, No. 1257, Madison, April 1961.

114. Forest Products Laboratory, "Charring Rate of Selected Woods - Transverse to Grain," Forest Service Research Paper, FLP 69, Madison, April 1967.

115. Bird, G. I., "Protection of Structural Steel Against Fire," Fire Note, No. 2, Dept. of Scientific and Industrial Research and Fire Offices' Committee, Joint Fire Research Organization, London, HMSO, 1961.

116. Robinson, W. C., The Parker Building Fire, Underwriters' Laboratories, Chicago, c. 1908.

117. Ferris, J. E., "Fire Hazards of Combustible Wallboards," Commonwealth Experimental Building Station Special Report, No. 18, Sydney, Oct. 1955.

118. Markwardt, L. J., Bruce, H. D., and Freas, A. D., "Brief Description of Some Fire-Test Methods Used for Wood and Wood Base Materials," Forest Service Report, No. 1976, Forest Products Laboratory, Madison, 1976.

119. Foster, H. D., Pinkston, E. R., and Ingberg, S. H., "Fire Resistance of Walls of Gravel-Aggregate Concrete Masonry Units," Building Materials and Structures, Report, BMS 120, National Bureau of Standards, Washington, March 1951.

120. Foster, H. D., Pinkston, E.R., and Ingberg, S. H., "Fire Resistance of Walls of Lightweight-Aggregate Concrete Masonry Units," Building Materials and Structures, Report BMS 117, National Bureau of Standards, Washington, May 1950.

121. Structural Clay Products Institute, "Structural Clay Tile Fireproofing," Technical Notes on Brick & Tile Construction, vol. 1, no. 11, San Francisco, Nov. 1950.

122. Structural Clay Products Institute, "Fire Resistance Ratings of Clay Masonry Walls - I," Technical Notes on Brick & Tile Construction, vol. 3, no. 12, San Francisco, Dec. 1952.

123. Structural Clay Products Institute, "Estimating the Fire Resistance of Clay Masonry Walls - II," Technical Notes on Brick & Tile Construction, vol. 4, no. 1, San Francisco, Jan. 1953.

124. Building Research Station, "Fire: Materials and Structures," Digest, No. 106, London, HMSO, 1958.

125. Mitchell, N. D., "Fire Hazard Tests with Masonry Chimneys," NFPA Publication, No. Q-43-7, Boston, Oct. 1949.

126. Clinton Wire Cloth Company, Some Test Data on Fireproof Floor Construction Relating to Cinder Concrete, Terra Cotta and Gypsum, Clinton, 1913.

127. Structural Engineers Association of Southern California, Fire Ratings Subcommittee, "Fire Ratings, a Report," part of Annual Report, Los Angeles, 1962, pp. 30-38.

128. Lawson, D. I., Fox, L. L., and Webster, C. T., "The Heating of Panels by Flue Pipes," Fire Research, Special Report, No. 1, Dept. of Scientific and Industrial Research and Fire Offices' Committee, London, HMSO, 1952.

129. Forest Products Laboratory, "Fire Resistance of Wood Construction," Excerpt from 'Wood Handbook - Basic Information on Wood as a Material of Construction with Data for its Use in Design and Specification,' Dept. of Agriculture Handbook, No. 72, Washington, 1955, pp. 337-350.

130. Goalwin, D. S., "Properties of Cavity Walls," Building Materials and Structures, Report BMS 136, National Bureau of Standards, Washington, May 1953.

131. Humphrey, R. L., "The Fire-Resistive Properties of Various Building Materials," Geological Survey Bulletin, 370, Washington, 1909.

132. National Lumber Manufacturers Association, "Comparative Fire Test on Wood and Steel Joists," Technical Report, No. 1, Washington, 1961.

133. National Lumber Manufacturers Association, "Comparative Fire Test of Timber and Steel Beams," Technical Report, No. 3, Washington, 1963.

134. Malhotra, H. L., and Morris, W. A., "Tests on Roof Construction Subjected to External Fire," Fire Note, No. 4, Dept. of Scientific and Industrial Research and Fire Offices' Committee, Joint Fire Research Organization, London, HMSO, 1963.

135. Brown, C. R., "Fire Tests of Treated and Untreated Wood Partitions," Research Paper, RP 1076, part of Journal of Research of the National Bureau of Standards, vol. 20, Washington, Feb. 1938, pp. 217-237.

136. Underwriters' Laboratories, "Report on Investigation of Fire Resistance of Wood Lath and Lime Plaster Interior Finish," Publication, SP. 1. 230, Chicago, Nov. 1922.

137. Underwriters' Laboratories, "Report on Interior Building Construction Consisting of Metal Lath and Gypsum Plaster on Wood Supports," Retardant, No. 1355, Chicago, 1922.

138. Underwriters' Laboratories, "An Investigation of the Effects of Fire Exposure Upon Hollow Concrete Building Units," Retardant, No. 1555, Chicago, May 1924.

139. Moran, T. H., "Comparative Fire Resistance Ratings of Douglas Fir Plywood," Douglas Fir Plywood Association Laboratory Bulletin, 57-A, Tacoma, 1957.

140. Gage Babcock & Association, "The Performance of Fire-Protective Materials Under Varying Conditions of Fire Severity," Report 6924, Chicago, 1969.

141. International Conference of Building Officials, Uniform Building Code (1979 ed.), Whittier, CA, 1979.

142. Babrauskas, V., and Williamson, R. B., "The Historical Basis of Fire Resistance Testing, Part I and Part II," Fire Technology, vol. 14, no. 3 & 4, Aug. & Nov. 1978, pp. 184-194, 205, 304-316.

143. Underwriters' Laboratories, "Fire Tests of Building Construction and Materials," 8th ed., Standard for Safety, UL263, Chicago, 1971.

144. Hold, H. G., Fire Protection in Buildings, Crosby, Lockwood, London, 1913.

145. Kollbrunner, C. F., "Steel Buildings and Fire Protection in Europe," Journal of the Structural Division, ASCE, vol. 85, no. ST9, Proc. Paper 2264, Nov. 1959, pp. 125-149.

146. Smith, P., "Investigation and Repair of Damage to Concrete Caused by Formwork and Falsework Fire," Journal of the American Concrete Institute, vol. 60, title no. 60-66, Nov. 1963, pp. 1535-1566.

147. "Repair of Fire Damage," 3 parts, Concrete Construction, March-May, 1972.

148. National Fire Protection Association, National Fire Codes; a Compilation of NFPA Codes, Standards, Recommended Practices and Manuals, 16 vols., Boston, 1978.

149. Ingberg, S. H. "Tests of Severity of Building Fires," NFPA Quarterly, vol. 22, no. 1, July 1928, pp. 43-61.

150. Underwriters' Laboratories, "Fire Exposure Tests of Ordinary Wood Doors," Bulletin of Research, no. 6, Dec. 1938, Chicago, 1942.

151. Parson, H., "The Tall Building Under Test of Fire," Red Book, no. 17, British Fire Prevention Committee, London, 1899.

152. Sachs, E. O., "The British Fire Prevention Committee Testing Station," Red Book, no. 13, British Fire Prevention Committee, London, 1899.

153. Sachs, E. O., "Fire Tests with Unprotected Columns," Red Book, no. 11, British Fire Prevention Committee, London, 1899.

154. British Fire Prevention Committee, "Fire Tests with Floors a Floor by the Expended Metal Company," Red Book, no. 14, London, 1899.

155. Engineering News, vol. 56, Aug. 9, 1906, pp. 135-140.

156. Engineering News, vol. 36, Aug. 6, 1896, pp. 92-94.

157. Bauschinger, J., Mittheilungen de Mech.-Tech. Lab. der K. Tech. Hochschule, München, vol. 12, 1885.

158. Engineering News, vol. 46, Dec. 26, 1901, pp. 482-486, 489-490.

159. The American Architect and Building News, vol. 31, March 28, 1891, pp. 195-201.

160. British Fire Prevention Committee, First International Fire Prevention Congress, Official Congress Report, London, 1903.

161. American Society for Testing Materials, Standard Specifications for Fire Tests of Materials and Construction (C19-18), Philadelphia, 1918.

162. International Organization for Standardization, Fire Resistance Tests on Elements of Building Construction (R834), London, 1968.

163. Engineering Record, vol. 35, Jan. 2, 1897, pp. 93-94; May 29, 1897, pp. 558-560; vol. 36, Sept. 18, 1897, pp. 337-340; Sept. 25, 1897, pp. 359-363; Oct. 2, 1897, pp. 382-387; Oct. 9, 1897, pp. 402-405.

164. Babrauskas, Vytenis, "Fire Endurance in Buildings," PhD Thesis. Fire Research Group, Report, No. UCB FRG 76-16, University of California, Berkeley, Nov. 1976.

165. The Institution of Structural Engineers and The Concrete Society, Fire Resistance of Concrete Structures, London, Aug. 1975.

INDEX